Theory and Applications of Computability

In cooperation with the Association Computability in Europe

Books published in this series will be of interest to the research community and graduate students, with a unique focus on issues of computability. The perspective of the series is multidisciplinary, recapturing the spirit of Turing by linking theoretical and real-world concerns from computer science, mathematics, biology, physics, and the philosophy of science. The series includes research monographs, advanced and graduate texts, and books that offer an original and informative view of computability and computational paradigms.

Rodney G. Downey • Alexander Melnikov

Computable Structure Theory

A Unified Approach

Springer

Rodney G. Downey
School of Mathematics and Statistics
Victoria University
Wellington, New Zealand

Alexander Melnikov
School of Mathematics and Statistics
Victoria University
Wellington, New Zealand

ISSN 2190-619X ISSN 2190-6203 (electronic)
Theory and Applications of Computability
ISBN 978-3-031-92432-3 ISBN 978-3-031-92433-0 (eBook)
https://doi.org/10.1007/978-3-031-92433-0

Cover illustration: Triangulated Tuatara by Alex Downey

This Springer imprint is published by the registered company Springer Nature Switzerland AG
The registered company address is: Gewerbestrasse 11, 6330 Cham, Switzerland

Contents

Dedicated to Kristin and Tanya

Preface

It is striking that Turing's fundamental paper [487], which proposed the modern model of computation, focused on processes involving real numbers. That is, this seminal paper did not concern itself only with discrete problems coded by integers. Rather, it had continuous problems (such as the calculation of integrals by better and better approximations) hand in hand with the Entscheidungsproblem, which is the problem of algorithmic decidability for first-order logic. Thus, at the dawn of the theory of computation, there was no real distinction between the continuous and the discrete.

However, subsequent events saw a divergence in the study of computational processes in these two domains; the discrete and the analytic were somehow treated as being quite different. They had their own communities with computability and complexity theorists in the case of the discrete, and people concerned with computable analysis and effective topology in the analytic settings.

In recent years, it has become noticeable that the ideas in the two areas are re-converging. Analytic structures, particularly those coded by dense sequences, share much in common with countable computable discrete structures.

> In this book, we present a unified theory of computably presented structures combining both countable and separable uncountable structures.

Our main objects of study are *computable* algebraic and topological structures such as linear orderings, graphs, Boolean algebras, groups, separable Banach and Polish spaces. The theories of these computable algebraic structures and computably presented spaces are now well-established and wide ranging. It is our belief that the theories of computable separable spaces and computable discrete algebraic structures are very tightly connected, and no firm line can be drawn between them. In the book, we present results and techniques that establish direct links between these theories.

It is essentially impossible to cover all major topics of the theory in one book, so we will have to be selective. The basic themes we address include the following:

(a) What does it mean for an infinite structure to be computable?

(b) How can we compare different (computable) presentations of a structure?

(c) Can we classify computably presentable structures in a given class?

As we will see, such questions are directly related to the problems that lie in the foundations of mathematics:

(i) Is it possible to show that a given problem is undecidable?

(ii) How do we measure the complexity of a problem?

(iii) Can we classify structures in a given class?

Note that Question (iii) does not seem to be related to computability at all. Nevertheless, computable structures can sometimes provide insights into problems of this nature, as will be explained in the second part of the book. Roughly speaking, the first half of the book addresses Questions (a)-(c), and the second half is mainly devoted to (i) - (iii).

In the light of our unifying theme, we will be introducing techniques in context. So, for example, when we introduce the finite injury priority method, we will more or less immediately show how it works in the context of computable structure theory. Similarly when a technique is introduced in either a discrete setting or an uncountable one, we will endeavour to also show how it resonates in the other setting.

The text provides a unified introduction to computability theory, computable algebra, and computable Polish space theory. Some topics, particularly those in computable abelian group theory and effective topological group theory, are presented in detail for the first time.

We had to limit the scope of the book to keep its length reasonable. Nevertheless, we believe that we have captured the major themes of development for the last few decades, at least from our biased point of view. We have also given references for further reading when we felt that including the material would have upset the cost/benefit ratio. In this spirit, throughout the book we have starred material. If a section is labelled with a star, this means that it can be safely ignored, as it is not essential to the rest of the book. Some further material will be included in the form of exercises, some of which will be given with hints.

Notes on exercises: The exercises in this book range from material we believe is suitable for self-guided students to material that is either not essential to the rest of the book, relies on topics not covered in the book, or is genuinely difficult—often at the level of a significant research paper. Sometimes we have added hints, but for many exercises, we have simply provided a reference, especially if the result is readily accessible in a paper available online. Exercises we consider particularly suitable for students are marked with $\circ$, those that are either not essential or rely on external material are marked with $*$, and genuinely difficult ones are marked with $**$.

Acknowledgements: We are grateful to our colleagues Keng Meng Ng, Denis Hirschfeldt, Steffen Lempp, Mars Yamaleev, D. Reed Solomon, Daniel Turetsky, Noam Greenberg, Mathieu Hoyrup, Nikolay Bazhenov, Paul Shafer, Luca San Mauro, Maxim Zubkov, Ruslan Kornev, as well as to our students Xavier Enright, Alibek Iskakov, and Sapir Ben-Shahar, who suggested many useful corrections. Thanks to Matthew Askes and Brendan Harding who helped with the diagrams. We gratefully acknowledge the support of the Rutherford Discovery Fellowship (RDF-VUW1902) and Marsden Fund (20-VUW-105), awarded by the Royal Society Te Apārangi, which allowed us to prioritise the writing of this book above all other commitments.

Part I

Computable presentability and effective duality

Chapter 1

Introduction

In the first part of the book, we investigate the problem of computable presentability for spaces and algebraic structures. Much of the theory revolves around a few basic definitions of computability that are tested and compared in various natural classes of mathematical structures. In the next few sections we will meet the main players in this book, and also give some of the main motivating themes. We will always bear in mind our central theme of unifying the countable and uncountable.

> In Part 1 of the book, we focus on results that establish a *direct connection* between algorithmic procedures in countable algebra and topology through *effective dualities.*

These effective dualities include the recently established computable Stone duality and computable Pontryagin duality. In the first part of the book we use these effective dualities to transfer several classical results from effective countable algebra (to appear as Theorem A) to the topological setting. These effective topological results, to be stated in Theorem B, appear to be fundamental. In the second part we will use these dualities to transfer classification-type results about countable algebraic structures to the uncountable setting.

The plan of this introductory chapter is as follows:

1. Section 1.1 motivates the considerations of the first half of the book.
2. Section 1.2 contains the main definitions of the book, the statements of Theorems A and B, and also states the effective dualities mentioned above.
3. In Section 1.3 we review conflicting terminology as presented in the literature.

R. G. Downey, A. Melnikov, *Computable Structure Theory*, Theory and Applications of Computability, https://doi.org/10.1007/978-3-031-92433-0_1

1.1 A motivating example: fields

To motivate the considerations of our book, and especially the first half of it, we consider several more specific questions and sample results that naturally arise in the class of countable fields. As we also see, effective processes in uncountable structures also arise naturally and perhaps even inevitably.

We adopt the Turing-Church Thesis that says that a process is algorithmically computable if and only if under some appropriate coding it becomes computable by a Turing machine. If the reader is not familiar with these concepts, they should rely on their intuition and wait until Section 2.1 where we give the precise definitions.

We first consider the questions analysed in a classic paper of Metakides and Nerode [384] who studied the "effective content of field theory." The following examples are drawn from this paper. Matakides and Nerode analysed the extent to which the following classical results were *effective* or *algorithmic*: Consider the well-known algebraic results for countable structures given below, ignoring the * symbols.

I (Steinitz) Every * field has a * algebraic closure.

II (Steinitz) Every * algebraically closed field has a * transcendence base.

III (Steinitz) Any two * algebraic closures of a * field F differ by an * F-automorphism.

IV (Artin-Schreier) Any * formally real field can be * ordered.

V (Artin-Schreier) Any two * real closures of a * ordered field F differ by a * F-automorphism.

VI (Krull) Any * proper Galois extension F' of a * field F has a * F-automorphism other than the identity.

If the $*$ symbols are removed, then these are classical results [15, 483, 319]. The question is:

> To what extent are these theorems algorithmic? What is their *effective, or algorithmic content*?

If we restrict ourselves to finite dimensional fields, then all of them are algorithmically true by Kronecker [318] in the sense that from the finite input data, we can *build* the relevant object, for example the transcendence basis in Steinitz (II). But what if we wish to consider arbitrary countably infinite fields? For instance, in the result of Steinitz (I), can we *construct* an algebraic closure for any countably infinite field? What do we even mean by this?

The point here is that classical computability theory was developed to answer questions like the *Entscheidungsproblem*[1] where algorithms take *finite descriptions* of instances to *finite descriptions* of their solutions. For instance, for the Entscheidungsproblem, a positive solution would need to be an algorithm A which takes as input a description (or a code) for a formula of predicate logic. This formula will commonly be given by its (Gödel) number $n \in \mathbb{N}$, although any coding of the formula into a number would suffice. The algorithm A on input n will say 1 or 0 as to whether or not the corresponding formula is true.

However, for question such as whether we can construct an algebraic closure of a countably infinite field, *the input itself is infinite.* To give this question meaning, we surely demand that the

[1] The decision problem for first order logic.

input data, here being an infinite field or an infinite group, be given algorithmically. Then we would ask that from the algorithm giving the input data we produce an algorithm for the solution. In particular, it would be natural to guess that the $*$ symbols in (I) - (VI) above can be replaced with "*computable*" or "*computably presented*". So now the effective version of Steinitz (I) might read:

> Every *computable* field has a *computable* algebraic closure.

A computable field is a tuple $(F, \cdot, +, =, ^{-1}, 0, 1)$ with F being a computable set, upon which the field operations are computable functions with input and output in $\mathbb{N}$. That is, we present the data via algorithms and then seek solutions we can obtain from this data. Using this formal notion of computable presentability, we can *formally illustrate* that for (I) the answer is "yes", as proven by Rabin [440] (see §2.2.2). Moreover, as we will see, this effectivisation can be obtained as part of a *general basic result* in computable structure theory: a computable version of the Henkin construction for predicate logic.

However sometimes the answer to the effective version is "no", such as for (IV), as proven by Ershov [157]: There is a computable real closed field with no computable ordering. Only (I) and (V) hold computably: (I) by Rabin [440], and (V) by effectivising Artin's proof in [15]. The others are also false for computable fields: the effective version of (III) was proven to be false by Frölich and Shepherdson [185], and the rest can be found in Metakides and Nerode [384].

In the case of a "no" answer, we are then led to further questions such as:

> Given that there is a computable formally real field with no computable ordering, and the classical theorem says that there is *some* ordering, how complicated must such an ordering be? For instance, does it always have some relatively simple ordering? Can we effectively describe the collection of all compatible orderings?

Metakides and Nerode [384] showed that for a computable formally real field, the space of orderings can be computably described as a Π^0_1 class (Exercise 4.2.62). Such a class can be thought of as the collection of infinite paths through a computable binary tree; a *Stone space.* Then we can appeal to the general theory of Π^0_1 classes (see Subection 2.1.6) to conclude, for example, that for each computable real closed field, there must be an ordering L of low Turing degree. The ordering L being *low* means that although L is not necessarily computable, we have $L' \equiv_T \varnothing'$. This means that the halting problem for Turing machines with oracle L is not computationally harder than the halting problem for machines without an oracle. These notions will be later explained in detail in Section 2.1.6.

We could also address very similar questions in other classes of structures, such as groups. For the class of abelian groups, the natural analogy of (IV) is:

IV* (Mal'cev, after Levi) Any * torsion-free abelian group can be * ordered.

Again we interpret * as "computable". Can every computable torsion-free abelian group be computably ordered? In its simplest form the answer to this question is "no", as shown by Downey and Kurtz [135]. This negative answer to this question can be easily derived as a consequence of the general results and methods developed in Section 5.1 for the class of torsion-free abelian groups (see Exercise 5.1.49).

The reader will note that such questions can also have other interpretations. Classically, we regard objects as being the same if they are isomorphic, but for computable structures, an isomorphism type might have many computable *presentations* (*computable copies*) of the same structure.

If a certain construction or property is not algorithmic for an effective version of a field or a group, is there *another computable copy* where the construction or property becomes algorithmic? We arrive at the following variation:

IV** Any orderable * group is (classically) isomorphic to an * ordered group.

In other words, taking * to be the property of being computable, if G is a computable group which is orderable (but not necessarily computably orderable), is G (classically, not computably) isomorphic to a computable group with a computable ordering?

Darbinyan [107] has used an elaborate construction to show that the answer is "no" in general. But in contrast with (IV*), the answer to (IV**) is "yes" if G is abelian (Exercise 5.1.55). The abelian case follows easily from a seemingly unrelated Dobrica's Theorem 5.1.37 about computable torsion-free abelian groups with computable bases. Quite interestingly, perhaps the best way to apply the result of Dobrica to (IV**) is to use the natural order in the uncountable field of real numbers.

The reader might note that we have only considered *countable* structures in our treatment of (I)-(VI) above. However, uncountable objects (such as Π^0_1 classes and the field of reals) naturally occur when we study effective presentations of countable structures. Also, Galois correspondence characterises effective algebraic field extensions in terms of "recursive" profinite groups which will be discussed in §4.2.6. Thus even if we are only interested in countable algebra, we also invariably have to deal with algorithmic processes in uncountable spaces. Additionally we might also be interested in studying uncountable structures on their own. For example, can we ask similar questions about, e.g., separable Banach spaces and Polish groups? Natural questions of this type arise in analysis. For example, given a "computable" continuous function on $[0, 1]$ which is bounded below, can we compute a minimum point? Can we compute its Riemann integral? As with $[0, 1]$, for a wide class of topological spaces there are natural countable descriptions of them given by a computable dense subset. Generally speaking, most (but not all) natural separable spaces coming from the classical literature have computable presentations in this sense.

Most of the connections made between the algorithmic questions in the countable and uncountable realms that we discussed earlier were primarily technical in nature. Uncountable objects may appear as technical tools inside effective algebraic proofs, e.g., Π^0_1 classes help to effectively linearly order a countable formally real field. Conversely, there is also a large body of literature that applies the ideas and methods in computable countable algebra to uncountable spaces. For sample results see [369, 360, 268]. These results should also be viewed as technical generalisations from the countable to the separable uncountable case.

As we have stated earlier, in the first part of the book we develop a unified theory that establishes a *direct connection* between algorithmic procedures in countable algebra and topology through *effective dualities*. These connections usually follow the principle: a countable structure in a certain class is algorithmically presented if and only if its uncountable "dual" is effectively presented. Furthermore, we will see that statements similar to the examples presented above sometimes remain true for weaker notions of algorithmic presentability, such as c.e. presentability and even computability relative to the halting problem (which is the canonical example of an undecidable problem). Examples of such properties include the effective versions of Stone and Pontryagin dualities established in the first part of the book.

Precisely calibrating the level of algorithmic effectiveness required for a classical fact to work *computably* often necessitates the development of relatively sophisticated machinery, which may not be directly related to the classical result at hand. Further, questions that arise in such studies may have no direct analogues in the respective fields of topology or algebra. In the second part of the book, we will see that such investigations lead to methods for addressing classical (non-algorithmic) mathematical questions using algorithmic means. However, in the first part of the book, we will focus on a thorough examination of the computable presentability of structures and spaces, as well as the associated machinery. We will defer the applications of these methods to classification problems until Part 2.

1.2 Computable structures and spaces

In this section we give several formal definitions that will be used throughout the book, and then we state Theorem A and Theorem B that are central to Part 1.

1.2.1 Computable countable structures

We assume that the reader is familiar with the notion of an algebraic structure. We adopt the Turing-Church Thesis, which states that a process is algorithmically computable if and only if, under some appropriate coding, it becomes computable by a Turing machine. Our algorithms are *abstract* in the sense that we do not impose any time or resource bounds on them. In other words, if a procedure is intuitively computable, then it can be taken to be formally computable.

A set X of natural numbers is *computably enumerable* (c.e.) if its elements can be listed by an algorithm, but the order in which its elements appear could be unpredictable or unnatural. A set is *computable* if we can algorithmically decide membership $x \in X$. It is clear that any finite set is computable, and any computable set is c.e., but the converse fails in general. For example, the *halting problem* is the standard example of a c.e. set that is not computable (Proposition 2.1.5). The halting problem encodes the set of all programs (for the universal Turing machine) that eventually halt their computation. (All these notions will be formally introduced and clarified in Section 2.1.) No further background in computability theory is assumed in this section.

Computable and c.e. presentations

Rabin [440] and, independently, Mal'cev [345, 346] proposed the following notion of computability for a countable algebraic structure. For simplicity, we restrict ourselves to structures with only finitely many operations and relations.

Definition 1.2.1 (Rabin, Mal'cev). An algebraic structure A is *computably presentable (or constructivisable)* if there exists an algebraic structure B isomorphic to A whose elements form a computable set of natural numbers, and the operations and relations on B are computable (as functions and relations on the natural numbers).

Such an isomorphic copy B upon natural numbers is usually called a *computable presentation*, a *computable (isomorphic) copy*, or a *constructivisation* of A. The notion above is now widely accepted as the standard "base" notion of computable presentability for countable algebraic structures. There are notions stronger than the definition given above, e.g., primitive recursive, decidable, polynomial-time, punctual, and automatic presentations. However, in many situations an algebraic structure satisfies the following weaker condition.

Definition 1.2.2. An algebraic structure A is *computably enumerably (c.e.) presented* (Σ^0_1-presented, positively presented) if it is isomorphic to the factor of a computable structure by a c.e. congruence.

A congruence is an equivalence relation that "respects" the operations of the structure. This means that for any operation f, whenever $x_i \cong y_i$ we have that $f(x_0, \ldots, x_n) \cong f(y_0, \ldots, y_n)$. For now, assume our structure is an algebra with no relations on it. We postpone the case of relations until §3.2.1. A standard example of a c.e. presented structure is the factor of a computable group G by a computably enumerable normal subgroup H. Such presentations are quite common in group theory, where they are (somewhat confusingly) called "recursive" in the literature; more about this in Section 1.3.

Example 1.2.3. Let G be a group. A *finite presentation* of G is a tuple $\langle x_1, \ldots, x_n \mid y_1, \ldots, y_m \rangle$ such that

$$G \cong \langle x_1, \ldots, x_n \rangle / \langle y_1, \ldots, y_m \rangle,$$

where $\langle x_1, \ldots, x_n \rangle$ is the free group F generated by $x_1, \ldots, x_n$, and $\langle y_1, \ldots, y_m \rangle$ is the normal subgroup of F generated by the "relations"

$$y_1, \ldots, y_m \in \langle x_1, \ldots, x_n \rangle$$

upon the "generators" $x_1, \ldots, x_n$. A finite presentation is evidently also a c.e. presentation of the group. Famously, the "word problem" (i.e., the congruence modulo $\langle y_1, \ldots, y_m \rangle$) might be undecidable in a finitely presented group ([51, 416]). It is not hard to see that such a G with undecidable word problem cannot have a computable presentation; for a detailed explanation see §2.2.1 (this is Theorem 2.2.10).

Thus, the two main notions of effective presentability defined above (being c.e. presented and having a computable presentation) already differ up to isomorphism for finitely presented groups. However, in some broad classes, every c.e. presented structure admits a computable presentation, but this typically requires a non-trivial proof. There has been considerable research in this direction for finitely generated groups; for instance, refer to the somewhat dated but excellent survey [390].

The book is dedicated to the study of structures that are *not* finitely generated and are often not even countable. Examples of computably and c.e. presented structures will be given in Section 2.2.

Relativisation and low presentations

Suppose that we have a structure that has no computable presentation, such as a c.e. presented structure. It still makes sense to try to understand what is the best way to present such structures, possibly through effective approximations or perhaps using additional computational power. In §3.1.1, we will show that these two ideas typically lead to equivalent definitions.

A common approach to this is through relativisation. Using oracle Turing machines, we can *relativise* Definition 1.2.1 to an oracle X, where X is a non-computable set written on the extra oracle tape of the machine computing a presentation of the structure. For example, since every c.e. set is computable relative to the aforementioned halting problem, a c.e. presented structure is computably presentable *relative to the halting problem.*

Perhaps the most interesting case is when X is, in some sense, close to being computable. We define X to be *low* if the halting problem for machines with oracle X is computationally no harder than the halting problem for machines without an oracle (see Section 2.1.6 for details). It is well known that there are low sets that are not computable. Nevertheless, low sets are usually viewed as being close to computable. Thus, when X is low, the definition below can be viewed as being somewhat effective.

Definition 1.2.4. An algebraic structure A is X-computable if the domain, the operations, and the relations on A are X-computable, i.e., computable by a Turing machine with oracle X.

Richter [451] was perhaps the first to systematically investigate X-computable structures. We will look at some of Richter's results in §3.2.3. In some instances, having a low presentation implies that a structure also has a computable presentation. In other circumstances, such as graphs or abelian groups, this fails to be true. Results asserting that low presentability implies computable presentability are very rare. We mention that recently, Marker and Miller [349] showed that every low differentially closed field has a computable presentation. Another well-known result will soon be stated.

Separating the notions: Theorem A

Linear orderings, Boolean algebras, and abelian groups are important players in this book. Not only are they excellent vehicles for displaying techniques, but understanding them also yields important results about Polish spaces. To illustrate the interplay of lowness, c.e. presentability, and computability, we will give detailed proofs of the following.

Theorem A.

1. There is a c.e. presented Boolean algebra not isomorphic to any computable Boolean algebra.
2. Every low Boolean algebra has a computable presentation.
3. Every c.e. presented torsion-free abelian group is isomorphic to a computable group.

We will define all terms used in Theorem A in due course. The first fact is due to Feiner [162, 161], the second to Downey and Jockusch [129], and the third result was established by Khisamiev [288]. It takes one more observation on top of Theorem A to fully separate low, c.e. and computable presentability; this will be explained in Section 2.2.1.

Of course, our exposition of the theory of computable structures will not be limited to these well-known results, but the three facts stated above should be viewed as being central to the first few chapters of the book, which are naturally restricted to countable structures. However, these fundamental results about Boolean algebras and torsion-free abelian groups will find applications in later chapters when we prove analogous results about Polish spaces.

1.2.2 Computable separable spaces

What happens when a structure is not countable? In his famous papers [487, 488], Turing introduced the concept of a step-by-step computation for real numbers. Paraphrasing Turing, we have the following notion.

Definition 1.2.5. A real ξ is:

- *right-c.e.* if $\{r \in \mathbb{Q} : \xi < r\}$ is computably enumerable (c.e.);
- *left-c.e.* if $\{r \in \mathbb{Q} : \xi > r\}$ is c.e.;
- *computable* if it is both left-c.e. and right-c.e.

Equivalently, a real ξ is computable if there is a computable procedure that, on input n, outputs a rational number r (represented as a fraction) such that $|\xi - r| < 2^{-n}$. This notion of effectiveness essentially relies on the density of the rationals in $\mathbb{R}$. In other words, $\mathbb{R}$ is "computable" in the sense that it is the completion of a computable countable set, the rationals. Various authors extended this idea beyond the reals to cover classical separable Banach spaces and, more generally, complete separable metric spaces (also called Polish spaces). For instance, one of the fundamental notions in the theory of computable structures considered in this book is the following definition. This definition was essentially due to Ceitin [82] and independently to Moschovakis [405].

Definition 1.2.6. A *presentation* of a Polish space M is given by a sequence $(x_i)_{i\in\omega}$ and a complete metric d such that $(x_i)_{i\in\omega}$ is dense in (M, d). A presentation $(x_i)_{i\in\omega}$ is:

- *right-c.e.* if $\{r \in \mathbb{Q} : d(x_i, x_j) < r\}$ are c.e. uniformly in i, j;
- *left-c.e.* if $\{r \in \mathbb{Q} : d(x_i, x_j) > r\}$ are c.e. uniformly in i, j;
- *computable* if it is both left-c.e. and right-c.e.

The points x_i are usually called special, rational, or ideal points.

In the definition above, "uniformly in i, j" means that there is an algorithm that, when given i, j as inputs, produces a listing of the corresponding set of rationals represented as fractions. In the present book, we are mostly focused on compact spaces and the metric spaces associated with Banach spaces, so the metric d is always assumed to be complete. However, Definition 1.2.6 depends on the notion of isomorphism that we use. For instance:

- If we view spaces up to isometry, then the metric d in Definition 1.2.6 is *fixed.*

- If we view spaces up to homeomorphism, then the metric d in Definition 1.2.6 is merely *compatible* with the topology of the space.

We could also view spaces up to quasi-isometry, homotopy, linear isometry for Banach spaces, topological isomorphism for Polish groups, and so on. We say that a Polish space is *computable* (left-c.e., right-c.e.) if it has a computable (respectively left-c.e., right-c.e.) presentation, and we usually emphasise whether we use homeomorphism or isometry as the notion of equivalence. When spaces are viewed up to isometry, we may refer to computable Polish presentations of the space as "computable structures" or "computability structures" *on* the space. This terminology, coined by Pour-El and Richards [435], will be convenient in the last chapter of the book.

Natural examples of computable Polish spaces come from functional analysis and topological algebra: see Section 2.4 for many examples. Similar to the situation in computable countable algebra, in Polish groups and separable Banach spaces we usually additionally require the standard operations to be computable in some sense. The difficulty is that the input of such an operation is an infinite sequence rather than a single natural number. There are several possible ways to address this issue; some approaches turn out to be equivalent, but some are not. We will give a more in-depth analysis of the assorted approaches to defining computable continuous real-valued functions in §2.3, and will provide a general definition for Polish spaces in §2.4.1. For the purpose of the present section, we restrict ourselves to Polish spaces with no additional operators.

Computable compactness

In the first half of the book, we will be mainly concerned with compact Polish spaces and groups. Compactness plays a central role in classical analysis and topology. In the words of Hewitt [249],

> "A great many propositions of analysis are:
>
> - trivial for finite sets;
> - true and reasonably simple for infinite compact sets;
> - either false or extremely difficult to prove for noncompact sets."

Since we are concerned with *algorithmic* problems on effective structures, it is often desirable to use an algorithmic form of compactness. The following notion was first explicitly defined by Mori, Tsujii, and Yasugi [403], to the best of our knowledge.

Definition 1.2.7. A computable Polish space M is called *computably compact* if there exists a computable function that, given n, outputs a finite cover of M by open balls of radii $< 2^{-n}$ that cover M.

The definition and the many equivalent forms of it will be further elaborated in the relevant section. Of course, the definition is naturally restricted to compact Polish spaces. The notion can be extended to locally compact spaces and to spaces that are not necessarily computable Polish, but we will not include this in our discussion. Computably compact presentations are very common in the modern literature; we cite the two recent surveys [270, 139]. The notion is furthermore exceptionally robust; we will prove the equivalence of the definition above to several other definitions.

Separating the notions: Theorem B

The reader undoubtedly recognises that Definitions 1.2.6 and 1.2.7, along with their relativisations, give rise to various notions of effective presentability for Polish spaces—some appearing more natural, while others seemingly more artificial. For example: right-c.e. effectively compact spaces, low Polish spaces, and so on. It seems to us that the three most natural and commonly occurring presentations of (compact) Polish spaces are:

$$\begin{array}{c}\text{computably compact}\\ \downarrow\\ \text{computable Polish}\\ \downarrow\\ \text{right-c.e. Polish}\end{array}$$

with the obvious implications denoted by the arrows above. In computable algebra, theorems separating and comparing the various notions of presentability such as c.e. presentable, computably presentable, and low structures are at least half a century old. Results of this kind are typically viewed as fundamental results of the theory. Thus, the following question arises naturally:

Can we separate by counter-examples the three notions of computable presentability for compact Polish spaces defined above?

If we view spaces up to isometry, the question above is easy; see Exercise 2.4.30 for a hint. However, if we view spaces *up to homeomorphism*, the situation becomes much more complex, especially if we want our examples to come from a reasonably natural class of compact spaces. This is where computable algebraic techniques and results (and specifically Theorem A) will become very useful. Indeed, every item of Theorem A will be used to prove some item in the *second main result* of the first part of the book that separates the three main notions for spaces:

Theorem B.

1. There exists a right-c.e. Stone space not homeomorphic to any computable Polish space.
2. Every computable Stone space is homeomorphic to a computably compact space.
3. There exists a connected compact computable Polish space not homeomorphic to any computably compact space.

In contrast with Theorem A, Theorem B is very recent: 1. has been proven in [35], 2. in [245], and 3. in [341]. A variety of techniques are necessary to give a complete and detailed proof of Theorem B. The central tools will be several effective dualities, which we state in the next subsection.

1.2.3 Effective dualities

Theorem B will be derived from Theorem A using several dualities that we state below, but we delay the definitions until the relevant sections.

1. A Boolean algebra B has a computable presentation iff its Stone space $\widehat{B}$ has a computable Polish presentation (Theorems 4.2.79 and 4.2.80).
2. Same as above, but for computable B and computably compact $\widehat{B}$ (Theorems 4.2.78 and 4.2.80).
3. A Boolean algebra B has a c.e. presentation iff $\widehat{B}$ has a right-c.e. Polish presentation (Theorem 4.2.81).
4. A torsion-free abelian group G has a computable presentation iff the connected compact domain of its Pontryagin dual $\widehat{G}$ has a computably compact presentation (Theorem 5.2.1).
5. A torsion-free abelian q-divisible group G has a Δ^0_2-presentation iff $\widehat{G}$ has a computable Polish presentation (Theorem 5.2.25).

In some sense, these results are perhaps more interesting (or "useful") than Theorem B itself. These dualities provide a direct link between computable algebra and computable topology. The effective dualities will be employed in the second part of the book to investigate classification problems in natural classes of structures and spaces.

To establish these (and a few additional) effective dualities, we need first to build the foundational machinery. This spans several classical topics, including priority techniques, methods of computable linear orders and computable Boolean algebras, the calculus of computably compact spaces, basics of computable abelian group theory, and even elements of algebraic topology. While we will not delve too deeply into these topics, we will provide references to the relevant literature.

Additionally, we will incorporate a number of classical results that, although not directly connected to Theorems A and B, fit well within the broader narrative of the book.

Part I consists of four chapters (excluding the introduction). The algebraic complexity of the chapters increases "monotonically in their index". We begin Part I with a discussion of computably enumerable sets which can be viewed as structures in the empty signature, and we end Part I with theorems about topological groups.

1.3 Conflicting terminology*

*Recall that throughout the book, sections and subsections marked with * can be safely skipped.*

There are several traditions in computable mathematics, to name a few: the Turing-Markov style of computable analysis [1], Mal'cev's school in Russia and Kazakhstan [159], the Weihrauch style of computable analysis in Germany [505], the Ash-Knight tradition in computable algebra in the USA and Australia [20, 401, 402], and computable Banach space theory in the style of Pour-El and Richards [435]. We also mention the closely related topics of effective descriptive set theory [406], classical combinatorial group theory [343, 390]. Because of the Cold War, some of these traditions had been largely isolated from each other for many decades. As a result, each of these traditions developed quite distinctive terminology and notation. In this short section, we will discuss some of the related terminology from the literature.

The versions of the fundamental definitions developed by different traditions are either equivalent or very closely related. On the other hand, the exact same term can also correspond to two (or more) non-equivalent notions, and this can be highly confusing.

For example, "recursively presented groups" in combinatorial group theory are called "c.e. presented groups" in this book[2]. A "recursively presented group with solvable word problem" is just a "computable group" in our terms. Unfortunately, "recursive" and "computable" are often used synonymously in the literature, so the term "recursive group" could potentially mean two different things. To make matters worse, the term "recursive group" can mean something entirely different in the context of profinite groups; e.g., [473]. In view of all these, we shall avoid the use of "recursive" throughout the book whenever possible.

Fröhlich and Shepherdson [185] used the term "explicit presentation," which can be traced back to van der Waerden. Rabin [440] abandoned this old-fashioned terminology and coined the notion of a computable algebraic structure, which we adopt. Some sources also refer to computable structures as computable (or recursive) models, typically in the context of the effective content of model theory (e.g., [229, 388]). In this approach, operations on the structure are often replaced with relations coding their graphs. It is then required that the atomic diagram (the collection of quantifier-free formulae with parameters from the computable domain of the structure) forms a computable set. This definition is, of course, equivalent to our definition of a computable structure, up to a change of notation.

There has been a longstanding tradition in the former USSR to use explicit *numberings* of computable structures. This approach was suggested by Mal'cev [345]. For example, a *constructivisation* ν of a countable group $(G, \cdot)$ is a surjective map (a "numbering") $\nu : \omega \to G$ such that $\{(m, n) : \nu(m) = \nu(n)\}$ is a computable set and $\nu(m) \cdot \nu(n) = \nu f(m, n)$ for some computable $f : \omega^2 \to \omega$. If we require that $\{(m, n) : \nu(m) = \nu(n)\}$ is merely computably enumerable, then we have a "positive" numbering of G. It is not hard to see that a group is constructivisable if and only if it is computably presentable. Similarly, it admits a positive numbering if and only if it is c.e. presented. The term "constructivisation" is quite illustrative and appealing, but for the sake of standardising terminology, we shall also avoid this term in the book.

Similarly, the German school of computable analysis [505] also makes codings explicit, calling them *representations*. These are partial functions from Baire space to the objects being represented. In our case, we will be dealing with separable spaces, and hence such representations will simply be (fast) Cauchy sequences. Adding such explicit representations would only obfuscate matters for the issues we are interested in.

In contrast with the above two frameworks, the modern Ash-Knight style computable algebra uses a much more simplified notation. There, elements of a computable presentation (Definition 1.2.1) are natural numbers rather than names of elements in some "ideal" structure. A similar approach, but in the context of separable metric spaces, is usually taken in effective descriptive set theory (see e.g., p. 96 of [406]). There, a *recursive presentation* of a Polish space is a dense subset $(x_i)_{i \in \omega}$ of the space so that the relations

$$P(i, j, m, n) \text{ if and only if } d(x_i, x_j) < \frac{m}{n+1};$$

$$Q(i, j, m, n) \text{ if and only if } d(x_i, x_j) \leqslant \frac{m}{n+1}$$

are computable relations. Our definition of a computable Polish space is equivalent to saying that P and $\neg Q$ are merely computably enumerable relations. It is easy to construct a discrete computable

[2] Authors working in combinatorial group theory referred to these as recursive groups because they had a presentation of the form $\langle x_1, \ldots, x_n, \ldots | r_1, \ldots, r_m, \ldots \rangle$. They noted that by using a padding trick, the c.e. sets of relations $\{r_i \mid i \in \mathbb{N}\}$ could be replaced by a computable set of relations. For example, replace r_i with $x_1^{-s} x_1^s r_1$, if r_i is enumerated at a stage s.

Polish space (in the sense of Definition 1.2.6) that is not isometric to any recursive Polish space; see Exercise 2.4.31. In this book, we will be mainly looking at spaces up to homeomorphism, and it is not hard to show that every computable Polish space is computably homeomorphic to a recursive one; this is Exercise 2.4.33. As we noted above, we shall avoid the use of "recursive" throughout the book, justified by Exercise 2.4.33.

Chapter 2

Foundations

In this chapter we lay the theoretical foundation of the theory and support it with a number of examples and counter-examples. All results in this chapter are related to the fundamental problem of computable presentability of a structure, a function, or a space. Most results discussed in this chapter are historic and are not directly related to Theorems A and B (or to their proofs). The main result of the chapter is as follows:

Theorem (Melnikov [369]). There is a computable presentation of the space $(C[0,1], d_{\sup})$ in which the norm is computable, but the operation $+$ is not.

The theorem will re-appear as Theorem 2.4.20. The theorem and its consequences will be used in the final Chapter 10 of the book. The plan for this chapter is as follows:

1. Section 2.1 contains the background in computability theory sufficient for understanding all proofs later in this chapter.
2. Section 2.2 contains classical examples of computable and c.e. presented structures and historical "naive" computable constructions in algebra and model theory.
3. Section 2.3 compares various classical definitions of computability for functions $\mathbb{R} \to \mathbb{R}$. The old theorems and notions discussed in Section 2.3 lie in the foundations of computable Polish and Banach space theory, even though in this section we restrict ourselves to the real line.
4. In Section 2.4 we introduce the notions of a computable Banach space and a computable Polish group, and we observe that these notions are equivalent for real Banach spaces. We also compare the notion(s) to the definition of a computable Polish space, and give several relatively non-trivial examples and counter-examples, including Theorem 2.4.20.

Deeper results will require more advanced techniques which will be developed in later chapters.

R. G. Downey, A. Melnikov, *Computable Structure Theory*, Theory and Applications of Computability, https://doi.org/10.1007/978-3-031-92433-0_2

2.1 Elementary computability theory

We use $\mathbb{N}$ and ω interchangeably to denote the set of non-negative integers. Notations such as 2^ω for Cantor space, the collection of all infinite binary sequences, are standard in the literature.

2.1.1 Codings and computable functions

Our initial concern is with functions of the form $A \to B$ where $A, B \subseteq \mathbb{N}$; i.e. *partial* functions on $\mathbb{N}$. If $A = \mathbb{N}$ then the function is called *total.* Looking only at $\mathbb{N}$ may seem rather restrictive. Later we will be concerned with functions which take subsets of the rationals or subsets of 2^ω as their domains. From the point of view of classical computability theory, where resources and efficiency don't matter, the definitions naturally extend to such objects by coding. For example, if we consider the rationals $\mathbb{Q}$, these can be considered as coded in $\mathbb{N}$ as follows:

> Let $r \in \mathbb{Q} - \{0\}$; write $r = (-1)^\delta \left(\frac{p}{q}\right)$ with $p, q \in \mathbb{N}$ in lowest terms and $\delta = 0$ or 1. Then define the Gödel number of r, $\#(r)$, as $2^\delta 3^p 5^q$, with the Gödel number of $r = 0$ to be 0.

Then by the fundamental theorem of arithmetic, $\#$ describes an injection from $\mathbb{Q}$ into $\mathbb{N}$ and furthermore given $n \in \mathbb{N}$ we can decide exactly which $r \in \mathbb{Q}$, if any, has $\#(r) = n$. Similarly if σ is a finite binary string, say $\sigma = a_1, a_2, \ldots a_n$, then we can define

$$\#(\sigma) = 2^{a_1+1}3^{a_2+1}\ldots(p_n)^{a_n+1},$$

where p_n denotes the n-th prime. There are a myriad of other codings possible. One could code the string σ by representing it as the number 1σ so that the string 01001 would correspond to 101001. The above procedures are called "effective coding" since they give an algorithm for the relevant injection. The actual coding and the way we represent objects does not matter too much until we look at structures arising from computable analysis, where more care is needed.

Another common coding we will use is to code pairs of natural numbers via a pairing function, which is any computable injection from $\mathbb{N}^2 \to \mathbb{N}$ with computable range. The *standard pairing function* is the function $\langle n, m\rangle = \frac{1}{2}(n+m)(n+m+1)+m$. This extends naturally to finite tuples of natural numbers: $\langle n, m, p\rangle = \langle\langle n, m\rangle, p\rangle$, and so on.

Convention 2.1.1. Henceforth, unless otherwise indicated, we will always regard the objects under discussion as being effectively coded in some fixed way.

As we have already stated earlier, we adopt:

> **Church-Turing Thesis.** The collection of algorithmic partial functions on the positive integers are exactly those that can be simulated by Turing machines.

An excellent discussion of the subtleties of the Church-Turing thesis can be found in Odifreddi [421]. A typical use of the Church-Turing thesis might be as follows. Suppose we can explain, in sufficient detail, a certain algorithmic process and convince ourselves that our procedure is intuitively mechanical. Then the thesis implies that with sufficient effort (perhaps also with enough motivation) one could design a set of instructions for a Turing machine that would be able to compute the process. This is similar to designing a pseudo-code in computer science, or even just explaining how one could design such a pseudo-code, and then claiming that it can actually be implemented in the required programming language if necessary. Even though this thesis might seem a bit too relaxed, it will allow us to compress most proofs tenfold if not more, thus allowing the proof to focus on the mathematical ideas.

The following fundamental property of partial computable functions will be essential throughout this book.

Property 2.1.2 (Enumeration Theorem: Existence of the Universal Turing Machine). *There is an algorithmic way of enumerating all the partial computable functions. That is, there is a partial computable function $f(e, x)$ of two variables such that*

$$f(e, x) = \varphi_e(x)$$

where $\varphi_e(x)$ denotes the e-th partial computable function on input x.

The point of Property 2.1.2 is that we can pretend that we have all the machines $\varphi_1, \varphi_2 \ldots$ in front of us; to compute 10 steps in the computation of the 3rd machine on input 20, we can pretend to walk to the 3rd machine, put 20 on the tape and run it for 10 steps; we write this as $\varphi_{3,10}(20)$.

Notation 2.1.3. Let φ_e be the e-th partial computable function. The standard notations for the result of s steps in the computation of φ_e on input x are $\varphi_{e,s}(x)$ and $\varphi_e(x)[s]$.

We write $\varphi_{e,s}(x) \downarrow$ or $\varphi_e(x)[s] \downarrow$ to indicate that the computation halts in *at most* s steps; otherwise, we write $\varphi_{e,s}(x) \uparrow$ or $\varphi_e(x)[s] \uparrow$. The exact choice of notation will depend on the context; however, we will usually avoid placing the stage as a superscript. More generally, we adopt the following standard

Notation 2.1.4. If A is an object that we compute, effectively enumerate, or computably approximate, then the result of s steps in this computation is typically denoted either A_s or $A[s]$.

Given any partial computable function f, there are infinitely many different algorithms to compute it. If φ_e is one such algorithm for computing f, we say that e is an *index* for f. We also write $f(x) \downarrow$ to denote that x is in the domain of f, and therefore the computation of any algorithm computing f halts on input x.

In many ways, Property 2.1.2 is the platform that makes undecidability proofs work since it allows us to carry out diagonalisation arguments within the class of partial computable functions. For instance we remind the reader of the following basic result.

Proposition 2.1.5 (Unsolvability of the halting problem). *There is no algorithm which given e, x decides if $\varphi_e(x) \downarrow$. That is, there is no algorithm that returns* 1 *if the e-th machine on input x halts, and returns* 0 *otherwise.*

The following result, called the *s-m-n Theorem*, is due to Kleene.

Theorem 2.1.6 (The s-m-n Theorem). *Suppose that $g(x, y)$ is a partial computable function of two variables. Then there is a total computable $s(x)$ such that for all x, y,*

$$\varphi_{s(x)}(y) = g(x, y).$$

We omit the proof. The above also holds if x and y are tuples $x_0, \ldots, x_m$ and $y_0, \ldots, y_n$, and there is also a version of the theorem that works for indices of Turing functionals (to be defined).

A basic notion in computability theory is m-reducibility. We say that a set A is many-one reducible, or simply m-reducible to another set B, written as $A \leqslant_m B$, if there is a computable function f such that for every x, $x \in A$ iff $f(x) \in B$. We write $A \equiv_m B$ iff $A \leqslant_m B$ and $B \leqslant_m A$. The relation $\leqslant_m$ is a pre-ordering and the equivalence classes are called *m-degrees.* The idea is that an instance of asking whether "$x \in A$" can be effectively transformed into a instance of the question "$f(x) \in B$", so B is at least as computationally complicated as A.

We let $K_0 = \{\langle e, x\rangle : \varphi_e(x) \downarrow\}$ denote the set representing the halting problem. A different, diagonal version of the set is $K = \{x : \varphi_x(x) \downarrow\}$. These two sets are the same up to m-degree as we now show.

Fact 2.1.7. $K \equiv_m K_0$.

Proof. Clearly $K \leqslant_m K_0$ since $x \in K$ iff $\langle x, x\rangle \in K_0$. To see $K_0 \leqslant_m K$, define a partial computable function g such that for all p, z,

$$g(p, z) = \begin{cases} 1 & \text{if } \varphi_x(y) \downarrow \text{ and } p = \langle x, y\rangle, \\ \uparrow & \text{otherwise.} \end{cases}$$

Via the s-m-n Theorem, there is a total computable $s(p)$ such that for all p, z,

$$\varphi_{s(p)}(z) = g(p, z).$$

Then $\langle x, y\rangle \in K_0$ iff $s(\langle x, y\rangle) \in K$. □

2.1.2 Computable and computably enumerable sets

For the next definition which already appeared in the introduction, recall that the characteristic function χ_A of a set $A \subseteq \omega$ is the total function so that $\chi_A(x) = 1$ if $x \in A$, and $\chi_A(x) = 0$ otherwise.

Definition 2.1.8. A set $A \subseteq \omega$ is called

(i) *computably enumerable* (c.e) if $A = \text{dom } \varphi_e$ for some e, and

(ii) *computable* if the characteristic function of A is computable.

We will let W_e denote the e-th computably enumerable set. That is, we let $W_e = \text{dom } \varphi_e$ and let $W_{e,s} = \{x \leqslant s \mid \varphi_{e,s}(x) \downarrow\}$ constitute s steps in the enumeration of W_e. The name *computably enumerable* comes from the notion of being 'effectively countable' via the following characterisation. The proof is a straightforward exercise.

Proposition 2.1.9. *An infinite set A is computably enumerable iff there is a (total) computable injective function f with $f(\mathbb{N}) = A$.*

Thus we can think of a computably enumerable set as an infinite listing of its elements that can be produced by an effective procedure. Notice that the listing of a c.e. set is *not* required to be in increasing order. Proposition 2.1.5 tells us that $K \equiv_m K_0$ are c.e. sets which are not computable. Another classical result is:

Theorem 2.1.10 (Post [431]). *A set A is computable iff both A and its complement are computably enumerable.*

Proof. To decide whether a given number $x \in A$, simultaneously list both A and its complement and wait to see where x appears first. Clearly, x must be listed in A or its complement after finitely many steps. Here we appeal to the Church-Turing thesis to see that this decision procedure can be computed by a Turing machine and A is therefore computable. Without it the proof would be unnecessarily tedious. □

For instance, it follows that the complement of the halting problem, $\overline{K}$, is not c.e..

Suppose we are interested in some property of partial functions, such as being total, which is independent of their implementation. The collection of indices of all partial functions having this property forms an *index set*, as defined below.

Definition 2.1.11. An *index set* is a set $A \subseteq \mathbb{N}$ such that if $x \in A$ and $\varphi_x = \varphi_y$, then $y \in A$.

For example, the collection of all indices of total computable functions

$$Tot = \{e : \varphi_e \text{ is total}\}$$

is an index set, whereas the halting problem $K = \{e : \varphi_e(e) \downarrow\}$ is not, as we demonstrate in the next subsection. The s-m-n Theorem 2.1.6 can be used to prove the following classical result.

Theorem 2.1.12 (Rice's Theorem [448]). *An index set A is computable iff $A = \mathbb{N}$ or $A = \emptyset$. Furthermore if $A \neq \mathbb{N}$ and $A \neq \emptyset$ then either $K \leqslant_m A$ or $K \leqslant_m \overline{A}$.*

Proof. The proof of Rice's Theorem is very similar to the proof that $K_0 \leqslant K$. Let $A \neq \mathbb{N}$, $\emptyset$ be an index set and without loss of generality, assume that $e \in \overline{A}$ where dom $\varphi_e = \emptyset$. (If $e \in A$ instead, we swap the roles of A and $\overline{A}$).

Since $A \neq \emptyset$, fix some $z \in A$. Then for some q, $\varphi_z(q) \downarrow$. By the *s-m-n* Theorem, there is a computable $s(x)$ such that, for all $y \in \mathbb{N}$,

$$\varphi_{s(x)}(y) = \begin{cases} \varphi_z(y) & \text{if } \varphi_x(x) \downarrow \\ \uparrow & \text{if } \varphi_x(x) \uparrow \end{cases} .$$

Then $\varphi_x(x)\downarrow$ implies $\varphi_{s(x)} = \varphi_z$ and so $s(x) \in A$, and $\varphi_x(x)\uparrow$ implies $\varphi_{s(x)} = \varphi_e$ and so $s(x) \notin A$. Thus $K \leqslant_m A$. □

Of course many decision problems are not coded by index sets and so can have decidable solutions. Rice's Theorem says that any nontrivial problem which is given purely in terms of machine descriptions cannot be decidable. Index sets will play a central role in the second half of the book.

2.1.3 Kleene's Recursion Theorem

A fundamental result in classical computability is the Recursion Theorem. Its proof uses the s-m-n Theorem 2.1.6 and allows us to use the index of the set we are building in a construction, as part of the construction of that very same set. This apparent circularity is what causes the Recursion Theorem to appear counter-intuitive, but does not actually cause a problem because our functions can be partial (computable) in general.

Theorem 2.1.13 (Recursion Theorem; Kleene [298]). *Suppose that f is a computable function. Then we can compute (from an index of f) a number n (called a* fixed point *of f) such that*

$$\varphi_n = \varphi_{f(n)} \text{ and hence, } W_n = W_{f(n)}.$$

Proof. First define d via the s-m-n Theorem 2.1.6 as $\varphi_{d(u)}(z)$ equals $\varphi_{\varphi_u(u)}(z)$ if $\varphi_u(u) \downarrow$ and $\varphi_{d(u)}(z)$ is otherwise undefined. By the s-m-n Theorem, d is total. Given an index for f, find an index v such that

$$\varphi_v = f \circ d,$$

noting that φ_v is total as well. Now letting $n = d(v)$, the following calculation

$$\varphi_n = \varphi_{d(v)} = \varphi_{\varphi_v(v)} = \varphi_{fd(v)} = \varphi_{f(n)}$$

shows that n is a fixed point for f. □

There are many variations on the Recursion Theorem. For example, if $f(x, y)$ is computable, then there is a computable function $n(y)$ such that, for all y,

$$\varphi_{n(y)} = \varphi_{f(n(y),y)}.$$

This result is usually called the *Recursion Theorem with Parameters* (Exercise 2.1.16). We now give a very simple application of the Recursion Theorem.

Example 2.1.14. We show that $K = \{e : \varphi_e(e) \downarrow\}$ is *not* an index set. (This fact also follows from Rice's Theorem 2.1.12.) By the s-m-n Theorem 2.1.6, let f be a computable function such that $\varphi_{f(n)}(n) \downarrow$ and $\varphi_{f(n)}(z)\uparrow$ for all $z \neq n$. Suppose that K is an index set. By the Recursion Theorem 2.1.13, let n be a fixed point for f, so that $\varphi_n = \varphi_{f(n)}$. Choose $m \neq n$, another index for φ_n. Then $\varphi_n(n) \downarrow$ and hence $n \in K$, and yet $\varphi_m(m) \uparrow$ and $m \notin K$, a contradiction. Note that this example also shows that there is a partial computable function φ_n that halts only upon its own index.

Exercises

Exercise° 2.1.15. Show that if g is a computable function there exists an index n such that $W_n = \{0, \ldots, g(n)\}$.

Exercise° 2.1.16 (Recursion Theorem with Parameters; Kleene [298]). Suppose $f(x, y)$ is a computable function, then there is a computable function $n(y)$ such that, for all y,

$$\varphi_{n(y)} = \varphi_{f(n(y),y)}.$$

Exercise 2.1.17 (Double Recursion Theorem; Muchnik [410], Smullyan [475]). If f and g are computable functions of two variables, there exist a, b such that $\varphi_a = \varphi_{f(a,b)}$ and $\varphi_b = \varphi_{g(a,b)}$.

2.1.4 Oracle computability and Turing degrees

As we have seen, the key idea used in the proof of Rice's Theorem 2.1.12 is that of *reducibility.* Recall that the reduction used in Rice's Theorem 2.1.12 is called an *m-reduction,* which is the simplest reduction. More specifically, $A \leqslant_m B$ if there is a computable function f such that $x \in A$ iff $f(x) \in B$. If f is 1-1, then we might emphasise the special nature of the m-reduction by writing $A \leqslant_1 B$. This is called a 1-reduction. Of course, there is no reason why we should restrict ourselves and ask the "oracle" B only one question in order to decide if $x \in A$. What about finitely many queries, perhaps bounded in some computable way dependent on the input x? This idea of more a general oracle access was introduced in another classic paper of Turing [490]. In other words, we can regard one problem B as being at least as hard as another problem A, by attaching to our computer an infinite read-only memory tape that contains B. Following Turing [490], we formalise this idea as follows.

We can extend the notion of a Turing machine to one with an extra read-only tape with infinitely many cells. We call such Turing machines *oracle Turing machines.* We can regard a normal machine as an oracle machine which never reads the extra oracle tape. The extra oracle tape can contain an infinite sequence; this sequence can be identified with (the characteristic function of) a set B. That is, B can be viewed as a function with range $\{0, 1\}$, and $B(x) = 1$ iff $x \in B$. Of course, each such function can also be viewed as an infinite string of 0-s and 1-s, with the n-th position being 1 iff $B(n) = 1$. In the same way that we could enumerate all partial computable functions, we also get:

Proposition 2.1.18 (Turing [490]). *There is a uniformly computable enumeration of all oracle Turing machines* $\{\Phi_e : e \in \mathbb{N}\}$.

We may identify $\Phi_e^{\varnothing}$ with φ_e.

Notation 2.1.19. We write $\Phi_e^B(n)$ or $\Phi_e(B; n)$ to denote the computation of the machine Φ_e with B stored in the oracle tape and performed with input n.

Then $\Phi_{e,s}^B(n)$ or $\Phi_{e,s}(B; n)$ denote the result of this computation after s steps (if there is any). The other standard, related, but not always equivalent notations in the literature are $\Phi_e(B; n)[s]$ and $\Phi_e^B(n)[s]$, but these notations we will usually avoid. For example, if B is computably enumerable and B_s is the part of B listed by stage s, then $\Phi_e(B; n)[s]$ should likely be interpreted as $\Phi_{e,s}(B_s; n)$ rather than $\Phi_{e,s}(B; n)$, unless otherwise specified.

Since we identify a set A with its characteristic function, an oracle machine Φ_e can be viewed as a *functional* that maps (partial) characteristic functions to (partial) characteristic functions:

$$B \mapsto \Phi_e^B,$$

where both sides are viewed as functions. Having this interpretation in mind, we write $\Phi_e(B)$ to mean $(\Phi_e^B(n))_{n\in\mathbb{N}}$. We will use this notation even when Φ_e is partial, in which case it maps partial functions to partial functions. We thus refer to oracle machines as *Turing operators* or *Turing functionals* to emphasise that we are interested in the fact that the operators act on sets (as opposed to acting on numbers).

Definition 2.1.20 (Turing Reduction [490]). We say that A is *Turing reducible* to B, written $A \leqslant_T B$, if $A = \Phi_e^B$ for some $e \in \mathbb{N}$. That is, we can compute A using an oracle Turing machine with an oracle for B.

We stress that there is no limit on the number of queries that can be used in a computation, only that it is finite. Thus, $\leqslant_m$ is a simple variation of Turing reduction as only one question is asked by an m-reduction on each input x. We also adopt:

The Relativised Church-Turing Thesis: The partial functions that are algorithmically computable *relative to* B are exactly those that are computable by a Turing machine that has B stored in its oracle tape.

We write $A \equiv_T B$ to mean that $A \leqslant_T B$ and $B \leqslant_T A$. This gives rise to an equivalence relation, and the equivalence classes under $\equiv_T$ are of the form $\deg(A) = \{B : B \equiv_T A\}$. The equivalence classes encode the notion of "equicomputability" and are called *Turing degrees*, *degrees of unsolvability*, or simply *degrees* throughout the rest of this chapter and the majority of the book (unless specifically stated otherwise).

We always use boldface letters for degrees. We let $\mathbf{0}$ denote the degree of the computable sets. If a degree contains a computably enumerable set, we will call it a *computably enumerable degree.* Note that not every set in a c.e. degree is computably enumerable; for instance we have $K \equiv_T \overline{K}$ but $\overline{K}$ is not c.e.. Additionally, we will often mix notation by writing, for example, $A \leqslant_T \mathbf{a}$, for a set A and a degree $\mathbf{a}$.

The use principle

Suppose $\Phi(A; x)\downarrow$. Let $u(\Phi(A; x))$ denote the use of this computation. Formally, $u(\Phi(A; x))$ at stage s is equal to

$$1 + max_{y\leqslant s}\{A(y) \text{ is used by } s \text{ stages of the computation } \Phi(A; x)\},$$

if this maximum exists, and 0 otherwise. The extra 1 is only to make the notation $A \upharpoonright u(\Phi(A; x))$ mean "the longest initial segment of A used in the computation of $\Phi(A; x)$", where $A \upharpoonright n = A \cap \{0, 1, \ldots, n-1\}$. If our notation gets a bit mixed up, this intended interpretation of the use should be prioritised over the formal details. Also, note that no parameter of the computation at stage s can be larger than s, so $max_{y\leqslant s}$ can be safely replaced with max_y in the definition of $u(\Phi(A; x))$. Also, there is a slight ambiguity in the interpretation of "s stages of the computation", i.e., of $\Phi(A; x)[s]$. If $A = \cup_s A_s$ is c.e. and is being listed, this will typically be interpreted as $\Phi_s(A_s; x)$. If A is written on the oracle tape all at once in advance, then s in A_s should be dropped. The intended interpretation will always be clear from the context. We identify sets with their characteristic functions, and we further identify characteristic functions with strings of 0-s and 1-s. For a finite string τ, $\Phi(\tau; n)\downarrow$ usually (implicitly) assumes that the use of the computation does not exceed the length of τ.

Lemma 2.1.21 (Use principle). *Suppose $\Phi(A; x)\downarrow$ and let $u = u(\Phi(A; x))$. Let B be any set such that $B \upharpoonright u = A \upharpoonright u$. Then $\Phi(A; x) = \Phi(B; x)$.*

Proof. Both A and B give the same answers to the oracle membership questions in the computations, hence the result must be the same. □

The use principle implies that Turing operators are continuous maps $2^\omega \to 2^\omega$. We will use this observation extensively throughout the book.

The jump operator

Many notions we have seen so far can be *relativised* to an oracle. In particular, if $A \leqslant_T B$ we say A is computable *relative to* B, or B*-computable*. A B-computably enumerable set is defined to be the domain of a partial B-computable function, and so on. Many notions and results can also be relativised to any set B by replacing "computable" with "B-computable" throughout. For instance, the following is the relativised version of the halting problem.

Definition 2.1.22 (Jump Operator). For any set A we define K^A to be the halting problem for machines with oracle A:

$$K^A = \{e : \Phi_e^A(e)\downarrow\}.$$

The set K^A is also denoted A' and is called *the jump of* A.

Since we can identify $\Phi_e^\varnothing$ with φ_e, we may assume that $\varnothing' = K$. Notice that K^A is c.e. relative to A. If the degree of A is $\mathbf{a}$ then we write $\mathbf{a}'$ for $\deg(A')$. Note that $\mathbf{a}'$ makes sense since $A \equiv_T B$ implies $A' \equiv_T B'$. In fact, $A \equiv_T B$ implies $A' \equiv_1 B'$; this is Exercise 2.1.26.

Notation 2.1.23. Noting the above equivalences, we will frequently identify $K^{(n-1)}$ with $\varnothing^{(n)}$, for any $n \geqslant 1$.

By relativising Proposition 2.1.5, we have $B <_T B'$ for all B. Consequently we can define a hierarchy of degrees

$$\mathbf{0}, \mathbf{0}', \mathbf{0}'', \dots, \mathbf{0}^{(n)}, \dots.$$

This hierarchy is does not collapse and is closely related to the arithmetic hierarchy that we discuss next. It can also be continued beyond ω iterations of the jump to form a transfinite hierarchy; we will discuss it in later chapters.

Note that it does not follow from the definition that $\mathbf{0} < \mathbf{a}$ implies $\mathbf{0}' < \mathbf{a}'$ (though of course $\mathbf{0} \leqslant \mathbf{a}$ does imply $\mathbf{0}' \leqslant \mathbf{a}'$). The following notion was already mentioned earlier; e.g., see Theorem A.

Definition 2.1.24. A degree $\mathbf{a}$ is *low* if $\mathbf{a}' = \mathbf{0}'$, and a set is low if its degree is low.

In the next chapter we will show that there exist low non-computable c.e. sets (Theorem 3.1.1). Low degrees are usually thought of as being somewhat close to $\mathbf{0}$; this can be made formal using the low_n hierarchy that we shall not define here.

Exercises

Exercise° 2.1.25. Prove that B is c.e. relative to A iff $B \leqslant_1 A'$.

Exercise° 2.1.26. Prove that if $A \leqslant_T B$ then $A' \leqslant_1 B'$.

2.1.5 The arithmetic hierarchy

In this chapter we need only the definition and a few basic facts about the arithmetical hierarchy. More about the hierarchy will be explained later. An n-ary relation is a set of tuples in $\mathbb{N}^n$. We define the classes Σ_n^0, Π_n^0, and Δ_n^0 as follows. A set B is Σ_n^0 if there is a computable relation $R(x_1, \ldots, x_n, x)$ with $x \in B$ iff

$$\underbrace{\exists x_1 \, \forall x_2 \, \exists \ldots Q_n x_n}_{n-1 \text{ alternations of quantifiers}} \qquad R(x_1, \ldots, x_n, x) \text{ holds,}$$

where Q_n is either $\forall$ or $\exists$ depending on whether n is odd or even.

A set B is Π_n^0 iff $\overline{B}$ is Σ_n^0. This means that a Π_n^0 set has a similar syntactical representation as above except we now have the leading quantifier $\forall$ followed by $n-1$ alternations of quantifiers. Note that we can always collapse two quantifiers of the form $\exists x_1 \, \exists x_2$ into a single $\exists x_3$ quantifier using the pairing function, which is why it makes sense to count only the number of *alternations* of quantifiers. Finally we say a set R is Δ_n^0 if it is both Σ_n^0 and Π_n^0.

Theorem 2.1.27 (Kleene [299]). *A set A is computably enumerable iff A is Σ_1^0.*

Proof. Suppose A is computably enumerable. Then $A = \operatorname{dom} \varphi_x$ for some x, and $y \in A$ iff $(\exists s)(\varphi_x^s(y) \downarrow)$, noting that the predicate "$\varphi_x^s(y) \downarrow$" is computable given x, y, s. Conversely, if A is Σ_1^0 then for some computable R we have $y \in A$ iff $(\exists z)(R(z, y))$. Define a partial computable function g by setting $g(y) = 0$ if we can find some number z such that $R(z, y)$ holds, and $g(y) \uparrow$ otherwise. Then $A = \operatorname{dom} g$. $\square$

By Theorem 2.1.10, $\Delta_1^0 = \Sigma_1^0 \cap \Pi_1^0$ consists of exactly the computable sets. It also follows from the undecidability of K that $K \in \Sigma_1^0 \backslash \Delta_1^0$ and its complement $\overline{K} \in \Pi_1^0 \backslash \Delta_1^0$. We get the arithmetical hierarchy of Kleene:

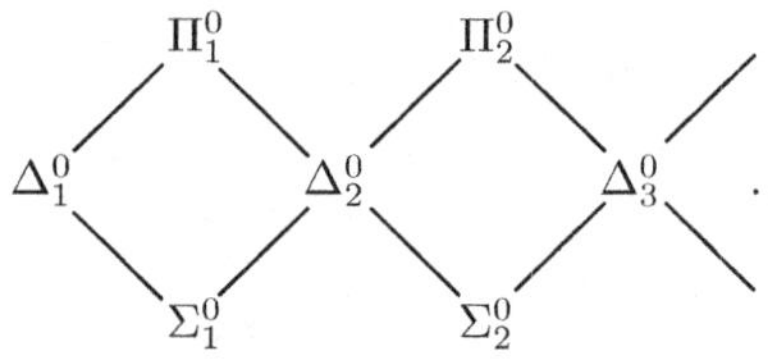

Here lines mean inclusion (rightward along the page). In Section 3.1.1, we establish all the inclusions and show that they are all proper. For instance, we will show that $\mathbf{0}^{(n)} \in \Sigma_n^0 \backslash \Delta_n^0$. However we will not need these facts in this chapter.

2.1.6 Π_1^0 classes

We write $2^{<\omega}$ to denote the set of all finite strings (tuples) of zeros and ones. It can be identified with the complete binary tree. Note that each infinite path through $2^{<\omega}$ is a (characteristic function of a) set. Every such infinite path is naturally an element of 2^ω.

A computable subtree of $2^{<\omega}$ is a computable collection S of strings of zeros and ones closed under initial segments. Then $P \in 2^\omega$ is a *path through* S if for all of its finite initial segments σ (written $\sigma \prec P$) we have $\sigma \in S$. We denote by $[S]$ the collection of paths through S.

Definition 2.1.28 (Π^0_1 class). A (binary) Π^0_1 class $C \subseteq 2^\omega$ is a collection of infinite paths through a computable subtree of $2^{<\omega}$.

An equivalent formulation is that C is a Π^0_1 class iff there is a computable relation R on finite strings such that

$$C = \{\alpha \in 2^\omega : \forall n R(\alpha \upharpoonright n)\}.$$

We remark that a more general definition of a computably bounded Π^0_1 class replaces 2^ω with a finitely and computably branching tree, meaning that at level n there are $f(n)$ many nodes for a computable function f. It is not difficult to show that for such a tree T and Π^0_1 class C defined on T, there is a Π^0_1 class $\hat{C}$ defined on 2^ω such that the members of C are in computable 1-1 correspondence with those of $\hat{C}$, and have the same many-one degrees. (See Excercise 2.1.31.)

If there in no computable bound on the level n branching then the theory of such Π^0_1 classes is quite different. For example, every path might code $\varnothing'$, whereas later we will show that a computably bounded Π^0_1 class always has a member of low Turing degree. We will usually say "Π^0_1-class" when we actually mean *computably bounded* Π^0_1 class, unless otherwise specified.

In this chapter we will only need the following well-known result. (More results will be presented in §4.2.3.) In some sense, the class constructed below is as far from being decidable as possible.

Theorem 2.1.29. *There exists a non-empty Π^0_1 class with no computable paths.*

Proof. Let A and B be disjoint c.e. sets. The the collection of *separating sets* $P = \{X : X \supseteq A \text{ and } X \cap B = \varnothing\}$ is a non-empty Π^0_1 class. Such a class of sets separating disjoint c.e. sets is called a *separating class.* A pair of c.e. sets A and B is effectively inseparable if there is no computable set C, such that $C \supseteq A$ and $\overline{C} \supseteq B$. To prove the theorem it is sufficient to come up with an effectively inseparable pair of c.e. sets. For example, by the proof of the incompleteness theorem, for Peano Arithmetic (PA), the c.e. sets of (Gödel codes for) provable formulae $A = \{\#\psi : PA \vdash \psi\}$ and $B = \{\#\psi : PA \vdash \neg\psi\}$ form an effectively inseparable pair. (For a more straightforward example, consider $A = \{e : \varphi_e(0) = 0\}$ and $B = \{e : \varphi_e(0) = 1\}$.) □

Exercises

Exercise° 2.1.30 (Folklore).

1. Show that, for every Π^0_2 predicate $\forall x \exists y R(x, y, z)$ where R is computable, we can uniformly replace R with a computable predicate P so that for all z,

   $$\forall x \exists y R(x, y, z) \text{ if and only if } \exists^\infty w P(w, z),$$

   where $\exists^\infty$ stands for "there exists infinitely many".
2. Same as above, but now uniformly replace P with an R.
3. In the notation above, prove that $\neg\exists^\infty w P(w, z)$ if and only if $\exists^{<\infty} w P(w, z)$, where $\exists^{<\infty}$ stands for "there exists at most finitely many". (Note the negation is not applied to P.)

4. Iterate (1) and (3) to uniformly derive a general form of any Π^0_n-predicate in terms of only $\exists^\infty$-quantifiers and at most one $\forall$-quantifier over a computable predicate.
5. Show that the naive analogue of (4) in terms of $\exists$ and $\exists^{<\infty}$ fails for Σ^0_n-predicates when $n \geqslant 4$.
6. Show that, for every Π^0_2 predicate $\forall x \exists y R(x, y, z)$ where R is computable, we can uniformly replace R with a computable predicate P so that for every x there exists at most one y with so that $P(x, y, z)$.

Exercise° 2.1.31. Suppose that f is a computable function and we make a computable tree T such that for all $\nu \in T$, ν has at most $f(n)$ many extensions. Show that elements in $[T]$ are in computable 1-1 correspondence to $[\hat{T}]$ where $\hat{T} \in 2^{<\omega}$.

Exercise° 2.1.32. Suppose $C = [T]$, $T \subseteq 2^{<\omega}$, is a Π^0_1.

1. Suppose C has only finitely many members (equivalently, there are only finitely many infinite paths through T). Show that in this case all these members are computable.
2. Suppose $\xi \in C$ is *isolated*, meaning that there is a $\sigma \in T$ so that ξ is the only infinite path through T that extends σ. Show that ξ is computable.
3. Suppose we are given an index e for $T \subseteq 2^\omega$ and $n \in \mathbb{N}$ so that $C = [T]$ has exactly n members ($n > 0$). Show that we can compute these members uniformly in e and n. (See also Fact 4.2.45.)

Exercise° 2.1.33. Show that if P is a non-empty Π^0_1 class $\subseteq 2^\omega$, then there exists $\alpha \in P$, such that α has c.e degree. (Hint: Think about the rightmost or leftmost path in the computable tree where $P = [T]$.)

2.2 Early effective algebra

This section contains examples of computable, low, and c.e. presented structures. For instance, we give examples of low and c.e. presented groups that have no computable presentation. We often give only extended sketches of these results, skipping most of the details not related to computability. Many of these examples will not be used in the rest of the book; they serve mainly as (historical) examples. The results here are mainly motivational, and hence the reader could skip the proofs of some of them without affecting their understanding of the details of the later results. However, at least skimming through the proofs is advised. An exception to this rule is Mal'cev's Theorem 2.2.16, as its proof contains fundamental ideas that will reappear frequently throughout the book. Familiarity with the elementary basics of group theory and field theory is assumed throughout this section.

2.2.1 Examples of computable and non-computable groups

According to Definition 1.2.1, a computable group $(G, \cdot)$ is one where the domain is a computable set and $\cdot$ is computable. Using brute-force search through all elements of G, we can compute x^{-1} for any given $x \in G$ in a computable group G.

Elementary example of computable groups

Clearly, every finite group is computable.

Example 2.2.1. The following groups are easily seen to be computable:

1. Finitely generated abelian groups.
2. The free group F_κ of rank $\kappa \leqslant \omega$. (When $\kappa = \omega$, we can additionally make the free generating set computable.)
3. The free abelian group of rank $\kappa \leqslant \omega$.
4. The invertible $n \times n$-matrices over $\mathbb{Q}$.
5. The $n \times n$-matrices over $\mathbb{Z}$ having determinant 1.
6. The additive group of any vector space of dimension $\leqslant \omega$ over $\mathbb{Q}$.
7. The additive group of any vector space of dimension $\leqslant \omega$ over any computable field (see Exercises 2.2.22 and 2.2.22 for examples of computable fields).
8. Any group isomorphic to F_ω/N, where F_ω is the computable presentation of the rank ω free group with a computable set of generators, and N is normal subgroup of F_ω that is computable (as a subset).
9. The subgroup of $(\mathbb{Q}, +)$ generated by $\{\frac{1}{p_i} : i \in S\}$, where p_i range over a computably enumerable set of primes S.
10. The subgroup G_S of $(\mathbb{Q}, +)$ generated by $\{1, \frac{1}{p_i^n} : \langle i, n\rangle \in S\}$, where S is a c.e. set.

We leave the verification of 1.-10. to Exercise 2.2.21.

Further examples are provided by the following, perhaps unexpected, result.

Theorem 2.2.2 (Rabin [440]). *Every finitely generated group of matrices over any field has a computable presentation. (Note that computability of the field is not assumed.)*

Proof. Suppose the group is generated by $A_1, \ldots, A_k$, and let U be the field. Consider the finitely many elements of the field that are mentioned in $A_1, \ldots, A_k$. If P is the prime field of the same characteristic as U, then all entries of any matrix from the group lie in a field of the form

$$U \cong P(x_0, \ldots, x_k, \alpha_0, \ldots, \alpha_m),$$

where the x_i are algebraically independent over P, and the α_j are algebraic over $P(x_0, \ldots, x_k)$. (This is because we have only finitely many entries, so simply adjoin them to the prime field if they are not already there.) By Exercises 2.2.22 and 2.2.23, the field U is computably presented. Using the computable presentation of U, we can begin with $A_1, \ldots, A_k$ and apply the (matrix) product and the (matrix) inverse operations iteratively to generate a computable presentation of the group. □

We cite [339] for an alternative proof of Theorem 2.2.2.

Remark 2.2.3. Rabin [440] points out that Theorem 2.2.2 has the consequence that every finitely generated group that is not computably presentable cannot have a faithful representation by matrices over a field (we omit the standard definition). In particular, not every finitely generated or even finitely presented group has a faithful presentation; e.g., Example 1.2.3 or Theorem 2.2.6 below. This consequence had been obtained before Rabin in 1940 by Fuchs–Rabinowitsch [196] using purely algebraic methods.

A low group with no computable presentation

We give the first elementary example of a low group with no computable presentation. The result is folklore.

Theorem 2.2.4. *There exists a low subgroup of* $(\mathbb{Q}, +)$ *that has no computable presentation.*

Proof. To establish the theorem, we give the first early example of a *characterisation* of computable presentability which, in its slightly stronger form (to appear as Theorem 5.1.16), is usually attributed to Mal'cev [346]. In the notation of Example 2.2.1(10), assume additionally that S is a set of pairs $\langle i, n\rangle$ such that $\langle i, n\rangle \in S$ implies $\langle i, k\rangle \in S$ for all $k \leqslant n$.

Proposition 2.2.5. *G_S is computably presentable iff S is computably enumerable.*

Proof. One implication is given by Example 2.2.1(10). For the other implication, let $A \cong G_S \subseteq \mathbb{Q}$ be a computable group. Fix the isomorphic image g of $1 \in G_S$ in A. To enumerate S, list all $x \in A$ and all positive integers m such that $mx = \underbrace{x + x + x + \ldots + x}_{x \text{ occurs } m \text{ times}} = g$. If $p_i^n x = g$ for some $x \in A$, then put $\langle i, n\rangle$ in S. This process is clearly effective. □

For example, if we take the complement $\overline{K}$ of the halting problem, and consider the set $S(\overline{K}) = \{\langle i, 1\rangle : i \in \overline{K}\}$, we see that $G_{S(\overline{K})}$ has no computable presentation, by Post's Theorem 2.1.10. Theorem 3.1.1 (to be proven later) states that there is a low non-computable c.e. set X. Let $S(\overline{X}) = \{\langle i, 1\rangle : i \in \overline{X}\}$. By Proposition 2.2.5, the group $G_{S(\overline{X})}$ is computable relative to X and thus is low. However, it has no computable presentation, again by Post's Theorem 2.1.10. □

A c.e. presented group with no computable presentation

In the context of groups, the definition of a c.e. presented structure (Definition 1.2.2) is equivalent to saying that

$$G \cong \langle a_i \mid r_i : i \in \mathbb{N}\rangle,$$

where a_i form a computable set of free generators of F_ω, and $r_i \in F_\omega$ is a c.e. set of *relations* that generate a normal subgroup that is c.e. as a subset of (this presentation of) F_ω; we leave this to Exercise 2.2.21. If the sets of a_i and r_j are both finite, we say that the group is *finitely presented.* Evidently, every computably presented group is c.e. presented. In [440], Rabin gives the following example which he attributes to Boone.

Theorem 2.2.6. *There exists a finitely generated, c.e. presented group without a computable presentation.*

Proof. Fix a set of natural numbers W and define

$$G_W = \langle x, y, u, t \mid u^i x u^{-i} = t^i y t^{-i}, i \in W\rangle.$$

Claim 2.2.7. *In G_W, $u^i x u^{-i} = t^i y t^{-i}$ holds iff $i \in W$.*

Proof of claim. Omitted. □

If W is c.e. then G_W is clearly c.e. presented. Let $W = K$, the halting problem.

Claim 2.2.8. *G_K is not isomorphic to any computable group.*

Proof of claim. Assume there is a computable copy of G_K, denote it A. Let $f : G_K \to A$ be an isomorphism. Fix elements $f(x), f(y), f(u), f(t)$. There are only finitely many such parameters, and we *non-uniformly* fix them; for more explanation see Remark 2.2.9 below. Then $u^i x u^{-i} = t^i y t^{-i}$ iff

$$f(u)^i f(x) f(u)^{-i} = f(t)^i f(y) f(t)^{-i},$$

and the latter has to be decidable in A, a contradiction. □

We conclude that G_K has the desired property. □

Remark 2.2.9. In the sketch above, we *non-uniformly* fixed a finite tuple of parameters. That is, we designed a computable procedure P that works when the parameters are correctly chosen. This can be viewed as follows: List all possible quadruples $\bar{x}_0, \bar{x}_1, \ldots, \bar{x}_i, \ldots$ and consider all possible procedures of the form $P(x_0), P(x_1), \ldots$. We know that one of these procedures does the job, but we do not necessarily know which one. We also used a non-uniform argument in the proof of Theorem 2.2.2 when we fixed x_i and α_j, and in the proof of Proposition 2.2.5 when we fixed the isomorphic image g of 1.

The proof of Theorem 2.2.6 can be pushed to a characterisation similar to Proposition 2.2.5. Such characterisations exist among finitely presented groups too; we cite [95, 49, 50]. Combined with the existence of non-computable low c.e. sets (Theorem 3.1.1), these results imply the existence of c.e. presented groups that are also low, yet not computably presented. Using completely different methods, in Corollary 9.3.23 we will give such examples among (non-finitely generated) abelian groups. Thus, we limit ourselves to a brief discussion below of the finitely presented case, sufficient to clarify Example 1.2.3 from the introduction.

Non-computable finitely presented groups

We outline the proof of the following fundamental result that appeared earlier as Example 1.2.3.

Theorem 2.2.10. *There exists a finitely presented group not isomorphic to any computable group.*

Proof sketch. The theorem can be derived from Theorem 2.2.11 and the two classical theorems that we state below.

Theorem 2.2.11 (Higman Embedding Theorem [250]). *Suppose that G is a finitely generated group. Then G can be isomorphically embedded into a finitely presented group iff G has a c.e. presentation.*

We omit the proof, but we note that it is much simpler than the original early proofs of the following classical:

Theorem 2.2.12 (Novikov [416], Boone [51]). *There is a finitely presented group with undecidable word problem*[1].

Proof of Theorem 2.2.12 using Theorem 2.2.11. Fix the finitely generated group G_K witnessing Theorem 2.2.6 and embed it into a finitely presented group H. Let $f : G_K \to H$ be the embedding. Arguing as in the second half of the proof of Theorem 2.2.6, we can deduce that $K = \{i : f(u)^i f(x) f(u)^{-i} = f(t)^i f(y) f(t)^{-i}\}$. □

We now explain why Theorem 2.2.10 follows from Theorem 2.2.12. Observe that f in the proof of Theorem 2.2.11 (and 2.2.6) was necessarily computable, but again non-uniformly (as explained in Remark 2.2.9). Of course, the same argument works for any finitely generated c.e. presented structure.

Proposition 2.2.13. *Suppose A is a finitely generated structure (in a language with no relational symbols). Then any two c.e. presented copies of A are computably isomorphic.*

Proof. Let X and Y be two c.e. copies of A. Non-uniformly fix any finite tuple of generators $\bar{x}$ in X and their isomorphic images $\bar{y}$ in Y. There is a unique extension of the map $\bar{x} \mapsto \bar{y}$ to an isomorphism between X and Y: just map $\bar{x}$-terms to the respective $\bar{y}$-terms. This extension is evidently computable. □

Corollary 2.2.14. *Suppose a finitely generated structure has a computable presentation A. Then in any c.e. presentation $B/_\sim$ of A, the c.e. equivalence relation $\sim$ is computable. In other words, every c.e. presentation of A is computable (up to notation change).*

Indeed, to decide $x \sim y$ in B, simply compute their isomorphic images in A and see whether the images are equal. As we noted earlier, any finitely presented group is evidently c.e. presented. Thus, to establish Theorem 2.2.10, it remains to combine the Novikov-Boone Theorem 2.2.12 with the corollary above. □

[1]That is, the equality modulo the relations is undecidable.

Groups with non-equivalent presentations

Mal'cev [345] was perhaps the first to propose that computable (presentations of) algebraic structures should be studied up to *computable* isomorphism. The inverse of a computable isomorphism is also computable (via a brute-force search argument). Thus, computably isomorphic structures exhibit identical computability-theoretic properties. We arrive at the important class of structures, ones where computable isomorphism type and isomorphism type coincide.

Definition 2.2.15 (Mal'cev). An algebraic structure is *computably categorical* or *autostable* if it has a unique computable presentation, up to computable isomorphism.

For example, finitely generated structures always possess computably unique presentations (Proposition 2.2.13). Thus, computable categoricity can be viewed as a generalisation of being finitely generated. But of course, there are many elementary examples that are not even close to being finitely generated. For example, it is also well-known and easy to see that the Rado graph (the random graph) is computably categorical, and so is the dense linear order $(\mathbb{Q}, <)$ (Ex. 2.2.20). But all these basic examples are relational structures. When a structure has operations, the analogy with being finitely generated sometimes becomes more direct. For instance, the following historical example is usually attributed to Mal'cev. Let $\mathbb{Q}_\alpha$ denote the additive group of the $\mathbb{Q}$-vector space of dimension $\alpha \in \mathbb{N} \cup \{\omega\}$ ($\mathbb{Q}_0 = \operatorname{span}_{\mathbb{Q}} \varnothing = \{0\}$). These groups are computably presented; this is Example 2.2.1(6).

Theorem 2.2.16 (Mal'cev). *$\mathbb{Q}_\alpha$ is computably categorical iff α is finite.*

The strategy used in this proof below will be extended in later chapters to prove Khisamiev's Theorem 5.1.41, which is (3) of Theorem A. Khisamiev's Theorem is not elementary, but it shares some key ideas with the proof below.

Proof. We would like to use scalar multiplication, however, we cannot use it directly since it is not in the language of groups. For an element x of an additive group and n a positive integer, write nx to denote

$$\underbrace{x + x + x + \ldots + x}_{x \text{ occurs } n \text{ times}}.$$

Also define $0g = 0$, and when n is a negative integer, set $nx = (-n)(-x)$. Clearly, the operation

$$(n, x) \mapsto nx$$

is computable in any computable presentation of $\mathbb{Q}_\alpha$. Further, given $m > 0$ and an element a, there exists a unique element y with the property $my = a$, and we can just brute-force search for y. Thus, more generally, the operation of scalar multiplication

$$(r, a) \mapsto ra,$$

where $r = \frac{m}{n} \in \mathbb{Q}$, is also computable in any computable presentation of $\mathbb{Q}_\alpha$. It follows that *the additive group structure on* $\mathbb{Q}_\alpha$ *uniquely and effectively determines the* $\mathbb{Q}$*-vector space structure* on it. For instance, it makes sense to say that two elements of the group are linearly independent. We are now ready to prove the theorem.

Assume $\alpha = n \in \mathbb{N}$. The case when $n = 0$ is trivial; assume $n > 0$. Given any two computable copies A and B of the group, *non-uniformly* (recall Remark 2.2.9) fix some finite bases $\bar{a} = a_0, \ldots, a_{n-1}$ and $\bar{b} = b_0, \ldots, b_{n-1}$ in A and B, respectively. Define $f : A \to B$ as follows. For $a \in A$, find integers $m, m_0, \ldots, m_{n-1}$ so that $m > 0$ and

$$ma = \sum_{i<n} m_i a_i.$$

In B, search of an element b so that

$$mb = \sum_{i<n} m_i b_i$$

and define $f(a) = b$. This map would be an isomorphism of the respective vector spaces over $\mathbb{Q}$ (it is linear), so it is evidently an additive group isomorphism between A and B. By the choice of n, $\bar{a}$ and $\bar{b}$, for any $a \in A$ such coefficients $m, m_0, \ldots, m_{n-1}$ and an element $b \in B$ *exist*. Thus, they will eventually be found. It follows that f is computable.

Now assume $\alpha = \omega$. We prove that the additive group $\mathbb{Q}_\omega$ admits two computable presentations that are not computably isomorphic. Let A be the "natural" computable presentation A of $\mathbb{Q}_\omega$, which is the collection of formal sums

$$\sum_{i \in \mathbb{N}} r_i a_i,$$

where almost all coefficients r_i are zero. (Alternatively, we can view each individual element as a finite tuple of rational numbers.) Given any two elements of this presentation, we can easily decide whether they are linearly independent as vectors over $\mathbb{Q}$.

We claim that it is sufficient to build a computable presentation B of $\mathbb{Q}_\omega$ in which there is no algorithm such that, given a pair of elements of B, decides whether they are linearly independent. Indeed, suppose such a B has been constructed, and assume $f : B \to A$ was a computable isomorphism. To decide whether $x, y \in B$ are linearly independent, calculate $f(x)$ and $f(y)$ and decide this property in A. (This is essentially an m-reduction.)

To this end, we construct a computable presentation $B = \bigcup_s B_s$ of $\mathbb{Q}_\omega$ in which linear independence for pairs of elements is undecidable. The idea is as follows. We reserve a sequence of elements $(a_i)_{i \in \mathbb{N}}$ that we initially keep linearly independent. Initially, we may think of these a_i as being the elements of the standard basis of the natural presentation of $\mathbb{Q}_\omega$. However, we will change this interpretation later for some (but not all) such a_i. If i enters the enumeration of the halting set K at stage t, declare $a_{2i+1} = m a_{2i}$, where m is a very large integer. Thus, we "discover" that a_{2i+1} is indeed dependent on a_{2i}, but the coefficient m witnessing this is so large that we have not seen it earlier. However, recall that we need to make sure that the group is computable, and if we are not being careful enough we may end up with a c.e. presentation. In other words, if we declare two elements unequal, we cannot possibly undo this at a later stage. The main subtlety is that we also need to preserve *inequalities* declared in B_s when we define B_{s+1}. This is why we need m to be very large.

Construction.

At stage 0, set $B_0 = \{0\}$ and $k(0) = 0$.

At the end of stage s, we have B_s that consists of all sums of the form

$$\sum_{i \notin K_s} (r_{2i}a_{2i} + r'_{2i+1}a_{2i+1}) + \sum_{j \in K_s} r''_j a_{2j}, \tag{2.1}$$

where $i, j \leqslant k(s)$ and the coefficients r_i, r'_i, r''_i range over the (reduced) fractions with numerators and denominators bounded by $k(s)$ in their absolute value. The operation $+$ is declared on these formal linear combinations naturally (i.e., $\sum_i r_i a_i + \sum_i q_i a_i = \sum_i (r_i + q_i) a_i$), provided that the result stays in B_s. Otherwise, $+$ is declared undefined yet.

At stage $s + 1$, consider two cases.

Case 1. No new number enters K at stage $s + 1$. Set $k(s + 1) = k(s) + 1$ and define B_{s+1} to consist of all sums of a similar form as we described above in (2.1), but with s replaced by $s + 1$ throughout. Declare $+$ on these elements naturally as well, as above. Clearly $B_s \subseteq B_{s+1}$. Let $g_{s+1} : B_s \to B_{s+1}$ be the subset embedding.

Case 2. Suppose j_0 is enumerated into K at stage $s+1$. Every element $b \in B_s$ can be expressed as

$$b = r_{2j_0}a_{2j_0} + r'_{2j_0+1}a_{2j_0+1} + \sum_{i \notin K_{s+1}} (r_{2i}a_{2i} + r'_{2i+1}a_{2i+1}) + \sum_{j \in K_s} r''_j a_{2j}. \tag{2.2}$$

Declare

$$a_{2j_0+1} = k(s)! a_{2j_0}$$

and define

$$g_{s+1}(b) = (r_{2j_0} + r'_{2j_0+1}k(s)!)a_{2j_0} + \sum_{i \notin K_{s+1}} (r_{2i}a_{2i} + r'_{2i+1}a_{2i+1}) + \sum_{j \in K_s} r''_j a_{2j}.$$

Set $k(s + 1) = (k(s) + 1)!$ and define B_{s+1} just as we did in the first case. Note however that in this case $g_{s+1} : B_s \to B_{s+1}$ is *not* the natural subset embedding.

Verification.

Claim 2.2.17. *For every s, g_s is injective.*

Proof. The map is clearly injective in the first case. In the second case, assume $g_{s+1}(b') = g_{s+1}(b)$, where b is as in (2.2), and

$$b' = q_{2j_0}a_{2j_0} + q'_{2j_0+1}a_{2j_0+1} + \sum_{i \notin K_{s+1}} (q_{2i}a_{2i} + q'_{2i+1}a_{2i+1}) + \sum_{j \in K_s} q''_j a_{2j}. \tag{2.3}$$

Then we immediately get that $q_{2i} = r_{2i}$ for any $i \notin K_{s+1}$, and that $q''_j = r''_j$ for each $j \in K_s$, and

$$r_{2j_0} + r'_{2j_0+1}k(s)! = q_{2j_0} + q'_{2j_0+1}k(s)!$$

Since $k(s)!$ is larger than both q_{2j_0} and r_{2j_0}, by considering this equality modulo $k(s)!$ we conclude that $r_{2j_0} = q_{2j_0}$ and, thus, $r'_{2j_0+1} = q'_{2j_0+1}$. (In fact, we should first turn this equality into an integer equality and then consider it modulo $k(s)!$; we leave the elementary details to the reader.) □

We now verify that g_{s+1} preserves the operation.

Claim 2.2.18. *Assume $b, b' \in B_s$ are so that $b + b' \in B_s$ was defined at stage s. Then $g_{s+1}(b + b') \in B_{s+1}$ and $g_{s+1}(b + b') = g_{s+1}(b) + g_{s+1}(b')$.*

Proof. The first case in the description of stage $s + 1$ is obvious, we focus on the second case. In the notation of (2.2) and (2.3) above, we clearly have

$$\begin{aligned} g_{s+1}(b + b') = {} & [(r_{2j_0} + q_{2j_0}) + k(s)!(r'_{2j_0+1} + q'_{2j_0+1})]a_{2j_0+1} + \\ & \sum_{i \notin K_{s+1}} ((r_{2i} + q_{2i})a_{2i} + (r'_{2i+1} + q'_{2i+1})a_{2i+1}) + \sum_{j \in K_s} (r''_j + q''_j)a_{2j} = \\ & g_{s+1}(b) + g_{s+1}(b'). \end{aligned} \tag{2.4}$$

Additionally,

$$|(r_{2j_0} + q_{2j_0}) + k(s)!(r'_{2j_0+1} + q'_{2j_0+1})| \leqslant k(s) + k(s)!k(s) \leqslant k(s)!(k(s) + 1) = k(s + 1),$$

and therefore $g_{s+1}(b + b') \in B_{s+1}$. □

We have defined a uniformly effective sequence of partial groups and partial embeddings

$$B_0 \rightarrow_{g_1} B_1 \rightarrow_{g_2} B_2 \rightarrow_{g_3} B_3 \rightarrow_{g_4} \cdots.$$

As further explained in Remark 2.2.19 below, the claims guarantee that as the limit of this process, we get a computable structure; denote it B.

Remark 2.2.19. From the two preceding claims, we deduce that we can safely take the union of the nested sequence of B_i, identifying each element of B_s with its g_{s+1}-image in B_{s+1}. In fancier terms, we can take the "direct limit" of this effective sequence. In the sequel, we will often identify elements of computable structures with their *indices* in ω; that is, the domain will often be thought of as either ω or its initial segment. If an element $b \in B_s$ receives an index $i \in \mathbb{N}$, then we declare that $g_{s+1}(b)$ also has index i, and so on. Another way to describe it is to say that $i \in \mathbb{N}$ "changes its interpretation" at stage $s + 1$ to be $g_{s+1}(b)$. The claims above guarantee that this interpretation change also preserves the operation, whenever it is defined. Also, the finite part of the open diagram (i.e., the quantifier-free atomic facts about the elements) defined at stage s is preserved at stage $s + 1$. This, in particular, includes both equality and inequality of elements (by the injectivity of g_{s+1}). In particular, in the limit we get a computable structure B in the language of one binary operation $+$, not just a c.e. presented structure.

We now argue that B is isomorphic to $\mathbb{Q}_\omega$.

Since for each individual i, i can enter K at most once, we have that the sequence $(g_s(a_{2i+1}))_{s \in \mathbb{N}}$ eventually stabilises at either a_{2i+1} or ma_{2i} for some integer m that depends on i. Also, $g_s(a_{2i}) = a_{2i}$, for every s and i, whenever it is defined. Every element b that ever enters B_s for some s can be expressed as a linear combination of finitely many elements a_j. It follows that the sequence $(g_s(b))_{s \in \mathbb{N}}$ stabilises for every such b, and its final value is an element in the $\mathbb{Q}$ -vector space spanned by linearly independent elements

$$\{a_{2i}, a_{2j+1} : i \in \mathbb{N}, j \notin K\}$$

using coefficients in $\mathbb{Q}$. Conversely, any such linear combination will eventually be added into B, because $k(s)$ is monotonically increasing. Thus, $B \cong \mathbb{Q}_\omega$.

Observe that in the resulting additive group B, a_{2i} and a_{2i+1} are dependent iff $i \in K$. As we explained earlier, this implies that A is not computably isomorphic to B. □

Can we describe computably categorical abelian groups? What about other structures? We will return to computable categoricity later when we have enough tools.

Exercises

Exercise° 2.2.20. Show that the dense linear order of the rationals is computably categorical.

Exercise° 2.2.21 (Folklore after Rabin and Mal'cev)**.**

1. Show that a (countable, discrete) group G is c.e. presented iff it is isomorphic to a factor of a computable free group F (w.l.o.g. upon a computable generating set) by a c.e. normal subgroup N.
2. Prove that G is computably presented iff N can be chosen to be computable.

The two exercises below are essentially due to Kronecker [318]; see also [185].

Exercise° 2.2.22. Let F be a computable field, and suppose α is algebraic over F. Show that $F(\alpha)$ has a computable presentation. (Hint: Consider $F[x]/\langle p(x)\rangle$, where $p \in F[x]$ is irreducible and $p(\alpha) = 0$.)

Exercise° 2.2.23. Let F be a computable field, and suppose x is transcendental over F. Show that the fraction field $F(x)$ has a computable presentation. (Hint: Use long division and the generalised Euclidean algorithm to produce an effective list of irreducible polynomial fractions over F.)

Exercise° 2.2.24 (Metakides and Nerode [384])**.** Show that for a computable formally real field, the space of orderings can be computably described as a Π^0_1 class.

Exercise° 2.2.25 (Folklore after Hatzikiriakou and Simpson [246])**.** Fix a computable presentation of a torsion-free abelian group. Show that the space of compatible linear orders on G can be (effectively) identified with elements of a certain Π^0_1-class $C \subseteq 2^\omega$.

2.2.2 The Henkin construction is computable

We assume that the reader is familiar with the Henkin construction from elementary model theory. If the reader is *not* familiar with this material, they can skip this subsection, since we won't need elementary model theory in the sequel, with the exception of several exercises. We will, however, occasionally encounter the following notion that we already mentioned at the beginning of the chapter.

Definition 2.2.26 (Decidable structure)**.** We call a structure A *decidable* if we can decide all first order statements about tuples of elements in A. (That is, the *full diagram* of A is computable.)

It is easy to find an example of a computable algebraic structure with no decidable presentation; e.g., Exercise 2.2.33. Examples of groups with this property will be encountered later (e.g., Theorem 5.1.18). In later chapters we will also see that such examples can be found among Boolean algebras and linear orders as well. Here, we present a general method of producing decidable structures using decidable theories.

A decidable theory has a decidable model

All our theories are first-order.

Definition 2.2.27. A theory T (in a computable language) is said to be *decidable* if there is an algorithm that, given any sentence φ in the language of T, determines whether $T \vdash \varphi$.

The following theorem is folklore and can be found in, e.g., [156].

Theorem 2.2.28. *A complete decidable theory T has a decidable model.*

Proof. We simply observe that the Henkin construction is computable. Let T be a decidable theory, and let $\{c_i : i \in \mathbb{N}\}$ be an enumeration of a computable set of new constants. Let $\{\sigma_i : i \in \mathbb{N}\}$ list all sentences in the language $L(T)$ of T together with $\{c_i : i \in \mathbb{N}\}$. We construct a model A and its complete decidable theory T_A expanding T in stages. We construct the theory T_A as $\{\tau_n : n \in \mathbb{N}\}$. As usual in a Henkin construction, at stage $2e + 1$, if τ_e is of the form $\exists x\theta(x)$, we add a Henkin witness c_i not occurring in

$$\phi_{2e} =_{\text{def}} \wedge_{i\leqslant 2e}\tau_i.$$

At stage $2e + 2$, we force the completeness of the diagram as follows. Let $\overline{x}$ be the first sequence of variables of the same length as $\overline{c}$ obtained from step $2e+1$, which does not occur in $\wedge_{i\leqslant 2e+1}\tau_i \rightarrow \sigma_e$. Using the decidability of T, check if

$$T \vdash \forall\overline{x}(\wedge_{i\leqslant 2e}\tau_i \rightarrow \sigma_e)[\overline{x}/\overline{c}].$$

If this holds, let $\delta_{2e+2} = \sigma_e$. Otherwise, let $\delta_{2e+2} = \neg\sigma_e$.

Define $c_i \sim c_j$ if and only if $c_i = c_j \in T_A$. To define the model, consider the collection of all equivalence classes modulo $\sim$, and declare that $\phi \in T_A$ holds on a tuple $\bar{a}$ of equivalence classes if it holds for some (equivalently, all) representatives of these classes. It is routine to verify that the resulting extension T_A of T is complete and computable, and that the structure $A/\sim$ is computable and is a model of T; this is Exercise 2.2.34. □

Deeper results in computable model theory require significantly more intricate techniques, but we will not cover these results in the book; see, e.g., [229] for a gentle introduction. We will only give a few elementary applications of the Henkin construction to algebra; see below.

An application to algebraic and real closures

The language of algebraically closed fields is $L = \{+, \cdot, -, 0, 1\}$, and the theory ACF has the field axioms together with, for each $n \in \omega$,

$$\varphi_n \equiv \forall x_1, \ldots, x_n \exists y(x_1 + x_2 y + \cdots + x_n y^n = 0).$$

Recall that a theory admits *elimination of quantifiers* if, for every $\varphi \in L(T)$, there is a quantifier-free $\psi(x_1, \ldots, x_n) \in L(T)$ such that $T \vdash \forall x_1, \ldots, x_n(\varphi \leftrightarrow \psi)$. We say that T admits *effective* elimination of quantifiers if we can compute ψ from φ uniformly. The following result is classical.

Theorem 2.2.29 (Tarski). *ACF admits effective elimination of quantifiers.*

Note that if T is a computable theory and is complete, then it is decidable.

Theorem 2.2.30 (Rabin [440]). *Every field can be computably embedded into its computable algebraic closure.*

Rabin's original proof constructed the algebraic closure using a quotient of a polynomial ring with infinitely many variables. We construct the algebraic closure by an effective Henkin-style construction. See, for example, [183, Theorem 2.5] where this construction is carried out in reverse mathematics.

Proof. Suppose F is a computable field. We first construct a computable algebraically closed field K and a computable embedding $\alpha\colon F \to K$. Let L_F be the language of fields with constant symbols for the elements of F, and let T be the theory of algebraically closed fields together with the atomic diagram of F. Since F has an algebraic closure, the theory is consistent.

By Theorem 2.2.29, it is possible to effectively replace every formula $\phi(\bar{a})$ in T (with parameters $\bar{a} \in F$) by a quantifier-free formula $\psi(\bar{a})$ such that

$$ACF \vdash \phi(\bar{a}) \leftrightarrow \psi(\bar{a}).$$

By our assumption, F is a computable field. Thus, we can effectively check whether $F \models \psi(\bar{a})$ and conclude that T is decidable. Using the computability of Henkin's construction (Theorem 2.2.28), we obtain a computable (indeed, decidable) K in which F is naturally listed as a computably enumerable subset named by the constants.

Computably list all elements of K that are algebraic over F. They form a computably enumerable subfield $alg(F)$ of K that is isomorphic to the algebraic closure $\overline{F}$ of F. To obtain a computable presentation of $\overline{F}$, re-enumerate the domain of $alg(F)$ by setting the first element that appears in its enumeration equal to index 0, the second to 1, and so on. (More generally, any computably enumerable substructure B of a computable structure A has a computable copy.) It is not difficult to organise this enumeration so that it gives a natural computable embedding from F to this new computable copy of $\bar{F}$; we leave the details to Exercise 2.2.35. The range of this embedding does not have to be computable, but it will of course be computably enumerable. □

Rabin proved that computability of the image of F in its closure is equivalent to F having a *splitting algorithm.* A field has a splitting algorithm if, given any polynomial in $F[x]$, we can decide whether it splits over the field. The reason why this might be necessary can be seen in, e.g., Exercise 2.2.22. One can easily construct a field without a splitting algorithm by a direct diagonalisation; for an explicit example, consider $\mathbb{Q}(\sqrt{p_i} : i \in K)$. This computable field will have non-computably isomorphic algebraic closures; see [185, 384].

Recall that an ordered field is real closed if every positive number has a square root, and all polynomials of odd degree have at least one root. Let RCF denote the first-order theory of real closed fields; it is well-known that $RCF = Th(\mathbb{R}, +, \cdot, <)$.

Theorem 2.2.31 (Tarski). *RCF admits effective quantifier elimination.*

The next theorem and its proof are very similar to what we had for algebraically closed fields.

Theorem 2.2.32 (Ershov [154], Madison [344]). *Every computable ordered field can be computably embedded into its computable real closure.*

Proof. This is the same as the proof for algebraic closure, but using quantifier elimination in RCF (instead of ACF). □

For related results about differential and difference fields that require more advanced methods, see [231, 242]. Analogous results are also known for abelian groups and their divisible hulls [474], as well as torsion-free locally nilpotent groups [159]. We will not develop this topic any further; [159] remains the standard reference for such constructions.

Exercises

Exercise° 2.2.33. An *equivalence structure* is an algebraic structure of the form $(X, \sim)$, where $\sim$ is an equivalence relation. Prove that there is a computable equivalence structure that has no decidable presentation.

Exercise° 2.2.34. Complete the proof of Theorem 2.2.28.

Exercise° 2.2.35. Complete the proof of Theorem 2.2.30.

Exercise° 2.2.36 (Folklore). Let V be a computable non-principal type in a complete decidable theory T. Show that T has a decidable model that omits V.

2.2.3 Computable vs. constructive ordinals

Historically, the ordinals (the well-orders) were the first broad class of relational structures to be studied thoroughly from the perspective of computable presentability. We assume that the reader is familiar with the notation and the elementary properties of ordinals. For instance, recall that for ordinals, $\beta \in \alpha$ (both viewed as sets) means that β is an initial segment of α (viewed as an order). Our definition of a computable structure, restricted to linear orders, gives the following notion of computability for ordinals.

Definition 2.2.37 (Computable ordinal). A *computable ordinal* is the order type of some computable well-ordering, that is $(A, \leqslant_A)$ where A is a computable set and $\leqslant_A$ is a computable well-ordering on the set.

It is clear that if L is a computable well-order and $R = \{\ell \in L : \ell < x\}$ is its initial segment, then R is computable as well. The first non-computable ordinal is called "omega-one CK", written ω_1^{CK}, where CK stands for "Church-Kleene".

Now we discuss "constructive" ordinals due to Kleene [300, 301]. The idea is to associate each ordinal with an algorithmically effective notation which carries information about immediate successors, predecessors, and limit points. For example, this information is necessary to design definitions by transfinite recursion. A computable presentation of a well-order does not have to have this additional information about its points.

Kleene's $\mathcal{O}$. Recall that φ_e stands for the partial recursive function with index e. Define a system of notations by specifying a set $\mathcal{O}$, a function $|\cdot|_{\mathcal{O}}$, and a (strict) ordering $<_{\mathcal{O}}$ on $\mathcal{O}$. Here $|a|_{\mathcal{O}} = \alpha$ means that $a \in \mathcal{O}$ is a *notation* for α. This is done as follows:

- 1 is the notation for 0.
- If a is the notation for α then 2^a is the notation for $\alpha + 1$.
- We now define $b <_{\mathcal{O}} 2^a$ if either $b <_{\mathcal{O}} a$ or $b = a$.
- For limit ordinals α we give notations $3 \cdot 5^e$, where

$$\varphi_e(0) <_{\mathcal{O}} \varphi_e(1) <_{\mathcal{O}} \varphi_e(2) \dots$$

  and α is the least upper bound for $|\varphi_e(n)|_{\mathcal{O}}$. (In particular, φ_e is total.)
- Define $b <_{\mathcal{O}} 3 \cdot 5^e$ if there exists an n with $b <_{\mathcal{O}} \varphi_e(n)$.

Kleene's $\mathcal{O}$ can be visualised as an infinitely branching, tree-like structure. It begins with

$$1, 2, 2^2, 2^{2^2} \dots,$$

which are the notations for $0, 1, 2, 3 \dots$, but then it becomes infinitely branching at every limit level. The branching occurs because there are, of course, infinitely many ways to list a sequence converging to a limit ordinal α. Thus, at level ω, the "tree" splits into infinitely many infinite chains of the form

$$3 \cdot 5^e, 2^{3 \cdot 5^e}, 2^{2^{3 \cdot 5^e}}, \dots,$$

where e ranges over the total functions such that $(\varphi_e(n))_{n \in \mathbb{N}}$ is a strictly increasing sequence (of notations) below ω. Elements from two different "chains" corresponding to distinct indices are incomparable under $<_{\mathcal{O}}$.

Definition 2.2.38 (Constructive ordinal)**.** The ordinals having notations in Kleene's $\mathcal{O}$ are called the *constructive* ordinals.

Note that there is a conflict of terminology with the post-Soviet tradition in computable mathematics. In that tradition, "constructive" was used as a synonym for "computable". Fortunately, these two notions of effective presentability for ordinals are actually equivalent.

Theorem 2.2.39 (Spector [481])**.** *A countable well-order (ordinal) is constructive iff it has a computable presentation.*

We outline the proof suppressing some details and emphasising the important steps and ideas. We refer the reader to Rogers [454] for a more detailed proof. The technique used in this proof will not be particularly useful in the sequel; perhaps the only exception is the proof of Theorem 2.3.7, which can also be skipped if necessary.

*Proof **. We show that every constructive ordinal has a computable presentation.

Lemma 2.2.40. *Suppose $|a|_{\mathcal{O}} = \alpha$, $a \in \mathcal{O}$. Then each $\beta \in \alpha$ receives exactly one notation $b <_{\mathcal{O}} a$. All such unique notations below a can be computably listed, and $<_{\mathcal{O}}$ restricted to these notations is computable.*

Proof. The first assertion of the lemma follows from the definition of $\mathcal{O}$. Essentially, this is because $\mathcal{O}$ is a (set-theoretic) tree. It is usually said that

$$p(a) = \{b : b \in \mathcal{O} \text{ and } b <_{\mathcal{O}} a\}$$

is a "path" through $\mathcal{O}$ (below a). Notations that lie on the same path are comparable under $<_{\mathcal{O}}$. Every ordinal $\beta \in \alpha$ receives a notation along $p(a)$.

We prove the second assertion of the lemma. Observe that the path below a can be enumerated (uniformly in a). This is because $b <_{\mathcal{O}} b'$ is equivalent to saying that there exist $a_0, \ldots, a_k \in p(a)$ so that $a_0 = b$, $a_k = b'$, and $a_i <_{\mathcal{O}} a_{i+1}$ according to one of the atomic cases in the definition of $<_{\mathcal{O}}$. (This follows by induction on $|a|_{\mathcal{O}}$.) We can therefore list both the path $p(a)$ and the order $<_{\mathcal{O}}$ restricted to the path. For $b \neq b'$ along the path, exactly one of the two possibilities $b <_{\mathcal{O}} b'$ or $b' <_{\mathcal{O}} b$ must occur. By Post's Theorem 2.1.10, we can decide $<_{\mathcal{O}}$ for numbers coming from $p(a)$. □

Given α with a notation a in $\mathcal{O}$, we use the lemma to produce a computable presentation of α. The case when α is finite is again trivial. Suppose α is infinite. Fix a computable function $f : \omega \to p(a) = \{b : b \in \mathcal{O} \text{ and } b <_{\mathcal{O}} a\}$ whose index can be computed uniformly in a. Define $L = (\omega, <)$ by the rule

$$i < j \text{ if and only if } f(i) <_{\mathcal{O}} f(j),$$

where $f(i), f(j) \in p(a)$ and therefore $<_{\mathcal{O}}$ will eventually be decided for these values. It follows that $L = (\omega, <) \cong \alpha$ is a computable presentation of α.

Now assume that α is a computable ordinal, and let L be its computable presentation. The obvious issue is that, in L, the property "x is a successor" can be undecidable. The idea is to turn L into a computable copy D of $\gamma > \alpha$ where these properties are decidable.

Lemma 2.2.41. *There is a computable presentation D of $\omega \cdot (1+\alpha)+1$ in which we can additionally decide whether $x \in D$ is a successor or a limit point. In the former case, we can additionally uniformly compute the predecessor of x, and in the latter case we can compute a computable monotonic sequence $(y_i)_{i \in \mathbb{N}}$ converging to x from below.*

Sketch. Recall that L is a computable presentation of α. Clearly, $1 + L + 1$ also has a computable presentation. Working effectively, we make progress towards replacing each point in $1 + L$ by longer and longer initial segments of the ω-chain. This way, we end up with a computable $D \cong \omega \cdot (1+\alpha)+1$.

In each ω-chain that we build, the first point is always a limit point, and the rest are successor points. In the latter case, we can always compute the predecessor. In the former case, unless it is the left-most point coming from the first ω-chain in $\omega \cdot (1 + \alpha) + 1$, it is a limit point. If x is like that, then we can evidently list $\{y \in D : y < x\}$. We can uniformly choose an increasing, first-found subsequence of $\{y \in D : y < x\}$ to be the desired monotonic sequence $(y_i)_{i \in \mathbb{N}}$ converging to x from below. □

We now show that the D constructed in the lemma above can be turned into a notation of γ in $\mathcal{O}$. If we succeed, then Lemma 2.2.40 will imply that $\alpha < \gamma$ also has a notation in $\mathcal{O}$. To turn D into a notation in $\mathcal{O}$, we would like to define $n(x)$ which, given $x \in D$, outputs a notation for $\{y \in D : y < x\}$ in $\mathcal{O}$. We will define a function $f(e, x)$ that imitates the behaviour of $n(x)$, and then we will use the Recursion Theorem 2.1.3 to argue that for some e, $f(e, x) = \varphi_e(x)$; i.e., f "knows" its own index. Thus, in the definition below, e should be understood as the (intended) index of the procedure that we define.

Define a partial computable function $f(e, x)$ by recursion as follows:

1. If x is the least point in D, set
$$f(e, x) = 1.$$

2. If x is a successor point and y is its immediate predecessor, call $f(e, y)$ (according to these instructions applied to inputs e, y). If it halts, set
$$f(e, x) = 2^{f(e,y)}.$$

3. When x is a limit point, let $(s(x, i))_{i \in \mathbb{N}}$ be a uniformly computable sequence converging to x. The function $s(\cdot, \cdot)$ is a computable function by our assumption about D. Using the s-m-n Theorem 2.1.6, define ψ by the rule
$$\varphi_{\psi(e,x)}(i) = \varphi_e(s(x, i)),$$
and then set
$$f(e, x) = 3 \cdot 5^{\psi(e,x)}.$$

This completes the definition of f. Even if f is partial, the instructions describing f in terms of e still make sense. By the s-m-n Theorem 2.1.6, there is a total computable function g such that
$$\varphi_{g(e)}(x) = f(e, x),$$
for all e and x. By Recursion Theorem 2.1.3, we can assume that e is so that
$$\varphi_e(x) = \varphi_{g(e)}(x) = f(e, x),$$
for every x. By transfinite induction, $n(x) = \varphi_e(x)$ outputs a notation for $\{y \in D : y < x\}$ in $\mathcal{O}$. In particular, when applied to the greatest element of D it gives a notation for γ in $\mathcal{O}$, proving that $\alpha < \gamma$ is constructive as well. □

Before we proceed, we note that a slight modification of the proof of Lemma 2.2.41 shows that any c.e. presented well-order is isomorphic to a computable one (Exercise 2.2.44). Thus, ordinals do not distinguish between computable and c.e. presentations, up to isomorphism. This observation can be pushed to show that every hyperarithmetical ordinal has a computable copy (Exercise 8.1.25); the hyperarithmetical hierarchy, which we briefly discuss next, extends the arithmetical hierarchy. We also note that the Fellner-Watnick Theorem 3.2.23, which is the main result of Section 3.2.6, can also be viewed as a generalisation of Lemma 2.2.41.

Extending the arithmetical hierarchy*

It is possible to extend the arithmetical hierarchy beyond ω to the computable ordinals. The resulting hierarchy is called the hyperarithmetical hierarchy. For instance, we can uniformly effectively define

$$\varnothing^{(\omega)} = \oplus_{n \in \mathbb{N}} \varnothing^{(n)} = \{\langle m, n \rangle : m \in \varnothing^{(n)}\},$$

and then define Δ^0_ω to be the class of all sets computable relative to $\varnothing^{(\omega)}$. In the first part of the book, we shall never actually go beyond ω. In Chapter 8 of the book we will see that this process can be iterated beyond ω to (all) computable ordinals $\alpha < \omega_1^{CK}$.

Exercises

Exercise° 2.2.42. Show that if $\{\alpha_i : i \in \mathbb{N}\}$ is a computable collection of computable ordinals then $\sup\{\alpha_i : i \in \mathbb{N}\}$ is a computable ordinal.

Exercise° 2.2.43. Produce a formal detailed proof of Lemma 2.2.41.

Exercise° 2.2.44. Show that there is a computable procedure which, on input a c.e. presentation L of a well-order, outputs a computable copy of $\omega \cdot L$. Conclude that every c.e. presented ordinal has a computable copy.

2.2.4 Historical remarks*

Neither of the authors is a historian of mathematics. With some trepidation, we offer some observations concerning the development of computable algebra.

The history of mathematics, and of algebra in particular, is deeply entwined with computation. Almost all of pre-20th century mathematics was fundamentally algorithmic. A notable exception to this is Hilbert's Basis Theorem. Hilbert's Basis Theorem [251] proves that every algebraic set over a field can be described as the set of common roots of finitely many polynomial equations. Famously, Hilbert's proof does *not* actually compute this finite basis, but shows that the basis must exist. This result was quite controversial at the time. Gordon, the supervisor of Emmy Noether, was an expert in calculating invariants. He is supposed to have said of Hilbert's proof:

> "This is not mathematics; this is theology."

Although this is likely a myth (since there is no record of it until 20 or so years after Gordon's death), it does reflect the mathematical thinking of the day. But it is also a salutary lesson in computable mathematics. Suppose that we want to actually calculate the invariants guaranteed by Hilbert's Theorem, something which turns out to be quite important in physics. How do we do this? To calculate a finite basis, we need not just an existence proof, but a computable version of the algebraic result. In the case of Hilbert's Basis Theorem, to do this calculation, we are led to the modern theory of Gröbner bases. (We cite Buchberger [69], who actually invented Gröbner bases.) It is certainly the case that authors such as Kronecker were quite sceptical of non-constructive methods. Much of the classical algebra of the early 20th century was indeed computable. For example, we will later give a modern interpretation of the Kronecker-Herrmann [318, 247] results about finite extensions of fields, which demonstrates this effectiveness. Early editions of van der Waerden's seminal books [493] (supposedly based on Emmy Noether's notes) had algorithmic proofs of various results on rings and fields. Interestingly, in later editions, many such proofs were replaced by slicker but less constructive proofs. Metakides and Nerode [385] gave an overview of the introduction of non-computable methods in algebra in the 20th century.

As with the case of Gröbner bases, answering algorithmic questions often yields a much deeper understanding of the mathematics where the question arises. One particularly fine example of this phenomenon comes from Dehn's work [109] from the early 20th century. Dehn analysed algorithmic questions about finitely presented groups. Dehn gave geometric algorithms for solving the "word problem" for certain kinds of finitely presented groups. He showed that certain classes of finitely presented groups were not only c.e. presented but had the equality relation being computable. In our terminology, they are computable groups. Dehn noted that the methods were specific to certain classes of groups and did not apply in general. Based on this observation, he articulated the three questions which provided a significant impetus to the huge area now called *combinatorial group theory*:

1. Is every finitely presented group computable? (The word problem.)
2. Given x and y in a finitely presented group, can we algorithmically decide if x is conjugate to y? (The conjugacy problem.)
3. Given two finitely presented groups, can we algorithmically decide if they are isomorphic? (The isomorphism problem.)

Recall that in Example 1.2.3 we already encountered the celebrated theorem of Novikov [416] and Boone [51] stating that there is a group that is finitely presented in which the word problem is not algorithmically decidable. In the terminology of this book, there is a finitely presented (thus, c.e. presented) group with no computable presentation. We cite the book [343] and the survey [390] for a detailed exposition of the subject. Such investigations are closely related to decidability problems for simplicial complexes in topology that we will discuss shortly.

In fact, we now know that the answers to all three of Dehn's problems are negative. The techniques developed to answer these *algorithmic* questions, such as small cancellation theory, HNN extensions, and Higman's Embedding Theorem 2.2.11, have proven to be enormously influential in group theory (see, for example, Lyndon and Schupp [343]).

The modern study of computable abstract structure theory begins with the work of Fröhlich and Shepherdson [185], Rabin [440], and Mal'cev [345, 346], particularly focusing on structures that are either not groups or are not finitely generated. Fröhlich and Shepherdson studied computable field extensions, and Mal'cev and Rabin also laid the foundations of the general theory of computable structures that applies to arbitrary algebraic structures, not just groups or fields.

In modern terminology, Fröhlich and Shepherdson showed that there exist two algebraic closures of a computable field with no computable isomorphism between the closures. It is worth noting that Fröhlich and Shepherdson's proof actually recycles a proof from van der Waerden [493]. It is quite clear that although the authors of the early 20th century did not have access to a formal theory of computation, which awaited the work of Turing and others in the 1930s, they definitely had a sharp intuitive idea of what an algorithmic procedure was. We will see this demonstrated again in the work of Borel in analysis.

Metakides and Nerode [384] later extended the Fröhlich-Shepherdson result. They showed that computable algebraic closures are computably unique if and only if the field has a (separable) *spliiting algorithm*. By a splitting algorithm, we mean that there is a uniformly computable procedure which decides if a polynomial over the field is irreducible. If there is such a splitting algorithm, the "usual" construction of an algebraic closure becomes computable.

A hallmark paper was Rabin's [440], where he proved the influential result that a computable field has a computable algebraic closure, *in spite of the fact that the usual "adjoining roots" method may not be computable, as there may not be a splitting algorithm.* Since a computable field may lack a computable method of determining whether a given polynomial is irreducible, the classical construction cannot be performed effectively. The hidden message is that there is some *other way to construct algebraic closures* than the one usually taught to students.

Early work on the non-computability of aspects of computable groups, fields, and other algebraic structures usually involved coding the halting problem into the question at hand. Since the 1960s, a variety of combinatorial techniques have been developed to understand the classical theory of computation, such as complex priority arguments (Soare [477]), Π^0_1 classes (Cenzer [84]), effective measure theory (Downey-Hirschfeldt [125]), along with a strengthened understanding of the model theory of algebraic structures. As a consequence, over the past 60 years, the study of computable algebraic structures has grown into a technically deep theory. Early books on the subject include [158, 20, 159]. The theory has many aspects, and it is essentially impossible to cover all major topics of the theory in one book, so we will have to be selective. We will put much emphasis on the aspects of the theory that are related to computable analysis and computable topology. For other aspects of the theory, such as definability, Ash-Knight style forcing, and the true stages techniques, see the books of Montalbán [401, 402].

2.3 Computability for real functions

This section presents several early examples of applications of computability in elementary analysis. Some of the key notions introduced in this chapter will be important throughout the rest of the book.

Recall that a real ξ is computable if, for every n, we can compute $y \in \mathbb{Q}$ such that $d(\xi, y) < 2^{-n}$, where every rational is given as (e.g.) an irreducible fraction. A sequence $(y_i)_{i\in\mathbb{N}}$ of rational numbers with the property $d(y_i, y_{i+1}) < 2^{-i-1}$ is called a *fast Cauchy sequence.* If such a sequence converges to ξ, then it is called a *fast Cauchy name* of ξ. Note that $d(\xi, y_i) < 2^{-i}$. We will sometimes omit "fast" and say simply "Cauchy name", and we sometimes omit $\mathbb{N}$ in the subscript and write simply $(y_i)_i$. Thus, a real is computable iff it has a computable (fast) Cauchy name.

Notation 2.3.1 ($\mathbb{R}_c$). Let $\mathbb{R}_c$ denote the set of all computable reals.

There are several potential notions of computability for a real function $f : \mathbb{R} \to \mathbb{R}$. Unfortunately, the two most common definitions used in the literature are not equivalent, so in the end we will have to make a choice.

Exercises

Exercise° 2.3.2. Take for granted that there exists a non-computable c.e. low set $A \subseteq \omega$ (Theorem 3.1.1). Prove that in any non-empty interval of $\mathbb{R}$ there exists a left-c.e. real that is low but not computable, and a right-c.e. real that is low but not computable.

Exercise° 2.3.3 (First stated in Rice [449], but likely known earlier). Show that the collection of computable real numbers forms a real closed field.

Exercise 2.3.4. Recall that a real z is called *left-c.e.* if its left cut $\{q \in \mathbb{Q} : q \leqslant z\}$ is c.e. A real z is called *d.c.e.* or *weakly computable* if there exist left-c.e. reals x and y such that $z = x - y$.

1. Show that a real z is left-c.e. if and only if there is a computable non-decreasing sequence of rational numbers $\{q_i : i \in \mathbb{N}\}$ such that $\lim_{i\to\infty} q_i = z$.
2. (Ambos-Spies, Weihrauch, and Zheng [10]). Prove that z is weakly computable if and only if there exists a computable sequence of rationals $\{d_i\}$ such that $d_i \to z$ and $\sum_{n=0}^{\infty} |d_{n+1} - d_n| < \infty$.
3. (Ambos-Spies, Weihrauch, and Zheng [10]). Prove that the d.c.e. reals form a field.
4. (Ng [415], Raichev [442]). Prove that the collection of d.c.e. reals forms a real closed field.

2.3.1 The constructive approach to functions

As usual, let $\varphi_1, \varphi_2, \ldots$ be a standard enumeration of the partial computable functions. We call $e \in \mathbb{N}$ an *index* of $x \in \mathbb{R}_c$ if e is the index of a function φ_e that lists a (rational, fast) Cauchy name of x. We now consider two definitions of computability for a real function that refer only to inputs in $\mathbb{R}_c$.

Definition 2.3.5. A function $f : \mathbb{R}_c \to \mathbb{R}_c$ is *Markov computable* if there exists a partial computable function $\nu : \omega \to \omega$ such that, given any index e of $x \in \mathbb{R}_c$, $\nu(e)$ exists, and is an index of $f(x)$. We call the function $\nu : \omega \to \omega$ the *index function* of f.

The uniform (functional) version of this definition is as follows.

Definition 2.3.6. We call a function $f : \mathbb{R}_c \to \mathbb{R}_c$ *Borel computable* if there exists an oracle Turing Machine Φ such that, for all $x \in \mathbb{R}_c$, any computable Cauchy name $(x_i)_{i \in \mathbb{N}}$ of x, and any $n \in \mathbb{N}$,

$$(\Phi^{(x_i)_i}(n))_{n \in \mathbb{N}}$$

is a fast Cauchy name of $f(x)$.

The definitions above are not due to Markov and Borel (more in §2.3.4); however, we shall use this terminology because it appears to be standard (e.g., [24, p.22]). Borel computability seems to be stronger than Markov computability. Nonetheless, these definitions are equivalent, as we now show.

Theorem 2.3.7 (Kreisel, Lacombe and Shoenfield [317], Markov [353]). *A function f is Markov computable iff it is Borel computable.*

Remark 2.3.8. In the literature, Theorem 2.3.7 is sometimes referred to as the Kreisel–Lacombe–Shoenfield–Ceitin Theorem. However, in his paper, Ceitin [82] merely extends the much earlier result of Markov (announced in [350] and published in [353]) to computable Polish spaces. We leave this more general version (for Polish spaces) to Exercise 2.4.35 since its proof is not really that different from the proof we present below. Exercise 2.4.35 will find an unexpected application in Part 2 of the book. In §8.2.4, it will be used to derive a theorem about effective reductions between classes of countable algebraic structures. See also Exercise 2.4.36 for an analogous notion for linear operators on computable Banach spaces.

Proof of Theorem 2.3.7. ($\Leftarrow$) Use the s-m-n Theorem 2.1.6 to compute an index for $(\Phi^{(x_i)_i}(n))_{n \in \mathbb{N}} = f(x)$ from the index for $(x_i)_{i \in \mathbb{N}}$. For that, replace the oracle Turing machine with a Turing machine in which the computable oracle becomes a part of its program.

($\Rightarrow$) This implication is quite neat. Assume f is Markov computable, and let ν be its index function, so if e is an index of a fast Cauchy name of a computable real x, then the function $\nu(e)$ is total and lists a fast Cauchy name of $f(x)$. Our task is to produce a uniform procedure that has to do something with sequences that are not necessarily computable Cauchy names. Since every finite partial sequence is extendible to a computable one, we still need to define what the functional does on them.

The naive idea is to use the first found computable Cauchy name, say an eventually constant one, that agrees with the input on a long enough initial segment. The obvious danger is that we may define the functional inconsistently. Indeed, the outputs for different computable fast Cauchy names of x must converge to the same $f(x)$. It seems that to achieve this, one needs to be able to see the future.

The actual idea is to "set a trap" using the Recursion Theorem 2.1.13. (In fact, we shall use the Recursion Theorem with Parameters presented as Exercise 2.1.16.) Given φ_e, we will slow it down to define $\varphi_{n(e)}$, which tests what ν (given by Def. 2.3.5) does on index e. It will attempt to copy φ_e for a long enough initial segment, and then wait for an extension of this initial segment that gives a different output under ν. If it has the opportunity, it will choose the values that make the disagreement occur. The paradoxical self-referential nature of the Recursion Theorem will imply that for computable fast Cauchy sequences, from some point on, the disagreement cannot possibly happen. This will allow us to conclude that, no matter how we modify φ_e beyond this long enough initial segment, all possible computable sequences extending this initial segment will yield the same computation. In the definition of the functional, we will refer to $\varphi_{n(e)}$ when we choose our computable approximations to a possibly non-computable oracle. This will resolve the issue informally discussed earlier.

The exact details are a bit tedious, though certainly not difficult. Unless the reader really wants to understand the technical details, the clever formal proof below can be skimmed through, as these techniques won't really be used anywhere in the sequel.

Formal proof. We will use the following notation:

- For a finite tuple σ of natural numbers, let σ' be the infinite string with prefix σ in which the last bit of σ is repeated infinitely often. We identify σ' with a function taking i to the ith element $\sigma'(i)$ of σ', and with some index of this function uniformly computable from σ.
- We also restrict ourselves to σ that are valid partial fast Cauchy, thus in particular making σ' a fast Cauchy for every such σ. Note that ν must halt on each such σ'.
- For a function g, we write $g \restriction_{t+1}$ to denote the partial function that is the restriction of g to inputs $\{0, \ldots, t\}$. We also identify strings with partial functions whose domains are initial segments of ω. For instance, $g \restriction_{t+1}$ is identified with the string $\langle g(0), g(1), \ldots, g(t)\rangle$, provided that the values $g(0), g(1), \ldots, g(t)$ are defined. Also, for a finite string σ, $g \restriction_{t+1} \subseteq \sigma$ means that $g \restriction_{t+1}$ is a prefix of σ (and, in particular, that $g(0), g(1), \ldots, g(t)$ are defined).

Using the Recursion Theorem with Parameters (Exercise 2.1.16), define

$$\varphi_{n(e)} = \begin{cases} \uparrow & \text{if } \nu(e)\uparrow; \\ \varphi_e & \text{if } \nu(e)\downarrow,\ \varphi_e \text{ is fast Cauchy} \\ & \text{and either } \nu(n(e))\uparrow \text{ or } \nu(n(e))\downarrow \neq \nu(e); \\ \sigma' & \text{if } \nu(e)[s]\downarrow = \nu(n(e))[s]\downarrow, \{0, \ldots, s\} \subseteq \operatorname{dom}(\varphi_e), \\ & \sigma \supseteq \varphi_e \restriction_{s+1}, \text{ and } \nu(\sigma') \neq \nu(e); \\ \varphi_e \restriction_{s+1} & \text{otherwise,} \end{cases}$$

where s is the first stage at which $\nu(n(e))$ halts (if it does). Note also that we set it equal to $\varphi_e \restriction_{s+1}$ if s is the first bit at which φ_e fails to be fast Cauchy. To see how the Recursion Theorem with Parameters (Exercise 2.1.16) is used, replace $n(e)$ by z on the right-hand side. We have that the index of the left-hand side depends uniformly on e, z, and thus the function we define can be expressed as $\varphi_{h(e,z)}$. There is a computable n such that $\varphi_{h(e,n(e))} = \varphi_{n(e)}$.

If φ_e is indeed a fast Cauchy sequence, then (in particular) $\nu(e)$ is defined. We cannot have $\nu(n(e))$ divergent or unequal to $\nu(e)$ because then $\varphi_{n(e)}$ would just copy φ_e, thus contradicting the choice of ν. Therefore, $\nu(e) = \nu(n(e))$. In particular, there can be no σ' extending the restriction of φ_e to $\{0, \dots, s\}$ such that $\nu(\sigma') \neq \nu(e)$. Since for every σ' necessarily $\nu(\sigma') \downarrow$, it follows that $\nu(\sigma') = \nu(e)$ for any σ' that agrees with φ_e on inputs $0, \dots, s$. (Note that we only really used that $\nu(e)$ and $\nu(n(e))$ were both defined to conclude this.) Since we will wait for a disagreement forever, $\varphi_{n(e)} = \varphi_e \upharpoonright_{s+1}$ is partial.

Suppose now that we have a fast Cauchy φ_i for which $\nu(i)$ and $\nu(n(i))$ are also defined, and such that φ_i agrees with φ_e on inputs $\{0, \dots, s\}$. Here $s = s(e)$ is the parameter defined above corresponding to e and $n(e)$. Then, as before, we have $\nu(i) = \nu(n(i))$. Suppose $\nu(i) \neq \nu(e)$, where $\nu(i)$ halts in $s' = s(i)$ steps, and additionally $\mathrm{dom}(\varphi_i) \supseteq \{0, \dots, s'\}$. Without loss of generality, suppose $s' \geqslant s$; the case when $s \geqslant s'$ is symmetric. Then, using the same argument as we had for $\nu(e)$, we conclude that $\nu(i) = \nu(\tau')$, where τ is a long enough finite string that agrees with φ_i up to s'. Since $s' > s$ and φ_i agrees with φ_e on inputs $\{0, \dots, s\}$, τ' will also agree with φ_e on the first s inputs. But since we assumed that $\nu(i) \neq \nu(e)$ and $\nu(i) = \nu(\tau')$, this contradicts the third clause in the definition of $\varphi_{n(e)}$ because τ' could be taken there to play the role of σ'.

Define a Turing functional Φ as follows. On input $(a_i)_{i \in \mathbb{N}}$ and an integer n, perform the following steps. Keep verifying that $|a_i - a_{i+1}| < 2^{-i+1}$; if an i is found for which it fails, then declare the functional divergent. Simultaneously, search for an index e such that $\nu(e)$ is defined, $\nu(n(e))$ halts in s steps, and φ_e (viewed as a sequence) agrees with $(a_i)_{i \in \mathbb{N}}$ on the first s bits. Output

$$\Phi^{(a_i)_i}(t) = \varphi_{\nu(j)}(t) \quad \text{for } t \leqslant s$$

where j is an index for $a_0, a_1, \dots, a_s, a_s, a_s, \dots$.

We verify that Φ is well-defined. Suppose some other e', s', and j' are found. Because φ_e and $\varphi_{e'}$ share an initial segment of length equal to either s or s', we have

$$\varphi_{\nu(j)} = \varphi_{\nu(e)} = \varphi_{\nu(e')} = \varphi_{\nu(j')}$$

by the same argument as we had above.

Now, assume $(a_i)_{i \in \mathbb{N}}$ is a computable fast Cauchy sequence with index j. Then φ_j will be among the functions that satisfy the properties listed in the definition of Φ, although perhaps j will not be the first index found that satisfies these conditions. But as we have just seen, this makes no difference, as the result will still be $\varphi_{\nu(j)}$. $\square$

We remark that a lot of classical elementary real analysis can be effectivised using Markov computability. This is the gist of the book by Aberth [1].

2.3.2 The uniform approach to computability

We now consider several definitions of a computable real-valued function that are not restricted to computable reals. All these notions turn out to be equivalent.

Kleene's approach

Recall that a fast Cauchy name of a real x is a sequence $(x_i)_{i \in \mathbb{N}}$ of rationals such that $|x_i - x| < 2^{-i-1}$.

Definition 2.3.9. We call a function

$$f : \mathbb{R} \to \mathbb{R}$$

Type II (type two) computable if there is a Turing functional that, given a fast Cauchy name of x, outputs a fast Cauchy name of $f(x)$.

Kleene (e.g., [302]) defined Type II computability for the Baire space ω^ω rather than for the reals $\mathbb{R}$. In the context of $\mathbb{R}$, the definition (in its equivalent form) is usually attributed to Lacombe [326, 327] and Grzegorczyk [227]; we will discuss these equivalent definitions shortly in §2.3.2. This notion of computability is Type II in the sense that it is a functional, i.e., it maps not natural numbers to natural numbers but sequences to sequences, and sequences can themselves be viewed as functions. If natural numbers are Type 0 objects and functions are Type I objects, then functionals are Type II. We remark that Markov computability, at least as stated in Definition 2.3.5, is Type I.

Notice that Definition 2.3.9 is simply an extension of Borel computability to arbitrary (fast) Cauchy names. Consider also the following definition. We are not sure who was the first to use it, but a specific uniform version of this notion due to Lacombe and Grzegorczyk will be defined in the next paragraph.

Definition 2.3.10 (Effective continuity, the ϵ-δ version)**.** A function $f : \mathbb{R} \to \mathbb{R}$ is *effectively continuous* if there is a c.e. family F of pairs (D, E) of (indices of) basic open intervals with rational endpoints such that

(C1) for every $(D, E) \in F$, we have $f(D) \subseteq E$;

(C2) for every real x and every basic open $E \ni f(x)$, there exists a basic open D with $(D, E) \in F$ and $x \in D$.

In the definition above, every interval is represented by its rational centre and its rational radius; the Gödel number i of this pair is the index of the interval U_i. It is easy to see that the definition above is equivalent to its topological version, as stated below; we leave the verification of this equivalence as an exercise.

Definition 2.3.11 (Effective continuity)**.** A function $f : \mathbb{R} \to \mathbb{R}$ is *effectively continuous* if there is a c.e. set W so that for every open interval U_i,

$$f^{-1}(U_i) = \bigcup_{(i,j) \in W} U_j.$$

If there is such a W which is c.e. relative to X, then f is said to be X-effectively continuous.

So we obtain the trivial but important result:

Lemma 2.3.12. *A function $f : \mathbb{R} \to \mathbb{R}$ is continuous iff it is X-effectively continuous for some oracle X.*

We remark that in the definition of Borel computability, we do not have to require the Cauchy sequences to be computable, though we still need their limits to be computable reals for the operator to work correctly. It follows that *Borel computability corresponds to effective continuity restricted to* $\mathbb{R}_c$, meaning that we need to replace all basic open E and D with $E \cap \mathbb{R}_c$ and $D \cap \mathbb{R}_c$ throughout Definition 2.3.10, and require x (and $f(x)$) to be in $\mathbb{R}_c$. But of course, a Markov computable function does not have to be continuous (or even defined) on all of $[0, 1]$.

The lemma below is also folklore and is similar to Theorem 2.3.7, but it is much easier to prove.

Lemma 2.3.13. *The following are equivalent for a function* $f : \mathbb{R} \to \mathbb{R}$*:*

1. f *is effectively continuous;*

2. f *is Type II computable.*

The proof relies on the use principle and is left to Exercise 2.3.20. A complete proof of the more general Lemma 4.2.9 will be presented later. The following properties of Type II computable functions are immediate from the definitions.

Proposition 2.3.14 (Folklore). *We write "computable" for "Type II computable".*

1. *(Effective Bolzano-Weierstrass) If* $f : [0, 1] \to \mathbb{R}$ *is computable, then the reals* $\sup_{x \in [0,1]} f(x)$ *and* $\inf_{x \in [0,1]} f(x)$ *are computable.*

2. *If* f, g *are computable on* $[0, 1]$ *then so are the functions* $f + g$, $f - g$, $f \cdot g$, $\sup\{f, g\}$, $\inf\{f, g\}$, *and* αg *for any computable real* α.

Computability via uniform convergence

Working independently, Lacombe [326, 327] and Grzegorczyk [227] gave the following definition of a computable real-valued function on the unit interval. This notion can be extended to the whole of $\mathbb{R}$ by, for example, considering intervals of the form $[-n, n]$, but for simplicity, we shall restrict ourselves to $[0, 1]$ throughout this subsection.

Definition 2.3.15. A function $f : [0, 1] \to \mathbb{R}$ is Lacombe-Grzegorczyk computable if:

1. f maps every computable sequence of points into a computable sequence of points, and

2. f is effectively uniformly continuous, i.e., there is a computable function $h : \omega \to \omega$ such that, for all x, y and all n, if we have $|x - y| < \frac{1}{h(n)}$ then $|f(x) - f(y)| < 2^{-n}$.

Another related notion was coined by Caldwell and Pour-El; e.g., [432].

Definition 2.3.16. A function $f : [0, 1] \to \mathbb{R}$ is uniformly computable if for every n we can compute (the coefficients of) a polynomial $p(x) \in \mathbb{Q}$ such that

$$\sup_{x \in [0,1]} |f(x) - p(x)| \leqslant 2^{-n}.$$

Recall Definition 1.2.6 of a computable Polish space.

Fact 2.3.17. *Let $C[0,1]$ be the set of all real-valued continuous functions $f : [0,1] \to \mathbb{R}$ under the metric of uniform convergence*

$$d_{\sup}(f, g) = \sup_{x \in [0,1]} |g(x) - f(x)|.$$

Then the collection $P_{\mathbb{Q}}$ of all polynomial functions $\mathbb{Q}[x]$ restricted to $[0,1]$ is a computable dense subset of $C[0,1]$, in the sense that, given (tuples of rationals describing) $p, q \in P_{\mathbb{Q}}$ and $n \in \mathbb{N}$, $d_{\sup}(p, q)$ can be uniformly computed to precision 2^{-n}.

It follows that $f : C[0,1] \to \mathbb{R}$ is uniformly computable iff f is a computable *point* in the computable Polish space $(C[0,1], d_{\sup}, P_{\mathbb{Q}})$. Further equivalences are contained in the following theorem.

Theorem 2.3.18. *For a function $f : C[0,1] \to \mathbb{R}$, the following are equivalent:*

1. *f is Type II computable;*
2. *f is Lacombe-Grzegorczyk computable;*
3. *f is uniformly computable.*

The proof is left to Exercise 2.3.21.

Exercises

Exercise° 2.3.19. Show that the functions $\cos x$, $\sin x$, $\sqrt{x}$ (for $x \geqslant 0$) and $\log x$ are all (Kleene) computable, but step-functions with rational parameters are not (Kleene) computable relative to *any* oracle.

Exercise° 2.3.20. Prove Lemma 2.3.13.

Exercise° 2.3.21. Prove Theorem 2.3.18.

Exercise° 2.3.22. Show that if $f : [0,1] \to \mathbb{R}$ is (Type II) computable, and $f(0) < 0$ and $f(1) > 0$ then we can compute a real $x \in [0,1]$ such that $f(x) = 0$.

Exercise° 2.3.23. Verify Proposition 2.3.14 (2) above: If $f : [0,1] \to \mathbb{R}$ is (Type II) computable, then the reals $\sup_{x \in [0,1]} f(x)$ and $\inf_{x \in [0,1]} f(x)$ are computable.

2.3.3 Type I computability vs. Type II computability

A Markov computable function does not necessarily have to be defined on non-computable points, let alone be continuous over the entire interval $[0,1]$. Since every Type II computable function is necessarily continuous, these two notions of computability are clearly different. It is possible to build a Markov computable function with no continuous extension; see Exercise 2.3.26. However, such results are not very satisfying.

Indeed, suppose $f : [0,1] \to \mathbb{R}$ is Markov computable and additionally continuous. Does it have to be Type II computable? Following the general pattern of the book, we prove:

Theorem 2.3.24 (Specker [480]). *There is a continuous, Markov computable $f : [0,1] \to \mathbb{R}$ that is not Type II computable.*

Proof. By Proposition 2.3.14, it is sufficient to construct a continuous Markov computable function whose supremum is not computable. The proof proceeds through the following steps:

First, interpret 2^ω as the Cantor (middle third) set in $[0,1]$, such that the set of strings extending a given finite string σ is in a natural 1-1 correspondence with the respective clopen set, which is the Cantor set. Under this interpretation, which is the standard homeomorphic embedding $g : 2^\omega \to [0,1]$, computable reals in the Cantor set clearly correspond to computable paths through 2^ω.

Using Theorem 2.1.29, fix a Π^0_1 class $C \subseteq 2^\omega$ without computable members. It follows, for instance, that the closed set $P = g(C)$ has no computable points. Additionally, we can computably enumerate a collection of basic open rational intervals such that they together make up $[0,1]\backslash P$. For that, initiate the effective list of the "middle thirds" straight away. If a clopen set X is enumerated in $2^\omega\backslash C$, then also adjoin $g(X)$ to the list.

Now fix a real α that is not computable but is left-c.e., in the sense that the left cut $\{q : q < \alpha\}$ is computably enumerable. (See Exercise 2.3.4.) For example, take

$$\alpha = \sum_{i \in K} 2^{-2^i}.$$

Let $\alpha = \lim_s \alpha_s$ be an effective approximation of α from below. We can set

$$\alpha_s = \sum_{i \leqslant K_s} 2^{-2^i},$$

where K_s is the part of K enumerated by stage s. Although we do not necessarily know how close α_s is to α, we know that $\alpha_s \leqslant \alpha_{s+1} < \alpha$ for every s, and that $\lim_s \alpha_s = \alpha$.

Define a function $f : [0,1] \to \mathbb{R}$ by the following construction.

Construction.

At stage s, consider each of the finitely many intervals of the form $I = \left[\frac{k}{3^{-s}}, \frac{k+1}{3^{-s}}\right]$ that have been used in the definition of the Cantor set. If such an interval has not yet been declared out of P, then set f_s equal to α_s on I. At each interval outside the Cantor set, define f_s to be linear, using the values of f_s at the endpoints of the interval[2]. This finishes the description of f_s.

Set $f(x) = \lim_s f_s(x)$ for each $x \in [0,1]$.

End of construction.

Verification. It is clear that f is continuous, simply because it is the limit of the continuous f_s with the rate of uniform convergence of the sequence computable relative to K. We argue that the function f constructed above is Markov computable.

[2]Note that f_{s-1} could have already been defined equal to α_{s-1} on I at the previous stage. In this case, the value of f on I needs to be updated. Otherwise, if such an interval has left P at or before stage s, do not update f on I.

Let β be a computable point. Then it must be that $\beta \notin P$, and (as we mentioned above) it must be inside a rational interval that is either outside the Cantor set or will eventually be declared out of P. We can eventually see which of the two possibilities occurs.

If we see $\beta \notin g(2^\omega)$, the function is defined to be linear using its values at the endpoints of the respective interval that contains β and is outside $g(2^\omega)$. If δ and γ are the left-most and right-most points of the Cantor set relative to β, then note that both δ and γ are computable. In particular, they cannot be in P, and therefore there will be a stage when some intervals containing these points will be declared 'out'. At the stage at which this happens, we use the values of f at these two points and linearity to calculate $f(\beta)$.

If $\beta \in g(2^\omega)$, it must be that $\beta \notin P$. At the stage at which β is declared out (with a whole ternary interval), we use the value of f on this interval according to the construction.

It follows that for every computable point β, we can uniformly calculate the final value of $f(\beta)$. The procedure does not have to halt if β is not computable, but this is fine. It follows that f is Markov computable.

We claim that $\sup f = \alpha$, which is a non-computable left-c.e. real. This follows from the fact that P is non-empty, so at every stage there will be an interval that is not declared out, and the definition of f will keep getting updated.

Since α is not computable, f cannot be a Type II computable function, by Proposition 2.3.14. □

When restricted to continuous functions $[0,1] \to \mathbb{R}$, Markov (Type I) computable functions are not too different from Type II computable functions, in the following sense.

Proposition 2.3.25. *Every Markov computable continuous function $f : [0,1] \to \mathbb{R}$ can be computed using the halting problem.*

Proof. By Theorem 2.3.18, it is sufficient to use the halting problem to calculate, given $\epsilon = 2^{-n}$, a $\delta = 2^{-m}$ such that

$$|x - z| < \delta \to |f(x) - f(z)| \leqslant \epsilon.$$

For a fixed $\delta = 2^{-m}$ and $\epsilon = 2^{-n}$, the collection of all x, z for which the above implication holds forms a closed subset of $[0,1]^2$. The condition holds for all x and y iff the open complement of this set is empty. Thus, if it *fails*, then it must be witnessed by some rational x and z. The careful choice of non-strict and strict inequalities, along with the Markov computability of f, implies that the failure of $|x - z| < \delta \to |f(x) - f(z)| \leqslant \epsilon$ is a c.e. condition. Indeed, the existence of rational x, y such that

$$|x - z| < \delta \,\&\, |f(x) - f(z)| > \epsilon$$

can be discovered at a finite stage by calculating f on more and more rational inputs. It follows that we can express that $\delta = 2^{-m}$ does not work for a fixed $\epsilon = 2^{-n}$ as an existential statement of the form

$$\exists k R(k, m, n),$$

where R is a computable relation. This implies that, given n, we can computably enumerate all m for which it fails. With the help of the halting problem, we can decide whether the procedure of searching for a counter-example halts. Therefore, given n, we can compute the least m that works for n. □

Recall that, for algebraic structures, if we had access to the halting problem, we could decide equality in a c.e. presented structure. In this sense, for continuous functions, Markov computability is related to Type II computability similarly to how c.e. presentations are related to computable presentations in algebra. Likewise, Type II computable functions are more "honestly" computable in the sense that they are (analytically) uniformly computable. In view of the equivalence of all uniform definitions we have seen so far (e.g., Theorem 2.3.18), we will adopt the following unified terminology:

A real function is *computable* if it is Type II computable.

Throughout the rest of the book, when we say that a continuous function is "computable", we mean that it is "Type II computable".

Exercises

Exercise° 2.3.26 (Specker [480]). Use the method of proof of Theorem 2.3.24 to construct a Markov computable function $f : [0,1] \to \mathbb{R}$ with no continuous extension on $[0,1]$.

Exercise 2.3.27 (Myhill [414]). Show that there exists a computable function $[0,1] \to \mathbb{R}$ which is differentiable, but does not have a computable derivative.

Exercise 2.3.28 (Pour-El and Richards [434]). Show that if the second derivative of a computable function $f : [0,1] \to \mathbb{R}$ exists and is continuous (but is not necessarily computable), then the derivative of f is computable.

Exercise* 2.3.29 (Pour-El and Richards [434]). Show that there exists a computable function $f : [0,1] \to \mathbb{R}$ such that, for every n, the nth derivative $f^{(n)}$ exists (and, thus, is computable by the previous exercise), but $f^{(n)}$ is not uniformly computable in n.

2.3.4 Historical remarks*

The theory of computable analysis can be traced back quite a long way. Nearly four thousand years ago, the Babylonians (see [174]) calculated $\sqrt{2}$ correctly as

$$1 + \frac{24}{60} + \frac{51}{60^2} + \frac{10}{60^3} \approx 1.414213.$$

There is a long history of mathematicians trying to calculate important real constants such as e and π explicitly, and we can see a direct line of reasoning to the work of Cauchy and other analysts of the 18th and 19th centuries. In 1912, Borel [53] gave an informal definition of a computable real number:

> "We say that a number α is computable if given a natural number n, we can obtain a rational number q that differs from α by at most $\frac{1}{n}$."

Borel had no formal definition of a computable procedure, so his definition of "we can obtain" was vague, although he said that "the operation can be executed in finite time with a safe method that is unambiguous."

In his famous papers [487, 488], Turing initiated the theory of computable real number functions. Indeed, he mentioned that one of his motivations for the invention of his famous machine model was to describe real number computation. The term "computable real number" first appeared in Turing [487]. However, Turing himself did not systematically develop the theory of computable real number functions, which was taken up by Banach and Mazur [29]. They used the following notion: A function $f : \mathbb{R}_c \to \mathbb{R}_c$ is said to be *Banach-Mazur computable* (*sequentially computable*) if f maps any given computable sequence $(r_i)_{i\in\mathbb{N}}$ of real numbers into a computable sequence of real numbers $(f(r_i))_{i\in\mathbb{N}}$. A function that is Markov computable must be Banach-Mazur computable, and while the converse holds in some cases, it is not true in general [248]. Banach-Mazur computability is perhaps too general for our purposes because it characterises functions as computable even if they may not be computed in the typical sense, i.e., by a Turing machine. Their investigations were unfortunately interrupted by the Second World War.

The subject was revived in the mid 1950s by Grzegorczyk and Lacombe (e.g., [226, 227, 225, 326, 327]) who laid out the foundations in detail. Around this time, there was a lot of material that might best be called "computable calculus", and in particular functions computable when restricted to the field of computable real numbers in $[0, 1]$.

The Russian/Soviet school also adopted Turing's ideas and further developed them (e.g., [351]), and today Turing's computable function is more commonly known as Markov computable. Much of this work was by authors such as Zaslavskii and Ceitin in papers including [514, 515, 81]. We cite the book of Aberth [1] for a detailed exposition of some of these early results. For many bibliographic references covering results obtained before 1998, see [63].

The Type II approach to computability is commonly attributed to Kleene (e.g., [302]), although many equivalent definitions were introduced by others around the same time. This approach to the computability of continuous functions has become standard in the modern literature [505, 435]. In this book, we exclusively focus on continuous functions, making this approach perfectly suitable for our purposes. One can extend this idea to Type III by considering maps between functionals, etc.; see Kleene [302] and Ershov [155] for a detailed study of computability in higher types.

2.4 Computable separable structures

In this section, we give the definitions of a computable Polish group and a computable Banach space, and we provide examples. Following the general pattern of the chapter, we mainly focus on establishing the foundations, i.e., we analyse and compare the definitions rather than apply them to prove some effective analytic results.

2.4.1 The basic definitions

Recall the definition of a computable Polish space (Definition 1.2.6). Let (M, d, X) be a computable Polish space, where X is a dense countable sequence in M that can be identified with ω. Elements of X are called special points.

A point $\xi \in M$ is computable if, for every n, we can compute $x \in X$ such that $d(\xi, x) < 2^{-n}$. A sequence $(x_i)_{i\in\mathbb{N}}$ in X with the property $d(x_i, x_{i+1}) < 2^{-i-1}$ is called a fast Cauchy sequence. If such a sequence converges to ξ, then it is called a (fast) Cauchy name of ξ. Note that $d(\xi, x_i) < 2^{-i}$. A point is computable if it has a computable fast Cauchy name.

It is not hard to check that if (M_i, d_i, X_i), for $i = 0, \ldots, k$, are computable Polish spaces, then so is

$$\left(\prod_{i\leqslant k} M_i, d, \prod_i X_i\right),$$

where, for $\bar{x} = (x_0, \ldots, x_k)$ and $\bar{y} = (y_0, \ldots, y_k)$, the metric is given by $d(\bar{x}, \bar{y}) = \sum_i d(x_i, y_i)$. Similarly, the product of uniformly computable sequences of computable Polish spaces is also a computable Polish space under the metric $\sum_{i\in\mathbb{N}} 2^{-i} d(x_i, y_i)$.

Convention 2.4.1. We will often identify a computable Polish space (M, d, X) with its domain M to simplify notation. However, the reader should keep in mind that M can have many different dense subsets, and we hope that this relaxed notation will not cause any confusion. We will emphasise which dense set is used when necessary. Additionally, the metric may or may not be fixed; this will also be clear from the context.

Definition 2.4.2. Let X and Y be computable Polish spaces. We say that

$$f : X \to Y$$

is *computable* if there is a Turing functional which, on input a fast Cauchy name of x in X, outputs a fast Cauchy name of $f(x)$.

The definition above is a direct generalisation of Definition 2.3.9 of Type II computability for real functions. Lemma 2.3.13 will be extended to cover arbitrary computable Polish spaces (Lemma 4.2.9). Thus, every computable f has to be (effectively) continuous.

Fix the standard computable presentation of the reals using $\mathbb{Q}$ and the usual metric. Also, fix the computable presentation of the complex numbers given by $\mathbb{Q} + i\mathbb{Q}$ and the complex norm.

Definition 2.4.3. Let M be a computable Polish space. A computable n-ary operation on M is computable function of the form

$$F : X_1 \times X_2 \ldots \times X_n \to X_{n+1},$$

where each X_i is either M or $\mathbb{R}$ or $\mathbb{C}$.

We also view constants as computable operations that, on any input, produce the same computable point. The definition below is essentially due to Stoltenberg-Hansen and Tucker [484].

Definition 2.4.4. A *computable Polish algebra* is a computable Polish space M together with a sequence of (uniformly) computable operations on M.

The definition above should be compared with the definition of a (discrete) computable structure due to Malcev and Rabin (Def. 1.2.1). In fact, it can be viewed as a generalisation of it in the following sense. Given a computable algebraic structure M upon $\mathbb{N}$, introduce the trivial discrete metric by the rule $d(x, y) = 1$ iff $x \neq y$. This metric turns the computable algebraic structure into a computable Polish algebra.

Perhaps the most important examples of computable Polish algebras are computable Banach spaces and computable Polish groups, which we discuss next.

2.4.2 Computable Polish groups

A computable Polish group is a computable Polish algebra of the form $(G, \cdot, ^{-1})$ that happens to be a group. In the computably compact case (Definition 1.2.7), we can drop the computability of $^{-1}$ from the definition without any loss of generality; this will appear later as Corollary 4.2.46. We will see shortly that the same can be said about the additive groups of real Banach spaces.

Examples of computable Polish groups

We give a few examples of computable Polish groups that are not associated with Banach spaces.

Lemma 2.4.5. *The following Polish groups (with their natural topologies) admit a computable Polish presentation.*

1. *The unit circle group* $\mathbb{R}/\mathbb{Z}$ *(under* $+$*).*
2. *The 3D rotation (special orthogonal) group* $SO(3)$*.*
3. *The additive group of* p*-adic integers.*
4. *The infinite symmetric group* S_∞ *of all permutations of* $\mathbb{N}$*.*

Proof. 1. is elementary, and the verification of 2. and 3. is left to Exercise 2.4.12.

We verify 4. This particular presentation was suggested in [220]. The obvious difficulty is that even if a map $f : \mathbb{N} \to \mathbb{N}$ is injective, we can never be sure if it is surjective, and thus we can never decide whether we should keep approximating it. In our presentation, an element of S_∞ is given as a path through a tree that consists of pairs (h, h^{-1}) where h is a permutation of $\mathbb{N}$. This ensures, via a local condition given by the tree, that the path encodes a permutation.

For strings σ_i, $i = 0, 1$ with natural number entries, and of the same length N, by $\sigma_0 \oplus \sigma_1$ we denote the string of length $2N$ which alternates between σ_0 and σ_1. That is, $(\sigma_0 \oplus \sigma_1)(2i+b) = \sigma_b(i)$ for $i < N$, $b = 0, 1$. The domain of our approximation structure for S_∞ is the computable tree of strings

$$\mathit{Tree}(S_\infty) = \{\sigma \oplus \tau \colon \sigma, \tau \text{ are 1-1} \wedge \sigma(\tau(k)) = k \wedge \tau(\sigma(i)) = i \text{ whenever defined}\}.$$

For functions f_0, f_1 on $\mathbb{N}$, we define a function $f_0 \oplus f_1$ on $\mathbb{N}$ by $(f_0 \oplus f_1)(2i + b) = f_b(i)$.

We view S_∞ as the group of objects of the form $h \oplus h^{-1}$ where h is a permutation of $\mathbb{N}$. Our concrete presentation of S_∞ is the group defined on the paths of $\mathit{Tree}(S_\infty)$. We view the paths through the tree as an ultrametric space (i.e., $d(x, z) \leqslant \max\{d(x, y), d(y, z)\}$), in which the (ultra)metric is the usual longest common prefix metric[3], and the dense set is given by permutations having finite support. If $f = f_0 \oplus f_1$ and $g = g_0 \oplus g_1$ in S_∞, we define $f^{-1} = f_1 \oplus f_0$ and $gf = (g_0 \circ f_0) \oplus (f_1 \circ g_1)$. It is immediate that these operations are computable. □

Further examples of computable Polish groups are provided by the following elementary lemma that first appeared in [313].

Lemma 2.4.6. *A discrete group admits a computable Polish presentation iff it is computably presentable (in the sense of Definition 1.2.1).*

Proof. If a discrete group is computably presented, then we use the trivial discrete metric

$$d(x, y) = 1 \text{ iff } x \neq y$$

to turn it into a computable Polish group.

Suppose the group G is computable Polish with respect to some metric compatible with its discrete topology and which makes the operations computable. All points in a discrete group are always special because all points are isolated. In particular, the identity e has to be special.

Consider the free group F (freely and formally) generated by the special points. Every finite "word" in the language of the group is a product of special points and their inverses[4].

Since the identity e is isolated, there is a rational $r > 0$ so that the basic open ball $B(e, r)$ contains only e,

$$B(e, r) = \{e\}.$$

To see whether $y = e$, compute the real $d(e, y)$ to precision $r/8$. If this approximation is $< r/4$, then $y = e$. Otherwise, it must be $> r/2$; in this case, $y \neq e$. Note that these properties are mutually exclusive and both are computably enumerable (c.e.). Thus, "$y = e$" is a decidable property for any computable point y in the group.

It follows from the argument above that we can decide when a word in F is equal to the identity in our group G. Then $G \cong F/R$, where R is a computable normal subgroup of F generated by the

[3] The distance is set equal to 2^{-n}, where n is the length of the longest common prefix of two strings.

[4] In particular, each such word can be viewed as a computable point uniformly in the indices of these special points used to form the word. For example, $x_{17}^{-1} x_{22} x_{17}$ is a computable point whose fast Cauchy name can be produced uniformly in the indices 17 and 22. Of course, $x_{17}^{-1} x_{22} x_{17}$ is equal to some special point x_i; but for our proof to work, we do not need to know which x_i it is exactly.

elements ("words") that are equal to e in G. As we discussed earlier, such groups admit computable presentations (Exercise 2.2.21). □

It can be shown that computable Polish groups are also closed under finite (or effectively countably infinite) direct products (Exercise 2.4.14). We remark that Lemma 2.4.6 has an analogue for right-c.e. Polish and c.e. presented Polish groups (Exercise 2.4.28); we leave the somewhat technical definition of a right-c.e. Polish group to Exercise 2.4.27. Thus, Theorem 2.2.6 implies that there is a right-c.e. Polish group that is not homeomorphic to any computable Polish group.

Computably compact groups

Recall that a computable Polish space M is called *computably compact* if there exists a computable function that, given n, outputs a finite cover of M by open balls centred on special points and having rational radii $< 2^{-n}$ that cover M.

Definition 2.4.7. A *computably compact group* is a computable Polish group that is additionally computably compact. That is, it is a computably compact space with computable group operations upon the space.

As we mentioned earlier, the computability of the inverse operation can be derived from the computability of the product; we delay the verification of this fact until Corollary 4.2.46. It is not hard to see that the groups from 1. and 3. of Lemma 2.4.5 are indeed computably compact groups (Exercise 2.4.13), but the verification of the computable compactness of $SO(3)$ is delayed until Exercise 4.2.64. Further examples can be obtained by noting that computably compact groups are closed under taking direct products (Exercise 2.4.15). Non-trivial examples of computably compact connected and profinite groups will be given in Chapters 5 and 9. In this subsection, we only briefly discuss another interesting related notion due to Turing.

In [489], Turing defines an *ϵ-approximation* to a Polish group $(G, \cdot)$ as a finite group $(X_\epsilon, \star_\epsilon)$ which is ϵ-dense in G, i.e.:

1. $X_\epsilon \subseteq G$ and for each $g \in G$, there exists an $x \in X_\epsilon$ with $d(g, x) < \epsilon$, and
2. for every $x, y \in X_\epsilon$,
$$d(x \cdot y, x \star_\epsilon y) < \epsilon.$$

Note that Turing did not require the operation on X_ϵ to be the same as the operation on G.

Definition 2.4.8 (Turing [489]). A Polish group is *approximable* if for every $\epsilon > 0$, it has an ϵ-approximation.

Every approximable Polish group is totally bounded and, thus, compact. While Turing did not define the following notion, his non-effective definition above was arguably motivated by algorithmic intuition.

Definition 2.4.9 (Essentially Turing [489]). A computable Polish group G is *computably approximable* if, for every $n > 0$, we can compute the strong index (i.e., a code of the finite tuple representing) of a 2^{-n}-approximation $(X_n, \star_n)$ to G, where X_n consists of special points.

It is easy to see that every computably approximable group is computably compact. In Chapter 4, we will see that every computably compact profinite group is also computably approximable (Corollary 4.2.108), and it follows from the materials of Chapter 5 that every connected computably compact abelian group is also computably approximable (Exercise 5.2.27). However, to draw these conclusions, we will have to develop quite a bit of machinery. Here, we only state the following neat result that should be compared to Theorem 2.2.2.

Theorem 2.4.10 (Essentially Turing [489])**.** *Every approximable Lie group*[5] *is* computably *approximable and, thus, admits a computably compact presentation. Indeed, such groups are exactly the compact abelian Lie groups.*

Proof sketch. In [489], Turing demonstrated that every approximable Lie group has to be compact and abelian. It is well-known ([429]) that the compact abelian Lie groups are exactly the groups of the form

$$F \times \mathbb{T}^k,$$

where F is finite abelian, and $\mathbb{T} \cong (\mathbb{R}, +)/(\mathbb{Z}, +)$ is the unit circle group from Lemma 2.4.5(1). It is routine to check that such groups are computably compact (Exercises 2.4.14 and 2.4.15) as witnessed by their ϵ-approximations by finite subgroups. □

In particular, the Lie group $SO(3)$ is computably compact (Exercise 4.2.64) but is not approximable (since it is not abelian). It follows that computable approximability is strictly stronger than computably compact presentability among compact Lie groups. We note that there is an approximable computable Polish group that is not isomorphic to any computably approximable one; see Exercise 9.5.17. We are not aware of any further work in computable analysis that would systematically study approximability of (compact) Polish groups.

Exercises

Exercise° 2.4.11 (Folklore)**.** Let $f : X \to Y$ be a computable surjective isometry between computable Polish spaces X and Y. Show that f^{-1} is also computable.

Exercise° 2.4.12. Finish the proof of Lemma 2.4.5.

Exercise° 2.4.13. Show that the groups in parts 1 and 2 of Lemma 2.4.5 are computably compact.

Exercise° 2.4.14. Show that the direct product $G \times H$ of computable Polish groups G and H is also a computable Polish group. Produce a computable presentation of $G \times H$ in which the projections onto the components are computable maps.

Exercise° 2.4.15. Show that the direct product $G \times H$ of computably compact groups is also computably compact.

[5] A Lie group is a smooth manifold with a differentiable group operation.

2.4.3 Computable Banach spaces

We now turn to Banach spaces. We are mainly interested in Banach spaces over $\mathbb{R}$, but of course, the definition below can be easily extended to cover $\mathbb{C}$-spaces.

Definition 2.4.16. A computable (real) Banach space is a computable Polish algebra of the form $\mathbb{B} = (B, +, \cdot, 0)$, where 0 is a constant, the metric on B induces the norm $||x|| = d(0, x)$, which, together with $+$ and the scalar product $\cdot : \mathbb{R} \times B \to B$, makes B a Banach space.

The definition above is a reformulation of the approach taken in, e.g., Pour-El and Richards [435].

Lemma 2.4.17 (Melnikov and Ng [376]). *For a real Banach space $\mathbb{B}$ upon a fixed computable Polish space induced by the norm, the following are equivalent:*

1. *$(B, +, -)$ is a computable Polish group;*
2. *$(B, +, \cdot, 0)$ is a computable Banach space.*

Furthermore, in 1. *computability of $+$ implies computability of the additive inverse $-$.*

Proof. The implication $2 \to 1$ is trivial because $-x = (-1)x$. We prove $1 \to 2$.

To see that 0 is a computable point, fix any special point x from the dense sequence in B and calculate $0 = x - x$. This makes the norm $||x|| = d(x, 0)$ computable. Clearly, for every $n \in \mathbb{N}$, we can uniformly compute the point

$$\underbrace{x + x + x + \ldots + x}_{x \text{ occurs } n \text{ times}},$$

and similarly when $n \in -\mathbb{N}$. This makes the operations $x \to nx$ uniformly computable when $n \in \mathbb{Z}$.

Note that, for an integer $m > 0$, $||my - x|| < \epsilon$ is equivalent to $||y - \frac{1}{m}x|| < \frac{\epsilon}{m}$. Thus, to compute $\frac{1}{m}x$ to precision 2^{-n}, we have to find a special point y such that my is $m2^{-n}$-close to x. Here we do not assume that x is a computable point. Instead, we assume that its fast Cauchy name is given to us (as an oracle). Since this search is uniform in x and m, we conclude that the operations $x \to \frac{1}{m}x$ are uniformly computable.

If a is a real and r is a rational approximation to a, we have

$$||rx - ax|| = |r - a|||x||.$$

So, to compute ax to precision 2^{-n}, we need to choose a rational r such that

$$|r - a| \leqslant \frac{2^{-n-1}}{\max\{||x||, 1\}}$$

and calculate rx to precision 2^{-n-1}. It follows that the uniform computability of scalar multiplication by rationals implies computability of scalar multiplication by reals. This finishes the proof of $1 \to 2$.

To see why $+$ additionally determines $-$, search for a point y such that $||x + y|| < 2^{-m}$, which is the same as to say that $||y - (-x)|| < 2^{-m}$, i.e., y is 2^{-n}-close to the point $-x$. □

In other words, assuming that we fix the metric associated with the norm on a separable (real) Banach space $\mathbb{B}$, we have:

$\mathbb{B}$ is computable Banach iff its additive group is computable Polish.

This fact simplifies the verification of computable presentability of Banach spaces.

Example 2.4.18. The following separable (real) Banach spaces are computable:

1. Separable Hilbert spaces.
2. The space $C[0,1] = C([0,1];\mathbb{R})$ of continuous functions on the unit interval, under point-wise operations and the supremum metric. The dense sequence is given by the polynomials with rational coefficients. Alternatively, we can use piecewise linear functions with rational parameters. (See also Fact 2.3.17.)
3. The spaces ℓ_p, where p is a computable real. Sequences with finite support form a computable dense set.
4. The spaces $L_p[0,1]$, where p is a computable real, and the dense sequence is given by, e.g., step functions with rational parameters.
5. The space $C^{(n)}([0,1];\mathbb{R})$ of n-times continuously differentiable functions $[0,1] \to \mathbb{R}$, under the norm
$$\sum_{i \leqslant n} \sup_{x \in [0,1]} |f^{(i)}(x)|$$
and the point-wise operations. The dense sequence is again given by polynomials with rational coefficients.

Some commonly encountered Banach spaces are not computably presentable. For example, ℓ_∞, the space of all bounded real sequences with the supremum norm, is not computably presentable as it is not even separable. In Corollary 4.2.114, we will see that there exists an $\varnothing'$-computable Banach space of the form $C(K;\mathbb{R})$, where K is compact, that is not isomorphic to any computable Banach space. Further, it can be shown that there is a low, right-c.e. presented Banach space not linearly isometric to any computable Banach space (Exercise 2.4.40).

Reconstructing the group operation from the norm

Mazur and Ulam (see, e.g., [469, 492]) showed that every surjective self-isometry of a Banach space is affine, and thus, when the norm is fixed, there is only one way to define $+$ to get (the additive group of) a Banach space. Is this effectively true? In other words, can we strengthen Lemma 2.4.17 by dropping the computability of $+$ as well?

We will answer this question in the affirmative for separable Hilbert spaces, where $+$ can be effectively recovered from the norm. As the central result of this section, we will prove that there is a computable dense sequence in $C[0,1]$ that computes the norm but does not compute $+$. These results will later allow us to draw conclusions about the effective uniqueness of computable presentations of these spaces.

Hilbert spaces

The proposition below states that in Hilbert spaces, the norm effectively determines the group structure (thus, also the linear space structure by Lemma 2.4.17).

Proposition 2.4.19 (Melnikov [369]). *Suppose a Hilbert space* $\mathbb{H}$ *has a computable Polish presentation in which* 0 *is a computable point. (This makes the norm computable.) Then the operation* + *is also computable with respect to this presentation.*

Proof. Fix a dense sequence $(\alpha_i)_{i\in\mathbb{N}}$ in which the vector 0 is a computable point. Recall that $d(x, y) = ||x - y||$. For instance, $||x|| = d(0, x)$ is computable for every computable point x. It is well-known that the parallelogram identity

$$||x + y||^2 + ||x - y||^2 = 2||x||^2 + 2||y||^2$$

characterises Hilbert spaces within the class of Banach spaces. We show that the operation + is computable with respect to (w.r.t.) $(\alpha_i)_{i\in\mathbb{N}}$.

Recall that points in the fixed dense sequence in a computable Polish or Banach space are called *special.* Given a positive rational $\epsilon < 1$ and (fast) Cauchy names for points x and y, find a special point z such that:

1. $|\,||z||^2 + ||x - y||^2 - 2||x||^2 - 2||y||^2\,| < \delta$,
2. $|\,||y - z|| - ||x||\,| < \delta$,
3. $|\,||x - z|| - ||y||\,| < \delta$,

where $\delta = \epsilon/(2||x|| + 2||y|| + 3)$. We may assume that $\delta < 1$. Applying the parallelogram identity, we obtain

$$|\,||z||^2 - ||x + y||^2\,| < \delta.$$

Using the well-known formula for the inner product, we get

$$||x + y - z||^2 = ||x + y||^2 + ||z||^2 - 2\langle x, z\rangle - 2\langle y, z\rangle.$$

Applying this formula again, we obtain

$$||y - z||^2 = ||y||^2 + ||z||^2 - 2\langle y, z\rangle$$

and

$$||x - z||^2 = ||x||^2 + ||z||^2 - 2\langle x, z\rangle.$$

We combine the three equations above:

$$||x + y - z||^2 = (||x + y||^2 - ||z||^2) + (||x - z||^2 - ||y||^2) + (||y - z||^2 - ||x||^2).$$

Taking into account $\delta < 1$, observe that

$$\begin{aligned}|\,||x - z||^2 - ||y||^2\,| &= |\,||x - z|| - ||y||\,| \cdot (||x - z|| + ||y||)\\ &< \delta(||y|| + \delta + ||y||)\\ &< \delta(2||y|| + 1),\end{aligned}$$

and similarly

$$|\,||y-z||^2 - ||x||^2\,| < \delta(2||x||+1).$$

Thus,

$$\begin{aligned}||x+y-z||^2 &\leqslant |\,||x+y||^2 - ||z||^2\,| + |\,||x-z||^2 - ||y||^2\,| + |\,||y-z||^2 - ||x||^2\,| \\ &< \delta + \delta(2||y||+1) + \delta(2||x||+1) \\ &= \delta(2||x|| + 2||y|| + 3) \\ &= \epsilon.\end{aligned}$$

It follows that we can produce a (fast) Cauchy name for $x+y$ uniformly in (fast) Cauchy names for x and y. □

The space $C[0,1]$

We write $C[0,1]$ to denote the space of continuous real-valued functions on the unit interval; see Example 2.4.18 for the definition. We fix the supremum metric

$$d(f,g) = d_{sup}(f,g) = \sup_{x\in[0,1]} |f(x) - g(x)|$$

associated with the supremum norm. We also fix the computable dense sequence $(l_i)_{i\in\mathbb{N}}$ consisting of piecewise linear functions with finitely many rational breaking points. The theorem below states that the supremum norm on $C[0,1]$ does not effectively determine the operation $+$. Therefore, the computability of $+$ cannot be dropped from the definition of a computable Banach space without consequences. This also means that the aforementioned theorem of Mazur and Ulam does not hold computably.

Theorem 2.4.20 (Melnikov [369]). *There is a computable presentation of the Polish space $(C[0,1], d_{sup})$ in which the constant function $\mathbf{0}$ is a computable point but the operation $+$ is not computable.*

In other words, we shall construct a computable presentation of $(C[0,1], d_{sup})$ in which the norm is computable, but $+$ is not. The computability of $\mathbf{0}$ will be used later in a corollary.

We remark that, in the proof below, we split the main task into a number of sub-tasks that are called *requirements*. We then prove the theorem by satisfying ("meeting") these requirements one by one. A special technique, called the priority technique, can sometimes help to meet all the requirements even if they conflict with each other. However, as we will explain in due course, in the present proof such conflicts can be completely avoided, and thus the proof is *injury-free*. Before we turn to the formal proof, we informally explain the main idea behind a typical injury-free proof.

Requirements and direct diagonalisation. We have already seen several proofs with relatively non-trivial constructions, e.g., Mal'cev's Theorem 2.2.16 and Specker's Theorem 2.3.24. However, the former proof coded K (the halting problem) into the linear independence relation in a vector space, and the latter coded K into the supremum of a continuous Markov computable function.

The present proof is a bit more complex than that, since it will rely on the direct diagonalisation technique rather than on coding.

Typically, in such arguments, one has some overall goal that one breaks down into smaller subgoals ("requirements") for which it is argued that they are all eventually met in the limit. As an archetype for such proofs, think of Cantor's proof that the collection of all infinite binary sequences is uncountable. One can conceive of the proof as follows.

Suppose we could list a collection of binary sequences $\mathcal{S} = \{S_0, S_1, ...\}$ with $S_e = s_{e,0}s_{e,1}...$. Our goal is to construct a binary sequence $U = u_0u_1...$ that is not on the list $\mathcal{S}$. This should be thought of as a game against our opponent who must supply us with $\mathcal{S}$. We shall construct U in stages, at stage t specifying only $u_0...u_t$, the initial segment of U of length $t + 1$.

Our *requirements* are the decomposition of the overall goal into subgoals of the form

$$R_e : U \neq S_e,$$

one for each $e \in \mathbb{N}$. Of course, we know how to satisfy these requirements. At stage e, we simply make $u_e \neq s_{e,e}$ by setting $u_e = 1$ or 0, and making $u_e = 1$ iff $s_{e,e} = 0$. Hence for all e, $U \neq S_e$; all the requirements are met. This is a contradiction to the fact that $\mathcal{S}$ supposedly lists all infinite binary sequences, as U is a binary sequence.

We now turn to the proof of Theorem 2.4.20. The proof that we present here is different from the proof in [369]; it can be found in [376].

Proof idea. We build a computable Polish presentation $X = (h_i)_{i\in\mathbb{N}}$ of $(C[0,1], d)$ which consists of points of the form $h_i = \lim_s h_{i,s}$, where $h_{i,s}$ is a computable double sequence of rational piecewise linear functions. We will have that for each fixed i, the sequence $(h_{i,s})_{s\in\mathbb{N}}$ eventually stabilises. Since computability of $+$ implies computability of scalar multiplication (see Lemma 2.4.17), it is sufficient to ensure that $(1/2)\cdot$ is not a computable operation with respect to $(h_i)_{i\in\mathbb{N}}$. We also need to ensure that the metric is computable, and so is the zero function.

Suppose we are diagonalising against the eth computable procedure Θ_e potentially representing $h_i \mapsto \frac{1}{2}h_i$. We will use a witness h_p which has constant value $16e$ on some small interval I_e (reserved exclusively for this requirement). The basic strategy will wait for $\Theta_e(p)$ to converge with high accuracy. We then adjust h_p on interval I_e by lowering its value $h_p(z)$ by $8e$ for some $z \in I_e$. This will ensure that Θ_e is "killed". To ensure that distances are preserved, we need to adjust h_m similarly on I_e for every h_m that takes on values larger than $8e$ on I_e; see Fig. 2.2. This leaves functions with norm $\leqslant 8e$ untouched after some stage.

The formal requirements. Let $L = (l_i)_{i\in\mathbb{N}}$ denote the effective sequence of all continuous piecewise linear functions with finitely many rational breakpoints (written rational p.l. functions) without repetition; this sequence in dense in the space (Example 2.4.18). We construct $X = (h_i)_{i\in\mathbb{N}}$ and meet the following global requirements:

(1) For every i, there is some s_i such that $h_{i,s_i} = h_{i,t}$ for every $t \geqslant s_i$.

(2) For every i, j and s, we have $d(h_{i,s}, h_{j,s}) = d(h_{i,s+1}, h_{j,s+1})$.

(3) For each m, there is exactly one k such that $\lim_s h_{k,s} = l_m$.

These three requirements clearly imply that $X = (h_i)_{i\in\mathbb{N}}$ is a computable presentation of $(C[0,1], d)$.

We fix an effective listing $(\Psi_e)_{e\in\mathbb{N}}$ of all partial computable functions of two arguments that satisfies the following conditions:

1. For every e, t, x, we have $d(\Psi_e(x,t), \Psi_e(x,t+1)) < 2^{-t-1}$, if $\Psi_e(x,t)$ and $\Psi_e(x,t+1)$ converge.
2. For every stage s and every e, t, x, we have $\Psi_{e,s}(x,t)\downarrow$ only if $\Psi_{e,s}(x,n)\downarrow$ for each $n \leqslant t$.

To see that $(\Psi_e)_{e\in\mathbb{N}}$ exists, we start with some universal listing of all partial computable functions of two variables and limit ourselves to only those which satisfy (1) and (2). For every e and x, set $\Theta_e(x) = \lim_{n\to\infty} \Psi_e(x,n)$ if the limit exists (where the limit is taken with respect to the metric on M), and set $\Theta_e(x)\uparrow$ otherwise.

Notation 2.4.21. At stage s we set $\Theta_{e,s}(x)$ equal to $\Psi_{e,s}(x,m)$ if m is the largest such that $\Psi_{e,s}(x,m)\downarrow$, and we set $\Theta_{e,s}(x)$ undefined otherwise. In the former case, we let $\theta_{e,s}(x) = m$. Thus, $\Theta_{e,s}(x)$ is our stage s guess about $\Theta_e(x)$, and $\theta_{e,s}(x)$ indicates the error between $\Theta_{e,s}(x)$ and $\Theta_e(x)$.

We need to satisfy, for every e, the requirements:

$$N_e : \Theta_e \text{ does not compute } x \mapsto \tfrac{1}{2}x \text{ in } X,$$

where Θ_e stands for the eth computable operator as defined in Notation 2.4.21.

In the following, $(I_e)_{e\in\mathbb{N}}$ stands for some effective listing of disjoint computable closed subintervals of $[0,1]$. We ensure that for each strategy N_e and each h_i, N_e is only allowed to modify h_i on the interval I_e. More specifically, when N_e requests for the interpretation of h_i to be changed at a stage s, we always ensure that $h_{i,s}(z) = h_{i,s+1}(z)$ for every $z \notin I_e$. The requirements N_e all act independently and at most once during the construction.

The detailed strategy for N_e is as follows. It will have its own witness, a rational p.l. function $w_e \in X$. The function w_e, when first defined at stage $2e$, is equal to $16e$ on the interval I_e, is equal to zero at the endpoints of $[0,1]$, and is linear outside I_e.

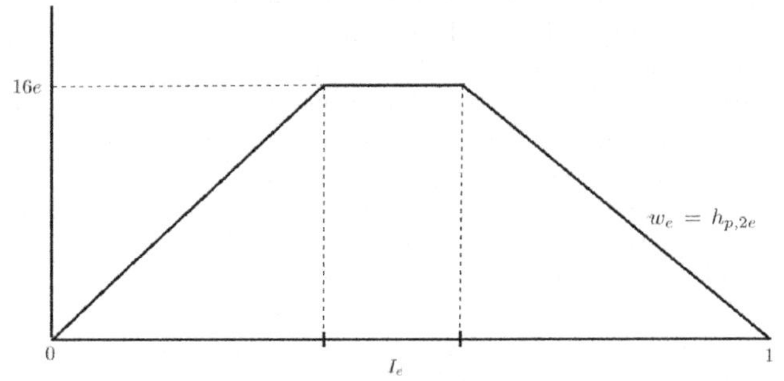

Figure 2.1: Function $h_{p,2e} = w_e$

Let p be such that $w_e = h_{p,2e}$. The strategy N_e does nothing until it sees a computation $\Theta_{e,s}(p)$ where $\theta_{e,s}(p) > e$ (see Notation 2.4.21). If we have

$$\sup_{z\in I_e} \left| \frac{1}{2} h_{p,s}(z) - h_{\Theta_{e,s}(p),s}(z) \right| = \sup_{z\in I_e} \left| h_{\Theta_{e,s}(p),s}(z) - 8e \right| > 2^{-e+1},$$

then the strategy does nothing for the rest of the construction, and we win N_e simply because

$$\sup_{z\in I_e} \left| h_{\Theta_{e,s}(p),s}(z) - f(z) \right| \leqslant d\left(h_{\Theta_{e,s}(p)}, f\right) < 2^{-e},$$

and thus

$$d\left(\frac{1}{2}h_p(z), f(z)\right) \geqslant \sup_{z\in I_e}\left|\frac{1}{2}h_p(z) - f(z)\right| = \sup_{z\in I_e}\left|\frac{1}{2}h_{p,s}(z) - f(z)\right| \geqslant 2^{-e},$$

where $f = \lim_{s\to\infty} h_{\Theta_{e,s}(p)}$. Thus we assume that at stage s we have

$$\sup_{z\in I_e}\left|h_{\Theta_{e,s}(p),s}(z) - 8e\right| \leqslant 2^{-e+1}. \tag{2.5}$$

The strategy N_e will then *act* as follows. Introduce a new interpretation $h_{p,t}$ as described below. (Notice that $h_{p,s}$ is equal to $h_{p,2e}$ on the interval I_e, but not necessarily outside this interval.) Choose a small sub-interval J of I_e satisfying the following: For all current interpretations $h_{i,s}$ and $h_{j,s}$ of X introduced so far, we have:

(i) $h_{i,s}$ is linear within J, i.e., $h_{i,s}$ has no breakpoints residing in J.

(ii) There is no pair $z_1, z_2 \in J$ such that $h_{i,s}(z_1) = 8e$ and $h_{i,s}(z_2) \neq 8e$.

(iii) If there is some $z \in J$ such that $h_{i,s}(z) = h_{j,s}(z)$, then $h_{i,s} \restriction J = h_{j,s} \restriction J$.

It is clear that J can be found effectively, since the construction has only looked at finitely many interpretations so far. Hence, each $h_{i,s}$, when restricted to J, is either strictly monotonic and does not take the value $8e$, or else it is constant on J. Furthermore, each pair $h_{i,s}$ and $h_{j,s}$ is either equal or non-intersecting in the interval J.

Now pick z to be the midpoint of J. For every interpretation $h_{i,s}$ such that $h_{i,s} \restriction I_e$ is strictly above $8e$, we set $h_{i,s+1}(z) = 8e$, $h_{i,s+1}(\min J) = h_{i,s}(\min J)$, and $h_{i,s+1}(\max J) = h_{i,s}(\max J)$. We linearly interpolate $h_{i,s+1}$ within J and keep $h_{i,s+1} = h_{i,s}$ unchanged outside J. This is illustrated by Figure 2.2.

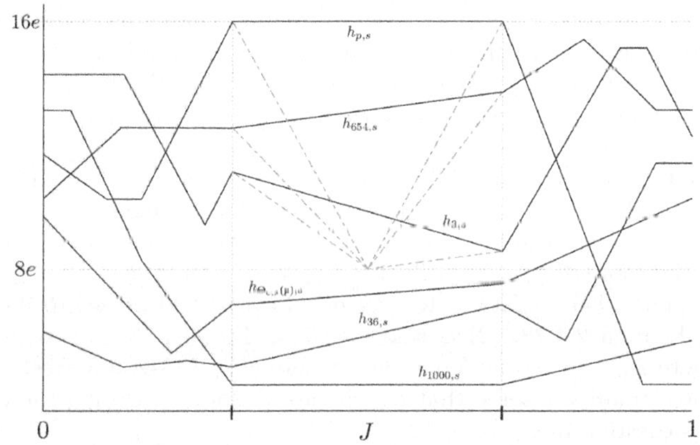

Figure 2.2: The modifications needed to get $h_{i,s+1}$.

Notice that this action only modifies each $h_{i,s}$ on the interval I_e. It is also straightforward to check the following:

Lemma 2.4.22. *Distances between the approximations are preserved.*

Proof. Fix i, j. We argue that $d(h_{i,s}, h_{j,s}) = d(h_{i,s+1}, h_{j,s+1})$. Let $m = |h_{i,s}(\min J) - h_{j,s}(\min J)|$ and $M = |h_{i,s}(\max J) - h_{j,s}(\max J)|$. Since there are no breakpoints of $h_{i,s}$ and $h_{j,s}$ in J, we clearly have

$$\sup_{v \in J} |h_{i,s}(v) - h_{j,s}(v)| = \max\{m, M\}.$$

Therefore, it is sufficient to see that

$$\sup_{v \in J} |h_{i,s+1}(v) - h_{j,s+1}(v)| = \max\{m, M\}.$$

If both $h_{i,s}$ and $h_{j,s}$ are modified, then this last equality follows easily from the fact that for every $\min J \leqslant v \leqslant z$, we have $|h_{i,s}(v) - h_{j,s}(v)| \leqslant m$, and for every $z \leqslant v \leqslant \max J$, we have $|h_{i,s}(v) - h_{j,s}(v)| \leqslant M$. Suppose, on the other hand, that $h_{i,s} \neq h_{i,s+1}$ and $h_{j,s} = h_{j,s+1}$. Then for every $v \in J$ we have

$$h_{i,s}(v) \geqslant h_{i,s+1}(v) \geqslant 8e \geqslant h_{j,s+1}(v) = h_{j,s}(v).$$

So we also have that $\sup_{v \in J} |h_{i,s+1}(v) - h_{j,s+1}(v)| = \max\{m, M\}$. □

Lemma 2.4.23. *N_e is satisfied.*

Proof. If N_e never acts, it is clearly satisfied, so we assume it acts at stage s as above. Since no approximation will ever be changed again within I_e after N_e acts, we have $h_p(z) = h_{p,s+1}(z) = 8e$, and

$$h_{\Theta_{e,s}(p)}(z) = h_{\Theta_{e,s}(p),s+1}(z) = \min\{8e, h_{\Theta_{e,s}(p),s}(z)\}.$$

By Equation (2.5) we have $h_{\Theta_{e,s}(p),s}(z) \geqslant 8e - 2^{-e+1}$ and so $h_{\Theta_{e,s}(p)}(z) \geqslant 8e - 2^{-e+1} > 7e$. Now since $\theta_{e,s}(p) > e$ we have $f(z) > 7e - 2^{-e} > 6e > \frac{1}{2}h_p(z)$. Hence $f \neq \frac{1}{2}h_p$. □

Construction. We fix an effective ordering of the N-requirements. At stage s of the *construction*, we simply let the strategies of the first s requirements act according to their instructions. Next, if we do not see l_m among $(h_{i,s})_{i \leqslant s}$ at stage $s \geqslant m$, we pick the least n such that h_n has no approximation so far and set $h_{n,s} = l_m$. This concludes the construction.

Verification. We first show that the global requirements are met. For (1), fix i, and let t be the first stage at which h_i receives its initial approximation (namely, $h_{i,t}$). Let D be such that $\|h_{i,t}\| = d(0, h_{i,t}) < 8D$. Only the strategies N_e where $e \leqslant D$ can possibly change the approximation at a later stage. Furthermore, if $t > s$ is such that $h_{i,t} \neq h_{i,s}$, then $\|h_{i,t}\| \leqslant \|h_{i,s}\|$. Each N-strategy acts at most once. Thus, there exists a stage after which h_i will be set to its final value, and so (1) is satisfied. By Lemma 2.4.22, (2) is also satisfied. Finally, (3) is satisfied because for each l_m, after a stage where $N_0, \ldots, N_D$ no longer act, where $\|l_m\| < 8D$, any fresh assignment of l_m to an h_i must be stable. Finally, observe that $h_{0,s}$ is never modified during the construction, so the interpretation of $\mathbf{0}$ is computable.

This finishes the proof of Theorem 2.4.20.

The computable Polish presentation $(h_i)_{i \in \mathbb{N}}$ of $(C[0,1], d_{sup})$ constructed above has several nice properties, which will be important later. A computable Polish presentation X of (M, d) is *rational-valued* if $d(x, y) \in \mathbb{Q}$ for every $x, y \in X$, and the distance d is represented by a computable function of two arguments mapping each pair of special points (x, y) to the corresponding rational number $d(x, y)$ (represented as a fraction). We also say that two computable Polish presentations L and L' of (M, d) are *limit equivalent* if there is a total computable function $g(x, s) : L \times \mathbb{N} \to L'$ of two arguments such that the sequence $(g(x, s))_{s \in \mathbb{N}}$ is eventually stable on every x, and

$$f(x) = \lim_{s \to \infty} g(x, s)$$

is an isometric bijection of L onto L', where the limit is taken with respect to the standard discrete metric on $\mathbb{N}$.

Recall that $L = (l_i)_{i \in \mathbb{N}}$ denotes the effective sequence of all continuous piecewise linear functions in $C[0, 1]$ with finitely many rational breakpoints. The proof presented above gives a slightly stronger version of Theorem 2.4.20.

Theorem 2.4.24 (Melnikov and Ng [376]). *There exists a rational-valued computable Polish presentation X on $(C[0, 1], d)$ which is limit equivalent to L, and such that the constant zero function* $\mathbf{0}$ *is computable in X but the operation $+$ is not.*

In Chapter 10, we will discuss computably categorical Banach spaces, where Theorem 2.4.24 will find an application.

Of course, $C[0, 1]$ does have a presentation that computes $+$, and the same can be said about all common separable Banach spaces that we are aware of. It is currently not known if there exists a pathological Banach space that has a computable Polish presentation but does not have a computable Banach presentation, up to isometry.

Question 2.4.25. *Is there a separable Banach space that has a computable Polish presentation (as a Polish space under $d(x, y) = ||x - y||$) but is not isometrically isomorphic to any computable Banach space?*

2.4.4 Exercises: Comparing presentations of spaces

Computable topological spaces

In several exercises, we will use the following classical definition found in [316, Def. 3.1] and [224, Def. 3.1]. (See also Kalantari and Weitkamp [278], Korovina and Kudinov [315], and Spreen [482].) We shall typically restrict the definition to Polish spaces; however, it makes sense for arbitrary countably based spaces.

Definition 2.4.26. A *computable topological presentation* of a topological space M is given by a sequence $(B_i)_{i \in \mathbb{N}}$ of non-empty basic open sets of M and a computably enumerable set W such that

$$B_i \cap B_j = \bigcup \{B_k : (i, j, k) \in W\},$$

for any $i, j \in \mathbb{N}$.

A computable topological presentation (of a compact space) is *effectively compact* if there exists a c.e. enumeration of finite tuples $\langle i_1, \ldots, i_k \rangle$ such that $B_{i_1}, \ldots, B_{i_k}$ is a cover of the space.

A computable topological presentation is *strong* if the relation $\{\langle i, j \rangle : B_i \cap B_j = \varnothing\}$ is computable.

Note that Definition 2.3.11 of an *effectively continuous map* makes sense for computable topological spaces if we interpret U_i as basic open sets. Whenever we say that a function or a homeomorphism between computable topological spaces is computable, we always interpret it using the natural generalisation of Definition 2.3.11. In Lemma 4.2.9, we will see that this approach is equivalent to the usual definition in the context of computable Polish spaces.

We will see that the notion of a computable topological space is, in general, too weak to be of any significant use. Surprisingly, any countably based T_0 space (effective or not in any sense whatsoever) admits a computable topological presentation; this will appear as Exercise 4.2.105. On the other hand, it appears that the most natural extra effectiveness conditions, such as effective regularity or normality, allow one to produce an effective compatible metric (to appear as Exercises 4.2.24 and 4.2.26), which is, of course, complete in the locally compact case. This includes the case of an effectively compact presentation (Exercise 4.2.41). For instance, in the effectively compact case, the notion turns out to be too closely related to the notion of a computably compact space to be of any significant advantage (Exercise 4.2.41). However, some of the elementary material related to computable topological spaces is viewed as common knowledge, so we include it here, and throughout the book, to inform the reader. The more interesting and difficult facts about such presentations will appear as exercises in later chapters (e.g., Exercises 4.2.24–4.2.26, 4.2.41, 4.2.104, 4.2.105, 4.2.112, 5.2.28 and 9.4.16).

Exercise° 2.4.27 (Folklore).

1. Show that for every right-c.e. Polish space, the collection of basic open balls

   $$\{B(x_i, r) : x_i \text{ spatial and } r \in \mathbb{Q}^+\}, \text{ where } B(x_i, r) = \{y \in M : d(x_i, y) < r\},$$

   forms a computable topological space (Def. 2.4.26). [Combined with Exercise 2.4.30, this allows one to define a *right-c.e. presentation* of a Polish group to be a right-c.e. Polish space together with effectively continuous operations upon the induced computable topological space. See also Exercises 2.4.29, 2.4.28, 4.2.112 and 5.2.28 for more about such presentations.]
2. Show that two non-homeomorphic spaces can have identical computable topological presentations. [Hint: Take any computable Polish space and consider its dense subset.]
3. Show that the product of computable topological spaces is again computable topological.
4. Let $f : X \to Y$ and $g : Y \to Z$ be computable (i.e., effectively continuous) maps between computable topological spaces. Show that their composition is also computable.
5. Say that a map $f : X \to Y$ between computable topological spaces is *effectively open* if, given a listing of an open set $U \subseteq X$ (given by some basic open sets making up U), we can uniformly produce a listing of $f(U)$. Prove that the composition of effectively open maps is again effectively open.

Exercise 2.4.28 (Koh, Melnikov, and Ng [313]). Show that for a discrete countable group G, the following are equivalent:

1. G is c.e. presentable.
2. G admits a right-c.e. Polish presentation under the discrete topology (in the sense of Exercise 2.4.27(1)).

Conclude that there exists a right-c.e. Polish group with no computable Polish presentation.

Exercise 2.4.29 (Koh, Melnikov, and Ng [313]). A *computable topological group* is a computable topological space (Def. 2.4.26) with group operations $\cdot$ and $^{-1}$ that are effectively continuous. Show that the operations are also effectively open, i.e., given enumerations of open sets U, V, we can list $U \cdot V$ and U^{-1}; see also Exercise 2.4.27(5). (Hint: For the inverse, note that $(U^{-1})^{-1} = U$. This also makes the map $(x, y) \to x^{-1}y$ effectively continuous. Enumerate all basic open B for which there is some basic open A satisfying

$$A \cap U \neq \varnothing \text{ and } A^{-1} \cdot B \subseteq V.$$

The union of all such B is actually equal to $U \cdot V$. If B is enumerated by the procedure above, let $a \in A \cap U$. For each $b \in B$, we have $b = a \cdot a^{-1}b \in U \cdot A^{-1} \cdot B \subseteq U \cdot V$, and so $B \subseteq U \cdot V$. Conversely, let $a \in U$ and $b \in V$. Since $a^{-1} \cdot ab = b \in V$, let A, B be basic open sets containing a and ab respectively, such that $A^{-1} \cdot B \subseteq V$. Thus, B will be enumerated by the procedure above and $ab \in B$.)

Computable and recursive Polish spaces

For the next two exercises, recall the definition of a "recursive" Polish space from Section 1.3: A *recursive presentation* of a Polish space is a dense subset $(x_i)_{i \in \mathbb{N}}$ of the space so that the relations

$$P(i, j, m, n) \text{ if and only if } d(x_i, x_j) < \frac{m}{n+1};$$

$$Q(i, j, m, n) \text{ if and only if } d(x_i, x_j) \leqslant \frac{m}{n+1}$$

are computable relations on ω^4.

Exercise° 2.4.30 (Folklore). Show that the three notions of computable presentability of a Polish space introduced in Section 1.2.2 differ up to isometry. (Hint: Note that there is a left-c.e. real that is not right-c.e., and a right-c.e. real that is not left-c.e. Consider the arc space $[0, \alpha] \subseteq \mathbb{R}$ under the usual metric. Show that $[0, \alpha]$ is isometric to a computable Polish space iff α is left-c.e., and $[0, \alpha]$ is isometric to a computably compact space iff α is computable.)

Exercise° 2.4.31 (Gregoriades, Kispéter, and Pauly [223]). Show that there exists a discrete computable Polish space not isometrically isomorphic to any "recursive" Polish space. (Hint: In a recursive space, we can decide whether the distance between a pair of points is equal to a given $n \in \mathbb{N}$. Build a discrete space. To diagonalise against the n-th potential recursive presentation, introduce a pair of points x_n and y_n so that $d(x_n, y_n) = n + \epsilon_n$, where ϵ_n is small and is controlled by us. To make sure it works up to isometry, make the pair 2^n-far from the points x_m, y_m, $m < n$.)

Exercise° 2.4.32 (Folklore; e.g., [223, 373]). Recall that in our definition of a computable Polish space we did not require that $x_i \neq x_j$ when $i \neq j$ in the dense sequence $(x_i)_{i \in \mathbb{N}}$. Show that every computable Polish space is computably isometric (via the identity map) to a computable Polish space in which there are no repetitions in the dense sequence. (Hint: Do not put x_i into your new refined sequence unless it looks sufficiently separated from the points you have already put in your sequence. If this never happens, then x_i must be in the completion.)

Exercise° 2.4.33 (Gregoriades, Kispéter, and Pauly [223]). Show that every computable Polish space is homeomorphic to a "recursive" space. (Hint: Assume the dense $(x_i)_{i \in \mathbb{N}}$ sequence has no repetitions and is infinite; see Ex. 2.4.32. It is sufficient to replace d with a new metric αd, where α is a positive computable real such that $\alpha d(x_i, x_j) \neq r$ for every rational r. Build α using a Cantor-style direct diagonalisation so that $\alpha \neq \frac{r}{d(x_i, x_j)}$.)

Exercise° 2.4.34 (Folklore; e.g., [197]). Let $\mathcal{K}$ be a class of at most countable structures in a computable language (signature). Assume the domains of all structures are either ω or its initial segment. Fix a computable Polish presentation of ω^ω given by the finite strings in $\omega^{<\omega}$ and the longest common prefix ultrametric (cf. proof of Lemma 2.4.5). Show that $\mathcal{K}$ can be associated with a subset $C(\mathcal{K})$ of "codes" in ω^ω in a way that the computable elements (paths) of $C(\mathcal{K})$ are in a uniformly effective 1-1 correspondence with the computable structures in $\mathcal{K}$. (Hint: Use specifically reserved levels of $\omega^{<\omega}$ to code the values of the operations and relations on various inputs.)

Exercise 2.4.35 (Ceitin [82]). Extend Kreisel-Lacombe-Shoenfield-Markov Theorem 2.3.7 to computable Polish spaces.

Computable Banach spaces

Exercise 2.4.36 (Pour-El and Richards [435], Chapters 4 and 5 therein[6]). Fix a computable Banach space $\mathbb{B}$. Say that a linear operator $T : \mathbb{B} \to \mathbb{B}$ is *effectively determined* if there exists a uniformly computable sequence of points $(e_i)_{i \in \mathbb{N}}$ such that the sequence $(e_i, T(e_i))$ is dense in the graph of T (viewed as a subset of $\mathbb{B} \times \mathbb{B}$).

1. Show that every (Type II) computable linear operator is effectively determined.
2. Show that every effectively determined operator that is furthermore bounded (equivalently, continuous) is computable.
3. Prove* that there is an effectively determined linear operator on a computable Hilbert space (of infinite dimension) that is *not* bounded[7].
4. Show* that for every self-adjoint effectively determined linear operator T on a computable Hilbert space, the eigenvalues of T are computable reals.
5. Show* that there exists a (Type II) computable self-adjoint linear operator on a computable Hilbert space whose eigenvalues do not form a uniformly computable sequence[8].

Exercise* 2.4.37 (Pour-El and Richards [435], Chapter 4). Show that there exists a computable (indeed, compact) self-adjoint operator $T : \mathbb{H} \to \mathbb{H}$ on computable Hilbert space $\mathbb{H}$ with the following properties.

1. The number $\lambda = 0$ is an eigenvalue of T of multiplicity one (i.e., the space of eigenvectors corresponding to $\lambda = 0$ is one-dimensional).

[6] While certain parts of this exercise are undoubtedly non-elementary, reworking these results based on [435] is appropriate for a student.

[7] We remind the reader that exercises marked with a $*$ are either challenging or require some material not covered in the book. Those marked with two stars are especially difficult and are included primarily to inform the reader.

[8] We remark that all these results do not actually depend on the choice of the computable presentation of the infinite-dimensional Hilbert space $\mathbb{H}$, since it is effectively unique up to computable linear isometry [435] (cf. Theorem 10.2.2). (The same can be said about any finite dimension too.) Further, any sequence $(e_i, T(e_i))$ witnessing that T is effectively determined gives a computable dense sequence $(e_i)_{i \in \mathbb{N}}$ in the space. Thus, at least in the case of $\mathbb{H}$, the choice of the sequence does not really affect the definition, and thus we are dealing with a very natural analogue of Type I (Markov) computability for *operators* on $\mathbb{H}$.

2. None of the eigenvectors corresponding to $\lambda = 0$ is computable.

Exercise 2.4.38 (Brattka [58]). 1. Show that if $T : X \to Y$ is a computable bijective (bounded) linear mapping between computable Banach spaces, then T^{-1} is computable.

2. Show that this is not uniform in general.
3. Prove that, indeed, the inversion mapping $T \mapsto T^{-1}$ is not computable in the case $X = Y = \ell_p$, where $p \geqslant 1$ is a computable real number.

Exercise* 2.4.39. Fix a Lebesgue space U whose dimension is at least 2 and whose exponent is p. Prove the following:

(1) Show that if p is a computable real, then the space has a computable Banach presentation. (Hint: Use the classification of separable L^p-spaces that can be found in, e.g., [83].)

(2) Assume U has a computable Banach presentation (i.e., U is linearly isometric to some computable Banach space). Show* that the exponent p is right-c.e. if it is smaller than 2, and otherwise it is left-c.e. (Brown, McNicholl, and Melnikov [68]).

(3) Extend** (2) to show that when $p \geqslant 2$ or when the space is finite-dimensional, then p has to be a computable real (McNicholl [361]).

Exercise 2.4.40. Prove that there exists a right-c.e. presented Banach space $\mathbb{B}$ (i.e., so that the norm is right-c.e.) which is additionally low, but so that $\mathbb{B}$ is not linearly isometric to any computable Banach space. (Hint: Recall that $\ell_p \subset \ell_q$ when $1 \leqslant p < q$, and this is true because the norms satisfy $||x||_p \leqslant ||x||_q$. Fix a right-c.e. low real $p > 2$ given by Exercise 2.3.2 and consider ℓ_p. It is given by the dense linear space of vector with rational coordinates and having finite support; this is a right-c.e. presentation which is indeed also low. By Exercise 2.4.39(3), ℓ_p is not linearly isometric to any computable Banach space. Note the same argument would work for a left-c.e. p as well.)

Exercise* 2.4.41 (Bosserhoff [54]). Let $\mathbb{X}$ be a Banach space. A sequence $(x_i)_{i\in\mathbb{N}} \in \mathbb{X}^{\mathbb{N}}$ is a *Schauder basis* of $\mathbb{X}$ if, for all $x \in \mathbb{X}$, there exists a unique sequence of coefficients $(a_i)_{i\in\mathbb{N}} \in \mathbb{R}^{\mathbb{N}}$ such that $\sum_{i=1}^{\infty} a_i x_i = x$. In his famous book [28], Banach asked if every separable Banach space has a Schauder basis. Banach's question was solved by Per Enflo [153], who proved that there exists a separable Banach space with no Schauder basis.

1. Show that Enflo's construction is, in fact, computable.
2. Prove that there exists a computable Banach space with a Schauder basis, but without a computable Schauder basis.

Exercise 2.4.42 (Qian [437]). Let $\mathbb{B}$ be a computable Banach space with a computable Schauder basis $(b_i)_{i\in\mathbb{N}}$ (Exercise 2.4.41). Show that the projection functions $P_i : \sum_i \lambda_i b_i \mapsto \lambda_i$ are computable, uniformly in i. [Hint: Take for granted the following classical theorem of Banach [28]: Let $\mathbb{B}$ be a Banach space and $(b_i)_{i\in\mathbb{N}} \subseteq \mathbb{B}$ a sequence of nonzero elements. Then $(x_i)_{i\in\mathbb{N}}$ is a basis of $\mathbb{B}$ iff: (1) there exists a constant $K \in \mathbb{R}$ such that for all $n, m \in \mathbb{N}$ with $m < n$, and for all sequences of scalars $(a_i)_{i\in\mathbb{N}}$, we have $\|\sum_{i=1}^{m} a_i b_i\| \leqslant K \|\sum_{i=1}^{n} a_i b_i\|$, and (2) the finite linear span of $(x_i)_{i\in\mathbb{N}}$ is dense in $\mathbb{B}$. Now suppose $M \geqslant K$, an upper bound on the basis constant of $(b_i)_{i\in\mathbb{N}}$. Fix some $x \in \mathbb{B}$. Search for some $(\beta_i^{s+1})_{i\leqslant m} \in \mathbb{Q}^{<\omega}$ such that $\|x - \sum_{i=1}^{m} \beta_i^{s+1} e_i\| < \varepsilon_{s+1}$ where ε_{s+1} is a value such that $\varepsilon_{s+1} < 4^{-s-1}/M$ and $2M\left(4^{-s-1}/M + \varepsilon_{s+1}\right) < 2^{-s-1}$. Noting that $\|P_i\| \leqslant 2M$, and assuming a finite sequence $(\beta_i^s)_i$ for s has already been found, we obtain that

$$\left\|\sum_i \beta_i^{s+1} b_i - \sum_i \beta_i^s b_i\right\|_\infty \leqslant \max_k \left\|P_k\left(\sum_i \beta_i^{s+1} b_i - \sum_i \beta_i^s b_i\right)\right\| \leqslant \max_k \|P_k\| \left(4^{-s}/M + \epsilon_{s+1}\right) < 2^{-s-1}.$$

In particular, we have that for each fixed i, $(\beta_i^s)_{i\in\mathbb{N}}$ is a fast Cauchy sequence converging to $P_i(x)$. See also Exercise 2.4.44 below.]

Exercise 2.4.43 (Downey, Greenberg and Qian [122]). Show that if B is a computable infinite dimensional Banach space, then there is a computable infinite dimensional subspace $\hat{B}$ of B with a computable Schauder basis (Exercise 2.4.41).

Exercise* 2.4.44 (Pour-El and Richards [435]). Let $\mathbb{X}, \mathbb{Y}$ be computable Banach spaces, and let $(e_i)_{i\in\mathbb{N}}$ be a computable sequence in $\mathbb{X}$ whose linear span is dense. Let $T : \mathbb{X} \to \mathbb{Y}$ be a linear operator with closed graph whose domain contains $\{e_i : i \in \mathbb{N}\}$ and such that the sequence $(T(e_i))_{i\in\mathbb{N}}$ is computable in $\mathbb{Y}$. Then T maps every computable element of its domain onto a computable element of $\mathbb{Y}$ iff T is bounded.

2.4.5 Historical remarks*

It was reasonably natural to extend early work on computable analysis (which was mostly computable calculus) to computable metric spaces, and Ceitin [82] is one early example. It is fair to say that the spaces with the most developed theory are computable compact spaces and computable Banach spaces [435, 56]. The study of computable separable spaces, especially Banach spaces, has attracted considerable interest, eventually culminating in books such as [1, 505, 435]. The systematic theory of computably compact spaces is a much more recent development, but it has been increasingly popular in recent years, recently culminating in the two large surveys [270, 139], which contain numerous results, proofs, and proof sketches. As assayed in Metakides and Nerode [385], many of these foundational results recycle some of the earlier theorems in constructive analysis presented in, e.g., Bishop [48]. However, more advanced results require more sophisticated techniques, including the priority method. The theory of computable Polish groups is a very recent development [375, 373], and this book is the first to present it comprehensively.

A historical curiosity is that for many decades, computable analysis had essentially been developing independently of the theory of countable computable structures. Only in the last decade, beginning with [369, 268], has there been a line of systematic investigations aiming to unite computable algebra and the theory of separable spaces; work in this direction is still ongoing. (See [94, 68] and the survey [138].) The main motivation of such investigations is that many aspects of these two theories are not really that different. As a result (as we will see in the present book), methods and results of one can be applied to the other and vice versa.

Later in the book, we will also use the classical notion of a simplicial complex. The algorithmic content of the topology of simplicial complexes is a classical field of study that is closely related to group theory. For instance, Adyan [3, 4] and Rabin [438] showed that it is undecidable to determine if two given finitely presented groups are isomorphic. Soon after, Markov [352] used these results to show that it is undecidable whether two compact manifolds of dimension $n \geqslant 4$, given as finite simplicial complexes, are homeomorphic. Markov computably transformed a finite presentation of a group $\langle X|R\rangle$ into a simplex $M_{\langle X|R\rangle}$ homeomorphic to an n-manifold with fundamental group $\langle X|R\rangle$, so that

$$\langle X|R\rangle \cong \langle Y|S\rangle \iff M_{\langle X|R\rangle} \cong_{hom} M_{\langle Y|S\rangle}.$$

This reduces the isomorphism problem for group presentations to the homeomorphism problem for simplices representing n-manifolds, showing that the latter problem is also undecidable. For $n \leqslant 3$, the homeomorphism problem for manifolds, presented as finite simplicial complexes, is decidable because of classification theorems; for $n = 3$, this uses the work of Perelman on Thurston's geometrisation conjecture (see [322]). For more results, see [430].

One of our goals in the book is to partially extend these ideas of Markov and others to Polish spaces that do not admit a triangulation, or perhaps are not even compact. To do so, we will use methods of algebraic topology, more specifically, Čech cohomology, in one of our proofs. This idea of using Čech cohomology first appeared in [341] and was then simplified and clarified in [139]. It is worth noting that much of classical algebraic topology is intrinsically computable. This is made explicit in [412] for simplicial (co)homology and in [31] for integral (co)homology of finitely generated groups. However, applications of algebraic topology in computable analysis are still rare. It was pioneered by Miller in [391] for the special case of the n-sphere. The effective content of algebraic topology and homological algebra has not yet been systematically explored.

2.5 What's next?

To proceed, we must develop additional methods, especially priority techniques. In the next chapter, we temporarily abandon computable analysis and the "actual" algebra (groups, fields, etc.) and develop a sufficient technical apparatus in the purely combinatorial context of Turing degrees and countable linear orders. We will return to separable spaces later, when we discuss Boolean algebras, where these techniques and results established for linear orders will find applications to Boolean algebras and their Stone spaces.

Chapter 3

Priority constructions and computable linear orders

The main goal of this chapter is to establish the following result, which will be used to prove Feiner's Theorem A (1) in the next chapter.

> **Theorem** (Fellner [164], Watnick [502]). There is a uniform procedure which, given a Δ^0_3-presentation of an infinite linear order L, outputs a computable copy of $\mathbb{Z} \cdot L$.

All definitions will be clarified in due course. A detailed proof of the Fellner-Watnick Theorem will be given in §3.2.6, where it will also be put into context. (Indeed, the original motivation behind proving the theorem was essentially unrelated to Feiner's Theorem A (1).)

We will need a lot more background in pure computability theory than was sufficient in the previous chapter. The chapter is divided into two sections:

1. Section 3.1 provides a brief introduction to the priority method. It includes the proofs of classical results in degree theory that are traditionally used to explain and demonstrate the priority techniques.
2. Section 3.2 applies the techniques explained in Section 3.1 to computably and c.e. presented linear orders. It culminates with two substantially different detailed proofs of the Fellner-Watnick Theorem.

The results presented in this chapter are many decades old and have been included in many excellent books and surveys. Thus, we often choose to give a friendly extended sketch (where appropriate) instead of giving a stage-by-stage construction and its verification in gory detail. We expect that the reader should be able to fill out the missing formal details if needed. But, of course, we shall give a complete and detailed proof of the main result of the chapter stated above.

R. G. Downey, A. Melnikov, *Computable Structure Theory*, Theory and Applications of Computability, https://doi.org/10.1007/978-3-031-92433-0_3

3.1 Priority techniques

Computably enumerable sets lie at the core of classical computability theory. The original impetus for computability theory was the study of effective processes in mathematics, seeking tools to show that, for instance, no algorithm can solve Hilbert's *Entscheidungsproblem* [252] (the problem of algorithmic decidability for first-order logic), nor is there one to solve Dehn's fundamental group theory problems, including the word, conjugacy, and isomorphism problems [109]. The original undecidability proofs tended to encode Turing machines into the problem at hand. Authors did not study properties of computability by abstracting away from a specific context. Nonetheless, it is evident that the use of effectively generated undecidable sets, such as the halting problem, was *implicit* in all the early foundational papers. Post's highly influential paper [431] pioneered the idea that computability could be studied separately by removing the context and examining the sets, as well as the processes generating those sets, themselves. In the words of Soare [477, p.ix]:

> "Post [431] stripped away the formalism associated with the development of recursive functions in the 1930's and revealed in a clear informal style the essential properties of r.e. sets and their role in Gödel's incompleteness theorems."

This idea of abstracting essential properties from applied contexts is not new and permeates mathematics; consider group theory, ideal theory, linear algebra, etc. As we can see, this idea is also essential to our book. We will take some time to develop the basic tools and machinery related to computably enumerable sets and degrees in isolation from algebra, analysis, or topology.

The main goal of this section is to accumulate enough techniques to prove two classical results that we will state shortly. For the first result below, recall that a set $A \leqslant \varnothing'$ is called low if $A' \leqslant_T \varnothing'$.

Theorem 3.1.1. *There exists a c.e. non-computable low set.*

The theorem is considered folklore; see [477, Theorem VII.1.1]. The proof of Theorem 3.1.1 is based on the methods used in the proof of Friedberg-Muchnik Theorem 3.1.21, which we also include. In Section 2.2, we already discussed one application of Theorem 3.1.1 to effective presentations of groups. Recall that Proposition 2.2.5 and Theorem 2.2.6 were established by "coding" a c.e. set W into a group G_W. This allowed us to separate the construction of the set from the definition of the respective group.

Unfortunately, it is often essentially impossible to separate the computability-theoretic combinatorics of sets and degrees from algebra using a coding technique. As a result, even in the context of linear orders, the use of more advanced priority techniques applied directly to the structure often seems inevitable. While certainly not *always* unavoidable, the infinite injury method is often the first line of attack, and the brute-force proofs using this method tend to be more flexible (i.e., easier to modify) than the "clever" proofs that avoid infinite injury.

It will be instructive to see the technique applied in the pure setting of sets, without any irrelevant combinatorics specific to the class of linear orders. Perhaps the most transparent and clear application of the infinite injury machinery known to us is the modern proof of the following result due to Lachlan [325] and Yates [513].

Theorem 3.1.2. *There exists a minimal pair of non-computable c.e. sets.*

(Non-computable c.e. sets A and B form a minimal pair if for all sets C, if $C \leqslant_T A, B$, then C is computable.)

We shall also present several classical results, such as Sacks' Splitting Theorem 3.1.34, that fit well into the story and will also help the reader to understand the priority technique. The finite injury of unbounded type used in the proof of Sacks' Splitting Theorem will be implemented in the description of computably categorical linear orders. Another classical result, the Friedberg Enumeration Theorem 3.1.43, which requires a "degenerate" infinite injury, will be important in Chapter 9 of the book.

3.1.1 The Limit Lemma and injury-free approximation

We use the notions and notation introduced in Section 2.1. For instance, we shall use the notation $\mathbf{0}^{(n)}$, $\varnothing^{(n)}$, K_0, $\leqslant_m$, $\leqslant_T$, Σ^0_n, Δ^0_n, Π^0_n, $\Phi(A; x)\downarrow$ without further clarification. We use $\Phi(A; x)$ and $\Phi^A(x)$ interchangeably. In this section we write $\Phi(A)$ to mean $(\Phi^A(n))_{n\in\mathbb{N}}$, in which case we usually (implicitly) assume $\Phi^A(n) \in \{0, 1\}$ whenever defined. We also write $A \upharpoonright m$ for $\{0, 1, \ldots, m-1\} \cap A$.

The Limit Lemma

We identify sets with their characteristic functions; for instance, $A(x) = 1$ is the same as saying that $x \in A$. Also, if g is a total function in two arguments and $|\{s : g(x, s) \neq g(x, s+1)\}| < \infty$, then we say that $\lim_s g(x, s)$ exists and is equal to the value z such that $z = g(x, s)$ for infinitely many s. Recall the use principle (Lemma 2.1.21).

Lemma 3.1.3 (Shoenfield's Limit Lemma [466]). *$A \leqslant_T K$ iff there is a computable function $g(\cdot,\cdot)$ such that, for all x,*

1. *$\lim_s g(x, s)$ exists;*
2. *$\lim_s g(x, s) = A(x)$.*

Proof. ($\Rightarrow$) Suppose $A \leqslant_T K$ so that for some procedure Φ_e, we have $\Phi_e(K) = A$. Define $g(x, s) = 0$ if $\Phi_{e,s}(K_s; x)\uparrow$ or $\Phi_{e,s}(K_s; x)\downarrow \neq 1$ and $g(x, s) = 1$ otherwise, where $K = \cup_s K_s$ is some computable enumeration of K. Given x, let $u = u(\Phi_e(K; x))$. Let $s = s(x)$ be any stage where $K_s \upharpoonright u = K \upharpoonright u$. Let $t \geqslant s$ be the least stage such that $\Phi_{e,t}(K; x)\downarrow$. As $K_t \upharpoonright u = K_s \upharpoonright u = K \upharpoonright u$, we have, by the use principle, $\Phi_{e,t}(K; x) = \Phi_{e,t}(K_t; x) = \Phi_{e,q}(K_q; x) = \Phi_e(K; x)$ for all $q \geqslant t$. Then for all $q \geqslant t$, $g(x, t) = g(x, q) = A(x)$ by definition.

($\Leftarrow$) Suppose such a function g exists. Without loss of generality, we can assume that $g(x, 0) = 0$ for all x. We construct a computably enumerable set B and a reduction Γ so that $\Gamma(B) = A$. Then $A \leqslant_T K$ since B is c.e. (see Section 2.1). Put $\langle x, n\rangle$ into B iff

$$|\{s : g(x, s) \neq g(x, s+1)\}| \geqslant n.$$

The reduction Γ is then defined as follows: on input x, compute the least n such that $\langle x, n\rangle \notin B$. Then $x \in A$ iff n is odd (since $g(x,0) = 0$, by assumption). □

Intuitively, in the ($\Leftarrow$) part of the above proof, K was used to decide whether

$$(\exists s)(g(x,s) \neq g(x,s+1))$$

and hence, computably in K, calculate the limit of $g(x,s)$. Such arguments can be made formal using the s-m-n Theorem 2.1.6. We will see a similar use of $\mathbf{0}'$ very shortly in the proof of Theorem 3.1.14, where the use of the s-m-n Theorem will be clarified. In the literature, similar details are often completely omitted, and we shall typically omit these details as well.

As we have seen, we can relativise results, definitions, and proofs in computability theory. For instance, the Limit Lemma above relativises to show that $A \leqslant_T K^B$ iff there is a B-computable function $f(x,s)$ such that $A(x) = \lim_s f(x,s)$ for all x. Combining this with induction, we get the following generalisation of the Limit Lemma 3.1.3.

Corollary 3.1.4. *Suppose that $n \geqslant 1$. Then, $A \leqslant_T \varnothing^{(n)}$ iff there is a computable function g of $n+1$ variables such that for all x,*

$$A(x) = \lim_{s_1}\lim_{s_2}\ldots\lim_{s_n} g(x, s_1, s_2, \ldots, s_n).$$

We also have the following useful fact characterising the arithmetical degrees.

Theorem 3.1.5 (Folklore). *Suppose that $n \geqslant 0$.*

(i) B is Σ^0_{n+1} iff B is c.e. in some Σ^0_n set if and only if B is c.e. in some Π^0_n set.

(ii) $\varnothing^{(n)}$ is Σ^0_n-complete, in the sense that $\varnothing^{(n)}$ is Σ^0_n and for all Σ^0_n sets B, $B \leqslant_m \varnothing^{(n)}$.

(iii) B is Σ^0_{n+1} iff B is c.e. in $\varnothing^{(n)}$.

(iv) B is Δ^0_{n+1} iff $B \leqslant_T \varnothing^{(n)}$.

Proof. (i) This is essentially Theorem 2.1.27. If B is c.e. relative to some Π^0_n set D, then for some e,

$$x \in B \text{ iff } \exists s \exists \sigma \subset D\, (x \in W^{\sigma}_{e,s}).$$

Now since $x \in W^{\sigma}_{e,s}$ is a computable property, the result follows since the property $\sigma \subset D$ is Δ^0_{n+1} and hence Σ^0_{n+1}, as it is a combination of Σ^0_n statements (asserting that certain things are in D) and Π^0_n statements (asserting things outside of D).

(ii) This follows by induction (exercise).

(iii) By (i) and (ii).

(iv) Note that B is Δ^0_{n+1} if and only if B and $\overline{B}$ are both Σ^0_{n+1}, and hence c.e. in $\varnothing^{(n)}$ by (ii). If a set and its complement are both c.e. in X, then they are computable in X, and hence $B \leqslant_T \varnothing^{(n)}$. □

The result above can be relativised to any oracle X. For that, define classes Σ^X_n and Π^X_n (also denoted $\Sigma^0_n(X)$ and $\Pi^0_n(X)$ or sometimes $\Sigma^{0,X}_n$ and $\Pi^{0,X}_n$) using X-computable relations in place of computable relations as the base. The jump operator correlates very nicely with levels of the relativised hierarchy. For instance, it follows from the theorem above that $\Delta^{\varnothing^{(m)}}_n = \Delta^0_{n+m}$ and $\Sigma^{\varnothing^{(m)}}_n = \Sigma^0_{n+m}$, and similarly for $\Pi^{\varnothing^{(m)}}_n$.

Remark 3.1.6. This correspondence between classes is also *uniform*, in the sense that given an index of a set in $\Sigma_n^{\varnothing^{(m)}}$, we can uniformly and effectively produce its index (its description) as a member of Σ^0_{n+m}, and vice versa. (Similarly for $\Pi_n^{\varnothing^{(m)}}$.)

For instance, if $S = \operatorname{dom}\Phi(\varnothing'; e, x)$, then

$$x \in S \text{ if and only if } \exists a \forall b\, R(f(e), a, b, x),$$

where R is a computable predicate and f is a total computable (indeed, primitive recursive) function. Informally, the primitive recursive function f simply turns a description of Φ into a description of R, but it does not actually need to have access to the oracle. The proof uses a careful analysis of the iterated Limit Lemma; we omit it and refer the reader to Rogers [454].

Movable markers and m-completeness

Note that in Theorem 3.1.5(ii), we can replace Σ^0_n and $\varnothing^{(n)}$ with Π^0_n and $\overline{\varnothing^{(n)}}$, and the result would still hold. We can further extend Theorem 3.1.5(ii). To do so, we shall define the notion of Σ^0_n- and Π^0_n-completeness for an arbitrary set, not just $\varnothing^{(n)}$ (respectively, $\overline{\varnothing^{(n)}}$).

Definition 3.1.7. If $\mathcal{C}$ is a class of sets, we say that B is $\mathcal{C}$-*complete* if B is in $\mathcal{C}$ and, for any A in $\mathcal{C}$, we have that $A \leqslant_m B$. If we do not require that B is in $\mathcal{C}$, then we say that B is $\mathcal{C}$-*hard.*

It is usually assumed that the reduction used to witness completeness (or hardness) is the m-reduction, but occasionally other reductions are used as well; for instance, one may wish to use $\leqslant_{wtt}$, as defined in Exercise 3.1.9.

The theorem below is folklore. The standard *movable markers* technique used in its proof will be quite useful in applications to structures (e.g., Theorem 3.2.11 in the next section). This technique will be useful in later chapters too, especially in Chapter 5. The movable markers technique is a precursor to the finite injury technique, and indeed, some of the simplest finite injury proofs are perhaps best viewed as dynamic approximation proofs using movable markers.

Theorem 3.1.8. *(i)* ***Fin*** *$= \{x : \operatorname{dom}\varphi_x$ is finite$\}$ is Σ^0_2-complete.*

(ii) ***Tot*** *$= \{x : \varphi_x$ is total$\}$ and* ***Inf*** *$= \{x : \operatorname{dom}\varphi_x$ is infinite$\}$ are both Π^0_2-complete.*

(iii) ***Cof*** *$= \{x : \operatorname{dom}\varphi_x$ is cofinite$\}$ is Σ^0_3-complete.*

Proof. We prove (i). We know that K' is Σ^0_2-complete. By the *s-m-n* Theorem 2.1.6, we have a computable function f such that

$$\varphi_{f(x)}(s)\downarrow \text{ iff } \exists t \geqslant s\, \widehat{\Phi}^{K_t}_{x,t}(x)\uparrow,$$

where $\widehat{\Phi}$ behaves exactly like Φ, except that it automatically $\uparrow$ for at least one step if the use on argument x changes from stage $t-1$ to t. Then $f(x)$ is in `Fin` iff $x \in K'$. The proof of (ii) is similar and is omitted.

We outline a proof of (iii) that uses the *movable markers* technique. Suppose that $S \in \Sigma^0_3$. Thus, there is a computable relation $R(x, y, s, t)$ such that for all x,

$$x \in S \text{ if and only if } \exists y \forall s \exists t R(x, y, s, t).$$

Using Exercise 2.1.30, replace this computable relation with another computable relation P such that

$$x \in S \text{ if and only if } \exists y \exists^\infty z P(x, y, z).$$

Given x, we shall (uniformly in x) build a computable function f_x such that

$$\operatorname{dom} f_x \text{ is cofinite if and only if } \exists y \exists^\infty z P(x, y, z).$$

Our construction will be effective in x, and we will appeal to the s-m-n Theorem 2.1.6 to conclude that, for some total computable g,

$$f_x = \varphi_{g(x)},$$

and this g will be the function witnessing the Σ^0_3-completeness of `Cof`.

At stage s, we will have $\mathbb{N} = \cup_{y \in \mathbb{N}} I_{y,s}$, where each $I_{y,s}$ is an interval of the form

$$I_{y,s} = [m_{y-1,s}, m_{y,s}],$$

so that $m_{-1,s} = 0$ for all s, and $m_{y,0} = y$ for all y. At stage s, the values of $m_{y,s+1}$ will be determined in the construction using P and the previous values $m_{y,s}$. These values will be the "movable markers" mentioned earlier. For every s, we shall additionally have

$$m_{y,s+1} \geqslant m_{y,s},$$

so the "markers" can only be "moved" to a larger value.

We shall move $m_{y,s}$ only if one more z is discovered for y (at stage s) such that $P(x, y, z)$ holds; if this happens, we say that P "fires" for y via z (at stage s). In this case, we shall "move" all the "markers" $m_{v,s}$ ($v \geqslant y$) by setting

$$m_{v,s+1} := m_{v+1,s}$$

for all $v \geqslant y$. We shall also declare $f_{x,s+1}(m_{y,s}) \downarrow= 1$, but we shall keep $f_{x,s+1}(m_{a,s+1}) \uparrow$ for all $m_{a,s+1}$, $a \in \mathbb{N}$. Note that, by induction,

$$(m_{y-1,s+1}, m_{y,s+1}) \subseteq \operatorname{dom} f_{x,s+1},$$

for every y and s.

If, for some y, the predicate "fires" infinitely often, i.e. $\exists y \exists^\infty z P(x, y, z)$, then assume y is the least such. In this case, we see that for all $v < y$, $\lim_s m_{v,s} = m_v < \infty$ exists, while for all $w \geqslant y$, $\lim_s m_{v,s} = \infty$. In particular, I_y will be cofinite, making $\operatorname{dom} f_x$ also cofinite.

Otherwise, if $\forall y \exists^{<\infty} z P(x, y, z)$, we will have that $\lim_s m_{v,s} = m_v < \infty$ for *all* $v \in \mathbb{N}$. In this case, we shall end up with infinitely many finite intervals I_y, one for each y. The function will remain undefined on all the boundary points m_y, making the complement of its domain infinite. $\square$

Exercises

Exercise° 3.1.9. We say that $A \leqslant_{wtt} B$ (where *wtt* stands for "weak truth table") if there is a Turing functional Φ and a computable f such that $\Phi^B = A$ and for each x, the use $\varphi^B(x) < f(x)$. Show that $A \leqslant_{wtt} \varnothing'$ iff there are computable functions h and f, such that, for all x, $A(x) = \lim_s h(x, s)$ and $|\{s : h(x, s+1) \neq h(x, s)\}| < f(x)$.

Exercise° 3.1.10. Prove that the index set $K_1 = \{e : W_e \neq \varnothing\}$ is Σ^0_1-complete.

Exercise° 3.1.11. Prove that the index set $\texttt{Disjoint} = \{\langle e, i\rangle : W_e \cap W_i = \varnothing\}$ is Π^0_1-complete.

Exercise° 3.1.12 (Rogers). Prove that the index set $\texttt{Comp} = \{e : W_e \text{ is computable }\}$ is Σ^0_3-complete.

Exercise° 3.1.13. A property P is called *computably invariant* if, given any computable bijection $f : \omega \to \omega$, $A \subset \omega$ has P iff $f(A)$ has P. For example, being c.e. is computably invariant. Show that being an index set is *not* computably invariant.

3.1.2 The finite extension method

The finite extension method is a somewhat blunt tool, but it is a good place to start. One can view this method as the purely set-theoretic version of the injury-free diagonalisation that we have already encountered in the proof of Theorem 2.4.20. The idea is that we build our set step by step, as a sequence of its initial segments, satisfying a list of requirements. The construction does not actually have to be computable, and thus the resulting set will also typically be non-computable.

Recall that Rice's Theorem 2.1.12 showed that all index sets are of degree $\geqslant \mathbf{0}'$. In 1944, Post [431] observed that all then known computably enumerable problems had the property that they were either of Turing degree $\mathbf{0}$ or $\mathbf{0}'$. He posed

Post's Problem. Does there exist a c.e. degree $\mathbf{a}$ with $\mathbf{0} < \mathbf{a} < \mathbf{0}'$?

As we see in the next section, Post's Problem was finally solved by Friedberg [181] and Muchnik [409] using a new and ingenious method called the priority method; this is Theorem 3.1.21. The method used to solve this question was an "effectivisation" of an earlier finite extension method discovered by Kleene and Post, the priority method.[1] Kleene and Post proved the following.

Theorem 3.1.14 (Kleene and Post [303]). *There exist degrees* $\mathbf{a}$ *and* $\mathbf{b}$ *both below* $\mathbf{0}'$ *and such that* $\mathbf{a}|\mathbf{b}$. *(That is, they are incomparable.)*

Proof of Theorem 3.1.14. We construct $A = \lim_s A_s$ and $B = \lim_s B_s$ in stages, and meet the requirements below for all $e \in \mathbb{N}$:

$$R_{2e} : \Phi_e(A) \neq B, \quad R_{2e+1} : \Phi_e(B) \neq A.$$

Note that if $A \leqslant_T B$, then there must be some procedure Φ with $\Phi(B) = A$. Hence, if we meet all the R_n, then $A \nleqslant_T B$ since we meet all the R_{2e+1} requirements. Similarly, $B \nleqslant_T A$ since we meet all the R_{2e} requirements. Thus, A and B will have incomparable T-degrees. The fact that $A, B \leqslant_T \varnothing'$ will come from the construction and will be observed at the end.

[1] It is possible to solve Post's problem without a priority argument; see Downey and Hirschfeldt [125]. The methods used there are *more difficult* than the priority method.

The argument is a finite extension one in the sense that at each stage s we specify a finite portion of A, namely A_s, and a finite portion B_s of B. Both A_s and B_s will be specified as strings of length $\leqslant t_s$ for some parameter t_s inductively defined in the construction. *The key invariant is that for all stages $u \geqslant s$, $A_s \preceq A_u$ and $B_s \preceq B_u$*, where $X \preceq Y$ means that (the finite binary string coding the characteristic function of) X is an initial segment of Y.

Thus, in a finite extension argument, after stage s we can only *extend* the portion of A (or B) we have specified so far. Or, to put it another way, at a later stage, we cannot change the sets on anything we have so far specified.

Construction.

Stage 0. Set $A_0 = B_0 = \lambda$ (the empty string). Set $t_0 = 0$.

Stage $2e+1$. (Attend R_{2e}.) We will have specified A_{2e}, B_{2e}, and t_{2e} at stage $2e$. Pick some number x, called a *witness*, with $x > t_{2e}$, and see if there is a string τ extending A_{2e} such that

$$\Phi_e(\tau; x)\downarrow.$$

If such a τ exists, choose the first such τ and set $A_{2e+1} = \tau$. For all q with $q \leqslant x$, set

$$B_{2e+1}(q) = \begin{cases} B_{2e}(q), & \text{if } q \leqslant \max\{z : z \in B_{2e}\}, \\ 1 - \Phi_e(\tau; x), & \text{if } q = x, \\ 0, & \text{otherwise.} \end{cases}$$

Set $t_{2e+1} = \max\{x, |\tau|\}$. If no such τ exists, then set $A_{2e+1} = A_{2e}$, $B_{2e+1} = B_{2e}$, and $t_{2e+1} = x$ (which is $\geqslant t_{2e} + 1$).

Stage $2e+2$. (Attend R_{2e+1}.) Proceed as we did in stage $2e+1$, except with the roles of A and B reversed.

End of Construction.

Verification. To verify the construction, we prove that we meet R_j for all j, and in fact, we meet R_j at stage $j+1$. This is proven by induction on j. First, note that for all n, $t_{n+1} > t_n$. We suppose that we have met R_j for all $j < n$ by stage n, and the parameter t_n is so large that it will protect all the computations for $j < n$. Without loss of generality, $n = 2e$. Now, at stage $n+1$, there are two cases to consider. If there is a τ extending A_n with $\Phi_e(\tau, x)\downarrow$, our action is to adopt τ as our next A_{n+1} and cause $\Phi_e(A_{n+1}; x) \neq B_{n+1}(x)$. We will then set t_{n+1} to be large enough that for all $n' \geqslant n$, for all z with $z \in u(\Phi_e(A_{n+1}; x))$, $A_{n'}(z) = A_n(z)$, and hence $\Phi_e(A; x) = \Phi_e(A_{n+1}; x) \neq B_{n+1}(x) = B(x)$. The other case is that no such τ exists. Then, since A is an extension of A_n, it can only be that $\Phi_e(A; x)\uparrow$, and hence in either case, we meet R_n.

Finally, we argue that $A, B \leqslant_T \varnothing'$. Notice that the construction is in fact fully computable except for the decision as to which case we are in at stage n. There, we must decide if there is some τ with a convergent computation. For instance, at stage $2e+1$, we must decide

$$\exists \tau, s[A_{2e} \preceq \tau \wedge \Phi_{e,s}(\tau; x)\downarrow].$$

This is a Σ_1^0 question uniformly in x, and hence can be decided by $\varnothing'$. Specifically, use the *s-m-n* Theorem 2.1.6 to construct a computable function $\varphi_{s(\langle f,\sigma,x\rangle)}$ such that for all z, $\varphi_{s(\langle f,\sigma,x\rangle)}(z) = 1$ if $\exists \tau, s[\sigma \preceq \tau \wedge \Phi_{f,s}(\tau; x)\downarrow]$ and $\varphi_{s(\langle f,\sigma,x\rangle)}(z)\uparrow$ otherwise. Then

$$\exists \tau, s[\sigma \preceq \tau \wedge \Phi_{f,s}(\tau; x)\downarrow]$$

if and only if $s(\langle f, \sigma, x\rangle) \in \varnothing'$. □

We again remark that the reasoning at the end of the above proof is quite common: $\varnothing'$ can answer any $\Delta_2^0 = \Sigma_2^0 \cap \Pi_2^0$-question and hence any Σ_1^0- and Π_1^0-question. We also note that neither A nor B constructed above can possibly be finite, even though we did not explicitly ensure that they were infinite. (But of course, we could always extend the sets at the stages when no τ exists arbitrarily.)

A slightly more subtle application of the finite extension argument is the following. We say that a pair of degrees $\mathbf{a}, \mathbf{b} \neq \mathbf{0}$ form a *minimal pair* if $\mathbf{c} < \mathbf{a}, \mathbf{b}$ implies $\mathbf{c} = \mathbf{0}$. That is, the infimum $\mathbf{a} \wedge \mathbf{b} = \mathbf{0}$.

Theorem 3.1.15 (Kleene and Post [303])**.** *Given* $\mathbf{a} \neq \mathbf{0}$*, there exists* $\mathbf{b} \neq \mathbf{0}$*, such that* $\mathbf{a}$ *and* $\mathbf{b}$ *form a minimal pair.*

Extended sketch. We will be given A of degree $\mathbf{a}$. We need to meet the requirements

$$R_{e,i} : \Phi_e^A = \Phi_i^B = f \text{ total, implies } f \text{ computable.}$$

We construct B using the finite extension method. Suppose that we have constructed B_s and dealt with $R_{e',i'}$ for $\langle e', i'\rangle < s$, and $s = \langle e, i\rangle$. We ask,

"Do there exist t, x, and σ with σ extending B_s, and $\Phi_{e,t}^A(x)\downarrow \neq \Phi_{i,t}^\sigma(x)\downarrow$?"

If so, we let $B_{s+1} = \sigma$. If not, $B_{s+1} = B_s$. If we can find such parameters, we force $\Phi_e^A \neq \Phi_i^B$. If we cannot and $\Phi_e^A = \Phi_i^B$ is total, then it must be computable, since for any x, we simply start computing $\Phi_{i,t}^\sigma(x)$ for all σ of length $\leqslant t$ and $t \geqslant s$ and $B_s \preceq \sigma$. We know one will halt by totality and it must be the correct answer. This is a computable method of evaluating $\Phi_i^B(x) = \Phi_e^A(x)$. □

The reader should note that in the proof above, $B \leqslant_T A'$.

Exercises

Exercise° 3.1.16. A set is called *immune* if it is infinite and has no infinite c.e. subsets.

1. Use the finite extension method to construct an immune set $A \leqslant_T \varnothing'$.
2. Suppose that A is an immune set and $a \in A$. Show that there is no computable permutation p of ω with $p(A) = A \setminus \{a\}$.
3. A set B is called *bi-immune* if both B and $\overline{B}$ are immune. Construct a set $B \leqslant_T \varnothing'$ which is bi-immune.

Exercise° 3.1.17. A set A is called *low* if $A' \equiv_T \varnothing'$. Use the finite extension method to construct $A \leqslant_T \varnothing'$ such that A is non-computable and low.

Exercise° 3.1.18. A set A is called *autoreducible* if there is a Turing functional Φ such that for all x,

$$A(x) = \Phi^{(A\setminus\{x\})}(x) = A(x).$$

That is, for each x, A can determine if $x \in A$, without using x in any query to itself. For example, a complete theory T has this property since, for each (code of) a sentence x, $x \in T$ iff $\neg x \notin T$. Using the finite extension method construct a set $A \leqslant_T \varnothing'$ which is *not* autoreducible.

Exercise 3.1.19 (Friedberg Completeness Criterion and low$_n$ sets; Friedberg [180]).

1. Suppose $X \geqslant_T \varnothing'$. Show that there exists a set A such that $A' \equiv_T X \equiv_T A \oplus \varnothing'$.
2. Use the first part to construct a set B which is low$_2$ and not low. That is, $B'' \equiv_T \varnothing''$, and $B' >_T \varnothing'$.

Exercise 3.1.20 (Jockusch and Shore [272]). Using the techniques of the previous question, show that for any $e \in \mathbb{N}$ and any $X \geqslant_T \varnothing'$, there is a set A such that $A \oplus W_e^A \equiv_T A \oplus \varnothing' \equiv_T X$. (Note that the previous question is the special case with $W_e^A = A'$.)

3.1.3 Post's problem and the finite injury method

A more subtle generalisation of the finite extension method is the *priority method.* We begin by looking at the simplest incarnation of this elegant technique, the *finite injury priority method.* This method is somewhat like the finite extension method but with backtracking.

The idea behind this method is the following. As an illustration, let's reconsider Post's Problem that asks for incomplete c.e. degrees. In the result below, we will construct *c.e.* sets A and B with incomparable Turing degrees. We need to satisfy the requirements

$$R_{2e} : \Phi_e(A) \neq B,$$

$$R_{2e+1} : \Phi_e(B) \neq A,$$

but this time we are not allowed to use an oracle in the construction.

Each requirement picks a *follower* x which it intends to use for diagonalisation and initially keeps it out of the respective set (say, B). It seems that we need to know the answer to

$$\text{“Does } \tau \text{ with } \Phi_e^\tau(x)\downarrow \text{ exist or not?”}$$

to decide which strategy to pursue. But the idea is that we first guess that no such τ exists for our witness x. This means that nothing is really done for the sake of R_{2e}, save keeping $x \notin B_s$, unless we see a stage where some $\tau \preceq A_s$ exists. If such a stage occurs, then we will try to make A extend τ and win as before by putting x into B if necessary. So whatever case occurs, we will win.

However, this change can potentially affect the computation of some other $R_{2e'+1}$. The requirement $R_{2e'+1}$ may have already found its computation, put y into A, and wants to preserve the computation of $\Phi_{e'}$ on B_s. However, it is possible that $x \leqslant u(\Phi_{e'}(B_s; y))$. If R_e enumerates x into B_s, it will “injure” $R_{2e'+1}$, in the standard terminology.

To make sure that every requirement is eventually met, we put a “priority ordering” on them and will allow R_j to injure R_i if R_j has higher priority than R_i. In our case, if $j < i$. If R_i is injured at stage s, then we will “initialise” the requirement R_i; i.e., we will restart its strategy by picking a new large follower.

We now turn to the formal proof of the Friedberg-Muchnik Theorem.

Theorem 3.1.21 (Friedberg [181], Muchnik [409]). *There exist c.e. sets A and B such that A and B have incomparable Turing degrees.*

Proof. We build c.e. sets $A = \cup_s A_s$ and $B = \cup_s B_s$ and satisfy the requirements below.

$$R_{2e} : \Phi_e(A) \neq B,$$

$$R_{2e+1} : \Phi_e(B) \neq A.$$

The strategy for a single R_j. We begin by looking at the strategy for a single R_j. Without loss of generality let $j = 2e$.

(i) Initially, we will pick a new fresh number $x = x(j)$ to *follow* R_j. This number is targeted for B, and of course, we have $x \notin B_s$.

(ii) We wait for a stage $t > s$ to occur with $\Phi_{e,t}(A_t; x) \downarrow = 0 = B_t(x)$. If stage t does not occur then we must have $\Phi_e(A; x) \neq B(x)$.

(iii) Should stage t occur, set $A_{t+1} = A_t$, and put x into $B_{t+1} - B_t$, causing

$$\Phi_{e,t+1}(A_{t+1}; x) = 0 \neq 1 = B_{t+1}(x) = B(x).$$

In the construction, we will protect this computation with priority $j = 2e$.

Note that when we take action (iii), we might injure $R_{2e'+1}$ if $x \leqslant u(\Phi_{e',t}(B_t; x'))$, which is the use of the computation $\Phi_{e',t}(B_t; x')$.

Definition 3.1.22. We say that R_j *requires attention* at stage s if j is least such that one of the following pertains:

(1) R_j has no follower at stage s.

(2) R_j has a follower $x(j, s)$ at stage s and furthermore, supposing that $j = 2e$,

$$\Phi_{e,s}(A_s; x(j, s)) \downarrow = 0 = B_s(x(j, s)).$$

(If $j = 2e + 1$, then we reverse the roles of A and B.)

Construction.

Stage 0. Set $A_0 = B_0 = \varnothing$.

Stage $s > 0$. Find the least j with R_j requiring attention. We suppose that $j = 2e$ without loss of generality. Adopt the appropriate case below.

(1) Pertains. Find a number x larger than any number seen so far in the construction and appoint $x(j, s) = x$. Initialise all $R_{j'}$ with $j' > j$. That is, cancel all followers associated with $R_{j'}$.

(2) Pertains. Initialise all $R_{j'}$ for $j' > j$. Set $A_s = A_{s-1}$ but set $B_s = B_{s-1} \cup \{x(j,s)\}$.

End of construction.

Verification. We prove by induction on j that

(a) each R_j receives attention only finitely often,

(b) $\lim_s x(j,s) = x(j)$ exists,

(c) R_j is met.

For an induction, assume (a), (b), and (c) hold for all $j' < j$. Let s_0 be a stage good for j: that is, for all $s \geqslant s_0$ and all $j' < j$,

$(a)'$ $R_{j'}$ does not require attention at stage s, and
$(b)'$ $x(j',s) = x(j',s_0)$.

Choose s_0 to be minimal such. (Indeed, it would be enough to fix s_0 so that $R_{j'}$ $(j' < j)$ will not require attention after s_0.) It can only be that R_j receives attention via (i) at stage s_0+1, and is appointed a large fresh follower $x = x(j, s_0+1)$. By the choice of s_0, x is never cancelled: the only requirements that could cancel x are $R_{j'}$ for $j' < j$. There are two possibilities.

The first is that (2) never pertains (with x). In this case, there is no stage $t > s_0 + 1$ with $\Phi_{e,t}(A_t; x) \downarrow = 0 = B_t(x)$. This means that either $\Phi_e(A; x) \uparrow$ or $\Phi_e(A; x) \neq 0 = B(x)$. In either case, we win.

The second case is that (2) pertains to R_j at some stage $s > s_0 + 1$. In this latter case, we act to cause a disagreement at stage s, namely

$$\Phi_{e,s}(A_s; x) = 0 \neq 1 = B_s(x).$$

Now since we initialise all R_k for $k > j$, and new followers are always appointed large and fresh, it follows that this stage s disagreement is permanent.

In either case, R_j only receives attention at most twice more after stage s_0, and is met. Furthermore, $x(j, s_0+1) = x(j,t) = x(j)$ for all $t > s_0$. This concludes the induction and, hence, the proof of the Friedberg-Muchnik Theorem. □

We remark that the above proof is an instance of the simplest of all finite injury arguments, as it is an example of what is called "bounded injury" in the sense that we can put a computable upper bound in advance on the number of injuries that R_j can possibly have. In this case, the bound is $2^j - 1$.

3.1.4 Low c.e. sets

We have accumulated enough techniques to prove the main result of this subsection. Recall that a set A is low if $A' \leqslant_T \varnothing'$. (Since $A \leqslant_T A'$, then necessarily $A \leqslant_T \varnothing'$.) We are ready to prove Theorem 3.1.1 that says that there exist c.e. non-computable low sets. We give two proofs. As far as we know, the first, indirect proof was first suggested by Soare [476].

Indirect proof of Theorem 3.1.1. We show that the two sets constructed in the proof of the Friedberg-Muchnik Theorem 3.1.21 are low. For that, using (the functional version of) the *s-m-n* Theorem 2.1.6, we define a computable h such that

$$\Phi_{h(e)}(A;y) = \begin{cases} 0 & \text{if } \Phi_e(A;e)\downarrow; \\ \uparrow & \text{otherwise.} \end{cases}$$

If A and $x(2e,s)$ are as in the proof of Theorem 3.1.21, then we define

$$g(e,s) = \begin{cases} 1 & \text{if } \Phi_{h(e),s}(A_s;x(2e,s))\downarrow = 0; \\ 0 & \text{otherwise.} \end{cases}$$

We argue that $\bar{g}(e) = \lim_s g(e,s)$ is well-defined; this is essentially because R_{2e} can be injured only finitely many times. Indeed, if $\exists^\infty s\, g(e,s) = 1$, then actually $\lim_s g(e,s)$ exists and is equal to 1. This situation corresponds to the existence of infinitely many s such that $\Phi_{h(e),s}(A_s;x(2e,s)) = 0$. However, if s is large enough so that R_{2e} is never injured again, then the computation $\Phi_{h(e),s}(A_s;x(2e,s))$ and the value $x(2e,s)$ are final. It follows that $\bar{g}$ is the characteristic function of A'. Hence, by the Limit Lemma 3.1.3, we have $A' = \{e : \lim_s g(e,s) = 1\} \leqslant_T K$. □

The proof above is clever, however, it is not really any easier than the direct construction that we outline below.

Sketch of a direct proof of Theorem 3.1.1. We construct a c.e. set A in stages. To make A noncomputable, we need to meet the requirements

$$P_e : \overline{A} \neq W_e.$$

To make A low, we meet the requirements

$$N_e : \exists^\infty s\ \Phi_{e,s}(A_s;x)\downarrow \implies \Phi(A;x)\downarrow.$$

If we define $g(x,s) = 1$ iff $\Phi_{e,s}(A_s;x)\downarrow$ and $g(x,s) = 0$ otherwise, then, provided that the construction is stage-by-stage computable, $g(x,s)$ will be computable. Moreover, if we meet every N_e, then $\lim_s g(e,s) = g(e)$ exists for each e, because A is c.e., and $\Phi(A;x)\downarrow$ implies that for some s, $\Phi(A;x) = \Phi_{e,s}(A_s;x)$. As in the previous proof, this will guarantee that $A' = \{e : \lim_s g(e,s) = 1\} \leqslant_T K$.

The strategy to meet N_e is as follows. If $\Phi_{e,s}(A_s;x)\downarrow$, then try to ensure that $A_s \upharpoonright u(\Phi_{e,s}(A_s;x)) = A \upharpoonright u(\Phi_{e,s}(A_s;x))$ by initialising all lower-priority requirements. This forces the lower-priority strategies to choose new large numbers as followers. The new followers will be too large to injure the computation $\Phi_{e,s}(A_s;x)$ after stage s.

The strategy for P_e is as follows. While $A_s \cap W_{e,s} = \varnothing$, P_e picks a follower x larger than any number seen so far. Then, if we see $x \in W_{e,s}$, put x into A_{s+1}.

We arrange the requirements in priority order according to their index:

$$N_0 > P_0 > N_1 > P_1 > N_2 > \ldots.$$

In the construction that we omit, the priority method sorts the actions out. Note that since P_e picks fresh numbers, P_e does not injure any N_j for $j < e$. It is easy to see that any N_e can be injured at most e times, and P_e is met as it is initialised at most 2^e times. □

Exercises

Exercise° 3.1.23 (Soare [477]). A set X is called *semi-low* if $H_X = \{e : W_e \cap X \neq \emptyset\} \leqslant_T \emptyset'$. Let A be a c.e. set. Prove that the following are equivalent:

(a) $\overline{A}$ is semi-low;

(b) There is an enumeration $A = \cup_{s\in\mathbb{N}} A_s$ of A such that for all e, $(\exists^\infty s)\ [W_{e,s} - A_s \neq \emptyset] \rightarrow W_e - A \neq \emptyset$.

(c) There exists a computable function f such that for all e:

(1) $W_e \cap \overline{A} = W_{f(e)} \cap \overline{A}$, and

(2) $W_e \cap \overline{A} = \emptyset \rightarrow W_{f(e)}$ is finite.

Exercise° 3.1.24 (Soare [477]). Show that every low set is semi-low. Let C be a c.e. set. Show that there is a c.e. set A with $A \equiv_T C$ and $\overline{A}$ is semi-low.

Exercise° 3.1.25 (Soare [477]). Let $D_0, D_1, \ldots$ be the standard enumeration of finite sets, where (say) a set is represented by a finite tuple in $2^{<\omega}$ with index n. Show that for every (infinite) c.e. set A, the c.e. set $X = \{n : D_n \cap A \neq \emptyset\}$ has the property $\{e : W_e \cap \overline{X} \neq \emptyset\} \equiv_T A'$. (In particular, every non-low c.e. degree contains a c.e. set X whose complement is not semi-low.)

Exercise° 3.1.26 (Ladner [328]). 1. Use the finite injury method to construct a c.e. set which is not autoreducible (see Exercise 3.1.18).

2. Show that a c.e. set A is autoreducible iff there exist c.e. sets $A_1 \sqcup A_2 = A$ with $A \equiv_T A_1 \equiv_T A_2$. (Such sets are called *mitotic*.)

Exercise° 3.1.27 (Friedberg's Splitting Theorem [182]). Show that if A is any non-computable c.e. set, then there exist two disjoint c.e. sets A_1, A_2 such that $A_1 \sqcup A_2 = A$, and both A_1 and A_2 are non-computable.

Exercise° 3.1.28. A c.e. set A is called *simple* if A is coinfinite and $\overline{A}$ is immune (Ex. 3.1.16). A *strong array* is a computable collection of finite sets $\mathcal{D} = \{D_{f(n)} : n \in \mathbb{N}\}$ with f computable. We say that a set X is hyperimmune if X is infinite and for every infinite strong array $\mathcal{D}$ there is some n with $D_{f(n)} \nsubseteq X$. We say that A is hypersimple if $\overline{A}$ is infinite and hyperimmune. Construct a c.e. set A which is simple but not hypersimple.

Exercise° 3.1.29. A set X is called *effectively immune* if it is infinite and there is a computable function g such that $W_e \subset X$ implies that $|W_{g(e)}| < g(e)$. Prove that if A is c.e. and $\overline{A}$ is effectively immune, then $A \equiv_T \emptyset'$.

3.1.5 Using (semi-)lowness*

In this subsection, we will examine how lowness can be used in constructions, particularly in the context of c.e. sets. This idea can be extended to handle semi-low sets (defined in Exercise 3.1.23), which will be used in Section 7.2 of Chapter 7 to describe Δ^0_2-categoricity of a broad class of groups. We will not need this technique until Chapter 7, and the impatient reader may skip this section and return to it later.

Whilst it is not easy to pinpoint the very first use of lowness, one early example was due to Robinson [453] (see Exercise 3.1.33). Lowness allows for a kind of "verification" of configurations of a low c.e. set, building on the theme that low sets resemble computable sets. We can use lowness to test whether some computation involving the low set X should be trusted, as follows. For example, suppose $X = \cup_s X_s$ is c.e. and low, and for some s, we have

$$\Phi_{e,s}^{X_s}(n) \downarrow= 0$$

with use $u(e,n,s) = u(\Phi_{e,s}(X_s;n))$. For the sake of some diagonalisation, perhaps we wish to enumerate n into some other set Y if this is a correct computation. However, at a later stage t, some $y \leqslant u(e,n,s)$ might enter $X_t \backslash X_s$, and then the computation $\Phi_{e,s}^{X_s}(n) \downarrow= 0$ might later change to $\Phi_{e,t}^{X_t}(n) \downarrow= 1$ (if the computation converges at all). Note that this can only happen if $\overline{X_s} \upharpoonright u(e,n,s) \neq \overline{X} \upharpoonright u(e,n,s)$. The following is an example of a lemma which can be used to test whether $\overline{X_s} \upharpoonright u(e,n,s) = \overline{X} \upharpoonright u(e,n,s)$. In the lemma, the finite set D_m is our way to approximate $\overline{X_s} \upharpoonright u(e,n,s)$; this will be further explained after we prove the lemma.

Lemma 3.1.30. *Suppose that X is low. Then*

$$Y = Y_j = \{j \mid \exists m \in W_j(D_m \subseteq \overline{X})\} \leqslant_T \varnothing',$$

where D_m is the mth canonical finite set.

Proof. By the definition of Y, we have $Y \in \Sigma_1^X$. By Theorem 3.1.5, relativised to X, this implies $Y \leqslant_T X'$. Since X is low, $X' \equiv_T \varnothing'$, and thus $Y \leqslant_T \varnothing'$. □

The way that this is used is as follows. Consider the situation described above, where we wish to test

$$\text{“Is } \Phi_{e,s}^{X_s}(n) \downarrow= 0 \text{ final?”}$$

at every stage s. This will be associated with some requirement R_e that monitors Φ_e. For the sake of R_e, we will build a c.e. set $W_{g(e)}$ whose index is given by the Recursion Theorem 2.1.13. Initially, $W_{g(e)} = \varnothing$. We know that $Y = Y_{g(e)} \leqslant_T \varnothing'$. By the Limit Lemma 3.1.3, there is a computable function h with the property $Y(i) = \lim_s h(i,s)$ for all i.

Now, assume that $h(g(e),s) = 0$ and $\Phi_{e,s}^{X_s}(n) \downarrow= 0$ with use $u(e,n,s)$. The idea is to put m into $W_{g(e),s}$, where $D_m = \overline{X_s} \upharpoonright u(e,n,s)$, and use it as a “test”. One of two things can happen:

- $h(g(e),t)$ changes to 1 for some $t \geqslant s+1$, or
- some $y \leqslant u(e,n,s)$ enters $X_t - X_s$ for some $t > s$.

Note that, knowing that one of the two *must* happen, we can pause the construction to wait and see which occurs. In the former case, we will believe that $\Phi_{e,s}^{X_s}(n) = 0$ is correct, and we then act accordingly in the construction. In the latter case, we will know that the computation at stage s is *not* correct and will act accordingly. Note that if we believe the former case, it is possible that some $y \leqslant u(e,n,s)$ might enter $X_v - X_s$ at some $v \geqslant s$. However, note that $D_m \nsubseteq \overline{X}$ and, hence, $m \notin Y$. Therefore, at some stage $t' \geqslant t$, we will have $h(g(e),t') = 0$. We can then repeat the action. However, since we know that $\lim_s h(g(e),s)$ exists, from some point onwards, we must eventually get the correct answer.

More generally, we might have the opportunity to use the Recursion Theorem 2.1.13 and the Limit Lemma 3.1.3 if some query to X is (uniformly) Δ_2^0. If we can somehow argue that *either* the oracle must change *or* our guess about the monitored computation must change, then we can imitate the method described above by (uniformly) retrying until our guess is finally correct.

This idea might work under an assumption weaker than lowness. For instance, in Exercise 3.1.23, we defined a set Z to be *semi-low* if

$$\{e : W_e \cap Z \neq \emptyset\} \leqslant_T \emptyset'.$$

In Exercises 3.1.24 and 3.1.25, we observed that semi-lowness is not a degree-invariant property, and that it is generally weaker than lowness. However, the condition $\{e : W_e \cap Z \neq \emptyset\} \leqslant_T \emptyset'$ might be sufficient to implement the general methodology described above in specific cases. We illustrate this technique with a simple application that only requires the (complement of a) set to be semi-low.

Theorem 3.1.31 (Downey and Jockusch [128]). *Show that if C is a low c.e. set, then there exists a c.e. set $A \leqslant_T C$, such that $A \nleqslant_m C$. Indeed, assuming that $\overline{C}$ is semi-low is enough.*

Sketch. We give a sketch that emphasises the use of (semi-)lowness and omits the combinatorics related to the finite injury technique.

We construct a c.e. set $A \leqslant_T C$ and meet the requirements

$$R_e : A \nleqslant_m C \text{ via } \varphi_e.$$

Let $C = \cup_s C_s$ be an enumeration of C so that at most one element enters $C_{s+1} - C_s$ at every stage s.

To ensure $A \leqslant_T C$, we will allow x to enter $A_{s+1} - A_s$ only if some $y \leqslant x$ enters $C_{s+1} - C_s$; this is called *simple permitting.*

For the sake of R_e, we will build an auxiliary c.e. set $V_e = W_{g(e)}$ whose index is given by the Recursion Theorem 2.1.13. As C is (semi)low,

$$S = \{j \mid W_j \cap \overline{C} \neq \emptyset\} \leqslant_T \emptyset'.$$

By the Limit Lemma 3.1.3, we have a computable function h with $S(j) = \lim_s h(j, s)$ for all j. Initially, we have $W_{g(e)} = \emptyset$ and (without loss of generality) $h(g(e), s) = 0$. If $y \in C_{s+1} \backslash C_s$, then look for the least e such that R_e seems unsatisfied by diagonalisation, and there is some $x \notin A_s$ such that $\varphi_{e,s}(x) \downarrow$ and $y \leqslant x$. If $\varphi_{e,s}(x) \in C_{s+1}$, then keep x out of A_t at stages $t > s$. If $\varphi_{e,s}(x) \notin C_{s+1}$, then we can enumerate $\varphi_{e,s}(x)$ into $W_{g(e),s}$. Now we can delay our decision as to whether to put $x \in A$ by waiting to see whether $h(g(e), t) = 1$ for some $t \geqslant s$, or $\varphi_{e,s}(x)$ enters C_t. As above, one of the two events must occur. In the latter case, we win again. In the former case, we have a certification that $\varphi_{e,s}(x) \notin C$. Then we put x into A_{s+1} and believe we have won. Notice that this second case can only occur finitely many times as $\lim_s h(g(e), s)$ exists. The argument then works by finite injury. □

Exercises

Exercise° 3.1.32. (Downey and Jockusch [128]) We say that $A \leqslant_{tt} B$ if there is a Turing functional Φ such that Φ^X is total for all X and $\Phi^B = A$. Show that if C is a low c.e. set, then there exists a c.e. set $A \leqslant_T C$, such that $A \nleqslant_{tt} C$.

Exercise 3.1.33. (Robinson [453]) Let $\mathbf{b}$ and $\mathbf{e}$ be c.e. degrees such that $\mathbf{c} < \mathbf{b}$, and $\mathbf{c}$ is low. Then there exist incomparable low c.e. degrees $\mathbf{a_0}$ and $\mathbf{a_1}$ such that $\mathbf{b} = \mathbf{a_0} \cup \mathbf{a_1}$ and $\mathbf{a_i} > \mathbf{c}$ for $i = 0, 1$.

3.1.6 Finite injury arguments of unbounded type

There are examples where the number of injuries is finite, but not bounded by any computable function. One such classical example is Sacks' Splitting Theorem below. We write $X \sqcup Y$ for the disjoint union of X and Y.

Theorem 3.1.34 (Sacks [457]). *Let B, C be c.e. sets, and assume C is non-computable. Then there exist disjoint c.e. sets A_0, A_1 such that $B = A_0 \sqcup A_1$ and $C \not\leqslant_T A_i$, for $i = 0, 1$.*

Corollary 3.1.35. *If B is a non-computable c.e. set, then there is a c.e. set A such that $\varnothing <_T A <_T B$.*

Proof of Corollary 3.1.35. Note that if $B = A_0 \sqcup A_1$, then $B \equiv_T A_0 \oplus A_1$ (Exercise 3.1.36). By setting $C = B$ in the theorem, we see that the only possibility is that $A_0 \equiv_T A_1$, and that A_0 and A_1 are both incomplete and below B. □

Proof sketch of Theorem 3.1.34. Let $C = \cup_s C_s$ and $B = \cup_s B_s$ be computable enumerations of C and B, respectively. Without loss of generality, we will assume that we are given an enumeration of B such that exactly one number enters $B_{s+1} - B_s$ at stage $s+1$. We build $A_i = \cup A_{i,s}$ in stages and meet the requirements

$$N_{e,i} : \Phi_e(A_i) \neq C,$$

for every $e \in \mathbb{N}$ and $i \in \{0, 1\}$. Also, for every s, we put the unique $b_s \in B_{s+1} - B_s$ into exactly one of $A_{0,s+1} - A_{0,s}$ or $A_{1,s+1} - A_{1,s}$. This causes $B = A_0 \sqcup A_1$.

To meet $N_{e,i}$, we define the *length of agreement function*

$$\ell^i(e, s) = \max\{x : \forall y < x[\Phi_{e,s}(A_{i,s}; y) = C_s(y)]\}$$

and the *maximum length of agreement function,*

$$m^i(e, s) = \max\{\ell^i(e, t) : t \leqslant s\}.$$

We also define the *restraint function*

$$r^i(e, s) = \max\{u(\Phi_{e,s}(A_{i,s}; x)) : x \leqslant m^i(e, s)\},$$

where u is the use of the respective computation. Note that we used $<$ in the definition of ℓ^i and $\leqslant$ in the definition of r^i, which reflects that we will try to preserve at least one disagreement in the Φ_e-computation by restraining elements to enter the respective A_i below $r^i(e, s)$. This is sometimes called *Sacks' preservation strategy.*

In the construction, if $x \in B_{s+1} - B_s$, see if there is a $\langle e, i\rangle \leqslant s$ least such that $x \leqslant r^i(e, i)$. If yes, then choose $\langle e, i\rangle$ least such and put x in $A_{1-i,s+1} - A_{1-i,s}$. If no such $\langle e, i\rangle$ exists, then put x in $A_{0,s+1} - A_{0,s}$.

We now sketch the verification. Suppose $N_{0,0}$ is not met, i.e., $\Phi_0(A_0) = C$. By induction, we see that the definition of r^0 ensures that the Φ_0-computation with use below $r^0(0, s)$ is final. This is because all numbers that appear in B below $r^0(0, s)$ will be put into A_1, for every s. But this would imply that A_0 has to be computable, and thus so would be C, contradicting the choice of C. It thus follows that $\Phi_0(A_0) \neq C$. Let x be the least such that either $\Phi_0(A_0)(x) \uparrow$ or $\Phi_0(A_0)(x) \downarrow\neq C(x)$. In the former case, it must be that we never see a divergent $\Phi_0(x)$-computation, for otherwise we'd preserve it. (This is where it is important that we preserve at least one disagreement.) In the latter case, the definition of r^0 ensured that we preserve the $\Phi_0(A_0)(x)$-computation as soon as it is discovered at some stage s. In either case $r^0(0,\cdot)$ cannot possibly increase after some stage. Thus, $\lim_s r^0(0,s)$ exists.

By induction on $\langle e,i\rangle$, we argue that (for every i and e) $\lim_s r^i(e,s)$ exists, and that every $N_{e,i}$ is met. For that, repeat the argument above for $N_{e,i}$ assuming the stage is large enough so that, for all $\langle e',i'\rangle$, the values $r^{i'}(e',\cdot)$ have reached their limits. Further, assume the stage is so large that all elements of B below $\max_{\langle e',i'\rangle<\langle e,i\rangle} r^{i'}(e',\cdot)$ have been listed in B by that stage. Using this stage and the initial segment of B as a non-uniform parameter, repeat the analysis above to conclude that $\lim_s r^i(e,s)$ exists and $N_{e,i}$ is met.

In contrast with the previous two theorems, the number of times $r^i(e,s)$ can increase, and thus the number of times $N_{e,i}$ can be "injured", cannot be computably bounded. □

Exercises

While none of the exercises below are marked with a star, some of them are not straightforward. The reader might want to look at the cited papers for the details. (The same is true about the exercises after the next subsection.)

Exercise° 3.1.36. Show that if A_0, A_1 are c.e. sets and $B = A_0 \sqcup A_1$, then $B \equiv_T A_0 \oplus A_1$.

Exercise° 3.1.37 (Ambos-Spies [9]). Use a movable marker construction to show that there is a Turing-complete computably enumerable set A such that if $A_1 \sqcup A_2 = A$ is a c.e. splitting of A, then one of A_1 or A_2 is low.

Exercise° 3.1.38. A set Y is called high$_n$ if $Y^{(n)} \equiv_T \varnothing^{(n+1)}$. If $n = 1$, we say that Y is high. Use Exercise 4.2.59 to construct a set $A <_T \varnothing'$ which is high. (Hint: Theorem 3.1.1 constructs a c.e. set W such that for each D, $D <_T W^D \equiv_T D'$.)

Exercise° 3.1.39 (Jockusch and Soare [272]). Prove the Pseudo-Jump Theorem:

(i) For each e such that $X <_T X \oplus W_e^X$ for all X, construct a c.e. set A such that $A \oplus W_e^A \equiv_T \varnothing'$. (See Ex. 3.1.20.)

(ii) Use part (i) to construct a high c.e. set $A < \varnothing'$. (Hint: Consider W_e^X to be the construction of a non-computable set c.e. relative to X and low relative to X.)

(iii) Use part (ii) to construct a low$_2$ c.e. set.

(iv) Show that there are c.e. sets A_i, B_i for $i \in \mathbb{N}$, such that A_n is properly low$_n$ and B_n is properly high$_n$ (i.e., not low$_{n-1}$ and not high$_{n-1}$, respectively).

Exercise° 3.1.40. Use the Recursion Theorem to show that the *proof* of Sacks' Splitting Theorem shows that if A is a c.e. non-computable set and $\varnothing \not\leqslant_T C$, then we can find a c.e. splitting $A_0 \sqcup A_1 = A$, such that $C \not\leqslant_T A_i$ for $i \in \{0,1\}$ and also that A_0 and A_1 are low.

Exercise 3.1.41 (Downey and Stob [149]). Show that if A is c.e. and $A \equiv \varnothing'$, then there is a c.e. splitting $A_0 \sqcup A_1 = A$ such that $\varnothing <_T A_0 \leqslant_T A_1$. (Hint: Use the Recursion Theorem 2.1.13 to force two elements into A.)

Exercise 3.1.42 (Downey and Stob [149]). We say that a coinfinite c.e. set A is *promptly simple* if there is a computable function f such that, for all e, if $|W_e| = \infty$, then $\exists^\infty s \exists x((x \in W_{e,s+1} - W_{e,s}) \wedge x \in A_{f(s+1)})$. Show that the splitting theorem in the previous exercise holds if we assume that A is promptly simple.

3.1.7 Constructions involving infinite injury

We now give two examples of priority methods involving infinite injury to the requirements.

Enumerating of all c.e. sets without repetition

Recall that there is a uniform enumeration $(W_e)_{e\in\mathbb{N}}$ of all c.e. sets; however, this uniformly c.e. list contains many repetitions. The theorem below states that the standard enumeration can be replaced with another one, $(V_i)_{i\in\mathbb{N}}$, in which every W_e will appear as some V_i, and furthermore, $i \neq j$ implies $V_i \neq V_j$.

Theorem 3.1.43 (Friedberg Enumeration Theorem [180]). *There exists a uniformly c.e. collection $\{V_i : i \in \mathbb{N}\}$ of c.e. sets that mentions each c.e. set* exactly once.

We call the enumeration $\{V_i : i \in \mathbb{N}\}$ a *Friedberg enumeration* of the c.e. sets. We will return to 1-1 enumerations in Chapter 7, where a good understanding of the proof below will be very helpful.

Proof. We have that $\{W_e : e \in \mathbb{N}\}$ lists all c.e. sets. An obvious strategy is to choose *minimal* indices, SMIN= $\{e_0, e_1, \ldots\}$ meaning that e_{i+1} is the first $j > e_i$ such that $W_j \notin \{W_e : e \leqslant e_i\}$. The problem is that this is a complicated set, which is certainly not computable (Exercise 3.1.53). However, the idea of the proof below is to try to guess such minimal indices and send wrong guesses to the *garbage.*

For the construction, we will first construct a sequence $\{V_i : i \in \mathbb{N}\}$ listing all c.e. sets *which are not* ω exactly once. Then we can add ω as V_{-1}. We let $V_0 = \varnothing$. In the construction, at each stage s we will have a list $\{V_i : i \leqslant p(s)\}$, for some $p(s) \geqslant s$. These are designated as either *active* or *garbage.* If V_i is active, it will be simulating exactly one $W_{e(i)}$. If it is declared as garbage at some stage s, we will make it equal to $\{0, \ldots, n_i\}$ where n_i is some large fresh number, hitherto unseen in the construction.

At stage $s+1$ of the construction, each active V_i has an associated *test* $t(i)$. Perform the following steps.

(i) For each active V_i, see if there is some $e' < e(i)$ such that $W_{e'} \upharpoonright t(i) = W_{e(i)} \upharpoonright t(i)[s+1]$, which means that this equality holds at stage s (i.e., $W_{e',s+1} \upharpoonright t(i) = W_{e(i),s+1} \upharpoonright t(i)$, and $t(i) = t_{s+1}(i)$ is the test for V_i at stage $s+1$). If so, then we declare that V_i is garbage and find a (least unused) fresh number i', declare that $V_{i'}$ is active, and $e(i') = e(i)$ and $t(i')$ to be large and fresh.

(ii) For each active V_i, see if there is some garbage V_j with $j < i$, and $V_j \upharpoonright t(i) = V_i \upharpoonright t(i)[s+1]$. If so, then we declare that V_i is garbage and find a fresh number i', declare that $V_{i'}$ is active, and $e(i') = e(i)$ and $t(i')$ to be large and fresh.

(iii) If $W_e \neq \varnothing[s+1]$ and it is least with no V_i simulating W_e, then for the least unused j, let V_j simulate W_e, define $j(e) = e$, and $t(j)$ to be large and fresh.

(iv) Finally, for all active V_i, make $V_i = W_{e(i)}[s+1]$.

We argue that this construction works. It is evident that if V_i is garbage, then $V_i = \{0, \dots, n_i\}$ and n_i is unique to i. This means that all the garbage V_i are distinct. We say that V_i is *permanently active* if it is never made into garbage. Suppose that $W_e \neq \omega$ is given and e is its minimal index. Then we claim that there is some permanently active V_i with $e(i) = e$. If $W_e = \varnothing$, then there is nothing to prove as $V_0 = \varnothing$. If $W_e \neq \varnothing$, then by step (iii) of the construction, we will associate some V_i with e at some stage s. Now, if this is not permanent, then V_i must get turned into garbage either by (i) or (ii). We have that (i) pertains only if some $W_{e'} \upharpoonright t(i) = W_e \upharpoonright t(i)[s]$. If this happens, then we will reset V_i to a new $V_{i'}$, but with a larger $t(i')$. Call this new $V_{i'}$ and others generated by its transitive closure under (i) or (ii) *heirs* of V_i. There can only be finitely many heirs to V_i generated by (i), since e is a minimal index, so $W_e \neq W_j$ for all $j < e$. There can also only be finitely many heirs by (ii). Each time (ii) is invoked, $W_e \upharpoonright t(i) \supseteq \{0, \dots, n_j\}$ for some $n_j \to \infty$, meaning that $W_e = \omega$, a contradiction. Thus for each e there is an i with $W_e = V_i$. If e is not a minimal index, and e' is some other index for W_e, then each V_i with $e(i) = e'$ will be turned into garbage by clause (i) and W_e. If $W_e = \omega$, this is also true by clause (ii). Thus each $W_e \neq \omega$ has a unique V_i simulating it. □

The above construction can be viewed as having a sort of "infinite injury". If we think of the requirements being

$$R_e : \text{If } e \text{ is a minimal index, then there is some } i \text{ such that } V_i = W_e,$$

then the requirement R_k can be "injured" by the requirement R_e at some stage s if $e < k$ and W_e agrees with W_k up to some large testing threshold $t(i)$ at stage s. If, in fact, k is not a minimal index, then R_k will be "injured" infinitely often by a higher priority R_e. Of course, in this case, R_k is also satisfied since k is not a minimal index.

A minimal pair of c.e. sets

In the next theorem, we give another example of an infinite injury construction, where more coordination is required between the different requirements. In that construction, we shall introduce the notion of a *priority tree* to help organise the process. Other terminology introduced there, such as *expansionary stages*, is now standard in the modern literature.

> **Theorem 3.1.44** (Lachlan [325], Yates [513]). *There exist non-computable c.e. sets A and B such that for all sets C, if $C \leqslant_T A, B$ then C is computable.*

As we have seen earlier, the degrees of such sets are called a minimal pair. In Theorem 3.1.15, we constructed a minimal pair by finite extension and forcing a disagreement where we could. That method is not possible for c.e. degrees, and something new is needed.

Proof. We construct $A = \cup_s A_s$ and $B = \cup_s B_s$ in stages to satisfy the requirements below for $e \in \mathbb{N}$:

$$\begin{aligned} R_e :& \quad \overline{A} \neq W_e; \\ Q_e :& \quad \overline{B} \neq W_e; \\ N_{i,j} :& \quad \Phi_i(A) = \Phi_j(B) = f \text{ and } f \text{ is total} \implies f \text{ is computable.} \end{aligned}$$

We meet the R_e and the Q_e by a Friedberg-Muchnik type strategy. That is, we shall pick a (fresh) follower x, targeted for A in the case of R_e, and wait until x enters $W_{e,s}$. Of course, should x never enter $W_{e,s}$ for any s, then $x \notin (W_e \cup A)$ and hence $A \neq \overline{W_e}$. Should x enter $W_{e,s}$ for some stage s, then we can win forever by putting x into A_t at some $t \geqslant s$.

The tricky requirements are the $N_{i,j}$. We will first discuss how to meet a single $N_{i,j}$ in isolation and then look at the coherence problems between the various requirements and the solution to these provided by the use of a tree of strategies.

One N-requirement in isolation. For a single $N_{i,j}$, we will need the auxiliary functions

$$\ell(i,j,s) = \max\{x : \forall y < x(\Phi_{i,s}(A_s;y)\downarrow = \Phi_{j,s}(B_s;y)\downarrow)\}$$

and

$$m\ell(i,j,s) = \max\{\ell(i,j,t) : t < s\}.$$

We call $\ell(i,j,s)$ the *length of agreement function* and call $m\ell(i,j,s)$ the *maximum length of agreement function.* As with the Sacks Splitting Theorem 3.1.34, note that the maximum length of agreement function is a sort of "high water mark" for lengths of agreement seen so far. We shall call a stage *$\langle i,j\rangle$-expansionary* if the current length of agreement exceeds the previous high water mark. That is, s will be called $\langle i,j\rangle$-expansionary if $\ell(i,j,s) > m\ell(i,j,s)$. Also, we let $mu(i,j,s)$ denote the maximum element used in any computation below $\ell(i,j,s)$.

The key idea for a single $N_{i,j}$ is the following. We allow some element x to enter A below the use of Φ_i, but nothing to enter B below the respective use of Φ_j that (used to) make the computations equal below $\ell(i,j,s)$. We do not allow any further elements $\leqslant mu(i,j,s)$ to enter B until the next $\langle i,j\rangle$-expansionary stage u. It can only be that $B_u(z) = B_s(z)$ for all $z \leqslant mu(i,j,s)$, and hence, by the use principle,

$$\Phi_{j,s}(B_s;y) = \Phi_{j,u}(B_u;y),$$

for any $y \leqslant \ell(i,j,s)$. But since u is expansionary, this means that

$$\Phi_{i,u}(A_u;y) = \Phi_{j,u}(B_u;y) = \Phi_{j,s}(B_s;y) = \Phi_{i,s}(A_s;y).$$

That is, even though the $\Phi_{i,u}(A_u;y)$-computations might have changed because x entered A, the result of the computation on y below the previous length of agreement remains the same. This, in particular, implies that $N_{i,j}$ will be met, since to compute f we just need to know the result of this computation. Finally, we remark that if no expansionary stage $u > s$ is ever found then $\Phi_i(A) \neq \Phi_j(B)$. Figure 3.1 might be helpful here.

Coherence. Consider two N-requirements N and N'. Now suppose that N has higher priority than N'. Now N requests us to only put numbers into A or B during its expansionary stages $\{s_1, s_2, \ldots\}$. Similarly, N' might request us to only put numbers in during stages $\{t_1, t_2, \ldots\}$. The problem is that these sets of stages might be *disjoint.* Then N blocks us from putting numbers into A or B at stages t_i, and N' blocks us from putting numbers in during stages s_i. Hence, the pair might block us from ever putting numbers into A or B.

This problem can be fixed as follows. We will have two versions of N', one believing that N has infinitely many expansionary stages, and the other will think that N has only finitely many expansionary stages. The latter will be initialised every time a new N-expansionary stage is found; it will respect the restraints imposed by N but will ignore N otherwise. The other version of N' will believe that N has infinitely many expansionary stages. This version will become active only

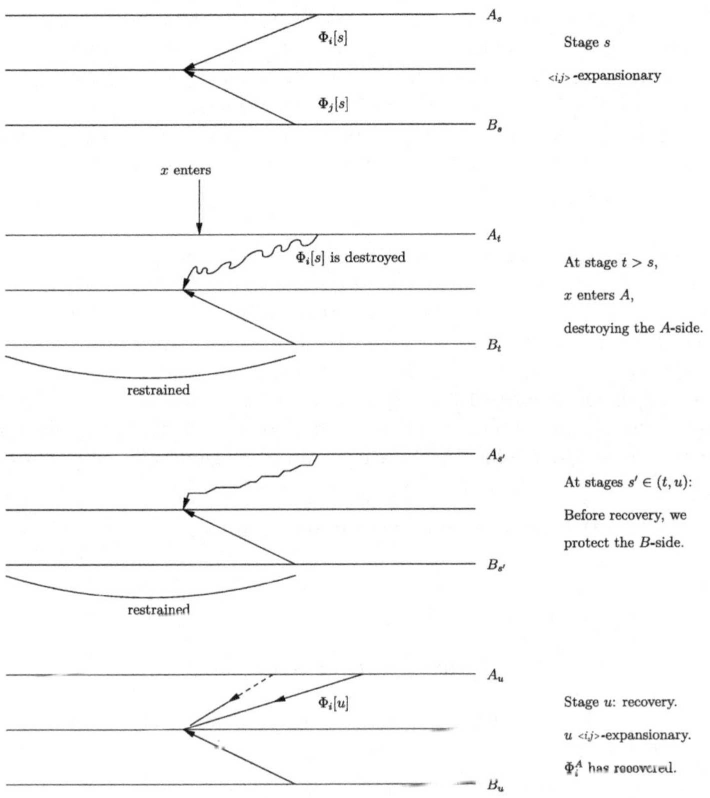

Figure 3.1: Recovery of computation.

at these expansionary stages. It will slow down its computation by making moves only at the stages $s_1, s_2, \ldots$ and will define its expansionary stages to be a subset of the expansionary stages of N. It will also coordinate its restraint with N, making sure that the A-side and the B-side are never simultaneously blocked. This version of N' strategy will remain dormant between $s_1, s_2, \ldots$, and indeed, it may be forever abandoned in case N has only finitely many expansionary stages. However, the restraint (on A or B) imposed by this version of N' will have to be respected by weaker priority strategies, including the other version of N'. Indeed, there is no way to know which version of N' has the right guess about N, so we must keep all possibilities open.

Now, if we also had to consider N'', it would need four clones, two for each version of N'. Note that this case analysis starts to resemble the full binary tree. The standard, modern way to organise such constructions is to make this idea explicit by introducing the tree of strategies. This tree helps to manage the case of multiple N-strategies, not just the two or three we considered above. Every strategy will have multiple versions, each associated with a node in a tree of "guesses" that aids in predicting the behaviour of other (clones of) strategies in the construction.

We now turn to the formal details.

The Priority Tree. We use the tree $\mathcal{T} = \{\infty, f\}^*$, i.e., the full binary tree consisting of finite strings of ∞-s and f-s, including the empty string λ. We assign $N_{i,j}$ to σ in $\mathcal{T}$ iff $|\sigma| = \langle i, j\rangle$, where $|\sigma|$ denotes the length of σ. Also, if $|\sigma| = 2e$ then we also assign R_e to σ and if $|\sigma| = 2e+1$ we assign Q_e to σ. For a requirement M, we will write M_σ for the version of M at σ.

We use lexicographical ordering $<_L$ on finite strings in $\mathcal{T}$ induced by $\infty < f$. We will say that M_σ has higher (or stronger) priority than M_τ if $\sigma <_L \tau$. As before, we also write $\sigma \preceq \tau$ if σ is an initial segment of τ; we slightly abuse notation and also include the case when τ is infinite. We write $\sigma\hat{\ }a$ to denote the string τ which is the extension of σ using symbol a adjoined to the end of σ.

Definition 3.1.45. (a) We define the notions σ-stage, $m\ell(\sigma, s)$, and σ-expansionary stage by induction on $|\sigma|$.
(i) Every stage s is a λ-stage.
(ii) Suppose that s is a τ-stage with $|\tau| = \langle i, j\rangle$. Let $\ell(\tau, s) = \ell(i, j, s)$. Define

$$m\ell(\tau, s) = \max\{\ell(\tau, t) : t \text{ is a } \tau\text{-stage } < s\}.$$

We say that s is τ-expansionary if $\ell(\tau, s) > m\ell(\tau, s)$ and declare s to be a $\tau\hat{\ }\infty$-stage. If $\ell(\tau, s) \leqslant m\ell(\tau, s)$, we declare that s is a $\tau\hat{\ }f$-stage. We define TP_s to be the unique σ of length s with s a σ-stage.

Definition 3.1.46. (a) We say that R_σ *requires attention* at stage s if $W_{e,s} \cap A_s = \emptyset$ (where $2e = |\sigma|$), s is a σ-stage and one of the following holds.
(i) R_σ currently has no follower.
(ii) R_σ has a follower $x \in W_{e,s}$.

We define Q_σ to require attention similarly.

Construction.

Step 1. Compute TP_s. Initialise all versions of requirements at guesses $\tau \nleq_L TP_s$.
Step 2. Find the R_σ or Q_σ of highest priority that requires attention at stage s. Without loss

of generality, we will suppose this to be R_σ. Initialise all requirements at guesses τ with $\tau \not\leq_L \sigma$. Adopt the appropriate case below.

- Definition 3.1.46 (i) holds. Appoint $x(\sigma, s) = s$ to follow R_σ. (Remember, s is larger than all computations seen so far by convention.)

- Definition 3.1.46 (ii) holds. Enumerate x into A_{s+1}.

End of Construction.

Verification. Let TP be the leftmost path visited infinitely often: That is, $\lambda \preceq TP$ and for all τ, if $\tau \preceq TP$, then $\tau\hat{}\infty \preceq TP$ iff $\exists^\infty s(\tau\hat{}\infty \preceq TP_s)$. Otherwise, $\tau\hat{}f \preceq TP$.

Lemma 3.1.47. *All the R_e and Q_e have versions that are met, and for all $\tau \preceq TP$, R_τ (or Q_τ) acts only finitely often.*

Proof. This lemma is proven by induction on e. We consider R_e. Let $\sigma \preceq TP$ with $|\sigma| = 2e$. Fix a stage s_0 such that for all $\tau <_L \sigma$ and $s > s_0$:
(i) if $\tau \not\leq \sigma$, s is not a τ-stage;
(ii) if M is a Q or R requirement assigned to τ, then M will not act at stage s.

Assuming s_0 to be least and a σ-stage, we can assume that either $W_{e,s} \cap A_{s_0} \neq \emptyset$ (in which case we are done), or R_σ receives attention via Case (i) getting a follower x at stage s_0. This follower is immortal by the choice of s_0 and the induction hypothesis. It will succeed in meeting R_e as in the basic module since it has priority at each σ-stage. □

Lemma 3.1.48. *All the $N_{i,j}$ have versions that are met.*

Proof. Again, we prove this by induction. Let $\sigma \preceq TP$ with $|\sigma| = \langle i, j\rangle$. Choose s_0 as in Lemma 3.1.47, so that no higher priority action can cause grief to N_σ. Now, if $\sigma\hat{}f \preceq TP$, we are done since $\liminf \ell(i, j, s) < \infty$ and hence $\Phi_i(A) \neq \Phi_j(B)$.

So we suppose that $\sigma\hat{}\infty \preceq TP$. To compute $\Phi_i(A; x)$, find the least $\sigma\hat{}\infty$-stage $s = s(x) > s_0$ such that $\ell(\sigma, s) > x$. Note that this can be computed from the parameters s_0 and σ. We claim that $\Phi_i(A; x) = \Phi_{i,s}(A; x)$. To see this, note that by Step 1 of the construction we will initialise all $\tau \not\leq_L TP_s$ at stage s. In particular, at stage s by choice of s_0 and since we appoint new followers to be large, and we are never above or left of $\sigma\hat{}\infty$ after stage s_0, the only numbers which are *below* s and can enter A or B *after* stage s are followers associated with $\gamma \succcurlyeq \sigma\hat{}\infty$. Such followers can enter their target sets only at $\sigma\hat{}\infty$ stages $s \geqslant s_0$. That is, they can only enter at σ-expansionary stages. In particular, as with the basic module, we can argue that for any $\upsilon\hat{}\infty$-stage $t \geqslant s$, at most one of A or B can change below $mu(\upsilon, t)$ before the next $\sigma\hat{}\infty$-stage $t' > t$.

Thus, exactly as in the basic module, we have that

$$\Phi_{i,s}(A_s; x) = \Phi_{i,t}(A_t; x) = \Phi_{j,s}(B_s; x) = \Phi_{j,t}(B_t; x) = \Phi_i(A; x) = \Phi_j(B; x),$$

for all $\sigma\hat{}\infty$-stages $t > s$. □

The minimal pair theorem is proved. □

3.1.8 Further reading*

A more extended introduction to priority techniques can be found in [125]. For finite injury techniques, the old-fashioned but thorough book by Rogers [454] is a good reference. The reader interested in c.e. sets and degrees is referred to Soare [477] for many further results and a large number of exercises. We remark that [477, Chapter VIII] presents alternative methods for handling combinatorics in infinite injury proofs, such as the Window and Thickness Lemmas, and also the Pinball Machine Model. The former approach is now typically considered old-fashioned and will not be used in this book, while the latter is also unnecessary for our purposes. For historical notes, see [478, part V].

Exercises

Since all the exercises below seem to require infinite injury, none of them is particularly easy. However, most of these results are old classics, and their proofs can be easily found in the literature.

Exercise° 3.1.49 (Downey and Welch [150], Ambos-Spies [9]). Construct a non-computable c.e. set A such that if $A = A_1 \sqcup A_2$ is a c.e. splitting of A, then the degrees of A_1 and A_2 form a minimal pair. (Hint: Modify the tree from the minimal pair argument. Use requirements $P_e : \overline{A} \neq W_e$, and $N_{\langle i,j,k\rangle} : W_i \sqcup W_j = A \wedge \Phi_k^{W_i} = \Phi_k^{W_j} = f$ total $\rightarrow f$ computable.)

Exercise° 3.1.50. A set A is called *piecewise computable* if $A^{[e]} =_{\text{def}} \{\langle e,x\rangle : \langle e,x\rangle \in A\}$, the e-th *slice* (column) of A is computable for every e.

1. Construct a c.e. set B such that, for each e, $B^{[e]}$ is an initial segment of $\omega^{[e]}$, and $B^{[e]}$ is finite iff φ_e is not total. We will say that B *codes* $\varnothing''$.
2. Show that if B codes $\varnothing''$, then $B' \equiv_T \varnothing''$. That is, B is high.

Exercise 3.1.51. In the notation and terminology of the previous exercise, $A \subseteq B$ is called a *thick* subset if for every e, $A^{[e]} =^* B^{[e]}$, meaning that $|B^{[e]} \backslash A^{[e]}| < \infty$. Show that if A is a thick subset of a set B coding $\varnothing''$, then A is also high.

Exercise 3.1.52 (Thickness Lemma – Shoenfield [467]). This is a weak form of the full Thickness Lemma. We refer the reader to Chapter VII of Soare [477] for more details. Thick subsets were defined in the previous exercise. Use the infinite injury method to prove that if $\varnothing <_T C$ is c.e. and B is a c.e. set coding $\varnothing''$, then there is a thick subset $A \subseteq B$ with $C \not\leqslant_T A$. This shows that there is an incomplete high c.e. degree.

Exercise* 3.1.53 (Meyer [387]). Let $MIN = \{e : \forall j < e\, \varphi_j \neq \varphi_e\}$, and $SMIN = \{e : \forall j < e\, W_e \neq W_j\}$. Show that for any acceptable enumeration of the partial computable functions, MIN $\equiv_T$ SMIN $\equiv_T \varnothing''$.

3.2 Computable linear orderings

In this section, we use the computability-theoretic techniques described in the previous section to prove several classical results about computable linear orders. We begin with the main definitions restricted to the class of linear orders. A detailed proof of the Fellner-Watnick Theorem, which is the main result of the present chapter, will be given in Section 3.2.6. Similarly to the previous section, we organise the results according to their combinatorial complexity. The most notable results include:

Theorem 3.2.1 (Feiner [161]). *There is a c.e. presented linear order not isomorphic to any computable one.*

Theorem 3.2.2 (Goncharov and Dzgoev [211], Remmel [446]). *A computable linear ordering A is computably categorical (Definition 2.2.15) iff A has only a finite number of adjacencies.*

In fact, Dzgoev announced Theorem 3.2.2 in 1978, but it was published only locally as a report[2].

3.2.1 The basic definitions, revisited

Recall that a linear order is just a partial order in which every two elements are comparable under $\leqslant$.

Definition 3.2.3. A *computable linear ordering* (or more precisely, a *computable presentation* of its order type) is a linear ordering $(A, \leqslant)$ where A is a computable set, and the ordering relation $\leqslant$ is a computable relation.

We typically assume that our order is countably infinite, as there is not much to say about the finite ones. Standard linear orderings such as ω, $\mathbb{Z}$, and $\mathbb{Q}$ are computably presentable.

Definition 3.2.4. We say that a linear ordering $(L, \leqslant_L)$ is c.e. presented if L is a computable set, and $\leqslant_L$ is a c.e. relation.

The way to think about a c.e. presentation is that we will discover if $x \leqslant_L y$, but may not know whether $x <_L y$ since we can later discover $x =_L y$. Thus, a c.e. presented linear ordering is $(A, <_A)/\equiv$, where $\equiv$ is a c.e. equivalence relation[3].

[2]The report is available at the library of the Sobolev Institute of Mathematics. According to Goncharov's MathSciNet review of [446], the reference is: Dzgoev "On constructivizations of some structures" (Russian), Akad. Nauk SSSR, Sibirsk. Otdel., Novosibirsk, 1978 (manuscript deposited at VINITI on July 26, 1978, Deposition No. 1606–79).

[3]Note that each $\equiv$-class has to be convex in the sense that if $a <_A b <_A c$ and $a \equiv c$, then $a \equiv b \equiv c$. Conversely, every convex equivalence relation on A can be used to define a quotient linear order.

3.2.2 Injury-free approximation. Feiner's Theorem

In this subsection we construct a c.e. presented order with no computable isomorphic copy; this is Feiner's Theorem 3.2.1 (for linear orders). The rest of the subsection is devoted to accumulating enough lemmas and propositions to prove the theorem.

Recall that a structure is computably categorical if any two computable copies (presentations) of the structure are computably isomorphic. For example, the usual back-and forth method shows that $(\mathbb{Q}, \leqslant)$ is computably categorical. Therefore, from the perspective of computable mathematics, all computable presentations of $(\mathbb{Q}, \leqslant)$ are essentially the same. Thus, there is no ambiguity in the statement of the following lemma.

Lemma 3.2.5. *$(A, \leqslant)$ is a computably presentable iff there is a computable subset of $(\mathbb{Q}, \leqslant)$ isomorphic to $(A, \leqslant)$.*

Sketch. Suppose $(A, \leqslant)$ is a computably presented linear order. Use the "forth" step in the usual back-and-forth proof of categoricity for $(\mathbb{Q}, \leqslant)$, but apply it to elements of $(A, \leqslant)$. Of course, even if A is itself dense, we do not have to make f onto. In fact, by the density of $\mathbb{Q}$, we can additionally ensure that $f(A)$ is actually a computable subset of $\mathbb{Q}$, not merely a c.e. subset. To achieve this, note that at any stage s, it is safe to declare finitely many rationals currently outside the range of f_s to be *permanently outside* of $f(A)$. By the density of $\mathbb{Q}$, we always have enough points to further extend the embedding.

The other implication in the statement of the lemma is obvious. □

Evidently, the lemma can be relativised to any oracle. Suppose L is Δ^0_2. Relativise Lemma 3.2.5 to $\mathbf{0}'$. This gives a Δ^0_2 subset of $\mathbb{Q}$ so that the induced order is computable. Thus, we can think of L as being a *Δ^0_2 set* $L = \lim_s L_s$ with a *uniformly computable* order on each L_s. Additionally, we can further exploit the density of $\mathbb{Q}$ to make sure that, once x is declared out of L, it never comes back:

Lemma 3.2.6. *If L is a Δ^0_2 ordering then $L \cong \hat{L}$ for some Π^0_1 subset of $\mathbb{Q}$.*

Sketch. Suppose $L \in \Delta^0_2$. So $L = \lim_s L_s$. When $x \in L_{s+1} - L_s$, use the density of $\mathbb{Q}$ to choose a point $\hat{x} \in \mathbb{Q}$ with Gödel number bigger than any number seen so far in the construction and map $f_s : x \to \hat{x}$ consistently with f_{t-1}. We will continue with this map unless x leaves L_t at some least $t > s$, which means that the order has been redefined on x. If $x \in L_t - L_{t+1}$, then we throw $\hat{x}$ out of the range of f forever. We let $f = \lim_s f_s$. If $x \in L$ then from some point onwards, $x \in L_t$. If $x \notin L$, then from some point onwards, $x \notin L_t$. Thus $f : L \cong \hat{L}$ where $\hat{L}$ is the ordering formed by $\hat{x}$ which are added to $\hat{L}$ and never leave. □

We will need the following folklore lemma that, in particular, implies that every Δ^0_2-presented linear order admits a c.e. presentation.

Lemma 3.2.7. *Every Π^0_1 subset of $\mathbb{Q}$ is isomorphic (as an order) to a c.e. presented linear order.*

Proof. We can of course assume that $\hat{L} \subseteq \mathbb{Q}$ is non-empty, say $x \in \hat{L}$. (Without loss of generality and up to a notation change, we could assume $x = 0$.) We build a c.e. presentation L of $\hat{L}$, where the latter is of course viewed as a sub-order of the rationals, and the former consists of computably ordered natural numbers $\mathbb{N}$ modulo a c.e. equivalence relation $\equiv$. At a stage s we will have defined only finitely many equivalence classes in L_s, each class currently consisting of finitely many numbers:

$$[\ell_0], [\ell_1], \ldots, [\ell_{c(s)}].$$

The classes are linearly ordered, and one of these classes contains x. At a later stage some of these classes may be declared equal, i.e., united into a bigger class. We also construct a map $\phi : \mathbb{Q} \to \mathbb{N}$ such that its restriction to $\hat{L}$ induces an isomorphism from $\hat{L}$ onto $L = (\mathbb{N}, \leqslant)/\equiv$. At a stage s we will have defined a finite partial map ϕ_s that induces an order-isomorphism between finitely many points in $\hat{L}_s \cap [0, \dots, s]$ and finitely many classes in L_s.

When an element r is enumerated into $\mathbb{Q}\backslash\hat{L}$ we say that r *leaves* $\hat{L}$. Assume that at every stage at most one element leaves $\hat{L}$.

The idea is that, when $r > x$ leaves $\hat{L}$ at stage $(s + 1)$, we want to declare it equal to the right-most point $q < r$ of the order that is still there. Similarly, when $r < x$, we declare it equal to the left-most point $q > r$ that has not yet left the order. Of course, we cannot quite do it in $\mathbb{Q}$ itself; indeed, think about the points in-between r and q. However, we can do this in L which currently has no classes between $[\phi_s(r)]$ and $[\phi_s(q)]$.

More formally, if $r > x$ leaves $\hat{L}$ at stage $(s + 1)$, search for the right-most point $q < r$ in L_s (which must exist) and declare $\phi_s(q) \equiv \phi_s(r)$. (Define $\equiv_{s+1}$ by modifying the current approximation $\equiv_s$ of $\equiv$ in the obvious way so that $\equiv_{s+1}$ is an equivalence relation; we omit the details.) The case when $r < x$ is symmetric. Then extend ϕ_s to a map ϕ_{s+1} defined on $\hat{L}_{s+1} \cap [0, \dots, s + 1]$ naturally.

The map $\psi = \bigcup_s \psi_s$ is onto by the construction, and $\equiv$ is also approximated via $\equiv_s$, which results in a c.e. congruence inducing a c.e. presentation L of some linear order. If $x, y \in \hat{L}$ then $[\phi(x)] \leqslant [\psi(y)]$ in L iff $x \leqslant y$ in $\hat{L}$. □

Corollary 3.2.8. *Every Δ^0_2 presentable linear order is isomorphic to a c.e. presentable one.*

Remark 3.2.9. In the proof of the lemma above, we in fact showed that a Π^0_1-subordering $\hat{L}$ of $\mathbb{Q}$ is computably isomorphic to a c.e. presented linear order, in the following sense. There is a computable $f : \mathbb{Q} \to \mathbb{N}$ so that its restriction to $\hat{L}$ induces an isomorphism between $\hat{L}$ and $L = \mathbb{N}/\equiv$. Conversely, it can be shown that a c.e. presented linear ordering L is computably isomorphic (in the same sense) to a Π^0_1 subset $\hat{L}$ of $\mathbb{Q}$. To see why, assume we have already defined $f(x)$ and $f(y)$, but now we have discovered $f(x) \equiv f(y)$. Since one of the two elements (x or y) are now out of the Π^0_1 set, the definition of f does not have to be adjusted. Thus, usually, *c.e. presented orders can be computably identified with Π^0_1 subsets of the rationals* without any loss of generality.

In a linear ordering $L = (L, \leqslant)$, an n-block (or a complete block of size n) is a set $x_1 < x_2 < \dots < x_n$ such that (x_i, x_{i+1}) is an adjacency for all $i < n$, x_1 is a left limit point or the first point of the ordering, and x_n is a right limit point, or the last point of the ordering.

Definition 3.2.10. If $L = (L, \leqslant)$ is a linear ordering, then $B(L) = \{n : L \text{ contains an } n \text{ block}\}$.

Note that $B(L)$ is a classical invariant. We have the techniques to show that it is also an effective invariant, in the following sense. If L is a computable ordering then $B(L)$ is Σ^0_3.

Theorem 3.2.11 (Lerman [336]). *If $S \in \Sigma^0_3$, then there is a computable L with $B(L) = S$.*

Proof. The proof resembles that of Theorem 3.1.8 (iii). Let S be Σ^0_3 and hence we have computable R such that

$$n \in S \text{ iff } \exists x \forall s \exists t R(n, x, s, t).$$

The construction begins with the sum of infinitely many computable copies of $\mathbb{Z}$

$$\mathbb{Z} + \mathbb{Z} + \mathbb{Z} + \dots.$$

Between the $\langle n, x\rangle$-th copy and $\langle n, x\rangle + 1$-st copies we will use a construction to add some new points in stages. At each stage s, between copy $\langle n, x\rangle$ and copy $\langle n, x\rangle + 1$ we will have constructed a finite ordering A_s consisting of three suborderings $A_s = C_s + D + E_s$. We will have a counter $s(n, x)$ to count the firings of $\langle n, x\rangle$. Initially, $s(n, x) = 0$. At stage 1 we set D to have exactly n elements, and C_1, E_1 to be empty. At each stage $s + 1$ for $s \geqslant 1$, we *always* add one element to the left end of C_s, and to the right end of E_s. If n, x has fired for $s(n, x)$ we set $s(n, x) = s(n, x) + 1$. In this firing case we will also add some new points, one to the right end of C_s and one to the left end of E_s. No such points are added if no firings for $s(n, x)$ happen.

Now the observation is that if $\langle n, x\rangle$ fires for all of its $s(n, x)$ (that is, $n \in S$ is witnessed by x) then C and E both grow into copies of $\mathbb{Z}$, so that $C + D + E = \mathbb{Z} + n + \mathbb{Z}$. Then $n \in B(L)$. If only finitely many firings, then the second part of the construction for $\langle n, x\rangle$ is invoked only finitely often, and hence $C + D + E = \mathbb{Z}$, for every x. That is, $n \notin B(L)$. That is $n \in S$ if $n \in B(L)$. □

Theorem 3.2.11 combined with the lemmas proved earlier allows us to construct a c.e. presented linear order not isomorphic to any computable one; this appeared as Theorem 3.2.1 earlier.

Proof of Feiner's Theorem 3.2.1. Applying Theorem 3.2.11 in relativised form, given any set S which is Σ_3^X, there is a X-computable linear ordering L with $B(L) = S$. Letting $X = K = \varnothing'$, and choosing $S = 0^{(4)}$ that is in $\Sigma_3^{\varnothing'} \setminus \Sigma_3^0$, (by Post's Theorem), we obtain a Δ_2^0 such order. By Lemma 3.2.6, the order is isomorphic to a Π_1^0 subset of $\mathbb{Q}$. By Lemma 3.2.7, it has a c.e. presentation. If L so constructed was isomorphic to a computable $\hat{L}$, then $B(\hat{L}) \in \Sigma_3^0$ (as $\hat{L}$ is computable), but this contradicts $B(\hat{L}) = B(L) = S \notin \Sigma_3^0$. □

Finally, we remark that in a decidable linear order L the block invariant $B(L)$ becomes computable. Consequently, Theorem 3.2.11 implies that there is a computable linear order with no decidable presentation.

3.2.3 Finite extension method. Richter's Theorem and the Frolov-Montalbán Theorem

Recall that in Section 2.2 we presented a way to "encode" a c.e. set into a presentation of a group. In Theorem 3.2.11, which was the key step in the proof of Theorem 3.2.1, a Σ_3^0 set was "coded" into the block relation of an order. However, this coding depended on the fixed Σ_3^0-approximation of the given set:

$$n \in S \text{ iff } \exists x \forall s \exists t\, R(n, x, s, t),$$

where R was computable. The same set can have many different Σ_3^0-approximations, i.e., many different potential such R. Thus, this transformation from Theorem 3.2.11 was not really well-behaved from the algebraic standpoint. Although this weak, notation-dependent coding was sufficient for our purposes, it has its limitations. For a set S to be an actual "Σ_3^0-invariant" of the respective order, Theorem 3.2.11 has to be slightly modified, as follows.

Lemma 3.2.12 (Folklore). *Given a set S, we can produce a linear order $L(S)$ such that S is Σ_3^X iff $L(S)$ has an X-computable presentation.*

Sketch. Use the shuffle sum of order-types $\mathbb{Z}$ and $\mathbb{Z} + n + \mathbb{Z}$. The *shuffle sum* $\biguplus_{i \in \mathbb{N}} L_i$ of order-types $(L_i)_{i \in \mathbb{N}}$ is obtained by replacing any point in $\mathbb{Q}$ with a copy of L_i for some i, so that each L_i gets densely distributed in the resulting order[4]. Note that any permutation ρ of ω results in the same order-type:

$$\biguplus_{i \in \mathbb{N}} L_i \cong \biguplus_{i \in \mathbb{N}} L_{\rho(i)}.$$

It is not difficult to see that, given a computable sequence $(L_i)_{i \in \mathbb{N}}$ of uniformly computable linear orders, we can computably turn it into their shuffle sum $\biguplus_{i \in \mathbb{N}} L_i$. In the proof of Theorem 3.2.11, we essentially produced a computable sequence of orders $(L_i)_{i \in \mathbb{N}}$ such that the order-type $\mathbb{Z} + n + \mathbb{Z}$ appeared among the L_i iff $n \in S$. Clearly, different Σ^0_3-approximations of S will give different such L_i. Define $L = \biguplus_{i \in \mathbb{N}} L_i$ and note that the resulting order-type depends only on S, and not on its Σ^0_3-approximation. □

The transformation described above is much better behaved. For example, if we replace S with the join of S with its complement (denoted $S \oplus \overline{S}$), then $L(S \oplus \overline{S})$ has an X-computable presentation iff $S \oplus \overline{S}$ is Σ^X_3 iff S is Δ^X_3 iff $S \leqslant_T X''$. In modern terminology, a linear order can have an arbitrary "second jump degree".

Can we do better than that? What about a Σ^0_1-coding of a set into a linear order? That is, can we find a construction of $L(S)$ such that $L(S)$ is computably presented if and only if S is Σ^0_1, and so that this could also be relativised?

We shall not attempt to define exactly what a Σ^0_1-coding of a set into a structure is, since issues such as uniformity will lead to a multitude of definitions. However, we should expect from any such "coding" that, given any set A, we can produce a structure $M(A)$ so that $M(A)$ has an X-computable copy if and only if $X \geqslant A$, via an argument similar to the one we had above for the "Σ^0_3-coding" given by Lemma 3.2.12.

We now use the finite extensions technique to illustrate that no such sufficiently well-behaved Σ^0_1-coding is possible in the setting of linear orderings.

Richter's Theorem

In the theorem below, **a** and **b** stand for Turing degrees.

Theorem 3.2.13 (Richter [450]). *Suppose* $\mathbf{a} \neq \mathbf{0}$ *and* $(A, \leqslant)$ *is an* **a***-computable presentation of a linear ordering. Then there is another* $\mathbf{b} \neq \mathbf{0}$ *and a presentation* $(B, \leqslant)$ *which is* **b***-presentable, such that* $\mathbf{a}, \mathbf{b}$ *form a minimal pair.*

Proof. We modify the technique of Theorem 3.1.15. In view of Lemma 3.2.5, we might as well regard A as a subset of $\mathbb{Q}$. We need to build another subset B of $\mathbb{Q}$ to meet the same requirements: $R_{e,i} : \Phi^A_e = \Phi^B_i = f$ total, implies f computable. The only extra condition is that $A \cong B$ as orderings. The only problem is the following. In the proof of Theorem 3.1.15, we could consider *any* σ with $B_s \preceq \sigma$ as a potential oracle for an x-computation. Now, our idea of making $A \cong B$ is that at stage s of the construction, we will have fixed a finite part of A, $A_s = \{a_1, \ldots, a_{n(s)}\}$ $(a_i <_{\mathbb{Q}} a_j)$ and $B_s = \{b_1, \ldots, b_{n(s)}\}$ with $a_i \mapsto b_i$ as our partial isomorphism. Now, it might be that $[a_i, a_{i+1}]$ is an adjacency of $(A, \leqslant)$, so that a σ which includes a rational between b_i and b_{i+1} cannot be used in a potential computation, as the stage s partial isomorphism cannot be extended to one including an element between b_i and b_{i+1}. Thus, we need to distinguish between finite subintervals of A_s and infinite ones. By adding enough points between elements, we can assume that rational s is in the domain of A, and that if $[a_i, a_{i+1}]$ is finite, it is an adjacency. This means

[4]That is, between any pair of blocks of the form L_i and L_j, and for any k, there is a block of the form L_k.

that we can correspondingly break B_s into a finite number of blocks which are infinite, for example $[\infty, b_1], [b_4, b_5], [b_6, b_7]$, say would correspond to the fact that A has infinitely many elements left of a_1, between a_4 and a_5, and between a_6 and a_7. Note that $[b_i, b_{i+1}] \cap \mathbb{Q}$ is a computable set. Then we call σ an *acceptable* string if $\sigma(n) = 1$ implies that $n = b_i$ for some i, or $n \in (b_i, b_{i+1})$ for some $[b_i, b_{i+1}]$ where $A \cap [a_i, a_{i+1}]$ is infinite. The proof is more or less identical to that of Theorem 3.1.15, and we ask for x, t and an acceptable σ with $\Phi_e^A(x) \downarrow \neq \Phi_i^{\sigma}(x) \downarrow$, and if so, let $B_{s+1} = \sigma$, and extend the partial isomorphism by adding enough elements of A to make $A_{s+1} \to \sigma$. □

Actually, Richter proved a more general model-theoretical result generalising Theorem 3.2.13; see Exercises 3.2.58-3.2.62 at the end of the chapter. What about a Σ_n^0-coding for other choices of n? We shall return to this question in §3.2.6.

The Frolov-Montalbán Theorem

Let *adj* denote the adjacency relation, that is, $adj(x, y)$ if $x < y$ and there is no z so that $x < z < y$. The following neat result was independently proven by Montalbán [399] and Frolov [190].

Theorem 3.2.14. *For a linear order L, the following are equivalent:*

1. *L has a low presentation;*
2. *L has a Δ_2^0-presentation in which the adjacency relation is also Δ_2^0.*

Proof. For $1 \to 2$, recall that $A \leqslant_T A' \equiv_T \varnothing'$, and that the X-computability of L implies $adj \leqslant_T X'$.

The proof of the harder implication, $2 \to 1$, combines ideas from the proof of Theorem 3.2.13 with Exercise 3.1.17. We will build a copy B of $(L, <)$ with domain $\{b_0, b_1, \ldots\}$. Indeed, we will build a bijection $f : \{b_0, b_1, \ldots\} \to L = \{\ell_0, \ell_1, \ldots\}$ and then define the order on B by "pulling back" the order on L along f. To clarify what this means, we need to introduce a notation.

For each $\sigma \in 2^n$, let $\psi_\sigma(x_0, \ldots, x_n)$ be the formula in the language of linear orders that completely describes the order on $x_0, \ldots, x_n$, depending on the values of σ. That is,

$$\bigwedge_{i:\sigma(i)=1} \psi_i \wedge \bigwedge_{i:\sigma(i)=0} \neg\psi_i,$$

where each ψ_i is a conjunction of atomic facts of the form $x_i < x_k$ and $x_i > x_k$, for $k < n$. Let $D(B) \in 2^\omega$ be the diagram of B; that is, we have that

$$B \models \psi_\sigma(b_0, \ldots, b_n) \text{ if and only if } \sigma \subseteq D(B).$$

At each stage s we define a finite one-to-one partial map $p_s : B \to L$ with domain $\{b_0, \ldots, b_{n_s}\}$, and then we will let $f = \bigcup_s p_s$. Given a finite one-to-one partial map p that maps $b_0, \ldots, b_n$ to $\ell_0, \ldots, \ell_n$, let $D(p)$ be the $\sigma \in 2^n$ such that $\psi_\sigma(\ell_0, \ldots, \ell_n)$ holds in L. Then we will have $D(B) = \bigcup_n D(p_n)$.

Construction:

- Let p_0 map b_0 to ℓ_0.

- At stage $s+1=2e$, extend p_s to p_{s+1} in any way so that b_e is in the domain of p_{s+1} and ℓ_e is in the image.
- At stage $s+1=2e+1$, we want to use $\varnothing'$ to decide the jump of $D(B)$. Suppose p_s maps $b_0,\ldots,b_{n_s}$ to $\ell_0,\ldots,\ell_{n_s}$. We will argue that, using $\varnothing'$, we can decide whether

$$\exists q \supseteq p_s \text{ such that } \{e\}^{D(q)}(e)\downarrow;$$

for now, take this property for granted. If the answer is positive, $\varnothing'$ can search for witnesses σ and $\overline{y}$ and use them to define p_{s+1}, adding $\overline{y}$ to the range of p_s. In this case, $\varnothing'$ knows that $e \in D(B)'$. Otherwise, we let $p_{s+1}=p_s$, and then $\varnothing'$ knows that $e \notin D(B)'$.

We have built B. To check that $D(B)' \leqslant_T \varnothing'$, it remains to verify that $\varnothing'$ can decide whether there exists an extension $q \supseteq p_s$ such that $\{e\}^{D(q)}(e)\downarrow$, where p_s maps $b_0,\ldots,b_{n_s}$ to $\ell_0,\ldots,\ell_{n_s}$. This is where we use that bot $<$ and *adj* are Δ^0_2 in L. Indeed, since $\varnothing'$ computes *adj* in L, it knows whether

$$\exists \sigma \supseteq p_s \ \{e\}^{\sigma}(e)\downarrow \ \& \ L \models \exists \overline{y}\, \psi_\sigma(\ell_0,\ldots,\ell_{n_s},\overline{y}).$$

To see why, use $\varnothing'$ to see which of the ℓ_j are in the same block, and which are not. Then see if there is a σ that obeys these restrictions and $\sigma \supseteq p_s \ \{e\}^{\sigma}(e)\downarrow$. Using $\varnothing'$, list the order on L and see if σ corresponds to a sub-order. It could be that σ claims the existence of 5 points between (e.g.) ℓ_1 and ℓ_2. Using that *adj* is computable in $\varnothing'$, we can see whether there are enough points between ℓ_1 and ℓ_2. Indeed, either a pair $\ell_1 < x < y < \ell_2$ is an adjacency or not. In the latter case we can $\varnothing'$-computably search for a point between x and y. It could be that we discover that ℓ_1 and ℓ_2 are in one block which is too small. In this case we try again, and search for a σ' that obeys this new restriction. It remains to note that there are only finitely many intervals between $\ell_0,\ldots,\ell_{n_s}$. Thus, eventually, we either find some extension or conclude that no such extension exists. □

It is natural to ask whether every low linear order is isomorphic to a computable order. While for Boolean algebras the answer to the analogous question is positive (to appear as Theorem 4.1.25), we will see that for linear orders, the answer is negative (Theorem 3.2.45).

3.2.4 Finite injury. Tennenbaum's Theorem

There are many illustrations of the finite injury method in the theory of computable linear orderings. One where the conflicts between the requirements are quite apparent is the following theorem. We write L^* to denote the linear order anti-isomorphic to L, i.e., the order in which $<$ is reversed to $>$. For example, ω^* is the order of negative integers.

Theorem 3.2.15 (Tennenbaum). *There is a computable copy of $\omega+\omega^*$ with no infinite computable ascending or descending suborderings.*

Extended Sketch. We will build the ordering $(A, \leqslant_A)$ in stages. The domain of A will be $\mathbb{N}$, so suborderings correspond to subsets of $\mathbb{N}$. Recall that W_e denotes the e-th c.e. set. We meet the requirements:

$$R_{2e} : |W_e| = \infty \rightarrow W_e \text{ is not an ascending subordering of } A,$$

$$R_{2e+1} : |W_e| = \infty \rightarrow W_e \text{ is not a descending subordering of } A.$$

Additionally, we must ensure that the order type of $(A, \leqslant_A)$ is $\omega + \omega^*$. To this end, we will build A as $B + C$, where we will refer to the members of B as *blue* and those in C as *red*.

At each stage s, we will let:

$$B_s = b_{0,s} <_A b_{1,s} <_A \cdots <_A b_{n,s},$$

and similarly,

$$C_s = c_{m,s} <_A c_{m-1,s} <_A \cdots <_A c_{0,s}.$$

Thus, we need to ensure that for all i, $\lim_i c_{i,s} = c_i$ exists, and similarly $\lim_s b_{i,s} = b_i$ exists. (We could explicitly add this as a new requirement, but it is unnecessary, as we will see.)

Now, the blue part is, of course, the ω part, and the red part is the ω^* part of A. At any stage $s + 1$, a red element can become blue and vice versa. If $b_{i,s}$ becomes red, then every element $x \in B_s$ with $x \geqslant_A b_{i,s}$ will also become red. If $b_{i,s}$ is the $\leqslant_A$-least element that becomes red and k elements become red, then at stage $s + 1$, the blue elements are now $b_{0,s}, \ldots, b_{i-1,s}$, and the red ones are $c_{m+k,s}, c_{m+k-1,s}, \ldots, c_{0,s}$. That is, $b_{j,s+1} = b_{j,s}$ for $j \leqslant i - 1$, and $c_{j,s+1}$ are the same for $j \leqslant m$, while for $j > m$, they are defined as the erstwhile blue elements.

We wish to ensure that W_e is not an ascending (blue) ω sequence. If W_e contains a blue element, then it cannot be such a sequence. So the obvious strategy for R_{2e} is to wait until we see some red $b_{i,s} \in W_{e,s}$ and make it red, as indicated. To ensure we don't do this for all elements, we will only consider $b_{i,s} > 2e$, so that R_{2e} has no authority to recolour elements $\{0, 1, \ldots, 2e\}$.

On the other hand, R_{2q+1} is trying to prevent W_q from being an infinite (red) ω^* sequence, and similarly, it wants to make red elements blue. If we allowed R_{2q+1} to undo the work we just did for R_{2e} by making the erstwhile $b_{i,s}$ blue again, we would undo the work needed to meet R_{2e}. Thus, when we act on R_k, we will do so with priority k, making some element the colour demanded by R_k, unless $R_{k'}$ for $k' < k$ wishes to change its colour.

So, we say R_{2e} *requires attention at stage* s if it is not currently declared satisfied, and we see some $b_{i,s} \in W_{e,s}$ not protected by any R_k for $k < 2e$ and $b_{i,s} > 2e$. Similarly, R_{2q+1} requires attention if some $c_{i,s}$, instead of $b_{i,s}$, satisfies the analogous conditions.

The construction is as follows: if any k requires attention, take the smallest such k, and perform the re-colouring demanded by R_k. At each stage, we will add one more blue element to the right of B_{s+1} and one more red element to the left of C_{s+1}. The remaining details involve letting the requirements "fight it out" by priorities.

An induction on k shows that we meet R_k. Once the higher priority requirements have ceased activity, if R_k requires attention via some $n \in W_{d,s}$, whatever colour R_k assigns to n will remain fixed, as R_k has priority. Finally, if W_d is infinite, such an n will occur. It should be clear that the resulting order is isomorphic to $\omega^* + \omega$. □

Note that R_k can only be injured at most $O(2^k)$ many times.

3.2.5 Unbounded finite injury. Computable categoricity

Recall that a structure is computably categorical if any two computable presentations of the structure are computably isomorphic. Since computable isomorphisms preserve essentially all computability-theoretic properties, as a consequence of Theorem 3.2.15, we have:

Corollary 3.2.16. *The linear order $\omega + \omega^*$ is not computably categorical.*

We have already mentioned that $(\mathbb{Q}, \leqslant)$ is computably categorical. Recall that Theorem 3.2.2 states that this is more or less the only such ordering. More specifically, it says that a computable linear ordering A is computably categorical if and only if A has only a finite number of adjacencies.

Proof sketch of Theorem 3.2.2. One direction is clear: if we non-uniformly map the adjacencies, then between them the orderings are dense and we can use Cantor's argument.

So suppose that $(A, \leqslant_A)$ has infinitely many adjacencies. We need to build a computable $B \cong A$ to meet the following requirements:

$$R_e : \varphi_e \text{ is not an isomorphism from } B \text{ to } A.$$

To make $B \cong A$, we will build an isomorphism $f : B \to A$. This will be built as $f(x) = \lim_s f_s(x)$ with f_s a partial map from B_s to A_s. Thus, for simplicity of notation, we have $f_s : b_{j_i} \mapsto a_{j_i}$ for $i \leqslant s$.

At stage $s + 1$, a new element a_{s+1} enters $A_{s+1} - A_s$, and perhaps $a_i <_A a_{s+1} <_A a_j$. Then we would need to add an element b_{s+1} with $b_i <_B b_{s+1} <_B b_j$, and $f_{s+1}(b_{s+1}) = a_{s+1}$. Clearly, this will build an isomorphism, but it is merely copying. This will change when we discuss how we meet the requirements, as we'll need to "move" the isomorphism.

Because we will later be possibly making $f_t(b) \neq f_s(b)$, we will need to make sure that we also meet the requirements:

$$N_b : \lim_s f_s(b) = f(b) \text{ exists},$$

and, additionally, in the construction, we will make sure that for every $a \in A$ there is a $b \in B$ such that $f(b) = a$. The ordering of the priorities will be:

$$R_0, N_0, R_1, N_1, \ldots.$$

We discuss how we meet a single requirement R_0 in isolation. Then we observe that we can let the priority method sort the construction so that all requirements are met.

One R_0 in isolation. Clearly, we need to do nothing unless φ_0 is total, so we will be monitoring the behaviour of φ_0, waiting for it to halt on more and more inputs. Suppose that we *knew* that $[a_i, a_j]$ was an adjacency in A. We could wait for $\varphi_{0,t}(b_p) \downarrow= a_i$, $\varphi_{0,t}(b_q) \downarrow= a_j$ to occur. If $[b_p, b_q]$ is not a B_t-adjacency, it never will be; so if we simply preserve the current $f_t : B_t \to A_t$. This will guarantee that φ_0 cannot be an isomorphism from B to A, as it takes a non-adjacency to an adjacency. If $[b_p, b_q]$ really is an adjacency, our action will be to split it in B_{t+1}, manufacturing the destruction of φ_0 as a possible isomorphism.

Since R_0 has the highest priority, we proceed as follows. We know that A is infinite, so new elements must occur either to the left of $f_t(b_p)$ or to the right of $f_t(b_q)$. Suppose the former. Let $b_{i_0} <_B b_{i_1} <_B \cdots <_B b_{i_k}$ denote the points of B_t left of b_p, and suppose that the new element a_{s+1} is between $f(b_{i_v})$ and $f(b_{i_{v+1}})$ (if it occurs to the left of b_{i_0} or to the right of b_{i_k}, it is easier, but similar). We add a new point b_{t+1} between b_p and b_q, and redefine $f_{t+1}(b_{t+1}) = f_t(b_p)$, and define $f_{t+1}(b_p) = f(b_{i_k})$, and similarly shift $f_{t+1}(b_{i_d})$ left for $b_{i_v} <_B b_{i_d} <_B b_p$, but leave $f_{t+1}(b_{i_e}) = f_t(b_{i_e})$ for $b_{i_e} \leqslant_B b_{i_v}$. We call this an R_0-*attack* on $[a_i, a_j]$. Fig. 3.2 and Fig. 3.3 might be helpful to see what is happening.

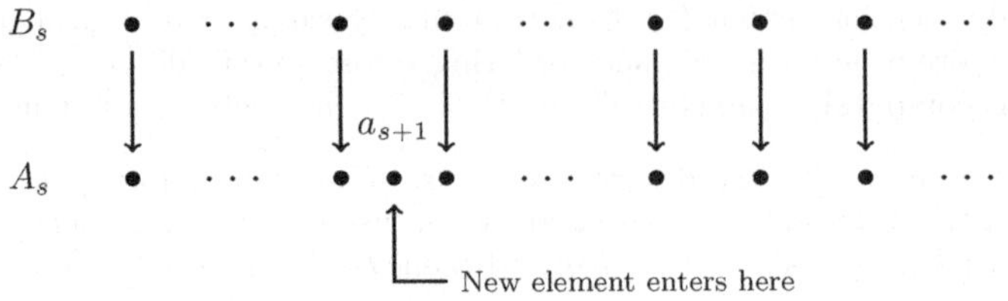

Figure 3.2: Stage s

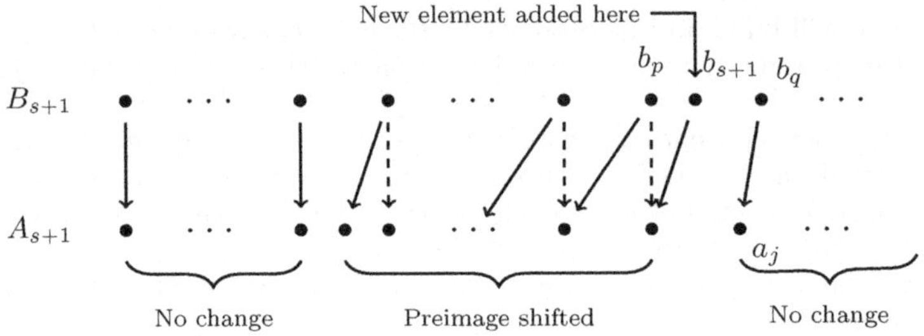

Figure 3.3: Stage $s+1$

Clearly, R_0 cannot know which pairs in A are adjacencies, so we will R_0-attack the current adjacency with the smallest Gödel number. Once we have attacked, we regard R_0 as satisfied, unless the adjacency $[a_i, a_j]$ used in the attack is split in A at a later stage s'. In that case, we will allow R_0 to become active again and try to attack the adjacency of $A_{s'}$ having the smallest Gödel number.

Since A has adjacencies, we will eventually meet R_0, after a finite (but unknown in advance) number of attacks. Note that this will potentially injure each N_j for $j > 0$, as each time we attack, we do keep a partial isomorphism at stage $t+1$, but we shift f_t on some elements.

The general case is similar but has some further difficulties. In particular, we clearly need to use the fact that the order has infinitely many adjacencies, not just one.

An outline of the general case.

We would like to attack R_e in a similar way, by splitting the pre-image of some adjacency $[a_i, a_j]$ in A_t. But now R_e must respect N_j for $j < e$. The strategy will attempt to preserve $f_t(j) = f(j)$ (with its respective priority). In particular, R_e cannot shift f on these few points.

These few points split B_t (and, hence, A_t) into finitely many intervals. Let $a_{i_0} <_A a_{i_1} <_A a_{i_2} <_A \cdots <_A a_{i_{e-1}}$ list these points in A, and let the corresponding points be labelled as $f^{-1}(a_{i_j}) = b_{i_j}$. Then A is split into

$$\{z : z <_A a_{i_0}\} \cup \bigsqcup_{p=0}^{e-2} [a_{i_p}, a_{i_{p+1}}) \sqcup \{z : a_{i_{e-1}} \leqslant_A z\},$$

and this induces the corresponding partition of B_t into sub-intervals (retracting via f_t).

The idea is simple: at least one of these sub-intervals has infinitely many adjacencies. Thus R_e pursues its strategy in each of the sub-intervals, noting that shifting can occur in an infinite subinterval without disturbing the endpoints. This allows N_j to be met for $j < e$ while still meeting R_e.

The remaining details are a routine application of the finite injury method; this is left as Exercise 3.2.18. □

Exercises

Exercise° 3.2.17. Construct a copy of ω such that the adjacency relation has degree $\mathbf{0}'$.

Exercise° 3.2.18. Give a detailed formal proof of Theorem 3.2.2.

Exercise° 3.2.19. A computable *partial* ordering $(P, \leqslant)$ is a computable set P and a partial ordering $\leqslant$ which is a computable relation on $P \times P$. A *linear extension* of a partial ordering $(P, \leqslant)$ is a linear ordering with the same domain P, $(P, \preccurlyeq)$ such that if $x \leqslant y$ then $x \preccurlyeq y$. A classical Szpilrajn extension theorem states that every partial ordering has a linear extension. Prove that a computable partial ordering has a computable linear extension.

Exercise 3.2.20 (Schwarz [460])**.** Let L be a computable linear ordering. Prove that the following are equivalent.

(i) L contains a dense interval.

(ii) Every computable linear ordering $\hat{L}$ isomorphic to L has a non-identity computable automorphism.

Exercise° 3.2.21. The well-known Dushnik-Miller ([151]) theorem states that every infinite linear ordering has a nontrivial (order-preserving) self-embedding. That is, $f : L \to L$ is order-preserving and for some $x \in L$, $f(x) \neq x$. Show that there is a computable copy L of $(\omega, <)$, such that L has no nontrivial computable self-embedding.

3.2.6 The Fellner-Watnick Theorem

A slight modification of Lerman's Theorem 3.2.11, namely Lemma 3.2.12, shows that any Σ_3^0-set S can be realised as a computable isomorphism invariant of a linear order $L(S)$, in the sense that $S \in \Sigma_3^0$ iff $L(S)$ is computably presented (and this can be relativised). On the other hand, Richter's Theorem 3.2.13 gives very strong evidence that there is no reasonable coding of Σ_1^0-sets into computable linear orders. In this section, we will prove that, for $n > 3$, a Σ_n^0-set can also be realised as an "effective invariant" of a computable linear order. To prove the result, we shall establish two standard "jump inversion" theorems for linear orders.

Of course, linear orders can have much more complicated invariants than just sets. For instance, given any linear order A, replace every point of the order by the order-type $(\mathbb{Q} + 2 + \mathbb{Q})$, which is the dense order of $\mathbb{Q}$ followed by two points and by another copy of $\mathbb{Q}$. Denote the resulting order by $Q(A) = (\mathbb{Q} + 2 + \mathbb{Q})A$. It should be clear that $A \cong B$ iff $Q(B) \cong Q(A)$, and thus A can be viewed as the isomorphism invariant of $Q(A)$. Similarly, we can define $\mathbb{Z}A$ by replacing every point of A by a copy of the order-type of the integers. We begin with the much simpler transformation $A \mapsto Q(A)$.

Theorem 3.2.22 (Downey and Knight [134])**.** *A linear ordering A is Δ_2^0-presentable iff $Q(A) = (\mathbb{Q} + 2 + \mathbb{Q})A$ is computably presentable.*

The proof is left as an exercise (Exercise 3.2.48), but the idea is clear. We can assume A is Π^0_1, so simply build a copy of $\mathbb{Q} + 2 + \mathbb{Q}$ around a point $z \in A$ unless z leaves A, in which case absorb the "junk" into the other copies of $\mathbb{Q}$. A far more complex theorem is the following:

Theorem 3.2.23 (Fellner [164], Watnick [502]). *A linear order A has an $\varnothing''$-computable presentation iff $\mathbb{Z}A$ is computably presentable.*

Before we turn to the proof, we will put it in context. There is a standard operator in the theory of linear orderings called the *condensation* operator, defined as follows.

Definition 3.2.24. Let B be a linear order. The (finite) condensation $C_F(B)$ of B is the result of identifying any two elements of B which are finitely far apart.

A *discrete* linear ordering is one where every element has an immediate predecessor and immediate successor (save perhaps the first point with no predecessor, and the last with no successor). If B is discrete and has neither first nor last points, then it is easy to see that it must be of order type $\mathbb{Z}A$ for some ordering A, so that $C_F(B) \cong A$. In some sense, $B = C_F^{-1}(A)$. It is not hard to see that if B is computable, then $C_F(B) \leqslant_T \varnothing''$ since it takes two quantifiers to ask if x and y in B are finitely far apart. This gives one implication in Theorem 3.2.23. Rosenstein [456] asked if the reverse implication held. This was answered in the affirmative by Fellner [165] and independently Watnick [502], and then independently rediscovered by Downey (unpublished).

The first proof of Theorem 3.2.23

The first proof that we present splits the construction of a computable copy of $\mathbb{Z}L$ into two lemmas. As far as we know, Zubkov was the first to note that the proof of Theorem 3.2.23 can be split into two finite injury proofs, but he never published his new proof. Montalbán takes a very similar approach in his book [401], but using (slightly) different methods.

Let *adj* denote the adjacency relation, i.e., $adj(x, y)$ holds if $x < y$ and there is no z such that $x < z < y$.

Lemma 3.2.25. *Let L be a non-empty Δ^0_2 linear order. Then the order $\mathbb{Z}L$ has a computable presentation in which adj is a computable relation. This is uniform.*

Proof. The proof does not actually need L to be infinite, it merely requires that $L \neq \varnothing$. Nonetheless, suppose the domain of L is (indexed by) $\mathbb{N}$:

$$|L| = \{\ell_i : i \in \mathbb{N}\},$$

and $<_L$ is Δ^0_2; the case of an initial segment of ω is essentially the same, up to a minor adjustment.

Notation 3.2.26. For a linear order Γ and $x, y \in \Gamma$, define $x \ll y$ if for the equivalence classes of x and y in $C_F(\Gamma)$, $[x]$ and $[y]$, we have $[x] < [y]$ in $C_F(\Gamma)$. This is the same as to say that $x < y$ and there are infinitely many points between x and y.

We build a computable linear order Γ with computable adjacency relation, in which every block (aka equivalence class in $C_F(\Gamma)$) is isomorphic to $\mathbb{Z}$. Additionally, we construct a Δ^0_2-map $\psi : L \to C_F(\Gamma)$ and for every $i \neq j$, we meet the requirements

$R_{i,j}: \ \ell_i < \ell_j$ if and only if $\psi(\ell_i) \ll \psi(\ell_j)$.

The map ψ will range over representatives of classes in $C_F(\Gamma)$. At every stage we will have defined $\psi_s(\ell_j) = m_{i,s}$. We will then prove that, for every i,

$$m_i = \lim_s m_{i,s} = \lim_s \psi_s(\ell_i) = \psi(\ell_i)$$

exists, and that the requirements are met, that each m_i lies in a $\mathbb{Z}$-block, and that every block has some m_j in it. We will not make ψ_s explicit in the construction, and will instead work only with $m_{i,s}$. Our strategies will work with ℓ_i (and m_i) rather than with the requirements; the requirements will be met somewhat indirectly, as a result of our actions.

Strategy for ℓ_0. In the construction, we will place $m_{0,s} = m_0$ and will begin building a $\mathbb{Z}$-block $[m_0]$ around m_0. We will denote the finite portion of $[m_0]$ at stage s by $M_{0,s}$. For every s, we will have $m_{0,s} = m_0$.

Strategy for ℓ_i, $i > 0$. The strategy assumes that $<_L$ restricted to $\ell_0, \ldots, \ell_{i-1}$ and approximated using the Limit Lemma 3.1.3, has not changed since the previous stage. If the approximation to $<_L$ has changed on $\ell_0, \ldots, \ell_{i-1}$, then the strategy is instantly *initialised* (to be clarified). If $M_{j,s}$, $j < i$, are the finite portions of $\mathbb{Z}$-blocks around $m_{j,s}$ $(j < i)$ built so far, and $m_{i,s}$ is undefined, then the strategy:

1. places $m_{i,s}$ between (or to the left or to the right of) the blocks $M_{j,s}$ so that

$$\ell_k \mapsto m_{k,s}, \ k \leqslant i$$

is an isomorphism between the current approximation to $<_L$ on $\ell_0, \ldots, \ell_i$ and $m_{0,s}, \ldots, m_{i,s}$, and

2. initiates the construction of $M_{i,s}$, which is (an attempt to build) a copy of $\mathbb{Z}$ around $m_{i,s}$ disjoint from $M_{j,s}$ $(j < s)$.

In $m_{i,s}$ is defined, then make progress in building the $\mathbb{Z}$-block around $m_{i,s}$ by placing a few more points around $m_{i,s}$, let the resulting block be $M_{i,s+1}$, and set $m_{i,s+1} = m_{i,s}$.

Initialisation. If a strategy ℓ_i $(i \leqslant s)$ needs to be initialised, then let k be a strategy so that $M_{k,s}$ is adjacent to $M_{i,k}$. In particular, $M_{k,s}$ is defined and thus the ℓ_k-strategy has not been initialised yet.

1. Set $m_{i,s}$ undefined.
2. Incorporate all points of $M_{i,s}$ into $M_{k,s}$ and set $M_{i,s}$ undefined.

Construction.

At stage 0, only the strategy for ℓ_0 acts.

At stage s, see if any strategies for ℓ_i $(i < s)$ need to be initialised. If so, initialise them one by one, until no strategies that need to be initialised are left. Then let the strategies for ℓ_i $(i \leqslant s)$ act according to their instructions.

Verification. By induction, we prove that $\lim_s m_{i,s}$ exists. The case when $i = 0$ is trivial, since the strategy working with ℓ_0 is never initialised. The case $i > 0$ follows by a straightforward induction, since our approximation to $<_L$ on $\ell_0, \ldots, \ell_{i-1}$ will eventually settle. Recall $\psi_s(\ell_i) = m_{i,s}$. Let $m_i = \lim_s m_{i,s}$ and $\psi(\ell_i) = \lim_s \psi_s(\ell_i) = m_i$.

Claim 3.2.27. *For every $i \neq j$, $R_{i,j}$ is met.*

Proof. Let s be so large that $m_i = m_{i,s}$ and $m_j = m_{j,s}$. Without loss of generality, assume $\ell_i < \ell_j$. The instructions of the strategies working with ℓ_i and ℓ_j guarantee that $m_i < m_j$. Further, $M_i = \bigcup_{t \geqslant s} M_{i,t}$ and $M_j = \bigcup_{t \geqslant s} M_{j,t}$ do not intersect and are isomorphic to $\mathbb{Z}$ each. It follows that $\psi(\ell_i) \ll \psi(\ell_j)$. □

Note we also showed that for every i there is an s such that $M_i = \bigcup_{t \geqslant s} M_{i,t}$. We must also have that $M_i \cong \mathbb{Z}$. It remains to note that for every s,

$$\Gamma_s = \bigsqcup_{i \leqslant s} M_{i,s},$$

and that every point x ever placed in Γ will eventually find itself in one of the M_i-blocks. We conclude that $\Gamma \cong \mathbb{Z}L$, as witnessed by ψ, the elements m_i, and their $\mathbb{Z}$-blocks M_i.

Finally, it remains to observe that no points will be put between an adjacent pair $x, y \in M_{i,s}$, for any i, s, even if this block will be later incorporated into some other block, due to initialisation. Thus, the adjacency relation is computable in Γ. □

The second lemma may look a bit less interesting, but it will in fact be more useful in the sequel. We state it in a slightly more general form than is needed to prove the theorem, because we shall need the stronger lemma in the next chapter. Also, compare the lemma with Theorem 3.2.14.

Lemma 3.2.28 (Downey and Jockusch [129]). *Suppose L is a Δ_2^0-linear order in which the adjacency relation is Δ_2^0. Then there exists a computable linear order $\widehat{L}$ which is, up to isomorphism, L except for an adjacency in L may be replaced by a finite block. If the linear order has infinitely many adjacencies and has no greatest and no least element, then we can produce $\widehat{L}$ uniformly.*

Proof. Without loss of generality, we may assume that $L \subseteq \mathbb{Q}$ having least and greatest elements and infinitely many adjacencies. We prove that there is a computable linear ordering $\widehat{L} \subseteq \mathbb{Q}$ and a function $h : L \to \widehat{L}$ such that the following conditions hold:

(i) h is 1–1 and order preserving and maps the least (greatest) element of L to the least (resp. greatest) element of $\widehat{L}$.

(ii) If $[a, b]$ is an adjacency of L, then $[h(a), h(b)] \cap \widehat{L}$ is finite.

(iii) If $c \in \widehat{L} - range(h)$, then

$$(\exists a, b)[a, b \in L \text{ and } adj(a, b) \text{ and } h(a) < c < h(b)].$$

By the Limit Lemma, L has a recursive approximation L_s (so that for almost all s, $a \in L$ if and only if $a \in L_s$). We may further assume that

$$\text{adj}(a, b) \text{ implies } \{s : (\exists c)(a < c < b \ \& \ a, b, c \in L_s)\} \text{ is finite.}$$

Using the Δ_2^0–ness of adj, we can "speed up" any approximation to L to get one with this property. We may also assume that the least and greatest elements of L are in L_0. We will construct $\widehat{L} = \cup_s \widehat{L}_s$, and $h = \lim_s h_s$ in stages.

At the initial stage $s = 0$ we map the least (greatest) element of L to 0 (respectively, 1) with highest priority and never change h on these arguments. At stage s, we are allowed to add new elements to $\widehat{L}$ between the current values of $h(a)$ and $h(b)$, $(a < b)$, only if there exists $c \in L_s$ with $a < c < b$. Thus (ii) will hold.

As before, let $q_0, q_1, \cdots$ be an effective enumeration of $\mathbb{Q}$. We require that any element added to $\widehat{L}$ at stage s be of the form q_m with $m \geqslant s$. Hence $\widehat{L}$ will be computable.

We have the following requirements:

$$
\begin{aligned}
R_{2m} \quad &: \quad q_m \in L \Longrightarrow h(q_m)\downarrow \ \&\ h(q_m) \in \widehat{L};\\
R_{2k+1} \quad &: \quad q_k \in \widehat{L} - h(L) \Longrightarrow (\exists\, i, j)\\
&\qquad \text{adj}\ (q_i, q_j)\ \&\ h(q_i) < q_k < h(q_j).
\end{aligned}
$$

Assign priorities as usual (the argument is finite injury). The construction is arranged so that $\mathrm{dom}(h_s) \subseteq L_s$ for every stage s. When we set $h_s(q_m) = q_k$, this assignment has the priority of R_p, where $p = \min\{2m, 2k+1\}$. If $q_m \in L_t$ for all $t \geqslant s$ and no requirement of higher priority than R_p acts after stage s, we will then have $h_t(q_m) = q_k$ for all $t \geqslant s$ (and hence $h(q_m) = q_k$).

Strategy for R_{2m}. If $q_m \in L_{s+1}$ and $h_s(q_m)$ is not defined, define $h_{s+1}(q_m) = q_t$ where t is chosen so that $t > m$ and this definition keeps h_{s+1} order preserving. Add q_t to $\widehat{L}$. If q_m leaves L or a higher priority requirement R_{2k+1} acts, cancel this value of h and start over. If $q_m \in L$, this will happen only finitely often, and $h(q_m) = \lim_s h_s(q_m)$ will exist and be in $\widehat{L}$.

Strategy for R_{2k+1}. Suppose $q_k \in \widehat{L}_s - h_s(L_s)$. (This situation arises when q_k is put into $\widehat{L}$ by some R_{2m}, but its apparent h-preimage seems to leave L or q_k is cancelled as an h-image by higher priority action.) Further, assume that there do not exist i and j with $h(q_i)$ and $h(q_j)$ defined with stronger priority than that of R_{2k+1} such that $\text{adj}_s\,(q_i, q_j)$ and $h_s(q_i) < q_k < h_s(q_j)$. Cancel all lower priority values of h. Choose i and j with $h_s(q_i)$ and $h_s(q_j)$ defined with stronger priority than R_{2k+1} such that $h(q_i) < q_k < h(q_j)$ and $(h(q_i), h(q_j))$ contains no values of h defined with stronger priority than that of R_{2k+1}. (Such i and j exist because we initially defined h on the least and greatest elements of L with highest priority.) By hypothesis, $\text{adj}_s\,(q_i, q_j)$ does not hold. As long as it continues not to hold, search for t such that $q_t \in L$ and $q_i < q_t < q_j$ by effective approximation. Set $h(q_t) = q_k$ (without changing $\widehat{L}$). If the candidate for t changes, start over by undefining all lower priority values of h. Also, if $\text{adj}\,(q_i, q_j)$ starts to hold, or q_i or q_j appears to leave L, start over.

These strategies combine by a standard finite injury argument. We omit further details, which are routine. The construction of $\widehat{L}$ is uniform. Recall however that we assumed that L had the greatest and the least element. We shall remove the least and the greatest element from $\widehat{L}$ if necessary. In case when L had no least or greatest element, these extra elements could not be a part of an adjacency. $\square$

Apply Lemma 3.2.25 relativised to $\varnothing'$ to obtain a Δ^0_2-copy of the order $\mathbb{Z}A$ in which the adjacency relation is also Δ^0_2. Now apply Lemma 3.2.28 to produce a computable presentation of $\mathbb{Z}A$; note that replacing an adjacency by a finite block does not change the isomorphism type of $\mathbb{Z}A$. Note this is all uniform.

The first proof of the Fellner-Watnick Theorem 3.2.23 is complete.

The second proof of Theorem 3.2.23 using the tree of strategies*

We now present a very detailed proof of Theorem 3.2.23 that uses the techniques from the proof of the Minimal Pair Theorem 3.1.44. The impatient reader may skip this subsection, as we will not need to use the tree of strategies until Chapter 9.

As we have seen, infinite injury can be replaced with two finite injury constructions in the specific case of Theorem 3.2.23. However, it appears that such iterated proofs are not always possible. It seems that in many situations, using the tree of strategies is a much more flexible approach, even if it may result in longer arguments.

The following proof was taken from Downey [119].

Infinite injury proof of Theorem 3.2.23. Recall we are given a $\varnothing''$-computable presentation of a linear order A, and we must produce a computable presentation of $\mathbb{Z}A$.

We assume A is non-empty, and indeed, we shall assume A is infinite, as the case when A is finite is elementary. We will also see that the proof is uniform, in the sense that given the index e for $\Phi_e^{\varnothing''} \cong A$, we can computably produce an index for a computable copy of $\mathbb{Z}A$.

Recall that in Lemma 3.2.6 we established that any Δ_2^0 order is isomorphic to a Π_1^0-subset of the rationals. Relativising this to $\varnothing'$, we obtain that we only need to deal with Π_2^0 subsets of $(\mathbb{Q}, \leqslant)$. This relativisation does not affect uniformity (see Remark 3.1.6).

The proof involves an infinite injury argument, similar to the construction of a minimal pair, as detailed in Theorem 3.1.44.

Before presenting the formal construction, we will informally explain how our construction works.

A nice representation of A. Let A be a Π_2^0-copy of the order. We let $\mathbb{Q} = \{x_0, x_1, x_2, \ldots\}$ be a computable enumeration of $\mathbb{Q}$.

First, we take a nice representation of A as first suggested by Jockusch, as follows. By the standard representation of a Π_2^0 set, we may suppose, without loss of generality, that $A = \{f(i) \mid \varphi_{f(i)} \text{ is total}\}$, where $\{\varphi_i : i \in \mathbb{N}\}$ lists the partial computable functions; see Theorem 3.1.8. We replace this representation with a better one, given by a tree that controls strategies. One nice property of this better representation is that if $j_1, \ldots, j_n$ is any finite subset of A, then $j_1, \ldots, j_n$ all "appear to be in A" *together* infinitely often. This is done as follows.

Define a stage s to be a *σ-stage* (for $\sigma \in 2^{<\omega}$) by induction on the length of σ, $|\sigma|$, as follows:

1. Every stage s is a λ-stage (λ is the empty string).
2. If s is a τ-stage and $|\tau| = j = f(i)$, then if $\varphi_{f(i),s}(y)\downarrow$ where
$$y = \mu z\{z \notin \operatorname{dom} \varphi_{f(i),t} : t \text{ is a } \tau\text{-stage and } t < s\},$$
we say s is a $\tau\widehat{\ }0$-stage. Otherwise, s is a $\tau\widehat{\ }1$-stage.

Then A is the "true path" of the above tree, in the following sense. Let β be the leftmost path visited infinitely often so that $\lambda \preceq \beta$, and if $\sigma \preceq \beta$, we have $\sigma\widehat{\ }0 \preceq \beta$ iff $\exists^\infty s$ (s is a $\sigma\widehat{\ }0$-stage); otherwise, $\sigma\widehat{\ }1 \preceq \beta$. Then
$$A = \{j : |\sigma| = j \text{ and } \sigma\widehat{\ }0 \preceq \beta\}.$$

We say that x_j *appears to be in A at stage s* if s is a σ-stage and $|\sigma| = j$. We define σ_s to be the unique string with $|\sigma_s| = s$ and s is a σ_s-stage. Then we say that A *appears to be* $\{j : \tau\widehat{\ }0 \preceq \sigma_s$ *and* $|\tau| = j\}$ *at stage* s, and define A_s to be this set.

Remark 3.2.29. Notice that $\forall s\ (x_0 \in A_s)$ because every stage is a λ-stage. This seemingly non-uniform assumption does not actually make the proof non-uniform. This is because when we realise an infinite (or indeed, non-empty) Δ_2^0 order A as a Π_2^0-subset of $\mathbb{Q}$, we can always assume that x_0 (or any fixed point) is in the subset. To ensure this, fix the first element in the domain ω of A and immediately map 0 to x_0. This will never be changed.

Recall that ω^* denotes the order of the negative integers.

A good model for $(B, <) = (C_F^{-1}(A), \leqslant)$. We must perform three basic tasks:

1. For points $x_i \in A$, we must build $\omega^* x_i \omega$.
2. We must "incorporate" all $x_j \notin A$, as well as all the auxiliary points from our attempts to build $\omega^* x_j \omega$ into blocks of the form $\omega^* x_i \omega$ for some $x_i \in A$.
3. We must ensure that nothing else is built.

A good model for B is given as follows. At stage s, we have a set of balls with various markings on them, arranged in a line. We have a supply of new balls we must add to this line, either inserted or added to the ends. These new balls will be z-balls, y-balls, or x_i-balls. The intention is that z-balls are attempting to be a part of an ω^*-block, y-balls part of an ω-block, and x_i-balls part of A. Later, we may change our mind and convert y to z, or x_i to y or z. *However, if an x_i-ball turns into a y-ball, it can't change back.*

The line of balls is referred to as the *surface*. An x_i-ball on the surface will be marked with a *guess* $\sigma \in 2^{<\omega}$. If there is a stage s where this guess proves wrong, we turn this x_i into a y- or a z-ball (with no guess).

The x_i-balls we must place on the surface at stage s are simply those that appear to be in A at stage s. Roughly speaking, we must place the y's and z's around the x_i's that appear in A at s, according to our definition of A_s above.

The strategies. To satisfy our three aims, we must have a strategy dictating where to place our new balls at each stage. This strategy must overcome several problems, whose solutions we outline below.

Incorporations. As a first approximation, let us suppose that we have three points which appear in A at stage s in the order

$$x_j < x_i < x_k.$$

Now suppose that *really* $x_j, x_k \in A$ and $x_i \notin A$. What we shall know is that x_j and x_k appear in A together infinitely often, but x_i only appears to be in A finitely often. For simplicity, let us suppose that x_j and x_k are successors in A. Thus, $\forall p\ (x_j \leqslant p \leqslant x_k \text{ and } p \in A \rightarrow p = x_j \text{ or } p = x_k)$. We see that locally $B = C_F^{-1}(A)$ should be one copy of $\omega^* + \omega$ at x_j and $\omega^* + \omega$ at x_k, with x_i not there. To achieve this, we shall incorporate x_i into the block of x_j. (The entire construction is "left justified".) We do this as follows. At stage s, when x_j appears to be in A, we must add one z before x_j and one y following x_j. As it stands, we put z immediately before x_j, but put y as far right as possible to be consistent with our current picture of A at stage s.

That is, to place y for the sake of x_j, we go as far right as possible until we see an x_p also appearing in A (or reach the end of A_s), and then repeat the process for x_p. For example, a typical situation might be

$$\begin{array}{ll} zzz\ x_j\ yz\ x_i\ yz\ x_k\ \cdots x_p \cdots & \text{at stage } s-1 \\ \underbrace{zzzz\ x_j\ yz\ x_i\ yz\ x_k\ \cdots x_p}_{\text{no change}}\ yz\ x_p \cdots & \text{at stage } s \end{array}$$

At stage s, it appears that $x_j \in A_s$ and x_p is the next member of A_s. This strategy works because x_j and x_k appear in A *together* infinitely often, so we build infinitely many z's before x_j, and similarly before x_k. (In this case, $p = k$.) Additionally, since x_i appears in A only finitely often, we build only finitely many y's and z's between x_j and x_k, and after that, we almost always incorporate x_i into x_j's ω-block whenever x_j and x_k appear in A together. Hence, we end up with the following structure:

$$\omega^* x_j \underbrace{yy \cdots yzz \cdots z}_{\text{finite}} x_i \underbrace{yy \cdots y \cdots}_{\omega} \omega^* x_k \cdots$$

Thus, we achieve $\omega^* x_j \omega \omega^* x_k$, as required. We refer to this strategy as *incorporation*, since we ensure that if x_i doesn't appear in A infinitely often, it gets incorporated into some block.

The problem. So far, we have explained how to place the y- and z-balls. The crucial property that allows this strategy to succeed is ensuring that between any two *successive* points of A (e.g., (x_j, x_k)), the wrongly placed "false points" (like x_i) can be rearranged into an ω-ordering.

Now, suppose that infinitely often, a new x_t appears so that $x_j < x_t < x_k$ appears in A at some stage s_t. It seems reasonable to place x_t between x_j and x_k. However, it is critical to determine how we should relate such x_t to x_i (in the previous notation).

Eventually, it will no longer appear that $x_i \in A$, so placing anything between x_j and x_i seems unnecessary. That is, although in the $\mathbb{Q}$-order we may have $x_j < x_t < x_i$, if it does not seem that $x_i \in A$ when we need to place an x_t-ball, we shall place x_t beyond x_i in the following order:

$$x_j < x_i < x_t < x_k.$$

Note that x_k appears in A at the same time.

Thus, assuming we are working to the right of x_0, our guiding principle is to always try to place new x_i's as far right as possible.

This brings us to a minor point.

We shall assume that x_0 is the least member of A.

This simplifies the presentation. When a new x_i appears, we first determine if $x_i < x_0$ or $x_0 < x_i$. If $x_i < x_0$, we work similarly, mutatis mutandis, so that everything proceeds to the left.

Summarising so far, our key idea is that we don't build between any x_p and x_q if x_q is x_p's current successor and x_p doesn't appear to be in A at the stage. This approach helps us overcome the problems induced by our strategy above.

The next situation we must consider is when x_t, x_j, x_i, x_k above are *all in A*, but infinitely often it appears that $x_j \in A$, $x_i \notin A$, and $x_k \in A$, and infinitely often it appears that $x_j \in A$, $x_i \notin A$, and $x_t \notin A$.

Following our strategy above, at stage s_0, when x_t appeared for the first time, we placed the balls in the order

$$x_j \; x_i \; x_t \; x_k \qquad \text{(at stage } s_0\text{)}$$

when their actual order (in A) is $x_j \; x_t \; x_i \; x_k$. We did this since it appeared that $x_i \notin A$ at stage s_0. But later, we see that $x_i \in A$ and also x_j, x_t, and $x_k \in A$ at some stage $s_1 > s_0$. We now realise that our initial guess regarding x_t's position was incorrect and now place

$$x_j \; x_t \; x_i \; \hat{x}_t \; x_k.$$

Here, $\hat{x}_t$ denotes a group of balls around x_t's old position. We can't remove these balls but also don't wish to build an $\omega^* + \omega$ block around $\hat{x}_t$, so our solution is to relabel the $\hat{x}_t$-balls as y- or z-balls.

This then gives rise to another problem. Later, we again see $x_t \in A$, but now $x_i \notin A$. We can't use the x_t position between x_j and x_i since we originally placed it there only because x_j and x_i appeared in A. Perhaps in another scenario, $x_i \notin A$ and $x_t \in A$, but infinitely many such x_t get inserted between x_j and x_i. We really should place new x_t as far right as possible consistent with our current picture of A_s. If, later, it appears that x_j and x_i are in A, we again build around the x_t we placed between x_j and x_i last time and cancel its current position. This leads to the fundamental idea of the proof.

Labelling. Whenever we place an x_t on the surface into a new position for the first time, we give it the *label* $\sigma \subseteq \sigma_s$ with $|\sigma| = t + 1$. Thus, we are indicating where x_t's correct position should be, assuming σ is correct. If we must move x_t, we do so because some x_i encoded in this guess, which appeared to be *out* of A, now appears to be *in* A, and $i < t$. (Of course, in σ_t this will appear as $\tau\hat{\ }1$ with $|\tau| = i$.) The condition $i < t$ is simply to determine which ball to move. When we move x_t, we give its new position a "better" label. The correct position for x_t is a stable position *corresponding to the leftmost label visited infinitely often.* The reader should note that, at any particular time, x_t might have several positions labelled on the surface; only (at most) one is correct. Here we need another notion. Let $\leqslant_L$ denote the lexicographic ordering on the tree. The reader should interpret $\sigma \leqslant_L \tau$ as $\sigma = \tau$ or σ is *stronger* than τ. Note that only those $\sigma \leqslant_L \tau$ will always have their apparent positions uncancelled. Those $\tau \leqslant_L \sigma$ only get visited finitely often and so only move σ finitely often. Hence, we shall argue that σ reaches a stable position and so does x_t. We now give the formal details of the construction, although we hope that the reader can see them for themselves.

Construction of $C_F^{-1}(A) = B$, stage $s + 1$.

Step 1 (Cancellation): Compute σ_s. Cancel all positions marked τ for $\tau \nleqslant_L \sigma_s$. Regard these now as y-balls and similarly, any z-balls associated with them become y-balls.

Step 2 (Placing x_j-balls): In order of j, for each τ with $|\tau| = j$ and $\tau\hat{\ }0 \subseteq \sigma_s$, proceed as follows:

If there is currently an (uncancelled) position marked $\tau\hat{\ }0$ on the surface, do nothing. If there fails to be such a position, establish one by placing an x_j-ball marked $\tau\hat{\ }0$ as far right as possible. This will be on the right of B_{s-1} unless there is an x_i-ball marked $\gamma\hat{\ }0 \subseteq \tau\hat{\ }0$ and $x_j < x_i$. In this case, for the $<_{\mathbb{Q}}$-least such x_i, we place x_j immediately left of x_i and put s z-balls immediately preceding x_i, i.e., $x_j zzz \cdots zx_i$.

Step 3 (Placing y, z-balls): Now place y- and z-balls as indicated in the discussion. That is, place another z-ball before x_0. Now go right and find the first x_j-ball, if any, marked $\gamma\hat{\ }0 \subseteq \sigma_s$ for some

γ. If there are no y- or z-balls between x_j and its predecessor, add a y-ball followed by a z-ball. If there are already blocks of y's and z's preceding x_j, add one further y to the y-block and one further z to the z-block. Continue until we get to the right end. Here, add one y-ball.

End of construction.

Verification. For the sake of the following lemmata, we shall adopt the following definitions:

Definition 3.2.30. *We say a* ball n appears to be in an x-ball g's ω^*-block at stage s *if n is a z-ball and, if m is any ball with m between n and g, then m is a z-ball.*

Definition 3.2.31. *We say a* ball n appears to be in an x-ball g's ω-block at stage s *if g has guess σ for some $\sigma \preceq \sigma_s$ (and g is not cancelled at s), $g \leqslant n$, and one of the following conditions is satisfied:*

1. *(a) there exists an x-ball g in $\operatorname{dom} B_s$ with $g \leqslant n < \hat{g}$ such that $\hat{g}$ has guess $\preceq \sigma_s$; and*

   (b) there does not exist an x-ball in $\operatorname{dom} B_s$ with guess $\preceq \sigma_s$ and $g < r \leqslant n$; and

   (c) there exists a y-ball p in $\operatorname{dom} B_s$ such that $n \leqslant p < \hat{g}$; or

2. *there does not exist an x-ball $\hat{g}$ in $\operatorname{dom} B_s$ with $g < \hat{g}$.*

Remark 3.2.32. *The intuition here is that either n occurs beyond the largest apparent member of A at s (in 2), or n occurs before the z-block of two apparently consecutive (at s) elements of A at stage s.*

We also say that position g (an x-ball) *receives attention at stage* s if the number of elements in g's apparent ω^*-block at s increases. Let β denote the leftmost path. That is, as in the discussion, β is the leftmost path visited infinitely often.

Lemma 3.2.33 (Stable position lemma). *Every x_i reaches a stable position. That is, either $x_i \notin A$, in which case x_i receives attention finitely often (in total), or $x_i \in A$, in which case there is a* unique *x_i-ball that receives attention infinitely often (and is, of course, never cancelled). This ball is marked $\sigma^\frown 0 \preceq \beta$ with $|\sigma| = i$.*

Proof. By induction. Suppose for all $j < i$, the lemma holds. Let s_0 be a stage such that $\forall s_0\ (\sigma^\frown 0 \leqslant_L \sigma_s)$. Let $s_1 > s_0$ be the least $\sigma^\frown 0$-stage exceeding s_0. At stage s_1, choosing s_0 minimal, we may suppose that we place an x_i-ball h, marked $\sigma^\frown 0$. We claim that this is x_i's stable position.

First, this position cannot be cancelled. We only cancel positions when their guess appears wrong (in step 1 of the construction). By 1 above, this cannot occur. Hence, this position is never cancelled.

Second, this position receives attention infinitely often since $\sigma^\frown 0 \preceq \beta$, and so $\exists^\infty s(\sigma^\frown 0 \preceq \sigma_s)$ by the definition of β.

Finally, h_i is unique. To see this, let g be another x_i-ball that receives attention infinitely often. This ball g must have guess γ, for some γ with $|\gamma| = i + 1$. There are three possibilities: either $\gamma \leqslant_L \beta$ and $\gamma \npreceq \beta$, $\gamma \preceq \beta$, or $\beta \leqslant_L \gamma$. In the first case, there are only finitely many γ-stages. We only add to this x_i-ball's z-block at γ-stages (by construction), and so such an x_i-ball can receive attention at most finitely often.

If $\gamma \preceq \beta$, then $\gamma = \sigma\hat{}0$. There are two possibilities: either g was appointed at a stage t *before* x_i (i.e., before s_1), or g was appointed *after* h, so at a stage $s_2 > s_1$. In the first case, our assumptions concerning the minimality of s_1 mean that there is a η-stage $\hat{t}$ with $t \leqslant \hat{t} \leqslant s_0$, with $\eta \leqslant_L \gamma$ and $\eta \neq \gamma$. Such a stage cancels g. If g is appointed after stage s_1, then g was appointed at a $\sigma\hat{}0$-stage. We only appoint new positions if there is not already a $\sigma\hat{}0$-position available. There is, of course, one—namely, the one occupied by h. Thus, we wouldn't have appointed g after all.

Finally, if $\beta <_L \gamma$, then $\sigma\hat{}0 \npreceq \gamma$. By step 1, at each $\sigma\hat{}0$-stage, we cancel any x-balls marked γ. Hence, g gets cancelled and so wouldn't have received attention after all. □

Lemma 3.2.34 (True z-ball lemma). *Let $x_i \in A$. Let $\hat{x}_i$ denote x_i's stable position (given by Lemma 3.2.33). Suppose n is an ω^*-ball of $\hat{x}_i$ at stage s_1. Then n is a z-ball of $\hat{x}_i$ at every $\sigma\hat{}0$-stage where $\sigma\hat{}0 \preceq \beta$ and $|\sigma| = i$. Also, if $z(n, s)$ is the number of balls between n and $\hat{x}_i$ at stage s, then for all $s \geqslant s_1$, $z(n, s) = z(n, s_1)$ ($= z(n)$, say).*

Proof. Let n and $\hat{x}_i$ be as above at stage s_1. By the definition of an ω^*-ball, n is a z-ball, and there are no balls between n and $\hat{x}_i$ save for z-balls. The construction (in step 2) specifically ensures that when we place new x-balls $x_j < \hat{x}_i$, we do not disrupt any currently placed z-balls. In particular, no new γ_j balls can be placed between n and $\hat{x}_i$. This, of course, means that $z(n, s) = z(n, s_1) = z(n)$ since new z-balls are always placed on the *left* end of $\hat{x}_i$'s apparent ω^*-block. □

Lemma 3.2.35 (True ω^*-block lemma). *Let $x_i \in A$ and $\hat{x}_i$ be as in Lemma 3.2.34. Then in B there appears an ω^* $\hat{x}_i$-block.*

Proof. By Lemma 3.2.34 and the fact that $\hat{x}_i$ receives attention infinitely often, we thus add infinitely many z-balls before $\hat{x}_i$. □

Definition 3.2.36. Suppose n is a z-ball of some $\hat{x}_i$. Then we say n is a stable ω^*-ball, and we say n adheres to $\hat{x}_i$ as an ω^*-ball.

Lemma 3.2.37 (ω-adherence lemma). *Let m be any ball, and suppose m is not a stable x_i-ball or a stable ω^*-ball. Then there exists a stable x_i-ball $\hat{x}_i$ and a stage $s(m)$ such that:*

1. *m appears to be an ω-ball of x_i at each $\sigma\hat{}0$-stage $> s(m)$, where $\sigma\hat{}0 \preceq \beta$ and $|\sigma| = i$.*
2. *If $d(m, \hat{x}_i, s)$ denotes the number of balls between m and $\hat{x}_i$ at stage s, then $\forall s > s(m)$ $(d(m, \hat{x}_i, s) = d(m, \hat{x}_i, s(m)) = d(m, \hat{x}_i))$.*

In this case, we say m adheres to x_i as an ω-ball.

Proof. This is the main lemma that our machinery—discussed before the construction—is meant to achieve. Either m is an unstable z-ball, an unstable x_j-ball for some j, or a y-ball. All of these cases are similar and can, roughly speaking, be treated simultaneously. Let s be the stage when m was placed on the surface. If m was an x_j-ball, set $c(m) = m$. If m was a y-ball, find the $\leqslant$-greatest x-ball x_j alive at stage s with $x_j \leqslant m$ and set $x(m) = x_j$. (Note: x_j does *not* need to be apparently in A at s. Also, x_j may later die or x_j may be $\hat{x}_j$.) Finally, if m was a z-ball, find the $\leqslant$-least x-ball x_j such that $m \geqslant x_j$, and set $c(m) = x_j$. Note that m appears to be in x_j's z-block at stage s. Since there are no dormant x-balls between m and $c(m)$, as in Lemma 3.2.34, we cannot place more balls between $c(m)$ and m after stage s due to the way we place balls. It therefore suffices to argue the lemma for $x(m) = x_j$ instead of m.

Without loss of generality, assume $x_j = c(m)$ is unstable, and either x_j is eventually cancelled or x_j is never cancelled but only appears in A finitely often.

Now let $\hat{x}_j$ be the $\leqslant$-greatest stable position $\leqslant x_j$ in B *at stage* s.

Claim 3.2.38. *After stage* s *we cannot add a stable position* $\hat{x}_t$ *with* $\hat{x}_i < \hat{x}_t < x_j$.

Suppose not, and $\hat{x}_t$ is such. Let $\sigma\hat{\ }0$ be the guess of x_t. The only time we place such a ball $\hat{x}_t$ between balls already on the surface is because the guess forces us to. That means there must be some ball x_q with guess $\gamma\hat{\ }0$ (say) and $\hat{x}_i < \hat{x}_t < x_q < x_j$ forcing $\hat{x}_t < x_j$. Since this is so, by priorities of movement it must be that $q < t$.

For suppose otherwise, and $t < q$. Now x_q must be already present on the surface at the stage s when x_i enters. Since this is a *new* position for x_t there must be no position marked $\sigma\hat{\ }0$ on the surface at stage s. Hence it cannot be that the guess $\gamma\hat{\ }0$ of x_q extends $\sigma\hat{\ }0$, because if this was so we would already have put down a x_t-ball g marked $\sigma\hat{\ }0$ by the time we put $\gamma\hat{\ }0$ down for z_q. But then this ball g cannot have been cancelled in the intervening stages since if g were cancelled, so too x_q would have been cancelled. (Note g would be cancelled since $\sigma\hat{\ }0 \not\leq_L \sigma_s$ but then as $\sigma\hat{\ }0 \preceq \gamma\hat{\ }0,\quad \gamma\hat{\ }0 \not\leq_L \sigma_s$). Hence we see $q < t$.

Since $q < t$ and $\hat{x}_t$ is a stable position, we must have that x_q is also a stable position. (Remember, in this case, $\sigma\hat{\ }0 \preceq \gamma\hat{\ }0$ and $\sigma\hat{\ }0 \preceq \beta$. If x_q's position is cancelled, so too is anything, in particular $\hat{x}_t$, marked $\sigma\hat{\ }0$.) In either case, we see that no such x_q (and hence $\hat{x}_t$) can exist, giving (3.2.38). Thus, we have Claim 3.2.38 that there are no stable positions between $c(m) = x_j$ and $\hat{x}_i$. We now show that x_j adheres to $\hat{x}_i$.

Let $x_{i_1}, \ldots, x_{i_n}$ list those x_0-balls alive at stage s with $\hat{x}_i < x_{i_1} < \cdots < x_{i_n} = x_j$. Let $s(m)$ be the least stage such that for all j with $1 \leqslant j \leqslant n$ we have

1. either x_{i_j} is cancelled at stage $s(m)$, or
2. $\forall s > s(m) \quad (\gamma_j \not\leq \sigma_s$ where γ_j is the guess of $x_{i_j})$.

Such a stage must exist by Claim 3.2.38 and the definition of stability. As the notation suggests, we claim that this $\hat{x}_i$ $(s(m))$ is correct, namely that

Claim 3.2.39. *1.* x_j *appears to be an* ω*-ball of* $\hat{x}_i$ *at each* $\sigma\hat{\ }0$*-stage* $> s(m)$.

2. $d(x_j, \hat{x}_i, s(m)) = d(x_j, \hat{x}_i, s)$ *for all* $s > s(m)$.

By construction, $2 \Rightarrow 1$ because of the way we place y-balls and the fact that $\hat{x}_i$ is stable. We argue that 2 holds in a similar way to our argument that there are no stable x-balls between x_j and $\hat{x}_i$. Suppose $d(x_j, \hat{x}_i, s(m)) < d(x_j, \hat{x}_i, s)$ for some least $s > s(m)$.

There are two ways we might insert new balls between x_j and $\hat{x}_i$. Either the new ball n is a y- or a z-ball placed between $\hat{x}_i$ and x_j because we see that x_q appears in A at s for $\hat{x}_i < x_q < x_j$ (notice that $x_q < x_j$ and $q \neq i_k$ for any $1 \leqslant k \leqslant n$ by choice of $s(m)$), or the ball is a new x-ball x_d, say, placed there because again we see x_q whose guess appears correct with $\hat{x}_i < z_q < z_j$. Again, $x_q \neq x_j$ or x_{i_k} for any $1 \leqslant k \leqslant n$. We claim that in either case, no such z_q can exist.

Using the reasoning of Claim 3.2.38, x_q, which did not exist at stage s, must have appeared at a stage where some of the x_{i_k} for $i \leqslant k \leqslant n$ looked correct, since otherwise we would have placed it beyond x_j. Also, using the same reasoning as Claim 3.2.38, it must be that q exceeds those x_{i_k} that forced it in (otherwise, x_{i_k} would be cancelled first). Thus, since x_q's guess *extends* such x_{i_k}'s guess, x_q's guess appears correct at best only when x_{i_k}'s does too. But the choice of $s(m)$ means that x_{i_k}'s guess never again looks correct. Thus, x_q's guess also never again looks correct. Hence stage s can't exist after all. This clinches 2 and hence Claim 3.2.39, and the lemma follows. ☐

Lemma 3.2.40 (Truth of outcome lemma). *Let n be any ball. Then either n is a stable x-ball, or n adheres to a stable x-ball.*

Proof. By Lemmas 3.2.35 and 3.2.37. ☐

Lemma 3.2.41. $C_F(A) \cong B$.

Proof. The desired isomorphism is induced by the injection $A \to B$ given by $\hat{x}_j \mapsto x_j$. The lemmata and the addition of y points achieve the rest. ☐

The second proof of the Fellner-Watnick Theorem 3.2.23 is complete. ☐

Corollary 3.2.42 (Watnick [502]). *For $n \geqslant 1$, a linear ordering L has an $\varnothing^{(2n)}$-computable presentation iff $\mathbb{Z}^n L$ is computably presentable*

Recall that in §3.2.3 we discussed the possibility of producing a linear order in which a set serves as its Σ^0_n-invariant. The well-known application of the Fellner-Watnick Theorem 3.2.23 and Theorem 3.2.22 is the following:

Corollary 3.2.43. *Fix any $n \in \mathbb{N}$, $n > 2$, and let S be an infinite set. There is a linear order $L(S)$ such that $L(S)$ has an X-computable presentation if and only if S is Σ^0_n.*

Proof. The case when $n = 3$ is covered by Lemma 3.2.12; let $\hat{L}(S)$ denote the order corresponding to the case when $n = 3$. If $n = 2k + 1$, where $k > 1$, then let $L(S) = \mathbb{Z}^{k-1}\hat{L}(S)$. If $n = 2k + 2$, where $k \geqslant 1$, then consider $(\eta + 2 + \eta)\mathbb{Z}^{k-1}\hat{L}(S)$. ☐

Recall also that no nice Σ^0_1-coding of a set into a linear order can possibly exist, as follows from Richter's Theorem 3.2.13. What about the case when $n = 2$? It follows from a result of Knight [304] that the case when $n = 2$ is also impossible; we omit the proof of this result.

3.2.7 Low linear orders. The Jockusch-Soare Theorem

An interesting application of the Fellner-Watnick Theorem 3.2.23 is the following result about low$_2$ discrete linear orders. Recall that a set X is low if $X' \equiv_T \varnothing'$, and it is low$_n$ if $X^{(n)} \equiv_T \varnothing^{(n)}$. Recall also that a *discrete* linear ordering is one where every element has an immediate predecessor and an immediate successor, except perhaps the first point with no predecessor and the last with no successor, if they exist. The corollary below was first established by Downey and Moses [147] for discrete low orders, and then Frolov [189] noted that the result clearly holds for low$_2$ discrete orders as well.

Corollary 3.2.44. *Every low$_2$ discrete linear order has a computable copy.*

Proof. We use the machinery accumulated for linear orders in the previous subsections without explicit reference. Suppose X is low$_2$, and suppose A is X-computable. Relativising to X, we conclude that $D = C_F(A)$ has to be an X''-computable copy. Since $X'' \equiv_T \varnothing''$, the copy is also Δ^0_3 (hence, Π^0_2). It could be that A starts with ω or ends with ω^*, or both. In each of the four possible cases, we could remove the greatest or the least point of D (or both) if necessary and produce a Δ^0_3 copy of the resulting $\tilde{D}$. Using the Fellner-Watnick Theorem 3.2.23, we can produce a computable presentation of $\mathbb{Z}\tilde{D}$. To get a computable presentation of A, we may have to adjoin ω to the left of $\mathbb{Z}\tilde{D}$ and ω^* to the right of it. □

For many years, it was not known whether every low linear order had to be isomorphic to a computable one. The question was resolved by Jockusch and Soare [273], who proved the following:

Theorem 3.2.45 (Jockusch and Soare [273]). *There is a low linear order not isomorphic to any computable linear order.*

The original proof of the theorem found in [273] involved infinite injury. We provide a fairly elementary, *injury-free* proof of this theorem, closely following the approach in [191] (see Theorem 1.21) and in [188]. The injury will be completely eliminated using Theorem 3.2.14.

Proof. We identify computable linear orders with c.e. subsets of the rationals; we slightly abuse notation and let $(W_e)_{e\in\mathbb{N}}$ be the uniform listing of all such subsets. We construct a Δ^0_2 linear order L with a Δ^0_2 adjacency relation and meet:

$$R_e : L \not\cong W_e.$$

Theorem 3.2.14 will guarantee that L has a low presentation. The linear order L will have the form:

$$\sum_{e\in\mathbb{N}} \eta + (2\langle e,0\rangle + 1) + \eta + L_e + \eta + (2\langle e,1\rangle + 1) + \eta,$$

where $\eta \cong (\mathbb{Q}, <)$, and each L_e will have no blocks of odd length. Each L_e will have the form

$$\sum_{i\in I_e} \eta + B_{e,i} + \eta,$$

where the sizes of the finite blocks $B_{e,i}$ will be even and increasing monotonically in i, and I_e will be either ω or finite, depending on the outcome of the R_e-strategy.

Note that, in W_e, it is $\Pi^0_1(\varnothing')$ to tell whether a given collection of points constitutes a block of a given size. Since we will be working relative to $\varnothing'$, if $W_e \cong L$, then there will be a stage at which the blocks $(2\langle e,0\rangle + 1)$ and $(2\langle e,1\rangle + 1)$ are finally located. If $L_e \not\cong W_e$, then of course we may have infinitely many unsuccessful $\Pi^0_1(\varnothing')$-attempts to locate these blocks. All activities of the strategy will be restricted to the sub-order of W_e which is between the current best candidates for $(2\langle e,0\rangle + 1)$ and $(2\langle e,1\rangle + 1)$ (if there are any).

The *idea* is that, in W_e, we can use $\varnothing'$ to ask if there are at least m points between a pair of points x, y, while in L we do not have to declare this immediately. We shall use this as leverage to diagonalise against W_e using L_e. There will be no interactions between different strategies.

Strategy for R_e. Initially, when it first becomes active, the strategy proceeds as follows:

(1) Create two blocks, $B_{e,0}$ and $B_{e,1}$, in L_e and keep $|B_{e,0}| = 2$ and $|B_{e,1}| = 4$.

(2) Wait for W_e to reveal two blocks of size (at least) 2 and 4. Meanwhile, make progress in building $L_e \cong \eta + 2 + \eta + 4 + \eta$ by placing points between the blocks and declaring them to be non-adjacent.

(3) Suppose W_e responds at stage s by showing blocks of similar size, say C and D. Let m be the number of points between $B_{e,0}$ and $B_{e,1}$ in $L_e[s]$. Using $\varnothing'$, check whether there are $2m + 2$ points between C and D.

(3.a) If there are $\leqslant 2m+2$ points, then proceed to build $L_e \cong \eta + 2 + \eta + 4 + \eta$.

(3.b) Otherwise, suppose there are more than $2m+2$ points between the blocks. Then adjoin $B_{e,1}$ to $B_{e,0}$ to create one n-block, where

$$n = 2m + 2 + |B_0[s]| + |B_1[s]|$$

is of course even. To achieve this, place one extra point into every non-adjacency empty interval between B_0 and B_1, and declare all successor pairs to be adjacent. (Place two points in one such interval to make sure n is even.) Proceed to building $L_e \cong \eta + n + \eta$.

In (1) and (2) we diagonalise trivially. In (3.a), L_e has no blocks of size > 4, while the respective part of W_e does. In (3.b), C and D cannot be part of one block of size n, while in L_e we will end up with one block of size n.

Thus, W_e must respond by changing its guess about the location of the $(2\langle e,0\rangle + 1)$- and $(2\langle e,1\rangle + 1)$-blocks in W_e. In this case, the R_e-strategy is restarted.

If i is the number of times the strategy has been restarted, then the strategy initiates building two blocks, $B_{e,2i}$ and $B_{e,2i+1}$ in L_e, located to the right of all other blocks $B_{e,j}$, $j < 2i$, which were created by the previous diagonalisation attempts.

Initially, the size $|B_{e,2i}|$ of $B_{e,2i}$ is declared to be a very large even number, and also $|B_{e,2i+1}| = 2|B_{e,2i}|$. Thus, the strategy initially attempts to build

$$L_e \cong \left(\sum_{j<2i} \eta + B_{e,j} + \eta \right) + B_{e,2i} + \eta + B_{e,2i+1} + \eta,$$

where j ranges over all blocks produced by the previous diagonalisation attempts of R_e.

Then the strategy proceeds to its new diagonalisation attempt (1)-(3), but with $B_{e,2i}$ and $B_{e,2i+1}$ playing the roles of B_0 and B_1, respectively. Since the sizes of $B_{e,2i}$ and $B_{e,2i+1}$ are much larger than all other blocks currently in L_e, the analysis of (1)-(3) remains the same.

So either R_e is met at an i-th iteration of the strategy, or we proceed to build infinitely many blocks inside L_e. Of course, this infinitary outcome is possible only if W_e does not have an interval isolated by blocks $(2\langle e,0\rangle + 1)$ and $(2\langle e,1\rangle + 1)$, and in this case R_e is also met. □

3.2.8 Further related results*

Beyond ω

Corollary 3.2.42 has a transfinite extension due to Ash, Jockusch and Knight [19] and Ash [18]; we omit the statement. The proof of this corollary is rather tedious, and goes beyond the material covered in the book. There are other books written which carefully present these methods. These methods introduced by Harrington [232, 233] are historically called "worker arguments" and "true stage" arguments. For detailed presentations, we refer the reader to Ash and Knight [20], and Montalbán [401, 402]. For worker arguments in the context of linear orders specifically, see [19].

Analogues of the Fellner-Watnick Theorem

With some modifications, an analogue of the Fellner-Watnick Theorem works for ω or ω^* in place of $\mathbb{Z}$.

Theorem 3.2.46 (Ash [18]). *Let X be a $\varnothing''$-computable linear ordering. Then the following linear orders are computably presented.*

(i) $\omega \cdot X$, *and*

(ii) $\omega^* \cdot X$.

While the transformations $X \mapsto \omega \cdot X$ and $X \mapsto \omega^* \cdot X$ in the theorem above are uniform for some order-types X, it is not uniform in general; see Exercise 3.2.57.

Relative to any non-computable oracle

Structures (that are not linear orders) can have the property that they are X-computably presentable iff $X >_T \varnothing$; this surprising theorem was established by Slaman [471] and, independently, Wehner [503]; see Exercises 3.2.63 and 3.2.64. It may seem surprising, but the following is a longstanding open question.

Question 3.2.47 (Downey). *Is there a linear ordering L which has presentations of every nonzero degree, but no computable presentation?*

In his PhD Thesis [393], Russel Miller showed that there is a linear ordering which has a presentation of every nonzero Δ_2^0-degree, yet has no computable copy. It is also known (though unpublished) that this can be an ordering with presentations in every hyperimmune degree, where $\mathbf{a}$ is hyperimmune if it computes a function not dominated by any computable function (see also Exercise 3.1.28); we omit the details. Also, for any $n \geqslant 2$ there exists a linear order that has an X-computable copy iff X is *not* low_n; see [186].

Further reading

For a more complete introduction to the theory of linear orders, we recommend reading the somewhat dated technical survey [119]. A more recent technical survey is [191], though it covers a more limited range of topics. For a detailed discussion of various results related to Question 3.2.47 but not restricted to linear orders, we cite the relatively recent (but rapidly ageing) survey [167], which contains no proofs but includes over 300 bibliographic references to the relevant literature.

Historical remarks

En passant, effective linear orderings have been studied since the 1940's and 50's, both in computable analysis and in the context of computable ordinals and their use in hierarchies in computability theory. We have also seen that Kleene [300] used computable well-ordered sets to give meaning to Turing degrees above $\mathbf{0}^{(\omega)}$. Specker [480], gave the first explicit example of a computable linearly ordered ascending sequence of rationals whose limit is a (left-c.e.) noncomputable real.

As far as we know, the first studies of the general class of computable linear orderings (and Boolean algebras) for their own sake began in the late 1960's with Feiner's thesis [161]. This thesis was the first to systematically focus on the subtle distinction between c.e. presentations and computable presentations in structures that are neither groups nor rings, and are *not* finitely generated. Effective linear orderings and Boolean algebras witnessed a lot of study in the 1970's and early 1980's particularly by Remmel, Lerman, LaRoche, Goncharov, and Dzgoev. We refer the reader to [119] for more on such developments. One significant feature of these results is the necessity for tools that enable much more indirect coding than had previously been seen in groups and fields. Most natural codings in linear orderings seem to use two or three quantifiers, and methods involve infinite injury arguments. Boolean algebras are even worse necessitating difficult techniques such as the Feiner's hierarchy as we will see in the next chapter.

Exercises

Exercise° 3.2.48. Prove Theorem 3.2.22.

Exercise° 3.2.49 (Folklore)**.** Prove that there is a c.e. presentable linear ordering not isomorphic to a low_2-presentable one.

Exercise° 3.2.50. Use Theorem 3.2.22 to show that there is a computable linear order L which is not isomorphic to any computable linear order with computable adjacency relation.

Exercise° 3.2.51 (Downey and Moses [147])**.** Show that if L is a semi-low (see Exercise 3.1.23) discrete linear ordering then it has a computable copy.

Exercise° 3.2.52 (Rosenstein [456])**.** Show that if $X \in \Sigma^0_2$, then there is a computable linear ordering of order type

$$\eta + n_0 + \eta + n_1 + \eta + \dots$$

where $n_0 < n_1 < n_2 < \dots$ lists X in order. (Recall that this is called a *strong* η-representation of X.)

Exercise 3.2.53 (Fellner [165])**.** Show that if $Y \in \Pi^0_2$, then Y has a strong η-representation (see the previous exercise).

Exercise* 3.2.54 (Lerman [336])**.** Show that if $S \in \Sigma^0_3$, then there is a computable linear ordering L of order type

$$\mathbb{Z} + n_0 + \mathbb{Z} + n_1 + \mathbb{Z} + \dots$$

where $n_0 < n_1 < \dots$ lists S in order of magnitude. This is called a *strong* $\mathbb{Z}$-representation of S.

Exercise 3.2.55 (Moses [408])**.** Show that for every n, there exists an n-decidable linear order with no $(n+1)$-decidable presentation. (Recall that a computable structure is n-decidable ($n \geqslant 1$) if we can uniformly decide first-order statements with n-1 alternations of quantifiers in the structure. For example, in a 2-decidable structure we can decide $\forall\exists$-statements.)

Exercise* 3.2.56 (Chisholm and Moses [91])**.** Show that there is a linear order that is n-decidable for all $n \in \mathbb{N}$, but has no decidable presentation.

Exercise° 3.2.57. Show that the procedure in Theorem 3.2.46 cannot possibly be uniform[5]. (Hint: Fix a uniform sequence of $\varnothing''$-computable linear orderings $(L_i)_{i\in\mathbb{N}}$ of the isomorphism types ω and $\mathbb{Z}$, so that $L_i \cong \mathbb{Z}$ iff the Π^0_4-predicate holds. Conclude that ωL_i witnesses the Π^0_4-completeness of "L_i has the least element", which contradicts the natural complexity of this property.)

See also Exercises 9.1.30 and 9.1.32.

[5]We are thankful to Maxim Zubkov for pointing this non-uniformity to us.

Exercises about degree spectra

The degree of the isomorphism type of a structure of A to be the least Turing degree $\mathbf{a}$, such that A has an $\mathbf{a}$-computable copy, if such least degree exists. The exercises below are based on the results of Richter that can be found in [450, 451].

Exercise° 3.2.58. Show that if $\mathbf{a}$ is the degree of an order-type (in the sense defined above), then $\mathbf{a} = \mathbf{0}$.

Exercise° 3.2.59. Suppose (i) and (ii) below hold for a degree $\mathbf{a}$ and a theory T over a finite language L.

(i) There is an infinite computable sequence of finite structures $\{A_i : i \in \mathbb{N}\}$ such that A_i is not embeddable into A_j for $i \neq j$.

(ii) For each $S \leqslant \omega$, there is a structure A_S such that:

(iia) A_S is a countable structure of T.

(iib) $A_S \leqslant_T S$.

(iic) A_i is embeddable into A_S iff $i \in S$.

Then there is a structure of L whose isomorphism type has degree $\mathbf{a}$.

Exercise° 3.2.60. Let $\mathbf{a}$ be any degree.

(i) There is an abelian group whose isomorphism type has degree $\mathbf{a}$.

(ii) There is a lattice whose isomorphism type has degree $\mathbf{a}$.

(iii) There is a graph whose isomorphism type has degree $\mathbf{a}$.

Exercise 3.2.61. We define the *computable embedding condition* as follows. Given a structure A, a finite structure C and an embedding $f : C \to A$, define the class $A_{C,f}$ to be

$$\{D : D \text{ is a finite structure extending } C \text{ embeddable into } A \text{ via a map extending } f\}.$$

Then A satisfies the computable extension property iff for all structures C isomorphic to a finite substructure of A, and for all functions f embedding C into A, the class $A_{C,f}$ is infinite and computable. Show that for any $\mathbf{a}$-computable structure A satisfying the computable extension property, there is Turing degree $\mathbf{b}$ and a $\mathbf{b}$-computable presentation of A so that $\mathbf{a}$ and $\mathbf{b}$ form a minimal pair. (Hint: Generalise the method used in the proof of Theorem 3.2.13.The key fact is that the $A_{C,f}$ is a computable collection, and this accords with the notion of acceptable string in the proof of Theorem 3.2.13.)

Exercise 3.2.62. Derive Theorem 3.2.13 (equivalently, Exercise 3.2.58) as a consequence of the previous exercise.

Exercise* 3.2.63 (Wehner [503]). A computable enumeration (a numbering) of a family of sets $\mathcal{S}$ is a c.e. set W such that $\mathcal{S} = \{W^{[n]} : n \in \mathbb{N}\}$, where $W^{[n]} = \{x : \langle n, x \rangle \in W\}$. Note we allow repetitions in the uniform enumeration of $\mathcal{S}$. Show that there exists a family of sets that has an X-computable enumeration iff $X >_T \varnothing$. (Hint: Consider the family $\{F \oplus \{n\} : F \text{ finite} \subseteq \omega \text{ and } F \neq W_n\}$. Given $X >_T 0$, use initial segments of X to extend finite sets in the family while avoiding W_n.)

Exercise° 3.2.64 (Slaman [471], Wehner [503]). Deduce from the previous exercise that there is a structure that has an X-computable presentation iff $X >_T \varnothing$. (We remark that Slaman gave a direct proof of this result that did not use Wehner's family.)

3.3 What's next?

In the next chapter, we will use linear orders to study Boolean algebras and their Stone spaces from the computability-theoretic standpoint. Results and techniques developed for linear orders will be rather useful, though certainly not unavoidable. Nonetheless, for the sake of exposition, we shall often choose to give a proof that uses linear orders over a proof that uses other methods.

Chapter 4

Boolean algebras and computable compactness

In this chapter we establish (1) and (2) of Theorem A and (1) and (2) of Theorem B; we state these results as separate theorems below:

Theorem (Feiner [162]). There is a c.e. presented Boolean algebra not isomorphic to any computable one.

Theorem (Downey and Jockusch [129]). Every low Boolean algebra has a computable presentation.

Theorem (Bazhenov, Harrison-Trainor, Melnikov [35]). There exists a right-c.e. Stone space not homeomorphic to any computable Polish space.

Theorem (Harrison-Trainor, Ng, Melnikov [245]). Every computable Polish Stone space is homeomorphic to a computably compact one.

In order to prove these results, we give a brief introduction to the theory of computable Boolean algebras and then to the theory of computably compact spaces. The chapter is subdivided into two sections:

1. Section 4.1 contains a brief introduction to the theory of effectively presented Boolean algebras. It includes the proofs of the theorems of Feiner, and Downey and Jockusch.

2. Section 4.2 lays the foundations of the theory of computably compact spaces, which will be useful throughout the rest of the book. It also provides detailed proofs of the remaining two results stated above.

We relate these subjects and connect them to the materials from the previous chapter using several *effective versions of Stone duality* between Boolean algebras, totally disconnected compact Polish spaces, and the interval algebras of linear orders. Although we do not restrict ourselves to the techniques and facts necessary to prove the results stated above, our exposition of these topics is very far from being complete.

R. G. Downey, A. Melnikov, *Computable Structure Theory*, Theory and Applications of Computability, https://doi.org/10.1007/978-3-031-92433-0_4

4.1 Computable Boolean algebras

Boolean algebras were introduced by George Boole in his book, *The Mathematical Analysis of Logic* (1847). Boole sought to develop a calculus of logical truths, building on Leibniz's earlier ideas. These algebraic structures are particularly important in logic and much less so outside of it. Consequently, the set-theoretic and model-theoretic aspects of Boolean algebras are very well-studied. The algorithmic aspects of computable Boolean algebras are also well-understood, with the field having an extensive theory, numerous results, and powerful techniques. As a consequence, we will need to be selective. In our book, Boolean algebras will be predominantly used as a tool to study the algorithmic aspects of totally disconnected compact spaces. Our introduction to the theory will therefore be rather brief and mostly restricted to the results that we need for proving the stated above theorems of Feiner and Downey-Jockusch. For a much more detailed introduction, we refer to Goncharov's book [207] and Remmel's survey [447]. Many results concerning computable Boolean algebras proven before 2000 can be found in [207, 447]. We will provide the reader with further references in due course.

4.1.1 Countable Boolean algebras

The definition of a Boolean algebra

Fix a set S, the collection of all its subsets $\mathcal{P}(S)$, and some $U \subseteq \mathcal{P}(S)$ closed under union, intersection, and complement relative to S, i.e., $\bar{X} = S \backslash X$. We can view U as an algebraic structure in which the individual elements are the subsets and the algebraic operations are $\cup$, $\cap$, and $\overline{X} = \neg X = S \backslash X$. This algebraic structure can also be viewed as a partial order under the subset relation $\subseteq$, in which the greatest element is S and the least is $\varnothing$; this order is a distributive lattice with "relative complements"; we omit the formal definitions. Note that $A \subseteq B$ iff $A \cap B = A$ iff $A \cup B = B$, so we could include the order in our signature if necessary.

Now suppose we are given an algebraic structure B in the language $\{\vee, \wedge, ^-, 0, 1\}$, where $\vee$, $\wedge$, $^-$ are simply algebraic operations that do not necessarily carry any set-theoretic interpretation. Can we put down a comprehensive list of axioms that would *guarantee* that the structure can be realised as $U \subseteq \mathcal{P}(S)$ for some S? In other words, can we find a set S and an interpretation that matches elements of B with elements of $\mathcal{P}(S)$ so that $\vee$, $\wedge$, $^-$ become $\cup$, $\cap$, and $\neg$ under this interpretation? It is well known that the answer is affirmative, and the theorem asserting this is known as Stone duality. To make sense of Stone duality formally, we first define the algebraic structures associated with it that are called Boolean algebras.

Definition 4.1.1. A *Boolean algebra* is an algebraic structure equipped with two binary operations $\wedge$ and $\vee$, a unary operation $^-$ (also sometimes denoted $\neg$ or $'$), and two distinguished elements 0 and 1 that satisfy the following axioms that hold for all elements a, b, c from the domain of the structure:

1. $a \vee (b \vee c) = (a \vee b) \vee c$
2. $a \wedge (b \wedge c) = (a \wedge b) \wedge c$
3. $a \vee b = b \vee a$
4. $a \wedge b = b \wedge a$
5. $a \vee (a \wedge b) = a$
6. $a \wedge (a \vee b) = a$
7. $a \vee (b \wedge c) = (a \vee b) \wedge (a \vee c)$
8. $a \wedge (b \vee c) = (a \wedge b) \vee (a \wedge c)$
9. $a \vee 0 = a$
10. $a \wedge 1 = a$
11. $a \vee \overline{a} = 1$
12. $a \wedge \overline{a} = 0$

Familiar examples include the simplest Boolean algebra $\mathbf{2} = \{0, 1\}$ consisting of two elements 0 and 1, and the Boolean algebra consisting of finite and co-finite subsets of the non-negative integers, under the set-theoretic operations.

We could use the axioms above to establish various expected properties such as $\bar{0} = 1$ and $\bar{1} = 0$, and that 1 is the unique element satisfying axiom 10, etc. We certainly do not claim that the axioms above are optimal in any sense. In fact, some of the axioms can be removed from the list above since they can be derived from the rest of the axioms. Unlike groups or fields, it seems that there is no fixed collection of axioms that would be accepted as 'standard' in the literature. For instance, in his book [207], Goncharov views Boolean algebras as structures in the language $\wedge, \vee, \bar{}$ (i.e., without 0 and 1) that satisfy the following axioms:

i. $a \vee b = b \vee a$
ii. $a \vee (b \vee c) = (a \vee b) \vee c$
iii. $a \vee (b \wedge c) = (a \vee b) \wedge (a \vee c)$
iv. $\overline{a \vee b} = \overline{a} \wedge \overline{b}$
v. $a \vee a = a$
vi. $(a \vee \bar{a}) \vee b = b$
vii. $\bar{\bar{a}} = a$

Then Goncharov derives the existence of the uniquely defined $0 = a \wedge \bar{a}$ and $1 = a \vee \bar{a}$ that do not depend on the choice of a. (We suggest that the reader verify this claim and also derive some of the axioms *i. – vii.* from 1. – 12. and vice versa.) Goncharov's choice of axioms is a bit more optimal, in the sense that his definition has fewer axioms. It can be shown using automated reasoning (see [359]) that Boolean algebras can be defined using *just one axiom*:

$$(((x \vee y)' \vee z)' \vee (x \vee (z' \vee (z \vee u)')')')' = z,$$

where x' stands for $\overline{x}$ (writing it down using $\bar{}$ is a challenge). Even though this is definitely a peculiar result, it is not practical in the sense that this axiom is neither natural nor convenient.

There are many other ways to define Boolean algebras using alternative operations. For example, we can view a Boolean algebra as a complemented distributive lattice with greatest and least elements. Alternatively, it can be viewed as a Boolean ring. The list goes on. In each case, one can reconstruct the operations $\wedge$ and $\vee$ from the operations used in these definitions in a nice, finitistic way. We will not provide any further analysis of the axiomatic approach to Boolean algebras, as it seems to be irrelevant to our story; for more details, see [207].

Ideals, filters, and Stone duality

Before we discuss Stone duality in more detail, we should mention two other notions central to the theory of Boolean algebras. An *ideal* $\mathcal{I}$ in a Boolean algebra B is a non-empty subset that is closed under $\vee$ and also has the property that $b \wedge i \in \mathcal{I}$ for any $i \in \mathcal{I}$ and $b \in B$. In particular, it follows that $0 \in \mathcal{I}$: since $\mathcal{I} \neq \varnothing$, we can take $i \in \mathcal{I}$ and conclude that $\bar{i} \wedge i = 0 \in \mathcal{I}$. Following the general pattern in commutative algebra, it is usually additionally assumed that $1 \notin \mathcal{I}$. The reason is that ideals should correspond to kernels of homomorphisms between Boolean algebras, and in the literature, it is almost uniformly assumed that $1 \neq 0$. For instance, if $\mathcal{I}$ is an ideal in B, we can define the factor-algebra $B/\mathcal{I}$ following the usual procedure of forming congruence classes modulo $\mathcal{I}$. (We can define $a = b \mod \mathcal{I}$ if $x \triangle y = (x \wedge \bar{y}) \vee (\bar{x} \wedge y) \in \mathcal{I}$, where $\triangle$ denotes the operation of "symmetric difference".) If we allowed $1 \in \mathcal{I}$, in which case, of course, $\mathcal{I} = B$, then we would say that $\mathcal{I}$ is *proper* if $1 \notin \mathcal{I}$. We assume $1 \notin \mathcal{I}$ for any ideal $\mathcal{I}$. Under this assumption, an ideal is *maximal* if, well, it is maximal (among all ideals under inclusion). An ideal is *prime* if for any element a, either a or $\bar{a}$ is in the ideal. The following lemma is not hard to show. We include its proof merely as an illustration of a typical use of the axioms. As we will explain shortly after the proof of the lemma, we will be able to avoid such formal arguments throughout the rest of the chapter.

Lemma 4.1.2. *In a Boolean algebra, an ideal is maximal iff it is prime.*

Proof. Suppose $\mathcal{I}$ is prime. Any other ideal $\mathcal{J} \supset \mathcal{I}$ must have $a \notin \mathcal{I}$. By the assumption on $\mathcal{I}$, $\bar{a} \in \mathcal{I} \subseteq \mathcal{J}$, and thus

$$1 = a \vee \bar{a} \in \mathcal{J},$$

showing that $\mathcal{I}$ must be maximal.

Conversely, fix a maximal $\mathcal{I}$. Suppose a is such that $a, \bar{a} \notin \mathcal{I}$. Extend the sets $\mathcal{I} \cup \{a\}$ and $\mathcal{I} \cup \{\bar{a}\}$, closing them under $\vee$ and under $\wedge$ with any element of the Boolean algebra. Denote the resulting ideals $\mathcal{J}_a$ and $\mathcal{J}_{\bar{a}}$. It is clear that both $\mathcal{J}_a$ and $\mathcal{J}_{\bar{a}}$ properly extend $\mathcal{I}$ and thus must contain 1. The ideal $\mathcal{J}_a$ consists of elements of the form $(x \vee a) \wedge b$, where $x \in \mathcal{I}$ and $b \in B$. The description of $\mathcal{J}_{\bar{a}}$ is similar, mutatis mutandis. In particular, for some $x, y \in \mathcal{I}$ and $b, c \in B$, we have

$$1 = (x \vee a) \wedge b \text{ and } 1 = (y \vee \bar{a}) \wedge c.$$

It is not difficult to derive from the axioms that whenever $1 = z \wedge w$, it must be that $z = w = 1$. So we conclude that

$$1 = x \vee a \text{ and } 1 = y \vee \bar{a}.$$

But then we have

$$x \vee y = (x \vee y) \vee 0 = (x \vee y) \vee (a \wedge \bar{a}),$$

and applying the distributivity axioms, we obtain

$$x \vee y = (x \vee y \vee a) \wedge (x \vee y \vee \bar{a}) = (1 \vee y) \wedge (1 \vee x) = 1 \wedge 1 = 1 \in \mathcal{I},$$

and this contradicts $\mathcal{I} \neq B$. □

It is also not too difficult to show that if $\mathcal{I}$ is maximal in B, then the factor $B/\mathcal{I}$ is just the two-element Boolean algebra $\{0, 1\}$. Additionally, given any non-zero element $a \in B$, there exists a maximal ideal $\mathcal{I}$ such that $a \notin \mathcal{I}$. The construction of such an $\mathcal{I}$ is fairly standard in commutative algebra. It is not hard to show that when we have $a \neq b$, for non-zero a and b, there exists a maximal $\mathcal{I}$ such that either $a \in \mathcal{I}$ and $b \notin \mathcal{I}$, or $b \in \mathcal{I}$ and $a \notin \mathcal{I}$. In other words, *maximal ideals separate points in B*.

If we swap $\wedge$ and $\vee$ in the definition of an ideal, we obtain the dual notion of a *filter* (closed under $\wedge$, stable under $\vee$ with any element of B). We also usually assume 0 is not in our filter. (Indeed, if we swap $\wedge$ and $\vee$ and 1 with 0 in the Boolean algebra, in the resulting Boolean algebra, ideals will become filters and filters will become ideals.) A maximal filter is called an *ultrafilter*. Similarly to the situation with maximal ideals, a filter $\mathcal{F}$ is an ultrafilter iff either $a \in \mathcal{F}$ or $\bar{a} \in \mathcal{F}$, for any $a \in B$. Indeed, $\mathcal{F}$ is an ultrafilter iff $\mathcal{I} = B \backslash \mathcal{F}$ is a maximal ideal, and vice versa. Recall that if $\mathcal{I}$ is maximal, then $\mathcal{I}$ is the kernel of a homomorphism $B \to \{0, 1\}$. Under this map, the complement of $\mathcal{I}$ is the ultrafilter corresponding to the pre-image of 1 under this homomorphism. Of course, ultrafilters also separate points in B.

We now return to Stone duality. One version of Stone duality is stated below.

Theorem 4.1.3. *Every Boolean algebra is isomorphic to some subalgebra $U \subseteq \mathcal{P}(S)$, where S is the set of its ultrafilters (alternatively, the set of its maximal ideals).*

We will give a complete proof of a different version of Stone duality later in Theorem 4.1.6. Since we will never actually use this particular version of the duality, we omit the formal details that are easy to find in the literature; e.g., [207].

Theorem 4.1.3 proof idea. Using slightly tedious but not difficult arguments (cf. Lemma 4.1.2), we find that our Boolean algebra B is isomorphic to some subset of $\mathcal{P}(S)$ in a certain canonical way:

$$a \in B \text{ is associated with } \{\mathcal{F} : a \in \mathcal{F}\},$$

where $\mathcal{F} \in S$ ranges over ultrafilters in B. Since ultrafilters separate points, the induced map is an isomorphic embedding from B into $\mathcal{P}(S)$; let U be the range of this embedding. □

Assuming Stone duality, we suggest the following:

> Do not memorise the axioms of Boolean algebras. Instead, think of the elements of the algebra as being (some) subsets of a fixed set S (identified with 1) and of $\wedge$, $\vee$, and $\bar{x}$ as being $\cap$, $\cup$, and $S \backslash x$, respectively.

This informal approach is obviously not always satisfactory, but it can be used to guide the reader's intuition.

Atoms

Define $a \leqslant b$ if $a \wedge b = a$; this partial order corresponds to the subset relation under the duality. We say that an element $a \neq 0$ *splits* if there are two non-zero elements x, y such that

$$a = x \vee y \text{ and } x \wedge y = 0.$$

This is, in fact, equivalent to saying that there is a non-zero $b < a$. In this case, we have that a splits into b and $\bar{b} \wedge a$:

$$a = b \vee (\bar{b} \wedge a).$$

The element $\bar{b} \wedge a$ is sometimes called *the complement of b relative to a*.

Note that the set

$$\hat{a} = \{b : b \leqslant a\}$$

of all elements below a can be viewed as a Boolean algebra in which a plays the role of 1. When $a \neq 1$, it is actually *not* a subalgebra in the usual model-theoretic sense (if we put 1 into the language). It is a *principal ideal* in the sense that it is the smallest ideal that contains $a \neq 1$. A finitely generated ideal can be defined similarly: it is the smallest ideal $\mathcal{I}$ that contains some finite collection of elements $a_0, \ldots, a_k \in \mathcal{I}$, which are then called its generators. Of course, the generators are not unique for such an $\mathcal{I}$. Indeed, it is easy to see that any finitely generated ideal is principal; consider $\hat{a}$, where $a = a_0 \vee a_1 \vee \ldots \vee a_k$ is the supremum/union of the finitely many generators $a_0, \ldots, a_k$.

More generally, we can take any collection of elements of a Boolean algebra and define the ideal generated by these elements. This is done by closing the set under finite unions and under intersections with arbitrary elements of the algebra. One such ideal that we shall refer to later is the ideal generated by all *atoms* in the algebra.

Definition 4.1.4. A non-zero element a of a Boolean algebra is called an *atom* if it does not split, i.e., there is no $b \neq 0$ such that $b < a$.

The most elementary case of Stone duality is when the Boolean algebra is finite. Indeed, if we could keep splitting an element, then the algebra would be infinite. It follows that every element of the algebra has to be the union of finitely many atoms. In this case, we can take the set of atoms to be S, and the algebra to be $\mathcal{P}(S)$. This, in particular, shows that two Boolean algebras having the same finite cardinality must be isomorphic, and that it has to be of the form $\mathcal{P}(\{1, \ldots, m\})$, and thus has to have size 2^m. This gives a complete classification of finite Boolean algebras.

Unfortunately, countably infinite Boolean algebras are much more complicated and are essentially *unclassifiable up to isomorphism*; we will discuss this in detail in Chapter 7. Researchers in this area have developed various tools to study broad subclasses of computable Boolean algebras. For instance, we will soon explore another version of Stone duality that relates Boolean algebras to linear orders. We will then discuss a version of Stone duality in terms of trees, and finally, in terms of abstract Stone spaces. This last version of duality will be used to derive corollaries in computable topology.

Interval representation

Fix a linear order L. Since all our orders will be countable, we can view L as a subset of the rationals. Consider the new linear order $L^* = L \cup \{\ell, \infty\}$, where ∞ is a new element larger than any element of L, and ℓ is also a new element which is smaller than any element of L. Consider the set $\text{Intalg}(L)$ consisting of finite unions of *left-closed right-open intervals* of the ordering L^*:

$$\text{Intalg}(L) = \{[a_0, b_0) \cup [a_1, b_1) \cup \ldots \cup [a_k, b_k) : a_i, b_i \in L \cup \{\ell, \infty\},\ k \in \mathbb{N}\},$$

where

$$[a, b) = \{x : a \leqslant x < b\}$$

for each fixed $a, b \in L \cup \{\ell, \infty\}$. We can additionally assume that $a_0 < b_0 < \ldots < a_k < b_k$ when $k > 0$. When $k = 0$, we allow the possibility $a_0 = b_0$ to ensure that we always have

$$[a_0, a_0) = [\ell, \ell) = [\infty, \infty) = \varnothing \in \text{Intalg}(L)$$

even in the extreme pathological case when $L \subseteq \mathbb{Q}$ is empty. Note we also always have $[\ell, \infty) = \{\ell\} \cup L \in \text{Intalg}(L)$.

Lemma 4.1.5 (Folklore). *Under the set-theoretic operations, Intalg(L) is a Boolean algebra. Furthermore, if L is computable, then so is the Boolean algebra Intalg(L).*

Proof. It is clear that the process described above is computable, so we only need to check that $\text{Intalg}(L)$ is indeed a Boolean algebra. Observe that $\text{Intalg}(L) \subseteq \mathcal{P}(\{\ell\} \cup L)$ and is closed under intersection, union, and complementation. Since $\{\ell\} \cup L$ has a least element, namely ℓ, $\text{Intalg}(L)$ also has the greatest element under inclusion, namely

$$[\ell, \infty) = \{\ell\} \cup L.$$

Thus, we have that $\text{Intalg}(L) \subseteq \mathcal{P}(\{\ell\} \cup L)$ contains both the greatest and the least element of the Boolean algebra $\mathcal{P}(L \cup \{\ell\})$. Since Boolean algebras are defined by universal axioms, the lemma follows. □

The following result is also folklore.

Theorem 4.1.6 (The Interval Representation Theorem). *Every Boolean algebra is isomorphic to an interval algebra Intalg(L) of a linear ordering L.*

Proof. Let $B = \cup_s B_s$ with $B_0 = \{0, 1\}$, and $B_{s+1} - B_s = \{b_s\}$. We will define at each stage a subalgebra $\widehat{B_s}$ containing B_s.

Define L_0 as the ordering with two points labelled 0 and 1; in the notation above, these elements will play the roles of ℓ and ∞. Thus we have the induced mapping g_0 with $0 \mapsto \varnothing$ and $1 \mapsto [0, 1)$. At stage $s+1$, we will have a set $\text{Atoms}_s = \{a_{s_1}, \ldots, a_{s_n}\}$ listing the atoms of the subalgebra of $\widehat{B_s}$, together with the linear ordering $L_s = 0, x_{s_1}, \ldots, x_{s_n} = 0$, so that $g_s(a_{s_j}) = [x_{s_{j-1}}, x_{s_j})$ induces an isomorphism from the subalgebras $\widehat{B_s}$ to $\text{Intalg}(L_s)$.

At stage $s+1$, if b_s is in $\widehat{B_s}$, we need do nothing. Otherwise, for each $a = a_{s_i}$ such that b_s *splits* a (i.e., both $x \wedge b_s$ and $x \wedge \overline{b_s}$ are non-trivial), add a new point y to L_{s+1} between $x_{s_{j-1}}$ and x_{s_j} to split the interval

$$[x_{s_{j-1}}, x_{s_j})$$

into $[x_{s_{j-1}}, y) \cup [y, x_{s_j})$. Map $a_{s_j} \wedge b_s$ to one of them, say, $[x_{s_{j-1}}, y)$, and map $a_{s_j} \wedge \overline{b_s}$ to $[y, x_{s_j})$. Note that this generates two new atoms for $\widehat{B_{s+1}}$.

Let $c_1, \ldots, c_m$ denote the atoms of $\widehat{B_{s+1}}$ below b_s. Clearly, we have ensured that the induced map

$$g(b_s) = g(c_1) \cup \cdots \cup g(c_m)$$

works. The result follows. □

The construction in the proof of Theorem 4.1.6 was restricted to $[0, 1] \cap \mathbb{Q}$. Thus, we will occasionally denote the extra elements ℓ and ∞ used to define $\text{Intalg}(L)$ by 0 and 1, respectively. It should be clear from the context when 0 corresponds to an element of the linear order and when it refers to the least element of a Boolean algebra, and similarly for 1. We remark that non-isomorphic linear orders can have isomorphic interval Boolean algebras; a sufficient condition will be presented in Theorem 4.1.12.

Sums of interval algebras

If we have a sequence of Boolean algebras $A_n = \text{Intalg}(L_n)$, $n \in \mathbb{N}$, then we write

$$\sum_{n\in\mathbb{N}} A_n$$

to denote

$$\text{Intalg}(L_0 + 1 + L_1 + 1 + L_2 + \ldots).$$

It is, of course, straightforward to define $\sum_{i\in I} A_i$, where I is some other set of indices. We remark that we can define $\sum_{i\in I} A_i$ to be $\sum_{i\in I} \text{Intalg}(L_i)$ for any choice of L_i such that $A_i \cong \text{Intalg}(L_i)$. Up to isomorphism, the result will not depend on the choice of L_i as long as $A_i \cong \text{Intalg}(L_i)$ for all i. In this subsection, a Boolean algebra will always be given as an interval algebra, thus for the purposes of this section, $\sum_{i\in I} A_i$ is merely a notational convenience.

Before we proceed, we should perhaps justify the use of the extra "ones" in the sequence

$$L_0 + 1 + L_1 + 1 + L_2 + \ldots.$$

If we take just two orders, say L_0 and L_1, then

$$[\ell, \infty) = [\ell, 1) \cup [1, \infty),$$

where ℓ is the extra element that we had to adjoin to the left of L_0, the point ∞ is an extra point added to the right of L_1, and the 1 is the point in-between L_0 and L_1 (which should not be confused with the greatest element $[\ell, \infty)$ of the Boolean algebra $\text{Intalg}(L_0 + 1 + L_1)$). Now it should be clear that the principal ideal generated by $[\ell, 1)$ is essentially $\text{Intalg}(L_0)$, in which the extra point 1 plays the role of ∞. Similarly, the principal ideal of $\text{Intalg}(L_0 + 1 + L_1)$ generated by $[1, \infty)$ is essentially $\text{Intalg}(L_1)$, but this time the extra point 1 plays the role of ℓ. Furthermore, $\overline{[\ell, 1)} = [1, \infty)$ in $\text{Intalg}(L_0 + 1 + L_1)$. In this sense, we have that

$$\text{Intalg}(L_0 + 1 + L_1) = \text{Intalg}(L_0) + \text{Intalg}(L_1),$$

and this sum is direct or disjoint in the sense explained above.

In the case of infinitely many L_i, $\sum_{i\in I} \text{Intalg}(L_i)$ can also be viewed as the infinite "disjoint sum" of the respective $\text{Intalg}(L_i)$. (It is not quite the same as the direct sum of infinitely many vector spaces because the algebras $B_i = \text{Intalg}(L_i)$ sort of "accumulate below ∞". This process can be described topologically in terms of Alexandroff compactification of the disjoint union of the dual spaces of B_i; more about topology later.)

Cantor-Bendixson derivative

Notice also that in the proof of the Representation Theorem 4.1.6, we get a 1-1 correspondence between *atoms* in B and *adjacencies* in L. This correspondence implies the alignment of the condensation derivative for linear orderings (where adjacencies are identified) with the Cantor-Bendixson derivative defined below.

Definition 4.1.7 (Cantor-Bendixson derivative). For a Boolean algebra B and $x, y \in B$, define $x =_1 y$ iff x and y differ by finitely many atoms. Then the α-th Cantor-Bendixson derivative of B, denoted by $D^\alpha(B)$, is defined via $D^0(B) = B$, $D^{(\beta+1)}(B) = D^{(\beta)}/_{=_1}$, and for limit α, $D^{(\alpha)}(B) = \cap_{\lambda<\alpha} D^{(\lambda)}(B)$.

For example, the interval algebra of ω becomes the two-element Boolean algebra after taking the derivative. This is because any two elements that differ by finitely many atoms are identified modulo the ideal generated by the atoms. Similarly,

$$B = D^{(1)}(\mathrm{Intalg}(\omega^2)) = \mathrm{Intalg}(\omega),$$

and indeed

$$B = D^{(m)}(\mathrm{Intalg}(\omega^n)) = \mathrm{Intalg}(\omega^{n-m}),$$

whenever $m < n$. This process can be iterated over ordinals α. Let $D^{(\alpha)}$ also denote the natural homomorphism from B to its α-th derivative $D^{(\alpha)}(B)$; we will only need the case when α is finite.

Definition 4.1.8. The Cantor-Bendixson rank of B is defined as the least α such that $D^{(\alpha)}(B) = D^{(\alpha+1)}(B)$.

That is, the algebra either vanishes after α iterations or becomes atomless. In the case when B is countable, which is the only case we ultimately care about, α has to be countable. Since all countable atomless Boolean algebras are isomorphic via the usual back-and-forth argument, if α is the rank of B and $D^{(\alpha)}(B)$ is non-trivial, then $D^{(\alpha)}(B)$ has to be isomorphic to Intalg($\mathbb{Q}$).

Definition 4.1.9. If $b \in B$ is such that $D^{(n)}(b)$ is an atom in $D^{(n)}(B)$, then we say that b is an *n-atom.*

For example, if the principal ideal generated by b is isomorphic to Intalg(ω) (after we declare $b = 1$), then b is a 1-atom. In fact, this gives a complete description of 1-atoms; we omit the proof. Similarly, an n-atom is characterised by the respective principal ideal being a copy of Intalg(ω^n); the transfinite analogy also works. Also, it should be clear that $D^{(\alpha)}(\mathrm{Intalg}(\mathbb{Q})) = \mathrm{Intalg}(\mathbb{Q})$ for any α. This is simply because Intalg($\mathbb{Q}$) has no atoms at all.

We will be using the Cantor-Bendixson derivative when we look at Feiner's Theorem. Our proof of Feiner's Theorem will use the Fellner-Watnick Theorem 3.2.23, and thus we will be using interval algebras of the form Intalg($\mathbb{Z}^n L$). Observe also that $\mathrm{Intalg}(\mathbb{Z}) = \mathrm{Intalg}(\omega^*) + \mathrm{Intalg}(\omega)$, where ω^* is the order of the negative integers. It is easy to see that $\mathrm{Intalg}(\omega^*) \cong \mathrm{Intalg}(\omega)$, so Intalg($\mathbb{Z}$) is just the sum of two 1-atoms, in the sense that its greatest element 1 splits into two 1-atoms. In particular, multiplying a discrete order L by $\mathbb{Z}$ (from the left) results in replacing every atom in the respective interval algebra by a pair of 1-atoms. It may not be immediately obvious, but Intalg($\mathbb{Z}^2$) is the sum of two 2-atoms, and so on. The situation is a bit different in the interval algebra $\mathbb{Z} \cdot \mathbb{Q}$; we have $D^{(1)}(\mathrm{Intalg}(\mathbb{Z} \cdot \mathbb{Q})) = \mathrm{Intalg}(\mathbb{Q})$. It does not contain any 1-atom, but it has plenty of atoms "all over the place". For instance, it satisfies the following property for $n = 0$.

Definition 4.1.10. Fix $n \geqslant 0$. We say that an element x of a Boolean algebra B is *atomic* (or 0-*atomic*) if for each non-zero $y \leqslant x$, there is some atom $z \leqslant y$. We say that x is *n-atomic* (for $n > 0$) if $D^{(n)}(x)$ is atomic in $D^{(n)}(B)$.

The definition has a transfinite extension, but we will not need it.

Definition 4.1.11. Fix $n \geqslant 0$. An element b of an algebra is 0-*atomless* or simply *atomless* if the principal ideal generated by the element is isomorphic to Intalg($\mathbb{Q}$) (when we declare $b = 1$). More generally, b is n-atomless if $D^{(n)}(b)$ is atomless in $D^{(n)}(B)$.

According to the definition above, the least element 0 is *not* atomless. Similarly, if an element is a finite union of atoms, then it is *not* 1-atomless, and so on.

The two definitions above are related as follows: Since all countably infinite atomless Boolean algebras are isomorphic, a is atomic iff no (non-zero) $b \leqslant a$ is atomless. In particular, α-atoms are atomic, but any algebra of the form $\text{Intalg}(L + \mathbb{Q})$ is not.

The Remmel-Vaught Theorem

We can ask many questions about the relationships between orderings and successivities, algebras and atoms, and order-types of L and isomorphism types of $\text{Intalg}(L)$. To prove the Downey-Jockusch Theorem about low Boolean algebras, we will need the following important result of this kind. Vaught [496] proved this result for atomic Boolean algebras, and Remmel [445] extended it to all countable Boolean algebras.

Theorem 4.1.12 (Remmel-Vaught). *(i) Suppose that B is any Boolean algebra having infinitely many atoms, and let $\widehat{B}$ be the algebra obtained from B splitting each atom of B a finite number of times. Then $\widehat{B}$ is isomorphic to B.*

(ii) (Rephrasing (i) in terms of linear orderings). Suppose that L and $\widehat{L}$ are linear orderings with infinitely many adjacencies, and $g : L \to \widehat{L}$ is an order-preserving embedding such that

(a) if $[x, y)$ is finite in L then $[g(x), g(y))$ is finite in $\widehat{L}$, and

(b) if $z \in \widehat{L}$ is not in the image of L, then there are x, y in L such that $[x, y)$ is finite and $z \in [g(x), g(y))$.

Then $\text{Intalg}(L) \cong \text{Intalg}(\widehat{L})$.

Proof. We use the form (i). Let $\langle Y \rangle$ denote the subalgebra generated by Y in $\text{Intalg}(\mathbb{Q})$. Here and henceforth, $\leqslant$ means $\leqslant_{[0,1]}$ since we view L and $\widehat{L}$ as suborderings of $\mathbb{Q} \cong \mathbb{Q} \cap [0, 1]$. For a subalgebra X of $\text{Intalg}(\mathbb{Q})$, let $I(X)$ denote the ideal generated by X, and $A(X)$ denote the set of atoms of X. It is relatively easy to see that $B/I(A(B)) \cong \widehat{B}/I(A(\widehat{B}))$.

To demonstrate that B is isomorphic to $\widehat{B}$, let $B = \{b_i : i \in \mathbb{N}\}$ and $\widehat{B} = \{c_i : i \in \mathbb{N}\}$. We build the isomorphism in stages, at each stage n, specifying a subalgebra B_n of B, a subalgebra $\widehat{B}_n$ of $\widehat{B}$, and an isomorphism $f_n : B_n \to \widehat{B}$. Inductively, we suppose that these parameters satisfy the following conditions.

(i) $a \in B_n$ is the union of exactly n atoms of B iff $f_n(a)$ is the union of exactly n atoms of $\widehat{B}$.

(ii) If $a \in B_n$ and $a \notin I(A(B))$ then

- $f_n(a) \notin I(A(\widehat{B}))$,
- $f_n(a) \equiv a \bmod I(A(\widehat{B}))$, and
- $|\{b \in A(B) : b \leqslant a\}| = |\{c \in A(\widehat{B}) : c \leqslant f_n(a)\}|$.

Since $0_B = 0_{\widehat{B}} = 0_{\mathbb{Q}}$ and $1_B = 1_{\widehat{B}} = 1_{\mathbb{Q}}$, at stage 0 the identity map can play the role of f_0 and will satisfy (i) and (ii). At stage 1, we do nothing else. Let $n \geqslant 1$.

Stage 2n. Assume that at stage $2n-1$ we have (i) and (ii) and additionally, $\langle\{b_0, \ldots, b_{n-1}\}\rangle \subseteq B_{2n-1}$, and $\langle\{c_0, \ldots, c_{n-1}\}\rangle \subseteq \widehat{B}_{2n-1}$. Let $a_0, \ldots, a_s$ list $A(B_n)$ so that $f_{2n-1}(a_0), \ldots, f_{2n-1}(a_s)$ list $A(\widehat{B})$. If $b_n \in B_{2n-1}$, do nothing. Otherwise, let $B_{2n} = \langle B_n \cup \{b_n\}\rangle$. We can renumber the list of atoms so that the atoms of B_{2n} are

$$a_0 \wedge b_n, a_0 - b_n, \ldots, a_j \wedge b_n, a_j - b_n, a_{j+1}, a_{j+2}, \ldots, a_s.$$

We define f_{2n} on the atoms of $A(B_{2n})$ and then extend it via the naturally induced map. For i with $j+1 \leqslant i \leqslant s$, let $f_{2n}(a_i) = f_{2n-1}(a_i)$. For the remaining i with $0 \leqslant i \leqslant j$, we define $f_{2n}(a_i \wedge b_n)$ and $f_{2n}(a_i - b_n)$ depending on one of the cases below.

Case 1. *a_i is the union of exactly m atoms of B.* Then for some k, $a_i \wedge b_n$ is the union of k atoms of B. Now $f_{2n-1}(a_i)$ is the union of exactly m atoms of $\widehat{B}$. Thus we let c be such that c is the union of exactly k atoms $\leqslant f_{2n-1}(a_i)$, and define $f_{2n}(a_i \wedge b_n) = c$ and $f_{2n}(a_i - b_n) = f_{2n-1}(a_i) - c$.

Case 2. $a_i \notin I(A(B))$. By hypothesis, $a_i \equiv f_{2n-1}(a_i) \bmod I(A(B))$. Thus $a_i \wedge b_n \equiv f_{2n-1}(a_i) \wedge b_n \bmod I(A(B))$ and $a_i - b_n \equiv f_{2n-1}(a_i) - b_n \bmod I(A(B))$. There are now 3 subcases.

Subcase 2a. $|\{b \in A(B) : b \leqslant a_i \wedge b_n\}| = |\{b \in A(B) : b \leqslant a_i - b_n\}| = \infty$. Then we can let $f_{2n}(a_i \wedge b_n) = f_{2n-1}(a_i) \wedge b_n$, $f_{2n}(a_i - b_n) = f_{2n-1}(a_i) - b_n$.

Subcase 2b. $|\{b \in A(B) : b \leqslant a_i \wedge b_n\}| = m < \infty$, *and* $|\{b \in A(B) : b \leqslant a_i - b_n\}| = k < \infty$. Then there are exactly $m = k$ atoms in $\widehat{B}$ below $f_{2n-1}(a_i)$, say, $c_{i_1}, \ldots, c_{i_{m+k}}$. Then let $d = [f_{2n-1}(a_i) \wedge b_n - \wedge_{j=1}^{m+k} c_{i_j}] \wedge \wedge_{j=1}^{m} c_{i_j}$. Then d has exactly m atoms of $\widehat{B}$ under it, and $d \equiv f_{2n-1} \wedge b_n \bmod I(A(\widehat{B}))$. Then we let $f_{2n}(a_i \wedge b_n) = d$ and $f_{2n}(a_i - b_n) = f_{2n-1}(a_i) - d$.

Subcase 2c. *Exactly one of $a_i - b_n$ or $a_i \wedge b_n$ has infinitely many atoms under it.* Without loss of generality, suppose that $|\{b \in A(B) : b \leqslant a_i \wedge b_n\}| = m < \infty$. Since $a_i \wedge b_n \equiv f_{2n-1}(a_i) \wedge b_n \bmod I(A(\widehat{B}))$, it follows that $f_{2n-1}(a_i) \wedge b_n$ has only finitely many atoms under it, say $c_{i_1}, \ldots, c_{i_k}$. By renumbering, let $g_1, \ldots, g_m$ denote the first m atoms under $f_{2n-1}(a_i)$. Let

$$d = [f_{2n-1}(a_i) \wedge b_n - \wedge_{j=1}^{k} c_{i_j}] \wedge \wedge_{j=1}^{k} g_j.$$

Then d has exactly m atoms of $\widehat{B}$ under it and $d \equiv f_{2n-1}(a_i) \wedge b_n \bmod I(A(\widehat{B}))$. We can now define $f_{2n}(a_i \wedge b_n) = d$ and $f_{2n}(a_i - b_n) = f_{2n-1}(a_i) - d$.

Finally, we let $\widehat{B}_{2n} = \langle\{f_{2n}(x) : x \in A(B_{2n})\}\rangle$. At odd stages, we do essentially the same thing except we use f^{-1} and go from $\widehat{B}_{2n+1}$ back to B_{2n+1}. In this way, one can easily see that $\cup_n f_n$ defines an isomorphism from B to $\widehat{B}$. □

We remark that the isomorphism constructed in the proof of (i) of the Remmel-Vaught Theorem above is not computable in general; see Exercise 4.1.18. In his thesis [485], Thurber proved an extension of the Remmel-Vaught Theorem. Specifically, Thurber replaced "atoms" in the above with "n-atoms", which are the atoms of the n-th Cantor-Bendixson derivative of the Boolean algebra. See also Knight and Stob [308]. We will not need these more general results.

Exercises

Exercise° 4.1.13. A Boolean algebra is atomic if every non-zero element of it bounds an atom. Show that if B is an infinite atomic Boolean algebra, and $B/\langle \mathrm{At}(B)\rangle \cong \mathbf{2}$, then $B \cong \mathrm{Intalg}(\omega)$. Here $\langle X\rangle$ denotes the ideal generated by X, $\mathrm{At}(Y)$ denotes the atoms of Y, and $\mathbf{2}$ is the two-element algebra.

Exercise* 4.1.14 (Thurber [485]). Let $n \geqslant 1$. Let B_1 and B_2 be Boolean algebras such that

- B_1 has infinitely many n-atoms, and
- B_2 results from B_1 by splitting its n-atoms a finite number of times.

Show that $B_2 \cong B_1$.

4.1.2 Effective presentations of Boolean algebras

Recall that a Boolean algebra is computable if its domain and its operations are computable. The Interval Representation Theorem 4.1.6 is clearly computable. Thus we obtain the following immediate

Corollary 4.1.15. *Every computable Boolean algebra is isomorphic to an interval algebra Intalg(L) of a computable linear ordering L. Furthermore, this is also uniform.*

We can use the corollary to apply methods and results developed in Chapter 2 to Boolean algebras. For instance, the corollary below follows from the corresponding results for linear orderings from Chapter 3. As far as we know, this was first observed by (cf., Feiner [163, Remarks 1-2]) for $n = 2$ and by Odintsov and Selivanov [423] for arbitrary $n \in \mathbb{N}$.

Corollary 4.1.16.

(i) Every Σ^0_n-(resp. Δ^0_n-, Π^0_n-) presentable Boolean algebra is representable as Intalg(L) with L Σ^0_n-(resp. Δ^0_n-, Π^0_n-) presentable. Furthermore, every Boolean algebra is isomorphic to a subalgebra of the free Boolean algebra Intalg($\mathbb{Q}$) of the same degree.

(ii) Consequently, every Π^0_{n+1}-presentable Boolean algebra is isomorphic to a Σ^0_n-presented one.

Here, Σ^0_n-presentability is understood in the sense of a factor $\mathrm{Intalg}(\mathbb{Q})/I$, where I is a Σ^0_n ideal, and similarly for Δ^0_n-, Π^0_n-presentability. In the case of Δ^0_n-presentability, this can lead to some confusion since it can stand for X-computable presentability for some $X \in \Delta^0_n$. Fortunately, in the case of countable Boolean algebras, these notions of Δ^0_n-presentability are equivalent (Exercise 4.1.20).

Boolean algebras can be constructed directly to satisfy a list of requirements. However, it is sometimes useful to instead build L in stages,

$$L = \bigcup_s L_s,$$

and consider

$$\mathrm{Intalg}(L) = \bigcup_s \mathrm{Intalg}(L_s) = \bigcup_s B_s,$$

where we can pass from L_s to $B_s = \mathrm{Intalg}(L_s)$ with all possible uniformity.

To illustrate this technique, we give one important application of the Remmel-Vaught Theorem 4.1.12. Recall that a structure is computably categorical if any two computable presentations of the structure are computably isomorphic. It is quite easy to see that $\beta = \mathrm{Intalg}(\mathbb{Q})$ is computably categorical. For that, use a straightforward back-and-forth procedure similar to that used for the dense linear order $\mathbb{Q}$. It follows that, similarly to linear orders, there are no interesting examples of computably categorical Boolean algebras, and indeed, L is computably categorical iff $\mathrm{Intalg}(L)$ is.

Theorem 4.1.17 (Goncharov and Dzgoev [211], LaRoche [331]). *A Boolean algebra is computably categorical iff it has only finitely many atoms.*

We give only an extended sketch that emphasises the role of the Remmel-Vaught Theorem 4.1.12 and omits some of the standard combinatorics related to priority constructions. A complete formal proof of the result can be found in [445], where it is derived from more general facts about the complexity of the set of atoms in a Boolean algebra. (See also Exercise 4.1.23.)

Extended Sketch. Suppose B is a computable Boolean algebra that has only finitely many atoms. Then it is either finite or is the sum of a finite algebra and the atomless algebra $\beta = \mathrm{Intalg}(\mathbb{Q})$. In each of these cases, one could easily construct a computable isomorphism between any two computable copies of B.

Now suppose B has infinitely many atoms. The idea is as follows. We build a computable copy A of B, and we meet:

$$P_e : \varphi_e : B \to A \text{ is not an isomorphism,}$$

where $(\varphi_e)_{e\in\mathbb{N}}$ is the effective list of all partial computable functions, as usual. To meet R_e, we will search for the next available atom $b \in B$, wait for $\varphi_e(b)\downarrow = a \in A$, and make sure a is not an atom by splitting it, if necessary. We also have to meet the global requirement

$$A \cong B.$$

This is done by making sure that A is obtained from B by splitting some atoms of B into finitely many atoms. Then we will have $A \cong B$ by the Remmel-Vaught Theorem 4.1.12.

If we choose to use linear orders and interval representations, we could view $B = \mathrm{Intalg}(L)$. In this case, we have $A = \mathrm{Intalg}(L')$, where L' is obtained from L by putting finitely many points inside some of the adjacencies in L. We can build A "around" B and essentially assume $B \subseteq A$ and $L \subseteq L'$. (Formally, we would have to define a computable isomorphic embedding $\psi : B \to A$, but this formalism would only obscure the main idea.) Under an appropriate effective indexing of the domain, we can view elements of B as natural numbers. We say that an element $b \in B$ has its index smaller than the index of $c \in B$ if the number associated with B is smaller than the number associated with c. The order of indices has order-type ω and should not be confused with the linear order L.

At stage s, we will have a finite algebra $B_s \subseteq B$ and a finite part $A_s \supseteq B_s$ of A. We also have $B_s \subseteq B_{s+1}$ and $A_s \subseteq A_{s+1}$ for every s. We then set $A = \bigcup_s A_s$ and $B = \bigcup_s B_s$. If it makes things easier, think of $B_s = \mathrm{Intalg}(L_s)$ and $A_s = \mathrm{Intalg}(L'_s)$, where $L_s \subseteq L'_s$ are finite linear orders, and where $L = \bigcup_s L_s$ and $L' = \bigcup_s L'_s$.

The strategy to meet P_e is as follows.

1. Fix a witness $b \in B$ with the smallest index that is an atom in B_s and has never been used before by P_e or P_j with $j < e$.

2. Wait for $\varphi_e(b)$ to converge.

3. Meanwhile, monitor b and see if it splits in B_t for some $t > s$. If it ever happens, initialise the strategy by picking a new witness.

4. If $a = \varphi_{e,t}(b) \in A_t$ is not an atom in A_t, then do nothing unless b later splits in B; in the latter case, go to 3.

5. If $a = \varphi_{e,t}(b) \downarrow$ is an atom in A_t, then consider the subcases:

   (a) If a is *restrained* from splitting (to be clarified shortly in (b)) by some P_i with $i < e$, then initiate the strategy by picking a new witness.

   (b) Otherwise, split a in A_{t+1} into a_0 and a_1 and *restrain* a_0 and a_1 from being split by strategies P_j, $j > e$.

6. From now on, monitor the element $c \in B$ naturally identified with $a \in A$ (i.e., $\psi(c) = a$ under the inclusion map $\psi : B \to A$). If at some later stage it splits in B into (say) c_0 and c_1, then:

   (a) Associate c_0 with a_0 and c_1 with a_1 under the inclusion relation[1].

   (b) Remove the restraint from a_0 and a_1 that was earlier imposed by P_e.

Think of P_i as having its priority higher than P_j if $i < j$.

Construction. To define A_s, we will mainly just copy B_s into A_s unless some extra atoms have to be adjoined to $A_s \backslash B_s$ due to the actions of the P_e-strategies. At stage s, let strategies P_e, $e < s$, act according to their instructions.

Verification (sketch). Instead of giving a dry formal verification by induction, we shall give a detailed informal explanation that emphasises the algebraic nature of the strategies.

We first discuss *one strategy in isolation*. Suppose first that P_e is the only requirement, and that we do not have to worry about any other P_j for $j \neq e$. We claim that in this case the strategy described above guarantees that φ_e cannot be an isomorphism from B to A. The strategy will be initialised only finitely many times. This is because we always search for a follower having the *smallest index* among the available atoms in B_s. Eventually we will find an element that is indeed an atom in B even though we will never be sure that it actually is an atom. Thus, we can assume that b never splits, P_e is never initialised, and b is the permanent witness for P_e. In this case, the outcomes are:

- $\varphi_e(b)$ never converges;

[1] In other words, we recycle a_0 and a_1 by setting $\psi(c_0) = a_0$ and $\psi(c_1) = a_1$. This is done to guarantee that we will end up with $A \cong B$ by the Remmel-Vaught Theorem 4.1.12.

- φ_e is not injective or fails to respect the algebraic operations[2];
- $\varphi_e(b) \downarrow= a$ and a is already not an atom in A_t;
- $\varphi_e(b) \downarrow= a$ is artificially split by the strategy in A_{t+1}.

In each of these cases, φ_e cannot possibly be an isomorphism, because any isomorphism has to be a total injective homomorphism that maps atoms to atoms.

Now suppose we have only two strategies, say P_0 and P_1. The only case when P_1 needs to worry about P_0 is when P_0 restrained some element, say a_0, due to its actions in 5(b), and then later we discover

$$\varphi_1(\tilde{b}) = a_0,$$

where $\tilde{b}$ is the current witness of P_1. Since the restraint imposed by P_0 has not been lifted from a_0, it means that a_0 still looks like an 'extra' atom in $A\backslash B$. In this case, we cannot possibly afford to split a_0 further, because in the presence of infinitely many strategies it may potentially result in a_0 bounding infinitely many elements, and this is very bad for the Remmel-Vaught Theorem 4.1.12, at least in its weakest form. We therefore initialise P_1 and resume searching for a witness whose image under φ_1 is not restrained by P_0. If we fail to find such a witness, then φ_1 cannot be onto, and thus it cannot be an isomorphism.

Now consider P_0 and its interactions with P_1. The strategy for P_0 completely ignores P_1. In particular, it is allowed to split an element restrained by P_1. As a result, we may end up with (say) $a_0 \in A\backslash B$ restrained by P_1, which is further split into $\tilde{a}_0$ and $\tilde{a}_1$ that are restrained by P_0. Note that the priority of the restraint imposed on atoms in $A\backslash B$ increases as we further split restrained elements. Consequently, in the presence of all strategies, every element that has ever been introduced in $A\backslash B$ can be further split only finitely many times.

The construction is a standard finite injury argument in which every strategy can be initialised only finitely many times. Thus, by induction, P_e will never be initialised after some stage t. Note that we cannot possibly have all elements of A permanently restrained by strategies P_j, $j < e$, simply because B is infinite. Also, B has infinitely many atoms, and therefore (as was already explained above) P_e will eventually find a stable witness. Finally, as we have already argued above, some atoms of B will perhaps be split into finitely many atoms in A, and otherwise, A is not really different from B. By the Remmel-Vaught Theorem 4.1.12, we conclude that $A \cong B$. □

Exercises

Exercise° 4.1.18 (Folklore). Suppose that computable Boolean algebras B and $\tilde{B}$ satisfy the assumptions of (i) in the Remmel-Vaught Theorem 4.1.12.

1. Show that B and $\tilde{B}$ do not have to be computably isomorphic.
2. Calculate the optimal upper bound Δ^0_n on the complexity of isomorphism between such B and $\tilde{B}$. In particular, for this n, show that there are such B and $\tilde{B}$ that are not Δ^0_{n-1}-isomorphic.

[2] In other words, it is not even an injective homomorphism. The possibility is often completely omitted in the literature perhaps because it is viewed as 'completely obvious'. The same can be said about some other 'trivial' outcomes that are often not even mentioned.

Exercise 4.1.19 (Remmel [443]). Recall that an ideal M of a Boolean algebra B is *maximal* if $1 \notin M$ and for all $b \in B$, either $\overline{b}$ or b are in M. Show that if B is a computable Boolean algebra and I is a computable proper ideal of B, then there is a computable maximal ideal M of B with $I \subseteq M$.

Exercise° 4.1.20 (Folklore).

1. Show that a Boolean algebra B is computable (or computably enumerable) iff $B \cong_{\text{comp}} \text{Intalg}(\mathbb{Q})/I$ for some computable (or computably enumerable) ideal I.
2. Extend this result to Π^0_n, Σ^0_n, and Δ^0_n presentations and ideals, respectively, for any $n \in \mathbb{N}$.

Exercise 4.1.21.

1. (Folklore). Show° that if $\{P_i : i \in \mathbb{N}\}$ is a computable list of propositional symbols, then the free Boolean algebra $\mathbb{P}$ generated by taking propositional formulae on this list generates a computable copy of Intalg($\mathbb{Q}$).
2. (Essentially Martin and Pour-El [354]). Show* that there is a c.e. copy of Intalg($\mathbb{Q}$) whose only c.e. filters are finitely generated.

Exercise° 4.1.22 (Remmel [445]). Recall that a set is *immune* if it is infinite and has no infinite computably enumerable subsets. Show that every computable Boolean algebra with infinitely many atoms is isomorphic to a computable Boolean algebra in which the set of atoms is immune.

Exercise* 4.1.23. Let At(B) denote the set of atoms of a Boolean algebra B.

1. (Remmel [444]). Show that if B is a computable Boolean algebra with infinitely many atoms, then B has a computable copy C, such that $\text{At}(C) \geqslant_T \varnothing'$. (Hint: Use the Remmel-Vaught Theorem 4.1.12.)
2. (Downey [118]). Show that if B is a computable Boolean algebra with infinitely many atoms, then B has a computable copy C, such that $\text{At}(C) \not\geqslant_T \varnothing'$.

Exercise 4.1.24** (Montalbán [398]). Suppose that B is a computable Boolean algebra having infinitely many atoms, and let $\mathbf{a}$ be the Turing degree of the set of atoms in B. Show that for every $\mathbf{d}$ so that $\mathbf{d}''' \geqslant \mathbf{a}'''$ there exists a computable copy of B in which the set of atoms has degree $\mathbf{d}$.

4.1.3 Low Boolean algebras. The proof of Theorem A(2).

Recall that Theorem 3.2.45 states that there is a low c.e. presented linear order *not* isomorphic to a computable one. We have seen low groups that are not isomorphic to computable ones too; e.g., Proposition 2.2.5. It is natural to ask whether the analogue of Theorem 3.2.45 holds for Boolean algebras.

Theorem 4.1.25 (Downey and Jockusch [129]). *Suppose that B is a low Boolean algebra. Then B is isomorphic to a computable Boolean algebra. Indeed, if B is $\varnothing'$-presentable and the atom relation for B is $\varnothing'$-computable, then B has a computable copy.*

Theorem 4.1.25 appeared earlier as Theorem A(2), but it states a bit more than was stated in Theorem A(2). Also, Theorem 4.1.25 should be compared with the Frolov-Montalbán Theorem 3.2.14.

Proof. We can view a low Boolean algebra as Intalg(L), the interval subalgebra of a low linear subordering of the rationals, $\mathbb{Q}$. Without loss of generality, we can suppose L has infinitely many adjacencies.

In light of Theorem 4.1.12(ii), to establish the desired result, it suffices to show that for any low linear ordering $L \subseteq \mathbb{Q}$ having infinitely many adjacencies, there is a computable linear ordering $\widehat{L} \subseteq \mathbb{Q}$ which is the same as L except that some adjacencies are replaced with finite blocks. But since L is low, it satisfies the premises of Lemma 3.2.28; thus such a $\widehat{L}$ indeed exists. Observe that the argument actually shows that any Δ_2^0-presented Boolean algebra with a Δ_2^0 set of atoms is isomorphic to a recursive Boolean algebra. □

The following conjecture is longstanding.

Conjecture 4.1.26 (Downey). *Every low$_n$ Boolean algebra is isomorphic to a computable one.*

The most recent progress on this question is the following.

Theorem 4.1.27 (Knight and Stob [308]). *Every low$_4$ Boolean algebra is isomorphic to a computable one.*

There is not much further evidence to support this conjecture. Harris and Montalbán [235] demonstrated that there is a genuine problem with the case $n = 5$, so new ideas will be needed. Another result indicating that the problem might be exceptionally combinatorially hard is [396].

4.1.4 Superatomic Boolean algebras*

We now very briefly discuss one important and well-studied class of Boolean algebras arising from ordinals. The class has many equivalent definitions (characterisations). We give the definition that involves the Cantor–Bendixson derivative (Definition 4.1.7).

Definition 4.1.28. The Boolean algebra B is *superatomic* if $X^{(\alpha)} = \mathbf{2}$ for some ordinal α, where $\mathbf{2}$ is the trivial two-element Boolean algebra.

Theorem 4.1.29 (Goncharov [199]). *For a Boolean algebra B, the following are equivalent:*

1. *B is computable and superatomic.*
2. *$B \cong$ Intalg(α) for some computable ordinal (well-order) α.*
3. *$B \cong$ Intalg($\omega^\beta \cdot k$), for some computable ordinal β and positive $k \in \mathbb{N}$.*

The non-effective version of the theorem is folklore; we cite [207, Proposition 1.5.7] and [207, Corollary 1.6.1]. The proof of the result is omitted, but we note that $3 \to 2$ and $2 \to 1$ are essentially obvious. We refer to p. 57 of [207] and to [199] for the details, and we refer to Exercise 8.1.26 for a proof sketch of $1 \to 3$.

In particular, if ω_1^{CK} denotes the least non-computable ordinal, then Intalg(ω_1^{CK}) does not have a computable copy; indeed, it does not even have a (hyper)arithmetical copy (this follows from Exercise 8.1.25). On the other hand, it is easy to see that a c.e. presented superatomic Boolean algebra has a computable presentation: combine Exercise 2.2.44 with Corollary 4.1.16. (Indeed, every hyperarithmetical superatomic Boolean algebra is isomorphic to a computable one, as follows from Exercise 8.1.25.) Thus, Feiner's Theorem 4.1.30 that we discuss next cannot be witnessed by a superatomic Boolean algebra.

4.1.5 Feiner's Theorem. The proof of Theorem A(1)

We would like to obtain a result for Boolean algebras analogous to Feiner's Theorem 3.2.1 for linear orders. However, we claim that the methods used in Chapter 1—specifically, coding some arithmetical set such as a Σ_3^0-set and then relativising, as we did for Theorem 3.2.1 using Lerman's Theorem 3.2.11—cannot be directly applied here.

Suppose we attempt to use $\mathrm{Intalg}(L)$, where L is constructed as in Theorem 3.2.11. Notice that $\mathrm{Intalg}(L) \cong \mathrm{Intalg}(\widehat{L})$, where $\widehat{L}$ replaces each block of size n in L with one of size 2. (Here, we are using the Remmel-Vaught Theorem 4.1.12.) It is not difficult to build a computable copy of $\mathrm{Intalg}(L)$.

We could attempt to code a more complicated arithmetical set using n-atoms. However, as mentioned earlier, Thurber proved an extension of the Remmel-Vaught Theorem 4.1.12 to n-atoms, leading to similar challenges.

Feiner's idea was to use n-atoms for all $n \in \mathbb{N}$ to "code" a uniformly Σ_n^0 relation $S(n)$ into a computable (or c.e. presented) Boolean algebra. That is, the membership $n \in S$ is uniformly Σ_n^0. To achieve this, he defined a new measure of complexity for sets computable from $\varnothing^{(\omega)} = \oplus_{n\in\mathbb{N}}\varnothing^{(n)}$:

$$\varnothing^{(\omega)} = \{\langle x, n\rangle : x \in \varnothing^{(n)}\},$$

that reflects the degree of uniformity necessary to compute the set.

Theorem 4.1.30 (Feiner [161, 162]). *There exists a c.e. presentable Boolean algebra not isomorphic to any computable Boolean algebra.*

The result appeared earlier as Theorem A(1). Our proof is similar to the version given by Thurber in his PhD Thesis [485].

The plan of the proof

For a Boolean algebra B, we will examine

$$S_B = \{n : B \text{ satisfies } \Gamma_n\},$$

where $\{\Gamma_n : n \in \mathbb{N}\}$ is a family of certain algebraic properties (predicates) that can be uniformly effectively described. These properties can be viewed as infinitary computable sentences in the language of Boolean algebras, but to keep things simple, we will avoid using infinitary logic.

Our algebra B will be set equal to the sum of principal ideals of the form

$$\mathrm{Intalg}(\mathbb{Z}^{n+1} + 1 + \mathbb{Q}) \text{ and } \mathrm{Intalg}(\mathbb{Z}^n\mathbb{Q} + 1 + \mathbb{Q}),$$

but we will not use the block of the second kind for some of the n. The property Γ_n will say that $\mathrm{Intalg}(\mathbb{Z}^n\mathbb{Q} + 1 + \mathbb{Q}) = \mathrm{Intalg}(\mathbb{Z}^n\mathbb{Q}) + \mathrm{Intalg}(\mathbb{Q})$ is among the blocks that we used. Here, $\mathrm{Intalg}(\mathbb{Q})$ serves as a "separator" between different coding locations of the form $\mathrm{Intalg}(\mathbb{Z}^{n+1})$ and $\mathrm{Intalg}(\mathbb{Z}^n\mathbb{Q})$.

If B is computable, then we will see that the complexity of checking Γ_n is Σ_{2n+3}^0 *uniformly in* n. We will clarify what this means later. It is therefore sufficient to produce a c.e. presented B in which checking whether Γ_n holds is *not* uniformly Σ_{2n+3}^0. This will be done using a variety of techniques and tools, including the Fellner-Watnick Theorem 3.2.23.

The property Γ_n

Our plan is to use algebras (actually, principal ideals) of the form

$$\mathrm{Intalg}(\mathbb{Z}^{n+1} + 1 + \mathbb{Q}) \text{ and } \mathrm{Intalg}(\mathbb{Z}^n\mathbb{Q} + 1 + \mathbb{Q})$$

as coding locations.

Recall that an element is atomic if every non-zero element below it bounds an atom, and that an element is atomless if it generates an infinite atomless principal ideal; in particular, 0 is *not* atomless. Recall also that x is n-atomic (for $n > 0$) if $D^{(n)}(x)$ is atomic in $D^{(n)}(B)$, and that b is n-atomless if $D^{(n)}(b)$ is atomless in $D^{(n)}(B)$.

Definition 4.1.31. For a Boolean algebra B, $a \in B$ and a natural number $n > 0$, define the following properties:

- $\gamma_n(a)$ holds when:
  1. a is $(n-1)$-atomic, and
  2. a is n-atomless.
- Define Γ_n to be the property saying that

$$\exists x\ \gamma_n(x).$$

Example 4.1.32. When $n = 1$ the property says that there is an element a that is atomic (0-atomic) and that it becomes atomless after taking the derivative once. It is clear that Γ_1 holds in $\mathrm{Intalg}(\mathbb{Z}\mathbb{Q} + 1 + \mathbb{Q})$ as witnessed by any element coming from the $\mathbb{Z}\mathbb{Q}$-part. (For instance, take $x = [\ell, a)$, where a is the separating element between $\mathbb{Z}\mathbb{Q}$ and the extra copy of $\mathbb{Q}$.)

Example 4.1.33. We argue that Γ_1 fails in $\mathrm{Intalg}(\mathbb{Z}^2 + 1 + \mathbb{Q})$. Fix $x \in \mathrm{Intalg}(\mathbb{Z}^2 + 1 + \mathbb{Q})$ and assume that $\gamma_1(x)$ holds. The element x is a finite union of half-open intervals $I_0, \ldots, I_k$. None of these intervals can come from the dense part of the order, for in this case x would not be atomic. This means that x comes from the principal ideal isomorphic to $\mathrm{Intalg}(\mathbb{Z}^2)$, which is the sum of two 2-atoms. If all of the intervals $I_0, \ldots, I_k$ making up x contain only finitely many points, then it means that $D^{(1)}(x) = 0$. This is impossible because x has to be 1-atomless and, in particular, non-zero in $D^{(1)}(B)$. On the other hand, if at least one of the intervals, say I_j, contains infinitely many points, then this means it has a copy of a 1-atom in it. This is the same as saying that $D^{(1)}(x)$ bounds an atom, so in this case x cannot be 1-atomless either. We conclude that Γ_1 fails in $\mathrm{Intalg}(\mathbb{Z}^2 + 1 + \mathbb{Q})$. Of course, there is nothing special about $\mathbb{Z}^2$ in this example, it could just as well be $\mathbb{Z}^m$ for any m.

Example 4.1.34. We argue that Γ_1 fails in $B = \mathrm{Intalg}(\mathbb{Z}^2\mathbb{Q} + 1 + \mathbb{Q})$. For suppose $x \in B$ is such that $\gamma_1(x)$ holds. If $I_0, \ldots, I_k$ make up x, then all of these half-open intervals must have empty intersection with $\mathbb{Q}$. (Otherwise, not every element below x will bound an atom; this is similar to Example 4.1.33.) If each of these I_j is fully contained in some copy of $\mathbb{Z}^2$, then either x is the sum of finitely many atoms or bounds a 1-atom. But this means that in this case x cannot be 1-atomless. The only case that is left is when at least one of these I_j contains infinitely many copies of $\mathbb{Z}^2$. But after taking the derivative, these will turn into infinitely many atoms, and thus x cannot be 1-atomless in this case either.

In the proposition below, fix any

$$B = \sum_{i \in \mathbb{N}} B_i,$$

where for each $i \in \mathbb{N}$, B_i is either Intalg($\mathbb{Z}^n\mathbb{Q} + 1 + \mathbb{Q}$) or Intalg($\mathbb{Z}^{n+1} + 1 + \mathbb{Q}$) for some $n \in \mathbb{N}$. For the purposes of the proof, we may assume that $n > 0$.

Proposition 4.1.35. *Then Γ_n holds in B iff there is an i such that B_i is isomorphic to Intalg($\mathbb{Z}^n\mathbb{Q}+ 1 + \mathbb{Q}$).*

Proof. The proof is essentially a generalisation of the examples above. If some B_i is isomorphic to Intalg($\mathbb{Z}^n\mathbb{Q} + 1 + \mathbb{Q}$), then take a equal to any subinterval of $\mathbb{Z}^n\mathbb{Q}$ in B_i and see that $\gamma_n(a)$ holds because a is $(n-1)$-atomic and a is n-atomless.

Now assume none of the B_i has the form Intalg($\mathbb{Z}^n\mathbb{Q} + 1 + \mathbb{Q}$), but Γ_n holds on B. So fix a such that $\gamma_n(a)$ holds. It takes n derivatives to make a atomless, and after taking the derivative $(n-1)$ times, a becomes atomic. So in particular, it takes *exactly* n derivatives to make a atomless, which means $D^{(n)}(a)$ generates Intalg($\mathbb{Q}$).

As in the examples preceding the proposition, let $a = I_0 \cup I_1 \cup \ldots \cup I_k$, where the I_j are non-empty half-open intervals in the linear order L of $B =$ Intalg(L). We identify each I_j with the respective element of the algebra. The linear order L looks as follows:

$$L = L_0 + 1 + \mathbb{Q} + 1 + L_1 + 1 + \mathbb{Q} + 1 + L_2 + \ldots,$$

where each L_i is either of the form $\mathbb{Z}^m\mathbb{Q}$ or of the form $\mathbb{Z}^m$ for some m. None of the I_j can possibly intersect the $\mathbb{Q}$-components, for in this case, a would not be n-atomic.

Some of the I_j can possibly intersect some of the $\mathbb{Z}^m\mathbb{Q}$ or $\mathbb{Z}^m$ for $m > n$. If I_j intersects an interval of this sort, then it intersects only one interval of this sort because of the $\mathbb{Q}$-separators. Also, we claim that such an I_j has to vanish after taking n derivatives. For if it does not, then it has to have an atom below it after taking n derivatives, and thus a cannot be n-atomless.

If I_j overlaps with some of the $\mathbb{Z}^m$ for $m < (n-1)$, then there is only one such $\mathbb{Z}^m$. Then I_j vanishes after taking $(n-1)$ derivatives. Similarly, if I_j intersects some interval of the form $\mathbb{Z}^m\mathbb{Q}$ for $m < n$, then it has to vanish already after taking $(n-1)$ derivatives. This is because if it becomes dense after taking $(n-1)$ derivatives, then a cannot be $(n-1)$-atomic.

The only case left is when I_j intersects $\mathbb{Z}^m$ for $m = (n-1)$. Since it cannot overlap with any of the $\mathbb{Q}$-parts, $I_j \subseteq \mathbb{Z}^{n-1}$. Now, it could certainly be $(n-1)$-atomic, since $\mathbb{Z}^{n-1}$ is the union of two $(n-1)$-atoms. However, it completely vanishes after taking n derivatives.

We conclude that every I_j in $a = I_0 \cup I_1 \cup \ldots \cup I_k$ has to vanish after taking n derivatives. But a was supposed to be n-atomless, so it should generate an ideal isomorphic to Intalg($\mathbb{Q}$) after taking n derivatives. This is a contradiction. □

Complexity analysis

Recall that Γ_n states that $\exists x\ \gamma_n(x)$, where $\gamma_n(a)$ holds when a is $(n-1)$-atomic and n-atomless.

Lemma 4.1.36. *Suppose B is an infinite computable Boolean algebra. Then the complexity of checking Γ_n in B is Σ^0_{2n+3}* uniformly in n.

A discussion before the proof. Formally, "Σ^0_{2n+3} uniformly in n" means that there is a computable binary predicate

$$R(\langle x_0, \ldots, x_{2n+3}\rangle, n)$$

such that

$$B \text{ satisfies } \Gamma_n \text{ iff } \exists x_0 \forall x_1 \exists x_2 \forall x_3 \ldots \exists x_{2n+3} R(\langle x_0, \ldots, x_{2n+3} \rangle, n),$$

where $\langle \ldots \rangle$ is the computable indexing of all finite tuples, e.g., $\langle x_0, x_1, x_2 \rangle$ is defined to be $\langle \langle x_0, x_1 \rangle, x_2 \rangle$ and so on. Equivalently, we could instead say that there is an oracle (partial) procedure U such that

$$B \text{ satisfies } \Gamma_n \text{ iff } U^{\varnothing^{(\omega)}}(n) \downarrow,$$

where, on input n, the procedure is allowed to use only the $\varnothing^{(2n+2)}$-section of

$$\varnothing^{(\omega)} = \{\langle x, m \rangle : x \in \varnothing^{(m)}\}.$$

In the proof below, the upper bounds on the complexity are obtained using induction in n. In the process of establishing these upper bounds, we will also establish the required uniformity. Recall that the elements of B are natural numbers, and that the operations in B are computable. (For simplicity and without loss of generality, we can assume that the domain of B is the whole of $\mathbb{N}$.) This also makes the induced order

$$a \leqslant b \text{ if and only if } a \wedge b = a$$

computable as well. Since we interpret elements of B as natural numbers, and the operations and relations on these elements also live within $\mathbb{N}$, the formulae that we shall write in the argument below can be viewed as first-order formulae in arithmetic augmented by the finitely many symbols for the operations of B.

The complexity of the subsets of $\mathbb{N}$ that these formulae define is derived based on counting the number of alternating number-quantifiers over a computable property. Only the alternations of quantifiers matter, since (e.g.) $\exists x \exists y \ldots$ can be replaced with $\exists w\, (w = \langle x, y \rangle \,\&\, \ldots)$. So we put the formula into its normal form and count the alternations of quantifiers in the quantifier prefix. Although this is a standard technique, so far in the book we have not yet seen an argument of this form that would require induction.

Proof. When $n = 0$, it is Π^0_1 to say that an element is an atom; just say that $a \neq 0$ and it does not split. An element a is atomless if every element below it splits, i.e.,

$$a \neq 0 \;\&\; \forall x \leqslant a\, [x \neq 0 \implies (\exists\, v, w \neq 0)\, (v \vee w = x \,\&\, v \wedge w = 0)],$$

which is Π^0_2. Also, an element a is atomic if

$$\forall \text{ non-zero } x \leqslant a\, \exists b \leqslant x \; b \text{ is an atom},$$

which is Π^0_3.

When $n = 1$, the congruence

$$x \sim_1 y \text{ if and only if } D^{(1)}x = D^{(1)}(y)$$

requires checking whether $x = y$ or $x \vartriangle y = (x \wedge \bar{y}) \vee (\bar{x} \wedge y)$ is a finite join of atoms. The latter is a search for a k and atoms $a_0, \ldots, a_k \in B$ so that $x \vartriangle y = a_0 \vee a_1 \vee \ldots \vee a_k$; this is Σ^0_2. It also follows that

$$x \leqslant_1 y \text{ if and only if } x \wedge y \sim_1 x,$$

which is the analogy of $\leqslant$ in $D^{(1)}(B)$, and is also Σ^0_2.

An element a is a 1-atom if $D^{(1)}(a)$ is an atom in $B/\sim_1$, and $\sim_1$ is Σ^0_2. An element a is a 1-atom iff $a \not\sim_1 0$ and it does not split, i.e., it is *not* the case that for some v, w,

$$v, w \not\sim_1 0 \;\&\; v \vee w \sim_1 x \;\&\; v \wedge w \sim_1 0.$$

This is a conjunction of Π^0_2 and $\neg\exists\Pi^0_2$, which is Π^0_3. This makes $\sim_2$ Σ^0_4.

By induction, we obtain that being n-atom is a Π^0_{2n+1}-property, and that $\sim_n$ and $\leqslant_n$ are Σ^0_{2n}-relations.

Recall that a is n-atomic when $D^{(n)}(a)$ is atomic in $D^{(n)}(B)$. This is the same as to say that $a \not\sim_n 0$ and for any $x \leqslant_n a$ such that $x \not\sim_n 0$, there exists an n-atom $b \leqslant_n x$. Since being n-atom is a Π^0_{2n+1}-property, we obtain that being n-atomic is Π^0_{2n+3}. Recall that a is n-atomless if $D^{(n)}(a)$ is atomless in $D^{(n)}(B)$. This is the same as expressing the property of being atomless but using the Σ^0_{2n} relations $\sim_n$ and $\leqslant_n$ instead of $=$ and $\leqslant$. This analysis gives the upper bound of Π^0_{2n+2}.

According to its definition, $\gamma_n(a)$ states that a is $(n-1)$-atomic (this is Π^0_{2n+1}) and n-atomless (which is Π^0_{2n+2}). We conclude that the upper bound for the complexity of

$$\Gamma_n \text{ if and only if } \exists a\, \gamma_n(a)$$

is Σ^0_{2n+3}.

It remains to discuss the uniformity. Given n, we can computably produce the first-order sentence (in arithmetic augmented with the operations of B) that "defines" the Γ_n. We elaborated this procedure in the inductive proof above; clearly, the proof was effective.

This definition of Γ_n requires only access to the operations of the Boolean algebra on top of the usual arithmetic. Since B is computable, these operations are given by (finitely many) computable functions on $\mathbb{N}$. If we fix the indices of these computable functions (which are now shared among all of these sentences representing Γ_n), we can uniformly Σ^0_{2n+3}-effectively verify these sentences. □

Constructing the coding blocks

Recall that our aim is to build a c.e. presented algebra of the form

$$B = \sum_{i \in \mathbb{N}} B_i,$$

where for each $i \in \mathbb{N}$, B_i is either Intalg($\mathbb{Z}^n\mathbb{Q} + 1 + \mathbb{Q}$) or Intalg($\mathbb{Z}^{n+1} + 1 + \mathbb{Q}$) for some $n \in \mathbb{N}$. In this algebra, we will have that "B satisfies Γ_n" is not uniformly Σ^0_{2n+3}. However, to keep B c.e. presented we need this invariant to be close enough to being uniformly Σ^0_{2n+3}. We are ready to state a technical fact that, after a bit of work, implies Feiner's Theorem.

Proposition 4.1.37. *Suppose "$n \in S$" is uniformly Σ^0_{2n+3}. Then there is a computable Boolean algebra B of the form*

$$B = \sum_{i \in \mathbb{N}} B_i,$$

where for each $i \in \mathbb{N}$, B_i is either Intalg($\mathbb{Z}^n\mathbb{Q} + 1 + \mathbb{Q}$) or Intalg($\mathbb{Z}^{n+1} + 1 + \mathbb{Q}$) for some $n \in \mathbb{N}$, and so that

$$S = \{n : \Gamma_n \textit{ holds in } B\}.$$

(By Proposition 4.1.35, we have that $S = \{n : \exists i\; B_i \cong$ Intalg($\mathbb{Z}^n\mathbb{Q} + 1 + \mathbb{Q}$)$\}$.)

Proof. We begin with constructing the elementary building blocks.

Lemma 4.1.38. *There is a uniform procedure which, given a Π^0_{2n+2}-predicate R and $i \in \mathbb{N}$, constructs a computable copy of $\mathbb{Z}^n\mathbb{Q}$ if $R(i)$ holds, and a computable copy of $\mathbb{Z}^{n+1}$, otherwise.*

Proof. It is sufficient to prove the case when $n = 0$ and then apply the Fellner-Watnick Theorem 3.2.23 n times. This is because our proof of the Fellner-Watnick Theorem was uniform assuming the input order was infinite. (Indeed, it was uniform when the input-order is just non-empty.) The Π^0_2-predicate is represented as $\forall x \exists y P(x, y, i)$, where P is computable. Start with building a copy of $\mathbb{Z}$, but every time the predicate "fires" make progress in turning $\mathbb{Z}$ into a copy of $\mathbb{Q}$. That is, if a witness y found for one more x, add a point to every current adjacency of $\mathbb{Z}$. □

Now fix a Σ^0_{2n+3}-predicate and view it as $\exists i R(i)$, where R is Π^0_{2n+2}. The lemma above gives a uniformly computable sequence of linear orders L_i, so that:

1. each L_i is either $\mathbb{Z}^n\mathbb{Q} + 1 + \mathbb{Q}$ or $\mathbb{Z}^{n+1} + 1 + \mathbb{Q}$, and
2. some L_i is $\mathbb{Z}^n\mathbb{Q} + 1 + \mathbb{Q}$ iff the Σ^0_{2n+3}-predicate holds.

Of course, the $(1 + \mathbb{Q})$-part can be added uniformly and essentially independently to the more complicated part of each such L_i. Our application of the Fellner-Watnick Theorem is uniform, and thus the construction of the L_i is uniform in i. Indeed, the whole module corresponding to Σ^0_{2n+3} is uniform in n.

Let R stand for the uniformly Σ^0_{2n+3}-instance "$n \in S$". Define the linear orders $U_{\langle n,i \rangle} = L_i$, where L_i are as described above for this specific R, and $\langle n, i \rangle$ is just the pairing function. Set $B_m = \mathrm{Intalg}(U_m)$ for every $m \in \mathbb{N}$, and define

$$B = \sum_{m \in \mathbb{N}} B_m.$$

It follows from Proposition 4.1.35 that B is exactly what we need. □

The final step in the proof

It is not difficult to construct a set S such that the relation $n \in S$ is uniformly Σ^0_{2n+4} but not uniformly Σ^0_{2n+3}. This fact can be stated in the following, more general terms. Fix a total strictly increasing computable function f. We write $\Sigma^0_{\langle f(n) \rangle}$ to denote the class of all sets X such that the relation "$n \in S$" is uniformly $\Sigma^0_{\langle f(n) \rangle}$. In our case, $f(n) = 2n + 3$.

Lemma 4.1.39. *In the notation above, there is a set $S \in \Sigma^0_{\langle f(n)+1 \rangle} \backslash \Sigma^0_{\langle f(n) \rangle}$.*

Proof. Let S_n be the effective uniform listing of all (uniform partial procedures that could potentially define) $\Sigma^0_{f(n)}$-sets. Declare $n \in S$ if $n \notin S_n$, and set $n \notin S$ otherwise. □

It is useful to view Σ^0_{n+m}-sets and sets Σ^0_n relative to $\varnothing^{(m)}$. Similarly, $\Sigma^0_{f(n)+1}$-sets can be viewed as $\Sigma^0_{f(n)}$-sets relative to $\varnothing'$.

In particular, if we fix $S \in \Sigma^0_{\langle 2n+4 \rangle} \backslash \Sigma^0_{\langle 2n+3 \rangle}$, we can apply Proposition 4.1.37 and produce a Δ^0_2 Boolean algebra B in which Γ_n codes $n \in S$. We can then appeal to Corollary 4.1.16 and conclude that B is c.e. presented. But note that in Corollary 4.1.16, Δ_2^0 stands for the complexity of the congruence. There are two ways to explain why Corollary 4.1.16 is sufficient.

The first explanation is algebraic. We can always reduce the complexity of any presentation of a Boolean algebra to the complexity of $=$ in this presentation, as follows. We can start with the free countable (atomless) Boolean algebra F generated freely by the elements of B, and then B is isomorphic to F/I, where I is generated by all elements of F that are equal to 0 when interpreted as elements of B. (This is Exercise 4.1.21(1).)

The second explanation is combinatorial. In the proof of Proposition 4.1.37, we will end up with a Δ_2^0 linear order, and we know from the previous section that it has a c.e. presentation. It should be clear now that the interval algebra of this order, which is exactly our B, also has a c.e. presentation.

If the reader is still uncomfortable with these explanations, in the next section we will give a *direct proof* of the c.e. presentability of any $\varnothing'$-computable Boolean algebra; this is Proposition 4.1.50. This proof will not rely on linear orders and will not refer to any of the previous results (including the present proof), so there will be no danger of circularity.

We conclude that B is c.e. presented, but it cannot possibly have a computable presentation. This is because Proposition 4.1.35 and Lemma 4.1.36 would then imply that $S \in \Sigma^0_{\langle 2n+3 \rangle}$, contradicting the choice of S.

The proof of Feiner's Theorem is complete.

Applications and consequences

As we remarked earlier, Feiner's technique has found numerous other applications. We refer the reader to Goncharov's book [207]. We give one illustration. Recall that an algebraic structure is decidable if its full first-order diagram is decidable.

Theorem 4.1.40 (Goncharov). *There is a computable Boolean algebra B not isomorphic to any decidable Boolean algebra.*

Sketch. In a decidable Boolean algebra, the set of atoms has to be computable. This is because being an atom is a first-order property in the language of Boolean algebras. It is therefore sufficient to construct a computable Boolean algebra that is not isomorphic to any computable algebra with a computable atom relation.

This is done using Feiner's coding technique. It is not hard to see that the uniformly Σ^0_{2n+3} properties Γ_n described in the proof of Feiner's Theorem (see Lemma 4.1.36) become uniformly Σ^0_{2n+2} in any computable algebra with a computable set of atoms (exercise). By Lemma 4.1.36 and Proposition 4.1.35, it is sufficient to fix a set $S \in \Sigma^0_{\langle 2n+3 \rangle} \backslash \Sigma^0_{\langle 2n+2 \rangle}$ and apply Proposition 4.1.37. □

The proof above does a bit more than claimed in the theorem. It is known (e.g., [207]) that the computability of the atom relation is equivalent to the computable algebra being 1-decidable, i.e., that we can decide the $\exists$-diagram of the algebra. (This is because every existential formula in the language of Boolean algebras can be computably rewritten into an equivalent quantifier-free statement in the language extended by the atom relation. This is folklore that can be traced back to Tarski.) An n-decidable presentation is defined similarly. Feiner's techniques can be used to show that there exist $(n+1)$-decidable Boolean algebras with no n-decidable presentations; we leave this to Exercise 4.1.41. For an in-depth study of n-decidable Boolean algebras, we cite Alaev [6].

Exercises

Exercise° 4.1.41 (Goncharov [201, 207]). Prove that, for every non-zero $n \in \mathbb{N}$, there is an n-decidable Boolean algebra that has no $(n+1)$-decidable presentation.

Exercise* 4.1.42 (Goncharov [201]). Show that for any natural number n, there exists a computable, n-atomic but not $(n+1)$-atomic Boolean algebra, which has no decidable presentation. In contrast, prove that every computable ω-atomic Boolean algebra has a decidable presentation.

Exercise* 4.1.43 (Goncharov [202]). Prove that there exists a Boolean algebra that is n-decidable for all $n \in \mathbb{N}$, yet has no decidable presentation.

4.1.6 Stone spaces and computable trees

We have used interval representations of Boolean algebras, but sometimes it is more convenient to use tree representations. This approach allows us to convert problems about Boolean algebras into problems concerning Π^0_1-classes, as defined in Section 2.1.6. This technique will also be very important in proving the two effective topological versions of Stone duality in the next section.

Stone spaces

Recall that one version of Stone duality, Theorem 4.1.3, says that a Boolean algebra is isomorphic to the set-theoretic sub-algebra of its ultrafilters (equivalently, maximal ideals).

Definition 4.1.44. The *Stone space* $\widehat{B}$ of a Boolean algebra B is defined to be the set X of all ultrafilters of B.

Such spaces are also occasionally called profinite spaces. One can equivalently define $\widehat{B}$ to be the set of all homomorphisms from B to $\mathbf{2}$. But, of course, ultrafilters in B are in a 1-1 correspondence with such $f: B \to \mathbf{2}$. Indeed, if f is a homomorphism of B onto $\mathbf{2}$, then $f^{-1}(1)$ is an ultrafilter in B, and $f^{-1}(0)$ is a maximal ideal in B. Conversely, given an ultrafilter $\mathcal{F}$, its complement $\mathcal{I} = B\backslash\mathcal{F}$ is a maximal ideal, and it follows that $B/\mathcal{I} \cong \mathbf{2}$ via f, so that $f^{-1}(1) = \mathcal{F}$.

We could view the set $\widehat{B}$ of all ultrafilters in B as a topological space. The space is topologised as follows: If $b \in B$, the family of all ultrafilters of B having b as an element is a typical basic clopen (closed and open) subset of $\widehat{B}$. (Equivalently, $\{f : f(a) = 1\}$.) Going in the other direction, the clopen subsets of $\widehat{B}$ form a Boolean algebra of sets which is isomorphic to B.

The crucial observation is that when we replace the internally defined notion of an ultrafilter with the external definition of topology, the duality becomes topological too. Specifically, we have the following folklore fact:

Theorem 4.1.45 (Topological Stone Duality). *For Boolean algebras B and C, B is isomorphic to C iff $\widehat{B}$ is homeomorphic to $\widehat{C}$.*

The topological version of Stone duality allows a great deal of flexibility in the way we represent the Stone space of B. This is because the algebra B is isomorphic to the algebra of the clopen sets of $\widehat{B}$. This exact same process would work for any space M provided that $M \cong_{\text{hom}} \widehat{B}$. For example, we state without proof the following well-known topological characterisation of Stone spaces.

Lemma 4.1.46 (Folklore). *For a Hausdorff space X, the following are equivalent (up to homeomorphism):*

1. *X is a Stone space;*

2. *X is the inverse (projective) limit of finite discrete spaces;*

3. *X is compact and totally disconnected.*

Since we will not need these characterisations in this section, we will not clarify the standard terminology used in the lemma above. We will return to the topological characterisations in the next section. Instead, we now describe some other, fairly explicit ways to represent Stone spaces using trees.

Tree representations

We want to represent the Stone spaces of c.e. Boolean algebras as Π^0_1 classes. Let T be a subtree of $2^{<\omega}$. Topologise $[T]$ by letting the basic open sets be those of the form $[\{\sigma\alpha : \alpha \in 2^\omega\}]$, where $\sigma \in T$. Then $[T]$ is a Polish space. The standard metric between $\xi, \eta \in [T]$ is defined to be

$$d(\xi, \eta) = 2^{-n},$$

where n is the length of the longest common initial segment of ξ and η, but one could also use Cantor's middle third construction and the metric inherited from $[0, 1]$. In particular, the space $[T]$ is compact (being a closed subset of $[0, 1]$). To see that $[T]$ is compact directly, use that T is finite-branching, i.e., it contains only finitely many strings of any given length.

The basic open sets are in fact clopen, i.e., open and closed. The collection of all clopen sets, which are exactly the finite unions of basic clopen sets, forms a Boolean algebra. Recall that $2^{<\omega}$ stands for the tree of finite strings of 0-s and 1-s, including the empty string, up to the prefix relation.

Lemma 4.1.47. *The Stone space of all ultrafilters of any countable Boolean algebra is homeomorphic to $[T]$ for some tree $T \subseteq 2^{<\omega}$.*

Proof. Let $\{b_n\}_{n\in\omega}$ be a listing of the universe of a countable Boolean algebra B. For any string $\sigma \in 2^{<\omega}$, let $b_\sigma \in B$ be defined as $\bigwedge\{b_n : \sigma(n) = 1\} \wedge \bigwedge\{\overline{b_n} : \sigma(n) = 0\}$, where $b_\lambda = 1_B$ by convention (λ is the empty string). Let $T = \{\sigma \in 2^{<\omega} : b_\sigma \neq 0_B\}$. If $f \in [T]$, let $U_f = \{b_n : f(n) = 1\}$. It is easily seen that U_f is an ultrafilter of B and that the mapping $f \mapsto U_f$ is a homeomorphism of $[T]$ onto the Stone space of B. □

We say that a tree $T \subseteq 2^{<\omega}$ *represents* the Boolean algebra B if $[T]$ is homeomorphic to the Stone space of B.

Remark 4.1.48. We note that there are at least two possible interpretations of the term "tree representation of a Boolean algebra" that can be encountered in the literature. In [207], Goncharov uses a different notion that he calls a "tree basis". Goncharov's approach is algebraic rather than topological. He views elements of the tree as independent generators, and under his approach dead ends of the tree correspond to atoms. In our approach, however, atoms correspond to isolated paths. Goncharov's approach is most convenient in the study of decidable Boolean algebras, while our approach is best suited for computable and c.e. presented Boolean algebras. These two tree representations are further compared in [423].

We see that for a countable Boolean algebra B, its Stone space is compact and separable. The following proposition gives connections between the computability properties of Boolean algebras and their representing trees. A string σ is called a *terminal node* of a tree T if $\sigma \in T$ and no string τ properly extending σ is in T. We write λ for the empty string.

Proposition 4.1.49 (Folklore; see [423]). *(i) Every computable Boolean algebra is represented by some computable tree $T \subseteq 2^{<\omega}$ with no terminal nodes. Conversely, every computable tree $T \subseteq 2^{<\omega}$ with no terminal nodes represents some computable Boolean algebra.*

(ii) Every c.e. presented Boolean algebra is represented by some computable tree $T \subseteq 2^{<\omega}$. Conversely, every computable tree $T \subseteq 2^{<\omega}$ represents some c.e. presented Boolean algebra.

Proof. The first statement of the first part follows immediately by effectivizing the proof above that every countable Boolean algebra is represented by some tree $T \subseteq 2^{<\omega}$. The converse is also easy to check. For the second part, assume that B is a c.e. Boolean algebra. The above argument produces a tree $T \subseteq 2^{<\omega}$ which represent B and is co–c.e., i.e. $2^{<\omega} - T$ is c.e. But then there is a computable tree $U \subseteq 2^{<\omega}$ such that $[U] = [T]$. (Let U consist of all strings $\sigma \in 2^{<\omega}$ such that no string $\tau \subseteq \sigma$ has been enumerated out of T by stage $|\sigma|$ in a fixed enumeration of $2^{<\omega} - T$.) The converse is easy to check because if T is a computable tree, $\{\sigma \in T : [T_\sigma] \neq \varnothing\}$ is co–c.e. by König's Lemma. □

This new correspondence allows us to give alternative proofs of some results about representations of Boolean algebras. For example, we can give an alternative proof of the instance of Corollary 4.1.16 that was necessary in the final step of our proof of Feiner's Theorem 4.1.30.

Proposition 4.1.50 (Feiner). *Each $\varnothing'$-computable Boolean algebra is isomorphic to some c.e. presented Boolean algebra.*

Proof. By Proposition 4.1.49 it suffices to show that for any $\varnothing'$–computable tree $T \subseteq 2^{<\omega}$ with no terminal nodes, there is a computable tree $[U] \subseteq 2^{<\omega}$ such that $[U] \cong [T]$. To each string $\sigma \in T$ we assign a string $f(\sigma) \in U$, where f will be a certain $\varnothing'$–computable partial function. We will have that, for $\sigma, \tau \in T$, that $f(\sigma) \preceq f(\tau)$ iff $\sigma \preceq \tau$. From this it follows easily (as T has no terminal nodes) that each string in the range of f is extendible to a path in $[U]$. Conversely, each string extendible to a path in $[U]$ will be extendible to a string in the range of f. From this it follows that f induces a homeomorphism φ from $[T]$ to $[U]$, i.e. $\varphi(g) = \bigcup_{\sigma \preceq g} f(\sigma)$, for $g \in [T]$.

To construct f and U, we use $\{T_s\}_{s \in \mathbb{N}}$, a computable approximation to T. By modifying this approximation if necessary, we may assume that for each s, the set T_s is a nonempty tree with no terminal nodes. Let s_0 be the least number s such that for all $t \geqslant s$ and all strings $\sigma \in 2^{<\omega}$ with $|\sigma| \leqslant 1$, $\sigma \in T$ iff $\sigma \in T_t$. Then $f(\lambda)$ will be a string in U of length s_0. Note that we may computably approximate s_0 in such a way that our initial approximation is 0, and if our approximation at $s+1$ differs from that at s, then our approximation to s_0 at stage $s+1$ is simply s. Thus our initial approximation to $f(\lambda)$ is λ, and we start the construction of U by letting it agree with T_0 on strings of length at most 1. (Note that T_0 contains λ and at least one string of length 1 since it is a nonempty tree with no terminal nodes.) In building U at stage $s+1$, we decide membership in U for all binary strings of length $s+1$, and we assume inductively that U contains at least one string of length s. If our approximation to s_0 changes at stage $s+1$, then we effectively choose a string τ of length s in U and let it be our new candidate for $f(\lambda)$. We want the rest of the construction of U to take place above τ, so we omit from U all strings of length $s+1$ which do not extend τ. Further, we act in the belief that the approximation T_{s+1} to T is correct on binary strings of length 1. Thus, for $i \leqslant 1$, we put $\tau^\frown i$ into U iff the string $\langle i \rangle \in T_{s+1}$. (This puts at least one string of length $s+1$

into U because T_{s+1} contains at least one string of length 1.) Obviously, this process must converge because our approximation to s_0 converges. The inductive step for defining $f(\sigma^\frown i)$ for $i \leqslant 1$ given $f(\sigma)$ is similar. We will have $f(\sigma^\frown i) \succeq f(\sigma)^\frown i$ and $|f(\sigma^\frown i)|$ will be the least number s such that $s > |f(\sigma)|$ and for all $t \geqslant s$ and all $j \leqslant 1$, $\sigma^\frown i^\frown j \in T$ iff $\sigma^\frown i^\frown j \in T_t$. □

In the final Boolean algebra subsection below we give a result that puts together all our methods developed so far, including tree representations, Downey-Jockusch Theorem 4.1.25, and Feiner's Theorem 4.1.30. We will not need this result in later chapters.

4.1.7 Rank 1 Boolean algebras*

Jockusch and Soare [274] essentially showed that, much in the spirit of Richter's Theorem 3.2.13, one has to use transfinite methods and uniformity to code any non-trivial information into a Boolean algebra (see Exercise 8.3.41). That is, there is no natural "coding" of Σ_n^0-sets into computable Boolean algebras, for any finite n. Also, the use of n-atoms for arbitrarily large $n \in \mathbb{N}$ seemed crucial in our proof of Feiner's Theorem 4.1.30. This leads to the following question.

Question 4.1.51. *Is there an arithmetical Boolean algebra of finite Cantor-Bendixson rank not isomorphic to a computable one?*

In this subsection, we use tree representations to answer this question in the affirmative.

If X is any Stone space, the Cantor–Bendixson derivative X' of X is the set of non–isolated points of X, with the subspace topology. In terms of trees, isolated points correspond to isolated paths. If X is a topological space, let $I(X)$ denote the closed set of points in X which are limit points of the isolated points of X. If X is a Stone space, then so is $I(X)$ (in the relative topology).

The following easy construction shows that every separable Stone space is homeomorphic to $I(X)$ for some separable Stone space X of Cantor–Bendixson rank $\leqslant 1$.

Definition 4.1.52. Let $T \subseteq 2^{<\omega}$ be a binary tree. Define a new tree $F(T) \subseteq 3^{<\omega}$ by starting with $2^{<\omega}$ and then attaching an isolated path to each node of T.

The paths through $F(T)$ are those infinite strings which either consist entirely of 0's and 1's or else consist of a string in T followed by an infinite string of 2's. The isolated paths in $[F(T)]$ are clearly those of the latter form, so $I([F(T)]) = [T]$. It also follows from these remarks that $[F(T)]' = [2^{<\omega}]$ which is a perfect space. Hence the Cantor–Bendixson rank of $[F(T)]$ is at most 1.

Theorem 4.1.53 (Downey and Jockusch [130]). *There is a c.e. presented Boolean algebra B of Cantor–Bendixson rank 1 which has no computable presentation.*

Proof. Let B_0 be a Boolean algebra which is $\varnothing^{(3)}$–c.e. but not isomorphic to any $\varnothing^{(3)}$–computable Boolean algebra. Such a Boolean algebra may be obtained by relativizing the proof of Feiner's theorem to $\varnothing^{(3)}$. Let $T \subseteq 2^{<\omega}$ be a $\varnothing^{(3)}$–computable tree which represents B_0. Such a tree T exists by Proposition 4.1.49, relativised to $\varnothing^{(3)}$. Finally, let B be a Boolean algebra represented by $F(T)$, where $F(T)$ is as defined in Definition 4.1.52. It is clear that B has Cantor–Bendixson rank at most 1. The following lemmas will show that B satisfies the rest of the conclusion of the theorem.

Lemma 4.1.54. *B is not isomorphic to any computable Boolean algebra.*

Proof. Suppose for a contradiction that B is isomorphic to B_1, a computable Boolean algebra. Let T_1 be a computable tree without terminal nodes which represents B_1. Such a tree T_1 exists by Proposition 4.1.49, and $[T_1] \cong [F(T)]$ where T is as chosen above. It follows that $I([T_1]) \cong I([F(T)]) = [T]$. We now define a Σ_2^0 tree $T_2 \subseteq T_1$ such that $[T_2] \cong I([T_1])$. First, let I be the set of strings σ on T_1 such that any two extensions of σ on T_1 are compatible. It is easily seen (using the fact that T_1 has no terminal nodes) that I is the set of nodes of T_1 which lie on a unique (necessarily isolated) branch of T_1. Let T_2 be the set of nodes $\sigma \in T_1$ such that there exist incompatible strings τ_1, τ_2 which are each in I and extend σ. It is easy to see that I is Π_1^0, and so T_2 is Σ_2^0, and that $[T_2] \cong I([T_1])$. Finally, let T_3 be the set of strings $\sigma \in T_2$ such that there exists $f \in T_2$ with $f \supseteq \sigma$, i.e. T_3 is the set of extendible nodes of T_2. By König's Lemma, T_3 is also the set of nodes σ of T_2 which have extensions in T_2 of each length $\geqslant |\sigma|$, so $T_3 \in \Pi_3^0$. Thus T_3 is a $\varnothing^{(3)}$–computable tree, and it clearly has no terminal nodes. But $[T_3] = [T_2] \cong I([T_1]) \cong [T]$, so T_3 represents B_0. Thus by Proposition 4.1.49, B_0 is isomorphic to some $\varnothing^{(3)}$–computable Boolean algebra, in contradiction to our choice of B_0. □

Since B is not isomorphic to any computable Boolean algebra, it does not have rank 0, so its rank is exactly 1. It remains to show that B is isomorphic to some c.e. Boolean algebra. The following lemma (relativised to $\varnothing'$) is the main step in showing that B is isomorphic to some c.e. Boolean algebra.

Lemma 4.1.55. *Let $U \subseteq 2^{<\omega}$ be a $\varnothing^{(2)}$–computable tree. There is a computable tree $V \subseteq 3^{<\omega}$ with no terminal nodes such that $[V] \cong [F(U)]$.*

Proof. By Theorem 4.1.25, it is sufficient to build a $\varnothing'$-computable V with a $\varnothing'$-computable set of atoms (isolated paths). Using Proposition 4.1.49 (relativised to $\varnothing'$), we can fix a Σ_2^0-tree $\hat{U}$ (and indeed, even a Δ_2^0-tree) such that $[U] = [\hat{U}]$.

The proof is therefore reduced to the following situation, which is then relativised to $\varnothing'$. We are given a c.e. $U \subseteq 2^{<\omega}$, and we have to construct a computable $V \subseteq 3^{<\omega}$ with no terminal nodes such that $[V] \cong [F(U)]$, and so that V has a computable set of atoms.

This is quite straightforward. Put an atom (an infinite isolated path) of the form $\sigma 2^\omega$ whenever σ is enumerated into $U \subseteq 2^{<\omega}$. We also declare it an atom (or isolated) immediately. Since the 2^ω-part does not contain isolated paths, we end up with a computable tree with a computable set of isolated paths. If σ is not on a path in 2^ω, then (by compactness) it bounds only finitely many nodes, so only finitely many extra atoms will be added below σ when compared to U. We can assume without loss of generality that $\tau = \sigma^-$, the predecessor of σ, is in U. In particular, τ already bounds an atom $\tau 2^\omega$ in V.

Without loss of generality, our Boolean algebra is infinite, and therefore the procedure described above will result in a (tree representing) a Boolean algebra having infinitely many atoms. By the Remmel-Vaught Theorem 4.1.12, we can *completely remove* the finite collection of atoms (paths) in V bounded by each such $\sigma \notin U$, and end up with a (tree representing) an isomorphic algebra. This is possible because $\tau = \sigma^-$ already bounds an atom, so we can identify this atom with the finite set of atoms below σ and apply the Remmel-Vaught Theorem 4.1.12. We conclude that $[F(U)] \cong [V]$. □

Lemma 4.1.55 (relativised to $\varnothing'$) shows that there is a $\varnothing'$–computable tree $V \subseteq 2^{<\omega}$ with no terminal nodes such that $[V] \cong [F(T)]$. Then, by relativising Proposition 4.1.49 to $\varnothing'$, it follows that V represents some $\varnothing'$–computable Boolean algebra B_2, and $B_2 \cong B$ since they are represented by V and $F(T)$, respectively. By Proposition 4.1.50, since B_1 is $\varnothing'$-computable, it is isomorphic to some c.e. Boolean algebra, and thus so is B. □

Exercises

Exercise° 4.1.56. Prove Feiner's Theorem 4.1.30 using generating trees instead of interval algebras.

Exercise 4.1.57. A an algebraic structure A is said to be *primitive recursive* if the domain and the operations of the structure are (uniformly) primitive recursive. A primitive recursive structure is *fully primitive recourse* or *punctual* if the domain of the structure is ω (or an initial segment of ω). A primitive recursive structure is *punctually* 1*-decidable* if there is a primitive recursive procedure that, given an existential sentence with parameters from the structure, outputs -1 if this sentence fails, and otherwise returns the tuple of elements of the structure witnessing the existential quantifiers. A bijection $f : \omega \to \omega$ is *punctual* if both f and f^{-1} are primitive recursive.

1. Prove° that every punctually 1-decidable structure is punctual.
2. Show that every computable Boolean algebra has a punctual presentation (Kalimullin, Melnikov, and Ng [282]).
3. Show that a Boolean algebra has a punctually 1-decidable presentation iff it has a computable copy in which the set of atoms is computable. Conclude that every 1-decidable Boolean algebra has a punctually 1-decidable presentation (Alaev [7], also rediscovered by Downey and Askes [23]).
4. Prove that a Boolean algebra has a unique 1-decidable presentation up to punctual isomorphism iff it is computably categorical (Alaev [7]).
5. Show that for a computable Boolean algebra B with a computable set of atoms, the following are equivalent (Dorzhieva et al. [115]):
   - Every 1-decidable presentation A of B is computably isomorphic to some punctually 1-decidable $P \cong B$.
   - B splits into finitely many $C_0, ..., C_k$ such that each C_i is either atomless, an atom, or a 1-atom.

   (Compare this description with Exercise 10.1.95.)

4.1.8 Further related results*

As mentioned earlier, there are numerous papers and results concerning computable Boolean algebras. The standard references for results proven before 2000 are [207] and [447]. Another reference is Odintsov's survey [422], which, however, overlaps significantly with [207]. When it comes to more recent results, there are few comprehensive surveys available. One notable reference is [43], although it focuses on specific topics rather than providing a broad overview.

4.2 Computably compact spaces

In this section, we return to computable separable spaces. We prove that there exists a computable right-c.e. Stone space that is not homeomorphic to any computable Polish space, and that every computable Polish Stone space is homeomorphic to a computably compact one. These were stated as (1) and (2) of Theorem B in the first chapter.

To prove these results, we will establish two more versions of Stone duality, but this time not restricted to spaces of the form $[T]$, where $T \subseteq 2^{<\omega}$. To handle an arbitrary compatible metric, we will need to develop the foundations of the theory of computably compact and computable Polish spaces, which will also be important in the next sections.

This section is based on [139].

4.2.1 Definitions

All of our spaces are Polish (separable and completely metrisable) spaces. Such spaces are also sometimes called Polishable. All spaces in this section are non-empty and compact, unless stated otherwise. In all our definitions we fix some metric compatible with the topology of the space. Since we are mainly interested in compact spaces where any compatible metric is complete, *we always assume that the metric is complete.*

Computable Polish spaces

Recall Definition 1.2.5:

Definition 4.2.1. A real ξ is

- *right-c.e.* if $\{r \in \mathbb{Q} : \xi < r\}$ is computably enumerable (c.e.);
- *left-c.e.* if $\{r \in \mathbb{Q} : \xi > r\}$ is c.e.;
- *computable* if it is both left-c.e. and right-c.e..

A Polish space (M, d) is *right-c.e. presented* or *admits a right-c.e. metric* if there exists a sequence $(\alpha_i)_{i \in \mathbb{N}}$ of M-points which is dense in M and such that for every $i, j \in \mathbb{N}$, the distance $d(\alpha_i, \alpha_j)$ is a right-c.e. real, uniformly in i and j. The definition of a left-c.e. Polish space is obtained from the notion of a right-c.e. Polish space using the notion of a left-c.e. real, mutatis mutandis.

In this section we usually consider Polish spaces under homeomorphism, that is, a Polish space has a right-c.e. (Polish) presentation if it is *homeomorphic* to the completion of a right-c.e. metrised space. To emphasise that neither the metric nor the dense sequence is fixed, we may occasionally use the term "computably metrised" rather than "computable Polish". Since most of our spaces are compact, the metric is automatically complete.

Definition 4.2.2. A Polish space is *computably presented* if there is a (complete, compatible) metric on the space which is computable, i.e., is both right-c.e. and left-c.e..

Basic open balls

Fix a Polish space, a dense sequence in the space, and a complete compatible metric d. A basic open ball is a ball of the form $\{x : d(x,c) < r\}$, where c is the "centre" of B that comes from the fixed dense set, and $r \in \mathbb{Q}$ is its "radius". For a basic ball B, we write $r(B)$ for the radius of B, and we use $c(B)$ to denote its (distinguished) centre. For a basic open ball B, write B^c for the basic closed ball with the same centre and radius as B. The closure $\overline{B}$ of B does not have to be equal to B^c in general (think of an isolated point in $B^c \backslash B$). In this section, $\overline{B}$ and B^c should not be confused with the (set-theoretic) complement of B; if we ever need to consider the complement of B, we will write $M \backslash B$. The set-theoretic inclusion of basic open balls is not c.e. in a computable Polish space in general. The following stronger notion is c.e.; it will be very useful.

Definition 4.2.3. A basic open ball B is said to be *formally included* in a basic open ball D, written $B \subseteq_{form} D$, if $r(B) + d(c(B), c(D)) < r(D)$.

The definition works for closed (or closures of) basic balls as well. This notion has been around for many decades; see, e.g., [482], where it is called strong inclusion. Formal inclusion is transitive; this is because $d(x,y) + r_2 < r_1$ and $d(y,z) + r_3 < r_2$ (together with the triangle inequality) imply $d(x,z) + r_3 < r_1$. If the centres and the radii are computable (not necessarily special and rational, respectively), formal inclusion remains c.e.. The same can be said about formal disjointness defined as follows. Two basic open balls B and D are *formally disjoint* if $r(B) + r(D) < d(c(B), c(D))$. We note that strong inclusion remains c.e. in the context of right-c.e. metric spaces, while strong disjointness remains c.e. in left-c.e. metric spaces.

Definition 4.2.4. Let X be a computable Polish space. For a point $x \in X$, its *name* is the set

$$N^x = \{i \in \mathbb{N} : x \in B_i\},$$

where $(B_i)_{i \in \mathbb{N}}$ is an effective listing of all basic open balls in X.

Recall that a point $x \in X$ is computable if there is a computable *fast Cauchy sequence* $(x_i)_{i \in \mathbb{N}}$ such that $d(x_i, x) < 2^{-i}$. Of course, 2^{-i} can be replaced with 2^{-i+17} or even $2^{-f(i)}$ for a sufficiently nice computable f if necessary, and these will be (computably) equivalent notions. We therefore allow $d(x_i, x) < 2^{-i+1}$ and $d(x_i, x) < 2^{-i-1}$ for such a sequence when convenient. In fact, this slightly annoying index can be completely removed from consideration using names of points.

Fact 4.2.5. *A point x in a computable Polish space is computable iff N^x is computably enumerable.*

We leave the proof as an exercise (Exercise 4.2.20).

Definition 4.2.6. An *open name* of an open set $U \subseteq X$ is a set $W \subseteq \omega$ such that

$$U = \bigcup_{i \in W} B_i.$$

An open set is *c.e.* or *effectively open* if it has a c.e. open name. A *closed name* of a closed set $C \subseteq X$ is the open name of $X \backslash C$. A closed set is *effectively closed* if $X \backslash C$ is effectively open.

Open names can be used to "topologise" another standard notion that we encountered earlier. Recall that in Definition 2.4.2 we defined a function f to be computable if there is a Turing functional which, on input a fast Cauchy name of x in $\mathcal{X}$, outputs a fast Cauchy name of $f(x)$.

An *enumeration operator* is a Turing operator (see §2.1.4) that is allowed to use only positive information about its oracle, i.e., when it asks a question "$n \in X$?" it can only use the positive answer "yes" in its instructions. It can be made more formal by allowing only instructions of the form "if $n \in X$ then do" and never using instructions of the form "if $n \notin X$ then do". The formal definition is as follows: A c.e. set W can be associated with a map $B \mapsto A$ according to the rule $A = \{x : (\exists u)(\langle x, u\rangle \in W \wedge D_u \subseteq B)\}$, where $(D_u)_{u \in \mathbb{N}}$ is a listing of all finite sets (given as tuples or finite strings). An enumeration operator turns *enumerations* into *enumerations*. A formal treatment of such operators and the induced enumeration reducibility can be found in [454]. Definition 2.4.2 is actually equivalent to the following.

Definition 4.2.7. We say that $f : X \to Y$ is computable if there exists an enumeration operator that, given (any enumeration of) the name N^x of $x \in X$, outputs (some enumeration of) the name $N^{f(x)}$ of $f(x) \in Y$.

To see why Definition 2.4.2 is equivalent to the definition above, note that Fact 4.2.5 was indeed uniform. Thus, we can computably turn an enumeration of N^x into a fast Cauchy name and vice versa.

Effectively continuous maps

The following definition is a generalisation of Definition 2.3.10.

Definition 4.2.8. Let X and Y be computable Polish spaces. A function $f\colon X \to Y$ is *effectively continuous* if there is a c.e. family F of pairs of (indices of) basic open sets such that:

(C1) for every $(U, V) \in F$, we have $f(U) \subseteq V$;

(C2) for every point $x \in X$ and every basic open E in Y such that $f(x) \in E$, there exists a basic open D in X with $(D, E) \in F$ and $x \in D$.

As we now show, this is equivalent to saying that, for some c.e. set F, $f^{-1}(B_i) = \bigcup_{(i,j)\in F} B_j$. (This follows from the elementary lemma below, which is a generalisation of Lemma 2.3.13.)

Lemma 4.2.9. *Let $f\colon X \to Y$ be a function between computable Polish spaces. Then the following conditions are equivalent:*

1. *f is effectively continuous.*
2. *There is an enumeration operator Φ that, on input an open name of an open set V in Y, lists an open name of the set $f^{-1}(V)$ in X.*
3. *f is computable, that is, there is an enumeration operator Ψ that, given the name of a point $x \in X$, enumerates the name of $f(x) \in Y$.*

Proof. (1) $\to$ (2). Suppose $V = \bigcup_{i\in W} B_i$. Note that (C2) implies that

$$f^{-1}(V) = \bigcup\{D \in X : (D, E) \in F \,\&\, \exists i \in W\, (D, B_i) \in F\},$$

and thus the name of $f^{-1}(V)$ can be listed using only positive information about W, with all possible uniformity.

(2) $\to$ (3). Note that $B \in N^{f(x)}$ iff $f^{-1}(B)$ contains a basic open set in N^x.

(3) → (1). Define a collection F of pairs (D, E) of (indices of) basic open sets in $X \times Y$ as follows. Fix a basic open E in Y. Enumerate all basic open D in X, and for each such D, enumerate all finite collections $D, A_1, \ldots, A_k$ of basic open sets (in X) such that $D \subseteq_{form} \cap_{i \leqslant k} A_i$ (meaning that D is formally contained in each A_i). Feed these finite collections to Φ and wait for some E to be enumerated in the output. When E is enumerated (if ever), put (D, E) into F.

We claim that F defined above satisfies (C1) and (C2). We check (C1). If $(D, E) \in F$, then $f(D) \subseteq E$. Indeed, suppose $d \in D$. There exists a sequence $D, A_1, \ldots, A_k$ such that $\varphi^{\{D, A_1, \ldots, A_k\}}$ enumerates E. Recall $D \subseteq_{form} \cap_i A_i$ implies $D \subseteq \cap_i A_i$, thus for any $d \in D$ the sequence listed by φ^{N^d} will contain E, and therefore $f(D) \subseteq E$.

We now check (C2). Fix $x \in X$ and a basic open $E \ni f(x)$. We must show that for some basic open $D \ni x$, $(D, E) \in F$. By assumption, φ^{N^x} enumerates $N^{f(x)}$ that contains E. Suppose E is listed with use $A_1, \ldots, A_k$. Since the A_i all contain x, there exists a basic open $D \ni x$ that is formally included in their intersection. Since the operator uses only positive information about its oracle, it will list E on input $\{D, A_1, \ldots, A_k\}$ as well, and thus (D, E) will be enumerated into F. (Indeed, in this argument, D does not actually need to be given to φ.) □

The proof above also works for right-c.e. spaces. It also works for computable topological spaces with c.e. formal (strong) inclusion that can be defined abstractly without any reference to a metric; see, e.g., [375, 482].

The definition of computable compactness

Recall that a complete metric space M is compact iff it is totally bounded; that is, for every $\varepsilon > 0$, there exists a finite set F of points such that every point of the space has distance less than ε to at least one point from F. A straightforward effectivisation of this criterion is the following definition that already appeared in the first chapter.

Definition 4.2.10. A computable Polish space is called *computably compact* if there exists a computable function that, given n, outputs a finite tuple of basic open balls of radii $< 2^{-n}$ that cover M.

When we consider finite covers, we usually say that we can *compute* a finite cover by basic open balls if we can compute the index of a finite set that codes the indices of the finitely many centres and the rational radii of basic open balls in the cover. This should not be confused with *enumerating* a finite cover, i.e., listing one ball after another in a c.e. fashion.

Fact 4.2.11. *Every compact computable Polish space is Δ^0_2-compact.*

Proof. Let $\mathcal{M} = (M, d, (p_i)_{i \in \mathbb{N}})$ be a compact computable metric space. A *compactness modulus* of $\mathcal{M}$ is any function that bounds

$$h(n) = \min\{j \colon \forall i \, \exists k < j \, d(p_i, p_k) \leqslant 2^{-n}\}$$

from above. We call h the least modulus of compactness.

Note that if $h(n) = j$, then the 2^{-n+1}-basic open balls centred in $p_0, \ldots, p_j$ cover the space. Since $d(p_i, p_k) \leqslant 2^{-n}$ is a Π^0_1 condition, if it fails, then its failure is witnessed by a special point.

Since the quantifier $\exists k < j$ is bounded, and since the space is compact, h is computable relative to $\varnothing'$. □

We note that a compact computable Polish space is computably compact iff it has a computable modulus of compactness.

The other two standard definitions

Definition 4.2.10 says that for every n we can compute one finite cover of the space by basic 2^{-n}-balls. From the computability-theoretic perspective, this definition does not seem quite as good as having access to all finite covers.

Definition 4.2.12. We define a computable Polish space to be *$*$-computably compact* if the collection of *all* finite covers of M by basic open balls can be given as a c.e. collection of explicit finite sets.

As we explain next, Definition 4.2.12 is also equivalent to:

Definition 4.2.13. We say that a computable Polish space is *computably countably compact* if there is a partial computable operator that, on input of any potential c.e. open basic cover, halts if it is a cover and outputs some finite sub-cover.

Interestingly, the two potential definitions suggested above (and a few more) turn out to be equivalent to Definition 4.2.10.

Theorem 4.2.14 (Folklore). *For a computable Polish space M, the following are equivalent:*

1. *M is computably compact;*
2. *M is $*$-computably compact;*
3. *M is computably countably compact.*

Proof. The implication (2) → (1) is obvious, and the equivalence of (2) and (3) is also elementary (Exercise 4.2.22).

Assume (1); we prove (2). Take a finite collection (B_i) of basic open sets and assume it is a cover. We must argue that eventually we will be able to effectively recognise that it is indeed a cover. The idea is that there exists an $\epsilon = 2^{-n}$ so small that every ϵ-cover of M is formally contained in this given cover. (This will be the "Lebesgue number" of the cover, in particular.) This will also be true for the ϵ-cover that will be given to us according to the definition of computable compactness. Since formal inclusion is c.e., we will be able to recognise that this formal inclusion has occurred.

It remains to prove that such an ϵ exists. We argue non-computably. Let c_i be the center of B_i, and r_i be its radius. Define for every i a function $f_i(x) = r_i - d(x, c_i)$ if x is in the ball B_i, and 0 otherwise. Define $g(x) = \sup_i f_i(x)$, which is also continuous. If (B_i) indeed was a cover, then the function g would be strictly positive because each x is inside one of the B_i.

Let v be its infimum that is achieved somewhere, by compactness. Take a rational $\epsilon = 2^{-m}$ less than $v/2$. Then for every point y, for some i, we have $\epsilon < r_i - d(y, c_i)$; that is,

$$d(y, c_i) + \epsilon < r_i,$$

equivalently, $B(y, \epsilon) \subset_{form} B_i$. This inclusion will still hold if we replace ϵ with an even smaller ϵ'. Thus, in particular, every basic open ϵ'-ball is formally included in one of the B_i. Consequently, (1) implies (2). □

Remark 4.2.15. The proof of (1) → (2) above additionally tells us that, for any given finite basic cover, there is an ϵ small enough so that any ϵ-cover formally refines the given cover. Also note that to recognise formal inclusion in a c.e. way, we do not need the radii r_i to be rational numbers; uniformly computable (real) r_i will suffice.

In view of Lemma 4.2.14, henceforth we use computable compactness and $*$-computable compactness interchangeably, and without further comment.

Elementary examples

Examples of computably compact spaces are the unit interval $[0, 1]$, the unit circle that can be viewed as the set of complex numbers having norm one: $\{\xi \in \mathbb{C} : ||\xi|| = 1\}$, the Hilbert cube, Cantor space 2^ω, and also geometric realisations (with rational parameters) of finite simplicial complexes that are central to algebraic topology. Simplicial complexes will be used as a tool later in the book.

One fundamental example comes from the theory of Π^0_1 classes. Cantor space 2^ω is a computable Polish space under the longest initial segment metric defined in §4.1.6: the distance between $\xi, \eta \in [T]$ is defined to be

$$d(\xi, \eta) = 2^{-n},$$

where n is the length of the longest common initial segment of ξ and η. Recall that in §2.1.6 we gave the following definition. A Π^0_1 class C is decidable if the tree $T \subseteq 2^{<\omega}$ such that $[T] = C$ is computable and has no dead ends. The fact below is immediate:

Fact 4.2.16. *Any non-empty decidable Π^0_1 class can be viewed as a computably compact space.*

Explanation. Let $C \subseteq 2^\omega$ be a decidable Π^0_1-class. The computable dense set is given by the effective enumeration of all paths through T of the form ℓ_σ, where σ ranges over all finite strings in T and ℓ_σ is the left-most infinite extension of σ along T. This makes C computable Polish. Computable compactness follows from the computable compactness of 2^ω and the computability of T. Indeed, we can list all 2^{-n}-covers corresponding to the basic clopen sets centred in $\sigma \in T$ at level n of T. □

We shall give much more intricate examples of computably compact spaces in due course.

There are several properties of computably compact spaces that are immediate from the definitions. These, for instance, include those summarised in the following:

Proposition 4.2.17.

1. *Let $f : M \to \mathbb{R}$ be computable and M be computably compact. Then $\sup_{x \in M} f(x)$ and $\inf_{x \in M} f(x)$ are computable real numbers. Furthermore, this is uniform.*

2. *The class of (non-empty) computably compact spaces is closed under taking (finite or computably infinite) direct products. More specifically, if $(M_i)_{i \in I}$ is a uniformly computable sequence of spaces, where $I \in \mathbb{N} \cup \{\omega\}$, then the direct product*

$$\prod_{i \in I} M_i$$

under (say) the metric

$$\sum_{i \in I} 2^{-i} \frac{d(x_i, y_i)}{1 + d(x_i, y_i)},$$

where x_i denotes the ith projection of $x \in \prod_{i \in I} M_i$, is a computably compact metric space.

We omit the elementary proof and leave it to Exercise 4.2.23. We remark that in (2), the choice of a dense computable sequence is not canonical.

Exercises

Some of the exercises below would be marked with at least one $*$ if they did not include extended hints.

Exercise° 4.2.18. Prove that a (compact) computable Polish space is computably compact if, and only if, for every n we can computably produce a finite tuple of basic *closed* balls that cover the space.

Exercise° 4.2.19 (Folklore)**.** Recall that the Hilbert cube is the space $H = [0, 1]^\omega$. The metric on the Hilbert cube H is given by Prop. 4.2.17(2). Since the usual metric on $[0, 1]$ is bounded by 1, we can simply use $\sum_i 2^{-i} d(x_i, z_i)$ for two points $(x_i)_{i \in \mathbb{N}}, (z_i)_{i \in \mathbb{N}} \in H$. Show that for a computable Polish (compact) M, the following are equivalent:

1. M is homeomorphic to a computably compact space;
2. M is homeomorphic to a computable closed subset of H.

Exercise° 4.2.20. Prove Fact 4.2.5.

Exercise° 4.2.21. Prove Theorem 4.1.45.

Exercise° 4.2.22. Complete the proof of Theorem 4.2.14.

Exercise° 4.2.23. Prove Proposition 4.2.17.

Exercise 4.2.24 (Dyment [152], Schröder [459])**.** It is easy to extend the notion of a c.e. (effectively) open set and an effectively closed set to computable topological spaces (Def. 2.4.26). A computable topological space X is *effectively normal* if, given (names of) disjoint effectively closed sets C_0 and C_1, we can effectively produce (names of) disjoint effectively open sets U_0 and U_1 that separate C_0 and C_1, i.e., so that $C_0 \subseteq U_0$ and $C_1 \subseteq U_1$. Prove the Effective Urysohn Lemma: Let X be an effectively normal computable topological space. Given disjoint effectively closed sets A and B uniformly produce a computable (i.e., effectively continuous) function $f_{A,B} : X \to [0, 1]$ so that $f_{A,B} \upharpoonright_A = 0$ and $f_{A,B} \upharpoonright_B = 1$. (Hint: Follow the standard textbook argument (e.g., [413, Thm 33.1]) but with one minor modification. To define the map, instead of the set of all rationals, use the set of the dyadic rationals.)

Exercise 4.2.25 (Amir and Hoyrup [11])**.** Show that any effectively compact, strong computable topological space (Def. 2.4.26) is effectively normal (Ex. 4.2.24).

Exercise* 4.2.26 (Dyment [152], Schröder [459])**.** Suppose that in an effectively normal (computable topological) space X there exists an effective enumeration $(C_i, D_i)_{i \in \mathbb{N}}$ of all (computable indices of) disjoint effectively closed subsets in X, perhaps with repetition. Assume that, additionally, for every $x \in X$ and every open $U \ni x$ there exist disjoint effectively closed $C \ni x$ and $D \supseteq (X \setminus U)$. Show that there exists a compatible metric on the space that can be realised as a computable (Type 2) function $X^2 \to \mathbb{R}$. (Hint: Use Exercise 4.2.24 to produce a uniformly effective list of functions

$$f_{C_i,D_i} : X \to [0,1]$$

that map the respective C_i to 1 and vanish at D_i. Define $g : X \to [0,1]^\omega$ to be

$$g(x) = (f_{C_i,D_i})_{i\in\mathbb{N}},$$

where the computable metric on $[0,1]^\omega$ is given by $d((x_i)_{i\in\mathbb{N}}, (y_i)_{i\in\mathbb{N}}) = \sum_i 2^{-i}|x_i - y_i|$. The function is computable. Since effectively closed sets separate points, it follows that g is injective, and thus g is a computable homeomorphic embedding of X into $[0,1]^\omega$.)

Exercise 4.2.27** (Miller [392]). Recall that the name of a point x in a computable Polish space is

$$\{i : x \in B_i\},$$

where $(B_i)_{i\in\mathbb{N}}$ is an effective list of all basic open balls of M (Def. 4.2.4). An oracle Y computes some fast Cauchy name of a point iff Y can enumerate N^x; this is Fact 4.2.5 (relativised). The *degree spectrum* of a point $x \in M$ is

$$DSp_M(x) = \{Y \in 2^\omega : N^x \text{ is c.e. relative to } Y\}.$$

1. Prove° that for any computable Polish space M and any $x \in M$, there is some point $y \in H = [0,1]^\omega$ (see Exercise 4.2.19) such that $DSp_M(x) = DSp_H(y)$.
2. Prove that for any computable Polish space M and any $x \in M$, there is some point $f \in C[0,1]$ (Example 2.4.18(2)) such that $DSp_M(x) = DSp_{C[0,1]}(f)$.
3. Show** that there is an $f \in C[0,1]$ such that $DSp_{C[0,1]}(f)$ contains no least Turing degree.

4.2.2 Deciding the intersection

One standard way of using a (finite) cover of a compact space in dimension theory and algebraic topology is to use Alexandroff's notion of a *nerve*, which we will need (and will properly define) in the next chapter. The nerve of a cover is a simplicial complex in which the faces are the collections of basic open sets that have a non-trivial intersection; i.e., each basic open set is a 0-dimensional simplex (a node), and balls $\{B, C, D\}$ form a 2-dimensional face if $B \cap C \cap D \neq \varnothing$, and so on.

From the computability-theoretic standpoint, the issue with this definition is that, for a fixed finite open cover, the non-emptiness of each specific intersection is merely Σ^0_1, and this cannot be improved in general. However, if we choose our covers very carefully, we can use special covers where we can decide intersections. Our next result shows that in a computably compact space we can find one such nice ϵ-cover for every $\epsilon \in \mathbb{Q}$. To state the result formally, we push the notion of computable compactness to its limits.

Definition 4.2.28 (Downey and Melnikov [139]). A set of basic open balls is $\cap$-decidable if for every finite sequence of balls $B_0, \ldots, B_i$ from the set, we can computably decide whether $\bigcap_i B_i = \varnothing$.

Definition 4.2.29 (Downey and Melnikov [139]). A (compact) computable Polish M is *nerve-decidable*, or ***-computably compact*, if for every $n > 0$ we can computably find a finite 2^{-n}-cover K_n (represented as a finite tuple of basic open balls) of M so that K_n is $\cap$-decidable uniformly in n. (By Remark 4.2.15, we can additionally assume that K_{n+1} formally refines K_n.)

Theorem 4.2.30 (Downey and Melnikov [139]). *A computable Polish M is nerve-decidable ($**$-computably compact) iff it is computably compact.*

Proof. Obviously, $**$-computably compactness implies computable compactness. To this end, we assume computable compactness of M. Recall that, for a basic open B, we write B^c for the basic closed ball with the same centre as B, and that the closure $\overline{B}$ of B does not have to be equal to B^c in general.

Lemma 4.2.31. *Suppose M is computably compact. Then, for basic closed balls B_i^c and B_j^c, the property $B_i^c \cap B_j^c = \varnothing$ is c.e. uniformly in i, j. The same is true for any finite collection of basic closed balls.*

Proof. The open set $M \backslash B_i^c$ is c.e.. Indeed, we just list all the basic open balls that are formally disjoint from B_i^c. Thus, the union of the complements, which is the complement of the intersection $B_i^c \cap B_j^c$, is also c.e. open. It covers the space iff the intersection is empty. By computable compactness of M, this is c.e. The case of finitely many balls is similar. □

If S is a finite cover, then for each subset $\{B_1, \ldots, B_k\}$ of S, exactly one of the possibilities is realised:

(*a*) $\bigcap_{i \leqslant k} B_i^c = \varnothing$, or

(*b*) $\bigcap_{i \leqslant k} B_i \neq \varnothing$, or

(*c*) $\bigcap_{i \leqslant k} B_i^c \neq \varnothing$ but $\bigcap_{i \leqslant k} B_i = \varnothing$.

Note that there are only finitely many conditions like that in total. Also, (*b*) is c.e., and (*a*) is c.e. by Lemma 4.2.31. However, (*c*) is more complex. We argue that there is an ϵ-cover for which the third possibility (*c*) is never realised for any finite collection of balls from the cover. If we succeed in proving that such a cover always exists, we will just search for a cover such that each finite collection of basic balls in the cover satisfies either (*a*) or (*b*).

Fix a finite $\epsilon/2$-cover of the space by basic open balls, and replace each ball in the cover with a ϵ-ball with the same centre. Let S be this new ϵ-cover. The third alternative (*c*) can be witnessed by at most one choice of radii. If this is the case, shrink the radii of all $B \in S$ by a $\delta < \epsilon/2$. Then (*a*) and (*b*) will still hold. □

A stronger condition

It will be convenient to have a computable system of covers (K_n) so that not only each K_n is $\cap$-decidable but the whole collection $\bigcup_n K_n$ is $\cap$-decidable. We are not sure whether such covers can be uniformly constructed for basic open balls with rational radii (represented as a pair of integers), but we can contract such a system for balls with centres in special points and uniformly computable radii that are not necessarily rational. We call such balls *basic computable open.*

Definition 4.2.32 (Downey and Melnikov [139]). A computable Polish M is *strongly computably compact* if for every $n > 0$ we can computably find a finite 2^{-n}-cover K_n (represented as a finite tuple of basic computable open balls) of M so that $\bigcup_n K_n$ is $\cap$-decidable.

As before, by Remark 4.2.15, we can additionally assume that K_{n+1} formally refines K_n.

Theorem 4.2.33 (Downey and Melnikov [139]). *A computable Polish space is computably compact iff it is strongly computably compact.*

Proof. By slightly increasing the radii of all the balls in a cover, we can ensure their radii are rational. Thus, every strongly computably compact space is computably compact. To this end, we assume the space is computably compact.

Iterate the proof of the previous Theorem 4.2.30. Suppose we have come up with a $\cap$-decidable K_0 and need to find K_1 so that $K_0 \cup K_1$ is $\cap$-decidable. But to find such a cover, we might have to slightly shrink the radii of the balls that we have already put into K_0. But note K_0 must still satisfy the closed properties $\bigcap_{i \leqslant k} B_i^c = \varnothing$ and finitely many open properties $\bigcap_{i \leqslant k} B_i \neq \varnothing$. The former is not an issue since the radii will decrease. The latter however needs to be maintained more carefully. When we first discover the finitely many open relations of the form of finitely many strict inequalities (when K_0 is first introduced), we also *compute* a rational parameter $\delta_0 > 0$ such that the relations will still hold if we decrease the radii of the balls by at most δ_0. This is possible since the conditions are just finitely many strict inequalities.

We then define $\delta_{0,n} = 2^{-n-2}\delta_0$ and note that $\sum_i \delta_{0,i} < \delta_0$. We intend to shrink the radii of each ball in K_0 by at most $\delta_{0,s}$ at stage s. This will make the radii in the balls computable while maintaining the finitely many conditions that K_0 needs to satisfy.

We also iterate this. When we define K_1, we will have more open conditions to maintain for $K_0 \cup K_1$. We compute a $\delta_1 > 0$ and set $\delta_{1,n} = 2^{-n-2}\delta_1$. We also ensure that $\delta_{1,n} \leqslant \delta_{0,n}$, for every n. When we define (our first approximation to balls in) K_2 at stage t, we will allow balls in K_0 to shrink by at most $\delta_{1,t} \leqslant \delta_{0,t}$ and balls in K_1 by at most $\delta_{1,t}$. All the finitely many conditions will still be satisfied.

We iterate this process until, in the end of the construction, we finally get a collection of computable balls $\bigcup_n K_n$. We leave the formal details to Exercises 4.2.37-4.2.39. □

Remark 4.2.34. In the proof of Theorem 4.2.33, instead of using basic open covers, we could use basic closed covers. We can also come up with any combination of open, closed and closures (e.g., decide whether $B_i \cap \overline{B_j} \cap B_k^c = \varnothing$) for any computable balls in K constructed in Theorem 4.2.33. Also, by Remark 4.2.15, we can always assume that K_{n+1} is formally contained in K_n, and this will still be true if we choose working with closed covers.

Definition 4.2.35. If a computable sequence (K_n) of finite 2^{-n}-covers of computable balls satisfies the properties described in Theorem 4.2.33 and Remark 4.2.34, then we say that (K_n) is a *fully $\cap$-decidable system of covers* of the space[3].

Lemma 4.2.36. *Let $K = \bigcup_n K_n$ be a fully $\cap$-decidable system of covers of a computably compact M. Then, for every closed ball D^c in K we can enumerate all basic open B in M such that $B \cap D^c \neq \varnothing$.*

[3]These K_n are represented as a uniformly computable sequence of indices of radii and centres, and we can choose whether we want to consider open, closed, or closures of open balls that have these parameters. For instance, when we say "$B^c(r,q)$ is in K_n" or "$\overline{B(r,q)}$ in K_n", or the same for $B(r,q)$, we really mean that parameters $\langle r, q\rangle$ are listed in K_n (where q is given as an index of a computable real).

Proof. Suppose $B \cap D^c \neq \varnothing$ and let x be any (not necessarily special) point in the intersection. Suppose the radius of B is δ, and let c_1 be the centre of B, and r_1 its radius. For some positive δ, we have $d(x, c_1) = r_1 - \delta$. Fix n so that $2^{-n} < \delta/2$, and consider the finite set K_n. Since K_n is a (closed or open) cover of the whole space, there must exist some $C \in K_n$ such that $x \in C$. Since $x \in D^c$, it must be that $C \cap D^c \neq \varnothing$, and (by our assumption) this can be recognised computably. We claim that for this C, we have that C is formally included into B. Indeed, if c_2 is the centre of C and r_2 is it radius, then we have that $d(x, c_2) < r_2 \leqslant 2^{-n} < \delta/2$, and therefore

$$d(c_1, c_2) + r_2 \leqslant d(x, c_1) + d(x, c_2) + \delta/2 < r_1 - \delta + \delta/2 + \delta/2 = r_1,$$

which is the same as to say that C is formally included into B. It follows that B intersects D^c iff there is an $n > 0$ and a ball $C \in K_n$ such that $C \cap D^c \neq \varnothing$ and C is formally included into B. This is a Σ^0_1-property. □

Exercises

Exercise° 4.2.37. Give a complete formal proof of Theorem 4.2.33.

Exercise° 4.2.38 (Folklore; see [57])**.** Prove the Effective Baire Category Theorem: Let $(U_i)_{i \in \mathbb{N}}$ be a sequence of uniformly c.e. dense open subsets of a computable Polish M. Then for any basic open B in M there is a computable $\xi \in B \cap \bigcap_{i \in \mathbb{N}} M_i$.

Exercise° 4.2.39 (Hoyrup[4])**.** Derive Theorem 4.2.33 using the Effective Baire Category Theorem (Exercise 4.2.38).

Exercise* 4.2.40 (Downey and Melnikov [139])**.** The covering dimension of M is the least $n \in \mathbb{N} \cup \{\infty\}$ such that every open cover of M has a refinement of order $n+1$, i.e., each point belongs to at most $n+1$ sets. It is well-known that a compact space of covering dimension n can be homomorphically embedded into $\mathbb{R}^{2n+1}$. Use $\cap$-decidable covers to prove that any computably compact Polish space of covering dimension n can be computably homomorphically embedded into $\mathbb{R}^{2n+1}$.

Exercise* 4.2.41 (Hoyrup, Melnikov, and Ng [267])**.** The notions of computable topological, strong computable topological, and effectively compact topological presentations were introduced in Definition 2.4.26. Show that, for a compact Polish space X, the following are equivalent:

1. X has an effectively compact strong topological presentation.
2. X admits a computably compact Polish presentation.

Conclude that every effectively compact topological space admits a Δ^0_2-Polish presentation. (Hint: To see why $2 \to 1$, use a strongly $\cap$-decidable system of covers. For the other implication, check that every X satisfying 1. also has the properties necessary to produce the metric given in Exercises 4.2.24, 4.2.25 and 4.2.26. Note the resulting metric is necessarily complete. Produce a computable dense sequence and verify that the presentation is computably compact.)

Exercise* 4.2.42 (Downey and Melnikov [139])**.** This exercise establishes that the computably compact spaces are exactly those whose continuous diagram is decidable. While it is not difficult, it requires some background in continuous logic ([46]). Let (M, d) be a computable Polish compact

[4] Personal communication.

space of diameter $\leqslant 1$. Define *the (full) continuous diagram* of (M, d) as follows. The atomic formulae are just $d(x, y)$ and the constant functions 0 and 1. We close these formulae under finite iterations of $\sup, \inf, \wedge, \vee, \frac{1}{2}\cdot$, and $\dot{-}$. We say that M is *continuously decidable* if its continuous diagram is uniformly Type II computable. That is, given (the Gödel number of) a continuous formula $\phi(\bar{x})$ from the full continuous diagram of M, we can uniformly produce an index for a Turing operator that computes the function

$$[\phi(\bar{x})] : M^n \to [0, 1],$$

where $\bar{x} = x_1, \ldots, x_n$. Show the following are equivalent:

1. M is continuously decidable;
2. M is computably compact.

4.2.3 Calculus of effectively closed sets

The notion of an effectively closed set is a generalisation of a Π^0_1 class (§2.1.6). We have already seen this notion in Definition 4.2.6, but we state it again here.

Definition 4.2.43. A closed subset C of a computable Polish M is effectively closed (Π^0_1-closed, or simply Π^0_1) if $M \backslash C$ is c.e. open.

In this subsection we present some well-known basic results about effectively closed sets that will be important in the sequel.

Elementary properties of effectively closed sets

It should be clear that effectively closed sets are closed under finite unions and arbitrary computable intersections (meaning that the effective procedures listing the complements must be uniformly indexed). The following lemma is also an immediate consequence of the definition:

Lemma 4.2.44 (Folklore). *Suppose $f : A \to B$ is a computable surjection, and assume C is effectively closed in B. Then $f^{-1}(C)$ is effectively closed (in A).*

Proof. This is because $A \backslash f^{-1}(C) = f^{-1}(B \backslash C)$ is c.e. open by Lemma 4.2.9. □

Another observation is an easy generalisation of a well-known fact about Π^0_1 classes (Exercise 2.1.32).

Fact 4.2.45 (Folklore). *Suppose $P = \{p\}$ is effectively closed singleton in a computably compact space X. Then the only point p of P is computable, and this is uniform.*

Proof. Given n, search for a basic open ball D of radius 2^{-n} and basic open $B_1, \ldots, B_m \in X \backslash P$ such that $D, B_1, \ldots, B_m$ cover X. Then $p \in D$. □

Corollary 4.2.46. *Let G be a computably compact space with a computable (multiplicative) group operation upon the space. Then the operation of taking the inverse is also computable in G.*

Proof. To se why the identity element is computable, observe that

$$\{x \mid x \cdot x = x\} = \{e\},$$

which is effectively closed. Thus, e is computable by Fact 4.2.45. Given x, to compute x^{-1} consider the closed set

$$\{y \mid x \cdot y = e\} = \{x^{-1}\}$$

and observe that it is effectively closed *relative to x*. Thus, the computability of $x \to x^{-1}$ follows from Fact 4.2.45 (relativised). □

Recall that in Theorem 2.1.29 we saw that a non-empty Π^0_1 class (i.e., an effectively closed subset of 2^ω) may contain no computable points. We now prove a well-known theorem of Jockusch and Soare [271] that guarantees that every effectively closed subset of a computably compact space has a *low* point. This theorem will be used in §7.1.2 and in many exercises throughout the book.

Theorem 4.2.47 (The Low Basis Theorem; Jockusch and Soare [271]). *Let X be a computable compact space and suppose $C \subseteq X$ is effectively closed (Π^0_1-closed) and non-empty. Then C has a low point (i.e., a point computable relative to a low degree).*

The result is often stated for the special important case when $X = 2^\omega$ in the literature. An even stronger version can be found in [61], but we won't need this level of generality.

Proof of Theorem 4.2.47. We need a few observations:

Fact 4.2.48. *Let C, D be Π^0_1-closed subsets of a computably compact X.*

1. *The condition $C \cap D = \varnothing$ is Σ^0_1.*
2. *If C has no computable members then either $C \cap D$ is infinite or empty.*

Proof. $C \cap D = \varnothing$ is equivalent to saying that their complements cover X, and this is c.e.. $C \cap D$ is itself Π^0_1, and if it is finite then it has an isolated point. (To apply Fact 4.2.45, consider the intersection of the class with the basic closed ball of small enough radius that contains the isolated point.) □

If we identify the dense set in the space with ω, a (fast) Cauchy name $(x_i)_{i \in \mathbb{N}}$ can be treated as a function $f : \omega \to \omega$. This clarifies the notation $\Phi_e^{(x_i)_{i \in \mathbb{N}}}$ that we use throughout the rest of the proof.

Fact 4.2.49. *Fix e. The collection of all points $x \in X$ that have a fast Cauchy name $(x_i)_{i \in \mathbb{N}}$ such that $\Phi_e^{(x_i)_{i \in \mathbb{N}}}(e) \downarrow$ forms an effectively open set, and this is uniform in e.*

Proof. Any computation $\Phi_e^{(x_i)_{i\in\mathbb{N}}}(e)\downarrow$ is witnessed by a finite initial segment $(x_i)_{i\leqslant s}$. Every fast Cauchy extension of $(x_i)_{i\leqslant s}$ will share the same computation with $(x_i)_{i\in\mathbb{N}}$. This gives a basic open ball of radius 2^{-s} centred in x_s. Being an initial segment of a fast Cauchy name is Σ_1^0 (we use strict $<$ in $d(x_i, x_{i+1}) < 2^{-i-1}$). We can effectively list all such initial segments. This, all such balls 'forcing' the halting computation can be effectively listed. □

We use these facts implicitly throughout the rest of the proof.

If C has a computable point then there is nothing to prove. So suppose C has no computable points. For an index e, let H_e be the effectively open set of points h such that $\Phi_e^{(x_i)_{i\in\mathbb{N}}}(e)\downarrow$ for some fast Cauchy $(x_i)_{i\in\mathbb{N}}$ of h. Then $H_0 \cap C \subseteq C$ is Π_1^0 and thus is either empty or it is infinite. If it is empty then set $D_0 = C$, otherwise let $D_0 = H_0 \cap C \subseteq C$. If D_e has been defined, let D_{e+1} be D_e if $H_{e+1} \cap D_e = \varnothing$, and otherwise let $D_{e+1} = H_{e+1} \cap D_e$ (which has to be infinite).

A collection of sets has a *finite intersection property* if the intersection of any finite sub-collection of sets in the family is non-empty. One of the equivalent formulations of compactness is that a space is compact iff any family of closed sets having the finite intersection property has non-empty intersection.

The resulting family of closed sets has the finite intersection property, and thus they share at least one common point. Let $\xi \in \cap_{i\in\mathbb{N}} D_i$. Let $(y_i)_{i\in\mathbb{N}}$ be a fast Cauchy name of ξ. To see whether $\Phi_e^{(y_i)_{i\in\mathbb{N}}}$ halts, use $\varnothing'$ to check whether H_e intersects D_{e-1} (where $D_{-1} = C$). If the intersection is empty, then $\Phi_e^{(y_i)_{i\in\mathbb{N}}}(e)\uparrow$. Otherwise, $\Phi_e^{(y_i)_{i\in\mathbb{N}}}(e)\downarrow$. It follows that $(y_i)_{i\in\mathbb{N}}$ is low. □

Computable closed sets

Effectively closed sets, Π_1^0-classes, computable functions, and computably compact spaces are closely technically related. To make this relationship explicit, we need one more definition. As usual, we identify basic open balls with their indices.

Definition 4.2.50. A closed subset of a computable Polish M is c.e. if $\{B : B \text{ basic open and } B \cap C \neq \varnothing\}$ is c.e..

Sets that satisfy the definition above are sometimes called *computably overt* in the literature.

Lemma 4.2.51 (Folklore). *A closed subset C of a computable Polish space M is computably enumerable iff C possesses a uniformly computable (in M) dense sequence of points*[5].

Proof of Lemma 4.2.51. Suppose C possesses such a computable sequence $(\alpha_i)_{i\in\mathbb{N}}$. Then the density of the sequence in C implies that $B_i \cap C \neq \varnothing$ iff $\exists j \alpha_j \in B_i$, which is a uniformly Σ_1^0 statement.

Now suppose C is a computably enumerable closed subset of M. Our goal is to construct a uniformly computable (finite or infinite) sequence of points $(\alpha_i)_{i\in I}$ that is dense in C. The proof below does not have to be non-uniform, but for notational convenience we split it into two cases, namely, when C is finite or infinite.

If C is finite, then it clearly contains only computable points. To see why, assume it is not empty (in this case there is nothing to prove) and let x be any point in C. Take a ball small enough so that $\{x\} \in B \cap C$. To get an 2^{-n}-approximation to x, wait for a basic open B' of radius $< 2^{-n}$ so that $B' \cap C \neq \varnothing$ and additionally B' is formally contained in B.

[5]Observe that the dense sequence makes C a computable Polish space under the induced metric.

Without loss of generality, we assume C is infinite. We uniformly approximate a computable sequence by stages. Before we describe stage s, recall that two basic open balls U and W are formally s-disjoint if $r(U) + r(W) < d(c(U), c(W))$ and this can be seen after calculating the radii and the distance with precision 2^{-s}. Then U and W are formally disjoint if the are formally s-disjoint for some s.

At stage 0, search for a basic open ball $B_{0,0}$ of radius < 1 such that $B_{0,0} \cap C \neq \varnothing$. If such a ball is never found then do nothing. If it is every found, go to the next stage.

At stage $s > 1$ first check whether there exists a basic open ball with index $< s$ which is formally s-disjoint from $B_{0,s-1}, \ldots, B_{s-1,s-1}$. If such a basic open B exists, then choose the first fund $B_{s,s} \subseteq_{form} B$ and $B_{i,s} \subseteq_{form} B_{i,s-1}$, $i < s$ such that $B_{j,s} \cap C \neq \varnothing$, the $B_{j,s}$ are pairwise formally disjoint and $r(B_{j,s}) < 2^{-s}$, $j = 0, \ldots, s$. Otherwise, if no such B exists, fix the first found pairwise formally disjoint $B_{0,s}, \ldots, B_{s,s}$ that intersect C, have radii $< 2^{-s}$, and such that $B_{i,s} \subseteq_{form} B_{i,s-1}$ for $i < s$ (note there is no further restriction on $B_{s,s}$). This ends the construction.

Let α_i be the unique point of the Polish space such that $\{\alpha_i\} = \bigcap_{j \geqslant i} B_{i,j}$. Since the construction is uniform and the radii of balls are rapidly shrinking, the points α_i form a uniformly computable sequence. Since each of the $B_{i,j}$ $(j = i, i+1, \ldots)$ intersects C and C is closed, each $\alpha_i \in C$. It remains to check that $(\alpha_i)_{i \in \mathbb{N}}$ is dense in C. Let $\overline{(\alpha_i)_{i \in \mathbb{N}}}$ be the completion of $(\alpha_i)_{i \in \mathbb{N}}$.

Suppose $c \in C$. We claim that $c \in \overline{(\alpha_i)_{i \in \mathbb{N}}}$. Assume $c \notin \overline{(\alpha_i)_{i \in \mathbb{N}}}$, and there is a ball U centred in c which is outside $\overline{(\alpha_i)_{i \in \mathbb{N}}}$. There will be a basic open ball $B' \ni c$ of radius at most 2^{-n} and which is formally contained in U with precision 2^{-n}:

$$d(c(U), c(B')) + r(B') < r(U) + 2^{-n}.$$

Then at every stage $s > n+4$ the balls $B_{i,s-1}$, $i = 0, \ldots, s-1$ will be formally s-disjoint from B, as will be readily witnessed by the metric. At some late enough stage s' we will get a confirmation that $B \cap C \neq \varnothing$. There exist only finitely many basic balls that have their index smaller than the index of B. Therefore, eventually B will be used to define $B_{t,t} \subseteq_{form} B$, contradicting the assumption that $U \cap \overline{(\alpha_i)_{i \in \mathbb{N}}} = \varnothing$. □

Definition 4.2.52. A closed subset of a computable Polish M is *computable* if it is c.e. and effectively closed.

As we mentioned immediately after the statement of Lemma 4.2.51, a c.e. closed subset of a computable metric space M can be viewed as a computable Polish space under the induced metric. The proposition below is also folklore.

Proposition 4.2.53. *For a closed subset C of a computably compact M, the following are equivalent:*

1. *C is a computably compact subspace of M;*
2. *C is computable.*

Proof. Assume (1). It is clear that C is c.e. (by Lemma 4.2.51). To list its complement, fix $x \in M \backslash C$. Let $\delta = \inf_{c \in C} d(x, c)$. Then any $\delta/4$ -cover of C must be formally disjoint from any ball centred in y with $d(y, x) < \delta/4$. For every n, fix a finite 2^{-n}-cover K of C. It follows that $M \backslash C$ is equal to the union of the (uniformly) effectively open sets U_n, where

$$\{B : B \text{ basic open and } B \text{ is formally disjoint from every ball in } K\}.$$

It follows that (2) holds.

Now assume (2). C is computable Polish by Lemma 4.2.51; let (y_j) be the computable dense subsequence. Fix $\epsilon = 2^{-n}$. We need to find an ϵ-cover of C by basic open balls. Note that y_j does not have to be special in M, and thus basic open balls in C do not have to be basic open in M. Nonetheless, an open ball of M centred in a computable point y and having a computable (more generally, left-c.e.) radius r is effectively open:

$$B(y, r) = \bigcup\{B(x, q) : d(x, y) + q < r\},$$

that is, $B(y, r)$ it is the union of basic open balls formally contained in it. Note effective openness of $B(y, r)$ is uniform in y and r. Regardless of whether the balls involved are basic or not, as long as their centres and radii are computable, the relation of formal containment remains c.e.

If a finite collection K of basic open (in C) balls formally contains a cover K' by basic open (in M) balls, then clearly K is a cover of C. We claim that this condition is also necessary (for K to cover C).

From the proof of Lemma 4.2.14 we know that, for a given cover K of C by (basic or not) open balls there is a small enough ϵ such that every ϵ-cover of C will be formally contained in at least one ball of K.

Take $\delta = \epsilon/4$. Fix a finite δ-cover K' of C by balls that are centred in special points of M, not C. Every $B' \in K'$ intersects C at some point x, and by the choice of δ, $d(x, cntr(B')) + \delta < \epsilon$, thus $B(\epsilon, x) \supset_{form} B'$. By transitivity of formal inclusion, we have that B' must be formally contained in some ball in K.

By computable compactness of M and computability of C, we can produce at least one δ-cover K' of C by basic open balls of M, uniformly in δ. (To see why, replace every basic open ball in the c.e open name of $M\backslash C$ by the effective union of balls of radii at most δ that are formally contained in it. This gives a new c.e. enumeration of the complement of C but with balls of radii at most δ. Then take the c.e. collection of all basic δ-balls that intersect C. Together these sets of balls cover M. Initiate the combined enumeration of these two c.e. sets and wait until at some finite stage we discover that we have a cover of M.) Since formal inclusion is c.e., this gives a procedure of listing covers of C by basic balls (in C). □

An immediate consequence of the proposition above is as follows.

Proposition 4.2.54. *Suppose $K = \bigcup_n K_n$ is a fully $\cap$-decidable system of covers of a computable Polish M. Then each computable closed ball D^c in K is a computable closed set (thus, is a computably compact subspace of M), and this is uniform.*

Proof. Lemma 4.2.36 says that each such D^c is c.e. closed. If $x \in M\backslash D^c$, then x inside an open ball C that is formally disjoint from D^c, and we claim that such balls can be computably enumerated. To see why, let c be the centre of D^c and r its radius, and assume $d(c, x) = r + \delta$. There must be a special x_i such that $d(x_i, x) < \delta/2$. Take the basic open ball $C = B(x_i, \delta/2)$. Then the distance between their centres is $d(c, \alpha) > r + \delta - d(x_i, x) > r + \delta - \delta/2 = r + \delta/2$, which is the sum of their raddii. So the balls are formally disjoint. Thus, the c.e. union of such open balls formally disjoint from D^c makes up the complement of D^c. □

Maps between computably compact spaces

The lemma below is well-known; e.g., [506, Theorem 3.3].

Lemma 4.2.55. *Suppose $f : X \to Y$ is a computable map, and assume X is computably compact. Then $f(X)$ is computably closed (in Y) and computably compact.*

Proof. Let (x_i) be a computable dense sequence in X. Then $(f(x_i))$ is dense in $f(X)$. (Every $\alpha = \lim_j x_j$ for some subsequence (x_j) and, by continuity, $f(\alpha) = \lim_j f(x_j)$, so $f(X) \subseteq cl(f(x_i))$. Suppose $\xi \in cl(f(x_i))$, say $\xi = \lim_j f(x_j)$. By compactness, (x_j) has a convergent subsequence (x_{j_k}), so let $z = \lim_k x_{j_k} \in X$ be its limit. Then $f(z) = \lim_k f(x_{j_k}) = \lim_j f(x_j) = \xi$.)

Initiate the enumeration of $f^{-1}(C)$ for each such basic open C; note it could be that some of these $f^{-1}(C)$ will be undefined. At some stage the preimages must cover the whole X. This gives a way of producing at least one 2^{-n}-cover of $f(X)$ uniformly in n; now apply Lemma 4.2.14. □

Combining Lemma 4.2.55 with Proposition 4.2.53, we get:

Corollary 4.2.56. *Suppose $f : X \to Y$ is computable and X is computably compact.*

- *If f is surjective then $Y = f(X)$ must be computably compact.*
- *$f(X)$ is a computable closed subset of Y.*

In computable algebra, the inverse of a computable bijective map is clearly computable as well. In contrast, there is no reason why the inverse of a computable bijection between spaces has to be computable even if its inverse is continuous (we mention here that this is actually true for isometric maps). The theorem below is elementary and is folklore (e.g., [61, Corollary 6.7]), but it is rather important because it tells us that effectively continuous maps are the right morphisms in the category of computably compact spaces.

Theorem 4.2.57. *Suppose $f : X \to Y$ is a computable bijection between computable Polish spaces, and assume X is computably compact. Then Y is also computably compact, and f^{-1} is computable.*

Proof. Computable compactness of Y follows from the corollary above. Given a (not necessarily) special point $y \in Y$, act computably relative to y. The set $Y \setminus \{y\}$ is effectively open relative to y. Since f is computable, $f^{-1}(Y \setminus \{y\})$ is effectively open relative to y. Since f is bijective, we have that

$$X \setminus f^{-1}(Y \setminus \{y\}) = \{f^{-1}(y)\},$$

which is a y-effectively closed singleton in X; apply Fact 4.2.45. □

In a computably compact group, the uniformly g-computable translation maps $x \to x \cdot g$ and $x \to g \cdot x$ have g-computable inverses. For instance, for each computable point g both maps are *effectively open*, meaning that they uniformly map names of open sets into names of open sets. The effective openness of the group operation can be derived in absence of computable compactness (see [313]), but we won't need this fact.

Exercises

Exercise° 4.2.58. Fix a computable linear operator $T : \mathbb{B} \to \mathbb{B}$, where the dimension of the computable Banach space $\mathbb{B}$ is finite. Show that the eigenvalues of T are computable, in the sense that they must be of the form $\xi + i\eta$, where ξ, η are computable reals. (Compare to Exercise 2.4.37.) [Hint: Use Fact 4.2.45.]

Exercise* 4.2.59 (Jockusch and Soare [271]). Show that there is an infinite Π^0_1 class C (i.e., an infinite effectively closed $C \subseteq 2^\omega$) such that, for all $f, g \in C$, if $f \neq g$, then f and g have incomparable Turing degrees.

Exercise* 4.2.60 (Jockusch and Soare [271]). Prove the following: Given any nonempty Π^0_1 class $C \subseteq 2^\omega$ which has no computable members, and any countable sequence of non-computable Turing degrees $\{\mathbf{a_i}\}$, $\mathcal{C}$ has $2^{\aleph_0}$ members f_i, mutually Turing incomparable, such that the degree of f_i is incomparable with each $\mathbf{a_i}$.

Exercise 4.2.61 (Jockusch and Soare [271]). Show that any nonempty Π^0_1 class $C \subseteq 2^\omega$ which has no computable members contains f, g whose greatest lower bound in the Turing degrees in $\mathbf{0}$. (Use Ex. 4.2.60 noting that the lower cone of each degree is countable.)

Exercise 4.2.62 (Metakides and Nerode [384]). Recall that a field F is orderable iff it is formally real, i.e., -1 is not a sum of squares in F.

1. Let F be a computable formally real field. Show that the class of compatible orderings on F can be represented as a Π^0_1-class. Conclude that every computable, orderable field admits a low compatible ordering.

2. Conversely, let C be a Π^0_1-class. Show* that there is a computable formally real field so that the compatible orders on F are in effective 1-1 correspondence with elements of C. Conclude that there is a computable, orderable field which is not computably orderable. [Hint: Matakides and Nerode effectivise the earlier work of Craven [98, 99]. Take a computably presented real closed field R (e.g., the real closure of $\mathbb{Q}$ using Theorem 2.2.32) and note that R has a computable order which (ignoring the field operations) is dense. Consider the field $R(y)$. The space $X(R(y))$ of compatible orderings of $R(y)$ is in an 1-1 effective correspondence with the Cantor space, which can be thought of as the Stone space of the atomless Boolean algebra freely generated by half-open half-closed intervals in R. Further, each clopen set in this space has the form $U_f = \{o \in X(R(y)) : o \text{ makes } f(y) > 0\}$. If $R = \{b_0, b_1, \ldots\}$ then such an f can be chosen to be a product of linear factors of the form $y - b_i$ (up to a sign). This gives a way to effectively represent a Π^0_1-class as $X(R(y)) \setminus \bigcup_{i \in W} U_{f_i}$, where W is a c.e. set. Take $F = R(\{y, \sqrt[2^n]{-f_i(y)} : i \in W, n \in \mathbb{N}\})$, noting that $f_i(y)$ have to be negative with respect to any compatible order, while the algebraic elements $\sqrt[2^n]{-f_i(y)}$ must all be positive. A naturally constructed computable copy of F will have the property that its space of compatible orders $X(F)$ is effectively homeomorphic to C.]

Exercise* 4.2.63 (Miller [395]). Suppose A is a computable algebraic field (i.e., algebraic over its prime subfield) with a splitting algorithm (i.e., an algorithm that given a polynomial decides whether it is irreducible). Prove that for any computable presentation B of A there exists an isomorphism $f : B \to A$ so that f is low.

Exercise 4.2.64. Show that the group $SO(3)$ admits a computably compact presentation. (Hint: Recall $M_3(\mathbb{R})$ is an inner product space under Frobenius inner product $tr(A^TB)$. Noting that $Q^T = Q^{-1}$ when $Q \in SO(3)$, we see that $SO(3)$ is contained in the computably compact closed ball $B = B^c(\mathbf{0}, 3)$ of M_3 under the Frobenius norm. Since $SO(3)$ is defined by the equation $det(Q) = 1$, it is effectively closed in B. To list a dense set of $SO(3)$, use Rodrigues' formula.)

Exercise° 4.2.65 (Brattka [60])**.** Let X and Y be computably Polish spaces and $f : X \to Y$, and assume X is computably compact. Then f is computable iff $graph(f)$ is effectively closed and iff $graph(f)$ is computable closed.

Exercise° 4.2.66 (Folklore; e.g. [139])**.**

1. Show that the space of all compact (or closed) subsets $\mathcal{C}(H)$ of the computable presentation of $H = [0,1]^\omega$ described in Exercise 4.2.19 is a computable Polish space. (Hint: The metric is given by the Hausdorff distance $d_{\mathrm{H}}(X,Y) := \max\{\sup_{x\in X} d(x,Y),\ \sup_{y\in Y} d(X,y)\}$. The countable dense set is formed by finite discrete subsets of special points of H.)
2. Show that for a c.e. closed subset C of H, C is computable iff C is a computable point in the computable presentation $\mathcal{C}(H)$ described in the hint above.

Exercise* 4.2.67 (Brattaka, de Brecht, Pauly [61]; see also [139])**.** Show that a compact computable Polish space X is computably compact iff there is a computable surjective $f : 2^\omega \to X$.

Exercise 4.2.68 (Folklore; e.g., [269])**.** Let X be a computable Polish space and let $K \subseteq X$ be compact. Say that K is *semicomputable* if we can list all finite open covers of K by basic open balls, as tuples of their strong indices. (This resembles the notion of computable compactness, but we do not assume that K is computable Polish; we use the basic balls of the larger space to list all finite covers of K.) Show that K is computably compact iff K is c.e. closed and semicomputable.

Exercise 4.2.69 (Iljazović [268]; see also [139])**.** Show that any isometric computable Polish presentation of a computably compact space is also computably compact. (Note that we do not require that the isometry is computable.)

Exercise° 4.2.70 (Folklore)**.** Suppose X is a computably compact subspace of a computable Polish space Y. Show that the inf-distance $d_{\inf}(y, X) = \inf_{x\in X} d(y, x)$ from a point y to the subspace X is a computable function.

Exercise° 4.2.71 (Folklore; see Metakides and Nerode [385])**.** We say that a closed subspace C of a computable Polish space X is *located* if $y \mapsto d_{\inf}(y, C) = \inf_{c\in C} d(y, c)$ is a computable function $X \to \mathbb{R}$. Show that, for a finite-dimensional subspace C of a computable (real) Banach space, the following are equivalent:

1. C is located;
2. C has a basis consisting of computable points.

[Hint: We can assume $C \neq \varnothing$. If C is located, then use a straightforward iterative procedure to approximate a computable $b_0 \in C$. If $C - \mathbb{R}b_0 \neq \varnothing$, then (non-uniformly) fix a basic closed ball disjoint from $\mathbb{R}b_0$, and repeat the procedure within the respective *open* ball to find b_1. Repeat until a finite computable basis $\bar{b}$ of C is found. Conversely, suppose C has a basis $\bar{b}$ consisting of computable points. The basic case is when the dimension is 1 and $\bar{b} = \langle b\rangle$. Note $d_{\inf}(x, C) = \inf_{\alpha\in\mathbb{R}} ||\alpha b - x|| \leqslant ||x||$, as witnessed by $\alpha = 0$. On the other hand, $|\alpha|||b|| - ||x|| = ||\alpha b|| - ||x|| \leqslant ||\alpha b - x||$, which means that when $|\alpha|||b|| - ||x||$ gets larger than $||x||$, we cannot possibly achieve the infimum of $||\alpha b - x||$. In other words, if $|\alpha| > \xi_x = \frac{2||x||}{||b||}$ the infimum cannot be attained. Thus, $d_{\inf}(x, C) = \inf\{||\alpha b - x|| : \alpha \in [-\xi_x, \xi_x]\}$. Since $[-\xi_x, \xi_x]$ is computably compact relative to x, this value is also computable relative to x (uniformly in x). To understand the inductive step, consider the case $\bar{b} = \langle b_0, b_1 \rangle$. Let $\delta_{1,x} = d_{\inf}(x, \mathbb{R}b_1) = \inf_{\lambda \in \mathbb{R}} ||\lambda b_1 - x||$, which is uniformly computable in x. As before, $||\alpha_0 b_0 + \alpha_1 b_1|| - ||x|| \leqslant ||\alpha_0 b_0 + \alpha_1 b_1 - x||$, and assuming $\alpha_0 \neq 0$, $|\alpha_0|||b_0 + \frac{\alpha_1}{\alpha_0} b_1|| \geqslant |\alpha_0| \delta_{1,b_0}$. We can bound the range of α_0 to $[-2||x||/\delta_{1,b_0}, 2||x||/\delta_{1,b_0}]$. A similar bound, this time using δ_{b_1}, can be put on α_1. This bounds the choice of (α_0, α_1) to an x-computably compact subset of $\mathbb{R}^2$, and the case when $\dim C = 2$ follows. The general case of $\dim C = n > 1$ is not very much different. It uses the distances from b_i to the spans of $\bar{b} - \{b_i\}$ to calculate parameters (that are computable by the I.H.) which give a way to bound the search to an x-computably compact set in $\mathbb{R}^n$.]

Exercise° 4.2.72. Let X be a located closed subset (Ex. 4.2.71) of a computable Polish space M. Show that X is computably closed. Further, if M is computably compact, then every computably closed $X \subseteq M$ is located, but without the computable compactness of M, this fails in general (even for compact M).

Exercise° 4.2.73 (Folklore). Recall that two norms $|| \cdot ||$ and $|| \cdot ||'$ on a Banach space $\mathbb{B}$ are equivalent, written $|| \cdot || \sim || \cdot ||'$, if there exists $c > 0$ such that for all $x \in \mathbb{B}$, $\frac{1}{C}||x|| \leqslant ||x||' \leqslant C||x||$. Note that this is an equivalence relation on the collection of all norms on $\mathbb{B}$.

1. Prove that if $|| \cdot || \sim || \cdot ||'$, then the identity operator Id on $\mathbb{B}$ is a homeomorphism from $(\mathbb{B}, || \cdot ||)$ to $(\mathbb{B}, || \cdot ||')$. [Hint: For a scalar λ, define $\lambda B_r(v) = B_{|\lambda| r}(\lambda v)$ and observe $CB_\varepsilon(x) \subseteq B'_\varepsilon(x) \subseteq \frac{1}{C} B_\varepsilon(x)$, where the notation should be self-explanatory.]

2. Prove that all norms on $\mathbb{R}^n$ are equivalent. [This can be found in any textbook that covers Banach spaces.]

3. Show that if $|| \cdot || \sim || \cdot ||'$ and both norms turn $\mathbb{B}$ into a computable Banach space, then $Id : (\mathbb{B}, || \cdot ||) \to (\mathbb{B}, || \cdot ||')$ and id^{-1} are computable. [Hint: Use that $y \in B'_\varepsilon(x)$ implies $y \in CB_\varepsilon(y) \subseteq B'_\delta(y) \subseteq_{form} B'_\varepsilon(x)$ for some δ. By decreasing the radii of some of these balls slightly if necessary, we can assume that the condition is open, i.e., it is stable under a slight variation of the parameters involved. In particular, $y \in CB_\varepsilon(u) \subseteq B'_\delta(u) \subseteq_{form} B'_\varepsilon(x)$ for some special u. Thus, to calculate $id^{-1}(B'_\varepsilon(x))$, list all basic open $CB_\varepsilon(u)$ so that $B'_\delta(u) \subseteq_{form} B'_\varepsilon(x)$.]

Exercise° 4.2.74. Show that the closed unit ball in a finite-dimensional computable Banach space is computably compact. [Hint: The closed unit ball B is evidently c.e. closed, so it is sufficient to list all finite basic open covers of B. Assume the dimension is n; let $e_1, \ldots, e_n$ be linearly independent special points. (Note that such special points exist.) We can replace the original dense set with the dense *linear* set $\sum_{i \leqslant n} r_i e_i$, $r_i \in \mathbb{Q}^+$. Define (say) the usual $|| \cdot ||_\infty$-norm on $\sum_{i \leqslant n} r_i e_i$. This gives a computable Banach presentation of $\mathbb{R}^n$ under $|| \cdot ||_\infty$. The closed unit $|| \cdot ||_\infty$-ball D is evidently computably compact (with respect to $|| \cdot ||_\infty$), and so are the n-scaled $|| \cdot ||_\infty$-balls nD of radii n. If $|| \cdot ||$ is the original norm, then $\eta : (a_1, \ldots, a_n) \mapsto ||\sum_{i \leqslant n} a_i e_i||$ is a computable map $\mathbb{R}^n \to \mathbb{R}$ (and thus $nD \to \mathbb{R}$). Recall that all norms on $\mathbb{R}^n$ are equivalent and induce the same topology (Ex. 4.2.73). The closed unit $|| \cdot ||$-ball B will be a closed subset of nD for some n. The special points that lie in the c.e. open $\eta^{-1}((0,1))$ are dense in B, and we can use $\eta^{-1}(1,\infty)$ to list the open complement of B in nD. This makes B a computably closed subset of the computably compact nD. By Proposition 4.2.53, B is computably compact, but with respect to the $||\cdot||_\infty$-norm. We can now appeal to Theorem 4.2.57 and Exercise 4.2.73 to conclude that B is also computably compact with respect to $||\cdot||$. Alternatively, we could directly list finite $||\cdot||$-covers of B, as follows. If $||\cdot|| \leqslant m \cdot ||\cdot||_\infty$, then every $\frac{\delta}{m}$-basic $||\cdot||_\infty$-ball will be contained in a δ-basic $||\cdot||$-ball, and this gives a way of listing finite covers of B with respect to the original norm. We remark that this exercise can also be derived from Exercise 4.2.92; see the hint to Exercise 4.2.94.]

Exercise° 4.2.75. Suppose $T : \mathbb{B} \to \mathbb{D}$ is a computable linear map between computable Banach spaces, where $\mathbb{B}$ is a computable Banach space of finite dimension. Show that $||T||$ is computable uniformly in the index for T. [Hint: Use Exercise 4.2.74, Proposition 4.2.17(1), and the formula $||T|| = \sup\{||Tx|| : ||x|| \leqslant 1\}$.]

Exercise° 4.2.76 (Computable Banach–Alaoglu Theorem; Brattka [59]). Let $\mathbb{B}$ be a computable Banach space. Show that there is a computably compact space X such that the closed unit ball of the dual space $\mathbb{B}'$ can be computably embedded into X as a Π^0_1 closed (thus, compact) subset. [Hint: Without loss of generality, we may assume there are no repetitions among the special points $(x_i)_{i\in\mathbb{N}}$ of $\mathbb{B}$ (Exercise 2.4.32). Consider the space $\prod_{i\in\mathbb{N}}[-||x_i||, ||x_i||]$. Every linear operator can be associated with a point in the space, and by continuity, the property of being linear is Π^0_1. We note that here, the effective correspondence between the closed unit ball and the respective Π^0_1 is *not* in the Type II sense with respect to the norm on $\mathbb{B}'$. The induced topology is the weak* topology. Clearly, we cannot hope for a stronger result, as the unit ball in $\mathbb{B}'$ may not be compact.]

4.2.4 Computable Stone duality. Proofs of Theorem B(1,2)

One of the topological characterisations of Stone spaces stated in Lemma 4.1.46 was as follows. A (Polish) X is a Stone space if, and only if, X is a compact and totally disconnected, i.e., it has no non-trivial connected subsets. The latter is equivalent to saying that the compact Polish space X is totally separated, i.e., whenever $x \neq y$ are points in X, there has to be a clopen set U such that $U \ni x$ and $X \setminus U \ni y$ (folklore). Having this property in mind, we say that a *clopen split* of M is a pair of disjoint (cl)open sets X, Y such that $X \sqcup Y = M$. The dual Boolean algebra is then equal to the Boolean algebra of all clopen sets of the space under the set-theoretic operations.

Computable Stone duality

Various versions of the elementary lemma below can be found in, e.g., [266, 139, 245]. Fix a compact M and assume it is computably metrised. Suppose an oracle Z is powerful enough to uniformly list all basic finite covers of M. Then we say that M is Z-computably compact.

Lemma 4.2.77. *Suppose M is computable Polish. If M is Z-computably compact, then there is a uniformly Z-computable list all clopen splits of M.*

Proof. Recall that two basic open balls $B(c_1, r_1)$ and $B(c_2, r_2)$ are formally disjoint if $r_1 + r_2 < d(c_1, c_2)$. Two sets of basic open balls are formally disjoint if any pair of basic open balls, one coming from the first set and the other from the second, are formally disjoint.

Suppose $M = X \sqcup Y$ is a clopen split, and let δ be the infimum-distance between these compact open sets

$$\delta = \inf_{(x,y)\in X\times Y} d(x,y).$$

(Since $X \times Y$ is compact and d is continuous, it attains its infimum at some pair (x_0, y_0). In particular, $\delta > 0$.)

Suppose $0 < \epsilon < \delta/4$. Then every finite ϵ-cover will consist of two formally disjoint subsets of basic open balls. Indeed, every ball covering a point in X cannot contain a point in Y, and every ball covering a point in Y cannot contain a point in X. If a basic open B has its centre in X and D has its centre in Y, then the distance between their centres is at least δ, while the sum of their radii is at most $\delta/2 < \delta$, making them formally disjoint. The same can be said about the respective closed basic balls.

On the other hand, if a finite open cover of M consists of two formally disjoint subcovers, then these subcovers induce a split of M into clopen components. Since the property of being formally disjoint is a c.e. property, Z is able to list all such covers. □

Another way to state the lemma above is that any modulus of compactness of M (see the proof of Fact 4.2.11) can computably enumerate the clopen splits of M.

Theorem 4.2.78 (Hoyrup, Kihara, and Selivanov [266]). *Let M be a computably compact Stone space (a totally disconnected compact Polish space). Then the Boolean algebra of its clopen subsets admits a computable presentation.*

Proof. Fix a fully $\cap$-decidable system of covers $K = \bigcup_{n\in\mathbb{N}} K_n$. Using the previous Lemma 4.2.77, effectively list all clopen splits of M into (open, formally disjoint names of) pairs of clopen sets. Let (X_i, Y_i) be the enumeration of these clopen splits; both of them have to be non-empty since each of them has to at least contain the centre of at least one basic ball. Both X_i and Y_i are given by their open as well as their closed covers, whichever is more convenient (recall Remark 4.2.34 and the discussion after when we argued that closed and open balls can be used interchangeably). Write X_i^1 for the corresponding Y_i in a clopen split; and let X_i^0 be another notation for X_i.

The Boolean algebra is generated by the empty set and arbitrary finite *non-empty* conjunctions of the form $\bigwedge_i X_i^{\epsilon_i}$, where $\epsilon_i \in \{0,1\}$. Since the system of covers K is fully $\cap$-decidable, we can indeed decide whether such a finite intersection is empty. In other words, if F is the free Boolean algebra generated by the X_i, then the Boolean algebra of clopen sets of M is isomorphic to F/I, where I is a computable ideal generated by the empty conjunctions. This makes the Boolean algebra computable. □

This result allows us to establish the following representation theorem.

Theorem 4.2.79 (Harrison-Trainor, Melnikov, and Ng [245]). *Let M be a computable Polish Stone space. Then the Boolean algebra of clopen subsets of M admits a computable presentation.*

Proof. Let M be the computable space. Recall that $\varnothing'$ can list all finite covers of the space, by Fact 4.2.11. Use Lemma 4.2.77 and relativise the previous Theorem 4.2.78 to get $\varnothing'$-computable presentation of the Boolean algebra of clopen sets.

Suppose we are given a non-zero element of the Boolean algebra. It is a non-empty clopen set represented via a finite union of basic balls. By slightly increasing the radii, we can assume they are given by rational parameters while still keeping the complement of the component formally disjoint. We can ask whether there exist two unequal special points x, y that are contained in this clopen set. This element is an atom iff no such pair of points exists. This question is *uniformly* Σ_1^0 *in the finite parameter* representing the clopen set. Even though it takes $\varnothing'$ to list such finite parameters, when we are already given such a parameter, $\varnothing'$ can decide the property.

Thus, the atom relation is $\varnothing'$-computable in the resulting $\varnothing'$-computable Boolean algebra. By Theorem 4.1.25, the Boolean algebra of clopen sets has a computable copy. □

The proof of Theorem B(2)

Recall that (2) of Theorem B states that every computable Polish Stone space is homeomorphic to a computably compact Stone space. It follows from the theorem below.

Theorem 4.2.80 (Harrison-Trainor, Melnikov, and Ng [245]). *For a countable Boolean algebra B and its dual Stone space $\widehat{B}$, the following are equivalent:*

1. *B has a computable presentation;*
2. *$\widehat{B}$ has a computable Polish presentation;*
3. *$\widehat{B}$ has a computably compact presentation.*

Proof. In the previous section, we explained that the dual Stone space $\widehat{B}$ of a computable Boolean algebra B can be represented as the collection of infinite paths $[T]$ through a computably branching, computable tree T without dead ends. We also explained in Fact 4.2.16 that $[T]$ can be viewed as a computably compact Polish space. The remaining implication is given by Theorem 4.2.79. □

Right-c.e. spaces and the proof of Theorem B(1)

Recall that a Polish space is right-c.e. if there is a dense sequence $(\alpha_i)_{i\in\mathbb{N}}$ such that $d(\alpha_i, \alpha_j)$ are uniformly right-c.e. reals, i.e., reals approximable from above. The sequence $(\alpha_i)_{i\in\mathbb{N}}$ may contain repetitions; equivalently, it is possible that $d(\alpha_i, \alpha_j) = 0$ for some i, j. In contrast with the computable Polish case, these repetitions cannot be removed in general. We will see that Π_1^0 classes can be viewed as right-c.e. metrised spaces. The following effective version of Stone duality was established in [35].

Theorem 4.2.81. *Let B be a countable Boolean algebra, and let $\widehat{B}$ be the dual Stone space. Then the following conditions are equivalent:*

(a) B has a c.e. presentation,

(b) $\widehat{B}$ admits right-c.e. presentation.

Before we prove this version of Stone duality, we explain how to use it to prove (1) of Theorem B. Recall that (1) of Theorem B says that there exists a right-c.e. Stone space that is not homeomorphic to any computable Polish space.

Proof of Theorem B(1). For a c.e. presented Boolean algebra B given by Feiner's Theorem 3.2.1. Then B does not have computable copies. By Theorem 4.2.81, one can assume that the Polish space $\widehat{B}$ is right-c.e.. On the other hand, since B is not computably presentable, Theorem 4.2.80 implies that $\widehat{B}$ is not homeomorphic to any computable Polish space. □

In the remainder of the section, we prove Theorem 4.2.81.

Proof of Theorem 4.2.81

Let T be a subtree of $2^{<\omega}$. As usual, $[T]$ denotes the set of all infinite paths through T. We say that T is a *pruned tree* if for any $\sigma \in T$, there is a path $x \in [T]$, which goes through σ. We write 2^ω to denote the space of paths through the complete binary tree, i.e., $2^\omega = [2^{<\omega}]$.

In the final step of the proof of Feiner's Theorem 3.2.1 we observed that every countable Boolean algebra B is a factor of the atomless Boolean algebra $F(B)$ freely generated by the elements of B. If B is c.e. presented then it is of the form $F(B)/I$, where I is a c.e. ideal. The topological version of this observation is Proposition 4.1.49(2) which says that B has a c.e. presentation iff B is isomorphic to the algebra of clopen subsets of $[T]$ for some computable T.

We prove (a)⇒(b) of Theorem 4.2.81. Suppose that a Boolean algebra B has a c.e. presentation. By Proposition 4.1.49, fix a computable T such that B is isomorphic to the algebra of clopen subsets of $[T]$. Equivalently, we can fix a co-c.e. pruned tree T with this property (by compactness, we can remove the dead ends in a c.e. way).

We define a right-c.e. Polish presentation (M, d) for the space $\widehat{B}$. We put $M = [T]$, and the distance d is induced by the standard ultrametric on the Cantor space 2^ω (i.e., $d(\xi, \eta) = 2^{-n}$, where n is the length of the longest common prefix of two unequal elements $\xi, \eta \in 2^\omega$). Indeed, we prove the following, more general fact.

Fact 4.2.82. *A non-empty Π^0_1 class $\subseteq 2^\omega$ can be viewed as a right-c.e. Polish space given by a right c.e. ultrametric on the Cantor space 2^ω.*

We build a dense sequence $(\alpha_i)_{i\in\mathbb{N}}$ inside (M, d) — our construction needs to ensure that the distances $d(\alpha_i, \alpha_j)$ are uniformly right-c.e.. Note that in general, a special point α_i could be equal to α_j for $i \neq j$.

Fix an effective sequence of finite trees $(T_s)_{s\in\mathbb{N}}$ such that for any $\sigma \in 2^{<\omega}$:

- if $\sigma \in T_s$, then $|\sigma| \leqslant s$;
- if $|\sigma| \leqslant s$ and $\sigma \notin T_s$, then $\sigma \notin T_{s+1}$;
- $\sigma \in T$ iff $(\exists s_0)(\forall s \geqslant s_0)(\sigma \in T_s)$;
- for every s, there is at most one $\tau \in T_s \backslash T_{s+1}$;

The sequence $(T_s)_{s\in\mathbb{N}}$ is constructed as follows. Since the tree T is co-c.e., we choose an effective enumeration of its complement: $2^{<\omega} \backslash T = \bigcup_{s\in\mathbb{N}} V_s$. At a stage $s+1$, we proceed as follows. First, we add to T_{s+1} all nodes σ such that $|\sigma| = s+1$, $\sigma \notin V_s$, and the parent of σ belongs to $T_s \backslash V_s$.

After that, we consider a finite set $F = T_s \cap V_s$. We choose a node $\tau \in F$ with maximal length, and delete it from T_{s+1}.

Now we are ready to construct our dense sequence $(\alpha_i)_{i\in\mathbb{N}}$. From now on, we assume that the distance $d(\cdot,\cdot)$ is induced by the metric on 2^ω.

At a stage s, for every $i \leqslant 2^s$, we define $\alpha_i[s]$ as a string σ such that $|\sigma| = s$ and $\sigma \in T_s$. Some of these numbers i could be declared *inactive*. In the end, we will obtain α_i as $\lim_s \alpha_i[s]$.

At stage 0, we define $\alpha_0[0]$ and $\alpha_1[0]$ as empty strings.

Stage $s+1$. For each active $i \leqslant 2^s$, we proceed as follows. The string $\alpha_i[s]$ satisfies one of the following two cases.

Case 1: $\alpha_i[s] \in T_{s+1}$ and there is a child σ of $\alpha_i[s]$ such that $\sigma \in T_{s+1}$. We define $\alpha_i[s+1] = \sigma$. Note the following: if $x \in 2^\omega$ and $\alpha_i[s] \nsubseteq x$, then $d(x, \alpha_i[s+1]) = d(x, \alpha_i[s])$.

Case 2: otherwise, either $\alpha_i[s] \notin T_{s+1}$, or $\alpha_i[s] \in T_{s+1}$ and no child of $\alpha_i[s]$ belongs to T_{s+1}. Since the tree T is pruned, this implies that $\alpha_i[s] \notin T$.

We find the largest $n < s$ such that for the string $\tau := \alpha_i[s] \upharpoonright n$, there is a $\xi \supset \tau$ with $|\xi| = s+1$ and $\xi \in T_{s+1}$. (Recall that $\rho \upharpoonright n$ stands for the initial segment of ρ of length n.) Note that the nodes $\xi \upharpoonright (n+1)$ and $\alpha_i[s] \upharpoonright (n+1)$ are siblings. Since T is pruned, one can show that every $\sigma \supseteq \alpha_i[s] \upharpoonright (n+1)$ does not belong to T.

For the chosen n, there could be several ξ satisfying the conditions above. We choose the leftmost one and set $\alpha_i[s+1] = \xi$. We note that every $x \in 2^\omega$ satisfies one of the following three conditions:

1. Suppose that $\tau \nsubseteq x$. Then $d(x,\xi) = d(x, \alpha_i[s])$.

2. Suppose that $\alpha_i[s] \upharpoonright (n+1) \subseteq x$. Then x is not a path through T.

3. Otherwise, we have $\xi \upharpoonright (n+1) \subseteq x$. Then $d(x,\xi) \leqslant 2^{-n-2} < 2^{-n-1} = d(x, \alpha_i[s])$.

For each i such that $2^s < i \leqslant 2^{s+1}$, we search for the leftmost $\sigma \in T_{s+1}$ such that $|\sigma| = s+1$ and $\sigma \notin \{\alpha_j[s+1] : j < i\}$. If such σ exists, then put $\alpha_i[s+1] = \sigma$. Otherwise, for all $t \geqslant s+1$, we set $\alpha_i[t] = \alpha_0[t]$, and we declare this i *inactive*.

This concludes the description of our construction. It is not hard to show that every bit of $\alpha_i[s]$ can change only finitely many times. Hence, $\alpha_i = \lim_s \alpha_i[s]$ is well-defined. Furthermore, the sequence $\{\alpha_i\}_{i\in\mathbb{N}}$ is dense in $([T], d)$.

The properties of the construction ensure that

$$d(\alpha_i[s+1], \alpha_j[s+1]) \leqslant d(\alpha_i[s], \alpha_j[s]) \text{ for all } i, j, s.$$

Therefore, for a rational q, the condition $d(\alpha_i, \alpha_j) < q$ holds iff there is a stage s such that $d(\alpha_i[s], \alpha_j[s]) < q$. We deduce that the reals $d(\alpha_i, \alpha_j)$ are uniformly right-c.e., and the Stone space $\widehat{B}$ has a right-c.e. Polish presentation.

Proof of (b)⇒(a). Suppose that the Stone space $\widehat{B}$ has a right-c.e. Polish presentation (M, d). This presentation is, in particular, Δ^0_2 Polish. By Theorem 4.2.80 relativised to $\varnothing'$, B has a Δ^0_2 presentation. By Proposition 4.1.50, B has a c.e. presentation as well.

The proof of Theorem 4.2.81, and therefore of Theorem B(1), is complete. □

4.2.5 Computably categorical Stone spaces

We apply out techniques to completely describe Sone spaces that have a unique computably compact presentation, up to computable homeomorphism. As we showed in Theorem 4.2.57, the inverse of a computable bijection between computably compact spaces is necessarily computable as well. Therefore, the following definition first proposed in [35] makes sense.

Definition 4.2.83. A compact space is *computably categorical* if it possesses a unique computably compact presentation, up to computable homeomorphism.

We emphasise that the definition above uses computable *homeomorphisms* to compare spaces. Recall that all our Stone spaces are Polish.

Theorem 4.2.84 (Bazhenov, Harrison-Trainor, and Melnikov [35]). *A Stone space S is computably categorical iff the dual Boolean algebra $\widehat{S}$ is computably categorical. (Thus, by Theorem 4.1.17, these are exactly the Stone spaces having only finitely many isolated points.)*

Proof. By Theorem 4.2.80, S has a computably compact presentation iff the Boolean algebra $\widehat{S}$ has a computable presentation. We use it without explicit reference. We give a somewhat brute-force proof of Theorem 4.2.84 similar to the argument in [35]; it utilises various theorems and facts proved in this chapter[6].

Suppose the dual Boolean algebra $\widehat{S}$ is computably categorical. Then, by Theorem 4.1.17, $\widehat{S}$ has only finitely many atoms. Under the duality, this translates to S having only finitely many isolated points. Consequently, either S is finite or is homeomorphic to $2^\omega \sqcup F$, where F is finite. (Here $\sqcup$ stands for the operation of taking the disjoint union of spaces, so that in the resulting space each of the two components is declared clopen.) In both cases it is easy to show that S is computably categorical; this is Exercise 4.2.85.

Now assume that S is computably categorical. Let A and B be two computable presentations of the Boolean algebra $\widehat{S}$. Using Proposition 4.1.49, produce two computable, computably branching trees with no dead ends T and Γ so that

$$[T] \cong \widehat{A} \simeq S \simeq \widehat{B} \cong [\Gamma].$$

Recall that every node σ in T is labeled by some element a of A, so that each clopen set $[\sigma]$ of all paths extending σ naturally corresponds to the principal ideal $\{c \,|\, c \leqslant a\}$, in the sense that the elements labelling the nodes extending σ generate this ideal. If $a_0, \ldots, a_n$ is the complete list of all elements corresponding to (i.e., labelling) the nodes of T at some level m, then

$$a_0 \vee a_1 \vee \ldots \vee a_n = 1$$

and

$$a_i \wedge a_j = 0, \text{ for all } 0 \leqslant i < j \leqslant n.$$

[6]There are other ways to prove the theorem. For example, one could view the dual space of X (the dual algebra of X) to be the collection of all continuous homomorphisms $X \to \{0,1\}$. Then one could argue that the calculation of the 'dual maps' is effective too.

As a consequence of all these properties, every element $a \in A$ can be expressed as a finite $\vee$-combination of elements labelling some nodes in T; the same can be said about B and $[\Gamma]$. In this sense, the elements labelling the nodes of the tree form a 'basis' of the respective Boolean algebra. The same can be said about nodes in Γ and elements of B. Under the usual shortest common prefix ultrametric, both $[T]$ and $[\Gamma]$ are evidently computably compact (Fact 4.2.16).

Let $\psi : [T] \to [\Gamma]$ be a computable homeomorphism, which must exist by our assumption about $S \cong [T] \cong [\Gamma]$. Let $A \upharpoonright T$ denote the collection of elements of A that label the nodes in T (together with 0 which does not label any node in T); define $B \upharpoonright \Gamma$ similarly. Define $\hat{\psi}$ For every $a \in A \upharpoonright T$ as follows.

Set $\hat{\psi}(0) = 0$. To define $\hat{\psi}(a)$ for a non-zero $a \in A \upharpoonright T$, let $\sigma \in T$ be the node labeled by a. Use computable compactness of $[T]$ and $[\Gamma]$ to find non-zero elements $b_0, \ldots, b_n \in B \upharpoonright \Gamma$ that label $\tau_1, \ldots, \tau_n \in \Gamma$, respectively, such that

$$\psi([\sigma]) = [\tau_1] \sqcup [\tau_1] \sqcup \ldots \sqcup [\tau_n],$$

where (as before) $\sqcup$ stands for the operation of taking disjoint union. It is not hard to see that such strings τ_i can be found effectively and uniformly in the index of σ; this is Exercise 4.2.86. Define

$$\widehat{\psi}(a) = b_0 \vee b_1 \vee \ldots \vee b_n.$$

Since every $x \in A$ can be expressed as a union of nodes in $a \in A \upharpoonright T$, this map is naturally extended to all A by the rule

$$\widehat{\psi}(x) = \vee_i \widehat{\psi}(a_i),$$

where a_i range over the finitely many incompatible (under $\leqslant$) elements in $A \upharpoonright T$ that make up x.

It should be clear that the map $\widehat{\psi}$ is well-defined on all of A, because ψ is a well-defined homeomorphism. For the same reason ψ is also onto. It is routine to check that, additionally, $\widehat{\psi}$ respects the Boolean algebra operations, and that the only pre-image of 0_B is 0_A; this is Exercise 4.2.87. Therefore, $\widehat{\psi}$ is a computable isomorphism of Boolean algebras A and B. □

Very little is known about computably categorical compact spaces. We suspect that much stronger results can be established that will eclipse Theorem 4.2.84. Computably categorical profinite abelian groups will be described in Section 9.5. We discuss computable profinite groups next.

Exercises

Exercise° 4.2.85 (Bazhenov, Harrison-Trainor, Melnikov [35]). Show that Cantor space is computably categorical (in the sense of Definition 4.2.83). Conclude that any Stone space with only finitely many atoms is also computably categorical.

Exercise° 4.2.86. In the notation of Theorem 4.2.84, show that for every $\sigma \in T$ we can uniformly effectively find (incompatible) strings $\tau_1, \ldots, \tau_n \in \Gamma$ such that $\psi([\sigma]) = [\tau_1] \sqcup [\tau_1] \sqcup \ldots \sqcup [\tau_n]$.

Exercise° 4.2.87. Finish the proof of Theorem 4.2.84 by showing that $\widehat{\psi}$ is an isomorphism of A onto B.

Exercise 4.2.88 (Hoyrup [265]). A subset of a Polish space M is G_δ is it is of the form $\bigcap_{i \in \mathbb{N}} U_i$, where U_i are open. It is effectively G_δ if the U_i are uniformly c.e. open sets. Assume that X is effectively G_δ and contains a computable (in M) dense (in X) sequence of points. Prove the following effective version of Alexandrov's Theorem (see [286, (3.11) Theorem] or [262, Thm 2-76]): There exists a *complete* metric d_X that turns X into a computable Polish space upon the dense sequence. Furthermore, M-computable points in X uniformly correspond to computable points in (X, d_X). (Hint: Fix a sequence (α_i) of special points in M. For some computable g_i, U_i is an effective union of open balls $B_n = B(\alpha_{g_i(n)}, r_n)$ with rational radii r_n and centres $\alpha_{g_i(n)}$. Let $f_{i,n}(x) = \max\left\{0, \frac{r - d(x, \alpha_{g_i(n)})}{r}\right\}$ and define $f_i = \sum_n 2^{-n} f_{i,n}$. Then $f_i(x) = 0$ iff $x \notin U_i$. For $x, y \in U_i$, define the metric $d_i(x, y) = d(x, y) + |f_i(x)^{-1} - f_i(y)^{-1}|$. Together with the c.e. set of special points that lie in U_i, this metric turns U_i into a computable Polish space. Define

$$d_X(x, y) = \sum_i 2^{-i-1} \frac{d_i(x, y)}{1 + d_i(x, y)}$$

which, together with the computable dense sequence in X, turns X into a computable Polish space. Note that a d_X-fast Cauchy sequence is also d_0- fast Cauchy, and since $d \leqslant d_0$, implies it is d-fast Cauchy. On the other hand, f_i are uniformly d-computable and bounded by 1. This makes d_X computable relative to d, and so any d-fast Cauchy name can be uniformly turned into a d_X-fast one.)

Exercise 4.2.89 (Le Roux and Ziegler [333]). Show that there exists a connected Π^0_1 closed subset of $[0, 1]^3$ with no computable points.

Exercise* 4.2.90 (Miller [391]). Recall that an *arc* is a topological space homeomorphic to the unit interval $[0, 1]$. Prove the following facts:

1. There is a Π^0_1 arc in $\mathbb{R}^2$ which is not computable (as a closed set).
2. There is a computable closed arc in $\mathbb{R}^2$ with non-computable endpoints.

Exercise* 4.2.91 (Miller [392]). Identify the Hilbert cube with its natural computable presentation described in Exercise 4.2.19. A subset $S \subseteq [0, 1]^\omega$ is convex if, whenever $d \in [0, 1]$ and $\alpha, \beta \in S$, then $d\alpha + (1 - d)\beta \in S$, where the sum is defined component-wise. Prove that there exists a non-empty convex Π^0_1 closed subset of the Hilbert cube $[0, 1]^\omega$ containing no computable points.

Exercise* 4.2.92 (Miller [391]). Fix a natural computable presentation of $\mathbb{R}^m$ induced by $\mathbb{Q}^n$ under the Euclidean metric.

1. Suppose $X \subseteq \mathbb{R}^m$ is Π^0_1 (effectively closed) and homeomorphic to an n-sphere. Show that X has to be computable (as a closed set).
2. $Y \subseteq \mathbb{R}^m$ homeomorphic to an n-ball, and both Y and its boundary spere are Π^0_1 closed. Show that Y is computable.

Exercise* 4.2.93 (Burnik and Iljazović [70]). If X is a connected 1-manifold with boundary, then X is a topological line (i.e. $X \cong \mathbb{R}$), a topological ray (i.e. $X \cong \mathbb{R}^+$), a topological circle, or an arc. Show that each semicomputable (Exercise 4.2.68) 1-manifold with finitely many connected components is computable.

Exercise 4.2.94. Let $\mathbb{B}$ be a computable (real) Banach space of finite dimension, and assume $\mathbb{V} \subseteq \mathbb{B}$ is a Π^0_1 linear subspace. Prove that $\mathbb{V}$ is computably closed. [Hint: First, note that we can fix a finite basis of $\mathbb{B}$ consisting of computable points. For that, observe that every subset generated by a finite computable tuple over $\mathbb{R}$ is closed; iterate starting with $\{0\}$. As explained in Exercise 4.2.73, $\mathbb{B}$ is computably homeomorphic to the "standard" presentation of $\mathbb{R}^n$ in which we can use the Euclidean norm or the ∞-norm upon $\mathbb{Q}^n$. The computable homeomorphism is witnessed by the identity operator and it has computable inverse. Under this effective homeomorphism, the unit ball S of $\mathbb{B}$ becomes a Π_1^0 subset of $\mathbb{R}^n$ (indeed, computably closed by Exercise 4.2.74), and $S \cap \mathbb{V}$ becomes a Π_1^0 subset of $\mathbb{R}^n$ homeomorphic to an m-ball for some $m \leqslant n$. By Exercise 4.2.92, this set is computably closed, and thus so is $S \cap \mathbb{V}$ (in $\mathbb{B}$). By Lemma 4.2.51, we can fix a $\mathbb{B}$-computable dense sequence in $S \cap \mathbb{V}$, and its $\mathbb{Q}$-linear span is computably dense in $\mathbb{V}$.]

Exercise 4.2.95. Show that for a finite dimensional (linear) subspace $\mathbb{V}$ of a computable real Banach space, the following are equivalent:

1. $\mathbb{V}$ is c.e. closed;
2. $\mathbb{V}$ has a basis consisting of computable points;
3. $\mathbb{V}$ is located (defined in Exercise 4.2.71);
4. $\mathbb{V}$ is Π_1^0 assuming $dim\,\mathbb{B} < \infty$;
5. $\mathbb{V}$ is computably closed.

[Hint: Use Exercises 4.2.71 and 4.2.94.]

Exercise 4.2.96. Let $\mathbb{B}$ and $\mathbb{D}$ be computable real Banach spaces and assume the dimension of $\mathbb{B}$ is finite. Let $T : \mathbb{B} \to \mathbb{D}$ be a computable linear operator. Show that $\ker(T) = \{x : T(x) = 0\}$ is computably closed in $\mathbb{B}$. [Hint: Use Exercise 4.2.94. We remark that one could instead use linear algebra to derive this fact, as follows. Fix a computable finite basis $\bar{b}$ of $\mathbb{B}$; this can be done as explained in the hint to Exercise 4.2.94. Using this basis, produce the matrix representing the linear operator; we note that all the coefficients in the matrix are computable reals, though we do not necessarily promise that we can decide which ones are equal to zero or pairwise equal. We do, however, know that this matrix can be transformed into one in the reduced row echelon form in finitely many (not necessarily uniformly effective) steps. The resulting matrix will still have all its coefficients as computable reals, and we can non-uniformly determine which ones are zero and which are not. Using this row echelon form, we can produce the basis for the null space of T following the usual textbook steps. The elements of this basis of $\ker(T)$ will be represented as linear combinations of the points in the finite basis $\bar{b}$ that we fixed earlier, with coefficients being computable reals. The $\mathbb{Q}$-span of these elements is dense in $\ker(T)$.]

Exercise 4.2.97. Let $T : \mathbb{B} \to \mathbb{D}$ be a computable linear operator between computable Banach spaces, let $\mathbb{K} = \ker(T) = \{x : T(x) = 0\}$, and assume $\dim \ker(T) < \infty$. Show that $\mathbb{B}/\mathbb{K}$ is a computable Banach space. [Hint: Use Exercise 4.2.95 to conclude that the standard quotient norm $||x||_{\mathbb{B}/\mathbb{K}} = \inf_{v \in \mathbb{K}} ||x + v||$ is computable; use the same dense set as in $\mathbb{B}$. More generally, this works for any located linear subspace $\mathbb{K}$.]

Exercise 4.2.98 (Effective Hahn–Banach Theorem in finite dimension; Metakides and Nerode [385]). Let $\mathbb{B}$ be a computable Banach space of finite dimension (over $\mathbb{R}$), and let $\mathbb{Y} \subseteq \mathbb{B}$ be a c.e. closed linear subspace of $\mathbb{B}$. Show that every computable linear $T : \mathbb{Y} \to \mathbb{R}$ has a computable linear extension $L : \mathbb{B} \to \mathbb{R}$ such that $||T|| = ||L||$. [Hint: In view of Exercises 4.2.97 and 4.2.95, we may assume T is injective on $\mathbb{B}$, and $\mathbb{Y}$ is computably closed in $\mathbb{B}$. Further, by scaling, we may assume that $||T|| = 1$. Use the following well-known fact. Let $(\mathbb{X}, \|\cdot\|)$ be a normed space, $\mathbb{Y} \subseteq \mathbb{X}$ a linear subspace, $x \in \mathbb{X}$, and let $\mathbb{Z}$ be the linear subspace generated by $\mathbb{Y} \cup \{x\}$. Let $F : \mathbb{Y} \to \mathbb{R}$ be a linear functional with $\|F\| = 1$. A functional $G : \mathbb{Z} \to \mathbb{R}$ with $G|_{\mathbb{Y}} = F|_{\mathbb{Y}}$ is a linear extension of F with $\|G\| = 1$ iff (†) $\sup_{u \in \mathbb{Y}}(F(u) - \|x - u\|) \leqslant G(x) \leqslant \inf_{v \in \mathbb{Y}}(F(v) + \|x - v\|)$. Noting that $\mathbb{B}$ has a basis $\bar{y}\bar{x}$ consisting of finitely many computable points, where $\bar{y}$ spans $\mathbb{Y}$, extend T to one more point x in $\bar{x}$ as follows. Observe that both the sup and the inf in (†) are left- and right-c.e. reals, respectively; indeed, these values can be approximated using a dense computable sequence in the c.e. closed $\mathbb{Y}$. Thus, either (†) defines a computable real or a non-singleton interval. In the latter case, non-uniformly pick any rational in this interval. In both cases, we can use a computable real to define the extension of T to x. To obtain L, iterate this process finitely many times[7].]

Exercise 4.2.99 (Effective Hahn–Banach theorem fails in general; Metakides, Nerode, and Shore [386], based on Bishop [48]). Show that there exists a computable Banach space $\mathbb{X}$ and a computable linear functional $T : \mathbb{Y} \to \mathbb{R}$ on a c.e. closed (indeed, computably located) linear subspace $\mathbb{Y} \subseteq \mathbb{X}$ with computable norm $\|T\|$, such that every extension $L : \mathbb{X} \to \mathbb{R}$ to the whole space with the same properties has a larger norm. [Hint: The basic idea for the diagonalisation module for (1) is depicted in the figure below. Initially, define T to be the identity on the x-axis, and define the norm on $\mathbb{R}^2$ to be (say) $||\cdot||_{2^{-s}}$ at stage s. With respect to this norm, the shape of the unit ball is as depicted in the diagram. The key observation is that, when $a \neq 0$, the only linear extension of T having norm 1 is that which fixes the bottom-right edge of the parallelogram. The same can be said about the norm $||\cdot||^*_{2^{-s}}$, but this time the top-right edge needs to be fixed. As 2^{-s} approaches 0, both norms approach the L_1-norm $||\cdot||$ on $\mathbb{R}^2$. In particular, we can 'switch gears' arbitrarily late in the construction, and turn $||\cdot||^*_{2^{-s}}$ into $||\cdot||_{2^{-s}}$ or vice versa. This provides a way to diagonalise against one potential computable extension Φ_e. To obtain the theorem, put the resulting sequence of computable Banach spaces $\mathbb{H}_e$ together using the Hilbert space direct sum. That is, $\mathbb{X} = \bigoplus_{e \in \mathbb{N}} \mathbb{H}_e = \{h \in \prod_{e \in \mathbb{N}} \mathbb{H}_e : \sum_{e \in \mathbb{N}} \|h(e)\|^2_{\mathbb{H}_e} < \infty\}$, under the sum-of-squares norm.]

Exercise° 4.2.100. Prove that there is a Π^0_1 (effectively closed) subset of $[0, 1]$ that is not homeomorphic to any computable Polish space. (Hint: Use Theorem 4.2.80.)

Exercise 4.2.101 (Melnikov and Ng [380]). Show that every left-c.e. Stone space is homeomorphic to a computably compact space.

Exercise* 4.2.102 (Harrison-Trainor, Melnikov, and Ng [245]). Show that for any computable ordinal α there exists a computable Polish space M and Π^0_1 (effectively closed) subset X of M so that X is not homeomorphic to any Δ^0_α-Polish space. (The classes Δ^0_α will be defined formally in Chapter 10. For now, assume $\alpha \subset \mathbb{N}$.)

[7] Metakides and Nerode [385] also note that, following the ideas of Bishop [48], this non-uniformity can be avoided if we relax the condition and require that $||T|| \leqslant ||L|| + \epsilon$, where $\epsilon > 0$ is any fixed rational. Then such an extension always exists and can be found uniformly (in all parameters, including ϵ). For that, they use a relaxed version of condition (†): (†$_{\epsilon_n}$) $\sup_{u \in \mathbb{Y}}((1 + \epsilon_n)F(u) - \|x - u\|) \leqslant G(x) \leqslant \inf_{v \in \mathbb{Y}}((1 + \epsilon_n)F(v) + \|x - v\|)$. They note that this search can thus be restricted to a computably compact subset of the respective finite power of $\mathbb{R}^n$, much in the spirit of the hint to Exercise 4.2.71. Thus, in particular, if we take $\prod_n (1 + \epsilon_n) < 1 + \epsilon$, we can iterate this process infinitely many times as well. In other words, if we begin with a computable linear operator T on a located finite-dimensional subspace, we can effectively find a linear extension L of T so that $||T|| \leqslant ||L|| + \epsilon$. Also, as noticed by Brattka [59], if such a computable T defined on a c.e. closed (not necessarily finite-dimensional) subspace $\mathbb{Y}$ admits a *unique* extension of the same norm to the entire space, then this extension has to be computable. Indeed, the inf and the sup in (†) are still (uniformly) right- and left-c.e. if $\mathbb{Y}$ is c.e. closed, and this makes the *unique* real defined by (†) uniformly computable. It does not matter whether $x \in \mathbb{Y}$, as this process still works.

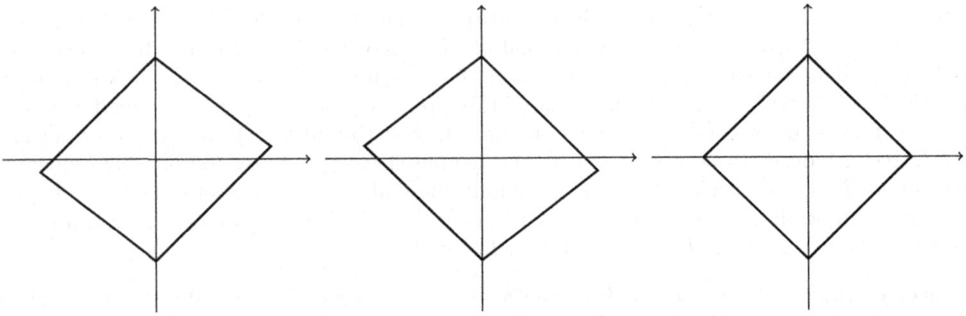

Figure 4.1: The idea for the diagonalisation module in Exercise 4.2.99. The parallelograms with corners $\{(0,1),(1+a,a),(0,-1),(-1-a,-a)\}$, $\{(0,1),(1+a,-a),(0,-1),(-1,a)\}$, and $\{(0,1),(1,0),(0,-1),(-1,0)\}$ serving as the unit balls for $||\cdot||_a$, $||\cdot||_a^*$, and $||\cdot||$ on $\mathbb{R}^2$.

Exercise 4.2.103 (Hoyrup, Melnikov, and Ng [267])**.** Prove that for a (separable) Stone space S and the dual Boolean algebra $\widehat{S}$, the following are equivalent:

1. S has an effectively compact computable topological presentation (Definition 2.4.26);
2. S has $0'$-compact computable topological presentation;
3. S has a Δ^0_2-compact Polish presentation;
4. S has a Δ^0_2-Polish presentation;
5. S has a right-c.e. Polish effectively compact presentation;
6. $\widehat{S}$ has a c.e. presentation;
7. $\widehat{S}$ has a Δ^0_2-presentation.

Recall that in Exercise 4.2.41 we saw that every effectively compact computable topological space has a Δ^0_2 Polish presentation (which is in fact Δ^0_2-compact). Conclude that this upper estimate (i.e., Δ^0_2) cannot be be improved to "computable" in general. Further, conclude that there exists an effectively compact topological space that is not homeomorphic to any effectively compact *strong* computable topological space (Definition 2.4.26).

Exercise* 4.2.104 (Bazhenov, Melnikov, and Ng [37])**.** Show that every Δ^0_2-Polish space (i.e., a Polish space computable relative to $\mathbf{0}'$) is $\mathbf{0}''$-computably homeomorphic to a computable topological space.

Exercise 4.2.105** (Hoyrup, Melnikov, and Ng [267])**.** Show that every countably-based T_0-space has a computable topological presentation.

4.2.6 Recursive profinite groups

In this subsection we apply the machinery of computably compact spaces to profinite groups. We assume that the reader is familiar with the notion of a projective (aka inverse) limit in the context of groups.

Definition 4.2.106 (La Roche [324], Smith [473]). A profinite group is *recursive* if it can be represented as the projective (aka inverse) limit of computable linear sequence of finite groups (F_i) under *surjective* $f_i : F_{i+1} \rightarrow_{onto} F_i$, where all these finite objects are uniformly computably represented by their strong indices (i.e., as indices of finite tuples).

In the theorem below, by a computably compact presentation we mean a computably compact Polish space with a computable group operation. By Corollary 4.2.46, the inverse operation is computable as well.

Theorem 4.2.107 (Downey and Melnikov [139]). *For a profinite group G, the following are equivalent:*

1. *G has a recursive presentation (in the sense defined above);*
2. *G has a computably compact presentation.*

Proof. Clearly, every recursive presentation can be viewed as a computably compact presentation (exercise). Now assume we are given a computably compact presentation of a profinite group. Using Lemma 4.2.77, computably list all clopen components of the group. Our plan is to list all normal clopen subgroups and then calculate their quotients to define a recursive inverse system representing the group.

To say that a clopen component is a normal subgroup, use the fact that every clopen component is a computable subspace of the group, and thus is computably compact, by Proposition 4.2.53. To see if a clopen C is a subgroup, search for a pair of finite covers, say (B_i) and (D_j), of C such that for every i, j there is a k with the property

$$B_i \cdot B_j \subseteq D_k$$

and for every i there is a k such that

$$B_i^{-1} \subseteq D_k.$$

We also search for a finite cover (U_n) of G such that for all n and i there is a k with

$$U_n^{-1} \cdot B_i \cdot U_n \subseteq D_k.$$

We argue that such a cover exists, and this will imply that we can computably list all clopen normal subgroups of G. Then we explain how to use these subgroups to build a recursive presentation of the group.

Since the clopen component C can be expressed as a (finite) union of open balls, the preimages of the clopen component under the computable maps $x, y \mapsto xy$, $x \mapsto x^{-1}$ and $z, x \mapsto z^{-1}xz$ in the respective product spaces can be uniformly listed. If C were not a normal subgroup then there will be special points witnessing this, and these would be witnessed together with sufficiently small basic open balls containing them. On the other hand, if C is a normal subgroup then every equation of the form, say,

$$z^{-1}xz = y,$$

where $x, y \in C$ and $z \in G$, would be witnessed by small enough basic open balls containing these points, i.e.,

$$U^{-1} \cdot B \cdot U \subseteq D,$$

where $z \in U$, $x \in B$, and $y \in D$. These products of these balls would give a cover of the respective compact product space. It follows that we can find a finite subcover.

We conclude that we can list all clopen normal subgroups of G. Note that, by the uniform computable compactness of each such clopen C, we can compute the diameter of C, which is $\sup_{x,y \in C} d(x, y)$. Using the techniques of Lemma 4.2.77 and Theorem 4.2.78 – that basically can be summarised by saying that we take the next cover by very small balls – we can furthermore produce a nested sequence of (finite open names of) clopen normal subgroups $\{C_i : i \in \mathbb{N}\}$ such that:

1. $C_{i+1} \subseteq C_i$ formally[8],
2. $diam\, C_i < 2^{-i}$.
3. For every i there exists a computable finite tuple $(x_{i,j})$ of special points (given by its strong index) such that $(x_{i,j}C_i)$ is a cover of G.
4. For every i, j, n, if $x_{i,j}C_{i+1} \subseteq x_{i+1,n}C_{i+1}$ then this inclusion is formal.
5. When $j \neq j'$, $x_{i,j}C_{i+1} \cap x_{i,j'}C_{i+1} = \emptyset$.

If we succeed, then $\bigcap_{i \in \mathbb{N}} C_i = \{e\}$, thus giving us a uniformly computable 'basis of identity' consisting of clopen normal subgroups of G. We will then use the cosets to calculate the finite G/C_i and the homomorphisms from G/C_{i+1} onto G/C_i.

More formally, we proceed by recursion. Assume C_{i-1} has been defined. We search for a C_i that satisfies all these five conditions. If we drop 'formal' in all these conditions, then it should be clear that such a C_i and $x_{i,j}$ must exist. Then fix such a C_i.

By Lemma 4.2.77 and the analysis of normality above, a normal clopen C_i will eventually be found, and furthermore both C_i and the finitely many cosets mod C_i will be represented as a finite collections of balls. Our task it to show that we can effectively recognise that these finite parameters describing the cosets define what we need. For that, we might need to adjust the finite covers by refining them so that, for instance, the inclusion is witnessed by formal inclusion of covers. This is done as follows.

We satisfy (1) by choosing the radii of a finite cover describing C_i small (see Remark 4.2.15), and we satisfy (2) by evaluating the computable diameter of the clopen set (this is again essentially done by further refining the cover). Here we use that C_i is indeed a computable closed set because of Lemma 4.2.77, so we can apply Proposition 4.2.53.

We elaborate why we will eventually find special points $(x_{i,j})$ and will eventually recognise that they satisfy (3). For that, note that each coset of C_i is open, and thus in particular contains a special point, say x. In particular, every coset mod C_i has the form xC_i. Since for every special x its coset xC_i is the image of C_i under the computable map $y \mapsto xy$ and C_i is computably compact with all possible uniformity, by Lemma 4.2.55 we conclude that xC_i is also computably compact, and with all possible uniformity. By refining the cover of xC_i (see Remark 4.2.15), we can ensure that all set-theoretical inclusions of xC_i into the clopen sets seen so far in the construction hold

[8] We can avoid using formal inclusions entirely in this case using that the sets are clopen. Similar for condition (4) below.

formally. We can also ensure that if two cosets do not intersect then this is also witnessed formally: take the radii of open balls much smaller than the pairwise distances between the finitely many clopen sets. This gives a way of computably recognising condition (5). We can also wait for finitely many such special points $x_{i,j}$ so that the respective cosets $x_{i,j}C_i$ cover the whole space.

To reconstruct the computable operation on G/C_i, calculate the product and the inverse on the special points $x_{i,j}$ with a sufficient precision until you see that the result is in one of the cosets modulo C_i. This is all computable because the cosets $x_{i,j}C_i$ are (uniformly) given by their finite open covers, and the operations on G are computable.

Finally, use effectiveness of condition (4) to calculate the surjective group-homomorphism $\phi_i : G/C_{i+1} \to G/C_i$ that maps every $x_{i+1,j}C_{i+1}$ to the unique coset $x_{i,j'}C_i$ that contains it. This gives a computable surjective inverse system $(G/C_i, \phi_i)_{i\in\mathbb{N}}$ the (inverse, projective) limit of which is topologically isomorphic to G. Since the system is uniformly computable (in the sense of strong indices of finite sets), this gives a recursive presentation of G. □

It is not difficult to produce an example of a profinite group that has a computable Polish presentation but has no computably compact presentation. This result follows from an example that can be found in [373]. We also state:

Corollary 4.2.108. *Every commutable compact profinite group is computably approximable (Definition 2.4.8).*

Proof. The 2^{-i}-approximations are uniformly given by the F_i in Definition 4.2.106. □

Exercises

For a computable compact space X, the space of all probability measures $\mathbb{P}(X)$ is a computable Polish space under the Wasserstein metric defined to be

$$d_w(\mu, \nu) = \sup \left| \int f d\mu - \int f d\nu \right|,$$

where the supremum is taken over all 1-Lipschitz functions upon X; that is, $|f(x) - f(y)| \leqslant d(x, y)$ for every $x, y \in X$. The dense set is given by Dirac measures which are the probability measures concentrated at finitely many special points of X. If X is a group than there exists a unique probability measure which is invariant under left translation by any element, called the (left) Haar measure. (There is one for the right translation too.)

Exercise 4.2.109 (Marcone and Valenti [347]). If X is computably compact then so is $\mathbb{P}(X)$.

Exercise* 4.2.110 (Pauly, Seon, and Ziegler [427]; see also [139]). For a compact computable Polish group G, the Haar measure in G is computable (as a point in $\mathbb{P}(G)$) iff G is computably compact.

Exercise* 4.2.111 (Bagaviev et al. [27]). Prove that every computable Polish space can be computably isometrically embedded into the natural presentation of $C[0, 1]$. (See [139] for a proof that uses computable compactness. We note, however, the the proof from [27] gives a uniformly primitive recursive embedding.)

Exercise 4.2.112 (Koh, Melnikov, and Ng [313]). We defined right-c.e. Polish presentations for groups in Exercise 2.4.27(1). In Exercise 2.4.28, we observed that there exists a discrete right-c.e. group not isomorphic to any computable Polish group. Prove that there exists an effectively compact (as a topological space according to Def. 2.4.26) right-c.e. Polish *profinite* abelian group not isomorphic to any computable Polish group.

4.2.7 Further related results*

As we mentioned earlier, this section is based on [139], which contains many more applications of computable compactness. Further excellent references are [270, 64]. When it comes to totally disconnected spaces specifically, the most closely related further reference is perhaps [380]. The standard references for computable Banach spaces are [56] and [435]. We also cite the recent survey [122].

We now very briefly discuss a few more recently established computable dualities.

Effective Banach–Stone duality

We state, without proof, another effective duality—this time between Stone spaces and Banach spaces. The classical Banach–Stone Theorem states that Banach spaces $C(X;\mathbb{R})$ and $C(Y;\mathbb{R})$ are isometrically isomorphic iff X and Y are homeomorphic.

Theorem 4.2.113 (Bazhenov, Harrison-Trainor, and Melnikov [35]). *Let X be a separable Stone space and let $C(X;\mathbb{R})$ be the Banach space of continuous functions $X \to \mathbb{R}$. Then the following are equivalent:*

(1) $C(X;\mathbb{R})$ has a presentation as a computable Banach space[9]*;*

(2) X has a computably compact presentation.

We emphasise that in (1) we consider $C(X;\mathbb{R})$ up to isometric linear isomorphism, but in (2) we view X up to homeomorphism. We omit the proof, but note that it uses the Downey-Jockusch Theorem 4.1.25. Koh, Melnikov, and Ng [311] have recently proven that there is a computable Banach space of the form $C(K;\mathbb{R})$ such that K is not homeomorphic to any computably compact space; see Exercise 9.1.36 for a hint. Thus, the effective Banach–Stone Theorem *fails* in general (at least in the strongest form). Theorems 4.2.113, 4.1.30, 4.2.80, and 4.2.81 combined give us the following corollary that we have already mentioned in Chapter 2 (§2.4.3):

Corollary 4.2.114. *There exists an $\varnothing'$-computable Banach space of the form $C(K;\mathbb{R})$, where K is a right-c.e. Stone space, that is not linearly isometric to any computable Banach space.*

As we noted in §2.4.3, there exist a *low* right-c.e. Banach space not linearly isometric to any computable Banach space (see Exercise 2.4.40 for a hint). Theorem 4.2.113, combined with Theorem 4.1.25 about low Boolean algebras, gives the following peculiar consequence.

Corollary 4.2.115. *Suppose $C(X;\mathbb{R})$ has a* low *Banach space presentation. If X is a Stone space, then $C(X;\mathbb{R})$ is isometrically isomorphic to a computable Banach space.*

In fact, it follows from Theorem 4.1.27 that the same can be said about *low*$_4$ Banach spaces of this form, by Theorem 4.1.27.

[9] Equivalently, as a computable Polish group, by Lemma 2.4.17.

Effective Gelfand duality

We also remark that it has recently been demonstrated in [71] that the Gelfand Duality Theorem between commutative unital C^*-algebras $C(K;\mathbb{C})$ and the respective compact domains K holds effectively. That is, the C^*-algebra $C(K;\mathbb{C})$ has a computable C^*-presentation (i.e., as a Banach space with additional operations $\times$ and $x \mapsto x^*$) if, and only if, K is homeomorphic to a computably compact space. The effective content of Gelfand Duality has been further investigated by McNicholl in [363]. Under some mild extra effectiveness conditions, it has been shown that (much like computable Stone duality) computable Gelfand Duality preserves computable categoricity in the right sense ([363]). Further algorithmic properties of C^*-algebras related to the general theme of the book have recently been established by Fox [175, 176] and Fox, Goldbring and Hart [177].

Duals of computable Banach spaces

If $\mathbb{B}$ is a computable and hence separable Banach space, then its dual (the space of bounded linear functionals $\mathbb{B} \to \mathbb{R}$) is not necessarily separable, let alone computable. The non-computability of the dual space is a significant impediment to the development of a theory of computable Banach spaces. Alternative methods must be found to replace classical arguments using the dual. For example, Brattka [56] showed that the dual space $\mathbb{B}'$ of a computable Banach space $\mathbb{B}$ is always a computable Banach space in a certain generalised sense. Just as effective Pontyragin and Stone dualities can be used to relate computable separable and discrete structures, this effective Banach space duality could potentially be used to develop a detailed and meaningful theory of non-separable Banach spaces. We also remark that the following question appears to be open:

Question 4.2.116. *Suppose $\mathbb{B}$ is a computable Banach space, and assume the dual $\mathbb{B}'$ of $\mathbb{B}$ is separable. Is it true that $\mathbb{B}'$ is linearly isometric to a computable Banach space?*

Some partial results were obtained in [62], where it was shown that if $\mathbb{B}$ has a computably shrinking effective Schauder basis (we omit the definitions), then $\mathbb{B}'$ has a natural computable Banach presentation. McNicholl conjectured that for a real $1 < p < 2$, if p is right-c.e., then the Lebesgue space $L^p[0,1]$ has a computable Banach presentation. In view of the well-known formula $\frac{1}{p} + \frac{1}{q} = 1$ for the exponent of the dual space and Exercise 2.4.39(2), a positive solution to the conjecture would imply that the answer to the question above is negative.

Computable t.d.l.c. groups and their dual ordered groupoids

A bit more is known about totally disconnected Polish groups, but not much more beyond the materials of §4.2.6 and Section 9.5 in Part II. The study of effective profinite groups began with Metakides and Nerode [384], La Roche [324], and Smith [473, 472]. Smith [473] showed that a profinite group is recursive iff it is topologically isomorphic to a decidable Π^0_1 class $[T]$ where the group operations are computable. Our Theorem 4.2.107 generalises his result to the case when the metric is not necessarily an ultrametric. As Smith [473] observed, Waterhouse's result [501] can be effectivised to prove a computable version of Galois correspondence between computable algebraic field extensions and profinite groups; we omit the statement.

In Lemma 2.4.5, we already encountered a non-compact totally disconnected group S_∞, which is the group of all permutations of $\mathbb{N}$. Various effective aspects of S_∞ were investigated in [220]. The special case of totally disconnected locally compact (t.d.l.c.) groups has been thoroughly investigated in [382, 341, 380]. In [382], it is also established that each t.d.l.c. group is effectively dual to the partially ordered groupoid of its clopen cosets, which is a countable structure. This is similar to the effective Stone duality established in this section, but the proof for t.d.l.c. groups is much more subtle. Pontryagin duality, which we discuss in the next chapter, is effective for computable abelian t.d.l.c. groups as well ([341]), but we will prove it only for profinite groups in Section 9.5.

4.3 What's next?

In the next chapter we prove another computable duality, this time between computable torsion-free abelian groups and certain computably compact spaces. Various fundamental results about computably compact spaces developed in this chapter will find direct applications in the next chapter. The tools developed for Boolean algebras will not be directly applied in the next chapter; however, working with Boolean algebras has hopefully prepared the reader for the somewhat more "truly algebraic" class of torsion-free abelian groups.

Chapter 5

Computable abelian groups and Pontryagin duality

In this chapter we prove the following results that appeared in the introduction as (3) of Theorem A and (3) of Theorem B, respectively.

Theorem (Khisamiev [288]). Every c.e. presented torsion-free abelian group is isomorphic to a computable one.

Theorem (Lupini, Melnikov, and Nies [341]). There exists a connected compact computable Polish space not homeomorphic to any computably compact space.

The key technical tool connecting these two theorems is a computable version of Pontryagin duality. Similarly to Stone duality between Boolean algebras and totally disconnected compact spaces, Pontryagin duality associates discrete countable abelian groups with compact connected Polish spaces of a certain kind. However, in contrast with Boolean algebras and Stone spaces, the relationship between the two theorems stated above is not quite as direct. It is more technical and involves a computable version of Čech cohomology, among other things. The chapter is (again) naturally split into two halves:

1. Section 5.1 contains a brief introduction to computable torsion-free abelian groups sufficient to prove theorems of Dobrica and Khisamiev which are required to prove effective Pontryagin duality.

2. Section 5.2 uses the results of Section 5.1 and Section 4.2 to prove two effective versions of Pontryagin duality necessary for establishing the second main theorem stated above. To prove these effective duality results, we will also need some elements of topological group theory and algebraic topology.

R. G. Downey, A. Melnikov, *Computable Structure Theory*, Theory and Applications of Computability, https://doi.org/10.1007/978-3-031-92433-0_5

5.1 Computable torsion-free abelian groups

5.1.1 Abelian groups

All groups in this section are additive and abelian. We assume that the reader is familiar with the standard notions of a factor-group, the order of an element, and the direct product of groups. The standard references for pure abelian group theory are Fuchs [194, 195] and Kaplansky [285]; we also recommend Kurosh [323] for a smooth and gentle introduction. We briefly review some basic notions specific to the field of abelian groups. Further notions will be introduced as needed.

Abelian groups basics

All our groups are additive and abelian. It is customary to use additive notation for the group operation in the abelian case. Recall the direct sum of abelian groups $(A_i)_{i \in I}$ is the group of all sequences $(a_i)_{i \in I}$, $a_i \in A_i$, that have finite support (i.e., are eventually 0). The standard notation is $\bigoplus_{i \in I} A_i$. This shouldn't be confused with the direct product of $(A_i)_{i \in I}$, which is not countable when I is infinite, unless almost all A_i are trivial.

Let A be an abelian group. Given a positive $n \in \mathbb{Z}$ and $a \in A$, define

$$na = \underbrace{a + a + a + \ldots + a}_{a \text{ repeated } n \text{ times}},$$

and also define $(-n)a = -(na)$ and $0a = 0$. We do not adjoin this operation to the language of groups and use it as an abbreviation. Using this notation, we list a few well-known standard notions and facts below.

Property 5.1.1. *Let A be an abelian group.*

(i) *A is* torsion-free *if $na \neq 0$ for any $n \neq 0$ and each non-zero $a \in A$.*

(ii) *A is* torsion *if for every a there exists an $n > 0$ such that $na = 0$. The least non-zero n such that $na = 0$ is the* order *of a.*

(iii) *The collection of all elements in A having finite order forms a subgroup $T(A)$, and $A/T(A)$ is torsion-free.*

(iv) *The torsion subgroup $T(A)$ further splits into a direct sum of maximal p-subgroups $T_p(A)$, in which the order of every element is some power of the respective prime p.*

Thus, in some sense, the study of abelian groups can be partially reduced to the theories of torsion-free and p-groups. Unfortunately, an abelian group does not necessarily split into a direct sum of its torsion and torsion-free subgroups. However, these two classes are traditionally viewed as central.

In this chapter we restrict ourselves to the class of torsion-free abelian groups.

The following definition will be rather important throughout the chapter.

Definition 5.1.2. Let A be an abelian group. Then $a_1, \ldots, a_k \in A$ are *linearly independent* ($\mathbb{Z}$-independent, Prüfer independent) if for each $m_1, \ldots, m_k \in \mathbb{Z}$, the equality

$$m_1 a_1 + \ldots + m_k a_k = 0$$

implies $m_i = 0$ for all $i \leqslant k$. We say that $a_1, \ldots, a_k$ are linearly dependent otherwise. A *basis* of A is a maximal linearly independent subset of A.

We will apply linear dependence mostly in the context of countable torsion-free abelian groups. We will see that countable torsion-free abelian groups are exactly the additive subgroups of $\bigoplus_{i \in \mathbb{N}} \mathbb{Q}$, the $\mathbb{Q}$-vector space of dimension ω that we have already encountered in Theorem 2.2.16. In this group, $\mathbb{Z}$-independence is equivalent to the usual linear independence over $\mathbb{Q}$. The *rank* of a (countable) torsion-free abelian group A is the smallest $\alpha \leqslant \omega$ such that $A \leqq \bigoplus_{i \in \alpha} \mathbb{Q}$. One can show that the cardinality of any basis of a torsion-free abelian A is exactly the rank of A. The rank of a finite subset of a group is defined similarly; we omit this material since it essentially repeats the standard notions and proofs from linear algebra, with only very minor adjustments. Some further details will be given later (e.g., Lemma 5.1.10, Exercise 5.1.44). Another intuitively clear property is stated below.

Lemma 5.1.3. *Suppose $\psi : A \to G$ is a homomorphism of torsion-free abelian groups that maps a basis B of A into a linearly independent set in G. Then:*

1. *ψ is injective, and*
2. *ψ maps any basis of A into a linearly independent set in G.*

Proof. We verify (1). If $\psi(g) = 0$, then for some integers n_b, almost all of which are zero, we have

$$\psi\left(\sum_{b \in B} n_b b\right) = \sum_{b \in B} n_b \psi(b) = 0.$$

Since $\psi(B)$ is independent in G, $n_b = 0$ for all $b \in B$.

Part (2) is left as an exercise. □

Free abelian groups

The free abelian group of rank $\alpha \leqslant \omega$ is the group of the form $\bigoplus_{i \in \alpha} \mathbb{Z}$, i.e., it is the direct sum of α copies of $(\mathbb{Z}, +)$. If $(f_i)_{i \in \alpha}$ are some fixed generators of these direct summands, then a typical element of this group has the form $\sum_i m_i f_i$, where $m_i \in \mathbb{Z}$ and $m_i = 0$ for almost all i. In this case, we may also write $\bigoplus_{i \in I} \mathbb{Z} f_i$ to emphasise the choice of generators. *Every generating set of a free abelian group that generates it freely is a basis (with respect to $\mathbb{Z}$-independence), but not every basis is necessarily a generating set.*

It is well known that a subgroup of a free abelian group is itself free abelian. We omit the proof. It follows that every abelian group is isomorphic to a factor of the form F/H, where F (and thus, H) are free abelian. The method of proof is similar to that used for Boolean algebras. If $A = \{a_i : i \in I\}$, then fix the free abelian group

$$F = \bigoplus_{i \in \alpha} \mathbb{Z} a_i$$

formally generated by $a_i \in A$. Then H is defined to be equal to $\sum_i m_i a_i$ for all formal linear combinations with $m_i \in \mathbb{Z}$ such that $\sum_i m_i a_i = 0$ in A. Note that in A, we can give each element $m_i a_i$ a meaning, while in F, $m_i a$ is just a formal expression. It is not hard to see that $A \cong F/H$.

The lemma below is also well known.

Lemma 5.1.4 (Rado [441]). *Let $G \leq F$ be free abelian groups. There exist linearly independent generating sets $g_1, \ldots, g_k$ and $f_1, \ldots, f_m$ ($k \leqslant m$) of G and F, respectively, and integers $n_1, \ldots, n_k$ such that for each $i \leqslant k$, we have $g_i = n_i f_i$.*

We omit the proof, which can be found in most textbooks that cover abelian groups; e.g., see [329, Theorem 7.8], where it is stated and proven in a slightly more general form. From the lemma and rank considerations, it follows easily that any finitely generated abelian group is a direct sum of cyclic groups. This consequence is known as *the classification of finitely generated abelian groups* in the literature. (Indeed, the lemma can be viewed as a reformulation of the classification of finitely generated abelian groups.) In the torsion-free case, these cyclic subgroups are infinite and, thus, are copies of $\mathbb{Z}$. In particular, every finitely generated torsion-free abelian group is free abelian.

Divisibility and pure subgroups

For $a \in A$ and a non-zero $n \in \mathbb{Z}$, the equation $nx = a$ does not have to be solvable in A. If there is such a solution, then we write $n|a$ and say that n *divides* a (in A). If for every $k \in \mathbb{N}$ we have $n^k|a$, then we write $n^\infty|a$ and say that n *infinitely divides* a. We will use the following standard notions and facts, which can be found in [194].

Property 5.1.5. Let A be an abelian group.

1. A subgroup B of A is *pure* or *serving* if for each $b \in B$ and $n \in \mathbb{Z}$, if $n|b$ in A, then $n|b$ already in B.

2. A group D is *divisible* if $n|d$ for every non-zero $n \in \mathbb{N}$ and every $d \in D$.

3. Every abelian group A can be isomorphically embedded into a divisible group.

4. For a *torsion-free* group A and a subset X of A, we can define the *pure closure* $(X)^*_A$ of X in A to be the least pure subgroup of A containing X.

5. A pure cyclic subgroup $C = \langle x \rangle$ of an abelian group A detaches as its direct summand: $A = B \oplus C$, for some $B \leq A$.

6. Every finitely generated subgroup H of A that is pure in A detaches as a direct summand of A, that is,

$$A \cong C \oplus H$$

for some $C \leq A$.

In a torsion-free group, there may exist at most one solution of $nx = a$ when $a \neq 0$; indeed, if there were two solutions $x_0 \neq x_1$, then we would have $n(x_0 - x_1) = 0$ for $x_0 - x_1 \neq 0$. Thus, 4. really makes sense only for a torsion-free group.

We explain 3. Fix $F/H \cong A$, where F is free abelian and H is a (free abelian) subgroup of F. If F is generated freely by $(f_i)_{i\in\mathbb{N}}$, then we can embed it isomorphically into the vector space $D(F)$ over $\mathbb{Q}$ upon the basis $(f_i)_{i\in\mathbb{N}}$. Then $D(F)/H$ makes sense because $H \leqq F \leqq D(F)$, and also $D(F)/H$ contains F/H as a subgroup. Since $D(F)$ is divisible, so is any of its homeomorphic images: this is because $nx = a$ becomes $n\phi(a) = \phi(x)$ under any homomorphism. Thus, $D(F)/H$ is divisible and contains an isomorphic copy of $F/H \cong A$.

Similarly to the algebraic closure of a field, a divisible group containing A can be chosen "minimal" in some standard sense that we will not define. It is unique up to isomorphism over A; it is called the *divisible hull* of A or the *divisible closure* of A. (We remark that this fact also has an effective analogue; see Exercise 5.1.36.) For example, if A is torsion-free and $(b_i)_{i\in I}$ is a basis of A, then the divisible closure of A is $\bigoplus_{i\in I} \mathbb{Q}b_i$, i.e., the (additive group of the formal) vector space over $\mathbb{Q}$ upon the basis $(b_i)_{i\in I}$. Note that A is naturally contained in $\bigoplus_{i\in I} \mathbb{Q}b_i$, because every element x of the *torsion-free* A can be represented as

$$\frac{1}{m}\sum_{i\in I} m_i b_i,$$

where almost all integers $m_i = 0$, the integer m is strictly positive, and $mx = \sum_{i\in\mathbb{N}} m_i b_i$. Such a linear combination must exist because $(b_i)_{i\in I}$ is a basis of A. If we additionally assume that $m > 0$ is the least positive integer such that $mx = \sum_i m_i b_i$ for some $m_i \in \mathbb{Z}$, then the choice of the integers m, m_i becomes unique for each $x \neq 0$. The torsion-freeness of A implies that the well-defined correspondence

$$x \mapsto \frac{1}{m}\sum_{i\in I} m_i b_i,$$

is a 1-1 homomorphism (Lemma 5.1.3). Thus, the most useful intuition that one could adopt when working with torsion-free abelian groups is as follows.

> Think of the elements of a torsion-free abelian group as "vectors" of the form $\frac{1}{m}\sum_i m_i v_i$, where the sum is finite (alternatively, almost all coefficients m_i are zero) and the v_i range over some basis in a vector space over $\mathbb{Q}$.

It will also be useful to avoid thinking of v_i as $\mathbb{Q}$-tuples, and rather, to think of them as formal sums. The reason is as follows. When considering formal sums, we can leverage the algebraic structure and properties of these sums more effectively. Also, a torsion-free abelian group is usually *not* closed under division by a non-zero integer. It is only closed under $+$ and $-$ and, thus, under multiplication by an arbitrary $m \in \mathbb{Z}$. In this sense, torsion-free abelian groups are generalisations of vector spaces over $\mathbb{Q}$. Nonetheless, this analogy with vector spaces can be extremely misleading. For instance, unlike vector spaces, a torsion-free group of rank > 2 does not have to split into a direct sum of non-trivial subgroups (folklore after Pontryagin and Levi). Torsion-free abelian groups may look tame, but in reality, they are much more poorly understood than, e.g., countable torsion abelian groups. No convenient invariants are known for countable torsion-free abelian groups of rank > 1, and thus we shall proceed with caution.

Linear span and pure subgroups

Fix a torsion-free abelian G.

Definition 5.1.6. For a set $S \subseteq G$, we write $\mathrm{span}(S)$ for the set of all elements that are "linearly spanned by S":

$$\mathrm{span}(S) = \{x \in G : \exists k \in \mathbb{N}\ \exists m, n_0, \dots, n_k \in \mathbb{Z}\ \exists c_0, \dots, c_k \in S\ mx = \sum_{0 \leqslant i \leqslant k} n_i c_i\}.$$

That is, $\mathrm{span}(S)$ it consists of all elements that can be expressed as a linear combination of some finite subset of S. (But notice the coefficient in front of x.) If $S = \varnothing$ we can agree that $\mathrm{span}(S) = \{0\}$.

Remark 5.1.7. In G, $\mathrm{span}(B)$ should not be confused with $\langle B \rangle$, which is the subgroup of G generated by B. (We have that B is linearly independent iff B generates $\langle B \rangle$ *freely*.)

In a torsion-free group, we have a nice description of linear span.

Lemma 5.1.8. *Let G be torsion-free and $S \subseteq G$. Then* $\mathrm{span}(S)$ *is equal to the least pure subgroup of G containing S (Property 5.1.5(4)):*

$$\mathrm{span}(S) = (S)^*_G.$$

Proof. The case when $S = \varnothing$ is trivial, so we can assume $S \neq \varnothing$. Since every element of $\mathrm{span}(S)$ satisfies $mx = \sum_i n_i c_i$, for some $c_i \in S$ and $m, m_i \in \mathbb{N}$, we have that $\mathrm{span}(S) \subseteq (S)^*_G$ simply because each such x is a solution of a linear equation with parameters from S. Conversely, suppose $x \in (S)^*_G$ but $x \notin \mathrm{span}(S)$, which means that x is independent of S. But we can build a pure subgroup of G starting with S, and then iteratively adjoining all solutions to linear equations (if there are any in G) with parameters in the set we defined so far, and then closing it under the group operations. By induction, at every step we will have only elements that are linearly dependent on S. This way we will construct a pure subgroup of G that contains S but does not contain x, contradicting the minimality of $(S)^*_G$. □

Observe that the procedure described above actually gives an algorithm that, given a subset S of a computable torsion-free G, enumerates $\mathrm{span}(S) = (S)^*_G$ with all possible uniformity.

Definition 5.1.9. Fix $S \subseteq G$, where G is torsion-free. We say that x and y are independent over S if $x \notin \mathrm{span}(S \cup \{y\})$ and $y \notin \mathrm{span}(S \cup \{x\})$.

In particular, we have that $x \notin \mathrm{span}(S) = (S)^*_G$, and the same can be said about y. The definition above essentially induces the notion of linear independence in the factor-group $G/(S)^*_G$ which itself is torsion-free; see Exercise 5.1.34. The following lemma is immediate.

Lemma 5.1.10. *In a torsion-free abelian group G, the span operator induced by linear independence satisfies the following properties for any $A, B \subseteq G$ and $a, b \in G$.*

1. $A \subseteq \mathrm{span}(A)$ *and* $\mathrm{span}(\mathrm{span}(A)) = \mathrm{span}(A)$,
2. $A \subseteq B \Rightarrow \mathrm{span}(A) \subseteq \mathrm{span}(B)$,
3. $\mathrm{span}(A)$ *is the union of the sets* $\mathrm{span}(F)$ *where F ranges over finite subsets of A, and*

4. *if* $a \in \text{span}(A \cup \{b\})$ *and* $a \notin \text{span}(A)$*, then* $b \in \text{span}(A \cup \{a\})$.

Proof. (1), (2) and (3) follow immediately from the definition of the span and Lemma 5.1.8. We therefore check only (4). If $a \in \text{span}(A \cup \{b\})$ and $a \notin \text{span}(A)$, it means that $ma = x + nb$, for some $x \in (A)^*_G = \text{span}(A)$ and non-zero $m, n \in \mathbb{N}$. But then $nb = ma - x \in \text{span}(A \cup \{a\})$, as required. □

The lemma above says that the *closure operator* $A \mapsto \text{span}(A)$ induced by linear independence is a "pregeometry" or a "Steinitz closure system"; we will not use it, but it can be useful [244]. For instance, the properties of the lemma suffice to show that every linearly independent set can be extended to a basis, and that the cardinalities of any two bases are the same; this is Exercise 5.1.44. (For the specific case of the closure operator *acl*, detailed proofs of these facts can be found in [348, Ch.6].)

Restricting independence

Let A be an abelian group. Recall that elements $a_1, \ldots, a_k \in A$ are dependent if

$$\exists m_1, \ldots, m_k \;\; m_1 a_1 + \ldots + m_k a_k = 0,$$

where not all m_i are equal to zero. These unbounded existential quantifiers correspond to unbounded search. A computable restricted version of this notion is defined as follows. Fix $s \in \mathbb{N}$. We write $\mathbb{Z} \restriction_{\leqslant s}$ to denote the set of integers of absolute value $\leqslant s$,

$$\mathbb{Z} \restriction_{\leqslant s} = \{m \in \mathbb{Z} : |m| \leqslant s\}.$$

If a is a non-zero element in a torsion-free group, then $\langle a \rangle \cong \mathbb{Z}$, and thus the notation $\langle a \rangle \restriction_{\leqslant s}$ also makes sense:

$$\langle a \rangle \restriction_{\leqslant s} = \{ma : |m| \leqslant s\}.$$

More generally, we write $\langle a_1, \ldots, a_k \rangle \restriction_{\leqslant s}$ to denote the collection of all sums of the form $\sum_{i \leqslant k} m_i a_i$, where $|m_i| \leqslant s$.

Definition 5.1.11. We say that $a_1, \ldots, a_k \in A$ are *s-independent* if

$$\forall m_1, \ldots, m_k \in \mathbb{Z} \restriction_{\leqslant s} \;\; m_1 a_1 + \ldots + m_k a_k = 0 \implies m_1 = m_2 = \ldots = m_k = 0,$$

and we say that they are s-dependent, otherwise.

We typically also assume $k \leqslant s$ in the definition, but this is optional. Of course, s-dependence implies dependence.

If G is indexed by natural numbers, $G = \{g_0, g_1, \ldots\}$, then we write $\text{span}_s(S)$ to denote the collection of all elements in the group having their indices $\leqslant s$ that are s-dependent on S. That is, $x \in \text{span}_s(S)$ if $x \in \{g_0, \ldots, g_s\}$, there is a $k \leqslant s$, elements $a_1, \ldots, a_k \in S$, and integers $m, m_1, \ldots, m_k \in \mathbb{Z} \restriction_{\leqslant s}$, $m \neq 0$, such that

$$mx = m_1 a_1 + \ldots + m_k a_k.$$

In various effective constructions, it is often convenient to assume $G_s = \{g_0, \ldots, g_{t(s)}\}$ rather than $\{g_0, \ldots, g_s\}$, where $t(s)$ depends on s. Then we restrict x to G_s.

Unfortunately, span_s is not nearly as well-behaved as span; for example, already (1) of Lemma 5.1.1 fails for span_s. Perhaps, the best we can say about span_s is

$$\text{span}(S) = \bigcup_{s \in \mathbb{N}} \text{span}_s(S),$$

and that for all s,

$$\text{span}_s(S) \subseteq \text{span}_{s+1}(S).$$

Partial groups

Since our groups are torsion-free, all their non-trivial finitely generated subgroups are infinite. However, we will need to deal with finite objects of the form $\text{span}_s(S)$ which are finite "segments" of subgroups of the group. More formally, we say that a $A \subseteq G$ is a *partial subgroup* if it satisfies the group axioms whenever the operation is defined.

A homomorphism of partial groups is a map that preserves the operations whenever they are defined. The (external) direct sum of partial groups $A \oplus B$ is defined as the partial group of tuples (a, b), where $a \in A$ and $b \in B$, under the partial component-wise operation.

We cannot usually apply algebraic group-theoretic techniques to partial groups directly. Every time we will have to find a way to extend a given partial group to an actual group. For instance, at stage s, we will typically have $C_s = \text{span}_s(B_s) \subseteq G$, for some finite set $B_s \subseteq G$. The set B_s will typically be s-independent, but perhaps not linearly independent in G. However, it is evidently independent in the free group $F = F(B_s)$ (formally) freely generated by B_s. When $B_s \subseteq G$, the formal free group F does *not* have to be a subgroup of G. For instance, we may later discover that B_s is t-dependent in G, for some large t. We will use F to decide some local properties about $\text{span}_s(B_s)$, one such application is described in the remark below.

Remark 5.1.12. Using, e.g., Rado's Lemma 5.1.4 and linear algebra applied to the free abelian group F formally generated by B_s, we can uniformly computably figure out the ranks of tuples in $C_s = \text{span}_s(B_s)$ as seen in F. (A different method that entirely avoids integer linear algebra will be presented in the proof of Lemma 5.1.22.) These ranks will not necessarily be equal to the actual ranks in G, however, if all coefficients witnessing the equalities that we need are sufficiently small, this will typically be enough to "switch gears" in the construction. For example, we may want to replace B_s with some other set $\tilde{B}_s$ so that $\tilde{B}_s$ and B_s are linearly interchangeable, i.e., $B_s \subseteq \text{span}(\tilde{B}_s)$ and $\tilde{B}_s \subseteq \text{span}(B_s)$. If all the coefficients witnessing $B_s \subseteq \text{span}(\tilde{B}_s)$ and $\tilde{B}_s \subseteq \text{span}(B_s)$ are smaller than t, where t is some parameter sufficient for our purposes, we could still replace B_s with $\tilde{B}_s$ even though B_s is actually not even linearly independent. We can perform all calculations in F. This is because we are really using $B_s \subseteq \text{span}_t(\tilde{B}_s)$ and $\tilde{B}_s \subseteq \text{span}_t(B_s)$. While we search for such a t and $\tilde{B}_s$, we may discover that B_s is not t-independent. In this case, we usually abandon the strategy.

5.1.2 Effective presentations of torsion-free abelian groups

All our groups are additive and at most countable. A group is computable if its domain is a computable set, and the group operations are computable. Of course, unless we forbid unbounded search, it is sufficient to assume that only $+$ is computable.

We say that $H \leq A$ is a computable subgroup of a (computable) group A if the domain of H is a computable subset of the domain of A. We define Σ^0_n- and Π^0_n-subgroups in a similar fashion.

The free abelian group on countably many generators clearly has a computable presentation with a computable generating set. We denote this presentation by $\mathbb{Z}_\omega$. In the context of abelian groups, the main definitions of the book can be formulated as follows.

Definition 5.1.13. Say that an abelian group A is Σ^0_n-presentable if $A \cong \mathbb{Z}_\omega/H$ for some Σ^0_n-subgroup H of $\mathbb{Z}_\omega$, and say that A is Π^0_n-presentable if $A \cong \mathbb{Z}_\omega/H$ for a Π^0_n-subgroup H of $\mathbb{Z}_\omega$. Define Δ^0_n-presentations similarly.

For instance, an infinite countable abelian group A has a computable (c.e.) presentation iff A is isomorphic to $\mathbb{Z}_\omega/H$, where $H \leqq \mathbb{Z}_\omega$ is a computable (respectively, c.e.) subgroup of $\mathbb{Z}_\omega$ (exercise). A group admits a Δ^0_n-presentation iff it has a $\mathbf{0}^{(n-1)}$-computable copy, so there is no danger of confusion.

Any c.e. subgroup of a computable group has a computable presentation. (A c.e. subgroup should not be confused with a c.e. presented group.) To see why, fix a computable function f such that range$(f) = H \leqq A$, where A is computable. Set $h_i = f(i)$. Given i, j, we can find the unique k such that $h_i +_A h_j = h_k$, and similarly for $-$. This gives a computable copy of H. Clearly, the range of f does not have to be computable in general. Also, evidently, this observation works for arbitrary computable structures and their c.e. substructures, not just groups.

Computable subgroups of the rationals

One of the earliest examples of a full description of computable groups in a given class belongs to Mal'cev [346]. Fix any element $a \in A$ of an additive group and any positive integer n. Recall that we say that $n > 0$ *divides* a (in A) and write $n|a$ if

$$\exists x \in A \ \ \underbrace{x + x + x + \ldots + x}_{n \text{ times}} = a.$$

Such an x (if it exists) is unique if A is a subgroup of $(\mathbb{Q}, +)$.

It should be clear that, up to isomorphism, the groups having their $\mathbb{Z}$-rank equal to 1 are exactly the non-null subgroups of $\mathbb{Q}$; we leave this to Exercise 5.1.29. Suppose $H \leqq (\mathbb{Q}, +)$ is a non-null group, and let $p_0, p_1, \ldots$ be the standard listing of all primes.

Definition 5.1.14. The *characteristic* of a non-zero element $h \in H$ is a sequence $(\alpha_0, \alpha_1, \ldots)$ where $\alpha_i = \infty$ in case $p_i^k|h$ for all k, and otherwise α_i is the largest $k \geqslant 0$ for which $p_i^k|h$ within H. We also say that α_i is the *p-height* of h.

Two characteristics $\chi = (\alpha_0, \alpha_1, \ldots)$ and $\xi = (\beta_0, \beta_1, \ldots)$ are *equivalent*, written $\chi \simeq \xi$, if

$$\sum_i |\alpha_i - \beta_i| < \infty,$$

which means that they can differ for finitely many i where furthermore the respective α_i and β_i must both be finite. The $\simeq$-equivalence class of h is called the (Baer) *type* of h, written $\mathbf{t}_H(h)$. It is not hard to see that any two non-zero elements of $H \leqq \mathbb{Q}$ are of the same type (Exercise 5.1.30). It thus makes sense to define the type of H, denoted by $\mathbf{t}(H)$, to be $\mathbf{t}_H(h)$ for some (equivalently, any) non-zero $h \in H$.

Theorem 5.1.15 (Baer [26], after Levi [337]). *Suppose $A, B \leqq \mathbb{Q}$ are non-trivial groups (equivalently, $rk(A) = rk(B) = 1$). Then $A \simeq B$ iff $\mathbf{t}(A) = \mathbf{t}(B)$.*

We leave the proof of the theorem to Exercise 5.1.30. To describe computable subgroups of $\mathbb{Q}$, we need to slightly adjust the standard invariants. Given a characteristic $\chi = (\alpha_0, \alpha_1, \ldots)$, define

$$S_\chi = \{\langle i, k\rangle : \alpha_i \geqslant k \geqslant 0\}.$$

Clearly, $\chi \simeq \xi$ iff $S_\chi =^* S_\xi$, i.e., the sets agree up to a finite difference. We say that a type $\mathbf{t}$ is computably enumerable (c.e.) if for some (equivalently, for all) $\chi \in \mathbf{t}$ the set S_χ is c.e.

Theorem 5.1.16 (Mal'cev [346]). *Suppose $A \leqq \mathbb{Q}$ is of type $\mathbf{t}$. Then A has a computable presentation iff $\mathbf{t}$ is c.e..*

Proof. If A has a computable presentation, then fix any non-zero element a of A and search through all other elements of the group and evaluate the operation on them to computably list $S_{\chi(a)}$. On the other hand, assume $S_\chi \in \mathbf{t}$ is c.e.. Then define the additive subgroup of the rationals H_χ generated by the set

$$\left\{\frac{1}{p_i^k} : \langle i, k\rangle \in S_\chi\right\}.$$

It follows from Theorem 5.1.15 that $H_\chi \cong A$. The group H_χ (under the additive group operations inherited from $\mathbb{Q}$) is evidently a c.e. subgroup of $\mathbb{Q}$, and, thus, is computably presentable. □

Recall that Theorem 3.1.1 states that there exists a c.e. non-computable low set A. If we "encode" A into a subgroup of $\mathbb{Q}$ via $p_i \mid 1$ iff $i \in \overline{A}$, then we obtain *a low group with no computable presentation*; we mentioned this already in Chapter 1, see also Exercise 5.1.31. The reader should also take a few moments to convince themselves that *every c.e. presented subgroup of* $(\mathbb{Q}, +)$ *has a computable copy* (Exercise 5.1.33).

Remark 5.1.17. Together with Theorem 2.2.6 and Theorem 3.1.1, the observations discussed above imply that the three notions of effective presentability defined in §1.2.1 (low, c.e., computable) are pairwise non-equivalent already in the class of groups. However, the historical example given in Theorem 2.2.6 was not commutative. It is also known that the notions differ in the class of abelian groups as well, as will be explained in Chapter 9 (see Corollaries 9.3.22 and 9.3.23).

Decidable subgroups of the rationals*

Recall that a group is *decidable* if it is computable and, furthermore, we can decide first-order statements about arbitrary tuples of elements in the group. Following the general theme of the book, we separate the notions of decidable and computable groups for subgroups of the rationals. Since the technical details related to the model-theoretic aspects of abelian groups will not be used in the sequel, we give only a sketch.

Theorem 5.1.18. *There exists a computable torsion-free abelian group that is decidable relative to a low oracle, but has no decidable presentation.*

Sketch. In model theory, one proves that abelian groups admit quantifier elimination down to Boolean combinations of existential formulae that involve only divisibility conditions of the form $m \mid x$, i.e., $\exists y\, my = x$. In the context of computable mathematics, a result of this sort is the following fact (see [291, Prop.1.1]).

Proposition 5.1.19. *For an abelian group G upon the domain $\mathbb{N}$, the following are equivalent:*

1. *$(G,+)$ is decidable;*

2. *$Th(G)$ is decidable, $(G,+)$ is computable, and the predicates $(p_i \mid \cdot)_{i\in\mathbb{N}}$ are uniformly computable, where $p_i \mid x$ is the predicate of divisibility by the i-th prime number.*

We skip the proof.

Let A be a low non-computable c.e. set (Theorem 3.1.1). Let G_A be the subgroup of $(\mathbb{Q},+)$ generated by

$$\left\{\frac{1}{p_i} : i \in A\right\}.$$

It is also known that, for *any* set $A \subseteq \omega$,

$$Th(G_A) = Th(\mathbb{Z}).$$

This can be concluded after calculating the Szmielew invariants of the groups (for the invariants, see, e.g., [291]). This calculation is relatively straightforward; see [198, Thm 17] where this result is stated in the required form. Finally, it is well-known that the theory of the integers with $+$ is decidable (e.g., [158]). It follows from Theorem 5.1.16 that, for a c.e. set A, the problem of decidable presentability of G_A can be completely reduced to the decidability of the divisibility predicates $(p_i \mid \cdot)_{i\in\mathbb{N}}$.

We claim that the decidability of G_A is equivalent to the computability of A. If H is any decidable presentation of G, then the predicates have to be uniformly computable by Proposition 5.1.19. This provides a method to decide $i \in A$, as follows. Fix any non-zero element of the group and appeal to Baer's classification Theorem 5.1.15. Conversely, if A is a computable set, then we can use the straightforward construction in Theorem 5.1.16 to build a computable copy of the group in which, additionally, the divisibility predicates are decidable. Thus, if A is low c.e. but not computable (Theorem 5.1.16), then G_A has a computable copy that is decidable *relative to the low oracle* A, but has no decidable presentation. □

The proposition above should be compared with Theorem 4.1.40 for Boolean algebras. Similarly to Theorem 4.1.40, we indeed separated 1-decidability and computable presentability for abelian groups.

Groups with linear dependence algorithm

Recall that in Theorem 2.2.16 we constructed a computable presentation of $\mathbb{Q}^{<\omega} = \bigoplus_{i\in\mathbb{N}} \mathbb{Q}$ that has no computable basis. But of course, $\mathbb{Q}^{<\omega}$ has a nice computable copy with a computable basis. Therefore, the notion defined below is *not* presentation-invariant.

Definition 5.1.20. We say that a computable torsion-free abelian group A has a *linear dependence algorithm* if, given any $a_1, \dots, a_k \in A$, we can uniformly decide if $a_1, \dots, a_k$ are linearly dependent.

Of course, in the definition above $k > 0$ is not fixed. We remark that this definition is an adaptation of the similar definition from the theory of computably vector spaces [383] and a special case of the more general notion of a computable pregeometry [148, 244]. It follows from Proposition 5.1.19 that having a linear dependence algorithm does not imply decidability or even 1-decidability. This is only to be expected, since linear (in)dependence is not a first-order property.

Remark 5.1.21. It can be shown that if G is a *homogeneous completely decomposable group* that is *not* divisible, then every decidable copy of G has an algorithm for linear independence [33] (to appear as Exercise 7.2.23). We delay the definition of (homogeneous) completely decomposable groups until Part 2, where they will play a significant role.

The fact below is well-known and holds much more generally in terms of effective pregeometries (e.g., [244]).

Fact 5.1.22. *For a computable torsion-free abelian group A, the following are equivalent:*

1. *A has a linear dependence algorithm;*
2. *A has a computable basis;*
3. *A has a c.e. basis.*

All implications are effectively uniform.

Proof. The implication (1) → (2) is an exercise, and (2) → (3) is straightforward. (For (1) → (2), build a basis B in stages. Also, specifically make sure that the element with index s is in span B_s.) We prove (3) → (1). Fix a c.e. basis B. To decide whether $x_1, \dots, x_n$ are dependent or not, find a finite set $A \subseteq B$ such that $\{x_1, \dots, x_n\} \subseteq \operatorname{span}(A)$. Then $x_1, \dots, x_n$ are independent iff there exist $a_1, \dots, a_n \in A$ such that

$$a_1, \dots, a_n \in \operatorname{span}(\{x_1, \dots, x_n\} \cup (A \backslash \{a_1, \dots, a_n\})).$$

The latter is a Σ^0_1-property. Since for $\{x_1, \dots, x_n\}$ being dependent is Σ^0_1, we conclude that the property is Δ^0_1, as desired. □

Is every computable torsion-free abelian group isomorphic to a computable group with a linear dependence algorithm? Below, we provide a positive answer to this question. But first, we give a characterisation of computable linear independence that will be useful later.

A useful characterisation of computable linear independence

In the next section, we will need to construct the dual of a torsion-free abelian group (to be defined). For that, we will need to uniformly computably access the finitely generated subgroups of our group (as opposed to merely its finite partial subgroups). More specifically, we will need to view a torsion-free abelian group as a "computable direct limit" (union) of finitely generated groups, where each such group also comes with its finite basis that generates it; cf. Definition 4.2.106.

Definition 5.1.23. We say that a computable G is *tractable* if there exists a uniformly computable ascending sequence of finitely generated abelian groups $(F_i)_{i \in \mathbb{N}}$ with the following properties:

1. $G = \bigcup_{i \in \mathbb{N}} F_i$.
2. $\{0\} = F_0 \subseteq F_1 \subseteq F_2 \subseteq F_3 \subseteq \dots$ is a uniformly computable sequence in which the set-theoretic embeddings are also computable.
3. For every F_i ($i > 0$) we can uniformly compute a (strong index for a) finite set of elements $h_0, \dots, h_{k(i)}$ such that
$$F_i = \langle h_0 \rangle \oplus \langle h_1 \rangle \oplus \dots \oplus \langle h_{k(i)} \rangle.$$

Recall that a subgroup H of an abelian A is pure (in A) if, for any $h \in H$ and each positive integer k, $\exists a \in A\ ka = h$ implies $\exists u \in H\ ku = h$. A finitely generated (f.g.) subgroup of A that is pure in A detaches in A (i.e., forms a direct summand of A); this appeared earlier as Property 5.1.5(6). Recall that we write $\langle h_1, \ldots, h_k \rangle$ for the f.g. subgroup of the given group generated by $h_1, \ldots, h_k$, and that $\langle h_1, \ldots, h_k \rangle \restriction_{\leqslant t}$ denotes the partial subgroup of it consisting of linear combinations $\sum_i n_i h_i$, where $|n_i| \leqslant t$, $i = 1, \ldots, k$.

Lemma 5.1.24 (Melnikov [373]). *Let G be a computable torsion-free abelian group. The following are equivalent:*

1. *G is tractable.*
2. *G has a computable basis.*

Proof of Lemma. Suppose G has a computable basis $B = \{b_1, b_2, \ldots\}$ (we include the possibility of B being finite or even empty). Enumerate B and G. Suppose at a stage we have effectively defined a f.g. partial subgroup

$$G_s = \langle h_1, \ldots, h_k \rangle \restriction_{\leqslant t},$$

where k, t depend on the stage, the h_i are linearly independent in G, and indeed $\mathrm{span}(h_1, \ldots, h_k) \subseteq \mathrm{span}(b_1, \ldots, b_k)$ in G. (It actually must be that $\mathrm{span}(h_1, \ldots, h_k) = \mathrm{span}(b_1, \ldots, b_k)$.) Furthermore, by linear independence we have that each h_i generates a finite initial segment of the infinite cyclic group, and also that

$$G_s = \langle h_1 \rangle \restriction_{\leqslant t} \oplus \ldots \oplus \langle h_k \rangle \restriction_{\leqslant t}.$$

Suppose a new element h enters the enumeration of the group. Before taking action, keep adjoining elements $b_{k+1}, b_{k+2}, \ldots$ from the basis B to $h_1, \ldots, h_k$ (noting that $\{h_1, \ldots, h_k, b_{k+1}, b_{k+2}, \ldots\}$ forms a basis of G). At a later stage we will have a f.g. partial group of the form

$$G'_u = \langle h_1 \rangle \restriction_{\leqslant u} \oplus \ldots \oplus \langle h_k \rangle \restriction_{\leqslant u} \oplus \langle b_{k+1} \rangle \restriction_{\leqslant u} \oplus \ldots \oplus \langle b_{k'} \rangle \restriction_{\leqslant u}$$

containing G_s. We keep doing so until we find a linear combination

$$mh = \sum_{i \leqslant k} n_i h_i + \sum_{k < i \leqslant k'} n'_j b_j,$$

where the integer coefficients m, n_i, and n'_j are reduced and $m > 0$ is smallest such. We are ready to define G_{s+1}. It will be a large enough finite partial subgroup approximating $\langle h, G'_u \rangle$. If $m = 1$, then we have $h \in G'_u$ and set $G_{s+1} = G'_u$. Otherwise, suppose $m > 1$. In this case, without loss of generality, $\langle h \rangle$ is pure in $X = \langle h, G'_u \rangle$; if it is not, replace h with h_0 so that $\langle h_0 \rangle = (h)^*_X$. (Using Rado's Lemma 5.1.4, fix linearly independent generating sets of $\langle h \rangle$ and X with the nice properties described in Lemma 5.1.4; it has to be that h is a multiple of some basic h_0 in X given by the lemma, since the only 1-element generating sets of $\langle h \rangle$ are $\{h\}$ and $\{-h\}$.) Since $\langle h \rangle$ is pure and cyclic, we have

$$\langle h, G'_u \rangle = \langle h \rangle \oplus H.$$

We choose any direct decomposition of H into cyclic summands, and we wait until a late enough stage v such that the generators of all these summands appear in the enumeration of G at stage v. Then we set $G_{s+1} = \langle h, G_v \rangle \restriction_{\leqslant v}$.

In both cases, we also record the information about the generators of G_{s+1} and the natural embedding of G_s into G_{s+1}. It is clear that this embedding is fully determined by how the generators of G_s are expressed in terms of the fixed generators of G_{s+1}.

Strictly speaking, $(G_s)_{s\in\mathbb{N}}$ is a sequence of finite partial subgroups, not a sequence of f.g. subgroups of G (as required). Otherwise, all the other properties that we need are satisfied by the sequence $(G_s)_{s\in\mathbb{N}}$. But note that $G = \bigcup_{i\in\mathbb{N}}\langle G_s\rangle$. Based on this observation, we claim that $\langle G_s\rangle$ is a uniformly computable subgroup of G. Indeed, for any $g \in G$, wait for $g \in G_v$ $(s \leqslant v)$, and then use the information about G_v, its generators, and how G_s is embedded into G_v to see if $g \in \langle G_s\rangle$. Finally, observe that the embedding of $\langle G_s\rangle$ into $\langle G_s\rangle$ is completely determined by the embedding of G_s into G_{s+1}. Thus, $G = \bigcup_{i\in\mathbb{N}}\langle G_s\rangle$ witnesses that G is tractable.

Conversely, suppose G is a tractable constructive group, and let $(F_i)_{i\in\mathbb{N}}$ be an ascending sequence of its subgroups witnessing its tractability. Suppose we have $g_1, \ldots, g_k \in G$. Then for some large enough m we must have $g_1, \ldots, g_k \in F_m$. The generators of F_m have to be linearly independent in F_m and, thus, in G. Since we can compute a full decomposition of F_m into infinite cyclic summands, we can decide whether $g_1, \ldots, g_k$ are independent using Fact 5.1.22. □

How to approach c.e. presented abelian groups

In this paragraph we discuss several useful ways to approach c.e. presented torsion-free abelian groups. We will need this in the proof of Khisamiev's Theorem 5.1.41.

Let A be our group and $U = L/E$ its c.e. presentation, where L is free abelian. A good way to think about U is to assume its domain is ω, the operations are computable, but the equality between elements is merely c.e.; equivalently, being equal to 0 is a c.e. unary relation. We have not yet dealt with c.e. presented structures that are not locally finite, i.e., where a finite subset does not necessarily generate a finite substructure. The difference is quite apparent: declaring $x = 0$ in a torsion-free abelian group results in setting $mx = 0$ for *all* $m \in \mathbb{Z}$. However, we can process only finitely many relations at any given stage. To make matters worse, at any finite stage we can really only examine a finite partial subgroup of U. While it is clear what this meant for a computable U, in the case of a c.e. presented group this needs to be further clarified.

Property 5.1.25. Assume $U = L/E$ is a c.e. presented abelian group, where L is computable free abelian, and E is its c.e. subgroup. Without loss of generality we may assume the following about L and E:

A) L is given together with a computable linearly independent set $(\ell_i)_{i\in\mathbb{N}}$ that generates it freely. In particular, we may assume the rank of L is ω.

B) We may assume that $E_s = \langle e_{0,s}\rangle\oplus\ldots\oplus\langle e_{k(s),s}\rangle$, where $e_0, \ldots, e_{k(s)} \in L$ are linearly independent. Note that E_s is an infinite object. It is given by the index of its independent generating set $e_{0,s}, \ldots, e_{k(s),s} \in L$.

C) Since U is torsion-free, we can further assume E_s is pure in L; equivalently, L/E_s is torsion-free (Exercise 5.1.34). Indeed, if $x \in E_s$ and $mh = x$, we know that for some $t \geqslant s$, $h \in E_t$, for otherwise $U = L/E$ would not be torsion-free. Thus, we can just put h into E_s straight away.

D) By taking $n(s)$ sufficiently large, we may assume

$$E_s \leqq L_s = \langle \ell_0\rangle \oplus \ldots \oplus \langle \ell_{n(s)}\rangle;$$

note L_s is also an infinite object and is given by the parameter $n(s)$.

E) Appealing to Rado's Lemma 5.1.4 (combined with brute force search), we see that E_s is a computable subgroup of L_s, uniformly in s.

F) Assuming U is torsion-free, the index of the free abelian group $\tilde{U}_s = L_s/E_s$ can be obtained uniformly from the indices of E_s and L_s. Further, using brute-force search combined with Rado's Lemma 5.1.4, we can uniformly compute a linearly independent subset $u_0, \ldots, u_{q(s)}$ of $\tilde{U}_s$ that generates it freely. We may set

$$U_s = (\langle u_0 \rangle \oplus \ldots \langle u_{q(s)} \rangle)|_{\leqslant s}.$$

To avoid various pathologies (e.g., the ℓ_j-coefficients of representatives of the cosets of u_i being too large), we may wish to replace $|_{\leqslant s}$ with $|_{\leqslant t(s)}$, where $t(s)$ is a monotonic function[1]. More generally, by increasing q and t when necessary, we can always assume that U_s is sufficiently large; for example, we may assume that it contains a linearly independent set of size s.

G) To relate U_{s+1} with U_s, we use the fact that $E_s \leqq E_{s+1} \leqq L_{s+1}$, where all subgroups are computable subsets, uniformly in s (by E). We have that $\tilde{U}_s = L_s/E_s$, and we can computably identify L_s/E_s with a (computable) subgroup of L_{s+1}/E_s. Further, by the Third Isomorphism Theorem,

$$L_{s+1}/E_{s+1} \cong \frac{L_{s+1}/E_s}{E_{s+1}/E_s}.$$

Using that all independent generating sets are uniformly computable in all these groups (as subsets of L), we can uniformly compute

$$\tilde{\eta}_s : L_{s+1}/E_s \to \frac{L_{s+1}/E_s}{E_{s+1}/E_s}$$

and its restriction η_s to $L_s/E_s \leqq L_{s+1}/E_s$,

$$\eta_s : L_s/E_s \to \frac{L_{s+1}/E_s}{E_{s+1}/E_s},$$

where the range of η_s is computable in $\dfrac{L_{s+1}/E_s}{E_{s+1}/E_s}$, uniformly in s. By F, we may further assume that the partial group U_{s+1} is large enough so that $\eta_s(U_s) \subseteq U_{s+1}$, and thus we arrive at

$$\eta_s : U_s \to U_{s+1}$$

which are uniformly computable homomorphisms of finite partial groups.

We leave further formal verification of the possibility of the assumptions summarised in A – G above as an exercise.

[1] For example, the function defined recursively via $t(0) = 2$ and $t(s+1) = 2(t(s)+1)!$ will usually suffice. It will significantly exceed all numbers mentioned in all parameters at stage s. This is because all these calculations can be performed in polynomial time instead of brute-force using linear algebra. Alternatively, we may first perform all our calculations using brute-force (or any method) and choose $t(s)$ to be large enough so that the partial subgroup $U_{t(s)}$ includes all elements of $\tilde{U}_s$ that we need; cf. Remark 5.1.12.

Remark 5.1.26. Of course, all of these assumptions can be made about computable groups as well, but in the computable case the maps $\eta_s : U_s \to U_{s+1}$ are additionally injective. (Recall Remark 2.2.19 in the proof of Mal'cev's Theorem 2.2.16.)

In summary, the sequence of U_s and η_s has pretty much every conceivable algorithmic property we could possibly hope for. To define U, at stage $s + 1$ monitor η_s to see which elements of U_s need to be identified (declared equal modulo E) in U_{s+1}. The domain of U can still be indexed by natural numbers, however, with repetition. Deciding whether i and j represent the same element of the abstract group is c.e., since we need to see whether for some s, these indices are declared equal in U_s. However, the operation $+$ on these indices is computable, since it can be computed inside the large enough U_s. Since the maps $\eta_s : U_s \to U_{s+1}$ are restrictions of homomorphisms $\tilde{\eta}_s : \tilde{U}_s \to \tilde{U}_{s+1}$, these calculations agree with the group operations throughout the entire sequence $(U_s)_{s\in\mathbb{N}}$. This presentation of U will be used in the proof of Khisamiev's Theorem 5.1.41.

Remark 5.1.27. The index of a coset in U is always assumed to be equal to the index of its smallest representative at every stage. It follows that for every coset $x + E$ in U there is a stage s large enough so that the index of $x + E_s$ in U_s is equal to the index of $x + E$ in U. In particular, we can use the smallest index representative technique in a c.e. presented group as well. (In the context of computable groups, we will use this technique shortly in the proof of Dobrica's Theorem 5.1.37.) However, we need to keep in mind that the index of an element (coset) in U_s may not be final at a stage, but it will settle at some stage.

Alternatively, we could use the uniformly computable sequence of finitely generated (free abelian) groups $\tilde{U}_s$ with distinguished independent generating sets, and uniformly computable homomorphisms $\eta_s : \tilde{U}_s \to \tilde{U}_{s+1}$. Then our presentation can be viewed as the *direct limit*

$$\varinjlim_{s\in\mathbb{N}}(\tilde{U}_s, \eta_s) = \left(\bigsqcup_{s\in\mathbb{N}} \tilde{U}_s\right) / \equiv \tag{5.1}$$

of the computable (linear) direct system $(\tilde{U}_s, \eta_s)_{s\in\mathbb{N}}$. (We shall write simply $\varinjlim_{s\in\mathbb{N}} \tilde{U}_s$ if there is no danger of confusion.). We give more details. For $u, v \in \bigsqcup_{s\in\mathbb{N}} \tilde{U}_s$, we set $u \equiv v$ if after several iterative applications of the η-homomorphisms, the images of v and u become equal in some $\tilde{U}_s$. Also, to define $+$ on classes $[\cdot]_\equiv$ modulo $\equiv$, fix $u \in \tilde{U}_s$ and $v \in \tilde{U}_t$; assume $t \leqslant s$. Iteratively apply the η-homomorphisms to u until the result of this process a is in $\tilde{U}_s$. Set

$$[u]_\equiv + [v]_\equiv = [a + v]_\equiv,$$

where $a + v$ is calculated in $\tilde{U}_s$. Since we've been avoiding explicit use of formalisms related to direct limits, we will leave the verification of $U \cong \varinjlim_{s\in\mathbb{N}} \tilde{U}_s$ to Exercise 5.1.35.

On the other hand, observe that $\equiv$ is clearly c.e. on $\sqcup_{s\in\mathbb{N}} \tilde{U}_s$. This gives the following useful fact. Recall that a finite presentation of a group is a tuple $\langle a_0, \ldots, a_k | r_0, \ldots r_n \rangle$ of generators and relations upon these generators. Since all our groups are abelian, we may assume that $a_0, \ldots, a_k$ are generators of the free abelian group of rank k, $\bigoplus_{i\leqslant k} \mathbb{Z}$, and $r_0, \ldots r_n \in \bigoplus_{i\leqslant k} \mathbb{Z}$. Note that the groups $\tilde{U}_s$ in (5.1) are uniformly finitely presented. Conversely, we clearly have:

Fact 5.1.28. *Let G be an abelian group. Suppose there is a uniformly computable directed sequence $(\tilde{U}_s, \eta_s)_{s\in\mathbb{N}}$ of finitely presented (abelian) groups and homomorphisms $\eta_s : \tilde{U}_s \to \tilde{U}_{s+1}$, where each*

$\tilde{U}_s$ *is given by a finite set of generators and relations, and whose direct limit (given by* (5.1)*) is isomorphic to* G*. Then* G *is c.e. presented, and the index of the c.e. presentation can be obtained uniformly from the index of the sequence* $(\tilde{U}_s, \eta_s)_{s \in \mathbb{N}}$.

Thus, for an abelian group, being c.e. presented and being presented via an effective direct limit (5.1) are synonymous, and this is uniform. This observation was certainly already known to Baumslag, Dyer, and Miller [31], though in a slightly different context and terminology. This approach won't help in the proof of Khisamiev's Theorem, but it will be useful in Section 5.2.

Exercises

Exercise° 5.1.29. Show that, up to isomorphism, the rank 1 torsion-free abelian groups are exactly the non-zero subgroups of $(\mathbb{Q}, +)$.

Exercise° 5.1.30. Check that any pair of non-zero elements in a given subgroup of $\mathbb{Q}$ have the same type. Prove Theorem 5.1.15.

Exercise° 5.1.31. Show that for any set A there exist subgroups G_A and H_A of $(\mathbb{Q}, +)$ such that G_A has an X-computable copy iff A is c.e. relative to X, and H_A has a computable copy iff $A \leqslant_T X$.

Exercise° 5.1.32. Generalise Theorem 5.1.16 to additive subgroups of $(\mathbb{Q}^n, +)$. (Hint: Consider a maximal $\mathbb{Q}$-independent subset of $G \leqslant (\mathbb{Q}^n, +)$. Produce an invariant for G generalising S_χ that uses the fixed basis as a parameter.)

Exercise° 5.1.33. Let G be an additive subgroup of $(\mathbb{Q}^n, +)$, for some finite n. Show that G has a c.e. presentation iff G has a computable copy. (Hint: Use Ex. 5.1.32.)

Exercise° 5.1.34. Let A be torsion-free abelian. Show that $H \leqq A$ is pure in A iff H/A is torsion-free.

Exercise° 5.1.35. Verify that the definition involving the direct limit (see (5.1)). Check that $\equiv$ is an equivalence relation and that the operation $+$ defined after (5.1) turns $\bigcup_i \tilde{U}_i / \equiv$ into an abelian group isomorphic to U.

Exercise* 5.1.36 (Smith [474])**.** The divisible closure (the divisible hull) of an abelian group A is "the smallest" divisible group D that contains A. Formally, if C is divisible and $A \to C$ is an isomorphic embedding, then there is an embedding of D into C over A. Show that every computable abelian group A can be computably embedded into its computable divisible closure.

5.1.3 Dobrica's Theorem

Theorem 5.1.37 (Dobrica [114])**.** *Every computable torsion-free abelian group is isomorphic to a computable group with a computable basis.*

Idea

We are given a computable (torsion-free abelian) A. If A has a finite basis, then there is nothing to prove. We therefore assume that the rank of A is infinite. We transform A into a computable B having a computable maximal linearly independent set $C = (c_i)_{i \in \mathbb{N}}$, as follows.

Build a Δ^0_2 isomorphism $\theta : B \to A$. Initially, let θ copy A into B without any change. Suppose we have $\theta(c_i) = a_i \in A$, $i = 0, 1, \ldots$. At a later stage, we may discover that, in A, a_0 and a_1 are linearly dependent. (The case of more than two a_i's is similar.) The idea is to use a modification of the strategy from the proof of Mal'cev's Theorem 2.2.16; however, it will need to be modified further. Choose the first found $d \in A$ which currently looks independent of a_0 and define

$$\theta(c_1) = a_1 + t!d,$$

where t is larger than any number mentioned so far in the construction. If previously

$$m\theta(x) = m_0 a_0 + m_1 a_1,$$

then we have to set

$$m\theta(x) = m_0 a_0 + m_1(a_1 + t!d) = (m_0 a_0 + m_1 a_1) + m_1 t!d,$$

where $(m_0 a_0 + m_1 a_1)$ is divisible by m as witnessed by the previous image of x, and $m_1 t!d$ is divisible by m because $m < t$ (recall t is large). *This is exactly why we used the factorial.* In particular, the relation $mx = m_0 c_0 + m_1 c_1$ will be preserved under θ.

If d is indeed independent of a_0, then so is $a_1 + t!d$; furthermore, $a_1 + t!d$ and d will have equal linear spans over a_0. Otherwise, the strategy will be repeated with a fresh d', and then perhaps d'' (etc.) until a $d^{(k)}$ truly independent over a_0 is found. In particular, it follows that this process of correcting mistakes will eventually stabilise[2].

Construction

We build a computable group B and its computable basis $C = (c_i)_{j \in \mathbb{N}}$. At every stage t, we also define a partial map $\theta_t : B_t \to A_t$. We also assume that for each t the partial group A_t is sufficiently large in the sense of Property 5.1.25 (F) and Remark 5.1.26.

At stage 0, begin with $C_0 = \varnothing$ in B_0, and let B_0 copy A_0 via $\theta_0 = Id$, i.e., without any nontrivial permutation.

Let $a_i = \theta_{t-1}(c_i) \in A_{t-1}$, $i < t$.

Stage t. We subdivide the stage into several phases:

(a) Choose $k < t$ largest such that $\theta_{t-1}(c_0), \ldots, \theta_{t-1}(c_k)$ are $2(t+1)!$-independent in A_t. Choose $d_{k+1}, \ldots d_t \in A_t$ such that

$$a_0 \ldots, a_k,\ a_{k+1} + t!d_{k+1},\ \ldots,\ a_{t-1} + t!d_{t-1},\ d_t$$

[2] The construction below can be viewed as a movable markers argument, where each "movable marker" $\boxed{i}$ corresponds to $\theta(c_i) \in A$ for some $c_i \in C$. When the value $\theta(c_i)$ needs to be changed, we "move" the "marker" to a new value. We will, however, not make movable markers explicit since the main complexity of the proof is not related to computability-theoretic combinatorics.

form a $2(t+1)!$-independent set, and $d_{k+1}, \ldots d_t \in A_t$ have the smallest possible indices (lexicographically).

(b) Define θ_t as follows.

(b.1) For each c_r, $k < r < t$, set

$$\theta_t(c_r) = \theta_{t-1}(c_r) + t!d_r = a_r + t!d_r.$$

(b.2) Declare $\theta_t(c_i) = \theta_{t-1}(c_i)$ for every $i \leqslant k$.

(b.3) Introduce c_t and declare $\theta_t(c_t) = d_t$.

(b.4) For each $x \in B_{t-1}$ such that $mx = \sum_{j<t} n_j c_j$, set

$$\theta_t(x) = \theta_{t-1}(x) + \sum_{t>r>k} \frac{n_r t!}{m} d_r.$$

(b.5) For every $a \in A_t$ that does not already have a θ_t-preimage, if

$$ma = \sum_{j \leqslant t} n_j \theta_t(c_j), \text{ where } m, |n_j| \leqslant t,$$

introduce a new element b in B_t, declare

$$mb = \sum_{j \leqslant t} n_j c_j$$

in B_t, and set $\theta_t(b) = a$.

Go to the next stage.

Verification

Claim 5.1.38. *Every stage of the construction eventually terminates its search.*

Proof. In (a), we search for elements in A_t which are $2(t+1)!$-independent. If $\theta_t(c_{i,0}), \ldots, \theta_t(c_{i,k})$ are indeed independent, then such elements must exist because the rank of A_t is infinite. Such elements (independent or not) will eventually be found.

In (b.4), for each x such that $mx = \sum_{j \leqslant t} n_j c_{i,j}$, we set

$$\theta_t(x) = \theta_{t-1}(x) + \sum_{t \geqslant r > k} \frac{n_r t!}{m} d_r;$$

such an element exists because $m < t$. □

Claim 5.1.39. *At the end of every stage t, each θ_t is an injective homomorphism of partial groups.*

Proof. By induction on t. The case when $t = 0$ is trivial. Suppose θ_{t-1} is an injective homomorphism of partial groups. Then, by induction, θ_t clearly respects the group operations whenever they are defined. Thus, we need only to verify that θ_t is injective.

Suppose $x, z \in B_{t-1}$ and therefore θ_{t-1} is defined on $x, z \in \operatorname{dom} \theta_{t-1}$, where

$$nz = \sum_j n_j c_j,$$

$$mx = \sum_j m_j c_j.$$

If θ_{t-1} and θ_t are equal on the domain of θ_{t-1}, then there is nothing to prove. Suppose θ needs to be redefined. Assume $\theta_t(z) = \theta_t(x)$, then

$$\theta_t(x) = \sum_{r>k} \frac{m_r t!}{m} d_r + \theta_{t-1}(x),$$

and

$$\theta_t(z) = \sum_{r>k} \frac{n_r t!}{n} d_r + \theta_{t-1}(x).$$

These values satisfy the equations:

$$m\theta_t(x) = \sum_{j \leqslant k} m_j \theta_{t-1}(c_j) + \sum_{r>k} m_r(\theta_{t-1}(c_r) + t! d_r),$$

$$n\theta_t(z) = \sum_{j \leqslant k} n_j \theta_{t-1}(c_j) + \sum_{r>k} n_r(\theta_{t-1}(c_r) + t! d_r).$$

Multiply the first equation by n and the second by m, and then subtract the first one from the second one. According to the instructions in (a) at stage t, the values $\theta_{t-1}(c_j)$ and $\theta_{t-1}(c_r) + t!d_r$ form a $2(t+1)!$-independent set. In particular, it must be that, for every $r > k$, $t!nm_r = t!mn_r$, and since both $m, n \neq 0$, we arrive at

$$\frac{m_r t!}{m} = \frac{n_r t!}{n}, \quad \text{for each } r > k.$$

Now recall that

$$\theta_t(x) = \sum_{r>k} \frac{m_r t!}{m} d_r + \theta_{t-1}(x)$$

and

$$\theta_t(z) = \sum_{r>k} \frac{n_r t!}{n} d_r + \theta_{t-1}(z).$$

Since $\sum_{r>k} \frac{n_r t!}{n} d_r = \sum_{r>k} \frac{m_r t!}{m} d_r$ by the above remarks, and $\theta_t(z) = \theta_t(x)$ by our assumption, we must have that

$$\theta_{t-1}(z) = \theta_{t-1}(x).$$

Since θ_{t-1} is injective by the inductive hypothesis, $z = x$.

It remains to argue that injectivity is maintained when we extend the domain of θ in (b.3) and (b.5). Recall that in (a) we chose $a_0, \ldots, a_k, a_{k+1} + t!d_{k+1}, \ldots, a_{t-1} + t!d_{t-1}, d_t$ to be $2(t+1)!$-independent, where $a_i = \theta_{t-1}(c_i)$, $i < t$. Thus, in (b.3) and, more generally, in (b.5), injectivity is maintained: this is because different choices of coefficients $n_j \leqslant t$ will result in different linear combinations in A_t. □

Claim 5.1.40. *For every $x \in B$, $\lim_t \theta_t(x)$ exists.*

Proof. Since $B = \mathrm{span}(C)$, it is sufficient to check the lemma for elements of C. In (a) we always choose $a_0, \ldots, a_k, a_{k+1} + t!d_{k+1}, \ldots, a_{t-1} + t!d_{t-1}, d_t$ to be $2(t+1)!$-independent, where $a_i = \theta_{t-1}(c_i)$ for $i < t$, so that the index of the tuple of d_i is lexicographically the smallest possible. Also, we make sure that $d_{k+1}, \ldots, d_t \in A$ have the lexicographically smallest possible indices. In particular, by induction, if θ has settled on $c_0, \ldots, c_k$ before stage t, then after finitely many attempts we will find a stable θ-image for c_{k+1}. □

Define $\theta(x) = \lim_t \theta_t(x)$. Since θ_t is injective at every stage, θ is injective too. It is a homomorphism because each θ_t is (of partial groups). It remains to check that θ is onto. Since in (a) we choose $d_{k+1}, \ldots, d_t \in A$ so that they have the least possible indices among all choices available, it follows that $\{\theta(c_i) : i \in \mathbb{N}\}$ is a maximal linearly independent subset of A. To see why, assume there exists $w \in A$ independent of $\{\theta(c_i) : i \in \mathbb{N}\}$. Go to a large enough stage of the construction to get a contradiction. Further, by (b.4) and (b.5), we have $\theta(B) = \mathrm{span}\, \theta(C)$. We conclude that the map $\theta(\cdot) = \lim_s \theta_s(\cdot)$ is a well-defined (Claim 5.1.40) surjective homomorphism $B \to A$, which is furthermore injective by Claim 5.1.39. Under θ, the computable subset C of B corresponds to a maximal linearly independent set in A.

The proof of Dobrica's Theorem is complete.

5.1.4 Khisamiev's Theorem A(3)

Theorem 5.1.41 (Khisamiev [288]). *Every c.e. presented torsion-free abelian group is isomorphic to a computable group. Furthermore, if the group is non-trivial, then this computable copy can be built uniformly in the index of the c.e. presentation.*

Khisamiev's Theorem is more subtle than Dobrica's Theorem 5.1.37. As far as we know, the only published proof of the result can be found in [288] where it is stated in more general terms. Our proof is different from the *non-uniform* proof in [288] (which is similar to the non-uniform proof of Dobrica's Theorem 5.1.37 presented above). In our proof, we do not assume that the rank of the group is infinite. But of course, if the rank is finite, an easy non-uniform argument shows that the group has a computable presentation; this is Exercise 5.1.46. We will need the extra uniformity in applications.

Proof. Let $U + L/E$ be a c.e. presentation of a non-trivial torsion-free abelian group. We assume that L and E in $U = L/E$ satisfy A)-G) of Property 5.1.25. In particular, by C), we assume that if x is declared equal to zero at a stage, then any element h such that $mh = x$ for some m is also immediately declared zero.

The setup

We build a computable presentation C and a Δ_2^0 isomorphism $f : C \to U$. At every stage, we have a finitely generated partial group C_s and a (finite, homomorphic) embedding f_s of C_s into the c.e. presentation U. We will also have

$$C_s = \text{span}_s(B_s),$$

where

$$B_s = \{b_0, b_1, \ldots, b_s\},$$

is s-independent. It is important to note that $\text{span}_s(B_s)$ should attempt to copy the span of $f_s(B_s)$ in U_s; of course, it could be later discovered (in U_t) that $f_s(B_s)$ is not linearly independent, and indeed, can even contain elements equal to zero. Our task, however, is to try to keep B_s independent. To achieve this, $f(b_i)$ of some of these b_i will likely need to be corrected. Unlike the previous proof of Dobrica's Theorem 5.1.37, we may have to also declare the current values of B_s to be dependent and redefine the values of some of these b_i in C by introducing new interpretations for such b_i in C.

Setting up the correction procedure

At some stage we may discover that the set $f(B_s)$ is linearly dependent. In general, $f_s(B_s)$ can become dependent in two different ways:

1. $f_s(B_s)$ is dependent in U because we discovered some non-trivial linear combination with large coefficients that we have not examined before.
2. Some previously non-zero $h \in \text{span}_s f_s(B_s)$ is declared equal to zero in U.

The former situation could happen in a computable U too, but the latter is specific to c.e. presented groups. Nonetheless, these two cases are not really that different. In the second case, we also have

$$mh = \sum_{i \leqslant s} n_i f_s(b_i),$$

where $m \neq 0$, and h is declared equal to 0. Because of our assumptions about the enumeration of U, $\sum_{i \leqslant s} n_i f_s(b_i) = 0$ in U. Thus, in the second case we also have $\sum_{i \leqslant s} n_i f_s(b_i) = 0$, but the coefficients n_i do not have to be "large".

The correction procedure

Recall that the s-independent $B_s = \{b_i : i \leqslant s\}$ is our attempt to approximate a basis of C at stage s, and $C_s = \text{span}_s(B_s)$. In the notation above, when $f_s(b_{j+1})$ is discovered to be in the span of

$$\{f_s(b_0), f_s(b_1), \ldots, f_s(b_j)\}$$

in U_{s+1}, we should be able to declare

$$b_k \in \text{span}(\{b_0, b_1, \ldots, b_j\}),$$

for $k = j + 1, \ldots, s$. Let j be smallest with this property. The case when $f_s(b_0)$ is declared 0 in U will be considered separately.

The algebraic recycling strategy. We can view

$$C_s = \mathrm{span}_s(\{b_0, b_1, \dots, b_s\})$$

at stage s as a finite initial segment of a free abelian group F freely generated by $b_0, b_1, \dots, b_s$; see, e.g., Remark 5.1.12 and the discussion preceding it. Using Lemma 5.1.4, fix $d_0, \dots, d_j$ and $c_0, \dots, c_s$ such that

$$\langle d_0, \dots, d_j \rangle = \langle b_0, \dots, b_j \rangle \text{ and } \langle c_0, \dots, c_s \rangle = F,$$

and furthermore $d_i = k_i c_i$ for some integers k_i, $i = 0, \dots, j$. Rank considerations imply that these generating sets must be linearly independent (in F), and thus each generates the respective group freely. Pick a very large natural number M and declare

$$c_k = M^{k-j} c_0, \ j < k \leqslant s.$$

Since M is very large, it defines an isomorphic embedding ξ of the partial group $\mathrm{span}_s(\{b_0, b_1, \dots, b_s\})$ into $\langle c_0, \dots, c_j \rangle$; the reasoning is exactly the same as in the old Mal'cev's Theorem 2.2.16.

Note $\langle c_0, \dots, c_j \rangle \supseteq \mathrm{span}_s\{b_0, \dots, b_j\}$. The map ξ fixes $c_0, \dots, c_j$ and, thus, also does not move $\mathrm{span}_s\{b_0, \dots, b_j\}$. However, it clearly puts $b_{j+1}, \dots, b_s$ into the span of $\{b_0, \dots, b_j\}$.

Declare f_{s+1} to be the linear extension of f_s to the initial segment of $\langle c_0, \dots, c_j \rangle$ that was sufficient to perform the manipulations described above (i.e., to define ξ). At the end of this process, declare $b_k, k \geqslant s$, undefined.

The spacial case when $f_s(b_0)$ is discovered to be in E (i.e., is 0 in U_{s+1}) is explained below.

Complete initialisation. Using the algebraic recycling strategy explained above with $j = 0$, put all b_k, $0 < k \leqslant s$, into $\langle c_0 \rangle$. (In particular, $b_1, \dots, b_s$ will be declared undefined and b_0 will be put in $\langle c_0 \rangle$.) Once this is done, pick a non-zero element $u \in U_{s+1}$ having the smallest index (in the sense of Remark 5.1.27) and declare $f_{s+1}(c_0) = u$. Extend f_{s+1} linearly to all other elements listed so far in $\langle c_0 \rangle$. In particular, $f_{s+1}(b_0) \in \langle u \rangle$. Note b_0 has not been declared undefined.

Redefining $b_k, k > j$. Suppose j, $0 \leqslant j \leqslant s$ is the largest index so that b_j is defined (see Remark 5.1.27). Pick fresh $a_{j+1}, \dots, a_{s+1}$ and declare

$$C_{s+1} = C_s \oplus (\langle a_{j+1} \rangle \oplus \dots \oplus \langle a_{s+1} \rangle) \mid_{\leqslant s+1} .$$

Define $f_{s+1}(b_i) = a_i$, $i = j+1, \dots, s+1$. Choose $u_{j+1}, \dots, u_{s+1} \in U_{s+1}$ with the least indices (lexicographically and in the sense of Remark 5.1.27) such that

$$f_{s+1}(b_0), \dots, f_{s+1}(b_j), u_{j+1}, \dots, u_{s+1}$$

are $(s+1)$-independent in U_{s+1}. Set

$$f_{s+1}(b_k) = u_k, \ \ k = j+1, \dots, s,$$

and then extend f_{s+1} linearly to C_{s+1}.

Finally, we need to make sure that $f = \lim_s f_s$ is surjective. This substage of stage $s+1$ will be performed after the "redefining b_k" substage described above, and so the (preliminary) values of f_{s+1} and C_{s+1} will already be defined.

Extending the range of f_{s+1}. If there is a $u \in cl_{s+1}(f_{s+1}(C_{s+1}))$ that is not in the range of f_{s+1},

$$u = \frac{1}{n} \sum_{i \leqslant s+1} n_i f_{s+1}(b_i),$$

then introduce a c in C_{s+1} (together with $kc, |k| \leqslant n$), set

$$c = \frac{1}{n} \sum_{i \leqslant s+1} n_i b_i,$$

and declare $f_{s+1}(c) = u$. Extend f_{s+1} linearly to C_{s+1}.

Construction

Build C, f and B in stages.

At stage 0, define $B_0 = \{b_0\}$, $C_0 = \{0, c\}$, where b_0 is interpreted as c ($c \neq 0$), and set $f_0(c) = u$, where $u \neq 0$ is the smallest index element of U_s that is not equal to 0.

At stage $s+1$, if $f_s(b_0)$ is declared 0 in U (i.e., is listed in E_{s+1}), then initialise the whole construction according to the instructions of complete initialisation. Otherwise, let j be largest such that $\{f_s(b_0), \ldots, f_s(b_j)\}$ are $s+1$-independent in U_{s+1}. Perform the algebraic recycling strategy to put b_k ($k \geqslant j$) into the span of $b_0, \ldots, b_j$, and then perform the "redefining b_k" procedure as described above to (re)define the values of $b_{j+1}, \ldots, b_{s+1}$ and to define f_{s+1}. Finish the step with the "extending the range of f_{s+1}" sub-step.

Proceed to the next stage.

Verification

Both the recycling strategy and the complete initialisation strategy allow us to preserve the finite open diagram of C_s when we define C_{s+1}, via the usual argument (this was explained in detail in Mal'cev's Theorem 2.2.16). We have two cases.

In the first case, we suppose the rank of U is infinite. Then, by induction, we argue that for every i, the value of b_i and also $f(b_i)$ eventually settle.

Indeed, b_0 is never redefined. Also, the complete initialisation strategy eventually locates a non-zero element in U, and once this happens $f(b_0)$ will never be redefined again. This is because we always use the smallest index element u; see Remark 5.1.27. (This is where we need $U \neq \{0\}$.)

The recycling strategy makes sure that, whenever the f-images of some elements are either discovered dependent or equal to zero, they are put into the span of $b_0, \ldots, b_j$ for the largest j so that $f_s(b_0), \ldots, f_s(b_j)$ still look independent. Furthermore, in this case $f_{s+1}(b_i) = f_s(b_i)$ for all $i \leqslant j$. Thus, since we always choose the new f-images for $b_{j+1}, \ldots, b_s$ to have the smallest possible indices lexicographically, we conclude that in the limit of the process $f(B)$ is maximal linearly independent in U. Since f is evidently a homomorphism, it follows that f is injective (Lemma 5.1.3). Further, $C = \text{span}(B)$ by construction, and furthermore in the very final phase of the construction we ensured that $f : B \to U$ is extended to the entire $U = \text{span}(f(B))$. It follows that in this case $C \cong U$ via the Δ^0_2 isomorphism f.

In the second case, assume the rank of U is n. Since U is non-trivial by our assumption, $n > 0$. In this case the values $b_k, k > n$ never settle in the construction. However, for each $i < n$, both b_i and $f(b_i)$ eventually settle, and thus $f(c)$ eventually settles for each $c \in \text{span}\{b_0, \ldots, b_{n-1}\}$. Further, C is built to be equal to $\text{span}\{b_0, \ldots, b_{n-1}\}$, and in the last phase of the construction we ensured that $f : C \to U = \text{span}(\{f(b_0), \ldots, f(b_{n-1})\})$, where $f = \lim_s f_s$, is surjective. Since f is a homomorphism and maps a basis to a basis, we conclude that f is a Δ^0_2-isomorphism of C onto U.

The proof of Khisamiev's Theorem is now finished. □

Combined with Dobrica's Theorem 5.1.37, we obtain:

Corollary 5.1.42. *Every c.e. presented torsion-free abelian group has a computable presentation with a computable basis.*

5.1.5 An application to categoricity*

Recall that an algebraic structure is called *computably categorical* if it has a unique computable presentation, up to a computable isomorphism.

Theorem 5.1.43 (Nurtazin [417]). *A computable torsion-free abelian group is computably categorical iff its rank is finite.*

Sketch. (For a complete proof, see, e.g., [203, 244].) The case when the group is the zero group is trivial. Suppose G has a finite basis $b_0, \ldots, b_k$, $k \geqslant 0$. Fix any other computable copy H and any isomorphism $f : G \to H$. We argue that f is computable. Non-uniformly fix $b_0, \ldots, b_k \in G$ and $f(b_0), \ldots, f(b_k) \in G$. Given $g \in G$, search for integers $m, n_0, \ldots, n_k$ such that

$$mg = \sum_{i \leqslant k} n_i b_i.$$

In H, search for h such that

$$mh = \sum_{i \leqslant k} n_i f(b_i).$$

It must be that $h = f(g)$, by torsion-freeness (as explained in §5.1.1).

Now assume the rank of G is infinite. By Dobrica's Theorem 5.1.37, we can assume that the group G has a computable basis. Our proof of Khisamiev's Theorem 5.1.41 gives a strategy of building a computable copy H of any given c.e. presented torsion-free group so that, when necessary, we can declare

$$b_k \in \text{span}\{b_0, b_1, \ldots, b_j\}, \; k > j,$$

where $B_s = \{b_i : i \leqslant s\}$ is our attempt to approximate a basis. Of course, the strategy works for computable groups as well.

The idea is to apply this strategy to build a computable copy H of G without a computable basis. If we succeed, then H and G cannot be computably isomorphic. This is because if we had a computable isomorphism $h : G \to H$, then the image of the computable basis of G would be a c.e. basis of H; apply Fact 5.1.22 to get a contradiction. (Compare this argument with the proof of Mal'cev's Theorem 2.2.16.)

By Fact 5.1.22, to build $H \cong G$ without a computable basis it is sufficient to diagonalise against all potential algorithms for linear independence in H. Interpret the e-th computable φ_e as the e-th potential algorithm for linear (in)dependence. We may assume $e > 0$. Build H and a Δ^0_2 isomorphism $f : H \to G$, as follows.

1. Let H copy G via f.
2. Wait for φ_e to declare $b_0, \ldots, b_e$ independent.
3. If this ever happens, use the algebraic strategy above to declare $b_k \in \operatorname{span}\{b_0, \ldots, b_{e-1}\}$ for $e \leqslant k \leqslant s$, introduce new interpretations for these b_k in H.
4. Then use elements linearly independent over $\{f_s(b_i) : i < e\}$ (i.e., in $G/\operatorname{span}(\{f_s(b_i) : i < e\})$) to define $f_{s+1}(b_k)$, $e \leqslant k \leqslant s$.

The rest of the proof is a standard finite injury argument; we omit it. □

Exercises

Exercise° 5.1.44. Using Lemma 5.1.10 prove that every linearly independent subset of a torsion-free abelian group can be extended to a basis of the group, and that all bases of the group have the same cardinality.

Exercise* 5.1.45 (Dobrica [114])**.** Prove Dobrica's Theorem 5.1.37 for arbitrary computable abelian groups.

Exercise* 5.1.46. Give a direct non-uniform proof that every c.e. presented torsion-free abelian group of finite rank has a computable presentation.

Exercise° 5.1.47. Give a complete and detailed proof of Theorem 5.1.43.

Exercise* 5.1.48. Prove Khisamiev's Theorem 5.1.41 using a modification of the combinatorial strategy from the proof of Dobrica's Theorem 5.1.37.

For the next exercises, recall that an (additive, abelian) group G is ordered if $a \leqslant b$ implies $a + c \leqslant b + c$, for any $a, b, c \in G$. It is well-known that an abelian group is orderable iff it is torsion-free; we cite books [314, 193] for a general exposition of the theory of ordered groups. Also, if $\leqslant$ is an order in an ordered abelian G, then we say that $a, b \in G$ lie in the same Archimedean class, or Archimedean equivalent (written $a \sim b$) if there exist non-zero integers n, m such that $a \leqslant nb$ and $b \leqslant ma$. The Archimedean classes are also ordered under the induced $[a]_\sim \preccurlyeq [b]_\sim$ iff $a \leqslant b$, which is well-defined on the classes. While none of the exercises below is marked with a star, some of them are certainly non-trivial, especially those based on the results of Solomon [479].

Exercise° 5.1.49 (Downey and Kurtz [135])**.** Show that there is a computable copy of the free abelian group that is not computably orderable. (Use a simplified version of the strategy from Khisamiev's theorem.)

Exercise 5.1.50 (Melnikov [366])**.** Show that, given a Δ^0_2 linear order L with a least element, we can produce an ordered group $G(L)$ so that the ordering $\preccurlyeq$ of the Archimedean classes of $G(L)$ is isomorphic to L. (Combine the strategy from the previous exercise with a Δ^0_2 approximation of L.)

Exercise° 5.1.51 (Solomon [479]). Show that if G is a computable presentation of a torsion-free abelian group of rank 1, then G has exactly two orders both of which are computable.

Exercise° 5.1.52 (Solomon [479]). Prove that if G is a computable presentation of a torsion-free abelian group with finite rank strictly greater than 1, then G has orders of every Turing degree.

Exercise 5.1.53 (Solomon [479]). Let G be a computable presentation of a torsion free abelian group with infinite rank. Show that G has orders of every degree $\mathbf{a} \geqslant_T \mathbf{0}'$.

Exercise 5.1.54 (Solomon [479]). Show that there exists a Π^0_1-class $C \subseteq 2^\omega$ so that for any computable presentation of a (torsion-free) abelian group G, C does not realise the orders on G in the sense of Exercise 2.2.25. (Hint: Use the previous three exercises and Exercise 4.2.59.)

Exercise° 5.1.55 (Solomon [479]). Use Dobrica's Theorem to prove that every computable torsion-free abelian G is isomorphic to a computably ordered, computable H. (Hint: Embed the group into the additive group of computable reals using, e.g., the reals $\sqrt{p_i}$.)

5.1.6 Some further remarks*

The strategy we utilised in Dobrica's Theorem was invented by Nurtazin [417], as far as we know. Of course, the key idea behind the strategy can be traced (at least) back to Mal'cev; see Theorem 2.2.16. Dobrica [114] adopted a modified version of this strategy. (We remark that the other standard spelling of Vyacheslav's family name is Dobritsa.) Notably, Dobrica's theorem does not require the group to be torsion-free in its full generality. Khisamiev [288] further extended this strategy to c.e. presented groups. Khisamiev's result answered a question of Baumslag, Dyer, and Miller [31], which was raised in the context of homologies of finitely presented groups; more about this in §8.3.2.

Goncharov, Lempp, and Solomon [215] applied this combinatorial strategy to ordered abelian groups, proving analogous versions of Dobrica's and Nurtazin's Theorems for such groups. However, as we have seen, this combinatorial strategy is not the ultimate method for handling computable torsion-free abelian groups. In fact, it appears that the strategy can often be circumvented if necessary. For instance, our proof of Khisamiev's Theorem avoids this combinatorial approach altogether, replacing it with an application of Rado's Lemma to (partial) groups. In [203], Goncharov bypasses the combinatorial strategy entirely when demonstrating that every abelian group of infinite rank has infinitely many computable copies, up to computable isomorphism. However, it is worth noting that he does reference Dobrica's Theorem when he fixes a copy with a computable base. Much more recently, Harrison-Trainor, Melnikov, and Montalbán [244] gave a simultaneous proof of Dobrica's Theorem and Nurtazin's Theorem, avoiding the combinatorial strategy using the notion of a c.e. pregeometry (see the discussion after Lemma 5.1.10).

The study of abstract computable pregeometries had been initiated long before [244]. Notions of independence play a central role in the study of the combinatorial properties of effectively presented vector spaces and of other structures with an appropriate notion of independence. Such studies were quite popular in the 1970s and 1980s; the standard reference is the fundamental paper of Metakides and Nerode [383], see also [110, 111, 112, 468, 117] and, for applications to reverse mathematics, [470]. Many results on subspaces of computable vector spaces remain true in the abstract setting of computable pregeometries (closure or span operators) – see the survey [148].

The main meta-theorem in [244] pushes this intuition even further. For instance, it follows from the meta-theorem that the natural analogies of Dobrica's Theorem and Nurtazin's Theorem hold for real closed fields and differentially closed fields, with respect to their natural notions of independence [244]. Also, [244] gives a combinatorially much simpler proof of the aforementioned results from [215] about ordered groups. The method from [244] has been applied in [373] to give a new "factorial-free" proof of Dobrica's Theorem for abelian groups that are not necessarily torsion-free. In our book, we do not need this degree of generality. However, [244, 148] could be worth looking at.

We shall return to discrete abelian groups in Part 2, where a much more in-depth analysis of (torsion-free, abelian) completely decomposable groups will be presented in Chapter 7. Computable torsion-free abelian groups (that are not completely decomposable) will be used as a technical tool in Chapter 8 to study connected spaces. Finally, the theory of computable abelian p-groups and their computable profinite duals will be presented in Chapter 9.

5.2 Effective Pontryagin duality

In this section, all groups are additive and abelian. Pontryagin duality is one of the main tools of abstract harmonic analysis (see textbook [173]), and it will also play a central role in this section. Let $\mathbb{T} = (\mathbb{R}, +)/(\mathbb{Z}, +)$, the unit circle group. Alternatively, $\mathbb{T}$ is the multiplicative group of the complex numbers having norm 1. For any locally compact topological group G, form its *dual group*

$$\widehat{G} = \{\chi \mid \chi \text{ is a continuous group homomorphism from } G \text{ to } \mathbb{T}\}.$$

It is easily seen that $\widehat{G}$ is itself a topological group under the operation $(\chi+\xi)(a) = \chi(a)+\xi(a)$ and the topology of uniform convergence. We will only need the case when G is either compact Polish or discrete countable. If G is compact Polish, then the topology of uniform convergence is exactly the topology given by the metric $\sup_{x\in G}(|\chi(x) - \chi(y)|)$ on the space of all continuous functions $G \to \mathbb{T}$, in which $\widehat{G}$ forms a closed set. We will soon explain (§5.2.1) an easy way to think about $\widehat{G}$ for a discrete countable G.

Pontryagin duality states that if G is compact or discrete abelian, then $G \cong \widehat{\widehat{G}}$. It is not hard to see that $\widehat{G}$ is discrete when G is compact. This means that the discrete dual $\widehat{G}$ of a compact abelian G contains all the information about G. Thus, the duality essentially reduces the study of compact abelian groups to the algebraic theory of abelian groups; a detailed exposition of this theory can be found in [340]. We note that van Kampen extended the duality to arbitrary locally compact abelian groups, but we will focus on the compact Polish/discrete countable case. As we noted earlier, a locally compact abelian G is discrete iff $\widehat{G}$ is compact abelian, and in this case, G is torsion-free (abelian) iff $\widehat{G}$ is connected compact. We take all these properties for granted and refer to the books [429, 404] for proofs.

The main goal of this section is to prove (3) of Theorem B. Recall that it states that the notions of effective compactness and computable Polish presentability differ in the class of connected Polish spaces, up to homeomorphism. Much of the work necessary to prove Theorem B(3) has been done in the previous section and in the previous chapter. In this section, we establish the following:

Theorem 5.2.1 (Melnikov [373], Lupini, Melnikov, and Nies [341]). *Let G be a (countable, discrete) torsion-free abelian group. Then G has a computable presentation iff the connected compact domain of its dual $\widehat{G}$ has a computably compact presentation.*

Theorem 5.2.1 does not require the group operations on $\widehat{G}$ to be computable. It will follow from the proof that, whenever the domain of the compact connected $\widehat{G}$ has a computably compact copy, it must also have a computably compact copy in which the group operations are computable. This theorem and its unexpected counterpart, Theorem 5.2.25, for Δ^0_2 groups and computable Polish spaces will be crucial in the proof of Theorem B(3).

5.2.1 From discrete to compact

IIn this subsection, we describe how to build a computable compact presentation of $\widehat{G}$ given a computable discrete torsion-free G. This provides one implication in Theorem 5.2.1.

Proposition 5.2.2. *Suppose G is computable torsion-free. Then $\widehat{G}$ admits a computably compact presentation. Furthermore, $\widehat{G}$ can be realised as a computable closed (thus, computably compact) subgroup of the computably compact group*

$$\mathbb{A} = \prod_{i \in \mathbb{N}} \mathbb{T}_i,$$

where each $\mathbb{T}_i$ is a copy of the natural computably compact presentation of the unit circle group $\mathbb{T} = \mathbb{R}/\mathbb{Z}$. If G has infinite rank, then this is uniform.

The proof below heavily relies on the material from the previous section and on the calculus of computably compact sets developed in the previous chapter. Modulo these results, the construction of a computably compact copy of $\widehat{G}$ and its verification are not too difficult.

Proof of Proposition 5.2.2. The direct product

$$\mathbb{A} = \prod_{i \in \mathbb{N}} \mathbb{T}_i$$

of infinitely many identical copies $\mathbb{T}_i$ of $\mathbb{T}$ carries the natural product metric

$$D(\chi, \rho) = \sum_{i=0}^{\infty} \frac{1}{2^{i+1}} d_i(\chi_i, \rho_i),$$

where each d_i denotes the shortest arc metric on $\mathbb{T}_i$.

Every compact abelian group can be realised as a closed subgroup of $\mathbb{A}$, as follows. Suppose $G = \{g_0 = 0, g_1, g_2, \ldots\}$ is a countably infinite discrete group. Let $\mathrm{Hom}(G, \mathbb{T})$ be the subset of $\mathbb{A} = \prod_{i \in \mathbb{N}} \mathbb{T}_i$, consisting of tuples $\chi = (\chi_0, \chi_1, \ldots)$, where each such tuple represents a group homomorphism $\chi : G \to \mathbb{T}$ such that $\chi(g_i) = \chi_i \in \mathbb{T}_i$. Since G is discrete, every group homomorphism $\chi : G \to \mathbb{T}$ is necessarily continuous. Thus, $\widehat{G} \cong \mathrm{Hom}(G, \mathbb{T})$. Since being a group homomorphism is a universal property, $\mathrm{Hom}(G, \mathbb{T})$ is closed in $\mathbb{A}$. Thus, Pontryagin duality implies that every separable compact abelian group is homeomorphic to a closed subgroup of $\mathbb{A}$. Our task is to show that this presentation can also be computably closed, though it is not guaranteed for an arbitrary computable presentation of G (Exercise 5.2.15). This will require some work.

We say that a point $x \in \mathbb{T}$ is *rational* if the respective point of the unit interval is a rational number. Then $\mathbb{T}$, equipped with rational points and the shortest arc metric, is evidently a computably compact Polish group.

Lemma 5.2.3. *Under the metric D and the component-wise operation, $\mathbb{A}$ is a computably compact Polish group.*

Proof. The underlying space $\mathbb{A}$ is a computable product of computably compact spaces; this was already stated in Proposition 4.2.17 earlier. The computable dense set is given by sequences $(a_i)_{i \in \mathbb{N}}$, where a_i is a rational point in $\mathbb{T}_i$, and almost all a_i are equal to zero. The group operation is clearly computable. □

Using Dobrica's Theorem 5.1.37, fix a computable presentation H of G that admits a computable maximal linearly independent set. By Fact 5.1.22, this is equivalent to saying that in H we can decide whether a given tuple of elements is dependent or not.

Recall that in Definition 5.1.23 we defined a tractable presentation as follows. A computable torsion-free H is *tractable* if there exists a uniformly computable ascending sequence of finitely generated abelian groups $(F_i)_{i\in\mathbb{N}}$ with the following properties:

1. $H = \bigcup_{i\in\mathbb{N}} F_i$.
2. $\{0\} = F_0 \subseteq F_1 \subseteq F_2 \subseteq F_3 \subseteq \ldots$ is a uniformly computable sequence in which the set-theoretic embeddings are also uniformly computable.
3. For every F_i $(i > 0)$, we can uniformly compute a finite set of elements $h_0, \ldots, h_{k(i)}$ such that
$$F_i = \langle h_0 \rangle \oplus \langle h_1 \rangle \oplus \ldots \oplus \langle h_{k(i)} \rangle.$$

By Lemma 5.1.24, H is tractable, and furthermore, the proof of Lemma 5.1.24 is uniform. Before we proceed, note that $\widehat{\mathbb{Z}} \cong \mathbb{T}$ (Exercise 5.2.13).

Lemma 5.2.4. *Suppose a computable torsion-free abelian group H is tractable. Then $Hom(H, \mathbb{T})$ is a computably closed subset of $\mathbb{A} = \prod_{i\in\mathbb{N}} \mathbb{T}$.*

Proof of Lemma 5.2.4. Suppose $C \subseteq \prod_{i\in\mathbb{N}} \mathbb{T}_i$ is a closed subgroup, then let $C_i \subseteq \mathbb{T}_i$ denote the projection of C onto $\mathbb{T}_i$. (By the torsion-freeness of H, we will actually have $C_i = \mathbb{T}_i$ when $i > 0$ and $C_0 = \{0\}$. We, however, stick with the notation C_i to emphasise that we are dealing with projections of C.) We say that C_i is *defined by a primitive relation* if one of the following possibilities is realised:

(1) there is a $j < i$ and a positive integer k such that every $\chi \in C$ satisfies $\chi_i = k\chi_j$,

(2) there is a $j < i$ and a non-negative integer k such that every $\chi \in C$ satisfies $k\chi_i = \chi_j$,

(3) there exist $u, v < i$ such that every $\chi \in C$ satisfies $\chi_i = \chi_u - \chi_v$,

(4) there exist $u, v < i$ such that every $\chi \in C$ satisfies $\chi_i = \chi_u + \chi_v$.

In each of the four cases, we assume that every finite sequence $(\chi_0, \ldots, \chi_i)$, where χ_j $(j \leqslant i)$ satisfies the respective linear conditions, can be extended to an infinite sequence $\chi \in C$. Note, if $j = k = 0$ in (2), then essentially there is no restriction on χ_i since $\chi_0 = 0$. We also say that $C \subseteq \prod_{i\in\mathbb{N}} \mathbb{T}_i$ is *effectively predictable* if each C_i is defined by a primitive relation that can be computed uniformly in i.

Claim 5.2.5. *Every effectively predictable (thus, closed) subgroup of $\prod_{i\in\mathbb{N}} \mathbb{T}_i$ is a c.e. closed subset of $\prod_{i\in\mathbb{N}} \mathbb{T}_i$.*

Proof of Claim. Recall that the dense sets in $\mathbb{T}_i$ are given by rational points. Fix the computable (in $\mathbb{A}$) dense (in C) sequence that is given by the sequences of rationals that satisfy the primitive relations. This makes C c.e. by Lemma 4.2.51. □

Claim 5.2.6. *Every effectively predictable subgroup of $\prod_{i\in\mathbb{N}} \mathbb{T}_i$ is effectively closed (Π^0_1) in $\prod_{i\in\mathbb{N}} \mathbb{T}_i$.*

Proof. A sequence χ is not in C if there is an i such that the i-th relation fails. This is an effectively open condition. □

By definition, computable closed sets are those Π^0_1 sets that are additionally c.e. closed. Since computable closed subsets of computably compact spaces are themselves computably compact by Proposition 4.2.53, it remains to prove the following claim.

Claim 5.2.7. *Suppose $(F_i)_{i \in \mathbb{N}}$ is a tractable computable presentation of a (discrete and torsion-free) abelian group G. Then $Hom(H, \mathbb{T}) \cong \widehat{G}$ is effectively predictable.*

Proof of Claim. Suppose $H = \{h_0 = 0, h_1, \ldots\}$. Our goal is to define $C_i \subseteq \mathbb{T}_i$ and list primitive relations defining each of the $C_i \subseteq \mathbb{T}_i$ in terms of some C_k, $k < i$. Using Rado's Lemma 5.1.4, we can uniformly effectively refine the sequence $(F_i)_{i \in \mathbb{N}}$ witnessing Definition 5.1.23 and assume that for every i, there exists an $a \in F_{i+1}$ such that either $F_{i+1} = \langle a \rangle \bigoplus F_i$ or $F_{i+1} = \langle F_i, a \rangle$ and for some m we have $ma \in F_{i+1}$, where m is assumed to be the least such. (See the proof of Lemma 5.1.24 where this was explained.)

We can effectively find full decompositions of F_i and F_{i+1} and the embedding, and thus we can effectively figure out which of the two possibilities is realised. Since we need to define characters, we need to specify the index of a in the enumeration of $H = \{h_0 = 0, h_1, \ldots\}$. So suppose $a = h_j \in H$.

In the former case, do not put any restriction on χ_i (formally, we declare that $0 \cdot \chi_i = \chi_0 = 0$ in (2)). For every n there exists an element $h_k = nh_j (= na)$; we will also set $C_k = \mathbb{T}_k$ and declare $\chi_k = n\chi_j$ as the respective primitive relation.

Otherwise, if we have $ma \in F_i$ and $a = h_j \in H$, then we check what is the k such that $h_k = mh_j \in F_i$, $k < j$, and declare $\chi_k = m\chi_j$. For each $0 < n < m$, there exists an element $h_k = nh_j (= na)$. As in the previous case, we will set $C_k = \mathbb{T}_k$ and declare $\chi_k = n\chi_j$ as the respective primitive relation.

We also keep working towards closing C under $+$ and $-$. Whenever we have $h_k = h_j + h_i$, declare $\chi_k = \chi_j + \chi_i$, and similarly for $-$. (We can assume that $k > i, j$, and we allow $i = j$.) The procedure described above witnesses that $C = Hom(H, \mathbb{T}) \cong \widehat{G}$ is effectively predictable. □

The proof of Lemma 5.2.4 is finished. □

Thus, we conclude that $Hom(H, \mathbb{T})$ is computable closed in $\mathbb{A}$ which is itself computably compact by Lemma 5.2.3. Proposition 4.2.53 implies that $Hom(H, \mathbb{T}) \cong \widehat{H} \cong \widehat{G}$ is computably compact. Together with the computable group operations inherited from $\mathbb{A}$, it forms a computably compact group. □

The proposition above can be restated in terms of computable direct and inverse limits, if necessary, and it essentially says that the well-known property

$$\widehat{\varinjlim_{i \in \mathbb{N}} G_i} \cong \varprojlim_{i \in \mathbb{N}} \widehat{G}_i$$

holds *computably* for discrete torsion-free $G = \varinjlim_{i \in \mathbb{N}} G_i$ (where $G_i \leq G$ are finitely generated) assuming G has an algorithm for linear independence. We remark that the torsion-freeness of G in Proposition 5.2.2 can be dropped; see Exercise 5.2.16. We will return to this in Chapter 9, where we will carefully verify this effective correspondence between inverse and direct limits of finite abelian groups.

Remark 5.2.8. The proof of Dobritsa's Theorem 5.1.37 is uniform in the case when the rank of G is infinite, and Fact 5.1.22 is uniform as well. The rest of the argument above is uniform without any restriction on the rank of G. In conclusion, provided that G has infinite rank or we have access to its basis, the index of the computably compact presentation of $\widehat{G}$ constructed in the proposition above can be obtained uniformly from the index of G (and the index of its basis, if it is given). This assertion could be true without any restriction on the rank, but it would require a much more careful proof (if true), and more importantly, we won't need full uniformity.

Turning a Δ^0_2 discrete group into a computable Polish space

Fix a prime number q.

Definition 5.2.9. We say that H is q-divisible if $q^\infty \mid h$ for every $h \in H$, i.e., for any $n \in \mathbb{N}$,

$$\forall h \in H \, \exists y \in H \;\; q^n y = h$$

holds in H.

The lemma below will be crucial in the proof of Theorem B(3). It is also somewhat unexpected.

Lemma 5.2.10 (Melnikov [374]). *Assume a Δ^0_2 group H is torsion-free abelian and q-divisible. Then its Pontryagin dual $\widehat{H}$ has a computable Polish presentation.*

Proof. We build $G \leqq \mathbb{A} = \mathbb{T}^\omega$, where, as before, $\mathbb{T}$ is the "natural" computably compact presentation of the unit circle group. Our goal is to ensure $G \cong \widehat{H}$.

Using Dobrica's Theorem 5.1.37 relativised to $0'$, fix a Δ^0_2 presentation with a Δ^0_2 basis B. Taking the retract of the domain under a suitable Δ^0_2 map, we can ensure that the indices of the basic elements form a computable set[3]. We can additionally assume that the index of the zero element is 0.

The basis freely generates a free abelian subgroup of H. Every non-zero element $h \in H$ satisfies a relation of the form

$$nh = \sum_{b \in B} m_b b,$$

where $n > 0$ and almost all coefficients m_b are zero. We can view H as a Σ^0_2 subgroup of $V_{|B|} = \bigoplus_{i<\mathrm{card}(B)} \mathbb{Q}$ so that the basis B is equal to the standard basis of $V_{|B|}$, where B is a computable set. To list H as a subgroup of $V_{|B|}$, it is sufficient to list elements $h \in H$ that satisfy relations of the form

$$p^v h = \sum_{b \in B} m_h b,$$

where p ranges over primes and v over positive integers. Note that if such an h exists, then it is unique. We computably guess whether the relation holds in the group in the spirit of the Limit Lemma 3.1.3. The relation holds in the group iff

$$\frac{\sum_{b \in B} m_b b}{p^v} \in H.$$

Since H is a Σ^0_2 subgroup in general, the relation holds if we eventually never see it to fail. However, if the relation fails, we can have infinitely many stages at which we believe that the relation might hold.

[3]The case of a finite basis is trivial. If the domain of H is ω and $B \subseteq \omega$ is infinite, fix an injective Δ^0_2 map $f : \omega \to \omega$ so that $\{f(2k+1) : k \in \mathbb{N}\} = B$ and $\{f(2k) : k \in \mathbb{N}\} = G \backslash B$. It is defined to be the map that lists the respective Δ^0_2 sets in natural increasing order. We define the new Δ^0_2 presentation of the group by setting $a_i = a_j + a_k$ if $f(i) = f(j) + f(k)$, which is $0'$-computable. In this presentation, $f^{-1}(B)$ is a computable set.

We reserve a special copy $\mathbb{T}_h$ of $\mathbb{T}$ for every such generator h, including $\mathbb{T}_b$ for each $b \in B$. In this notation, the elements of G are sequences of the form $(\chi_h)_{h\in H}$ going through the computably compact $\mathbb{A} \cong \prod_{h\in H} \mathbb{T}_h$. Similarly to the proof of Proposition 5.2.2, we would like to declare $m\chi_h = \sum_{b\in B} m_b\chi_b$ for all $\chi \in \mathbb{T}_h \cap G$ whenever we have a relation $mh = \sum_{b\in B} m_b b$. This, however, seems to require $\varnothing'$ in general, so we cannot just get away with relativising Proposition 5.2.2.

An informal outline of the basic strategy. The idea is to ensure that, in the dual group, $m\chi_h = \sum_{b\in B} m_b\chi_b$ holds up to (say) error 2^{-s} at stage s, and therefore it can be corrected later if necessary. If we decide that $m\chi_h = \sum_{b\in B} m_b\chi_b$ no longer holds for any h (in the spirit of the Limit Lemma 3.1.3), we will be able to turn it into a relation of the form $q^k h = \sum_{b\in B} m_b b$, where q is the fixed prime (so that the group is q-divisible) and k is very large.

We provide additional details. At each stage, we will place only finitely many rational points into each circle. Suppose we initially have a relation of the form

$$m\chi_h = \sum_{b\in B} m_b\chi_b, \tag{†}$$

which is approximated by finitely many intervals in the respective copies of the unit circle, with an error of 2^{-s} at stage s. If this relation is consistent for h and remains unchanged, we will continue to refine these intervals, making them smaller. Consequently, we will obtain a closed set where the relation indeed holds. The process of shrinking intervals will enable us to approximate a dense computable subset of this closed set.

Now suppose we no longer believe in the relation $m\chi_h = \sum_{b\in B} m_b\chi_b$ at stage s. At this stage, we have only finitely many intervals approximating the relation with precision 2^{-s}. Choose k sufficiently large such that each of these finitely many intervals in $\mathbb{T}_h$ contains at least one point of the form $\frac{r}{q^k-m}$, where $r \leqslant (q^k - m)$. Under the map $x \mapsto q^k x$, the point $\frac{r}{q^k-m}$ on the unit circle is mapped to

$$\frac{rq^k}{q^k-m} = \frac{r(q^k-m+m)}{q^k-m} = r + \frac{rm}{q^k-m} = \frac{rm}{q^k-m}. \tag{‡}$$

In other words, $x \mapsto mx$ and $x \mapsto q^k x$ agree on all points of this form. Consequently, for these points,

$$q^k\chi_h = \sum_{b\in B} m_b\chi_b \tag{*}$$

will also hold "up to 2^{-s}", which corresponds to

$$\exists h \; q^k h = \sum_{b\in B} m_b b$$

in H, and this holds vacuously since the group is q-divisible. Therefore, we can consistently *switch* the approximation of our closed set locally, perhaps making further necessary adjustments, as explained below.

Remark 5.2.11. For instance, if we are dealing with the relation $mh = b$ and changing it to $q^k h = b$, then we need to introduce 2^{-s}-approximations for many new pre-images of b under the (adjusted) map. For example, there are only two pre-images under the map $x \mapsto 2x$, but there are four pre-images under the map $x \mapsto 4x$. Such adjustments can make the c.e. closed set that we build not effectively closed (not Π^0_1) in general.

Then we introduce a fresh circle corresponding to the relation

$$mh = \sum_{b \in B} m_b b,$$

and repeat the process. If the relation indeed holds, then we will eventually find a "stable" circle realising it. Otherwise, we will continue adding circles to witness the q-divisibility of the group, and as a result, no circle will actually realise the relation. The verification is similar to Proposition 5.2.2, but, as explained in Remark 5.2.11, we cannot guarantee computable compactness.

Note there is essentially no interaction between strategies working with different generators of H in $V_{|B|}$. There are only two further subtleties that perhaps need to be discussed.

1. What if, at some stage, we mistakenly believe that $mh = b$ and $nh = b$ for $m \neq n$? Our presentation H of G, using $V_{|B|}$, automatically resolves all such issues, since every subgroup of $V_{|B|}$ is clearly torsion-free.
2. The relations of q^n-divisibility play a special role in the construction; in particular, we must ensure that all elements are q^∞-divisible. In the construction, we implement the basic strategies only for generators with relations of the form $nh = \sum_{b \in B} m_b b$, where n and q are co-prime. Meanwhile, we manually introduce more and more new circle-components for relations of the form $q^k h = b$, for each b. To ensure that no such relations are missing, we continue adding circles corresponding to elements g such that $g = mh$ (as in the proof of Proposition 5.2.2).

Now the reader hopefully sees how to organise the construction, but we include a (somewhat compressed) formal proof nonetheless. To make our proof a bit more transparent, we will actually split the relation $m\chi_h = \sum_{b \in B} m_b \chi_b$ into two relations: $\chi_r = \sum_{b \in B} m_b \chi_b$ and $m\chi_h = \chi_r$. Thus, we will apply the strategy outlined above to relations of the form $m\chi_h = \chi_r$. This is, of course, a mere notational convenience.

Formal proof. We first provide a "crude" construction that manipulates copies of the unit circle and relations of the form $n\chi_h = \sum_{b \in B} m_b \chi_b$, which seemingly yields a merely $0'$-computable Polish presentation of the dual group. We then combine this crude definition with the approximation technique outlined above to define a computable Polish group.

The crude construction. For every $b \in B$, reserve a copy $\mathbb{T}_b$ of the unit circle group. At stage s of the crude construction, introduce one more $\mathbb{T}_h$ and monitor one more relation of the form $p^v h = \sum_{b \in B} m_b b$. For every such relation that had not previously been considered, do the following:

(a.1) Introduce a copy of the unit circle group $\mathbb{T}_r$ and declare

$$\chi_r = \sum_{b \in B} m_b \chi_b.$$

(a.2) Introduce a new copy of the unit circle group $\mathbb{T}_{r,0}$ and declare it active.

(a.3) Declare $p^v \chi_{r,0} = \chi_r$. (Informally, $\chi_{r,0}$ is our initial attempt to define a "stable" circle for h.)

For every relation that has already been considered, check if it still holds according to the Σ^0_2-approximation that we fixed above. If it does not hold, then let u be largest such that $\mathbb{T}_{r',u}$ is currently active, and so that $\chi_{r'} = p^w \chi_{r',u}$ was declared at the stage when $\mathbb{T}_{r',u}$ was introduced.

(b.1) Declare $\mathbb{T}_{r',u}$ inactive and declare $\chi_{r'} = p^v \chi_{r',u}$ *dismissed.*

(b.2) Choose k very large (to be clarified) and declare $\chi_{r'} = q^k \chi_{r',u}$.

(b.3) Introduce a new copy $\mathbb{T}_{r',u+1}$ of the unit circle group and declare it active.

(b.4) Declare $\chi_{r'} = p^v \chi_{r',u+1}$.

The construction builds a system of circles and relations between them. If in step (b.2) we define k carelessly, we might end up with a crude construction that builds a Δ^0_2-closed subgroup of $\mathbb{T}^\omega$. We now explain how this crude construction can be approximated computably.

Approximating the crude construction computably. Note that, for intervals $I, J \subseteq \mathbb{T}$ with rational end-points and $m \in \mathbb{Z}$, both mI and $I + J$ are also intervals with rational end-points.

Definition 5.2.12. For a finite equation of the form $mh = \sum_{b \in B} m_b b$, its *$\epsilon$-name* is a finite collection of open basic intervals $U^i_h \in \mathbb{T}_h$ and $U^j_b \in \mathbb{T}_b$ of diameter ϵ with the following properties:

1. U^j_b cover $\mathbb{T}_b$;
2. U^i_h cover $\mathbb{T}_h$;
3. If $mx = \sum_{b \in B} m_b y_b$ holds for $x \in \mathbb{T}_h$ and $y_b \in \mathbb{T}_b$ then for some intervals $U^i_h \ni x$ and $U^{j_b}_b \ni y_b$,

$$mU^i_h = \sum_{b \in B} m_b U^{j_b}_b.$$

To define G, we monitor the crude construction. At each stage s, we perform the following steps:

1. Enumerate additional points into each copy of the unit circle that has been introduced by the crude construction.
2. For every declared relation that has not yet been dismissed, except for those declared in (b.4) of the current stage, refine its 2^{-s+1}-name to a 2^{-s}-name.
3. For every relation of the form $\chi_r = q^k \chi_{r,u}$ introduced at a (b.2)-type substage of the current stage for some r, replace the 2^{-s+1}-name of the dismissed relation $\chi_r = p^v \chi_{r,u}$ with a 2^{-s}-name of $\chi_r = q^k \chi_{r,u}$.

In particular, in step (b.2) we choose k sufficiently large so that substep (3) can be executed according to (‡) and the discussion preceding it. This ensures that we obtain a closed subset C of the constructed copy of $\mathbb{T}^\omega$, consisting of all characters that satisfy the relations introduced during the construction. (Note that this copy is not literally $\prod_{h \in H} \mathbb{T}_h$, because it requires $\varnothing'$ to identify the circle corresponding to $h \in H$.)

The verification. We first explain how to list a dense subset of C. At every stage, we list only finite tuples of special points in the constructed copy of $\mathbb{T}^\omega$ that satisfy the currently declared relations up to 2^{-s} (in the sense of Definition 5.2.12). During the construction, we will keep refining and extending each such finite tuple, so that the result will converge uniformly to an infinite tuple of points going through the whole of $\mathbb{T}^\omega$. This is done arbitrarily, say, by choosing the smallest available index among the special points that obey the current 2^{-s}-names (Definition 5.2.12).

It should be clear that C is equal to the completion of the uniformly computable list of points that we build according to the procedure described above. To finish the verification, note that C is topologically isomorphic to the character group of H, and therefore is a computable presentation of $\widehat{H} \cong \widehat{G}$ under the operations inherited from $\mathbb{T}^\omega$. □

Exercises

Exercise° 5.2.13 (Folklore)**.** Show that the unit circle group $\mathbb{T}$ and the group of integers $\mathbb{Z}$ are dual to each other. Show that cyclic groups are self-dual.

Exercise° 5.2.14 (Folklore)**.** Let A, B be either discrete or compact abelian groups. Show that $\widehat{A \oplus B} \cong \widehat{A} \oplus \widehat{B}$.

Exercise 5.2.15 (Melnikov [373])**.** Prove that the Π^0_1-closed subset $\mathrm{Hom}(G, \mathbb{T})$ of $\mathbb{A}$ defined in the proof of Proposition 5.2.2 does not have to be computably compact in general, i.e., without the extra assumption that the computable discrete group G has a computable basis.

Exercise* 5.2.16 (Melnikov [373])**.** Prove Proposition 5.2.2 for arbitrary computable abelian G. (Hint: Use Exercise 5.1.45.)

5.2.2 From compact to discrete

In this subsection, we establish the remaining implication of Theorem 5.2.1, and also its version for connected computable Polish spaces and Δ^0_2-groups. Before we proceed, we need to review the basic definitions from algebraic topology. The reason is that, for a compact connected abelian G, its torsion-free dual group is isomorphic to the first Čech cohomology group (to be clarified). Thus, to finish the proof of Theorem 5.2.1, we need to calculate the latter for a computably compact $\widehat{G}$. We explain all terms below.

Simplicial (co)homology

We very briefly review the classical notions of simplicial homology and cohomology; for a thorough introduction see [411].

Geometric simplicial complexes. Recall that a simplicial complex is essentially a space with a triangulation. A simplex is the higher-dimensional counterpart of a triangle. Thus, a point is a 0-simplex, a line segment is a 1-simplex, a triangle is a 2-simplex, and so on. Objects in the space composed solely of these simplices are called simplicial complexes. A (geometric) simplicial complex K in $\mathbb{R}^n$ is a collection of simplices in $\mathbb{R}^n$ such that:

1. Every face of a simplex in K is also in K, and
2. The intersection of any two simplices in K is a face of each of them.

We will only need the case where the complex is finite (i.e., composed of finitely many simplices).

An n-simplex on $(n+1)$ vertices in $\mathbb{R}^m$, where $m \geqslant n$, can be easily parameterised using linearly independent vectors. If we have $w_0, \ldots, w_n \in \mathbb{R}^m$ such that $v_1 = w_1 - w_0, \ldots, v_n = w_n - w_0$ are linearly independent, then the simplex is parameterised via

$$\left\{\sum_{i=1}^{n} \alpha_i v_i : \sum_{i=1}^{n} \alpha_i = 1 \text{ and } \alpha_i \geqslant 0,\ i = 1, \ldots, n\right\}.$$

It is not difficult to see that when v_i are computable, the resulting object is computably compact, but we will not need this. What we need is the cohomology group of a (finite) *abstract* simplicial complex. These notions are defined below.

Abstract simplicial complexes. An abstract simplicial complex is essentially a higher-dimensional generalisation of a graph. It consists of:

1. a set V of objects called *vertices*, and
2. a set S of subsets of V called *simplices* such that:
   (a) $\sigma \in S$ and $\tau \subseteq \sigma$ implies $\tau \in S$, and
   (b) for each $v \in V$, $\{v\} \in S$.

If $\tau \subseteq \sigma$, we say that τ is a face of σ. If σ has $k+1$ elements, then its dimension is k.

We are interested in finite simplicial complexes. Every finite abstract simplicial complex can be realised as a geometric simplicial complex, but we will not actually need it. In fact, all parameters in such a realisation can be chosen to be rational. Since we know there is one, we can simply search for it. Thus, the geometric realisation is also uniformly effective in the right sense. We omit further details. However, it is perhaps useful to have some geometric intuition when working with abstract simplicial complexes.

Computing homology. Suppose $K = (V, S)$ is a finite abstract simplicial complex. Orient every simplex in S by choosing the order of vertices. The orientation can, for instance, be defined using the sign of the determinant of $[v_1 - v_0, \ldots, v_k - v_0]$ in the geometric realisation; recall the vectors can be taken to have rational coordinates. However, the orientation can be done essentially arbitrarily (more details below); it is known that we always arrive at the same (co)homology groups as a result. Thus, we do not actually have to calculate a geometric interpretation.

If $\sigma \in S$, we assume it is already oriented and write $\sigma = (v_0, \ldots, v_k)$. We also write $(v_0, \ldots, \widehat{v_i} \ldots, v_k)$ to denote the (oriented) face of σ that has all vertices except for v_i, and which has the orientation from v_0 to v_k. A simplicial k-chain is a finite formal sum

$$\sum_i c_i \sigma_i,$$

where $c_i \in \mathbb{Z}$ and σ_i are oriented k-simplices, i.e., for each i, $\sigma_i = (v_{i,0}, \ldots, v_{i,k})$ for some $v_{i,j} \in V$.

Also, we would like to have $(v_0, v_1) = -(v_1, v_0)$. More generally, if π is a permutation of vertices in σ and τ is obtained from σ by applying this permutation, then we also declare $\sigma = \operatorname{sgn}(\pi)\tau$, where sgn is the sign of the permutation. Equivalently, swapping two adjacent vertices results in a change of the sign of σ in the group. Either way, we have finitely many generators and finitely many relations upon these generators that we define computably.

We arrive at a computable finitely generated abelian group

$$C_k = G_k/R_k,$$

where G_k is freely generated by oriented k-chains and R_k is freely generated by the aforementioned relations. By Rado's Lemma 5.1.4, we can uniformly computably find a set of generators of the resulting finitely generated abelian group so that these generators determine a (complete) direct decomposition of $C_k = G_k/R_k$ into cyclic summands. Also, define the *boundary operator* ∂_k by the rule

$$\partial_k(v_0, \ldots, v_k) = \sum_{i \leqslant k}(-1)^i(v_0, \ldots, \widehat{v_i} \ldots, v_k).$$

This can be extended to k-dimensional chains linearly, thus inducing a group homomorphism

$$\partial_k : C_k \to C_{k-1}.$$

(We can set ∂_0 equal to the zero map $\mathbf{0}$.) Note that, so far, all definitions were uniformly computable in the strongest sense possible.

It is well-known and easy to see that $\partial_k \circ \partial_{k+1} = \mathbf{0}$, and thus

$$Im\, \partial_{k+1} \subseteq Ker\, \partial_k.$$

Definition 5.2.17. Define the *k-th homology group* of a finite simplicial complex K to be

$$H_k(K) = Ker\, \partial_k / Im\, \partial_{k+1}.$$

The other standard notation is $H_k(K; \mathbb{Z})$ to emphasise that we used coefficients $c_i \in \mathbb{Z}$ when we defined k-chains. It could be any other abelian group, but we will not need this degree of generality.

We see that simplicial homology is computable.

Fact 5.2.18 (Folklore; see [411]). *Given a (strong index of an abstract) simplicial complex, we can uniformly compute its i-th homology group represented as $\bigoplus_{i \leqslant k}\langle a_i \rangle$, where $a_0, \ldots, a_k$ are the generators of the group such that the orders of the cyclic $\langle a_i \rangle$ are also uniformly computable.*

We leave the formal verification of this fact as an exercise. (Chapter 1 (Section 11) of [411] contains a rather detailed verification of Fact 5.2.18.)

Simplicial cohomology. Following the general pattern of the book, we will need a notion that is *dual* to homology. For the chain of groups and homomorphisms

$$\cdots \to C_{i+1} \overset{\partial_{i+1}}{\to} C_i \overset{\partial_i}{\to} C_{i-1} \to \cdots$$

define the dual "cochain"

$$\cdots \leftarrow C^*_{i+1} \overset{d_i}{\leftarrow} C^*_i \overset{d_{i-1}}{\leftarrow} C^*_{i-1} \leftarrow \cdots$$

where

$$C^*_i := \mathrm{Hom}(C_i, \mathbb{Z})$$

is the abelian group of all homomorphisms from C_i to $\mathbb{Z}$ and

$$d_{i-1} : C^*_{i-1} \to C^*_i$$

is the homomorphic map dual to ∂_i in the sense that

$$(d_{i-1}f)(c) = f(\partial_i c),$$

for any $f \in C_i^*$ and $c \in C_i$. The above relation also fully determines what the map d_{i-1} does on every such f. Also, each element f of C_i^* is fully described by where it maps the finitely many generators of C_i. This is all uniformly computable. It s not hard to show that $Im\, d_{i-1} \subseteq Ker\, d_i$. (When $i = 0$ assume $Im(d_{-1}) = \{0\}$.) We arrive at:

Definition 5.2.19. Define the *i-th cohomology group* of a finite simplicial complex K to be

$$H^i(K) = Ker\, d_i / Im\, d_{i-1}.$$

Again, $H^i(K;\mathbb{Z})$ is the other standard notation, because in our definitions $\mathbb{Z}$ can be replaced with some other abelian group G throughout, giving the notions "with coefficients in G". We will need only cohomologies "with coefficients in $\mathbb{Z}$". It is not difficult to see that a version of Fact 5.2.18 holds for the cohomology groups as well; we delay the verification of this claim until later (this is Claim 5.2.22).

Čech cohomology

We follow Section 73 of [411]. Given a compact M, let $\mathcal{N}$ be the directed set of all its finite open covers (under refinement). Since the covers by basic ϵ-balls, where ϵ ranges over positive rationals, are cofinal among all covers, without loss of generality we can restrict ourselves only to covers by basic open balls with rational radii. For instance, $\mathcal{N}$ could be a *nerve-decidable system* of covers (Definition 4.2.29) well-ordered under formal refinement instead of the usual refinement.

For each cover K from the system $\mathcal{N}$, recall that its *nerve* $N(K)$ is the collection of all subsets in the cover that intersect non-trivially. One can view $N(K)$ as a (finite, abstract) simplicial complex in which the n-dimensional faces are exactly the n-element subsets X of $N(K)$ such that $\bigcap\{Y : Y \in X\}$ is a non-empty set. For these finite simplicial complexes, we can define their *cohomology groups*

$$H^i(N(K)) = \ker(d_i)/\operatorname{im}(d_{i-1})$$

(with coefficients in $\mathbb{Z}$). If we have two covers, K and $\tilde{K}$, and $\tilde{K}$ refines K, then any refinement map

$$r : N(\tilde{K}) \to N(K), \text{ such that } r(B) \subseteq \tilde{B},$$

is *simplicial* in the sense that the images of the vertices of a simplex always span a simplex. Indeed, if several balls intersect, then the bigger balls that contain them must obviously intersect as well.

This map induces a map between k-chain complexes as well. It is well known that, furthermore, it induces *homomorphisms*

$$\psi^i_{K,\tilde{K}} : H^i(N(K)) \to H^i(N(\tilde{K}))$$

between the respective i-th cohomology groups; see Section 73 of [411] for a careful verification. Keep refining covers and keep "extending" the respective cohomology groups; note, however, that some elements of $H^i(N(K))$ can be declared equal to zero when we "extend" them to $H^i(N(\tilde{K}))$.

Consider the group obtained in the limit of this process. Recall the discussion after Property 5.1.25, where in (5.1) we (somewhat informally) introduced the notion of the direct limit in the specific context of well-ordered systems of finitely generated abelian groups. We arrive at:

Definition 5.2.20. In the notation above, let *the i-th Čech cohomology group $H^i(M)$ of M* be the (direct) limit of $H^i(N(K))$, $K \in \mathcal{N}$, induced by the (inverse) system $\mathcal{N}$.

For a finite simplicial complex, its Čech cohomology is isomorphic to its simplicial cohomology; see the last chapter of [411]. The i-th Čech cohomology groups are homeomorphism invariants of the given space. (In particular, the definition of $H^i(M)$ does not depend on the choice of covers, as long as every cover is eventually refined by an ϵ-cover, for any $\epsilon > 0$.) Proving this fact is actually not that difficult, but due to the complexity and length of the definitions involved, we omit it. We refer the reader to the last chapter of [411]. We take these properties of Čech cohomology for granted. Our next task is to analyse the computability-theoretic complexity of Čech cohomology.

Computing Čech cohomology

Recall that a (discrete, countable) group is c.e. presented if it is isomorphic to a factor of a computable free group by a computably enumerable subgroup. In other words, the operations of the group are computable but the equality is c.e..

Theorem 5.2.21 (Lupini, Melnikov, and Nies [341]). *For a computably compact space M, its i-th Čech cohomology group admits a c.e. presentation uniformly in i.*

The proof below first appeared in [139]; it is different from the proof in [341].

Proof of Theorem 5.2.21. As we noted above, we can assume that we are given a system of 2^{-n} covers $\mathcal{N}$ that is linearly nested under formal inclusion and is $\cap$-decidable (nerve-decidable); by Theorem 4.2.33 and Remark 4.2.34, this can be done uniformly computably.

Fix a $\cap$-decidable finite cover K and a positive $i \in \mathbb{N}$.

Claim 5.2.22. *The groups $H^i(N(K))$ are finitely presented, and the strong index of $H^i(N(K))$ can be obtained uniformly uniformly from K and i.*

Proof. The finite complex $N(K)$ is computable because the cover K is $\cap$-decidable. A close examination of the definitions shows that, given K (as a finite set of parameters) and i, we can compute the generators of $C^i = Hom(C_i, \mathbb{Z})$ and compute d_i. In the notation of Definition 5.2.19, we can uniformly computably find a finite set of independent generators (a_j) of $Ker(d_i)$ and a finite set of generators (b_s) of $Im(d_{i-1})$. □

Recall that a group admits a c.e. presentation if it is isomorphic to a factor of a computable group by a Σ^0_1 subgroup.

Claim 5.2.23. *The direct limit $\lim_{K \in \mathcal{N}} H^i(N(K))$ admits a c.e. presentation.*

Proof. Recall that $\mathcal{N}$ consists of $\cap$-decidable covers linearly ordered under the refinement relation. The refinement relation is c.e. and implies refinement. As explained earlier, if we have $K \subseteq_{form} K'$ in $\mathcal{N}$, the it (uniformly effectively) induces a simplicial map between the respective nerves $N(K)$ and $N(K')$. The latter in turn uniformly computably induces a homomorphism between the respective cohomology groups $\phi : H^i(N(K)) \to H^i(N(K'))$. By Claim 5.2.22, these abelian abelian groups are (uniformly) finitely presented. It follows that

$$\lim_{K \in \mathcal{N}} H^i(N(K)) = H^i(M)$$

can be viewed as a c.e. presentation of $H^i(M)$, by Fact 5.1.28. □

It is clear that the proofs of the claims above are uniform in i. This finishes the proof of the theorem. □

5.2.3 Effective dualities and the proof of Theorem B(3).

Recall that Theorem 5.2.1 states that, for a discrete torsion-free abelian group G, G has a computable presentation iff the compact connected domain of the dual group $\widehat{G}$ admits a computably compact presentation. Note that we do not need to assume that the operation of $\widehat{G}$ is computable. And here is why.

Theorem 5.2.24 (Folklore). *For a compact connected Polish abelian group H,*

$$H^1(H) \cong \widehat{H},$$

where as usual $\widehat{H}$ denotes the Pontryagin dual of H, and $H^1(G)$ stands for the first Čech cohomology group of the underlying space.

One does not need the operation of G to define $H^1(G)$; therefore homeomorphic connected compact abelian groups are necessarily isomorphic as topological groups. The proof of this theorem is essentially a direct calculation; it is omitted since it is not really related to our story. See Exercise 9.5.18 in Part 2 of the book for a hint. For a detailed proof of Theorem 5.2.24, see pp. 474–477 of [263]. For an even more detailed (and quite technical) general exposition of cohomology theories of compact abelian groups, we cite [264].

Proof of Theorem 5.2.1. If G is computable torsion-free, then its dual is computably compact by Proposition 5.2.2. Conversely, if the underlying space of the connected $\widehat{G}$ is computably compact, then by Theorem 5.2.21, we can calculate a c.e. presentation of $H^1(G)$. By Theorem 5.2.24, we obtain a c.e. presentation of the *torsion-free* abelian group $\widehat{\widehat{G}} \cong G$ (by Pontryagin duality). By Khisamiev's Theorem 5.1.41, G has a computable copy. □

We will also use the following version of effective Pontryagin duality established in [374]. Fix a prime q.

Theorem 5.2.25 (Melnikov [374]). *Suppose G is a q-divisible torsion-free abelian group. Then G has a Δ^0_2 presentation iff the compact dual $\widehat{G}$ of G has a computable Polish presentation (as a space).*

Proof. If G is Δ^0_2, then $\widehat{G}$ is computable by Lemma 5.2.10. Otherwise, if $\widehat{G}$ is computable, then $\widehat{G}$ is computably compact relative to $0'$, by Fact 4.2.11. Relativise Theorem 5.2.1 to obtain a Δ^0_2 presentation of G. □

We note that another version of effective Pontryagin duality, this time between torsion discrete and profinite abelian groups, will be established in Section 9.5.

Recall that to prove Theorem B(3), we need to find an example of a connected compact space that has a computable Polish presentation, but has no computably compact copy.

Proof of Theorem B(3)

Fix a non-computable c.e. set A; we can take A to be low (Theorem 3.1.1). Consider the group of the rationals G_A generated by the set

$$\left\{\frac{1}{2^k}, \frac{1}{p_i} : i \in \overline{A} \text{ and } k \in \mathbb{N}\right\},$$

where $(p_i)_{i\in\mathbb{N}}$ is the natural list of all prime numbers. By Malcev's old result (Theorem 5.1.16) and the remarks after Theorem 5.1.16, G_A has an X-computable copy iff A is computable relative to X. In particular, G_A has a Δ^0_2 presentation, but it has no computable presentation. Also, G_A is clearly torsion-free and 2-divisible. It follows from Theorem 5.2.25 that the connected compact domain of $\widehat{G}_A$ has a computable Polish presentation. However, Theorem 5.2.1 implies that the compact connected domain of $\widehat{G}_A$ cannot possibly have a computably compact presentation, as it would contradict the choice of A.

The proof of Theorem B(3) is complete.

Exercises

Exercise° 5.2.26. Let $(A_i)_{i\in\mathbb{N}}$ be discrete abelian groups. Show that

$$\widehat{\bigoplus_{i\in\mathbb{N}} A_i} \cong \prod_{i\in\mathbb{N}} \widehat{A_i}.$$

Exercise 5.2.27. Show that every computably compact abelian group admits a computably approximable presentation in the sense of Definition 2.4.8. (Hint: Given G, consider $\widehat{\widehat{G}}$ and use that the proof of Proposition 5.2.2 produces a computably approximable group.)

Exercise 5.2.28** (Effective Birkhoff-Kakutani Theorem; Koh, Melnikov and Ng [313]). A *computable topological group* is a computable topological space (Def 2.4.26) with group operations $\cdot$ and $^{-1}$ that are effectively continuous. (See also Exercise 2.4.29.) Let G be a computable topological group. Show:

1 There exists a computable left-invariant metric[4] realised as an effectively continuous functional $G \times G \to \mathbb{R}$.

[4]That is, a metric d compatible with the group topology that has the property $d(z \cdot x, z \cdot y) = d(x, y)$, for all x, y, z.

2 Further, in cases when the group is either abelian or locally compact, it is possible to produce an effective dense sequence, giving a *right-c.e. Polish presentation* of the group.

3 Conclude that for locally compact and for abelian groups, computable topological and right-c.e. presentability (Exercise 2.4.27) are *equivalent.*

4 Note that the result in (3) cannot be improved to "computable Polish" in general. (This follows from Exercises 2.4.28 and 4.2.112.)

Exercise* 5.2.29 (Melnikov and Ng [380], Lupini, Melnikov, and Nies [341]). Take for granted the well-known van Dantzig's Theorem, which says that a totally disconnected locally compact (t.d.l.c.) group always contains a compact open subgroup. Prove that the following are equivalent for an infinite t.d.l.c. (Polish) group:

1. $G \cong ([T], \cdot, ^{-1})$, where:

   (a) T is either the natural copy of $2^{<\omega}$ (in which case the group is profinite) or is obtained by joining ω-many natural copies of $2^{<\omega}$ below the common root;

   (b) the operations are computable upon the space $[T]$ of paths through T.

2. G is computable Polish and furthermore *computably locally compact* ([512, 507]): There is a uniform procedure which, given (the name N^x of) any point x, outputs a basic open B and a computable compact $K \subseteq G$ such that $x \in B \subseteq K$, where K is given by a sequence of finite open 2^{-n}-covers so that each ball in the cover is centred in a (computable) point in K and has a rational radius.

3. Show** that if G is additionally abelian, then (1) and (2) are equivalent to G being the limit of a computable inverse system of uniformly computable abelian groups:

$$A_0 \leftarrow_{\psi_0} A_1 \leftarrow_{\psi_1} A_2 \leftarrow_{\psi_2} \cdots,$$

   where the groups A_i are uniformly computable and the kernel of the surjective homomorphisms ψ_i are finite and are uniformly given by their strong indices (i.e., as finite sets).

Exercise* 5.2.30 (Lupini, Melnikov, and Nies [341]). Suppose that A is a computably locally compact t.d.l.c. group, as defined in Exercise 5.2.29.

1. Show that the dual $\widehat{A}$ of A is also computably locally compact. (This is not how it is stated in [341], but the proof in [341] gives computable local compactness.)

2. If additionally $\widehat{A}$ is also t.d.l.c., then A is computably locally compact iff $\widehat{A}$ is computably locally compact.

5.2.4 Further related results: comparing notions*

In this chapter, we completed the task of separating right-c.e., computable Polish, and computably compact spaces up to homeomorphism, and we did so using effective dualities. Both left- and right-c.e. Polish spaces form natural subclasses of Δ^0_2 Polish spaces. It has been shown in [37] that every Δ^0_2 Polish space admits a *computable topological* presentation, which is another classical notion of presentability in effective topology (Definition 2.4.26). The notions and the implications between

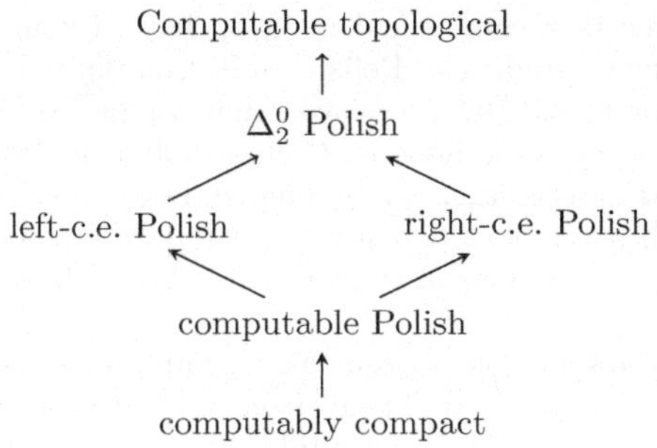

Figure 5.1: Notions of computable presentability for Polish spaces.

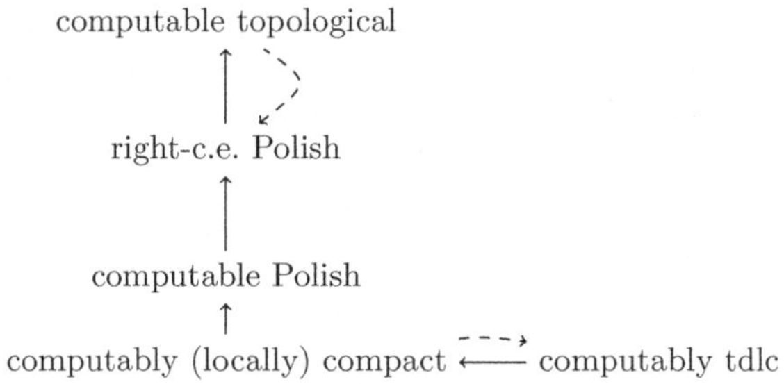

Figure 5.2: Notions of computable presentability for locally compact Polish groups.

them are summarised in the first diagram. Arrows illustrate the implications between these notions *up to homeomorphism*.

It has been shown in [37, 245, 266, 341, 35], culminating in [311], that the only implications between the notions in Fig. 5.1 are those depicted in the diagram. Some of these results appeared (or will appear) as exercises throughout the book; e.g. Exercises 4.2.24–4.2.26, 4.2.41, 4.2.104, 4.2.105, and 9.4.16. For example, in Exercise 4.2.104, we saw that every (not necessarily computable) Polish space admits a computable topological presentation, making such "presentations" essentially meaningless. On the other hand, in the compact case, this notion can be used to give (yet) another equivalent form of computable compactness (Exercise 4.2.41). In the context of closed surfaces, a few more notions were compared in [241]; it was discovered that all these notions are arithmetically equivalent to each other. Not all of the papers cited above rely on dualities to compare these notions up to homeomorphism. For example, the key combinatorial tool in [311] (see Exercise 9.1.36) is the notion of a *limitwise monotonic set*. We will introduce and utilise limitwise monotonic sets and functions in Chapter 9.

We now briefly discuss computable Polish groups. The second diagram illustrates implications between several natural notions of effective presentability for locally compact groups, up to topological group-isomorphism. Arrows illustrate the evident implications, and the dashed arrows represent the implications recently established in [313, 380]. Examples of compact and discrete groups separating the notions of "right-c.e. Polish" and "computable Polish" have been suggested in [313, Section 5]. In Theorem 4.2.107 we proved that for profinite groups, recursive and computably compact presentations are synonymous. Computably tdlc groups generalise the notion of a recursive profinite group; in Exercise 5.2.29 we extended Theorem 4.2.107 to such groups. See also Exercise 5.2.28 that establishes that locally compact "computable topological" groups are precisely the right-c.e. Polish ones, up to topological isomorphism. We will return to computably presented Polish groups in Part 2.

A special case of a computable Polish group is a computable Banach space (Lemma 2.4.17). Banach spaces are usually viewed up to linear *isometry* (equivalently, up to isometric group-isomorphism). For left-c.e., right-c.e., and computable Banach spaces compared up to linear isometry, see Exercise 2.4.40.

5.3 What's next?

We have finished proving Theorems A and B, and we have established several effective dualities in the process. In the second part of the book we will use these dualities to transform results about discrete structures into results about Polish spaces and groups. The second part will also be generally more technically challenging. For example, a generalisation of the infinite injury technique, the infamous $\mathbf{0}'''$ (zero-triple) technique, aka the "Lachlan monster", will be used to prove the main result of Section 9.

Part II

Computable classification

Chapter 6

Introduction

Classification theory is a hoary old theme in mathematics and indeed, science. In the second part of the book, we apply the methods and results developed in Part 1 to investigate the complexity of the classification problem in various classes of algebraic and topological structures.

> In Part 2 of the book, we focus on results that measure the complexity of the classification problem in common algebraic and topological classes.

Much of the theory revolves around a few basic definitions of what it means for a class to have an algorithmically useful classification. These definitions will be tested in various natural classes of mathematical structures, with a strong emphasis on discrete and compact abelian groups and compact metric spaces. We will always bear in mind our central theme of unifying the countable and the uncountable. Nonetheless, the second part of the book is generally more focused on countable structures, and especially discrete abelian groups. Their computable classification theory is much more developed than its separable counterpart. Another reason is that computability theory appears to integrate more easily with discrete algebra and combinatorics. Using the effective dualities established in Part 1, along with some further theorems, these results will find direct applications to separable structures.

R. G. Downey, A. Melnikov, *Computable Structure Theory*, Theory and Applications of Computability, https://doi.org/10.1007/978-3-031-92433-0_6

6.1 Classification via computation

Generally, in mathematics, we categorise objects into groups we regard as being the same. We typically have a notion of equivalence $\equiv$, and a class $\mathcal{C}$ of objects that we compare using $\equiv$. The equivalence classes will form a *classification* of the objects. In algebra and combinatorics, $\equiv$ would typically be isomorphism $\cong$. Examples of classifications up to isomorphism include the famous classifications of the finite abelian groups in terms of cyclic groups, compact 2-surfaces in terms of connected sums of tori or projective planes, and the celebrated classification of the finite simple groups. However, other equivalences are useful too. For example, the Turing degrees are a classification of the class of sets under $\equiv_T$, where $\equiv_T$ means "equi-computability". In metric spaces, we would often use either isometry or homeomorphism as classifiers, but one could also use bi-Lipschitz maps, homotopy equivalence, diffeomorphisms, and so on. We typically get quite distinct classifications with the different equivalence relations: for example, spaces might be homeomorphic but not isometric.

For a fixed class $\mathcal{C}$, the kinds of questions studied classically are:

1. How do various invariants compare?
2. How do we compare the complexity of different classifications?
3. Can we show there can be no reasonable classification?

Logic is a great tool to tackle such questions. Since this is a book about computable structure theory, we will concentrate on using computability theory for these tasks.

6.1.1 The three main approaches used in the book

We will examine the following three mutually related approaches:

(i) Classification up to X-computable isomorphism

Since we are interested in the effective content of mathematics, it is natural for us to replace isomorphism with computable isomorphism. Recall the definition:

Definition 6.1.1 (Mal'cev). A computable structure A is *computably categorical* if any other computable B isomorphic to A is *computably* isomorphic to A.

In Part I of this book, we classified computably categorical linear orderings, Boolean algebras, Stone spaces, and torsion-free abelian groups. Sometimes computable classification and classification coincide, such as in the case of finitely generated structures. But for most classes of structures, these are quite different. One recurrent theme in this area is to seek to understand how semantic notions (like computable isomorphism) relate to syntactic notions (such as definability); we will discuss this in detail in Chapter 10. Sometimes it will be natural to look at more complex classifiers, such as X-computable isomorphisms.

Definition 6.1.2. Fix $n \in \mathbb{N}$, $n > 0$. A computable structure A is *Δ^0_n-categorical* if for any computable $B \cong A$, there is a Δ^0_n isomorphism between A and B.

The notion has a transfinite analogue, but we won't really need it. This approach will be useful in other classification themes, such as *(ii)* below; these applications are mostly technical in their nature.

(ii) Classification via index sets

Index sets (Definition 2.1.11) have been used in computability theory and computable structure theory for many decades; e.g., [113], but in computable topology and analysis, this approach is relatively new (e.g., [138]).

Let $(M_e)_{e\in\mathbb{N}}$ be a uniform enumeration of all (partial) computable structures in the language of $\mathcal{K}$; we will never encounter the unnatural situation when such a list does not exist.

Definition 6.1.3 (Goncharov and Knight [213]). For a class $\mathcal{K}$ of structures (or spaces), define the following index sets.

1. $I(\mathcal{K}) = \{e : M_e \text{ is a member of } \mathcal{K}\}$ is called *the recognition problem for $\mathcal{K}$, the characterisation problem for $\mathcal{K}$*, or simply *the index set of $\mathcal{K}$*.
2. $E(\mathcal{K}) = \{\langle i,j\rangle : i,j \in I(\mathcal{K}), M_i \cong M_j\}$ is called *the isomorphism problem for $\mathcal{K}$. (Here "E" stands for "equivalent" under the fixed classifier.)*

Note that the isomorphism problem mimics the similar problem in combinatorial group theory mentioned in Part 1 of the book. The intuition is that these index sets usually accurately reflect the complexity of the classification problem for a given class $\mathcal{K}$. We will see that, for common classes, the estimates obtained for the complexity of $E(\mathcal{K})$ tend to always relativise, and thus reflect the complexity of classification of arbitrary members of $\mathcal{K}$, not just of the computable ones.

Because of Rice's Theorem 2.1.12, neither $E(\mathcal{K})$ nor $I(\mathcal{K})$ can be computable for any reasonable class. The next best estimate that we can hope for is that $E(\mathcal{K})$ and $I(\mathcal{K})$ are *arithmetical.*

Definition 6.1.4. We say that a class $\mathcal{K}$ is *arithmetical* if $E(\mathcal{K})$ and $I(\mathcal{K})$ are arithmetical sets, i.e., are both Σ^0_n for some $n \in \mathbb{N}$.

Goncharov and Knight [213] suggested that a class $\mathcal{K}$ should be viewed as tractable if both $I(\mathcal{K})$ and $E(\mathcal{K})$ are hyperarithmetical sets. However, after over 20 years of systematic research, the following phenomenon has been observed. In algebra, analysis, and topology, obtaining an index set of *transfinite* hyperarithmetical complexity seems to require a *transfinite* definition of the class. For example, we could fix some computable $\alpha \geqslant \omega$ and consider abelian p-groups of Ulm type $\leqslant \alpha$ or all countable compact spaces of Cantor-Bendixson rank at most α. Nonetheless, examples of transfinite index sets can be found in model theory. For instance, White [509] and, independently, Pavloskii [428] showed that the index set of computable homogeneous models is $\Sigma^0_{\omega+2}$-complete.

On the other hand, if a class is not arithmetical, then either $E(\mathcal{K})$ or $I(\mathcal{K})$ (or both) will typically be Π^1_1- or Σ^1_1-complete (or beyond). At this stage, it is enough to say that Σ^1_1-completeness means that the problem is as hard as checking through *all possible functions* $\mathbb{N} \to \mathbb{N}$; Π^1_1-sets are the complements of Σ^1_1 sets (the formal definitions will be given in Chapter 8). We arrive at the following dichotomy, which seems to universally hold in algebra, topology, and computable analysis.

The index set dichotomy. Let $\mathcal{K}$ be a "natural" class of algebraic or topological structures. Then one of the two alternatives holds:

- $\mathcal{K}$ is arithmetical, or
- either $E(\mathcal{K})$ or $I(\mathcal{K})$ is Σ^1_1- or Π^1_1-complete (or beyond).

Here, "natural" means "not ad hoc"—a standard class that can be found in the literature. Of course, counterexamples can be easily manufactured, but we have yet to discover a class in the literature that fails the dichotomy. Thus, it makes sense to agree that a class should be regarded as (potentially) "tractable" if it is arithmetical, and "difficult" or even "unclassifiable" if the other alternative holds.

The approach via index sets has its limitations, such as Rice's Theorem 2.1.12. Also, it is not applicable to finitely generated structures for which it tends to be too crude. It also depends on the exact choice of computable presentability that we use; however, as we will discuss in Chapter 7, this never seems to cause any issues.

There are other ways of using computability theory to measure the classification problem in a class, circumventing some of these obstacles. In the book, we will primarily focus on one more approach.

(iii) Enumerations with no repetition

Note that if both $I(\mathcal{K})$ and $E(\mathcal{K})$ are arithmetical, i.e., both are computable relative to $\varnothing^{(n)}$ for some $n \in \mathbb{N}$, then we can use $\varnothing^{(n)}$ to remove repetitions from $I(\mathcal{K})$, so that every computable structure from the class appears in the new list *exactly once*. We could argue that we "fully understand" a class $\mathcal{K}$ if we could make a *computable* listing $\{A_i : i \in \mathbb{N}\}$ of all the structures of $\mathcal{K}$ that mentions exactly one structure per isomorphism type. For example, all finitely generated abelian groups can be listed up to isomorphism without repetition. Similarly, there is a 1-1 list of all compact 2-surfaces up to homeomorphism. Both lists are algorithmically effective.

Recall that Friedberg [182] proved that there is a uniformly computably enumerable list of all c.e. sets with no repetition (up to the usual equality of sets); see Theorem 3.1.43. He produced such a list in spite of the fact that the index set $\{\langle i, j\rangle : W_i = W_j\}$ is Π^0_2-complete. This is known as a *Friedberg enumeration* of all c.e. sets. Motivated by this classical theorem, Goncharov and Knight [213] suggested the following definition.

Definition 6.1.5. Let $\mathcal{K}$ be a class of structures. We say that $\mathcal{K}$ admits a *Friedberg enumeration* (a *Friedberg numbering* or a *Friedberg listing*) if there is a uniformly computable listing of all computably presentable members of $\mathcal{K}$ without repetition, up to isomorphism.

In such a list, every member of $\mathcal{K}$ is usually represented by its index. For example, it could be the strong index of the group presentation in the list of all finitely presented abelian groups up to isomorphism, or the index of a computably compact presentation in the list of all orientable compact surfaces up to homeomorphism. But the surfaces can also be given by the strong indices of the finite simplicial complexes representing them. It will be always clear from the context which algorithmic presentations we use to produce a Friedberg enumeration.

This approach is restricted to computable members of the class, but the classifier (the notion of isomorphism) does not have to be effective. As usual, results of this sort can be relativised to any oracle. However, this approach seems more suitable for understanding computable members of $\mathcal{K}$ specifically.

There are several trivial reasons related to index set calculations that prevent many classes from having a Friedberg enumeration. As a result, there are very few positive results in the literature when it comes to Friedberg enumerations. Thus, even if the reader thinks that (iii) does not always accurately reflect the complexity of classifying structures in $\mathcal{K}$, *any result showing the existence of a Friedberg enumeration for a class $\mathcal{K}$ should be viewed as a very strong positive classification-type result about $\mathcal{K}$.* Some of these positive results require sophisticated machinery, such as $\mathbf{0}'''$-priority arguments, which we did not need in Part 1 of the book.

6.1.2 Other approaches

Before we discuss the main results of Part 2 of the book, we remark that the approaches above are not the only methods used to study the classification problem in a class of structures. For example, in descriptive set theory, one uses Borel reductions to compare classes of structures; see the book [197]. Another approach is due to Shelah; we cite [465] and [332] for the details. His idea is to consider the (cardinal) number of non-isomorphic models of a theory. This material is way beyond the scope of the present book, and we only mention it for interest. Instead, we briefly discuss a few further computability-theoretic classification themes, some of which will be used in the book as a technical tool.

Effective reductions between classes

This relatively popular approach (e.g., [170, 283]) effectivises the standard definition from descriptive set theory. Suppose $\mathcal{C}_1$ and $\mathcal{C}_2$ are classes of countable structures up to isomorphism, where every structure is identified with its atomic diagram. We say that $\mathcal{C}_1$ is *effectively reducible* to $\mathcal{C}_2$, written $\mathcal{C}_1 \leqslant_{EFF} \mathcal{C}_2$, if there exists a Turing functional Φ such that

$$M \cong N \text{ if and only if } \Phi(M) \cong \Phi(N),$$

where M, N range over $\mathcal{C}_1$, and $\Phi(M), \Phi(N)$ are structures in $\mathcal{C}_2$. The above definition has several natural variations (e.g., [170, 283]). We won't focus too much on this method, but we will use some known reductions between classes to make conclusions about index sets. More about this approach can be found in Section 8.2. This approach also has a natural "Type 1" analogue, which requires the reduction to work on indices of computable structures rather than for arbitrary presentations; it will also be defined and discussed in Section 8.2.

Effective universality

In Section 8.2, we will also discuss computably universal classes, which is a notion (essentially) coined in [257]. These are the most complex classes from a computability-theoretic perspective. Such classes can "effectively encode" any other class of countable algebraic structures, preserving important properties like classical and computable isomorphism types, and more. Like effective reductions between classes, computably universal classes will serve as a tool. However, we won't be developing a detailed theory of the effective universality of structures.

The two themes discussed above can also be extended to classes of separable structures. Natural examples of such generalised effective transformations are the computable dualities presented in Part 1 of the book; we delay further discussion until Section 8.2. One last theme is as follows.

Classification using infinitary formulae

In computable structure theory, another approach, discussed and tested already in [213], involves using *computable* infinitary $L_{\omega_1\omega}$ formulas. These formulas allow for computable infinite conjunctions and disjunctions. This approach explores the algorithmic content of the well-known fact that every countable structure is fully described by an infinitary sentence up to isomorphism (Theorem 10.1.2).

In Exercise 10.1.65, we will see that for discrete algebraic structures, this approach is very closely related to the approach via index sets, though these two frameworks are not equivalent even for subgroups of $(\mathbb{Q}, +)$ (Exercise 10.1.66). In the book, we shall avoid the use of infinitary logic when possible; we only use it in Chapter 10, and only in passing. Also, there has been very little progress in this direction in the context of *separable* structures (e.g., [381]). For more about this topic, we cite the recent survey [238].

6.2 The main results of Part II

In Chapter 7, we develop a systematic approach to the classification of structures and spaces using index sets and Δ^0_n-categoricity. We will see that the two themes are very closely related technically. The main result of Chapter 7 is as follows:

Theorem C (Lupini, Melnikov, and Nies [341], based on Downey and Melnikov [136]). Let $\mathcal{K}$ be the class of direct products of solenoid groups.

1. $\mathcal{K}$ admits an arithmetical characterisation among all Polish groups.
2. The isomorphism problem for $\mathcal{K}$ is arithmetical.

In both 1. and 2., the upper estimate is Σ^0_7. Solenoids are exactly the Pontryagin duals of rank 1 torsion-free abelian groups, and their topology uniquely determines the group operation on them. Solenoid spaces were introduced by Vietoris in [499]. Solenoids are important in the area of dynamical systems; e.g., [510]. These objects are also used throughout topology and topological group theory as a source of examples and counterexamples. The direct products of solenoid spaces are the Pontryagin duals of completely decomposable groups, introduced and first studied by Baer [26]. Completely decomposable groups traditionally play an important role in the theory of infinite abelian groups [195]. Their compact duals are also reasonably well-studied [340]. Among other tools, the proof of Theorem C relies on the effective Pontryagin duality from Part 1 of the book and on our technical result in [137], which states that completely decomposable groups are Δ^0_5-categorical. We will include a detailed proof of the result from [137], as well as some further related results about computable completely decomposable groups. In Exercise 7.2.49, we will see that it is possible to "computably transform" a linear order in a completely decomposable group and, thus, into the compact groups in Theorem C. This observation seems to suggest that $\mathcal{K}$ cannot possibly be an arithmetical class. However, as we will discuss at the end of Chapter 7, there is no contradiction because the "coding" hinted at in Exercise 7.2.49 is ill-behaved. Theorem C illustrates that the sketch in Exercise 7.2.49 cannot be modified to produce a sufficiently well-behaved transformation. This is certainly a non-trivial fact, as reflected in the complexity of the proof of Theorem C.

Chapter 8 lays the foundations of the non-classification theory, in the spirit of [213]. In this section, we are mainly focused on proving that the isomorphism problem in certain classes is Σ^1_1-complete. We develop sufficient machinery to prove:

Theorem D (Melnikov [373], based on Downey and Montalbán [146]). The homeomorphism problem for connected compact Polish spaces is Σ^1_1-complete.

Theorem D was proven in [373], but it was stated for connected compact groups. As noted in [139, 341], the result holds for spaces too, via the same proof. Just as with Theorem C, this result is derived from a similar result for discrete structures due to Downey and Montalbán [146], using effective Pontryagin duality. Chapter 8 is, of course, not limited to the proof of Theorem D. We will also show that many other classes of structures and spaces have Σ^1_1-complete isomorphism problems, including linear orders, Stone spaces up to homeomorphism, Banach spaces up to linear isometry, two-step nilpotent groups, and internal domains. To obtain some of these results, we will use the effective dualities established in Part 1. Other results, such as the Σ^1_1-completeness for integral domains, will follow from the *effective completeness* of the respective class. The notion of effective completeness has two versions, Type I (on indices) and Type II (for arbitrary members of the class). For example, in Theorem 8.3.10 we will establish that the isomorphism of computable torsion-free abelian groups is complete among all Σ^1_1-equivalence relations, and thus, the same is true for their connected duals too (Theorem 8.3.28). Theorem 8.3.28 implies Theorem D, but is harder to prove.

Chapter 9 is focused on the approach to classification via Friedberg enumeration. As we will see in Chapter 7, positive results stating the existence of a Friedberg enumeration in a class are very rare; the reason for this is often a straightforward index set calculation. Recall also that proving the original Friedberg Enumeration Theorem 3.1.43 for c.e. sets under equality requires infinite injury. Note that the index set $\{\langle i, j\rangle : W_i = W_j\}$ is merely Π^0_2-complete. In the main result of Chapter 9 stated below, the complexity of the isomorphism problem increases monotonically with the parameter n.

Theorem E (Downey, Melnikov, and Ng [145]). For any fixed $n > 0$, there is a Friedberg enumeration of all pro-p abelian groups of pro-Ulm type $\leqslant n$.

Here, pro-Ulm type is the Ulm type of the Pontryagin dual of the group; this will be clarified later. Chapter 9 presents a systematic exposition of computable abelian p-group theory, as it is necessary to prove Theorem E. Following the general pattern of the book, the main theorem is established through a sequence of transformations. First, we establish the existence of a Friedberg enumeration for equivalence structures [143]; this requires an application of the $\mathbf{0}'''$-machinery. Then we use methods of Khisamiev and Ash, Knight, and Oates to produce a Friedberg enumeration of abelian p-groups of Ulm type $\leqslant n$. To derive Theorem E, in Theorem 9.5.7, we establish another effective version of Pontryagin duality between torsion discrete and profinite abelian groups.

In Chapter 10, we attempt to classify computable structures up to computable isomorphism. We will see that the notion of computable categoricity admits several variations, the most popular being that of relative computable categoricity. Relative computable categoricity is a Type II analogue of computable categoricity (which itself is Type I), since we will show it is witnessed by Turing functionals. This Type II version of computable categoricity is well-behaved and admits a syntactical reformulation. The analogy with the situation in computable analysis is indeed striking, especially with the material of Section 2.3. For instance, we will prove theorems resembling the Kreisel-Lacombe-Shoenfield-Markov Theorem 2.3.7 and Specker's Theorem 2.3.24 from Part I. We will include several results about separable structures in Chapter 10, but these results work only for presentations viewed *up to isometry*. For example, in Theorem 10.2.9, we will give the description of (isometrically) relatively computably categorical Polish spaces in terms of approximate Scott families, and in Proposition 10.2.14, we apply approximate Scott families to the Urysohn space.

We will also outline the proof of the well-known theorem of Goncharov, stating that there is a structure with exactly two computable presentations, up to computable isomorphism (Theorem 10.3.2). Goncharov's theorem has consequences for structures in several standard classes, including metric spaces up to isometry.

We will finish the chapter, and the book, with a detailed proof of the following result:

Theorem F (Melnikov and Ng [376]). The space $(C[0,1], d_{sup})$ has infinitely many isometric, but not computably isometric, computable Polish presentations.

The main technical step in the proof of Theorem F, Theorem 10.3.20, generalises (and indeed, implies) another well-known theorem of Goncharov (Theorem 10.3.2) about discrete computable structures. Using Theorem 10.3.20, Theorem F will be derived from Theorem 2.4.20, which was the main result of Chapter 2.

Chapter 7

Classification theory

This chapter is mostly focused on establishing arithmetical upper bounds for index sets for various classes. The machinery of Δ_n^0-categoricity will be crucial in establishing many index set estimates, including the main result of the chapter, which is as follows.

Theorem C (Lupini, Nies, and Melnikov [341], based on Downey and Melnikov [136]). Let $\mathcal{K}$ be the class of direct products of solenoid groups.

1. $\mathcal{K}$ admits an arithmetical characterisation among all Polish groups.
2. The isomorphism problem for $\mathcal{K}$ is arithmetical.

The structure of the chapter is as follows:

1. Section 7.1 contains several basic results about index sets of structures and spaces. These index set estimates serve as examples, though some are not particularly straightforward and will be used in Section 7.3. We also briefly discuss Friedberg enumerations, but deeper results are postponed to Chapter 9.
2. Section 7.2 is devoted to computable completely decomposable groups. The main result of the section says that this class is arithmetical. We also include a complete classification of Δ_2^0-categoricity for homogeneous completely decomposable groups in our discussion.
3. Section 7.3 combines the results of Section 7.1 and Section 7.2 with effective Pontryagin duality (established in Part 1) to prove Theorem C.

This chapter relies heavily on the results and techniques established in Part 1, particularly on the content of Chapters 3 and 4.

R. G. Downey, A. Melnikov, *Computable Structure Theory*, Theory and Applications of Computability, https://doi.org/10.1007/978-3-031-92433-0_7

7.1 Calculus of index sets for structures and spaces

This section contains foundational material that will be important for the rest of Part 2 of the book. The results and examples discussed in the section are technically not difficult; however, some of them are quite neat.

7.1.1 Discrete countable structures

In this subsection, we go over some well-known examples of index sets. We also simultaneously establish upper estimates for Δ^0_n-categoricity (to be defined) in these classes. The elementary properties of arithmetical predicates summarised in Exercise 2.1.30 will be rather useful throughout this section.

Fix a class $\mathcal{K}$ of structures in a computable (typically, finite) language. We will be looking at

$$I(\mathcal{K}) = \{M_e : M_e \text{ represents a member of } \mathcal{K}\},$$

where $(M_e)_{e\in\mathbb{N}}$ is a uniformly effective list of all partial computable structures in the language of $\mathcal{K}$.

Proposition 7.1.1 (Folklore). *For the following classes, $I(\mathcal{K})$ is Π^0_2.*

(i) Linear orderings.

(ii) Graphs.

(iii) Groups.

(iv) Abelian groups.

(v) Torsion abelian groups.

(vi) Torsion-free abelian groups.

(vii) Equivalence structures.

(viii) Boolean algebras.

(ix) Vector spaces over a fixed computable field.

(x) Rings.

(xi) Trees.

(xii) Partial orderings.

Proof. They are all basically the same. All these classes are $\forall\exists$-axiomatisable, and some are $\forall$-axiomatisable. For M_e, being total is Π^0_2. □

Vector spaces

A computable structure is Δ_n^0-categorical if any two computable presentations of the structure are Δ_n^0-isomorphic.

Fact 7.1.2. *If every structure in a class $\mathcal{K}$ is Δ_n^0-categorical, and $I(\mathcal{K})$ is arithmetical, then $E(\mathcal{K})$ is arithmetical too.*

Proof. The complexity of saying that $A \cong_{\Delta_n^0} B$ is Σ_{n+2}^0. □

For example, the isomorphism problem for vectors spaces over a fixed field is arithmetical, since every vector space is Δ_2^0-categorical. This provides us with the upper estimate Σ_4^0 for the isomorphism problem. This upper bound can be improved. The following fact is folklore; e.g. [213, 72, 146].

Theorem 7.1.3. *$E(\mathcal{K})$ is Π_3^0-complete for the class of computable vector spaces over $\mathbb{Q}$. Indeed, if V_∞ is the $\mathbb{Q}$-vector space of dimension ω, then*

$$\{i : M_i \cong V_\infty\}$$

is Π_3^0-complete. The same is true about the additive group of V_∞.

Proof. Being a computable vector space over $\mathbb{Q}$ is Π_2^0, by Proposition 7.1.1 (vii). Let U, V be two computable vector spaces. To express that $\dim V \leqslant \dim U$, state that for all n, if there is a linearly independent n-tuple $\bar{v} \in V$, then there is a linearly independent n-tuple $\bar{u} \in U$. Being a linearly independent tuple is a Π_1^0-property, thus making $\dim V \leqslant \dim U$ a Π_3^0-property. Clearly, $\dim V \geqslant \dim U$ is also Π_3^0. Of course, $V \cong U$ iff $\dim U = \dim V$. Therefore $E(\mathcal{K}) \in \Pi_3^0$.

To see why the index set of V_∞ is also Π_3^0, observe that

$$\forall n\ \exists b_0, \ldots, b_n\ \{b_0, \ldots, b_n\} \text{ is linearly independent}$$

is a Π_3^0 property.

In Theorem 2.2.16 we proved that the additive group operation *effectively determines* the vector space scalar multiplication in V_∞. Being divisible, abelian, and torsion-free is Π_2^0. As a consequence, the upper bound for the index set of V_∞ remains Π_3^0 in the signature of additive groups.

We also note that the Π_3^0-completeness of the index set of V_∞ (among spaces) implies the Π_3^0-completeness of $E(\mathcal{K})$. To obtain the Π_3^0-completeness of $E(\mathcal{K})$, simply consider pairs of the form (V_e, V_∞), where V_∞ is identified with some computable presentation of this space. To this end, we outline the proof of Π_3^0-completeness of the index set of V_∞.

Fix a Π_3^0-complete set X. For each e, we will define a space V_e with

$$V_e \cong \begin{cases} V_\infty & \text{if } e \in X, \\ \text{a finite dimensional space} & \text{otherwise.} \end{cases}$$

Since our construction will be uniform, the index of V_e with respect to the enumeration of all structures in the language of $\mathbb{Q}$-vector spaces will also be computable:

$$V_e = M_{f(e)},$$

for some total computable f, thus illustrating the desired Π^0_3-completeness for V_∞. Let R be a computable predicate such that

$$e \in X \text{ if and only if } \forall n \exists^{<\infty} m\, R(e, n, m);$$

it exists by Exercise 2.1.30. Reserve elements $\{b, b_{n,m} : m, n \in \mathbb{N}\}$, and keep them linearly independent. If the predicate $R(e, n, m)$ "fires" on m, then use the strategy explained in detail in the proof of Theorem 2.2.16 to put the following elements in the span of b:

1. $b_{n,k}$, $k \geqslant m$;
2. $b_{n',m'}$, $n' > n$, $m' \in \mathbb{N}$.

It is easy to see that V_e is a non-trivial vector space (i.e., dim $V_e \geqslant 1$), and furthermore dim V_e is finite iff for some n, the predicate fires infinitely often. □

A similar argument works for any infinite computable field; see Exercise 7.1.13. In the case of a finite field the upper bound drops down to Π^0_2, we leave the proof to Exercise 7.1.14.

Theorem 7.1.4. *$E(\mathcal{K})$ is Π^0_2 complete for the class of computable vector spaces over a fixed finite field F.*

An equivalence structure is a structure of the form $(X, \sim)$, where $\sim$ is an equivalence relation on X. We will need the proposition below in Chapter 9.

Proposition 7.1.5 (Calvert, Cenzer, Harizanov, and Morozov [75]). *The isomorphism problem for computable equivalence structures is Π^0_4-complete. Indeed, there is an equivalence structure R so that*

$$\{i : M_i \cong R\}$$

is Π^0_4-complete.

Proof. The axioms of equivalence structures are Π^0_1, and totality of a presentation is Π^0_2. Given M_i and M_j, first check if the presentations are total and represent equivalence structures. To see whether $M_i \cong M_j$, it is sufficient to state that, for each n:

1. If M_i has (at least) n classes of size exactly $\lambda \in \mathbb{N} \cup \{\infty\}$, then M_j also has (at least) n classes of size exactly λ.
2. The same but with M_j and M_i interchanged.

Effectively in $\mathbf{0}''$ we can calculate the size of each individual class. This gives the upper bound Π^0_4.

To prove Π^0_4-completeness, let R be the structure with infinitely many infinite classes and with infinitely many finite classes of each finite size. Clearly, R has a computable presentation which we identify with R. Using Exercise 2.1.30, represent a Π^0_4 predicate P as follows:

$$P(e) \text{ if and only if } \exists^\infty i \exists^\infty j\, U(i, j, e),$$

where U is a computable predicate and "$\exists^\infty$" stands for "there exist infinitely many". Let also F be the structure that has no infinite classes but has infinitely many finite classes of each fixed size; this structure is also clearly computable.

We build an equivalence structure B_e to be $F \sqcup E$, where the i-th class of E has a size equal to $1 + \text{card}\{j : U(i, j, e) \text{ holds}\}$. The construction is uniform, so for some total computable f, we have that $B_e = M_{f(e)}$. Also, B_e has infinitely many infinite classes iff $P(e)$ holds. In this case, $B_e \cong R$, and otherwise $B_e \ncong R$. □

Ordinals

We fix the enumeration $(M_e)_{e\in\mathbb{N}}$ of all partial computable structures in the language of one binary relation. The theorem below can be extended to transfinite levels, but we omit the transfinite version and leave it to Exercise 8.1.27.

Theorem 7.1.6 (Folklore after Ash). *Fix $n \in \mathbb{N}$.*

(i) $\omega^{(n+1)}$ is (sharply) Δ^0_{2n+2}-categorical.

(ii) The index set $\{e : M_e \cong \omega^{(n+1)}\}$ is Π^0_{2n+3}-complete.

Sketch. For $n = 0$. Using $\varnothing'$, we can decide the adjacency relation in any computable copy of ω; this gives the upper bound Δ^0_2 in (i). This is sharp because of Exercise 3.2.17. The index set of ω is Π^0_3 since a linear order is ω iff it has the least element 0, and every other element of the order is in a finite block with 0. The Π^0_3-completeness uses the method of Theorem 7.1.3 to produce

$$A_e = \begin{cases} \omega & \text{if } e \in S, \\ k + \omega^* + \omega & \text{if } e \notin S, \end{cases}$$

where S is a Π^0_3-complete set.

The case when $n > 0$ follows by induction. The upper bound is routine. Iterate Theorem 3.2.46 (which is uniform for the linear orders that we use) to obtain

$$B_e = \begin{cases} \omega^n & \text{if } e \in S, \\ \omega^n(k + \omega^* + \omega) & \text{if } e \notin S, \end{cases}$$

where S is Π^0_{2n+3}-complete. □

The theorem above skips some levels in the arithmetical hierarchy. We can also obtain index sets of other levels using Theorem 3.2.22 and Watnick's Theorem 3.2.23; we leave this to Exercise 7.1.11.

Superatomic Boolean algebras

A (computable) Boolean algebra is superatomic if it is the interval algebra of a (computable) ordinal; see §4.1.4. The theorem below also has a transfinite version, but we leave it to Exercise 8.1.28.

Theorem 7.1.7 (Knight). *Fix $n \in \mathbb{N}$, Then*

(i) Intalg(ω^{n+1}) is Δ^0_{2n+2}-categorical but not Δ^0_{2n+1}-categorical.

(ii) The index set $\{e : M_e \cong$ Intalg$(\omega^{n+1})\}$ is Π^0_{2n+3}-complete.

Sketch. We have calculated the complexity of some properties needed to establish the upper bounds in the proof of Feiner's Theorem 4.1.30. We also note that, in the proof of Theorem 7.1.6, if $e \notin S$ (where $S \in \Pi^0_{2n+3}$), then we construct a linear ordering of Hausdorff rank one higher than ω^{n+1}, and hence the corresponding Boolean algebra will not be isomorphic to Intalg(ω^{n+1}). This is because, for example, the Stone space will have a higher Cantor-Bendixson rank. We leave the details to Exercise 7.1.12. □

Torsion-free abelian groups of rank 1

We use the materials and notation introduced in §5.1.2, and specifically the classification of subgroups of $\mathbb{Q}$ by their types (Theorem 5.1.15). Recall that a torsion-free abelian group has rank 1 iff it is isomorphic to a (non-zero) subgroup of $\mathbb{Q}$. Let $\mathcal{K}_1$ be the class of torsion-free abelian groups having rank $\leqslant 1$. (We also include the rank-zero trivial group $\{0\}$ in the class for convenience; this won't make any difference.) The following elementary (but neat) proposition is due to Calvert [73], and in the slightly stronger form given in Corollary 7.1.9, it appeared in [132].

Proposition 7.1.8. *1. The index set of $\mathcal{K}_1$ is Π^0_2-complete.*

2. The isomorphism problem for $\mathcal{K}_1$ is Σ^0_3-complete.

Proof. Being total and torsion-free abelian is Π^0_2. The rank is > 1 iff there exists $a, b \in G$ such that for all $m, n \in \mathbb{N}$,

$$ma + nb = 0 \rightarrow m = n = 0,$$

which is Σ^0_2. (The Π^0_2-completeness follows vacuously from the Π^0_2-completeness of Tot. Alternatively, simply stop building a copy of $\mathbb{Z}$ if Π^0_2 no longer fires.)

The upper bound is Σ^0_3 because any two groups in the class are isomorphic if, and only if, they are computably isomorphic. The Σ^0_3-completeness follows from Theorem 3.1.8(iii) combined with Theorem 5.1.15, but we outline a direct proof below.

Fix a Σ^0_3-predicate P and represent it as $\exists x \exists^\infty y R(x, y, i)$, where R is computable. Produce a computable group G_i, realised as a c.e. subset of $\mathbb{Q}$ that contains 1, with the following properties.

$$P(i) \text{ fails} \implies \mathbf{t}(G_i) \ni (1, 1, 1, 1, \ldots, 1, 1, \ldots)$$

and

$$P(i) \text{ holds} \implies \mathbf{t}(G_i) \not\ni (1, 1, 1, 1, \ldots, 1, 1, \ldots),$$

where the latter is achieved by keeping 1 non-divisible by infinitely many primes. If we succeed, Theorem 5.1.15 will guarantee Σ^0_3-completeness. Indeed, if H is some fixed computable presentation of the group having the same type as $(1, 1, 1, 1, \ldots, 1, 1, \ldots)$, then simply consider pairs (H, G_i).

We now explain the construction of G_i. We begin with $1 \in G_i$, and initially make progress in building $G_i \cong \mathbb{Z}$. Each x in $\exists x \exists^\infty y R(x, y, i)$ controls a marker which occupies some prime p_i. Initially, x occupies p_x. The goal of the marker is to keep 1 non-divisible by the p_i that it occupies.

If x "fires", all markers $y \geqslant x$ move. We move x to the next available p_j (currently occupied by $x+1$), and we move $x+1$ also either to the position of $x+2$ (if it is defined), and so on. The largest marker defined at the stage is moved to the next available prime larger than all primes occupied by any markers. We also introduce one more marker and place it on the next prime. We declare 1 divisible by the prime p_i (once), where p_i was the prime that previously carried x. If $P(i)$ fails, then every marker eventually settles, making 1 non-divisible by infinitely many primes. Otherwise, some marker will keep moving, making 1 divisible by almost all primes. □

Notice that the Σ^0_3-completeness is achieved using just one group. This is the group of type $(1, 1, 1, \ldots, 1, 1, \ldots)$.

Corollary 7.1.9 (Downey, Kach, Lempp, and Turetsky [132]). *There exists a computably categorical group H such that $I(H) = \{e : M_e \cong H\}$ is Σ^0_3-complete.*

We remark that essentially all completeness estimates we've seen have had a slightly stronger feature. For example, in the corollary above, the index set $I(H)$ was Σ^0_3-complete *within the class of rank 1 torsion-free abelian groups*, meaning that under both Σ^0_3- and Π^0_3-outcomes, we can produce a rank 1 torsion-free abelian group. As far as we know, the notion of "completeness within" for structures was coined by Calvert. Such results are slightly more insightful in the rare cases when the complexity of the index set of a class is not less complex than the (crude) estimate for the isomorphism problem in the class. We, however, usually won't concern ourselves too much with this minutiae, and will just aim for whatever is easier to prove.

Using other presentations*

We remark that in all results above we used computable presentations $(M_e)_{e \in \mathbb{N}}$, but we could have used c.e. presentations instead. Another possibility is to use computable presentations with some additional properties, e.g., vector spaces or abelian groups with a linear independence algorithm.

However, c.e. presentations and computable presentations are only one Turing jump apart, and the general intuition is that, at least in common classes, the complexity estimates for index sets will also be at most one quantifier apart. We cite [47] for many examples comparing index set complexity of various properties of groups using c.e., computable, and finite presentations; all these results confirm our intuition. Recall also that we are aiming at establishing *arithmetical* estimates, and typically it makes almost no difference which presentations we use. Indeed, we could perhaps use n-decidable or Δ^0_n presentations for some fixed $n \in \mathbb{N}$, and we would likely still get arithmetical estimates. (Of course, decidable presentations do not fall into this pattern.) In this sense, the index set approach is quite robust. Using computable presentations seems more convenient simply because it is usually easier to prove completeness results and because typically more is known about computable members of the class.

But, of course, we know that c.e. presentations differ from computable presentations in general (e.g., Theorem A in Part 1), so we must be careful. Pathological examples can be rather complex, and it is perhaps natural to wonder how complex they can be *exactly*. Among computably presented structures, the index sets of decidable, 1-decidable, automatic, primitive recursive, and polynomial-time presentable structures are all Σ^1_1-complete [237, 34] (see Exercise 8.1.42). Using a modification of the main construction in [34], one can show that this index set of computably presentable structures among all c.e. presented structures is Σ^1_1-complete as well (Exercise 8.1.43). We will discuss Σ^1_1-completeness in the next chapter.

In view of Feiner's Theorem and the effective Stone duality established in Part 1, we are especially interested in the complexity of this index set restricted to Boolean algebras.

Question 7.1.10. *What is the complexity of the index set of computably presentable Boolean algebras among all c.e. presented structures?*

The proof of Feiner's Theorem 4.1.30 can be easily modified to show that the index set in the question above is not arithmetical (i.e., not Σ^0_n for any n); this is left as Exercise 7.1.15. However, nothing is known beyond this observation.

Exercises

See also Exercises 9.3.59-9.3.61 for the index sets of subclasses of abelian p-groups, and see Exercise 10.1.31 for the index sets of decidably categorical structures.

Exercise° 7.1.11. Calculate the index sets of the orders of the form $\mathbb{Z}^{n+1}$, $(\mathbb{Q}+2+\mathbb{Q})^{n+1}$, $(\mathbb{Q}+2+\mathbb{Q})^{n+1}\mathbb{Z}^{m+1}$, and $\mathbb{Z}^{m+1}(\mathbb{Q}+2+\mathbb{Q})^{n+1}$; $m, n \in \mathbb{N}$.

Exercise° 7.1.12. Give details for Theorem 7.1.7.

Exercise° 7.1.13. Prove Theorem 7.1.3 for vector spaces over any fixed infinite computable field. (Note that a slight modification to the initialisation strategy is necessary in the case when the characteristic is finite.)

Exercise° 7.1.14. Prove Theorem 7.1.4.

Exercise 7.1.15. Show that the index set of c.e. presented Boolean algebras that admit a computable copy is not arithmetical. Conclude that the index sets of compact right-c.e. Polish spaces that admit a computable Polish presentation is also not arithmetical. (Hint: Use a uniform modification of the proof of Feiner's Theorem 4.1.30.)

Exercise 7.1.16 (Calvert [72]). Show that the isomorphism problem for the class of computable algebraically closed fields of any fixed characteristic is Π^0_3-complete.

Exercise 7.1.17 (Calvert [72]). Show that for the class of Archimedean real closed fields (i.e., those that have no "infinitely large" elements; equivalently, those embeddable into $\mathbb{R}$), the isomorphism problem is Π^0_3-complete.

Exercise 7.1.18 (Calvert, Harizanov, Knight, and Miller [76]). Define the class d-Σ^0_n to consist of all sets of the form $X\backslash Y$, where $X, Y \in \Sigma^0_n$. Let A be a computable Archimedean real-closed ordered field (see the previous exercise).

1. If the transcendence degree is 0 (i.e., A is isomorphic to the ordered field of algebraic reals), then $I(A) = \{i : A_i \cong A\}$ is Π^0_2-complete (within the class of Archimedean real-closed ordered fields).
2. If the transcendence degree is finite but not 0, then $I(A)$ is m-complete in the class d-Σ^0_2 (within the class of Archimedean real-closed ordered fields).
3. If the transcendence degree is infinite, then $I(A)$ is Π^0_3-complete (within the class of Archimedean real-closed ordered fields).

Exercise* 7.1.19 (White [509]; Pavlovsky [428]). Define the class $\Sigma^0_{\omega+2}$ to be the collection of sets that are Σ^0_3 relative to $\varnothing^{(\omega)}$.

1. Show that the index set of computable prime models is $\Sigma^0_{\omega+2}$-complete.
2. Show that the index set of computable homogeneous models is $\Sigma^0_{\omega+2}$-complete.

Exercise 7.1.20** (Carson et al. [80], McCoy and Wallbaum [358]). Let F_ω denote the free (non-abelian) group of rank ω. Show that the index set $\{i : G_i \cong F_\omega\}$ is Π^0_4-complete.

Exercise 7.1.21** (Boone and Rogers [52]). Show that the collection of all finite presentations having a decidable word problem forms a Σ^0_3-complete set[1].

[1] While the proof is not particularly difficult, it relies on techniques that are not covered in the book.

7.1.2 Compact spaces and groups

Fix an effective listing $(M_i)_{i\in\mathbb{N}}$ of all (partial) computable Polish spaces. Each such M_i is given by a dense sequence that can be identified with ω and a (partial) computable metric on it. We could list all partial computably compact spaces in a similar way; we will discuss this approach a bit later. The estimates that we would obtain in these two different frameworks would be at most one jump apart, and it will make little difference to us. However, it is not entirely obvious that in the case of compact Polish spaces we indeed get this robustness of the index set approach; establishing this foundation is one of the main goals of this subsection. As the main result of the subsection, we will prove that *the index set of solenoid groups is arithmetical.*

The space of isometries

In this subsection, we present sharp arithmetical calculations for the index sets of compact spaces up to isometry. For that, we need to accumulate enough information about computably isometric compact spaces.

An isometry is a metric-preserving map. It is clearly continuous and is always injective. If an isometry is surjective, then we say that it is an isometric isomorphism. Using a brute-force search, we can easily show that the inverse of a computable isometric isomorphism is always a computable map even if the spaces are not computably compact. In particular, we do not need to refer to Theorem 4.2.57 to compute the inverse of an isometric isomorphism. Further, to compute an isometry f between two computable Polish spaces X and Y, it is sufficient to uniformly compute the isometry on the special points of X. Indeed, if $(x_i)_{i\in\mathbb{N}}$ is a fast Cauchy name of a point x in X, then $(f(x_i))_{i\in\mathbb{N}}$ is also fast Cauchy in Y. We can use the computation of $f(x_i)$ to find a special $y_i \in Y$ so that $d(y_i, f(x_i)) < 2^{-i}$ and conclude that $(y_i)_{i\in\mathbb{N}}$ converges to $f(x)$ quickly. We write $X \cong_{iso} Y$ to mean that X and Y are isometrically isomorphic.

Theorem 7.1.22. *For computably compact metric spaces X, Y, $X \cong_{iso} Y$ is a Π^0_1-property (of the indices of X and Y).*

Proof. We modify the proof of an unpublished result of Nies and Melnikov stated in [381] (see Section 4.2 in [139] for a slightly different proof).

Proposition 7.1.23. *There is an informally effective procedure which, given two computably compact spaces X and Y, outputs a computably compact space $F(X,Y)$ and an index of an effectively closed subset $I(X,Y) \subseteq F(X,Y)$ whose members are exactly the (codes for) isometries $X \to Y$.*

Indeed, the computable compactness of X can be dropped. Also, $F(X,Y)$ is just a convenient computably compact presentation of 2^ω which, in view of Theorem 4.2.84 (which is uniform in the case of 2^ω), can be identified with 2^ω.

Proof of Proposition 7.1.23. Suppose the special points of Y are given by the sequence $(r_i)_{i\in\mathbb{N}}$, and let $(p_i)_{i\in\mathbb{N}}$ be the dense computable sequence in X. Using the computable compactness of Y, for each n, fix a finite 2^{-n}-cover of the space. Define the space F_n of all finite partial maps which assign each of the first n special points in X to one of the centres of the finitely many 2^{-n}-balls covering Y. It is computably compact under the discrete metric $d(a,b) = 1$ iff $a \neq b$. Define

$$F(X,Y) = \prod_{n \in \mathbb{N}} F_n,$$

which is also computably compact (Proposition 4.2.17). Each element of $F(X,Y)$ is a string of finite tuples

$$(\sigma_1, \sigma_2, \ldots),$$

where $\sigma_n = \langle r_{j_0}, r_{j_1}, \ldots, r_{j_{n-1}} \rangle$ is a tuple of special points in Y. These points are the images of the first n special points $p_0, \ldots, p_{n-1} \in X$, respectively, under the partial map coded by σ_n. To isolate the elements that code isometries, consider the following conditions:

1. $|d_Y(r_{j_i}, r_{j_k}) - d_X(p_i, p_k)| \leqslant 2^{-n+1}$ for each $i < k < n$,
2. $d_Y(\sigma_n, \sigma_{n+1} \restriction n) \leqslant 2^{-n}$,

where the distance between strings of equal length is the supremum of the distances of the respective coordinates. These conditions are Π^0_1 and define an effectively closed subset $I(X,Y)$ of $F(X,Y)$. Clearly, every point in $I(X,Y)$ codes an isometry. Further, π is an isometry iff the sequence $(\pi_n)_{n \in \mathbb{N}}$ lies in $I(X,Y)$, and this correspondence is computably uniform. □

We note that $\pi \in Iso(X,Y)$ can have more than one "name" in $I(X,Y)$. (It is Π^0_1 to tell whether two members of $I(X,Y)$ code the same isometry.)

Claim 7.1.24. *Suppose X is computably compact. For an isometry $f\colon X \to Y$, "being onto" is Π^0_1 relative to f.*

Proof. We have that $f(X)$ is compact and thus closed; therefore, f is not onto iff

$$\exists i \, d_{inf}(r_i, f(X)) = \inf_{y \in f(X)} d(r_i, y) > 0,$$

where the space $f(X)$ is computably compact relative to f by Lemma 4.2.55. In particular, the inf-distance to $f(X)$ is f-computable, by Exercise 4.2.70 (also relativised). This makes "f being not onto" Σ^0_1 relative to f. □

We return to the proof of Theorem 7.1.22. If a counter-example is found for $\pi \in I(X,Y)$ witnessing that (f represented by) π is not onto, then the non-surjectivity is witnessed by an "initial segment" of π, i.e., by some clopen set in $F(X,Y)$ containing π. By restricting $I(X,Y)$ to the effectively closed set $I^*(X,Y)$ of surjective isometries, we see that $X \cong_{iso} Y$ iff $I^*(X,Y) \neq \varnothing$. Since $F(X,Y)$ is computably compact, if $I^*(X,Y) = \varnothing$, then it will be effectively recognised at a finite stage: just wait for the complement of $I^*(X,Y)$ to cover $F(X,Y)$. It follows that $X \cong_{iso} Y$ is Π^0_1 in the indices of X and Y. Theorem 7.1.22 is proved. □

In the proof of the theorem above, we established that $Iso(X,Y)$ can be represented as a Π^0_1-class. Even though this representation does not have to be 1-1, we can now appeal to facts about Π^0_1 classes and effectively closed sets. By The Low Basis Theorem 4.2.47, we obtain:

Corollary 7.1.25. *For a computably compact space X, if $Y \cong_{iso} X$, then there is a low isometric isomorphism witnessing this*[2].

Of course, self-isometries of a compact Polish space form a topological group which (as we have essentially established above) is compact. In contrast with Stone spaces, a compact topological group cannot have isolated points unless it is discrete and, thus, finite; this is because translations by elements are self-homeomorphisms of the group. If the identity is isolated, then all points are isolated too, and if one point is not, then all points are not. In particular, a compact space has either finitely many or uncountably many self-isometries.

Corollary 7.1.26 (Iljazović [268]). *Suppose a computably compact X has only finitely many self-isometries. If for some computable Polish Y we have $Y \cong_{iso} X$, then there is a computable isometric isomorphism witnessing this.*

Proof. We use the notation from the proof of Theorem 7.1.22 throughout. We know that $Iso(X, Y)$ must contain an isolated point Θ. It can be tempting to refer to Fact 4.2.45 to establish the corollary. However, if $F : I^*(X, Y) \to Iso(X, Y)$ is the computable functional associating members of $I^*(X, Y)$ with isometries, then $F^{-1}(\Theta)$ does not have to be a singleton. (We remark that $I^*(X, X) = I(X, X)$ since every self-isometry of a compact space is surjective.)

Since every isometry is clearly continuous, we can view every isometry as an element of $C(X, Y)$ under the supremum metric

$$d_{sup}(f, g) = \sup_{x \in X} d(f(x), g(x)).$$

Take any $U \subseteq C(X, Y)$ that contains a unique member $\Theta \in Iso(X, Y)$.

Fix an n so that 2^{-n} is smaller than the diameter of U. Fix a sufficiently long $\sigma \in F(X, Y)$ that:

1. σ is extendible to an infinite path in $I^*(X, Y)$, and

2. the F-images of all its extensions in $I^*(X, Y)$ lie in U.

Note that for any such extension $\pi \in I^*(X, Y)$, we must have $F(\pi) = \Theta$.

Given $m > n$, wait for a late enough stage s so that all extensions of σ that have not yet been declared out of $I^*(X, Y)$ have their potential F-images at a distance of at most 2^{-m} from each other, in the sense of (2) of the proof of Theorem 7.1.22.

Since Θ is isolated in U and $I^*(X, Y)$ is a Π^0_1 class, it follows that such a stage must exist. Choose any (finite initial segment of) such ρ_m extending σ that has not yet been declared out of $I^*(X, Y)$. It determines a finite partial map that can be used to calculate Θ to precision 2^{-m+1}. It follows that Θ is computable. □

For example, geometric simplices are isometrically computably categorical with respect to computably compact presentations.

We are now ready to apply these techniques to calculate the complexity of the classification problem for compact spaces, up to isometry. Recall that $(M_i)_{i \in \mathbb{N}}$ is the uniform enumeration of all (partial) computable Polish spaces. We identify M_i with its completion, and we write $M_i \cong_{iso} M_j$ to mean that (the completions of) M_i and M_j are isometrically isomorphic.

[2] We do not assume that Y is computably compact, because this is guaranteed by Exercise 4.2.69. The same comment applies to the next corollary too.

Corollary 7.1.27 (Melnikov and Nies [381]). *1. The recognition problem for compact spaces*

$$I_{comp} = \{i : M_i \text{ is compact}\}$$

is Π^0_3-complete.

2. The isometric isomorphism problem for compact spaces

$$E_{iso} = \{\langle i, j\rangle : M_i \cong_{iso} M_j \text{ \& } M_i, M_j \text{ are compact}\}$$

is Π^0_3-complete[3].

Proof. For (1), say that the metric is total, is indeed a metric, and for every n there is a 2^{-n}-cover of the space by *closed* basic balls $D_0, \ldots, D_k$. The latter is the same as to say that for every special point x, $x \in D_0 \cup \ldots \cup D_k$; this is because the complement of $D_0 \cup \ldots \cup D_k$ is open. Since $x \in D_i$ is a Π^0_1-property, the overall complexity of

$$\forall n\, \exists k\, \forall x \,\text{special}\quad x \in D_0 \cup \ldots \cup D_k$$

is Π^0_3. The Π^0_3-completeness can be witnessed by closed c.e. subspaces of the standard computable copy of ω^ω; see Exercise 7.1.42.

For (2), recall that every compact computable Polish space is computably compact relative to $\mathbf{0}'$, by Fact 4.2.11. Thus, by Theorem 7.1.22, given $i, j \in I_{comp}$, it is Π^0_2 in i, j to tell that M_i and M_j are isometrically isomorphic. The Π^0_3-completeness of E_{iso} follows from the Π^0_3-completeness of I_{comp} vacuously; simply consider pairs $(C_i, C_i)_{i\in\mathbb{N}}$ for the sequence $(C_i)_{i\in\mathbb{N}}$ witnessing the Π^0_3-completeness of I_{comp}. (To see that the upper estimate Π^0_2 is optimal *provided that* M_i, M_j are both compact, fix some infinite compact C. Produce a finite subspace of C in the Σ^0_2-outcome, and C in the Π^0_2-outcome.) □

Compact Polish groups

In this subsection, we measure the complexity of the index sets of profinite and compact connected groups. The first step is to measure the complexity of being a compact group.

Fix an effective listing $G_0, G_1, \ldots$ of (partially) computable Polish spaces in which every (perhaps, partial) computable Polish space G_i is additionally equipped with a pair of c.e. sets that are interpreted as names of (partial, potential) group operations on G_i. We usually identify G_i with its completion $\overline{G_i}$. Recall Definition 4.2.8:

Definition 7.1.28. A function $f\colon G \to M$ is *effectively continuous* if there is a c.e. family F of pairs (D, E) of (indices of) basic open balls such that:

(C1) for every $(D, E) \in F$, we have $f(D) \subseteq E$;

(C2) for every $x \in G$ and every basic open $E \ni f(x)$, there exists a basic open D with $(D, E) \in F$ and $x \in D$.

For technical convenience, we will use the following uniform variation of Definition 7.1.28. If B is a basic open ball, let B^c denote the basic closed ball having the same radius $r(B)$ and centre $c(B)$ as B.

[3] Indeed, it is Π^0_3-complete within the Π^0_3-complete set I_{comp}, meaning that its Π^0_3-completeness can be witnessed using only indices from I_{comp} for both the Π^0_2 and the Σ^0_2-outcomes.

Definition 7.1.29. Let f be a continuous function between Polish metric spaces M and N. A $*$-name of f is any collection of pairs of basic open balls (B, C) such that $f(B) \subseteq C^c$, and for every $x \in M$ and every $\epsilon > 0$, there exists $(B, C) \in \Psi$ such that $B \ni x$ and $r(C) < \epsilon$.

We can uniformly pass from a $*$-name of f (Definition 7.1.29) to a name of f (Definition 7.1.28) and back, as follows. Suppose Ψ is a name of f. Since $f(B) \subset C$ implies $f(B) \subset C^c$, every $*$-name is a name. Now suppose Ψ is a $*$-name of f. Using $\epsilon/2$ instead of ϵ in Definition 7.1.29, fix (B, C) with $r(C) < \epsilon/2$ such that $x \in B$ and $f(B) \subset C^c$. Replace C with an equicentric $C' \supset C$ such that $r(C) < r(C') < \epsilon$. We have $f(B) \subset C^c \subseteq C'$ and $r(C') < \epsilon$.

The uniform procedure of passing from a name to a $*$-name can be applied to any c.e. set W which does not even have to be a name of any function. We denote the resulting c.e. set by W^*. Then W is a name iff W^* is a $*$-name (of the same function).

Having this uniform correspondence in mind, without loss of generality, *we may always assume that any (partial) computable group is given by the $*$-names of its potential operations.* We shall follow this convention throughout the rest of this subsection.

Proposition 7.1.30. *The index set $CPGr = \{i : G_i$ is a compact Polish group$\}$ is Π^0_3-complete.*

Proof of Proposition 7.1.30. We are given a triple $G_i = (G, W, U)$, where G is a (partial) computable Polish space and W, U are c.e. sets interpreted as potential $*$-names. We need to check whether G is compact and W and U are names of computable group operations on G. By Corollary 7.1.27(1), it is Π^0_3 to tell whether G is compact Polish.

Lemma 7.1.31. *Let G and M be computable Polish spaces that are also compact. Then*

$$\{e : W_e \text{ is a } * \text{-name of a computable } f : G \to M\}$$

is Π^0_3, uniformly in G and M.

Proof. Fix a c.e. set Ψ and interpret it as a set of pairs of basic open balls with rational radii:

$$\Psi = \{(C, B) : C, B \text{ basic open in } G, M \text{ respectively}\}.$$

To ensure that Ψ is a $*$-name of a computable operation, we require that Ψ additionally satisfies:

1. For every $(B_0, C_0), \ldots, (B_n, C_n) \in \Psi$, $\bigcap_i B_i \neq \varnothing$ implies $\bigcap_i C_i \neq \varnothing$.

2. For each rational $\epsilon > 0$, there exists a finite cover $B_0, \ldots, B_k$ of G and $(B_0, C_0), \ldots, (B_k, C_k) \in \Psi$ such that $r(C_i) < \epsilon$ for $i = 1, \ldots, k$.

We first check that (1) and (2) are (at most) Π^0_3, and then we prove that they capture the property of being a $*$-name.

Claim 7.1.32. *The properties (1) and (2) are Π^0_3.*

Proof of Claim. Since Ψ is c.e. and the intersection of open sets must be witnessed by special points from the respective computable structures, it is clear that (1) is Π^0_2. To see why (2) is Π^0_3, recall that every compact Polish space is $0'$-computably compact. Thus, "being a finite cover" is Σ^0_2. Therefore, (2) is Π^0_3. □

Clearly, if Ψ is a $*$-name of a computable operation $f : G \to M$, then Ψ satisfies (1) and (2) (recall G is compact).

Claim 7.1.33. *If Ψ satisfies (1) and (2), then it is a $*$-name of a computable operation.*

Proof. We define a map ψ as follows. For every $x \in G$, choose $(B, C) \in \Psi$ such that $x \in B$ and declare C^c *a (closed) Ψ-neighbourhood of $\psi(x)$*. (Note that if x is a computable point, then this process is effective.) Set $\psi(x)$ to be equal to any point in the intersection

$$\bigcap \{C : C \text{ is a } \Psi\text{-neighbourhood of } \psi(x)\}.$$

Property (1) implies that any two Ψ-neighbourhoods of $\psi(x)$ have a non-empty intersection. Let C_n be the intersection of the first n Ψ-neighbourhoods of $\psi(x)$ in any (not necessarily effective) list of such neighbourhoods. Then (C_n) is a nested sequence of non-empty compact sets, thus it has a non-empty intersection. Property (2) guarantees that for every ϵ there exists a Ψ-neighbourhood of $\psi(x)$ of size ϵ. Therefore, the intersection is a singleton. We conclude that ψ is a (total) function.

We claim that Ψ is a $*$-name for ψ. Property (2) implies that for every $\epsilon > 0$ there exists $(B, C) \in \Psi$ such that $B \ni x$ and $r(C) < \epsilon$. It remains to show that for each $(B, C) \in \Psi$ we have $\psi(B) \subseteq C^c$. Fix $x \in B$. Then

$$\{\psi(x)\} = \bigcap \{C : C \text{ is a } \Psi\text{-neighbourhood of } \psi(x)\}.$$

Since C is a Ψ-neighbourhood of $\psi(x)$, it follows that $\psi(x) \in C^c$. □

To finish the proof of the lemma, observe that our analysis was fully uniform in G and M. □

We return to the proof of Proposition 7.1.30. The product space $G \times G$ is compact. There is a uniform procedure that, given a computable Polish space G, outputs a computable presentation of $G \times G$. It follows from Lemma 7.1.31 that the index set of Polish spaces equipped with two well-defined computable operations is Π^0_3. To finish the proof of Proposition 7.1.30, recall that the group axioms are closed properties and thus can be checked only for special points (set $e = x \cdot x^{-1}$ for the first found x). The Π^0_3-completeness is Exercise 7.1.44. □

> **Theorem 7.1.34** (Melnikov [373]). *(1) For a computable Polish compact group, being connected is Π^0_2.*
>
> *(2) The index set of profinite Polish groups is Π^0_3-complete.*

Proof. Part (1) follows from Lemma 4.2.77 with $Z = \varnothing'$ because every compact computable Polish space is $\varnothing'$-computably compact. This upper bound is optimal; see Exercise 7.1.45.

We prove (2). We establish that the index set is Π^0_3; the Π^0_3-completeness is Exercise 7.1.45. It is well-known that a compact Polish group is profinite iff its neutral element 1_G has a basis consisting of normal clopen subgroups. Recall that a closed subgroup of a profinite group is itself profinite.

Consider the following procedure. Let $D_0 = G$. At stage $s > 0$, let D_s be the first found clopen normal subgroup such that the diameter of D_s is at most 2^{-s} and $D_s \subseteq D_{s-1}$ (if such D_s exists at all). To find a clopen subgroup, we use $\varnothing'$ to find a clopen split that works (Lemma 4.2.77). For that, search for finitely many closed balls witnessing that the group is disconnected, where the union of the first k of them together forms a normal subgroup. Since all involved sets are clopen, it suffices to check the inclusion, the diameter, normality, and the group operations only for special points. (Compare this to the proof of Theorem 4.2.107.)

It follows that $\varnothing'$ is capable of uniformly finding such a D_s (if it exists), and thus $\forall s(D_s$ is defined) is a Π^0_3-statement equivalent to profiniteness for a compact group G. □

Remark 7.1.35. The proof of 1. above of course also gives that being a connected compact space is arithmetical (Π^0_2).

Switching to computably compact presentations

In Part 1 of the book, we saw that computably compact presentations are more suited for compact Polish spaces than other effective presentations. Recall also that one of the many equivalent formulations of computable compactness involved covers by basic closed balls (Exercise 4.2.18).

Definition 7.1.36. A computably compact presentation is given by:

1. A computable Polish presentation M.
2. A (closed) modulus of compactness which, given n, outputs a finite tuple of *closed* basic balls of radii $\leqslant 2^{-n}$ that cover the space.

Every computable Polish space that is also compact is $\varnothing'$-compact. Also, being compact for a computable Polish space is arithmetical by Corollary 7.1.27. In view of the results established so far in this section (e.g., Proposition 7.1.30), if we only aim for arithmetical estimates and are only interested in compact objects, it makes sense to use computably compact presentations. It is immediate that some complexity estimates become one level lower if we use computably compact presentations; e.g., compare Corollaries 7.1.22 and 7.1.27.

It seems that switching between computable Polish and computably compact presentations usually does not make any difference, unless we worry about sharp estimates. But we also recognise that pathologies can be rather complex in general. The following question seems to be of considerable technical interest.

Question 7.1.37. *What is the complexity of the index set of computable Polish (compact) spaces that admit a computably compact presentation?*

Index sets of computably compact groups

Fix the enumeration of all (partial) *computably compact* groups $(G_i)_{i\in\mathbb{N}}$, where each computable Polish domain additionally comes with a (potential) computable modulus of continuity. In exchange, by Corollary 4.2.46, we can drop the (name for the potential) inverse operation and keep only the product.

Proposition 7.1.38. *With respect to computably compact presentations, the index set of compact groups is* Π^0_2.

Proof. This is a routine modification of Proposition 7.1.30. We only need to check the totality of the operation; compactness is automatic provided the Π^0_1 condition 2. of Definition 7.1.36 is satisfied. (We of course get Π^0_2-completeness since Tot can be easily coded into the index set.) □

The estimates for connectedness and being profinite in Theorem 7.1.34 become Π^0_1 and Π^0_2, respectively.

Proposition 7.1.39.
1. For a computably compact Polish group, being connected is Π^0_1.
2. The index set of profinite Polish groups with respect to computably compact presentations is Π^0_2*-complete.*

We leave the proof of Propositions 7.1.38 and 7.1.39 to Exercise 7.1.46. We illustrate the use of compact presentations on the index set of solenoid groups.

Solenoid groups

Recall that a class is arithmetical if both the recognition and the isomorphism problems are arithmetical for the class. Also, a solenoid group is the Pontryagin dual of a subgroup of $(\mathbb{Q}, +)$.

Remark 7.1.40. Why are such groups called "solenoids"? Every $H \leqq \mathbb{Q}$ can be represented as the direct limit of a sequence

$$\mathbb{Z} \rightarrow_{m_0} \mathbb{Z} \rightarrow_{m_1} \mathbb{Z} \rightarrow_{m_2} \mathbb{Z} \rightarrow_{m_3} \cdots,$$

where $\mathbb{Z} \rightarrow_{m_i} \mathbb{Z}$ maps $\mathbb{Z}$ isomorphically onto $m_i\mathbb{Z} \leqq \mathbb{Z}$. Under Pontryagin duality, $\widehat{H}$ is isomorphic to

$$\mathbb{T} \leftarrow_{m_0} \mathbb{T} \leftarrow_{m_1} \mathbb{T} \leftarrow_{m_2} \mathbb{T} \leftarrow_{m_3} \cdots,$$

where each $\mathbb{T} \leftarrow_{m_i} \mathbb{T}$ is just $m_i x \leftarrow x$. (These further properties of Pontryagin duality will be discussed in Section 9.5.1; in this informal explanation we take them for granted.) This map $\mathbb{T} \leftarrow_{m_i} \mathbb{T}$ can be visualised as $\mathbb{T}$ being "wrapped around" $\mathbb{T}$ exactly m_i times. As we zoom in, we see more and more copies of the unit circle wrapped around the first copy. This perhaps explains the name.

The theorem below was not stated explicitly in [341], but it follows easily from the methods developed in this paper (cf. [341, Corollary 1.4(3)]).

Theorem 7.1.41. *Solenoid groups form an arithmetical class.*

Proof. We use computably compact presentations. We will prove that, with respect to computably compact presentations, the recognition problem is Π^0_2-complete and the (topological) isomorphism problem is Σ^0_3-complete.

Propositions 7.1.38 and 7.1.39 guarantee that being a connected compact group is an arithmetical property (Π^0_2), and being abelian is Π^0_1 and can be checked only for special points. It is also Σ^0_1 to tell whether the group is non-zero. (If it is zero, we are done.) Given G that is already known to be non-zero and connected compact abelian, apply Theorem 5.2.21 (and Theorem 5.2.24) and produce a c.e. presentation of its discrete Pontryagin dual. Since G is non-zero, we can uniformly pass to its computable presentation using Khisamiev's Theorem 5.1.41. Let H be the resulting computable presentation of the dual. By Proposition 7.1.8, it is Π^0_2 to tell whether H has rank 1. This makes the index set of solenoid groups Π^0_2; and indeed, it is evidently Π^0_2-complete.

Given two groups, use the procedure described above to pass to their duals. (The case when at least one of them is the zero group is trivial and can be considered separately.) By Proposition 7.1.8, expressing that the duals are isomorphic is Σ^0_3. Further, the Σ^0_3-completeness for discrete groups established in Proposition 7.1.8 is witnessed by rank 1 groups, which of course come with a basis (being the element $1 \in \mathbb{Q}$, for example). Proposition 5.2.2 is uniform provided the discrete torsion-free groups are equipped with computable bases, as explained in Remark 5.2.8. This gives the Σ^0_3-completeness for their solenoid duals. □

Note that using computable Polish presentations would give estimates with one extra quantifier, but the estimates will of course remain arithmetical, and the result would still hold. The respective index sets must also be complete at their respective levels, but this we won't verify.

Exercises

Exercise° 7.1.42. Let P be a Π^0_3-predicate and identify ω^ω with its standard computable presentation by strings under the usual ultra-metric. Build a uniform sequence $(C_i)_{i \in \mathbb{N}}$ of c.e. closed subsets of ω^ω such that C_i is compact iff $P(i)$ holds.

Exercise° 7.1.43 (J. Miller (unpublished); see [139]). Let $A, B \subseteq \mathbb{N}$ be disjoint c.e. sets. There are isometric computably compact computable metric spaces L, R such that any representation of an isometry computes a set S such that $A \subseteq S$ and $B \cap S = \varnothing$.

Exercise° 7.1.44. Prove that the index set of compact Polish groups is Π^0_3-complete.

Exercise° 7.1.45. (1) Let P be a Π^0_2-predicate. Prove that there is a uniformly computable sequence of compact Polish groups $(H_i)_{i \in \mathbb{N}}$ so that

$$P(i) \text{ if and only if } H_i \text{ is connected.}$$

In other words, being connected is Π^0_2-complete within compact groups.

(2) Iterate the strategy from (1) above to prove that the index set of profinite groups is Π^0_3-complete within compact groups.

Exercise° 7.1.46. Prove Propositions 7.1.38 and 7.1.39.

Exercise 7.1.47 (Cenzer and Remmel [87]). Show that the index set of the computably bounded Π^0_1 classes in ω^ω is Σ^0_3-complete, and the index set of the computably bounded Π^0_1 classes which have infinitely many computable members is Π^0_4-complete.

Exercise 7.1.48 (Cenzer and Remmel [88]). Prove the following results about (Type II) computable functions $[0, 1] \to [0, 1]$:

1. The index set of functions having a computable derivative is Σ^0_3-complete.
2. The index set of functions having a continuous derivative is Π^0_3-complete[4].

3*. The index set of functions having a continuous derivative but no computable derivative is Π^0_3-complete.

Exercise* 7.1.49 (Qian [437]). Show that the index set of computable Banach spaces with computable Schauder bases (Exercise 2.4.41) is Σ^0_3-complete.

[4] We remark that Westrick [508] extended this result to establish sharp index set estimates for functions at arbitrary (computable) transfinite levels of the Kechris-Woodin differentiability hierarchy of functions.

Exercise* 7.1.50 (Xie [511]). Let $\mathbb{X}$ be a Banach space and $(x_i)_{i\in\mathbb{N}}$ be a Schauder basis of $\mathbb{X}$ (Exercise 2.4.41), with $\{S_i\}_{i\in\mathbb{N}}$ its associated sequence of projections. The *basis constant* of (x_i), denoted $\mathrm{bc}((x_i))$, is the value $\sup_i \|S_i\|$. The basis constant of the space $\mathbb{X}$, denoted $\mathrm{bc}(\mathbb{X})$, is the infimum of basis constants across all of its bases. A Banach space $\mathbb{X}$ is said to have the local basis structure (LBS) if there is some constant $K \in \mathbb{R}$ such that for any finite-dimensional subspace $\mathbb{B} \subseteq \mathbb{X}$, there exists a finite-dimensional space $\mathbb{L} \subseteq \mathbb{X}$ such that $\mathbb{B} \subseteq \mathbb{L}$ and $\mathrm{bc}(\mathbb{L}) \leqslant K$. Show that the index-set of computable Banach spaces with the local basis structure is Σ^0_3-complete.

Exercise 7.1.51** (Becher and Slaman [45]). Show that the set of indices for computable real numbers which are normal to at least one base is Σ^0_4-complete[5].

Exercise 7.1.52** (Harrison-Trainor and Melnikov [241]).

1. Show that every compact 2-surface is Δ^0_{26}-categorical (with respect to computable Polish presentations) up to homeomorphism[6].
2. Use the classification of all compact 2-surfaces to conclude that the index set and the homeomorphism problem of closed surfaces are arithmetical.

7.1.3 Index sets and Friedberg enumerations

Before we move on to more technical results, we also mention one elementary application of index sets to Friedberg enumerations. Recall that a Friedberg enumeration of a class K is a computable listing $\{V_e : e \in \mathbb{N}\}$ of all computable structures in K where every isomorphism type occurs precisely once. Whilst we treat this concept in more detail in Chapter 9, we are in a position to show that certain classes do not have Friedberg enumerations based on index sets.

The following observation is (essentially) due to Goncharov and Knight [213], but in this particular form was first stated in [330]. In the fact below, $\mathcal{K}$ can be either a class of algebraic structures or of topological objects.

Fact 7.1.53. *Fix a class $\mathcal{K}$ and a positive $n \in \mathbb{N}$. Suppose $I(\mathcal{K})$ is Δ^0_n and $I(\mathcal{K})$ is Σ^0_n-complete. Then $\mathcal{K}$ does not admit a Friedberg enumeration.*

Proof. Assume $(U_i)_{i\in\mathbb{N}}$ is a Friedberg enumeration of isomorphism types in $\mathcal{K}$. Let $(M_e)_{e\in\mathbb{N}}$ be the uniform enumeration of all structures in the language of $\mathcal{K}$ given by the universal Turing machine. There is a computable f such that $U_i = M_{f(i)}$. For every $m, n \in I(\mathcal{K})$, $M_m \not\cong M_n$ iff

$$\exists i \neq j \ (n, f(i)) \in E(\mathcal{K}) \ \wedge \ (m, f(j)) \in E(\mathcal{K}),$$

which is Σ^0_n. This makes $E(\mathcal{K})$ Δ^0_n, contrary to the hypothesis. $\square$

Of course, Σ^0_n-completeness in the fact above can be replaced with "Σ^0_n and not Δ^0_n". The following is a consequence of Proposition 7.1.8 and Fact 7.1.53.

Corollary 7.1.54 (Lange, Miller, Steiner [330]). *There is no Friedberg enumeration of all additive subgroups of $\mathbb{Q}$ up to isomorphism.*

The index set calculations in the proof of Theorem 7.1.41 imply:

---

[5]A real number is said to be normal to base b if, for every positive integer n, all possible strings of digits in base b of length n have asymptotic density b^{-n}. That is, all strings of digits occur with equal likelihood in the expansion of the real number to base b. We omit the formal definitions since they are irrelevant to the content of the book.

[6]It is perhaps not surprising that we do not know at present whether 26 is optimal.

Corollary 7.1.55. *There is no Friedberg enumeration of computably compact solenoid groups, up to topological group isomorphism.*

But of course, the sufficient uniformity in Effective Pontryagin Duality makes the corollaries above equivalent. Thus, using the proof of Theorem 7.1.41 was not really necessary.

Exercises

Exercise° 7.1.56. Prove that there is no Friedberg enumeration of all computably compact spaces up to isometry. (Give a direct diagonalisation argument.)

Exercise° 7.1.57. Prove that there is no Friedberg enumeration of all computably compact spaces up to computable homeomorphism.

Exercise° 7.1.58. A finite presentation (f.p.) of a group is a tuple

$$\langle x_1, \ldots, x_m | r_1, \ldots, r_m \rangle$$

where $x_1, \ldots, x_m$ are generators and $r_1, \ldots, r_m$ are relations upon the generators. A group is finitely presented (f.p.) if it has a finite presentation. Let $(F_i)_{i \in \mathbb{N}}$ be the uniform enumeration of all finite presentations given by their strong indices (i.e., as pairs of tuples). The *isomorphism problem* for f.p. groups is the collection $\{(i, j) : F_i \cong F_j\}$, where $F_i \cong F_j$ means that the groups represented by F_i and F_j are isomorphic. Take for granted that isomorphism of two finitely presented groups is an undecidable problem (Adyan [3, 4] and Rabin [438]).

(i) Show that the isomorphism problem for f.p. groups is Σ^0_1.

(ii) Conclude that there is no c.e. list of finite presentations (given by their strong indices) in which every f.p. group is mentioned exactly once, up to isomorphism.

7.2 Completely decomposable groups

In this section, we examine an important illustrative class of torsion-free abelian groups in much detail. Elements of the theory of computable torsion-free abelian groups were presented in Section 5.1. There, we proved Nurtazin's Theorem 5.1.43, which classified the computably categorical torsion-free abelian groups exactly as those having finite rank. In the present section, we will concentrate on a class of torsion-free abelian groups which are best understood *classically* beyond the rank one groups. For an important subclass of homogeneous completely decomposable groups, we will be able to completely describe the Δ^0_n-categorical members of the class, for all n. The techniques that we will accumulate in this section will be used in the next section to produce estimates for index sets.

In this section all our groups are discrete and at most countable.

Definition 7.2.1. A torsion-free abelian group is *completely decomposable* (c.d.) if it is isomorphic to

$$\bigoplus_{i \in I} H_i,$$

where H_i is a subgroup of the rationals $\mathbb{Q}$ under addition, for every $i \in I$. If all the elementary summands H_i are pairwise isomorphic, we say that the group is also homogeneous.

The subgroups H_i in $\bigoplus_{i \in I} H_i$ will be referred to as elementary (direct) components and elementary (direct) summands of the group. Algebraic properties of completely decomposable groups are quite well studied, especially in the countable case; see Fuchs [195]. Baer [26] was the first to systematically study this class of groups. He showed that, up to isomorphism, a countable completely decomposable group $\bigoplus_{i \in \mathbb{N}} H_i$ is fully determined by the isomorphism types of the elementary summands H_i, and that the complete decomposition is unique up to a permutation of the summands. Thus, the homogeneous case is completely described by the type (see Theorem 5.1.15) and the rank of the group. In this section we prove:

Theorem 7.2.2 (Downey and Melnikov [136]). *Every computable homogeneous completely decomposable group is Δ^0_3-categorical.*

We also completely describe Δ^0_2-categoricity in the class (Theorem 7.2.25). With enough machinery accumulated, we will prove the main result of this section:

Theorem 7.2.3 (Downey and Melnikov [137]). *Every computable completely decomposable group is Δ^0_5-categorical.*

As we will see, the theorem and its proof will allow us to give arithmetical estimates for the index set and the isomorphism problem of completely decomposable groups. To understand the effective categoricity of these groups, we will need both new uses of computability theory in the study of torsion-free abelian groups, and some new algebraic structure theory, as described in the next subsection.

7.2.1 Background, notation, and conventions

We refer the reader to Section 5.1 for the basic definitions of abelian group theory. We recall some of these definitions here. Also, we will slightly adjust our notation (e.g., Definition 7.2.5).

Recall that we write $k|g$ *in* G (or simply $k|g$ if it is clear from the context which group is considered) and say that k *divides* g *in* G if there exists an element $h \in G$ for which $kh = g$, and we say that h is a k*-root* of g. (The latter term is not standard in abelian group theory, but it will be convenient to us.) Clearly, $k|g$ is simply an abbreviation for the formula

$$(\exists h)(\underbrace{h + h + \ldots + h}_{h \text{ repeated } k \text{ times}} = g)$$

in the signature of abelian groups. If the group G is torsion-free then every $g \in G$ has at most one k-root, for every $k \neq 0$.

Definition 7.2.4. Suppose G is a torsion-free abelian group, g is an element of G, and $n|g$ some n. If $r = \frac{m}{n}$ then we denote by rg the (unique) element mh such that $nh = g$.

The definition below already appeared in §5.1.1, but we state it again here.

Definition 7.2.5 (Pure subgroups and [X]). A subgroup A of G is called *pure* if for every $a \in A$ and every n, $n|a$ in G implies $n|a$ in A. For any subset X of G we denote by $[X]$ the least pure subgroup of G that contains X to avoid conflict of notation (the pure closure of X in G).

The standard notations for the least pure subgroup of G containing X are $(X)^*_G$ and $(X)^*$ (when G is clear from the context); we used this notation in §5.1.1. However, in this section we will stick with $[X]$ which resembles the notation for the localisation of an integral domain. The reason we adjust our notation is to avoid conflict with the following convenient notation.

Notation 7.2.6. Let G be an abelian group and $A \subseteq G$. Suppose $\{r_a : a \in A\}$ is a set of (rational) indices. If we write $\sum_{a \in A} r_a a$ then we assume that $r_a a \neq 0$ for at most finitely many $a \in A$, and every element $r_a a$ is well-defined in G, according to Definition 7.2.4. We will use this convention without explicit reference to it. Now suppose $R \leqq \mathbb{Q}$, and $A \subseteq G$. We denote by $(A)_R$ the subgroup of G (if this subgroup exists) generated by $A \subseteq G$ over $R \leqq Q$, i.e.

$$(A)_R = \{\sum_{a \in A} r_a a : r_a \in R\}.$$

Finally, for $R \leqq Q$ and $a \in G$, we denote by Ra the subgroup $(\{a\})_R$ of G.

Let $R \leqq \mathbb{Q}$. If a set $A \leqq G$ is linearly independent then every element of $(A)_R$ has the unique presentation $\sum_{a \in A} r_a a$. Therefore, $(A)_R = \bigoplus_{a \in A} Ra$ for every linearly independent set A.

Computable homogeneous c.d. groups

We fix a computable presentation of the rationals $(\mathbb{Q}, +, \times)$. This structure is computably categorical, and thus we do not need to be more specific in our description of this copy. Recall Theorem 5.1.16:

Theorem 7.2.7 (Mal'cev [346]). *Let G be a torsion-free abelian group of rank 1. Then the following are equivalent:*

(1) The group G has a computable presentation.

(2) The type $\mathbf{t}(G)$ is c.e.

(3) The group G is isomorphic to a c.e. additive subgroup R of $(\mathbb{Q}, +)$. Furthermore, we may assume that $1 \in R$.

The (1) $\leftrightarrow$ (2) part of Theorem 7.2.7 can be easily generalised to the class of homogeneous completely decomposable groups:

Proposition 7.2.8. *A homogeneous completely decomposable group G has a computable presentation iff $\mathbf{t}(G)$ is c.e..*

Computable modules and categoricity. The definition of G_P.

We omit the standard definition of a module over a ring, but we recall that this is essentially a vector space, except the "scalars" range over a ring rather than over a field. We say that C is a computable presentation of a module M over a fixed computable presentation of a ring R if C is a computable presentation of M as an abelian group and the operation $\cdot : R \times C \to C$ is computable.

Fix a set of primes P. Let $\mathbb{Q}^{(P)}$ be the subgroup of the rationals $(\mathbb{Q}, +)$ generated by the set of fractions $\{\frac{1}{p^k} : k \in \mathbb{N} \text{ and } p \in P\}$, and let $G_P = \bigoplus_{i\in\mathbb{N}} \mathbb{Q}^{(P)}$, i.e., the direct sum of ω copies of $\mathbb{Q}^{(P)}$. Of course, $\mathbb{Q}^{(P)}$ can also be viewed as a ring. The proof of the following two facts are left as exercises.

Proposition 7.2.9. *The following are equivalent:*

1. *P is c.e.*
2. *$\mathbb{Q}^{(P)}$ is a c.e. subring of a computable presentation of $(\mathbb{Q}, +, \times)$.*
3. *G_P is computably presentable as an abelian group.*
4. *G_P is computably presentable as a module over $\mathbb{Q}^{(P)}$.*

Lemma 7.2.10. *For a c.e. set of primes P, the following are equivalent:*

1. *Every computable presentation of the group G_P has a Σ^0_n basis which generates this presentation as a module over $\mathbb{Q}^{(P)}$.*
2. *The group G_P is Δ^0_n-categorical.*
3. *The $\mathbb{Q}^{(P)}$-module G_P is Δ^0_n-categorical.*

Thus, from the computability-theoretic point of view, G_P may be alternatively considered as an abelian group or a $\mathbb{Q}^{(P)}$-module.

Exercises

Exercise° 7.2.11. Prove Proposition 7.2.9 and Lemma 7.2.10.

7.2.2 S-independence and excellent S-bases

The notion of p-independence (for a single prime p) is a fundamental concept in abelian group theory (see [194], Chapter VI). We introduce a certain generalisation of p-independence to *sets* of primes. The notion of S-independence below can also be viewed as a restriction of the notion of independence used in Pontryagin's freeness criterion (it can be found in [192]). It states that a countable abelian group is free if, and only if, it is the union of a countable chain of pure free groups of finite rank. To obtain Pontryagin's notion, just let S in Definition 7.2.12 below be the set of all primes.

Definition 7.2.12 (S-independence and excellent bases)**.** Let S be a set of primes, and let G be a torsion-free abelian group. If $S \neq \varnothing$, then we say that elements $b_1, \ldots, b_k$ of G are S-independent in G if

$$p \mid \sum_{i \in \{1,\ldots,k\}} m_i b_i$$

in G implies

$$\bigwedge_{i \in \{1,\ldots,k\}} p \mid m_i,$$

for all integers $m_1, \ldots, m_k$ and any $p \in S$. If $S = \varnothing$, then we say that elements are S-independent if they are simply linearly independent. Every maximal S-independent subset of G is said to be an S-*basis* of G. We say that an S-basis is *excellent* if it is a maximal linearly independent subset of G.

It is easy to check that S-independence in general implies linear independence. However, an S-basis does not have to be excellent; we leave this to Exercise 7.2.24.

Since "p-independence" and "P-independence" sound exactly the same, and since P can be confused with the set of all primes (rather than interpreted as some set of primes), we chose to use S instead of P in the definition above. Another reason is given in Notation 7.2.13 below.

Notation 7.2.13. In this section P stands for a set of primes (which is not necessarily the set of all prims) and $\tilde{P}$ for the complement of P within the set of *all* primes:

$$\tilde{P} = \{p : p \text{ is prime and } p \notin P\}.$$

The set $\tilde{P}$ will typically serve as the set S in "S-independence" throughout the rest of the chapter. We also will often consider the group

$$G_P \cong \bigoplus_{i \in I} \mathbb{Q}^{(P)},$$

where $\tilde{P}$-independence will play a special role.

Lemma 7.2.14. *Fix $B \subseteq G_P \cong \bigoplus_{i \in I} \mathbb{Q}^{(P)}$. Then B is an excellent $\tilde{P}$-basis of G_P iff $G = \bigoplus_{b \in B} \mathbb{Q}^{(P)} b$.*

Before we prove the lemma, we discuss the extreme cases. Let $\mathcal{P}$ be the set of *all* primes. Then $\tilde{\mathcal{P}} = \varnothing$. Recall that $\varnothing$-independence is simply linear independence, and $G_{\mathcal{P}} \cong \bigoplus_{i \in \mathbb{N}} \mathbb{Q}$. It is well-known that every maximal linearly independent set generates the vector space over $\mathbb{Q}$. If $P = \varnothing$ then $G_{\varnothing} \cong \bigoplus_{i \in \mathbb{N}} \mathbb{Z}$ is the free abelian group of the rank ω. As a consequence of the lemma, every excellent $\mathcal{P}$-basis of $G_{\varnothing}$ generates it as a free abelian group; this is a reformulation of the Pontryagin's criterion that we mentioned above.

Proof. ($\Rightarrow$). We assume that $P \neq \varnothing$ throughout, and we write G for G_P. Let B be an excellent $\tilde{P}$-basis of $G = G_P$. Suppose $g \in G$. By our assumption, B is a basis of G. Therefore, there exist integers m and m_b, $b \in B$, such that $mg = \sum_b m_b b$. Suppose $m = pm'$ for some $p \in \tilde{P}$. By Definition 7.2.12, $p | m_b$ for all $b \in B$. Therefore, without loss of generality, we can assume that $(m, p) = 1$, for every $p \in \tilde{P}$. By the definition of G, we have:

$$g = \sum_b \frac{m_b}{m} b \in (B)_{\mathbb{Q}^{(P)}} \leqq G.$$

The set B is linearly independent, therefore $(B)_{\mathbb{Q}^{(P)}} = \bigoplus_{b \in B} \mathbb{Q}^{(P)} b$ (see the discussion after Notation 7.2.6). We have $g \in (B)_{\mathbb{Q}^{(P)}} \leqq G$ for every $g \in G$. Thus, $G = (B)_{\mathbb{Q}^{(P)}}$.

($\Leftarrow$). Let $G = \bigoplus_{b \in B} \mathbb{Q}^{(P)} b$ for $B \subseteq G$, and $ph = \sum_{b \in B} m_b b$, where m_b is integer for every $b \in B$, and $p \in \tilde{P}$. We have $h \in G_P$ and thus $h = \sum_{b \in B} h_b$, where $h_b \in \mathbb{Q}^{(P)} b$ for each $b \in B$ (recall that $h_b = 0$ for a.e. b).

Therefore $ph = p \sum_{b \in B} h_b = \sum_{b \in B} ph_b = \sum_b m_b b$, and $ph_b = m_b b$ for every b (by the uniqueness of the decomposition of an element). Each elementary direct component of G in the considered decomposition has the form $\mathbb{Q}^{(P)} b$. In other words, the element b plays the role of $\mathbf{1}$ in the corresponding $\mathbb{Q}^{(P)}$-component of this decomposition. Now recall that $p \notin P$. Thus, $m_b \neq 0$ implies $p | m_b$ for every b, by the definition of $\mathbb{Q}^{(P)}$. □

In later proofs we will have to approximate an *excellent* basis stage-by-stage, using a certain oracle. Recall that not every maximal $\tilde{P}$-independent set is an excellent basis of G_P. Therefore, we need to show that, for a given finite $\tilde{P}$-independent subset B of G_P and an element $g \in G_P$, there exists a finite extension $B^\star$ of B such that $B^\star$ is $\tilde{P}$-independent and the element g is contained in the $\mathbb{Q}^{(P)}$-span of $B^\star$.

Proposition 7.2.15. *Suppose $B \subseteq G_P$ is a finite $\tilde{P}$-independent subset of G_P. For every $g \in G_P$ there exists a finite $\tilde{P}$-independent set $B^\star \subseteq G_P$ such that $B \subseteq B^\star$ and $g \in (B^\star)_{\mathbb{Q}^{(P)}}$.*

Proof. Pick $\{e_i : i \in \mathbb{N}\} \subseteq G_P$ such that $G_P = \bigoplus_{i \in \mathbb{N}} \mathbb{Q}^{(P)} e_i$. Let $\{e_0, e_1, \ldots, e_n\}$ be such that both $B = \{b_0, \ldots, b_k\}$ and g are contained in $(\{e_0, e_1, \ldots, e_n\})_{\mathbb{Q}^{(P)}}$. We may assume $k < n$. We will need the following well-known generalisation of Rado's Lemma 5.1.4.

Lemma 7.2.16. *Suppose $B = \{b_0, \ldots, b_k\} \subseteq \bigoplus_{i \in \{0, \ldots, n\}} \mathbb{Q}^{(P)} e_i$, is a linearly independent set. There exists a set $C = \{c_0, \ldots, c_n\} \subseteq \bigoplus_{i \in \{0, \ldots, n\}} \mathbb{Q}^{(P)} e_i$, and coefficients $r_0, \ldots, r_k \in \mathbb{Q}^{(P)}$ such that*
(1) $\bigoplus_{i \in \{0, \ldots, n\}} \mathbb{Q}^{(P)} e_i, = \bigoplus_{i \in \{0, \ldots, n\}} \mathbb{Q}^{(P)} c_i$, and
(2) $(\{r_0 c_0, \ldots, r_k c_k\})_{\mathbb{Q}^{(P)}} = (B)_{\mathbb{Q}^{(P)}}$.

Proof. It is a special case of a well-known fact ([329], Theorem 7.8) which holds in general for every finitely generated module over a principal ideal domain (note that $\mathbb{Q}^{(P)}$ is a principal ideal domain). □

We show that if B is $\tilde{P}$-independent (and not merely linearly independent) then we can set $B^\star = \{b_0, \ldots, b_k\} \cup \{c_{k+1}, \ldots, c_n\}$, where $C = \{c_0, \ldots, c_n\}$ is the set from Lemma 7.2.16. Suppose

$$p | \sum_{0 \leqslant i \leqslant k} n_i b_i + \sum_{k+1 \leqslant i \leqslant n} n_i c_i$$

for a prime $p \in \tilde{P}$. We have

$$\bigoplus_{i \in \{0,\ldots,n\}} \mathbb{Q}^{(P)} e_i = \bigoplus_{1 \leqslant i \leqslant k} \mathbb{Q}^{(P)} c_i \oplus \bigoplus_{k+1 \leqslant i \leqslant n} \mathbb{Q}^{(P)} c_i,$$

and $\sum_{1 \leqslant i \leqslant k} n_i b_i \in \bigoplus_{1 \leqslant i \leqslant k} \mathbb{Q}^{(P)} c_i$. By the purity of direct components, we have

$$p | \textstyle\sum_{1 \leqslant i \leqslant k} n_i b_i \text{ within } \bigoplus_{1 \leqslant i \leqslant k} \mathbb{Q}^{(P)} c_i$$

and

$$p | \textstyle\sum_{k+1 \leqslant i \leqslant n} n_i c_i \text{ within } \bigoplus_{k+1 \leqslant i \leqslant n} \mathbb{Q}^{(P)} c_i.$$

But the former implies $p | n_i$ for all $1 \leqslant i \leqslant k$ by our assumption, and the latter implies $p | n_i$ for all $k+1 \leqslant i \leqslant n$ by the choice of C and Lemma 7.2.14.

The set $B^\star$ is actually an excellent $\tilde{P}$-basis of $\bigoplus_{i \in \{0,\ldots,n\}} \mathbb{Q}^{(P)} e_i$, since the cardinality of $B^\star$ is $n+1$, which is exactly the rank of $\bigoplus_{i \in \{0,\ldots,n\}} \mathbb{Q}^{(P)} e_i$. Therefore, the set

$$B^\star = \{b_0, \ldots, b_k\} \cup \{c_{k+1}, \ldots, c_n\}$$

is a $\tilde{P}$-independent set with the needed properties. □

Suppose G is a torsion-free abelian group, and $a, b \in G$. Recall that $\chi(a) \leqslant \chi(b)$ iff $h_i(a) \leqslant h_i(b)$ for all i. In other words, $p^k | a$ implies $p^k | b$, for all $k \in \mathbb{N}$ and every prime p.

Definition 7.2.17. Let G be a torsion-free abelian group. For a given characteristic α, let $G[\alpha] = \{g \in G : \alpha \leqslant \chi(g)\}$.

We have $h_i(a) = h_i(-a)$ and $\inf(h_i(a), h_i(b)) \leqslant h_i(a+b)$, for all i. Furthermore, $\chi(0) \geqslant \alpha$, for every characteristic α. Therefore, $G[\alpha]$ is a subgroup of G.

Notation 7.2.18. Let $\alpha = (\alpha_0, \alpha_2, \ldots)$. The subgroup of $(\mathbb{Q}, +)$ generated by elements of the form $1/p_k^x$ where $x \leqslant \alpha_k$, will be denoted $\mathbb{Q}(\alpha)$.

Recall that the type is an equivalence class of characteristics. Thus, the type of $H \leqq \mathbb{Q}$ is simply the type of any nonzero element of H. We are ready to state and prove the main result of this subsection. Recall Notation 7.2.13.

Theorem 7.2.19. *Let $\mathcal{G} = \bigoplus_{i \in \mathbb{N}} H$, where $H \leqq \mathbb{Q}$, $\mathbf{t}(H) = \mathbf{f}$ and $\alpha = (\alpha_0, \alpha_1, \ldots)$ is of type $\mathbf{f}$. Then $\mathcal{G}[\alpha] \cong G_P$, where $P = \{p_i : h_i = \infty \text{ in } \alpha\}$. Furthermore, if B is an excellent $\tilde{P}$-basis of $\mathcal{G}[\alpha]$, then $\mathcal{G}$ is generated by B over $\mathbb{Q}(\alpha)$.*

Proof. We prove that $\mathcal{G}[\alpha] \cong G_P$.

Let g_i be the element of the i-th presentation of H in the decomposition $\mathcal{G} = \bigoplus_{i\in\mathbb{N}} H$ such that $\chi(g_i) = \alpha$. The collection $\{g_i : i \in \mathbb{N}\}$ is a basis of $\mathcal{G}$. Therefore, $\{g_i : i \in \mathbb{N}\}$ is a basis of $\mathcal{G}[\alpha]$. By the definition of P, $(\{g_i : i \in \mathbb{N}\})_{\mathbb{Q}^{(P)}}$ is a subgroup of $\mathcal{G}[\alpha]$. Furthermore, since $\{g_i : i \in \mathbb{N}\}$ is linearly independent,

$$(\{g_i : i \in \mathbb{N}\})_{\mathbb{Q}^{(P)}} \cong \bigoplus_{i\in\mathbb{N}} \mathbb{Q}^{(P)} g_i.$$

Thus, we have

$$\bigoplus_{i\in\mathbb{N}} \mathbb{Q}^{(P)} g_i \subseteq \mathcal{G}[\alpha].$$

We are going to show that every element $g \in \mathcal{G}[\alpha]$ is generated by $\{g_i : i \in \mathbb{N}\}$ over $\mathbb{Q}^{(P)}$. This will imply $\mathcal{G}[\alpha] \cong G_P$.

Pick any nonzero $g \in \mathcal{G}[\alpha]$. The set $\{g_i : i \in \mathbb{N}\}$ is a basis of $\mathcal{G}[\alpha]$, therefore $ng = \sum_{i\in\mathbb{N}} m_i g_i$ for some integers n and m_i, $i \in \mathbb{N}$. Since direct components are pure, $n | \sum_{i\in I} m_i g_i$ implies $n | m_i g_i$ for every $i \in \mathbb{N}$, and $g = \sum_{i\in I} \frac{m_i}{n} g_i$. After reductions we have $g = \sum_{i\in I} \frac{m_i'}{n_i} g_i$, where $\frac{m_i'}{n_i}$ is irreducible. It suffices to show that $\frac{m_i'}{n_i} \in \mathbb{Q}^{(P)}$.

Assume there is i such that $\frac{m_i'}{n_i} \notin \mathbb{Q}^{(P)}$. Equivalently, for some $p_k \in \tilde{P}$, we have $m_i' \neq 0$ and $n_i = p_k n_i'$, where n_i' is an integer (recall that $\frac{m_i'}{n_i}$ is irreducible).

We have $h_k(\frac{m_i'}{n_i} g_i) = h_k(\frac{m_i'}{n_i'} \frac{g_i}{p_k}) \leqslant h_k(\frac{g_i}{p_k})$, since m_i' is not divisible by p_k. But $h_k(\frac{g_i}{p_k}) < h_k(g_i)$ (recall that $h_k(g_i)$ is finite). It is straightforward from the definitions of h_k that $h_k(g) = \min\{h_k(\frac{m_i'}{n_i} g_i) : i \in I, m_i \neq 0\}$, since each g_i belongs to a separate direct component of $\mathcal{G}$. Therefore $h_k(g) \leqslant h_k(\frac{m_i'}{n_i} g_i) < h_k(g_i)$. But $\chi(g_i) = \alpha$. Thus, $\chi(g) \ngeqslant \alpha$ and $g \notin \mathcal{G}[\alpha]$, and this contradicts our choice of g. Therefore, $\mathcal{G}[\alpha] \cong G_P$.

We show that if B is an excellent $\tilde{P}$-basis of $\mathcal{G}[\alpha]$, then $\mathcal{G} = (B)_{\mathbb{Q}(\alpha)}$ (recall Notation 7.2.6).

For every $b \in B$ consider the minimal pure subgroup $[b]$ which contains b (recall Definition 7.2.5). Consider

$$H(B) = \sum_{b\in B}{}' [b] \leq \mathcal{G}.$$

In fact, $H(B) = \bigoplus_{b\in B} [b]$, because B is linearly independent within $\mathcal{G}[\alpha]$ and, therefore, within $\mathcal{G}$ as well.

By our choice, $b \in \mathcal{G}[\alpha]$. Thus, $\chi(b) \geqslant \alpha$ within $\mathcal{G}$. We show that in fact $\chi(b) = \alpha$. Assume $\chi(b) > \alpha$. We have $b = pa$ for some $a \in \mathcal{G}[\alpha]$ and $p \in \tilde{P}$. But B is $\tilde{P}$-independent. This contradicts the fact that $p | 1 \cdot b$ and 1 is evidently not divisible by p. Therefore, we have

$$[b] = \mathbb{Q}(\alpha) b,$$

and thus $H(B) = (B)_{\mathbb{Q}(\alpha)}$. It remains to prove that $\mathcal{G} \subseteq H(B)$.

Pick any nonzero $g \in \mathcal{G}$. There exist integers m and n such that $(m, n) = 1$ and $\chi(\frac{m}{n} g) = \alpha$. To see this we use the fact that $\chi(g) \in \mathbf{f}$. It is enough to make only finitely many changes to $\chi(g)$ to make it equivalent to α.

Equivalently, $\frac{m}{n}g \in \mathcal{G}[\alpha]$. Applying Lemma 7.2.14, we obtain

$$\frac{m}{n}g = \sum_{b \in B, r_b \in \mathbb{Q}^{(P)}} r_b b.$$

By our assumption, $\chi(b) = \chi(\frac{m}{n}g) = \alpha$, for every $b \in B$. Obviously, $m | \frac{m}{n}g$ in $\mathcal{G}$. Therefore, by the definition of α and B, we have $m|b$ *in* $\mathbb{Q}(\alpha)b$. Thus, there exist $x_b \in [b] = \mathbb{Q}(\alpha)b$ such that $mx_b = b$. We can set

$$g = \sum_{b \in B} nr_b x_b,$$

where $nr_b x_b \in [b]$. This shows $\mathcal{G} = (B)_{\mathbb{Q}(\alpha)}$. □

7.2.3 Δ^0_3-categoricity and the proof of Theorem 7.2.2

Recall that the main result of this section, Theorem 7.2.2, states that *every computably presentable homogeneous completely decomposable group is Δ^0_3-categorical.* The proof of the Theorem 7.2.2 is based on the lemma below.

Lemma 7.2.20. *Let $\mathcal{G} = \bigoplus_{i \in \mathbb{N}} H$, where $H \leqq \mathbb{Q}$, the type $\mathbf{t}(H)$ is $\mathbf{f}$, and $\alpha = (\alpha_0, \alpha_1, \ldots)$ is a characteristic of type $\mathbf{f}$. Let G_1 and G_2 be computable presentations of $\mathcal{G}$. Suppose that both $G_1[\alpha]$ and $G_2[\alpha]$ have Σ^0_n excellent $\tilde{P}$-bases. Then there exists an Δ^0_n isomorphism from G_1 onto G_2.*

We first prove Theorem 7.2.2, and then prove Lemma 7.2.20. We need to show that a given homogeneous completely decomposable group satisfies the hypothesis of Lemma 7.2.20 with $n = 3$.

Proof of Theorem 7.2.2. Let G be a computable presentation of $\mathcal{G} \cong \bigoplus_{i \in \mathbb{N}} H$, where $H \leqslant \mathbb{Q}$. Let α be a characteristic of type $\mathbf{t}(H)$ and

$$P = \{p_k : \alpha_k = \infty \text{ in } \alpha\}.$$

By Theorem 7.2.19 and Lemma 7.2.20, it suffices to construct a excellent $\tilde{P}$-basis of $G[\alpha]$ which is Σ^0_3.

We are building $C = \bigcup_n C_n$. Assume that we are given C_{n-1}. At step n of the procedure, we do the following:

1. Pick the n-th element g_n of $G[\alpha]$.
2. Find an extension C_n of C_{n-1} in $G[\alpha]$ such that:
(a) C_n is a finite $\tilde{P}$-independent set, and
(b) $C_n \cup \{g_n\}$ is linearly dependent.

Let $G = \bigoplus_{i \in I} Re_i$, where $\chi(e_i) = \alpha$ and $R \cong H$. Observe that at stage n of the procedure we have

$$g_n \cup C_{n-1} \subset (\{e_0, \ldots, e_k\})_{\mathbb{Q}^{(P)}},$$

for some k. By Proposition 7.2.15, the needed extension denoted by C_n can be found. It suffices to check that the construction is effective relative to $\mathbf{0}''$.

By Theorem 7.2.19, we have $G[\alpha] \cong G_P$, where $P = \{p : p^\infty | h\}$ is a Π^0_2 set of primes.

Claim 7.2.21. $G[\alpha]$ *is a* Π^0_2*-subgroup of* G.

Proof. Pick any $h \in G$ with $\chi(h) = \alpha$. By its definition, for every $g \in G$, the property $\chi(g) \geqslant \alpha$ is equivalent to

$$\forall p \text{ prime } \forall k \in \mathbb{N}\,((\exists x)p^k x = h \to (\exists y)p^k y = g).$$

Therefore, the group $G[\alpha]$ is a Π^0_2-subgroup of G. ☐

Claim 7.2.22. *There is a* $\mathbf{0}''$*-computable procedure which decides if a given finite set* $B \subseteq G[\alpha]$ *is* $\tilde{P}$*-independent, uniformly in the index of* B.

Proof. Note that in general $P \in \Pi^0_2$. By Claim 7.2.21, the group $G[\alpha]$ is a Π^0_2-subgroup of G. Thus, the condition "B is a $\tilde{P}$-independent set in $G[\alpha]$" seems to be merely Π^0_3:

$$\forall \overline{m} \in \mathbb{Z}^{<\infty}\, \forall p\,\text{prime} \left(\left[p \notin P \wedge (\exists x) \left(x \in G[\alpha] \wedge px = \sum_{b \in B} m_b b \right) \right] \to \bigwedge_b p \mid m_b \right).$$

The idea is to substitute the Σ^0_3 condition $(\exists x)(x \in G[\alpha] \wedge px = \sum_{b \in B} m_b b)$ by an equivalent Σ^0_2 one, using a non-uniform parameter $c \in G$ such that $\chi(c) = \alpha$. We are going to show that for every $p_v \notin P$, the property

$$(\exists x) \left(x \in G[\alpha] \wedge p_v x = \sum_{b \in B} m_b b \right)$$

is equivalent to

$$(\exists k)(\exists y \in G) \left(\alpha_v < k \wedge p_v^k y = \sum_{b \in B} m_b b \right),$$

where α_v is the v-th component of α corresponding to p_v and

$$\alpha_v < k \Leftrightarrow \neg(\alpha_v \geqslant k) \Leftrightarrow \neg(\exists \xi)(p_v^k \xi = c).$$

Suppose there is $x \in G[\alpha]$ such that $p_v x = \sum_{b \in B} m_b b$. Since $h_v(x) \geqslant \alpha_v$, we have $p_v^{\alpha_v} y = x$ and $p_v^{\alpha_v + 1} y = p_v x$, for some $y \in G$, so we can set $k = \alpha_v + 1$. For the converse, suppose there exist such k and y. Then $p_v x = p_v^k y$ for $x = p_v^{k-1} y$. We have $k > \alpha_v$, and therefore $(k-1) \geqslant \alpha_v$. But $h_v(x) \geqslant (k-1)$ because $x = p_v^{k-1} y$ is divisible by $k-1$, and thus $h_v(x) \geqslant \alpha_v$. The characteristic of x differs from the characteristic of y only at the position for the prime p_v. Thus, for every $w \neq v$,

$$h_w(x) = h_w(p_v^k y) = h_w \left(\sum_{b \in B} m_b b \right) \geqslant \alpha_w,$$

since $\sum_{b \in B} m_b b \in G[\alpha]$. Therefore, $\chi(x) \geqslant \alpha$ and $x \in G[\alpha]$. ☐

By Claim 7.2.21 and Claim 7.2.22, the procedure is computable relative to $\mathbf{0}''$. ☐

Proof of Lemma 7.2.20. Recall that G_1 and G_2 are computable presentations of $\mathcal{G}$ such that both $G_1[\alpha]$ and $G_2[\alpha]$ have Σ^0_n excellent $\tilde{P}$-bases. We need to show that there exists an Δ^0_n isomorphism from G_1 onto G_2. Let B_1 and B_2 be excellent $\tilde{P}$-bases of $G_1[\alpha]$ and $G_2[\alpha]$, respectively.

Observe that the group $\mathbb{Q}(\alpha)$ is isomorphic to a c.e. additive subgroup R of $(\mathbb{Q}, +, \times)$. Furthermore, we may assume that $1 \in R$. (Pick h with $\chi(h) = \alpha$ non-uniformly, and consider $[h]$.) By Theorem 7.2.19, we have

$$G_1 = \bigoplus_{b \in B_1} Rb \cong G_2 = \bigoplus_{b' \in B_2} Rb'.$$

To build a Δ_n^0 isomorphism from G_1 to G_2 first define the map from B_1 onto B_2 using a standard back-and-forth argument. Then extend it to the whole G_1 using the fact that $r \cdot b$ can be found effectively and uniformly, for every $r \in R$ and $b \in B_1$. □

Exercises

Exercise* 7.2.23 (Bazhenov, Goncharov, and Melnikov [33]). Let H be a decidable homogeneous completely decomposable group that is not a divisible group. Show that H admits a computable maximal linearly independent set.

Exercise° 7.2.24. Show that, for $S \neq \varnothing$, a maximal S-independent set does not have to be maximal linearly independent. (Hint: For example, Lemma 35.1 in [194] implies that the free abelian group of rank ω contains a $\{p\}$-basis which is not excellent.)

7.2.4 Semi-low sets, and Δ_2^0-categoricity

In this subsection we give a complete and detailed proof of Theorem 7.2.25 that describes Δ_2^0-categoricity of homogeneous c.d. groups in terms of semi-low sets. This result will not be used in the sequel. However, this is the only satisfactory description of Δ_2^0-categoricity in a non-trivial natural class that we are aware of. Even Δ_2^0-categorical equivalence structures do not seem to possess such a description, as we will briefly discuss in Chapter 9. Thus, the result presented in this section is rather unusual and surprising.

A set A is *semi-low* if the set

$$H_A = \{e : W_e \cap A \neq \varnothing\} = \{e : W_e \nsubseteq \overline{A}\}$$

is computable relative to $\varnothing'$. We have already met semi-low sets in Exercises 3.1.23, 3.1.24, and 3.1.25, and in §3.1.5 we saw that semilowness can be used to imitate lowness in some constructions.

This notion arose in the understanding of the automorphisms of the lattice of c.e. sets [477]. It is quite remarkable that this notion captures Δ_2^0-categoricity of a homogeneous c.d. group, as we show next. As before, G_P stands for the direct sum of infinitely many copies of the localisation of the integers by a set of primes P.

Theorem 7.2.25 (Downey and Melnikov [136]). *A computable homogeneous completely decomposable group A of rank ω is Δ_2^0-categorical iff A is isomorphic to*

$$G_P \cong \bigoplus_{i \in I} \mathbb{Q}^{(P)},$$

where P is a c.e. set of primes such that

$$\tilde{P} = \{p : p \text{ prime and } p \notin P\}$$

is semi-low.

Before we prove the theorem, we discuss its corollaries. Combined with Lemma 7.2.10, the theorem above gives:

Corollary 7.2.26. *For a c.e. set P, the following are equivalent:*

1. *G_P has a Σ^0_2 excellent $\tilde{P}$-basis;*
2. *G_P has a Σ^0_2-basis as a free $\mathbb{Q}^{(P)}$-module*[7]*;*
3. *G_P is Δ^0_2-categorical;*
4. *$\tilde{P}$ is semi-low.*

In particular, since every low set is semi-low (Exercise 3.1.23), we have:

Corollary 7.2.27. *If a c.e. set of primes P is low then G_P is Δ^0_2-categorical.*

In particular, the free abelian group of rank ω is Δ^0_2-categorical. It is known that every non-low c.e. degree contains a c.e. set whose complement is not semi-low; see Exercise 3.1.25. Thus, we conclude that the upper bound $n = 3$ for Δ^0_n-categorcity of homogeneous c.d. groups established in Theorem 7.2.2 cannot be improved in general.

Corollary 7.2.28. *There exists a c.e. set of primes P so that the homogeneous c.d. group G_P is not Δ^0_2-categorical. Indeed, any non-low c.e. degree contains a c.e. set of primes P with this property.*

In Chapter 10 we will study a "Type II" version of Δ^0_2-categoricity, relative Δ^0_2-categoricity. A computable structure H is relatively Δ^0_2-categorical if every X-computable copy of H is isomorphic to H via an X'-computable isomorphism. For the class of homogeneous completely decomposable groups, this notion corresponds exactly to G_P having a computable sets of primes P; this is Exercise 10.1.94. It follows from Theorem 3.1.1 that relative and "plain" Δ^0_2-categoricity differ for the class of homogeneous completely decomposable groups.

Proof of Theorem 7.2.25 and computable settling time*

The proof of the theorem is not particularly difficult; however, it is relatively long. The proof can be either skipped or skimmed through. However, the reader should note that the notion of a computable settling time strongly resembles the notion of a limitwise monotonic function that will be of central importance in Chapter 9.

Proof of Theorem 7.2.25. The proof is split into several parts. Each part corresponds to a different hypothesis on the isomorphism type of G. Different cases will need different techniques and strategies.

We need the following technical notion:

Definition 7.2.29 (Computable settling time)**.** Let $\alpha = (h_i)_{i\in\mathbb{N}}$ be a sequence where $h_i \in \omega \cup \{\infty\}$ for each i (in other words, let α be a characteristic). Also, suppose that there is a non-decreasing uniform computable approximation $h_{i,s}$ such that $h_i = \sup_s h_{i,s}$, for every i (in other words, the characteristic is c.e.).

[7] That is, $\{e_0, e_1, \ldots\}$ so that $G_P = \bigoplus_{i\in\mathbb{N}} \mathbb{Q}^{(P)} e_i$.

We say that α has *a computable settling time* if there is a (total) computable function $\psi : \omega \to \omega$ such that

$$h_i = \begin{cases} h_{i,\psi(i)}, & \text{if } h_i \text{ is finite,} \\ \infty, & \text{otherwise,} \end{cases}$$

for every i. We also say that ψ is a computable settling time for $(h_{i,s})_{i,s\in\mathbb{N}}$.

This is the same as saying that, given i, there exists an effective (and uniform) way to compute a stage s after which the approximation of h_i either does not increase, or increases and tends to infinity. Note that this is the property of a characteristic, not the property of some specific computable approximation. Indeed, given an approximation of α having a computable settling time, we can define a computable settling time for any other computable approximation of α. Furthermore, as can be easily seen, this is a type-invariant property. Thus, we can also speak of types having computable settling times.

If a homogeneous completely decomposable group G of type $\mathbf{f}$ is computable, then $\mathbf{f}$ is c.e. (see Proposition 7.2.8). Suppose that G is a computable homogeneous completely decomposable group of type $\mathbf{f}$, and let $\alpha = (h_i)_{i\in\mathbb{N}}$ be a characteristic of type $\mathbf{f}$. We consider the cases:

1. The type $\mathbf{f}$ of G has no computable settling time. In this case G is not Δ^0_2-categorical by Proposition 7.2.32. Observe that if $\mathbf{f}$ has no computable settling time then the set $Fin(\alpha) = \{i : 0 < h_i < \infty\}$ has to be infinite (see, e.g., Proposition 7.2.9). Thus, G can not be isomorphic to G_P, for a set of primes P.

2. The type $\mathbf{f}$ of G has a computable settling time, $Fin(\alpha) = \{i : 0 < h_i < \infty\}$ is empty (finite), and the set $\{i : h_i = 0\}$ is semi-low. In other words, the group G is isomorphic to G_P with P semi-low. In this case G is Δ^0_2-categorical, by Proposition 7.2.30 below.

3. The type $\mathbf{f}$ of G has a computable settling time, the set $Fin(\alpha) = \{i : 0 < h_i < \infty\}$ is empty (finite), and the set $\{i : h_i = 0\}$ is not semi-low. Here G is again isomorphic to G_P, but in this case G is not Δ^0_2-categorical, by Proposition 7.2.33 below.

4. The type $\mathbf{f}$ of G has a computable settling time, and the set $Fin(\alpha) = \{i : 0 < h_i < \infty\}$ is infinite and not semi-low. As in the above case[8], G is not Δ^0_2-categorical, by Proposition 7.2.33.

5. The type $\mathbf{f}$ of G has a computable settling time, and the set $Fin(\alpha) = \{i : 0 < h_i < \infty\}$ is infinite and semi-low. The group is not Δ^0_2-categorical, by Proposition 7.2.34 below.

We first discuss why case (3) and case (4) above can be collapsed into one case. First, define $Inf(\alpha) = \{i : h_i = \infty\}$ and $V = \{i : 0 < h_{i,\psi(i)} < \infty\}$, where ψ is a computable settling time for α. Note that V is c.e.. Evidently, $\overline{Inf(\alpha)} = \underline{Fin(\alpha)} \cup \{i : h_i = 0\}$ and $Fin(\alpha) = \overline{Inf(\alpha)} \cap V$. We claim that "$Fin(\alpha)$ is not semi-low" implies "$\overline{Inf(\alpha)}$ is not semi-low". We assume that $\overline{Inf(\alpha)}$ is semi-low and observe that

$$\{e : W_e \cap Fin(\alpha) \neq \varnothing\} = \{e : W_e \cap V \cap \overline{Inf(\alpha)} \neq \varnothing\} = \{e : W_{s(e)} \cap \overline{Inf(\alpha)} \neq \varnothing\}$$

for a computable function s. Therefore, $H_{Fin(\alpha)} \leqslant_m H_{\overline{Inf(\alpha)}} \leqslant_T \varnothing'$, as required.

[8]We distinguish these two cases only because these cases correspond to (algebraically) different types of groups. We discuss a bit later why these cases are essentially not different.

Therefore, cases (3) and (4) can be combined into

($3'$) If $\mathbf{f}$ has a computable settling time and $\overline{Inf(\alpha)}$ is not semi-low, then G is not Δ^0_2-categorical.

Now we state and prove the propositions which cover all the cases above.

Recall that, by Proposition 7.2.9, the group G_P has a computable presentation as a group (module) iff P is c.e..

Proposition 7.2.30. *If $\tilde{P}$ is semi-low (and co-c.e.) then G_P is Δ^0_2-categorical.*

Proof. The proof may be viewed as a simpler version of the proof of Theorem 7.2.2. Let $G = \{g_0 = 0, g_1, \dots\}$ be a computable copy of G_P. By Lemma 7.2.10, it is enough to build a Σ^0_2 excellent $\tilde{P}$-basis of G.

We are building $C = \bigcup_n C_n$. Assume that we are given C_{n-1}. At stage n of the construction, we do the following:

1. Pick the n-th element g_n of G.

2. Find an extension C_n of C_{n-1} in G such that (a) C_n is a finite $\tilde{P}$-independent set, and (b) $C_n \cup \{g_n\}$ is linearly dependent.

The algebraic part of the verification is the same as in Theorem 7.2.2 (and is actually simpler). Thus, it is enough to show that (a) in (2) above can be checked effectively and uniformly in $\varnothing'$. Given a finite set F of elements of G, define a c.e. set V consisting of primes which could potentially witness that F is $\tilde{P}$-dependent:

$$V = \Big\{p \,:\, \exists \overline{m} \in Z^{card(F)}\Big[p|(\sum_{g\in F} m_g g) \wedge (\bigvee_{g\in F} p \not| m_g)\Big]\Big\}.$$

The c.e. index of V can be obtained uniformly from the index of F. It can be easily seen from the definition of $\tilde{P}$-independence that

$$V \cap \tilde{P} = \varnothing \text{ iff } F \text{ is } \tilde{P}\text{-independent}.$$

By our assumption on $\tilde{P}$, this can be decided effectively in $\varnothing'$. $\square$

Fix a computable listing $\{\Phi_e(x, y)\}_{e\in\mathbb{N}}$ of all partial computable functions of two arguments; e.g., $\Phi_e(x, y) = \varphi_e(\langle x, y\rangle)$, where $\langle x, y\rangle = 2^x 3^y$. We say that $\lim_s \Phi_e(x, s)$ exists if $\Phi_e(x, s)\downarrow$ for every e and s and the sequence $(\Phi_e(x, s))_{s\in\mathbb{N}}$ stabilises. In the upcoming propositions we will use the following:

Notation 7.2.31. Fix an effective listing $\{\Psi_e(x, s)\}_{e\in\mathbb{N}}$ of total computable functions of two arguments satisfying the property:

$$(\lim_s \Phi_e(x, s) \text{ exists}) \Rightarrow (\lim_s \Phi_e(x, s) = \lim_s \Psi_e(x, s)),$$

for every x and e.

Proposition 7.2.32. *Suppose that the type $\mathbf{f}$ of a computably presentable $G = \bigoplus_{i\in\mathbb{N}} H$ has no computable settling time. Then G is not Δ^0_2-categorical.*

Proof of Proposition 7.2.32. In the construction below we identify elements of A and B and the corresponding elements of ω. It suffices to build two computable presentations, A and B, of the group $G = \bigoplus_{i \in \mathbb{N}} H$, and meet the requirements:

R_e : $\lim_t \Psi_e(b_e, t)$ exists $\Rightarrow$ $\lim_t \Psi_e(x, t)$ is not an isomorphism from B to A.

The nonzero element b_e is a witness for the R_e strategy below. More specifically, we enumerate $A = \bigoplus_{n \in \mathbb{N}} Ha_n$ and $B = \bigoplus_{e \in \mathbb{N}} C_e b_e$ in such a way that the sets $\{a_n : n \in \mathbb{N}\}$ and $\{b_e : e \in \mathbb{N}\}$ are computable. Let $(h_i)_{i \in \mathbb{N}}$ be a characteristic of type $\mathbf{f}$. Fix a computable approximation $(h_{i,s})_{i,s \in \mathbb{N}}$ of $(h_i)_{i \in \mathbb{N}}$ such that (1) $h_{i,s} \leqslant h_{i,s+1}$, and (2) $h_i = \lim_s h_{i,s}$, for every i and s.

We make sure $\chi(a_n) = (h_i)_{i \in \mathbb{N}}$, for every n, while the characteristic $\chi(b_e) = (d(e)_i)_{i \in \omega}$ of b_e will be merely equivalent to $(h_i)_{i \in \omega}$, for each e (thus, $C_e \cong H$, for each e).

The construction is injury-free, and we do not need any priority order on the strategies.

For every e, the strategy for R_e defines its own computable function ψ_e which is an attempt to define a computable settling time for $(h_i)_{i \in \omega}$. Since it will be clear from the construction at which stage ψ_e is defined (if ever), we omit the extra index t in $\psi_{e,t}$ and write simply ψ_e. We omit the index t for parameters $k_{e,i,t}$ as well. To define ψ_e the strategy uses the sequence $(k_{e,i})_{i \in \mathbb{N}}$ (to be defined in the construction).

Strategy for R_e: If at a stage s of the construction the parameter $k_{e,0}$ is undefined then:

1. Compute $\Psi_e(b_e, s)$. From this moment on, the strategy is always waiting for $t > s$ such that $\Psi_e(b_e, t) \neq \Psi_e(b_e, s)$. As soon as such a t is found, R_e initialises by making all its parameters undefined and also making $d(e)_{j,t} = h_{j,t}$ for every j we have ever seen so far.

2. Let $a \in A$ be such that $a = \Psi_e(b_e, s)$. Find integers c_n and c such that $ca = \sum_n c_n a_n$. Let j be a fresh large index such that (1) the prime p_j does not occur in the decompositions of the coefficients c and c_n, (2) $h_{j,s} > 0$, and (3) $d(e)_{j,s} < h_{j,s}$.

3. Once j is found, declare $\psi_e(j) = s$. We may assume that at stage s such an index j can be found, otherwise we speed up the approximation $(h_{i,s})_{i,s \in \mathbb{N}}$ during the construction. From this moment on, make sure $d(e)_{j,t} = h_{j,t} - 1$ for every $t \geqslant s$, unless the strategy initialises. Set $k_{e,0} = j$, and proceed.

Now assume that the parameters $k_{e,0}, \ldots k_{e,y}$ have already been defined by the strategy. We also assume that $\psi_e(i)$ has already been defined for each i such that $k_{e,0} \leqslant i \leqslant max\{k_{e,x} : 0 \leqslant x \leqslant y\}$. Assume also that $k_{e,y}$ was first defined at stage $u < s$. Then do the following:

I. Wait for a stage $t \geqslant s$ (of the construction) such that either (a) $h_{i,t} > h_{i,s}$ for some i such that $k_{e,0} \leqslant i \leqslant max\{k_{e,x} : 0 \leqslant x \leqslant y\}$ and $i \notin \{k_{e,0}, \ldots k_{e,y}\}$, or (b) $h_{i,u} < h_{i,t}$ for each $i \in \{k_{e,0}, \ldots k_{e,y}\}$. While waiting, make $d(e)_{j,r} = h_{j,r}$ (r is the current stage of the construction), where $j \leqslant r$ and $j \notin \{k_{e,0}, \ldots k_{e,y}\}$.

II. If (a) holds for some i, then set $k_{e,(y+1)} = i$. If (b) holds, then let i be a fresh large index such that (1) $h_{i,t} > 0$, and (2) $d(e)_{i,t} < h_{i,t}$, and set $k_{e,(y+1)} = i$. In this case also define $\psi_e(j)$ to be equal to the current stage for every j such that $max\{k_{e,x} : 0 \leqslant x \leqslant y\} < j \leqslant k_{e,(y+1)}$. Then proceed to III.

III. Set $d(e)_{i,t} = h_{i,t} - 1$ at every later stage t, where $i = k_{e,(y+1)}$, unless the strategy initialises.

End of strategy.

Construction. At stage 0, start enumerating A and B as free abelian groups over $\{a_n\}_{n \in \mathbb{N}}$ and $\{b_e\}_{k \in \mathbb{N}}$, respectively. Initialise R_e, for all e.

At stage s, let strategies R_e, $e \leqslant s$, act according to their instructions. If R_e acted at the previous stage, then return to its instructions at the position it was left at the previous stage.

Make $\chi(a_n) = (h_{i,s})_{i \in \mathbb{N}}$ in A_s for every $n \leqslant s$, and $\chi(b_e) = (d(e)_{i,s})_{i \in \mathbb{N}}$ in B_s for every $e \leqslant s$, by making a_n and b_e divisible by corresponding powers of primes.

End of construction.

Verification. For each e, the following cases are possible:

1. $\lim_s \Psi_e(b_e, s)$ does not exist. In this case the strategy initialises infinitely often. By the way the strategy is initialised, the characteristic of b_e is identical to α.

2. $\lim_s \Psi_{e,s}(b_e, s)$ exists and is equal to $\Psi_e(b_e, l)$. The domain of ψ_e should be co-infinite. For if it was co-finite, then α would have a computable settling time. Therefore, there is a parameter $k_{e,y}$ such that the $k^{th}_{e,y}$ position in α is finite. Thus, the strategy ensures $\lim_s \Phi_{e,s}(b_e, s)$ is not an isomorphism since the characteristic of b_e and α differ at the $k^{th}_{e,y}$ position. Therefore, α differs from $\chi(b_e)$ in at most finitely many positions, and the differences are finitary.

In both cases $\chi(b_e)$ is equivalent to α. By Theorem 5.1.15, $A \cong B \cong G$. □

Recall that cases (3) and (4) were both reduced to:

Proposition 7.2.33. *Let G be computable homogeneous completely decomposable abelian group of type $\mathbf{f}$, and suppose $\alpha = (\sup_s h_{i,s})_{i \in \mathbb{N}}$ in $\mathbf{f}$ has computable settling time ψ. Furthermore, suppose $\overline{Inf(\alpha)}$ is not semi-low. Then G is not Δ^0_2-categorical.*

Proof of Proposition 7.2.33. We build two computable copies of G by stages. Recall that the first copy $A = \bigoplus_i Ha_i$ is a "nice" copy with $\chi(a_i) = \alpha$, for every i. The second ("bad") copy $B = \bigoplus_{e \in \mathbb{N}} \bigoplus_{n \in \mathbb{N}} C_{e,n} b_{e,n}$ is built in such a way that $\chi(b_{e,n})$ is equivalent to α, for every e and n.

Recall Notation 7.2.31. It suffices to meet the requirements:

$R_e : (\forall n) \lim_t \Psi_e(b_{e,n}, t)$ exists $\Rightarrow \lim_t \Psi_e(x, t)$ is not an isomorphism from B to A.

The strategy for R_e initially attempts to define a total Γ such that $\Gamma(n) = 0$ iff $W_n \subseteq Inf(\alpha)$. If we succeeded, this would imply

$$H_{\overline{Inf(\alpha)}} = \{n : W_n \cap \overline{Inf(\alpha)} \neq \varnothing\} = \{n : W_n \nsubseteq Inf(\alpha)\} \leqslant_T \varnothing',$$

contradicting the hypothesis. In the following, we write I in place of $Inf(\alpha)$. We split R_e into substrategies $R_{e,n}$, $n \in \mathbb{N}$:

Substrategy $R_{e,n}$. Permanently assign the element $b_{e,n}$ to $R_{e,n}$. Suppose that the strategy becomes active for the first time at stage s of the construction. Then:

1. Start by setting $\Gamma_s(n, s) = 0$ (we may suppose that $\Gamma_j(n, j) = 0$, for every $j < s$). At a later stage t, we define $\Gamma_t(n, t)$ to be equal to $\Gamma_{t-1}(n, t-1)$, unless we have a specific instruction not to do so.

2. Wait for a stage $t > s$ and a number $j \in W_{n,t} \backslash I_t$.

3. We see $p = p_j$ with $j \in W_{n,t} \backslash I_t$ at a later stage t. Find $a \in A_t$ such that $a = \Psi_e(b_e, t)$ (recall that the enumeration of A is controlled by us). Find integers c_n and c such that $ca = \sum_n c_n a_n$. Let k be a fresh large natural number such that (i) the prime $p = p_j$ has power at most $[k/2]$ in the decompositions of the coefficients c and c_n, and (ii) $h_{j,\psi(j)} < [k/2]$, where ψ is the computable settling time. Note that (*i*) and (*ii*) imply k is so large that p^k does not divide $a = \Psi_e(b_{e,n}, t)$ within A, unless $j \in I_t$. Make $b_{e,n}$ divisible by p^k within B.

   Wait for one of the two things to happen:

   I. (*I* changes first). We see $j \in I_u$ at a later stage $u > t$, and $\Psi_e(b_{e,n}, v) = \Psi_e(b_{e,n}, t)$ for each $v \in (t, u]$. We return to (2) with u in place of s.

   II. (Ψ_e changes first). We see $\Psi_e(b_{e,n}, u) \neq \Psi_e(b_{e,n}, t)$ for $u > t$, and $j \in W_{n,v} \backslash I_v$ for each $v \in (t, u]$. Then set $\Gamma_u(n, u) = 1$ and start waiting for a stage $w > u$ such that $j \in I_w$. If such a stage w is found, then we set $\Gamma_w(n, w) = 0$ and go to (2) with w in place of s (and we do nothing, otherwise).

End of strategy.

Construction. At stage 0, start enumerating A and B as free abelian groups over $\{a_i\}_{i \in \mathbb{N}}$ and $\{b_{e,n}\}_{e,n \in \mathbb{N}}$.

At stage s, let strategies $R_{e,n}$, $e, n \leqslant s$, act according to their instructions. If $R_{e,n}$ acted at the previous stage, then return to its instruction at the position it was left at the previous stage.

Make $\chi(a_i) = \alpha = (h_j)_{j \in \mathbb{N}}$ in A for every i. For every $e, n \in \mathbb{N}$, make $\chi_j(b_{e,n}) = h_j$ in B for every j except at most one position, according to the instructions of $R_{e,n}$. We do so by making a_i and $b_{e,n}$ divisible by corresponding powers of primes.

End of construction.

Verification. By Theorem 5.1.15, $A \cong B \cong G$. Assume that $\lim_s \Psi_{e,s}(b_{e,n}, s)$ exists for every n (thus, *II* does not get visited infinitely often). Given n, consider the cases:

- $R_{e,n}$ eventually waits forever at substage (2). Then $\lim_s \Phi(n, s) = 0$ and $W_n \subseteq I$. Thus, we have a correct guess about $H_{\overline{Inf(\alpha)}}$.
- $R_{e,n}$ visits I of (3) from some point on. Then $\lim_s \Phi(n, s) = 0$ and $W_n \subseteq I$, and we again have a correct guess about $H_{\overline{Inf(\alpha)}}$.
- $R_{e,n}$ eventually waits forever at substage (3). Then $x_{e,n}$ witnesses that $\lim_s \Phi_{e,s}(x_{e,n}, s)$ is not an isomorphism from B to A.

There should be at least one n for which $\lim_s \Phi(n, s) \neq H_{\overline{Inf(\alpha)}}(n)$. Therefore, for at least one n, the strategy $R_{e,n}$ eventually waits forever at substage (3). Thus, R_e is met. □

Proposition 7.2.34. *If the type* **f** *of a computable homogeneous completely decomposable group* G *has a computable settling time, and* $Fin(\alpha) = \{i : 0 < h_i < \infty\}$ *is infinite and semi-low for* $\alpha = (\alpha_i)_{i \in \mathbb{N}}$ *of type* **f**, *then* G *is not* Δ_2^0*-categorical.*

Proof of Proposition 7.2.34. Let Γ be a computable function such that

$$Fin(\alpha) \cap W_n = \lim_s \Gamma(n, s).$$

As in the proof of Proposition 7.2.32, we are building two computable copies,

$$A = \bigoplus_{n \in \mathbb{N}} Ha_n \text{ and } B = \bigoplus_{e \in \mathbb{N}} C_e b_e,$$

of G. We make $\chi(a_n) = \alpha$ and $\chi(b_e) = (d(e))_{i \in \mathbb{N}} \simeq \alpha$, for every n and e. Recall Notation 7.2.31. The requirements are:

R_e : If $\lim_t \Psi_e(b_e, t)$ exists, then $\lim_t \Psi_e(x, t)$ is not an isomorphism from B to A.

For every e, the strategy for R_e will enumerate its own sequence of c.e. sets. The indexes for the sets are listed by a computable function g of two arguments:

$$\{W_{g(e,s)}\}_{s \in \mathbb{N}}.$$

Let $(h_{i,s})_{i,s \in \mathbb{N}}$ be a computable approximation of α such that, for every i, either $\alpha_i = h_{i,0}$ or $\alpha_i = \infty$. Also, let $n(0), n(1) \ldots$ be an effective increasing enumeration of the infinite computable set $N = \{i : h_{i,0} \neq 0\}$.

The strategy for R_e: Suppose $s = 0$ or $\Psi_e(b_e, s) \neq \Psi_e(b_e, s-1)$. Do the following substeps:

1. Make $\chi(b_e) = (d(e))_{i \in \mathbb{N}}$ and α equal at all positions seen so far.

2. Begin enumerating $W_{g(e,s)}$ by setting $W_{g(e,s)} = \varnothing$.

3. Wait for a stage u such that $\Gamma(g(e, s), u) = 0$.

4. Let $a \in A$ be such that $a = \Psi_e(b_e, s)$. If $a = 0$ do nothing. If $a \neq 0$, find integers c_m and c such that $ca = \sum_m c_m a_m$. Let $n(i) \in N$ be a fresh large number such that (1) the prime $p_{n(i)}$ does not occur in the decompositions of the coefficients c and c_m, (2) $h_{n(i),0} > 0$, and (3) $d(e)_{k,s} = 0$ for every $k \geqslant n(i)$.

5. Enumerate $n(i)$ into $W_{g(e,s)}$. Keep $d(e)_{n(i),l} = 0$ for $l \geqslant s$ (unless we have a specific instruction not to do so). *Restrain* the element b_e by not allowing the construction to make it divisible by any prime greater than $p_{n(i)}$.

6. Wait for one of the following three things to happen:

   I. $\Psi_e(b_e, s) \neq \Psi_e(b_e, t)$ at a later stage t. Then declare b_e not restrained and restart the strategy with t in place of s (go to (1); for instance, make b_e divisible by the corresponding power of $p_{n(i)}$).

   II. The number $n(i)$ enters the c.e. set $Inf(\alpha)$ at a stage $s > t$ (thus, $h_{n(i)} = \infty$). Make b_e infinitely divisible by $p_{n(i)}$ and return to (5) with $n(i+1)$ in place of $n(i)$ keeping b_e restrained.

   III. $\Gamma(g(e, s), t) = 1$ (thus, we believe $W_{g(e,s)} \cap Fin(\alpha) \neq \varnothing$ and $j \in Fin(\alpha)$). We remove the restraint from the element b_e allowing the construction to make b_e divisible by p_i with $i \notin W_{g(e,s)}$ if needed. We keep b_e not divisible by $p_{n(i)}$.
   If at a later stage r the number $n(i)$ enters $Inf(\alpha)_r$ (thus, $W_{g(e,s),r} \subseteq Inf(\alpha)_r$), then make b_e infinitely divisible by $p_n(i)$. In this case also wait for a stage $w \geqslant r$ such that $\Gamma(g(e, s), w) = 0$. Then return to (4) with a new fresh and large $n(j)$.

End of strategy.

Construction: At stage 0, start enumerating A and B as free abelian groups over $\{a_n\}_{n\in\mathbb{N}}$ and $\{b_e\}_{k\in\mathbb{N}}$, respectively.

At stage s, let strategies R_e, $e \leqslant s$, act according to their instructions. If R_e acted at the previous stage, then return to its instruction at the position it was left at the previous stage.

Make $\chi(a_n) = (h_{i,s})_{i\in\mathbb{N}}$ in A_s for every $n \leqslant s$, and $(h_{i,s})_{i\in\mathbb{N}} = (d(e)_{i,s})_{i\in\mathbb{N}}$ in B_s for every $e \leqslant s$ which is not restrained, unless R_e keeps $d(e)_{i,s} = 0$.

End of construction.

Verification. If $\lim_t \Psi_e(b_e, t)$ does not exist, then we reach I of (6) infinitely often and, therefore, $\chi(b_e) = \alpha$. Assume that $\lim_t \Psi_e(b_e, t)$ exists. Let s be a stage such that

$$\Psi_e(b_e, s) = \lim_t \Psi_e(b_e, t).$$

Let $u \geqslant s$ be a stage such that $\lim_t \Gamma(g(e,s), t) = \Gamma(g(e,s), u)$.

The set $W_{g(e,s)}$ is designed to make $\lim_t \Gamma(g(e,s), t) = 1$. If $\Gamma(g(e,s), u) = 0$ was the case, then we would add more elements to $W_{g(e,s)}$ at a stage $v \geqslant u$ and eventually put some $n(j) \in Fin(\alpha)$ into $W_{g(e,s)}$, a contradiction.

By the definition of Γ, if $\lim_t \Gamma(g(e,s), t) = 1$, then there is at least one $j \in W_{g(e,s)} \cap Fin(\alpha)$. Furthermore, the strategy guarantees that there is exactly one such a j, namely the last witness $n(i)$ which visits III of the strategy at some stage and stays there from this stage on. For instance, the element b_e will eventually be unrestrained (see the construction).

The algebraic strategy guarantees b_e is not divisible by $p_{n(i)}$ while the image is. Furthermore, b_e is declared not restrained as soon as we reach III with $n(i)$, meaning that the characteristic of b_e satisfies the property $d(e)_j = \alpha_j$ for each $j \neq n(i)$. It remains to apply Theorem 5.1.15. □

We note that in the proposition above the algebraic strategy from Proposition 7.2.33 would not succeed. Theorem 7.2.25 is proved. □

7.2.5 Arbitrary completely decomposable groups

As we mentioned earlier, the isomorphism type of a completely decomposable group is fully determined by the types of its elementary summands (elementary direct components), and each elementary summand can be described by its *type*. However, the collection of types of the elementary components may (in some sense) "encode" a countable linear order (see Exercise 7.2.49), and one may expect that there is no arithmetical upper bound on the complexity of isomorphisms between such groups. Nonetheless, in this section we prove Theorem 7.2.3 that states that *every computable completely decomposable group is* Δ^0_5*-categorical.*

The proof of Theorem 7.2.3 extends methods from §7.2. Using the algebraic machinery developed in the proof of Theorem 7.2.3, in the next section we will show that the index set of computable completely decomposable groups is arithmetical.

Proof of Theorem 7.2.3

We prove that every completely decomposable group is Δ^0_5-categorical. The proof of this theorem is divided into several parts. In the first part we state and prove algebraic facts about decompositions of completely decomposable groups, not all of the facts are well-known. In the second part we introduce an algebraic notion of a basic pair which is central to the proof, and prove the main algebraic lemma. In the third part we give the construction which builds an isomorphism between any two copies of the group, and in the forth part we verify that the construction is computable in $\mathbf{0}^{(4)}$.

Proof of Theorem 7.2.3. Let G be a completely decomposable group.

Decompositions of completely decomposable groups. Fix any complete decomposition of G into elementary summands. For a type $\mathbf{f}$, denote by $G_{(\mathbf{f})}$ the sum of all elementary summands of G having type $\mathbf{f}$. If the group G has no elementary summands of type $\mathbf{f}$, then we set $G_{(\mathbf{f})} = 0$. We have:

$$G = \bigoplus_{\mathbf{f}} G_{(\mathbf{f})},$$

where $\mathbf{f}$ ranges over all types. Whenever we are given a completely decomposable group, we usually fix a complete decomposition of it. Given two types $\mathbf{t}$ and $\mathbf{s}$, write $\mathbf{t} \leqslant \mathbf{s}$ if for some $\chi \in \mathbf{t}$ and $\rho \in \mathbf{s}$, we have $\chi \leqslant \rho$ (component-wise).

Definition 7.2.35. For a torsion-free abelian group A and a type $\mathbf{f}$, denote by $A_{\mathbf{f}}$ the subgroup generated by elements of having types $\geqslant \mathbf{f}$, and denote by $A^{\star}_{\mathbf{f}}$ the subgroup of A generated by the elements having types $> \mathbf{f}$.

Remark 7.2.36. Note that, in general, $A^{\star}_{\mathbf{f}}$ may contain elements of type $\mathbf{f}$. For example, consider a group having elementary components of only two types:

$$A = A_{(\mathbf{s})} \oplus A_{(\mathbf{t})},$$

where $inf\{\mathbf{s}, \mathbf{t}\} = \mathbf{f}$ and both $\mathbf{s}$ and $\mathbf{t}$ are strictly greater than $\mathbf{f}$. We have $A^{\star}_{\mathbf{f}} = A$. As can be easily seen, the group A contains elements of type $\mathbf{f}$. Every element having non-zero projections onto both summands has this property.

Fact 7.2.37. *Let G be a completely decomposable group, and let $G = \bigoplus_t G_{(t)}$ be its decomposition in homogeneous completely decomposable summands. For every type f,*

$$G_f = \bigoplus_{t \geqslant f} G_{(t)}$$

and

$$G^{\star}_f = \bigoplus_{t > f} G_{(t)}.$$

Proof. Clearly, $\bigoplus_{\mathbf{t} \geqslant \mathbf{f}} G_{(\mathbf{t})}$ is contained in $G_{\mathbf{f}}$. For every element g of G, let $g = \sum_{\mathbf{t}} g_{\mathbf{t}}$ be its decomposition into projections onto the homogeneous summands $G_{(\mathbf{t})}$. Here $\mathbf{t}$ ranges over all types, and $g_{\mathbf{t}} = 0$ for almost every $\mathbf{t}$. Note that the type of g is the infimum of the types of the projections. Therefore, only projections onto the components of types $\geqslant \mathbf{f}$ may occur if $\mathbf{t}(g) \geqslant \mathbf{f}$. This shows

$$G_{\mathbf{f}} = \bigoplus_{\mathbf{t} \geqslant \mathbf{f}} G_{(\mathbf{t})}.$$

The proof for $G^{\star}_{\mathbf{f}} = \bigoplus_{\mathbf{t} > \mathbf{f}} G_{(\mathbf{t})}$ is similar. □

As a consequence of the preceding fact,

$$G_{\mathbf{f}}/G_{\mathbf{f}}^{\star} \cong G_{(\mathbf{f})}.$$

We can not expect this group to be definable within G, and we have to deal with the quotient $G_{\mathbf{f}}/G_{\mathbf{f}}^{\star}$ isomorphic to $G_{(\mathbf{f})}$.

Let α be a characteristic of type **f**. Define $G[\alpha] = \{g \in G : \alpha \leqslant \chi(g)\}$.

Fact 7.2.38. *In the notation introduced above, $G[\alpha] = H[\alpha] \oplus C$, where $C \leqslant G_{\mathbf{f}}^{\star}$ and $H = G_{(\mathbf{f})}$ which is the sum of elementary components of G having type* **f**.

Proof. By Fact 7.2.37,

$$G_{\mathbf{f}} = G_{(\mathbf{f})} \oplus G_{\mathbf{f}}^{\star}.$$

By its definition, $G[\alpha] \subseteq G_{\mathbf{f}}$. For every $g \in G_{\mathbf{f}}$, $\chi(g) \geqslant \alpha$ implies the projection of g onto $G_{(\mathbf{f})}$ has characteristic $\geqslant \alpha$. Also, every element $H[\alpha]$ can be realised as a projection of a $g \in G_{\mathbf{f}}$ with $\chi(g) \geqslant \alpha$. The fact now follows. □

Let P be a set of primes which is not the set of all primes. As before, let $\mathbb{Q}^{(P)}$ be the additive subgroup of the rationals $(\mathbb{Q}, +)$ generated by fractions of the form $\frac{1}{p^m}$, where $p \in P$ and $m \in \mathbb{N}$. Let r be a cardinal number. Define

$$V_{P,r} = \bigoplus_{i<r} \mathbb{Q}^{(P)}.$$

Let $\alpha = (\alpha_i)_{i \in \mathbb{N}}$ be a characteristic. Consider the group $(G[\alpha] + G_{\mathbf{f}}^{\star})/G_{\mathbf{f}}^{\star}$. By Fact 7.2.38,

$$(G[\alpha] + G_{\mathbf{f}}^{\star})/G_{\mathbf{f}}^{\star} \cong H[\alpha],$$

where $H = G_{(\mathbf{f})}$. The group H is homogeneous completely decomposable of type **f**. An easy modification of the first part of Theorem 7.2.19 implies:

Fact 7.2.39. *For any characteristic α,*

$$H[\alpha] \cong V_{P,r},$$

where $P = \{p_i : \alpha_i = \infty\}$, and r is the rank of H.

Let P be a set of primes corresponding to a type **f** in the sense as above. Recall that, for a set of primes P,

$$\tilde{P} = \{p : p \text{ is prime and } p \notin P\}.$$

The second part of Theorem 7.2.19 gives:

Fact 7.2.40. *If a set B is an excellent $\tilde{P}$-basis of $(G[\alpha] + G_{\mathbf{f}}^{\star})/G_{\mathbf{f}}^{\star}$, then $G_{\mathbf{f}}/G_{\mathbf{f}}^{\star}$ is generated by B over $\mathbb{Q}(\alpha)$, where $\mathbb{Q}(\alpha)$ is the subgroup of $(\mathbb{Q}, +)$ containing 1 in which $\chi(1) = \alpha$.*

Recall that $(G[\alpha] + G_{\mathbf{f}}^{\star})/G_{\mathbf{f}}^{\star} \cong H[\alpha]$, where $H = G_{(\mathbf{f})}$, the homogeneous component of G having type **f**. For a collection $B \subset (G[\alpha] + G_{\mathbf{f}}^{\star})/G_{\mathbf{f}}^{\star}$, we say that $C \subset G$ is a set of representatives of B if each element from C belongs to a class from B, end each class from B has a unique representative in C. The definition of S-independence implies:

Fact 7.2.41. *Let $B \subset (G[\alpha] + G^\star_f)/G^\star_f$ be $\tilde{P}$-independent, and let C be any set of representatives of B. Then the projection of C onto $G_{(f)}$ is $\tilde{P}$-independent in $H[\alpha]$, where $H = G_{(f)}$.*

We should note that in Fact 7.2.40 "generated" clearly means "generated mod $G^\star_{\mathbf{f}}$". This is also one of the main difficulties in proving the theorem: we need to deal with representatives of classes mod $G^\star_{\mathbf{f}}$, not with elements of $H[\alpha]$. This difficulty is circumvented by using *basic pairs.*

Basic pairs. We will need a listing of characteristics representing types of the elementary summands of G such that this listing does not contain equivalent characteristics:

Notation 7.2.42. Let G be a completely decomposable group. In the following, $(\mathbf{f}_i)_{i\in I}$ stands for the listing of types of nonzero homogeneous components, and for every $i \in I$, $\alpha_i = (\alpha_{i,j})_{j\in\mathbb{N}}$ is a characteristic of type $\mathbf{f}_i$. Define also $P_i = \{p_j : \alpha_{i,j} = \infty\}$, where $p_0, p_1 \ldots$ is the standard listing of primes.

Definition 7.2.43. We say that a pair (σ, v) is *basic* if the following conditions hold:

1. σ is a finite tuple of elements of G;
2. $v : \sigma \to I$ is a function;
3. $i \neq j$ implies $\alpha_i \not\sim \alpha_j$, for every $i, j \in \mathit{range}\, v$;
4. for every i, if $v^{-1}(i) \neq \varnothing$ then $v^{-1}(i)$ is a set of representatives of $\tilde{P}_i$-independent classes in $(G[\alpha_i] + G^\star_{\mathbf{f}_i})/G^\star_{\mathbf{f}_i}$.

Notation 7.2.44. Given $R \subseteq \mathbb{Q}$ and $X \subseteq G$, denote by $[X]_R$ the set of sums

$$\sum_{x\in X} r_x x$$

where $r_x \in R$ for every x, and $r_x = 0$ for almost all x. We also assume G contains $r_x x$, for every $x \in X$.

Given a basic (σ, v), let $\mathit{Span}\,(\sigma, v) = \sum_{i\in\mathbb{N}}[v^{-1}(i)]_{\mathbb{Q}(\alpha_i)}$, where $[\varnothing]_{\mathbb{Q}(\alpha_i)} = 0$. By the definition of basic pairs, the sum above is in fact direct:

$$\mathit{Span}\,(\sigma, v) = \bigoplus_{i\in\mathbb{N}}[v^{-1}(i)]_{\mathbb{Q}(\alpha_i)},$$

and, furthermore, every homogeneous summand of this direct decomposition splits into elementary components, each elementary component being the span of an element of σ over the corresponding $\mathbb{Q}(\alpha_i)$.

Thus, for every basic (σ, v), the subgroup $\mathit{Span}\,(\sigma, v)$ is a completely decomposable group of rank $|\sigma|$ with homogeneous components $[v^{-1}(i)]_{\mathbb{Q}(\alpha_i)}$, where $v^{-1}(i)$ is an excellent $\tilde{P}_i$-basis of $[v^{-1}(i)]_{\mathbb{Q}(\alpha_i)}$.

The lemma below is central to the proof of the theorem.

Lemma 7.2.45. *For every basic pair (σ, v) and every element $g \in G$ there is a basic pair (τ, u) such that $\sigma \subseteq \tau$ and $g \in \mathit{Span}\,(\tau, u)$.*

Proof. Note that $Span(\sigma, v)$ is contained in $A \leq G$ which is a finite direct sum of elementary summands of G. By the definition of a basic pair, if $v^{-1}(i) \neq \varnothing$ then $v^{-1}(i)$ is a set of representatives of $\tilde{P}_i$-independent classes in $(G[\alpha_i] + G^{\star}_{\mathbf{f}_i})/G^{\star}_{\mathbf{f}_i}$. By Fact 7.2.39,

$$H[\alpha_i] \cong V_{P_i,k},$$

where $H = G_{(\mathbf{f}_i)}$ and k is the rank of H. By Fact 7.2.41, the projection of $v^{-1}(i)$ onto $H = G_{\mathbf{f}_i}$ is $\tilde{P}_i$-independent within $H[\alpha_i]$. Furthermore, the projection of $v^{-1}(i)$ onto H is contained in $A_{(\mathbf{f}_i)}[\alpha_i]$ which is isomorphic to $V_{P_i,k}$, where $k \in \mathbb{N}$. By Proposition 7.2.15, the projection of $v^{-1}(i)$ can be extended to an excellent $\tilde{P}_i$-basis of $A_{(\mathbf{f}_i)}[\alpha_i]$. Note that, considering the pre-image of this extension under the projection onto H, we may choose representatives C_i of an excellent $\tilde{P}_i$-basis of $(A[\alpha_i] + A^{\star}_{\mathbf{f}_i})/A^{\star}_{\mathbf{f}_i}$ so that these representatives are contained in A.

Let τ be the union of the C_i, where i ranges over the set $J = \{i : A_{(\mathbf{f}_i)} \neq 0\}$, and let u be a function which maps every element of τ into its characteristic. We prove by induction that

$$Span(\tau, u) = A.$$

The group A is of finite rank, and the partial ordering $\{\mathbf{f}_i : i \in J\}$ of the types of its elementary components is finite. We argue by induction on the number of types in this partial ordering, as follows. By Fact 7.2.40, for every $i \in J$ the factor-group $A_{\mathbf{f}_i}/A^{\star}_{\mathbf{f}_i}$ is generated over $\mathbb{Q}(\alpha_i)$ by the classes corresponding to C_i. (Recall the discussion after Fact 7.2.41.) Let $j \in J$ be such that $\mathbf{f}_j$ is maximal in $\{\mathbf{f}_i : i \in J\}$. By Fact 7.2.37,

$$A^{\star}_{\mathbf{f}_j} = 0.$$

Consequently, C_j generates $A_{\mathbf{j}}$ over $\mathbb{Q}(\alpha_i)$.

Let $\rho = \tau - C_j = \bigcup_{i \in J - \{j\}} C_i$, and let w be the restriction of u onto ρ. By the induction hypothesis,

$$A/A_{(\mathbf{f}_j)} = Span(\rho, w)/A_{(\mathbf{f}_j)}.$$

Therefore, every element of $\bigoplus_{i \neq j} A_{(\mathbf{f}_i)}$ is generated by elements of ρ and elements of $A_{(\mathbf{f}_j)}$. This shows $Span(\tau, u) = A$ and, by the choice of A, we have $g \in Span(\tau, u)$ □

Building an isomorphism. Given a computable completely decomposable group $G = \{g_0 = 0, g_1, \ldots\}$, we define stage-by-stage a sequence of basic pairs

$$(\sigma_0, v_0), (\sigma_1, v_1), \ldots$$

starting with $\sigma_0 = \varnothing$ such that $Span(\sigma_j, v_j)$ contains g_j, for each j. Without loss of generality, we may assume G has infinite rank (for otherwise it is computably categorical). By Lemma 7.2.45, we obtain an infinite sequence:

$$(\sigma_j, v_j)_{j \in \mathbb{N}}.$$

Consider $B = \bigcup_{j \in \mathbb{N}} \sigma_j$ and $U = \bigcup_{j \in \mathbb{N}} v_j$. The set B is a basis of G, and, by the definition of a basic pair,

$$G = \bigoplus_{i \in I} [U^{-1}(i)]_{\mathbb{Q}(\alpha_i)},$$

where $(\mathbf{f}_i)_{i \in I}$ is the listing of types of homogeneous components, and for every $i \in I$,

$$\alpha_i = (\alpha_{i,j})_{j \in \mathbb{N}}$$

is a characteristic of type $\mathbf{f}_i$. On the other hand,

$$G = \bigoplus_{b \in B} R_b,$$

where R_b is the span of b over $\mathbb{Q}[\alpha_j]$ with $j = U(b)$. Thus, if we are given (B, U), we can uniformly construct a "regular" decomposition of G into elementary components with U pointing the characteristic of a given "regular" element b in this decomposition.

It remains to observe that we may run this process on any other computable copy D of G and obtain a pair (T, V), where T is a basis and V is a function mapping elements of T into their characteristics. Given (B, U) and (T, V), we stage-by-stage map $b \in B$ to a rational multiple of $c \in T$ having the same characteristic as b, and then extend this map to an isomorphism of G onto D in the obvious way.

It remains to check which oracle is sufficient to build such a sequence of basic pairs in a given computable completely decomposable group G.

Calculating the complexity. Let $G = (g_0, g_1, \ldots)$ be a computable completely decomposable group, and let $G = \bigoplus_{i \in I} G_{(\mathbf{f}_i)}$ be its decomposition into homogeneous completely decomposable components. For every j, let $\beta_j = \chi(g_j)$. We need an enumeration of characteristics which correspond to different types.

Fact 7.2.46. *There exists a Σ^0_4 set $\mathcal{J} \subseteq \omega$ such that*

(a.) for every $i \in I$ there exists $j \in \mathcal{J}$ such that $\beta_j \in \mathbf{f}_i$;

(b.) $\beta_i \not\sim \beta_j$, for every $i, j \in \mathcal{J}$.

Proof. It is sufficient to show that the relation $\{(i, j) : \beta_i \sim \beta_j\}$ is Σ^0_3. Note that there is a 1-1 correspondence between β_i and the set of pairs

$$X_i = \{(j, k) : p_k^j | g_i\}.$$

The family of sets $(X_i)_{i \in \mathbb{N}}$ has a uniform enumeration. Also, every X_i is associated to the corresponding element g_i in an effectively uniform way. It remains to observe that $\beta_i \sim \beta_j$ if, and only if, $X_i =^* X_j$, for every i, j. □

Note that $(\mathbf{f}_j)_{j \in \mathcal{J}}$ are not necessarily exactly the types which correspond to non-zero $G_{(\mathbf{f}_i)}$ in the decomposition of G. Using $\mathcal{J}$ we establish a $\mathbf{0}'''$-computable uniform enumeration of pairwise non-equivalent c.e. characteristics $(\alpha_i)_{i \in I}$ covering all types of non-zero component present in the complete decomposition of G.

Fact 7.2.47. *For every j:*

(1.) $G[\alpha_j]$ is Π^0_2 uniformly in j;

(2.) $G^\star_{\mathbf{f}_j}$ is Σ^0_4 uniformly in j.

Proof. We have

$$g \in G[\alpha_j] \Leftrightarrow \chi(g) \geqslant \alpha_j \Leftrightarrow (\forall k)(\forall n)(p_k^n | g_j \to (n, k) \in \alpha_j)$$

which is Π^0_2. Also, $g \in G^\star_{\mathbf{f}_j}$ iff

$$(\exists k \in \mathbb{N})(\exists g_1, \ldots, g_k \in G)(\exists n \in \mathbb{N})\left(\chi(g_i) \not\sim \alpha_j \wedge \chi(g_i) \geqslant \alpha_j \wedge ng = \sum_{1 \leqslant i \leqslant k} g_i\right)$$

which is Σ^0_4, because $\chi(g_i) \not\sim \alpha_j$ is Π^0_3 as we have observed in the proof of Fact 7.2.46, and $\chi(g_i) \geqslant \alpha_j$ is Π^0_2. □

As a consequence of this fact, $(G[\alpha] + G^\star_{\mathbf{f}_j})/G^\star_{\mathbf{f}_j}$ has a Σ^0_5 set of representatives. We need more:

Fact 7.2.48. *Given i and elements $g_1, \ldots, g_k \in G[\alpha_j]$, the statement "the classes of $g_1, \ldots, g_k$ are $\tilde{P}_i$-independent in $(G[\alpha_j] + G^\star_{\mathbf{f}_j})/G^\star_{\mathbf{f}_j}$" is Π^0_4 uniformly in the indices of elements and in α_j.*

Proof. It is sufficient to require that, for every choice of coefficients $m_1, \ldots, m_k$ and for every prime p,

$$\left[(\exists y)(\exists x)\left(x \in G[\alpha] \wedge y \in G^\star_{\mathbf{f}_j} \wedge p \notin P_j \wedge px + y = \sum_{s \leqslant k} m_s g_s\right)\right] \Rightarrow \bigwedge_{s \leqslant k} p \mid m_s,$$

which is Π^0_4, by the preceding facts. □

Thus, $\mathbf{0}^{(4)}$ can build a sequence of basic pairs generating the whole group G. This finishes the proof. □

Exercises

Exercise 7.2.49. Given two types $\mathbf{t}$ and $\mathbf{s}$, write $\mathbf{t} \leqslant \mathbf{s}$ if for some $\chi \in \mathbf{t}$ and $\rho \in \mathbf{s}$, we have $\chi \leqslant \rho$ (component-wise). Let L be a computable linear order. Give an example of a computable completely decomposable group $G(L)$ in which the order on the types of the elementary summands is isomorphic to L. (Hint: Given a countable linear order L, associate every point $\ell \in L$ with a type $\mathbf{t}(\ell)$ so that

$$\ell \leqslant \ell_0 \text{ if and only if } \mathbf{t}(\ell) \leqslant \mathbf{t}(\ell_0).$$

Use a dichotomy-like process manipulating with infinite computable disjoint sets of primes. Then define $G(L) = \bigoplus_{\ell \in L} \mathbb{Q}(\mathbf{t}(\ell))$, where $\mathbb{Q}(\mathbf{t}(\ell)) \leqq \mathbb{Q}$ has type $\mathbf{t}(\ell)$.)

Exercise 7.2.50 (Melnikov [365, 374]). Using the family from Exercise 3.2.63, show that there exists a completely decomposable group that has an X-computable presentation iff $X' >_T \varnothing'$, i.e., iff X is non-low[9]. Using the material of Subsection 5.2.3, deduce that there exists a space that has an X-computably compact presentation iff X in non-low, and that there is a space that has an X-computable Polish presentation iff $X'' >_T \varnothing''$, i.e., X is non-low$_2$.

Exercise* 7.2.51 (Kach, Lange, and Solomon [276]). A computable completely decomposable group is effectively completely decomposable if it splits into the direct sum of its uniformly computable subgroups of rank 1. Let H be an effectively completely decomposable group of infinite rank. Show that there is a computable presentation G of H so that in G, the Turing degrees of orders compatible with the group operation (i.e., turning the group into an ordered group) are not closed upwards.

[9]This exercise is not difficult but requires a new idea.

Exercise* 7.2.52 (Downey et al. [120]). For a prime p, let $\mathbb{Q}^{(p)} = [\mathbb{Z}]_p = \langle\{\frac{1}{p^n} : n \in \mathbb{N}\}\rangle \leqq \mathbb{Q}$. Show that $\bigoplus_{p \in S} \mathbb{Q}^{(p)}$ is computably presentable iff S is a Σ^0_3 set of primes.

7.3 Applications to index sets

This section contains upper estimates for the characterisation problem and the isomorphism problem of completely decomposable groups and their Pontryagin duals, the products of solenoids.

7.3.1 Completely decomposable groups

Our machinery enables us to prove:

Theorem 7.3.1 (Downey and Melnikov [137]). *Completely decomposable groups form an arithmetical class. Both the recognition and the isomorphism problem for this class have complexity* Σ^0_7.

Proof. Recall that every characteristic $\alpha = (\alpha_i)_{i\in\mathbb{N}}$ can be viewed as a set of pairs $\{(k,i) : k \leqslant \alpha_i\}$ thst we call the corresponding *characteristic sequence.* The following fact is easy but helpful.

Fact 7.3.2. *There exists a uniform enumeration of all c.e. characteristic sequences.*

Proof. Given an enumeration of a c.e. set, effectively and uniformly transform it into an enumeration of a characteristic sequence by closing every column $\{s : (s,i) \in W_e\}$ downwards. □

Identifying characteristics and corresponding characteristic sequences, let $(\beta_j)_{j\in\mathbb{N}}$ be the uniform enumeration from Fact 7.3.2. The isomorphism type of a completely decomposable group $G = \bigoplus_{\mathbf{f}} G_{(\mathbf{f})}$ is uniquely determined by the set

$$\{(\mathbf{f}, k) : rank\,(G_{(\mathbf{f})}) = k\}.$$

We may replace every type in the set above by a characteristic of that type, and still get a full invariant describing G up to an isomorphism, modulo the equivalence of the characteristics. The proof of Theorem 7.2.3 enables us to re-formulate Theorem 7.2.3 as follows:

Fact 7.3.3. *For every computable completely decomposable group G, there is a $\mathbf{0}^{(4)}$ enumeration of a set of the form*

$$\{(j,s) : rank\,(G_{(\boldsymbol{t}(\beta_j))}) \geqslant s > 0\},$$

where $\beta_i \not\sim \beta_j$ and $G = \bigoplus_j G_{(\boldsymbol{t}(\beta_j))}$.

On the other hand, every uniformly computable family of characteristic sequences can be realised as one corresponding to a direct decomposition of a computable completely decomposable group:

Fact 7.3.4. *For every set $S = \{(j,s) : s \geqslant 1\}$ such that $\{\beta_j : (j,1) \in S\}$ is a set of pairwise non-equivalent characteristics in the uniform enumeration of all characteristics $(\beta_j)_{j\in\mathbb{N}}$, there exists a computable completely decomposable group of the form $\bigoplus_{j:(j,1)\in S}\left(\bigoplus_{k:(j,k)\in S}\mathbb{Q}(\beta_j)\right)$.*

Proof. The proof is not difficult and is left as Exercise 7.3.6. □

Summarising the above, every computable completely decomposable group has a corresponding Σ^0_5 family of characteristic sequences, and every such sequence can be associated to a $\mathbf{0}^{(4)}$-computable completely decomposable group in a uniform way. Also, Fact 7.3.2 can be relativised to $\mathbf{0}^{(4)}$. We obtain:

Fact 7.3.5. *There is a Σ^0_5 listing $(A_i)_{i\in\mathbb{N}}$ of $\mathbf{0}^{(4)}$-computable completely decomposable groups containing all isomorphism types of computable completely decomposable groups (possibly with repetitions).*

Note that every group A_i from the enumeration provided by Fact 7.3.5 and Fact 7.3.4 has a $\mathbf{0}^{(4)}$-computable complete decomposition algorithm: every group A_i is given together with a basis, where each element of the basis belongs to a separate component. By Fact 7.3.3, we may assume that characteristics of elements of the complete decomposition basis of A_i are c.e. (whereas the indexes of these characteristics are merely c.e. in $\mathbf{0}^{(4)}$). By the third part of the proof of Theorem 7.2.3, if a computable completely decomposable group G is isomorphic to A_i, then this isomorphism is in fact Δ^0_5: it suffices to build a sequence of basic pairs in G, take their union, and then map each element from the union to an element of the basis of A_i having an equivalent (c.e.) characteristic, maybe up to a rational multiple. Notice that we are implicitly using the uniqueness of the complete decomposition of G.

Thus, a computable structure is a completely decomposable group if, and only if, it is isomorphic to one of the groups $(A_i)_{i\in\mathbb{N}}$ from Fact 7.3.5 via a $\mathbf{0}^{(4)}$-computable isomorphism. Given a computable structure M_j, we ask "Is there i and a $\mathbf{0}^{(4)}$-isomorphism from A_i onto M_j?" which is uniformly Σ^0_7. It follows that the recognition problem is Σ^0_7.

It remains to check that the isomorphism problem is Σ^0_7. But this follows from the recognition problem being Σ^0_7 and from Fact 7.1.2. □

7.3.2 Products of solenoids. Proof of Theorem C

Recall that Theorem C states that direct products of solenoid groups form an arithmetical class (under topological isomorphism). Recall that a solenoid group is a group that is dual to some (non-nil) subgroup of $\mathbb{Q}$. It is well known that, under Pontryagin duality, direct sums become direct products, and vice versa. Thus, the duals of direct products of solenoid groups are exactly the completely decomposable groups.

In Theorem 7.3.1 we established that completely decomposable groups form an arithmetical class. The rest of the proof is very similar to the proof of Theorem 7.1.41. We use computably compact presentations; computable Polish presentations would give one more extra quantifier for the complexity.

Propositions 7.1.38 and 7.1.39 guarantee that being a connected compact group is an arithmetical property (Π^0_2), and being abelian is Π^0_1 and can be checked only for special points. It is also Σ^0_1 to tell whether the group is non-zero. (If it is zero we are done.) Given G that is already known to be non-zero and connected compact abelian, apply Theorem 5.2.21 (and Theorem 5.2.24) and produce a c.e.-presentation of its discrete Pontryagin dual. Since G is non-zero, we can uniformly pass to its computable presentation using Khisamiev's Theorem 5.1.41. Let H be the resulting computable presentation of the dual. By Theorem 7.3.1, it is Σ^0_7 to tell whether the resulting abelian group is completely decomposable. This makes the recognition problem for products of solenoid groups Σ^0_7.

Given two groups, use the procedure described above to pass to their duals. (The case when at least one of them is the zero-group is trivial and can be considered separately.) By Theorem 7.3.1, expressing that the duals are isomorphic is Σ^0_7.

This finishes the proof of Theorem C.

Exercises

Exercise° 7.3.6. Prove Fact 7.3.4.

Exercise* 7.3.7. Prove Theorem 7.3.11.

Exercise* 7.3.8. Prove Theorem 7.3.10.

7.3.3 Concluding remarks and further related results*

We conjecture that the estimate Σ^0_7 (for both the completely decomposable groups and their duals) is not optimal, i.e., the index sets are not Σ^0_7-complete. However, obtaining optimal estimates remains an open problem.

Problem 7.3.9 (Downey and Melnikov [137]). *Provide optimal complexity calculations for the isomorphism problem and the recognition problem for the class of completely decomposable groups (and for the class of their Pontryagin duals).*

In contrast with Theorem 7.3.1 and Theorem C, we do know that $n = 5$ in Theorem 7.2.3 is sharp:

Theorem 7.3.10 (Downey and Melnikov [137]). *There is a computable completely decomposable group which is not Δ^0_4-categorical.*

For instance, Theorem 7.2.2 fails beyond homogeneous completely decomposable groups. It could be that every Δ^0_4-categorical group was already Δ^0_3-categorical, similarly to well-orderings [16].

Theorem 7.3.11 (Downey and Melnikov [137]). *There is a computable completely decomposable group which is Δ^0_4-categorical but not Δ^0_3-categorical.*

It is perhaps worth noting that Δ^0_n-categoricity of torsion-free abelian groups that are not completely decomposable is very poorly understood (for $n > 1$). The best we know is the following technical result that solved an open problem posed by Goncharov.

Theorem 7.3.12 (Melnikov [372]). *For every $n > 0$ there exists a torsion-free abelian group $\mathcal{G}_n$ which is (relatively) Δ^0_{2n}-categorical, but not Δ^0_{2n-1}-categorical.*

The theorem has a transfinite extension to "even" successor computable ordinals; the formal statement will be given in Exercise 10.1.118. The reader may find it intriguing why the case of odd levels of the (hyper)arithmetical hierarchy in the theorem above is still open. For a detailed explanation of the technical difficulties and a further discussion of many related results, see [372].

Historical remark

The systematic study of effective properties of completely decomposable groups was initiated by Khisamiev and Krykpaeva [293] and then further developed by Khisamiev in [292]. The main notion investigated in these papers was that of an effectively completely decomposable group defined in Exercise 7.2.51. The earliest results about completely decomposable groups without any extra restrictions on effectiveness of the decomposition can be found [365, 120]. The main result of [120] (see Exercise 7.2.52) was further extended in [452]. We also cite [33] and [276] for further results about computable completely decomposable groups. See also [222] where uncountable free abelian groups were investigated using methods of higher recursion theory.

The "unsatisfactory" coding

Finally, recall that one can "computably code" an arbitrary countable linear order into the types of a completely decomposable group; see Exercise 7.2.49. We can further use effective Pontryagin duality to turn a computable linear order into a computably compact connected group. In the next chapter, we will see that computable linear orders are unclassifiable, while Theorem C suggests that the class of Pontryagin duals of completely decomposable groups is generally tame.

But clearly, there is no contradiction. The "transformation" hinted in Exercise 7.2.49 is heavily dependent on the specific presentation of the input linear order, with two different presentations very likely giving non-isomorphic groups. To get a contradiction with the results of the present chapter, one must find an effective transformation that is well-defined on isomorphism types, i.e., so that $L \mapsto G(L)$ has the property $L \cong \Gamma$ if and only if $G(L) \cong G(\Gamma)$. Theorem 7.2.3 and Theorem C illustrate that no such well-behaved transformation can possibly exist.

7.4 What's next?

In the next chapter we will develop enough techniques to illustrate that in many natural classes of structures and spaces, the index sets are Π^1_1- and Σ^1_1-complete. In the book, we put emphasis on positive results, and the "nonclassification" chapter that follows next is a bit more brief than this chapter. However, references to the (vast) relevant literature will be provided.

Chapter 8

Nonclassification theory

In this chapter we will prove that in many classes having an arithmetical recognition problem, the isomorphism problem is Σ^1_1-complete. These classes include Boolean algebras, separable Banach spaces, and integral domains. The main result of this chapter is as follows.

Theorem D (Melnikov [373], based on Downey and Montalbán [146]). The homeomorphism problem for connected compact Polish spaces is Σ^1_1-complete

The structure of this chapter is as follows:

1. Section 8.1 contains the basic definitions and results related to Σ^1_1- and Π^1_1-complete sets.
2. Section 8.2 gives a brief introduction to the recently emerged theory of effective reducibilities between classes of structures, with applications to index sets.
3. Section 8.3 presents the proof of the Downey-Montalbán Theorem, which states that the isomorphism problem for torsion-free abelian groups is Σ^1_1-complete. We also present a proof of a stronger result, Theorem 8.3.10, which establishes the completeness of the class of torsion-free abelian groups among Σ^1_1 equivalence relations. Combined with effective Pontryagin duality (established in Part 1) and the material of Section 8.1, these theorems will be used to derive results about separable spaces, including Theorem D.

R. G. Downey, A. Melnikov, *Computable Structure Theory*, Theory and Applications of Computability, https://doi.org/10.1007/978-3-031-92433-0_8

8.1 Foundations

This section contains a brief exposition of definitions and facts concerning Σ^1_1- and Π^1_1-completeness, and their immediate applications to the index sets of discrete and separable structures.

8.1.1 Definitions and notation

The classical theorems discussed in this subsection will be utilised as "black boxes" in applications without further modifications. This is in contrast with priority techniques, which had to be combined with algebra and combinatorics throughout much of Part 1 of the book. For a detailed exposition of this topic, see [454, Ch.16]. For a more in-depth study of higher recursion theory, refer to [458]. Additionally, [20] contains a detailed and comprehensive introduction to the hyperarithmetical and the analytical hierarchies.

We identify a (total) function $f : \omega \to \omega$ with a member of ω^ω. We say that a relation $R(f_1, \ldots, f_n; x_1, \ldots, x_m) \subseteq (\omega^\omega)^n \times \omega^m$ is computable if there is a Turing functional Φ_e with the property

$$\Phi_e^{f_1,f_2,\ldots,f_n}(x_1,\ldots,x_m) = \begin{cases} 1, & \text{if } R(f_1,\ldots,f_n;x_1,\ldots,x_m) = 1; \\ 0, & \text{otherwise.} \end{cases}$$

In $\Phi_e^{f_1,f_2,\ldots,f_n}(x_1,\ldots,x_m)$, we either use a multi-tape Turing machine of fix some computable coding of $f_1, f_2, \ldots, f_n$ into one function, e.g., $f_1 \oplus f_2 \ldots \oplus f_n(\langle i,k\rangle) = f_i(k)$, $1 \leqslant i \leqslant n$. The *analytical relations* are the members of the least class that includes computable relations (in the sense above) and is closed under complements and projections on individual and function variables. In other words, a relation is analytical if it can be written in the form

$$Q_1\xi_1 Q_2\xi_2 \ldots Q_n\xi_n R$$

where R is a computable relation, $Q_i \in \{\forall, \exists\}$, and each ξ_i either ranges over ω or over ω^ω. For example, if R is computable on $(\omega^\omega)^3 \times \omega^2$, then $\{(h,m) : \exists f \forall n \forall g R(f,g,h;n,m)\} \subseteq (\omega^\omega) \times \omega$ is analytical.

Theorem 8.1.1 (The Normal Form). *Any analytical relation P has a definition of the form*

$$(\dagger) \quad Q_1 f_1\, Q_2 f_2 \;\ldots, Q_k f_k\, Q_{k+1} z\, R(f_1, \ldots, f_k, \ldots; z, \ldots),$$

where R is a computable relation, the quantifiers Q_i alternate between $\exists$ and $\forall$, and all but the last one ranges over function variables.

Proof. This follows from Exercise 8.1.21 which summarises the rules of manipulating with quantifiers. □

Definition 8.1.2. Fix $k > 0$. An analytical relation P is:

1. Π^1_k if (there is a computable R so that) P satisfies ($\dagger$) with $Q_1 = \forall$;
2. Σ^1_k if P satisfies ($\dagger$) with $Q_1 = \exists$.

We also say that P is Δ^1_k if it is both Π^1_k and Σ^1_k.

We also use Σ^1_k and Π^1_k to denote the respective complexity classes. Thus, $\Delta^1_k = \Pi^1_k \cap \Sigma^1_k$. Clearly, $P \in \Pi^1_k$ if and only if $\neg P \in \Sigma^1_k$, where $\neg P$ is the complement of P in the respective domain.

We are mainly interested in relations on (subsets of) ω; thus, the reader may safely assume that all our analytic relations are subsets of ω unless otherwise specified.

Theorem 8.1.3 (Kleene). *For every k, there is a set in $\Sigma^1_k \backslash \Pi^1_k$ and a set in $\Pi^1_k \backslash \Sigma^1_k$.*

The proof of this fact can be derived from the existence of the universal Σ^1_k- and Π^1_k-relations combined with the usual diagonalisation; see Exercises 8.1.23 and 8.1.24.

8.1.2 Σ^1_1- and Π^1_1-complete sets

Explicit examples of sets in $\Sigma^1_1 \backslash \Pi^1_1$ are the Σ^1_1-complete sets.

Definition 8.1.4. A Σ^1_1 set $P \subseteq \omega$ is Σ^1_1-complete if any other Σ^1_1-set $\subseteq \omega$ is m-reducible to P. The notion of a Π^1_1-complete $P \subseteq \omega$ is defined similarly.

We give two examples of a natural Σ^1_1-complete and Π^1_1-complete *index* sets. As before, by a tree we mean a subset of $\omega^{<\omega}$ closed under prefix. An *infinite path* through a tree T, or sometimes just a *path* through T, is $p \in \omega^\omega$ such that all finite initial segments of p lie in T. A tree with no path is called *well-founded.* Otherwise, T is *ill-founded.*

Interpret every c.e. set W_e as a collection of indices of finite strings. We can uniformly modify every c.e. set into a c.e. set W^*_e that closes the listed strings under prefixes, i.e., whenever σ is a prefix of τ and τ is listed in W^*_e, we also put σ in W^*_e. Denote the resulting tree $\subseteq \omega^{<\omega}$ listed by W^*_e by T_e. Define the index sets

$$WF = \{e : T_e \text{ is well-founded}\}$$

and

$$IF = \{e : T_e \text{ is ill-founded}\} = \omega \backslash WF.$$

Theorem 8.1.5 (Folklore). *IF is Σ^1_1-complete and, thus, WF is Π^1_1-complete.*

Proof. It is clear that $WF \in \Pi^1_1$ and $IF \in \Sigma^1_1$. Fix a Σ^1_1-set S and a computable relation R such that $x \in S$ if and only if $\exists f \forall y R(f; y, x)$. Given x, define a c.e. set V_x as follows. Put the code for a string σ in V_x if for some y, $\forall z \leqslant y\, R(\sigma; z, x)$ holds, where σ is identified with $\sigma 0^\omega$, and also σ is assumed to be shorter than the use of (the computable functional witnessing the computability of) R on inputs x, z $(z \leqslant y)$.

Using the s-m-n Theorem, fix a computable function g such that $W_{g(x)} = V_x$, for every x. The tree $T_{g(x)}$ has an infinite path if, and only if, $x \in S$. Thus, g witnesses $S \leqslant_m IF$. Indeed, g also simultaneously witnesses the Π^1_1-completeness of WF. □

With each tree in Baire space, we associate an ordering.

Definition 8.1.6. The *Kleene-Brouwer ordering* $<_{KB}$ on strings of a tree $T \subseteq \omega^{<\omega}$ is defined as follows. Declare $\sigma <_{KB} \tau$ if one of the two conditions hold:

1. σ is a (proper) initial segment of τ, or

2. σ and τ agree up to, but not including, position n, and $\sigma(n) < \tau(n)$.

We let $\sigma \leqslant_{KB} \tau$ if $\sigma = \tau$ or $\sigma <_{KB} \tau$.

It is easy to see that $<_{KB}$ is a linear ordering.

Remark 8.1.7. Kleene used the ordering extensively in his work. However, originally, the ordering had been defined before Kleene by Lusin and Sierpinski [342], and Brower [66] in the 1920's. In keeping with the standard terminology, we will refer to the ordering as we have.

Theorem 8.1.8. *T is well-founded iff $<_{KB}$ is a well-ordering of nodes in T.*

Proof. If T is ill-founded and $\alpha \in [T]$, then the finite initial segments of α form an infinite descending $<_{KB}$-chain. Conversely, suppose that $\sigma_1, \sigma_2, \ldots$ is an infinite descending $<_{KB}$-chain. Then for each n, the set $\{k : \exists m\, \sigma_m(n) = k\}$ is bounded from above, and, thus, $\sigma_1, \sigma_2, \ldots$ belong to the restriction of T to a finitely branching subtree of $\omega^{<\omega}$. It follows that $[T] \neq \varnothing$ (e.g., by König's lemma). □

Further note that the transformation $T \mapsto KB(T)$ is effectively uniform. Let $(L_e)_{e \in \mathbb{N}}$ be a uniformly effective enumeration of all partial computable structures in the language of linear orders (one binary relation). We arrive at:

Corollary 8.1.9. *The index set $\{e : L_e$ is an ordinal$\}$ is Π^1_1-complete.*

Kleene's $\mathcal{O}$

Recall that in Section 2.2.3 we defined $\mathcal{O}$ to be the collection of all (notations for) constructive ordinals. In Section 2.2.3 we also proved that an ordinal is computable if, and only if, it is constructive, i.e., has a notation in $\mathcal{O}$. In view of Corollary 8.1.9 the classical theorem below is perhaps not too surprising. However, its proof is non-trivial and is omitted.

Theorem 8.1.10 (Kleene, Spector). *$\mathcal{O}$ is Π^1_1-complete.*

Kleene's idea was to extend the definition of $\varnothing^{(\omega)} = \oplus_{n \in \mathbb{N}} \varnothing^{(n)}$ as follows:

Definition 8.1.11 (Kleene). We define the sets $H(o)$ for notations o by effective transfinite recursion as follows.

- $H(1) = \varnothing$
- $H(2^a) = H(a)'$
- $H(3 \cdot 5^e) = \{\langle u, v\rangle : \exists n(u \leqslant_{\mathcal{O}} \varphi_e(n) \wedge v \in H(u))\}$.

The operator H is well-behaved under $\leqslant_T$. The result is certainly nontrivial, and was thought to be quite surprising when first proved; we omit the proof.

Theorem 8.1.12 (Spector [481]). *1. There is a partial computable function f such that for notations $a, b \in \mathcal{O}$, if $|a|_{\mathcal{O}} < |b|_{\mathcal{O}}$ then $f(a, b)$ is an index for $H(a)$ as a set computable from $H(b)$.*

2. Thus if a and b are notations for an ordinal α, $H(a) \equiv_T H(b)$.

Thus, for $a \in \mathcal{O}$ is a notation for α, then the "α-th Turing jump" can be well-defined up to $\equiv_T$ via

$$H(2^b) = H(b)'$$

if $a = 2^b$ and, when $a = 3 \cdot 5^e$,

$$H(3 \cdot 5^e) = \{\langle u, v\rangle : u <_{\mathcal{O}} 3 \cdot 5^e \wedge v \in H(u)\} = \{\langle u, v\rangle : \exists n(u \leqslant_{\mathcal{O}} \varphi_e(n) \wedge v \in H(u))\}.$$

Theorem 8.1.13 (Kleene [301]). *$A \in \Delta^1_1$ iff $A \leqslant_T H(a)$ for some $a \in \mathcal{O}$.*

For instance, every arithmetical set is Δ^1_1, and so is every $\varnothing^{(\omega)}$-computable set. For a computable ordinal $\alpha \geqslant \omega$, define Δ^0_α to be the complexity class of all sets computable relative to $H(a)$, where a is any notation for α.

Since $H(a) \equiv_T H(b)$ for any other notation b for α, the class is well-defined. (Note that for a finite $\alpha = n$, the Δ^0_{n+1}-sets are those computable in $0^{(n)}$, not in $0^{(n+1)}$.) It follows from the theorem above that

$$\Delta^1_1 = \bigcup_{\alpha < \omega_1^{CK}} \Delta^0_\alpha,$$

which is a remarkable fact. The classes Δ^0_α, $\alpha \geqslant \omega$, form the extension of the arithmetical hierarchy known as the hyperarithmetical hierarchy. If $X \subseteq \omega$, then we can define the classes $\Delta^0_\alpha(X)$ and $\Delta^1_1(X)$ by iterating $X^0, X', X'', \ldots$ over computable ordinals. It is also not too hard to show that if $X \in \Delta^0_\alpha$ and $Y \in \Delta^0_\beta(X)$, then $Y \in \Delta^0_{\alpha+\beta}$, i.e. "hyperarithmetical relative to a hyperarithmetical oracle is again hyperarithmetical"; e.g. [20, Proposition 5.21].

The Harrison order

For proofs of the following two facts, we refer to Lemma 2.1.III and Lemma 2.2.III of [458], respectively.

Theorem 8.1.14 (Kleene). *There exists a computable linear ordering with an infinite descending sequence but no hyperarithmetical descending sequence.*

Theorem 8.1.15 (Harrison). *Any computable linear order with the property as in Theorem 8.1.14 must have order type*

$$\omega_1^{CK}(1 + \mathbb{Q}) + \delta,$$

where ω_1^{CK} is the least non-computable ordinal, and $\delta < \omega_1^{CK}$ is a computable ordinal.

Note that if L satisfies Theorem 8.1.14 then so does $L \cdot \omega$. Thus, there is a computable linear order of the order-type $\omega_1^{CK}(1 + \mathbb{Q})$ which has no hyperarithmetical descending sequence.

Definition 8.1.16. The linear order $\omega_1^{CK}(1 + \mathbb{Q})$ will be called *the Harrison order* and denoted $\mathcal{H}$.

The linear order $\mathcal{H}$ is not computably categorical. However, we shall occasionally stretch our notation and identify $\mathcal{H}$ with *some* computable copy of $\mathcal{H}$; the exact choice of this copy will usually be clear from the context.

A useful lemma

Given a linear order L indexed by natural numbers, define the tree of descending sequences $DS(L)$ of L to be the collection of all tuples $\langle \ell_0, \dots \ell_k \rangle \in L^{<\omega} = \omega^{<\omega}$ such that $\ell_0 >_L \dots >_L \ell_k$ (together with the empty string). It should be clear that L is a well-order iff $DS(T)$ is well-founded.

It T, S are trees, then define their product $U = T * U = \{(\sigma, \tau) : |\sigma| = |\tau|, \sigma \in T, \tau \in U\}$ that can be viewed as a tree in $\omega^{<\omega} \times \omega^{<\omega} \cong \omega^{<\omega}$, under the coordinate-wise prefix relation on pairs. Note that $T * U$ is ill-founded iff *both* U and T are ill-founded. Further, any infinite path through $T * U$ computes an infinite path through T and an infinite path through U.

The following lemma will be used in applications to index sets. It can be found in [213]; see the proof of [213, Theorem 4.4(d)] where it is sketched. A variation of the lemma is [146, Lemma 3.1].

Lemma 8.1.17. *There is a computable operator R mapping computable trees to computable trees with the following properties:*

1. *$R(T)$ is well-founded iff T is well-founded, and*

2. *if $R(T)$ is not well-founded, then $R(T) \cong DS(\mathcal{H})$,*

where $\mathcal{H}$ is the Harrison order and $DS(\mathcal{L})$ be the tree of finite descending sequences of a linear ordering $\mathcal{L}$.

Proof. Identify $\mathcal{H}$ with a a computable linear order $L \cong \mathcal{H}$ given by Theorem 8.1.14 (and the remarks after Theorem 8.1.15)). Given T, define $R_0(T) = T * DS(\mathcal{H})$. If T is not well-founded, then $R_0(T)$ has no hyperarithmetical paths, and thus by Theorem 8.1.15, $KB(R_0(T)) \cong \omega_1^{CK}(1 + \mathbb{Q}) + \delta$. Define $I(T) = KB(R_0(T)) \cdot \omega$ and let $R(T) = DS(I(T)))$. By Theorem 8.1.8 and the properties of $L \mapsto DS(L)$, T is well-founded iff $R(T)$ is. If T is not well-founded, then $R(T) \cong DS(\mathcal{H})$. □

Remark 8.1.18. Note that in the proof above we also established that $I(T) \cong \mathcal{H}$ iff T is not well-founded, and otherwise $I(T)$ is a computable ordinal. (This is [92, Lemma 5.2].)

Tree rank

Definition 8.1.19. Let T be a subtree of $\omega^{<\omega}$. We define the *tree rank* of $x \in T$, denoted by $\mathrm{tr}(x)$, by induction:

1. $\mathrm{tr}(x) = 0$ if x has no successor;

2. For $\alpha > 0$, $\mathrm{tr}(x) = \alpha$ if α is the least ordinal greater than $\mathrm{tr}(y)$ for all successors y of x;

3. $\mathrm{tr}(x) = \infty$ if x does not have an ordinal tree rank.

The tree rank of the tree T is defined to be the rank of the top node 0.

The tree rank of the tree $S * T$ is the minimum of the tree ranks of S and T. In particular, $S * T$ has an infinite path iff both S and T have infinite paths. We note that if T has no infinite paths, then its tree rank must be a computable ordinal; the proof is similar to that outlined in the hint to Exercise 8.1.26. We always have

$$\mathrm{tr}((\sigma, \tau)) = \min(\mathrm{tr}(a), \mathrm{tr}(\tau)),$$

where $\sigma \in S$ and $\tau \in T$.

Recall that WF denotes the set of indices of computable well-founded trees on ω. For each computable ordinal α, let WF_α denote the set of indices of computable trees of tree rank less than α. This fact below a consequence (a reformulation) of the Bounding Principle.

Fact 8.1.20. *If f is a hyperarithmetical function from a hyperarithmetical subset of ω into WF, there exists a computable α such that the range of f is contained in WF_α.*

To see why Fact 8.1.20 would hold, first consider the case when f is computable; put the trees $T_{f(x)}$ together under the common root. The rank of this well-founded computable tree is a computable ordinal and bounds the rank of each $T_{f(x)}$. Now, if f is hyperarithmetical, then we can relativise this to f and use that "hyperarithmetical relative to hyperarithmetical is again hyperarithmetical". We omit the formal proof and refer the reader to, e.g., [458, Corollary II.3.4].

Exercises

Exercise° 8.1.21 (Kleene). Write $\forall^0$ and $\exists^0$ for quantifiers ranging over numbers, and $\exists^1$ and $\forall^1$ for quantifiers ranging over functions. In the context of analytic relations, prove that the following prefix transformations are permissible; i.e., in each case, for any predicate form with the given prefix, an equivalent predicate form with the new prefix can be obtained:

1. $\ldots\exists^0\exists^0\ldots \mapsto \ldots\exists^0\ldots$ and the same for $\forall^0$;
2. $\ldots\exists^1\exists^1\ldots \mapsto \ldots\exists^1\ldots$ and the same for $\forall^1$;
3. $\ldots\exists^0\ldots \mapsto \ldots\exists^1\ldots$ and the same for $\forall^0$ and $\forall^1$;
4. $\ldots\forall^0\exists^1\ldots \mapsto \ldots\exists^1\forall^0\ldots$;
5. $\ldots\exists^0\forall^1\ldots \mapsto \ldots\forall^1\exists^0\ldots$;
6. $\ldots \mapsto \ldots Q\ldots$, where Q is any quantifier (adding "dummy quantifiers").

[For a detailed proof, see [454, Theorem III, Ch.16]. Also, see [20, pp.75-77] for an extended sketch.]

Exercise° 8.1.22. If S, R are Σ^1_k-relations on $(\omega^\omega)^n \times \omega^m$ $(m+n>0)$, then so are $S \wedge R$, $S \vee R$. The same is true for Π^1_k-relations.

Exercise° 8.1.23 (Kleene). Based on the existence of the uniformly effective enumeration of all functionals (§2.1.4), prove the following. For each $k \geqslant 1$ and each tuple of variables $\bar{x}$ ranging over $(\omega^\omega)^n \times \omega^m$ $(m+n>0)$, there is a Σ^1_k-relation $U(\bar{x}, e)$ so that the list $\{U(\bar{x}, e) : e \in \mathbb{N}\}$ includes all Σ^1_k-relations. (The same for Π^1_k.)

Exercise° 8.1.24. Prove Theorem 8.1.3 assuming the previous exercise.

Exercise 8.1.25 (Spector). Show that any hyperarithmetical well-ordering is isomorphic to a computable one. (Hint. Let ω_1^X denote the first ordinal that is not X-computable. All results of this section and Section 2.2.3 can be relativised to X. Suppose L is hyperarithmetical, say X-computable for some hyperarithmetical X. It is sufficient to show that $\omega_1^X = \omega_1^{CK}$. Obviously, $\omega_1^X \geqslant \omega_1^{CK}$. Assume $\omega_1^X > \omega_1^{CK}$, and thus there exists an X-computable presentation $U \cong \omega_1^{CK}$. Then for a computable linear order L, being well-founded would be equivalent to saying that L is isomorphically embeddable into U, which is a Σ^1_1-property (because of, e.g., Theorem 8.1.13), contradicting Theorem 8.1.5. See [458, Cor.7.4.II] for a detailed proof.)

Exercise 8.1.26. Prove 1 → 3 of Theorem 4.1.29. [Hint: The Stone space X of such a Boolean algebra can be viewed as a computable closed subset of 2^ω. The operation of taking the Cantor-Bendixson derivative is a monotone Δ^1_1-operator on X, and it is well-known that this implies that the CB-rank of the space has to be $\leqslant \omega_1^{CK}$; see, e.g., [458, Theorem 8.9.III]. Since the space is countable, we obtain that each $\xi \in X$ is ranked, and thus by ([458, §8.11.III]), we obtain that the sequence stabilises at a computable ordinal. It follows that the computable superatomic Boolean algebras are exactly the algebras of the form $\mathrm{Int}(\omega^\alpha \cdot k)$, where α is a computable ordinal. See Exercise 10.1.64 for a different version of proof.]

Exercise* 8.1.27 (Ash). Prove an extension of Theorem 7.1.6 to ordinals of the form ω^α, where α is a computable ordinal.

Exercise* 8.1.28. ([20, Theorem 17.8]) State and prove an extension of Theorem 7.1.7 to Boolean algebras of the form $\mathrm{Intalg}(\omega^\alpha)$, where α is a computable ordinal.

8.1.3 Index sets of discrete structures

All results in this subsection can be found in [213].

The recognition problem

Recall that in Corollary 8.1.9 we established that the recognition problem for (the index set of) well-orders is Π^1_1 complete. Recall that the superatomic Boolean algebras are exactly the interval algebras of ordinals; see §4.1.4. We obtain:

Theorem 8.1.29. *The recognition problem for superatomic Boolean algebras is Π^1_1-complete.*

Proof. A Boolean algebra is superatomic iff $\mathcal{B} \cong \mathrm{Intalg}(L)$ for a well-ordered L, and indeed, any such L has to be well-ordered. Since we can uniformly calculate some such L from a given presentation of a Boolean algebra (Corollary 4.1.15), the result follows from Corollary 8.1.9. □

Another elementary consequence is as follows. Recall that an abelian group A is divisible if, and only if, for every $x \in A$ and every integer $m > 0$, we have that

$$\exists y \in A \ my = x.$$

An abelian group is reduced if it does not have non-trivial divisible subgroups. It is well-known that the maximal divisible subgroup of an abelian group detaches as its direct summand. Fix an effective enumeration $(A_i)_{i \in \mathbb{N}}$ of all partial computable structures in the language of groups.

Theorem 8.1.30. *The index set*

$$\{i : A_i \text{ is a reduced abelian } p\text{-group}\}$$

is Π^1_1-complete.

Proof. The foundations of the theory of abelian p-groups will be explained in detail in Section 9.3.1. There, in Corollary 9.3.4 we will show that for an abelian p-group, being not reduced is equivalent to the existence of a sequence of non-zero elements $(x_i)_{i\in\mathbb{N}}$ such that for all i, $px_i = x_{i+1}$. Thus, the property of being reduced is Π^1_1.

In Chapter 9.3.1 we will also describe a uniform transformation $T \mapsto G(T)$ that turns trees into abelian p-groups. The resulting p-group is defined by relations $x = py$ where y is the successor of x and x, y range over the nodes of T. The root of T corresponds to the element 0. It follows from Corollary 9.3.4 that the group $G(T)$ is reduced if, and only if, T is well-founded. (We can uniformly add at least one non-root node to T to make sure that $G(T)$ is never the trivial group if we define the trivial group to be divisible.) Being well-founded for T is Π^1_1-complete by Theorem 8.1.5. □

A single structure can have its index set maximally complicated.

Theorem 8.1.31. *The recognition problem for the Harrison order (Definition 8.1.16)*

$$\{i : L_i \cong \mathcal{H}\}$$

is Σ^1_1-complete. The same is true for the computable tree $DS(\mathcal{H})$.

Proof. For a fixed computable (countable, discrete) structure, its recognition problem is always Σ^1_1. The Σ^1_1-completeness for $\mathcal{H}$ follows from Corollary 8.1.9, Remark 8.1.18, and in the case of $DS(\mathcal{H})$ it is guaranteed by Theorem 8.1.5 and Lemma 8.1.17. □

Remark 8.1.32. In the theorem above, the Σ^1_1-completeness for $\mathcal{H}$ is witnessed by a uniformly computable sequence $(L_i)_{i\in\mathbb{N}}$ so that L_i is not an ordinal iff $L_i \cong \mathcal{H}$. Similarly, in the case of $DS(\mathcal{H})$, the Π^1_1-outcome is always witnessed by a well-founded tree.

The isomorphism problem

For a class of (countable, discrete) computable structures $\mathcal{K}$ with arithmetical recognition problem, the isomorphism problem

$$\{(i,j) : M_i, M_j \in \mathcal{K}, M_i \cong M_j\}$$

is always Σ^1_1. If it is Σ^1_1-complete then we say that the class is analytic complete. In the previous chapter we observed that the recognition problem for all classes from the theorem below are arithmetical. Thus, the theorem says that these classes are analytic complete.

Theorem 8.1.33 (Goncharov and Knight [213]). *For the following classes, the isomorphism problem is Σ^1_1-complete.*

1. *Trees.*
2. *Linear orders.*
3. *Boolean algebras.*
4. *Abelian p-groups.*

Proof. 1. is Lemma 8.1.17, and 2. is Remark 8.1.18. To prove 3., consider the functional that on input a linear order L outputs Intalg(L). Let $(L_i)_{i\in\mathbb{N}}$ be a uniformly computable sequence of computable linear orders such that $L_i \cong \mathcal{H}$ iff L_i is not an ordinal; see Remark 8.1.32. If L_i is a computable ordinal, then the Stone space of $Intalg(L_i)$ vanishes after taking ω_1^{CK}-many CB-derivatives. Otherwise, it stabilises at the perfect kernel which is non-trivial when $L_i \cong \mathcal{H}$. (Note that the Boolean algebra $Intalg(L_i)$ is superatomic iff L_i is an ordinal; see §4.1.4). It follows that the uniformly computable sequence of pairs

$$(Intalg(L_i), Intalg(\mathcal{H}))_{i\in\mathbb{N}}$$

witnesses the Σ^1_1-completeness in 3.

For 4., fix a uniformly computable sequence $(T_i)_{i\in\mathbb{N}}$ witnessing the Σ^1_1-completeness of the recognition problem for $DS(\mathcal{H})$; see Remark 8.1.32. In this sequence, T_i is well-founded iff $T_i \ncong DS(\mathcal{H})$. In the proof of Theorem 8.1.30 we defined a uniformly computable transformation $T \mapsto G(T)$ with the property that T is well-founded iff $G(T)$ is a reduced abelian p-group. Thus, the sequence

$$(G(T_i), G(DS(\mathcal{H})))_{i\in\mathbb{N}}$$

witnesses the Σ^1_1-completeness in 4. □

8.1.4 Index sets of separable structures

All results of this subsection are either folklore or can be found in [139] or [373].

Upper bounds

It is obvious that the isomorphism problem for discrete countable structures is Σ^1_1 provided that the recognition problem for the class is Σ^1_1. For separable structures the situation is more complex; however, in the two important cases of Banach spaces and compact spaces we also obtain the upper bound Σ^1_1, as we prove next.

Fix a uniformly computable list $(M_i)_{i\in\mathbb{N}}$ of all partial computable Polish spaces. We slightly abuse our notation and identify M_i with its completion $\overline{M_i}$.

Proposition 8.1.34. *The homeomorphism problem*

$$\{\langle i,j\rangle : M_i \cong_{hom} M_j \,\&\, M_i, M_j \text{ are compact}\}$$

for compact computable Polish spaces is Σ^1_1.

Proof. It is arithmetical to say that M_i is a (presentation of a) compact Polish space; see (1) of Corollary 7.1.27. To say that there is a homeomorphism $f : M_i \to M_j$, it is sufficient to state that there exist continuous surjective $f_1 : M_i \to M_j$ and $f_2 : M_j \to M_i$ such that $f_1 \circ f_2 = Id_{M_i}$. Every $g : X \to Y$ between compact X and Y can be represented by, e.g., a pair $(\tilde{g}, m)$ where $\tilde{g} : \omega^2 \to \omega$ and $m : \omega \to \omega$, and where the function $\tilde{g}(n,k)$ is interpreted as the image of the nth special point with precision 2^{-k}, and m as the modulus of uniform continuity.

We claim that it is arithmetical to say that $(\tilde{g}, m)$ represents a continuous function

$$\lim_k g(\cdot, k) : X \to Y.$$

Indeed, totality is arithmetical. Also one can express that m is a modulus of continuity that works for $\tilde{g}$ as a closed property. Thus, as before, if it fails then it must fail for some special points. Since the continuous image of a compact space is closed, it is arithmetical to say that $(\tilde{g}, m)$ represents a surjective function. (If it does not, then again there is a special point in the complement witnessing this.)

This allows to state the existence of f_1 and f_2 in a Σ_1^1 way. Finally, to say that $f_1 \circ f_2 = Id_{M_i}$, it is sufficient to say that it is true for special points because the property is (again) closed. This can be expressed arithmetically (in the presentations of) f_1 and f_2. □

Before we proceed, we note that the same result clearly holds if we used computably compact presentations instead of computable presentations.

Fix a computable list $(B_i)_{i\in\mathbb{N}}$ of all (partial) computable linear spaces over $\mathbb{Q}$ with a computable norm. (We again identify $\overline{B_i}$ with B_i.)

Proposition 8.1.35. *The linear isometric isomorphism problem* $\{\langle i,j\rangle : B_i \cong_{iso} B_j\}$ *for computable separable Banach spaces is* Σ_1^1.

Proof. Exercise 8.1.38. □

Finally, fix a uniformly computable list of all (partial) computable Polish groups $(G_i)_{i\in\mathbb{N}}$. Each G_i is given by a computable Polish space equipped with operators for the group operations. In §7.1.2 we proved that the index set of compact Polish groups is arithmetical.

Proposition 8.1.36. *The topological group isomorphism problem*

$$\{\langle i,j\rangle : G_i \cong G_j \text{ and } G_i, G_j \text{ are compact groups}\}$$

for compact groups is Σ_1^1.

Proof. Exercise 8.1.39. □

Similarly to compact spaces, the result will of course also hold true for computably compact groups. As we noted earlier, in the case of computably compact groups we do not even need to require the inverse operation to be explicitly included into a presentation; this is Corollary 4.2.46.

Completeness results

All structures in the theorem below are separable.

Theorem 8.1.37. *The isomorphism problem is* Σ_1^1*-complete in the following classes:*

1. *Stone spaces (Downey and Melnikov [139]).*
2. *Banach spaces (Downey and Melnikov [139]).*
3. *Profinite abelian groups (Melnikov [373]).*

Proof. It follows from Propositions 8.1.34,8.1.35 and 8.1.36 that the upper bound in each case is Σ^1_1.

To establish completeness in 1., use the Σ^1_1-completeness of the isomorphism problem for Boolean algebras (Theorem 8.1.33) combined with Effective Stone Duality (Theorem 4.2.80) established in Part 1. Note that Σ^1_1-completeness is witnessed by computably compact spaces for which Effective Stone Duality is uniform.

To prove 2., use 1. and the Effective Banach-Stone Duality (Theorem 4.2.113). But in fact, we do not need the full power of this effective duality. If $(S_i, B_i)_{i\in\mathbb{N}}$ are the computably compact Stone spaces witnessing the Σ^1_1-completeness in 1., then the sequence of pairs of Banach spaces $(C(S_i, \mathbb{R}), C(B_i, \mathbb{R}))$ is clearly uniformly computable and witnesses the Σ^1_1-completeness in 2.

To establish 3., recall that the isomorphism problem for abelian p-groups is Σ^1_1-complete (Theorem 8.1.33). In Section 9.5 we will establish that Pontyragin duality between computably discrete torsion and computably compact profinite abelian groups is uniformly computable. Taking this result for granted, we obtain the Σ^1_1-completeness in 3. □

Note that 1. and 3. of the theorem above are witnessed by *computably compact* Stone spaces and *computably compact* pro-p abelian groups, respectively. Thus, these completeness results remain true if we restrict ourselves to computably compact presentations. Notice also that in all three cases, there is a single structure so that the set of indices of other structures isomorphic to it is Σ^1_1-complete.

Exercises

Exercise° 8.1.38. Prove Proposition 8.1.35.

Exercise° 8.1.39. Prove Proposition 8.1.36.

Exercise° 8.1.40 (Melnikov [368, 366]). Show that the isomorphism problem for computable ordered abelian groups is Σ^1_1-complete. (Hint: Use the strategy from Exercise 5.1.50.)

Exercise 8.1.41 (Calvert [72]). Show that the isomorphism problem for real closed fields, and thus for formally real fields, is Σ^1_1-complete. Deduce that the isomorphism problem for computable fields of characteristic zero is Σ^1_1-complete. (Hint: We remark that the order in a real-closed field can be reconstructed using (or imitated by) the field operations: for an $a \neq 0$, $a > 0$ iff $\exists x\, x^2 = a$. Fix a linear order L. Consider the theory of real-closed fields together with a collection of new symbols $\{c_\ell : \ell \in L\}$ and sentences stating that:

1. for all $\ell \in L$, $0 < c_\ell$;
2. for each pair $\ell < \ell''$ in L, for every polynomial $p(c_\ell)$ in c_ℓ with coefficients involving $\{c_{\ell''} : \ell'' < \ell\}$, we have $p(c_\ell) < c_{\ell'}$.

Using the material of §2.2.2, uniformly in L produce a computable formally real field $F(L)$. The computable real-closed (or formally real, depending on the signature) field $F(L)$ codes L into the Archimedean classes of positive elements.)

The next few exercises are here mainly to inform the reader. Their proofs are too complex and go beyond the material covered in the book.

Exercise 8.1.42** (Bazhenov, Harrison-Trainor, Kalimullin, and Melnikov [34]). A computable structure in a finite language (signature) is:

1. *polynomial-time* if the domain, the operations, and relations on the structure are computable in polynomial time;
2. *primitive recursive* if the domain, the operations, and relations on the structure are primitive recursive;
3. *punctual* or *fully primitive recursive* if it is primitive recursive and the domain is ω;
4. *automatic* if there is a finite automaton computing its open diagram (the definition is sensitive to the exact choice of formal details, see [34]).

Show that the following index sets are Σ^1_1-complete among computable structures:

1. automatically presentable structures;
2. structures that admit a polynomial-time presentation;
3. primitive recursively presentable structures;
4. punctually presentable structures;
5. structures that admit a 1-decidable presentation;
6. structures that admit a decidable presentation (Harrison-Trainor [237]).

Exercise* 8.1.43 (Kalimullin (unpublished), based on [34]). Use the main result of [34] to prove that the index set of computably presented structures among all c.e. presented structures is Σ^1_1-complete. (Hint: First, define a transformation Ψ such that, for the structures witnessing the main result in [34], A has a computable presentation iff $\Phi(A)$ has a Δ^0_2 presentation, and A has a 1-decidable presentation iff $\Phi(A)$ has a computable presentation. For that, add predicates coding the $\exists$-diagram of the structure. This readily gives that the index set of computably presented structures among Δ^0_2-presented structures is Σ^1_1-complete. This is because the outcomes of the main construction in [34] give either an automatic (thus, decidable) structure or a structure that is not even 1-decidable. To get the claimed Σ^1_1-completeness among c.e. presented structures, restrict this idea to the diagonalisation modules. Modify the diagonalisation module and use a $\exists$-definable congruence $\sim$ on pairs of automorphic points in A, and consider $A/\sim$.)

Exercise 8.1.44** (Downey et al. [131]). Show that the index set of computably categorical structures is Π^1_1-complete.

8.2 Effective reductions between classes

In this section we discuss effectively universal and effectively complete classes. The notion of an effectively universal class is not easy to define formally. In this section the simpler notion of effective completeness and the even weaker notion of FF-completeness will generally suffice.

8.2.1 Effective transformations between structures

All our classes have computable languages (signatures). The following is an effectivisation of the standard Borel reduction from descriptive set theory that we briefly discussed in §6.1.2.

Definition 8.2.1 (Calvert, Cummins, Knight, and Miller [283]). Fix two classes of computable structures $\mathcal{K}_1$ and $\mathcal{K}_2$. Then $\mathcal{K}_1$ is effectively reducible to $\mathcal{K}_2$,

$$\mathcal{K}_1 \leqslant_{EFF} \mathcal{K}_2,$$

if there is a Turing functional Φ taking open diagrams of structures in $\mathcal{K}_1$ and outputting diagrams of structures in $\mathcal{K}_2$, such that

$$A \cong B \text{ iff } \Phi(A) \cong \Phi(B).$$

(The most common but somewhat unfortunate notation in the literature is $\mathcal{K}_1 \leqslant_{tc} \mathcal{K}_2$.) The definition above is usually used only for discrete structures in the literature, however, it can be useful for separable structures as well. For example, the effective dualities proved in Part 1 the book induce reductions between classes of discrete and separable structures, and the effective Banach-Stone duality induces an effective reduction between compact spaces K and the respective Banach spaces $C(K;\mathbb{R})$.

Definition 8.2.2. A class $\mathcal{U}$ is *effectively complete* with respect to discrete structures if for any class $\mathcal{K}$ of discrete countable structures, $\mathcal{K} \leqslant_{EFF} \mathcal{U}$.

It is immediate from the results in the preceding sections that every effectively complete class has its isomorphism problem Σ^1_1-hard. Thus, for example, the class of completely decomposable groups is not effectively complete, and the same can be said about the class of solenoid groups (or their products).

Preserving spectra and computable dimension

Many natural examples of effective transformations witnessing $\leqslant_{EFF}$ satisfy various further properties beyond Definition 8.2.1. Perhaps, the two most important properties of effective transformations that occur throughout the literature are as follows.

Definition 8.2.3. Assume $\mathcal{K} \leqslant_{EFF} \mathcal{U}$ via Φ. We say that:

1. Φ *preserves computable dimension* if it preserves the number of computable copies of structures, up to computable isomorphism.

2. Φ *preserves degree spectra* if A has an X-computable copy iff $\Phi(A)$ admits an X-computable copy.

That is, 1. says that if A has exactly $\kappa \in \mathbb{N} \cup \{\omega\}$ computable presentations up to computable isomorphism, then so does $\Phi(A)$. In particular, it maps computably categorical structures to computably categorical structures. (In Chapter 10 we will prove that for every $\kappa \in \mathbb{N} \cup \{\omega\}$ there exist an algebraic structure having exactly κ computable presentations, up to computable isomorphism.)

In the literature, in 2. it is often further assumed that the coded structures from $\mathcal{K}$ are "*automorphically non-trivial*", i.e., they do not become homogeneous after fixing finitely many constants. This assumption becomes unnecessary according to our definitions. Under the slightly more strict definition saying that A is $\mathbf{a}$-computable if $deg_T(A) = \mathbf{a}$ (rather than $deg_T(A) \leqslant \mathbf{a}$), this assumption is required to exclude the pathological case when the degrees of presentations are not closed upwards [304].

For the purposes of the book we usually restrict ourselves to the case when $\mathcal{K}$ ranges over discrete countable structures, while $\mathcal{U}$ could be a class of spaces. This restriction is of course not really necessary in general. For example, effective Stone duality between computably compact Stone spaces and Boolean algebras preserves computable dimension and degree spectra in the right sense; this easily follows from the proofs of Theorems 4.2.80 and 4.2.84, and from Exercise 10.4.8. The same can be said about effective Pontryagin duality mapping profinite abelian groups to their torsion abelian duals; this can be easily extracted from Theorem 9.5.7 that will be presented in the next chapter. We will usually further restrict our classes to their infinite members, since to establish Σ^1_1- or Π^1_1-completeness we do not need to consider finite structures. This assumption will sometimes be implicit in the proofs and proof sketches contained in the next two subsections. Nonetheless, most proofs that we present here can be adjusted to work for finite structures as well. For effective reducibilities between classes of finite structures, see [283].

Definition 8.2.4 (Hirschfeldt, Khoussainov, Shore, and Slinko [257]). An effectively complete class $\mathcal{U}$ is *complete with respect to computable dimension* if for every class of discrete structures $\mathcal{K}$, there is a transformation Φ witnessing $\mathcal{K} \leqslant_{EFF} \mathcal{U}$ that preserves computable dimension.

Similarly, we have:

Definition 8.2.5 (Hirschfeldt, Khoussainov, Shore, and Slinko [257]). An effectively complete class $\mathcal{U}$ is *complete with respect to degree spectra* if Φ witnessing $\mathcal{K} \leqslant_{EFF} \mathcal{U}$ can always be chosen to preserve degree spectra.

Definitions 8.2.4 and 8.2.5 are directly related to the two central problems of effective algebra: the existence and uniqueness of effective presentations for structures in the class. Since in this section we mainly look at index sets and their Σ_1^1-completeness, the much weaker notion of effective completeness will usually suffice. However, in this section our proofs of effective completeness will give transformations that are much stronger than just witnessing $\leqslant_{EFF}$, but we leave the verification of Definitions 8.2.4 and 8.2.5 for these transfoirmations to the exercises. These additional properties of the transformations will be useful in the last chapter.

Effectively universal classes*

Definitions 8.2.4 and 8.2.5 are just two examples of the many properties that natural transformations between structures can preserve. For a few examples of such additional properties, see Exercises 8.2.10 and 8.2.11. Seeking to define a transformation between classes that preserves as many properties as possible yet is not too restricted, we arrive at a "zoo" of definitions. For example, we may or may not allow our transformation to use a finite tuple of parameters in the structures. This approach was taken in Hirschfeldt, Khoussainov, Shore, and Slinko [257].

It is however worth noting that a formal and general definition of a strong effective reduction between classes *exists* and can be found in [243]. Essentially all known useful effective transformations in the literature that satisfy Definitions 8.2.4 and 8.2.5 satisfy this definition from [243].

Furthermore, this definition is *robust* in the following sense. It was illustrated in [243] that the notion in terms of functionals has an equivalent formulation using $L^c_{\omega_1\omega}$-definability. Both definitions are a bit too technical and we won't need them; thus, we omit them.

The canonical example of an effectively universal class is the class $\mathcal{G}$ of directed graphs. We will see that, for any other class $\mathcal{K}$ of countable structures, the transformation Φ witnessing $\mathcal{K} \leqslant_{EFF} \mathcal{G}$ satisfies much more than required in Definition 8.2.1 and even more than is stated in Definitions 8.2.4 and 8.2.5 above. Indeed, from the perspective of computable structure theory, the graph $\Phi(A)$ is "essentially" A, up to a change of the signature.

8.2.2 Simple codings

The results contained in this subsection are folklore, and their proofs and proof sketches follow Hirschfeldt, Khoussainov, Shore, and Slinko [257] closely[1]. The only exception is Theorem 8.2.9 which is trivial. As an immediate corollary of these results and Theorem 8.1.33, we obtain that *in all of these classes the isomorphism problem is Σ_1^1-complete.* As we already noted before, the transformations witnessing the effective completeness for these classes preserve computable dimension and degree spectra, but the verification of these claims will be left as an exercise.

Throughout the rest of this subsection, we usually assume that our structures are infinite and have domain ω. The case of finite structures is usually not very interesting to us since we seek to prove non-classification type results. (The case of finite structures was investigated in [283].) But at least for the simple codings in this section, the coding in the finite case is typically exactly the same as for the infinite case.

The effective completeness of directed graphs

The result below is folklore.

[1] We thank Denis Hirschfeldt for allowing us to use the diagrams from [257].

Theorem 8.2.6. *Directed graphs form an effectively complete class.*

Proof. Let A be a countable structure in a computable language with (possibly infinitely many) constants $c_0, c_1, \ldots$, function symbols $f_0, f_1, \ldots$, and relation symbols $R_0, R_1, \ldots$. Let k_i be the arity of f_i and let l_i be the arity of R_i. See Fig. 8.2.2 for the idea behind the transformation. We remark that there are many different ways to define $A \mapsto G(A)$, so the particular transformation from [257] which we adopt is certainly not canonical.

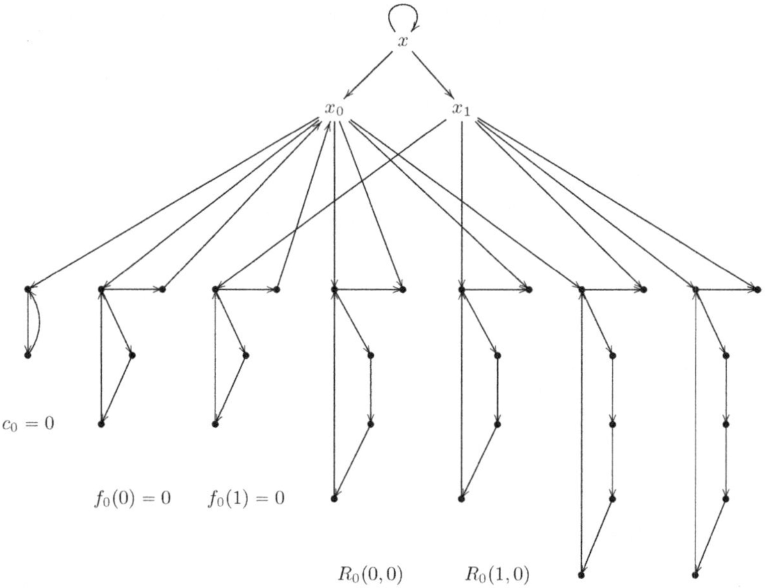

Figure 8.1: A portion of $G(A)$.

We give the detailed formal definition of $A \mapsto G(A)$. The corresponding directed graph $G = G(A)$ and the set of directed edges $E = E(A)$ consists of the following:

1. A unique node x with $(x, x) \in E$.
2. A node x_i for each element $i \in A$, with $(x, x_i) \in E$.
3. Suppose $j = c_i$ in A. Then there is a cycle[2] of length $4i + 2$ with an edge from x_j to one of the elements of this cycle.

[2] A cycle of length k is a sequence of elements $x_0, \ldots, x_{k-1}$ so that there is an edge from x_i to $x_{i+1 \, mod \, k}$, e.g., from x_{k-1} back to x_0.

4. For each function f_i and each tuple $(j_0, \ldots, j_{k_i-1})$ in A, we put:

   (a) a cycle C of length $4i + 3$;

   (b) a chain of elements $y_0, \ldots, y_{k_i}$, where y_0 is an element of C, with an edge from y_n to y_{n+1} for each $n < k_i$;

   (c) an edge from x_{j_n} to y_n for each $n < k_i$;

   (d) an edge from y_{k_i} to x_j, where $j = f_i(j_0, \ldots, j_{k_i-1})$ in A.

5. For each relation R_i and each tuple $(j_0, \ldots, j_{l_i-1})$ in A such that $R_i(j_0, \ldots, j_{l_i-1})$ holds in A:

   (a) a cycle C of length $4i + 4$;

   (b) a chain of elements $y_0, \ldots, y_{l_i-1}$, where y_0 is an element of C, with an edge from y_n to y_{n+1} for each $n < l_i - 1$;

   (c) and an edge from x_{j_n} to y_n for each $n < l_i$.

6. For each relation R_i and each tuple $(j_0, \ldots, j_{l_i-1})$ in A such that $R_i(j_0, \ldots, j_{l_i-1})$ does not hold in A:

   (a) a cycle C of length $4i + 5$;

   (b) a chain of elements $y_0, \ldots, y_{l_i-1}$, where y_0 is an element of C, with an edge from y_n to y_{n+1} for each $n < l_i - 1$; and

   (c) an edge from x_{j_n} to y_n for each $n < l_i$.

It is immediate that the transformation $A \mapsto G(A)$ is uniformly effective, and furthermore, there exist first-order existential formulae defining the relations and their complements of A inside $G(A)$, as well as the functions and the constant symbols. The verification of these properties is left as an exercise. In particular, it follows that $A \mapsto G(A)$ is injective on isomorphism types. □

Undirected Graphs

Recall that an undirected graph is simple if it does not have edges of the form (x, x). Since we view our graphs as predicate structures, multiple edges are automatically excluded.

Theorem 8.2.7. *The class of undirected simple graphs is effectively complete.*

Proof. By Theorem 8.2.6, it is sufficient to prove that the class of directed graphs is effectively reducible to the class of undirected graphs. Let G be a directed graph with edge relation E. We define the symmetric, irreflexive graph $H(G) = (H, F)$. See Fig. 8.2 for the idea behind the transformation $G \mapsto H(G)$. (Fig. 8.2 shows a part of $H(G)$ coding $E(0,1)$, $E(1,0)$, $E(1,2)$, $E(2,2)$, $\neg E(0,0)$, $\neg E(0,2)$, $\neg E(1,1)$, $\neg E(2,0)$, and $\neg E(2,1)$.)

The formal definition is as follows.

1. $H = \{a, \hat{a}, b\} \cup \{c_i, d_i, e_i : i \in G\}$.

2. $F(x, y)$ holds only in the following cases.

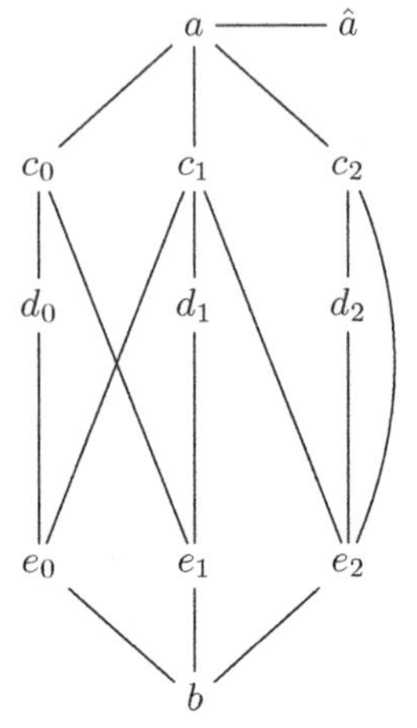

Figure 8.2: A part of $H(G)$.

(a) $F(a, \hat{a})$ and $F(\hat{a}, a)$.

(b) For all $i \in G$,

  i. $F(a, c_i)$ and $F(c_i, a)$,

  ii. $F(b, e_i)$ and $F(e_i, b)$,

  iii. $F(c_i, d_i)$ and $F(d_i, c_i)$,

  iv. $F(d_i, e_i)$ and $F(e_i, d_i)$.

(c) If $E(i, j)$ then $F(c_i, e_j)$ and $F(e_j, c_i)$.

Obviously, $G \cong G'$ implies $H(G) \cong H(G')$. To see why $H(G) \cong H(G')$ implies $G \cong G'$, we show that G is definable in $H(G)$ (indeed, it is first-order $\exists$-definable). Let

$$D = \{x \in H : x \neq \hat{a} \wedge F(a, x)\} = \{c_i : i \in G\}$$

and

$$R(x, y) = \{(x, y) : D(x) \wedge D(y) \wedge \exists d, e(F(b, e) \wedge F(y, d) \wedge F(d, e) \wedge F(x, e))\}.$$

The definable predicates D and R essentially imitate G and E. To see that D and R are invariant, it is enough to notice that $x = a$ is the only element of H_G that satisfies the formula

$$\exists^\infty y(F(x, y)) \wedge \exists z(F(x, z) \wedge \forall w(F(w, z) \rightarrow w = x)),$$

$x = \hat{a}$ is the only element of H_G that satisfies

$$F(x, a) \wedge \forall y(F(x, y) \rightarrow y = a),$$

and $x = b$ is the only element of H_G that satisfies

$$\exists^\infty y(F(x, y)) \wedge \neg F(a, x) \wedge \neg \exists z(F(a, z) \wedge F(x, z)).$$

We omit further elementary details. □

Exercises

Exercise° 8.2.8. Prove that the transformations in the proofs of Theorems 8.2.7 and 8.2.6 witness that the respective classes are effectively complete with respect to computable dimension and degree spectra.

Discrete metric spaces under isometry

We say that a discrete metric space M is rational-valued if $d(x, y) \in \mathbb{Q}$ for any $x, y \in M$. Such spaces can be viewed as algebraic structures in the language $(D_r)_{r \in \mathbb{Q}}$, where $D_r(x, y)$ holds iff $d(x, y) = r$. Further, two discrete metric spaces are isometrically isomorphic iff the respective algebraic structures are isomorphic. Thus, such as space is not really an honest metric space; it can be viewed as a countable algebraic structure. The fact below is trivial, but it will be useful in the last chapter. Recall that, according to Definition 8.2.2, an effectively complete class has to be $\leqslant_{EFF}$-above any class of *discrete* structures in a computable language.

Theorem 8.2.9. *Rational-valued discrete Polish spaces (viewed up to isometry) are effectively complete.*

Proof. We use the universality of undirected simple graphs established in Theorem 8.2.7. For a graph G, define $M(G)$ upon the set of vertices of G under the following metric:

1. $x \neq y$ and $(x, y) \in E$ implies $d(x, y) = 1$;
2. $x \neq y$ and $(x, y) \notin E$ implies $d(x, y) = 2$;
3. $d(x, x) = 0$, for all x.

The triangle inequality $d(x, y) + d(y, z) \geqslant d(x, z)$ follows from a straightforward case analysis, and the positivity of d is evident. Also,

$$(x, y) \in E \text{ if and only if } d(x, y) = 1,$$

so the graph relation is definable in $M(G)$. □

Of course, every space of the form $M(G)$ is automatically a computable Polish space. On the other hand, in every computable Polish space isometric to $M(G)$ we can uniformly decide whether $d(x, y) = 0$ or 1 or 2, for every pair of points. It should be clear that the transformation $G \mapsto M(G)$ preserves computable dimension and degree spectra, and indeed, perhaps any reasonable property one can possibly think of.

Exercises

Exercise° 8.2.10. Let Φ be one of the transformations described in Theorems 8.2.6, 8.2.7, and 8.2.9. Prove that A is X-computably isomorphic to B if, and only if, $\Phi(A)$ and $\Phi(B)$ are X-computably isomorphic.

Exercise° **8.2.11.** Let Φ be one of the transformations described in Theorems 8.2.6, 8.2.7, and 8.2.9. Show that there is a uniformly effective transformation $X \mapsto \mathcal{S}(X)$ such that whenever $X \cong \Phi(A)$, we have that $\mathcal{S}(X) \cong A$. Furthermore, and this reverse transformation $\mathcal{S}$ has the property described in the previous exercise, i.e., respects X-computable isomorphisms (whenever it is defined).

8.2.3 Integral domains and 2-step nilpotent groups*

Recall that an integral domain is a commutative ring with identity that has no zero divisors, i.e., $a \cdot b = 0$ implies that either $a = 0$ or $b = 0$. Every integral domain is contained in its field of fractions, and every finite integral domain is a field.

Theorem 8.2.12 (Hirschfeldt, Khoussainov, Shore, and Slinko [257]). *Let p be either 0 or a prime. The class of integral domains of characteristic p is effectively complete. Furthermore, it is complete with respect to degree spectra and computable dimension.*

Sketch. It is sufficient to effectively transform graphs into integral domains. The graphs constructed in §8.2.2 have the following property: for every finite set of nodes S there exist nodes $x, y \notin S$ that are connected by an edge. Let G be a symmetric, irreflexive, countably infinite graph with edge relation E, having the property mentioned above. We assume that the domain $|G|$ of G is ω.

The construction is essentially the same for the cases of infinite and finite characteristic p. Thus, it makes sense to adopt a unified notation. Let $\mathbb{Z}_0 = \mathbb{Z}$. We write $I_p = I$ be the set of invertible elements of $\mathbb{Z}_p$, including the case when $p = 0$. (Note that for each p, I is finite.) The integral domain $A(G)$ is defined to be

$$\mathbb{Z}_p\left[x_i : i \in \mathbb{N}\right]\left[\frac{y}{x_i x_j} : E(i,j)\right]\left[\frac{z}{x_i x_j} : \neg E(i,j)\right]\left[\frac{y}{x_i^n} : i, n \in \omega\right],$$

where, fractions are interpreted as elements of the enveloping fraction field of $\mathbb{Z}_p[x_i, z, y]$, and $R[X][Y]$ is identified with $R[X \cup Y]$ under the natural isomorphism.

It is easy to see that when G is computable, $A(G)$ is computable too, and this is uniform. To reconstruct G from an isomorphic copy A of $A(G)$, define

$$D = \{x \in A : x \notin I \wedge \exists r(x^2 r = z)\},$$

$$Q = \{(x, x') : D(x) \wedge \exists a \in I(x' = ax)\},$$

and

$$R = \{(x, x') : D(x) \wedge D(x') \wedge \neg Q(x, x') \wedge \exists r(rxx' = y)\}.$$

Using a relatively involved (but self-contained) combinatorial argument, it is possible to show that D and R are invariant under any automorphism of the ring. The graph G is isomorphic to (D, R) defined on the Q-classes of elements. The definability analysis indeed shows that both R and D are computable relative to A, thus giving universality with respect to degree spectra.

It takes a bit more effort to prove that the coding also gives completeness with respect to computable dimension, since any isomorphism between copies of $A(G)$ effectively induces an isomorphism between the respective definable copies of G, and vice versa. (We omit the details.) □

As noted in [257], the proof outlined above also illustrates that the multiplicative semigroup upon the generators

$$\{\pm 1\} \cup \{x_i : i \in \mathbb{N}\} \cup \left\{\frac{y}{x_i x_j} : E(i,j)\right\} \cup \left\{\frac{z}{x_i x_j} : \neg E(i,j)\right\} \cup \left\{\frac{y}{x_i^n} : i, n \in \omega\right\}.$$

also effectively codes the graph G preserving most effective properties of interest, including computable dimension and degree spectra.

Recall that every field is trivially an integral domain. Using algebraic geometry, the following stronger result was established by Miller, Poonen, Schoutens, and Shlapentokh [397]; we omit the proof.

Theorem 8.2.13. *Theorem 8.2.12 holds for the class of fields (of arbitrary characteristic).*

Let G be a group. The *center* of G is the set $\{x \in |G| : \forall y \in |G|\,(xy = yx)\}$. The *commutator* $[x, y]$ of $x, y \in |G|$ is the element $xyx^{-1}y^{-1}$. The group G is *2-step nilpotent* if $[x, y]$ is in the center of G for every pair of elements $x, y \in |G|$.

Theorem 8.2.14 (Harisanov et al. [230], based on [257]). *The class of 2-step nilpotent groups is effectively complete. Furthermore, it is complete with respect to degree spectra and computable dimension.*

Proof idea. This is an effectivisation of an old result that [257] attributes to Mal'cev. By Theorem 8.2.12, it is sufficient to define a transformation $R \mapsto G_R$ from integral domains of any fixed characteristic to 2-step nilpotent groups. Let $R = (|R|, +, \cdot, 0, 1)$ be a countably infinite integral domain of characteristic $p > 2$. Define G_R to be the group of upper triangular 3×3 matrices of the form

$$\begin{pmatrix} 1 & b & c \\ 0 & 1 & a \\ 0 & 0 & 1 \end{pmatrix},$$

$a, b, c \in |R|$, which is known as the *Heisenberg group* of R. The verification is omitted, but we note that [257] uses two non-uniform parameters to reconstruct the integral domain from G_R, and in [230] the use of these two parameters is eliminated. (Indeed, it is illustrated that any pair of non-commutative elements could be used to reconstruct R.) □

In all examples of transformations that we presented above the effective completeness was witnessed by a transformation that preserved lots of computability-theoretic properties of the transformed structures, including computable dimension. However, Definition 8.2.2 in its raw form does not assume any of these additional properties. If we take Definition 8.2.2 as the base of our theory, and do not put any additional restrictions we of course can obtain more effective completeness results, and often via much simpler proofs. For example, it is easy to see that the elementary embedding of graphs into fields of characteristic zero described in Exercise 8.2.32 is effective, thus showing that such fields lie on to under $\leqslant_{EFF}$. In contrast, establishing "effective universality" of fields (Theorem 8.2.13) requires a lot more effort.

8.2.4 A Type I version of effective completeness

Definition 8.2.2 uses a Turing functional Φ, perhaps with parameters, to transform arbitrary countable structures from the given class. The "Type I" analogue of this notion works on indices of computable structures. It is as follows. Note that the isomorphism problem for a class of computable discrete structures is typically Σ^1_1, and the same is true for a class of computably compact Polish spaces. Further, the induced equivalence relation is defined on a (hyper)arithmetical subset of ω which corresponds to the index set of the class. The definition below is abstract and does not mention classes of structures at all, and it is not restricted to isomorphism problems on structures.

Definition 8.2.15 (Friedman and Fokina [170]). Let E, R be a Σ^1_1-equivalence relations on (hyper)arithmetical subsets of ω.

1. We write $E \leqslant_{FF} R$ and say that E is FF-reducible to R if there is a (partial) computable function f such that
$$xEy \text{ if and only if } f(x)Rf(y),$$
where we of course assume f always halts on the support of E and outputs an index in $\mathcal{K}'$.

2. A Σ^1_1 equivalence relation on ω is *effectively complete among* Σ^1_1*-equivalence relations*, or simply FF-complete, if it is FF-reducible to any Σ^1_1 equivalence relation on ω.

3. We say that a class $\mathcal{K}$ of computable structures is FF-complete if the isomorphism problem for $\mathcal{K}$ is FF-complete.

We are mainly interested in the FF-reducibility between classes of structures.

Separating the Type I and Type II effective reducibilities

It is clear that $\mathcal{K}_1 \leqslant_{EFF} \mathcal{K}_2$ implies $\mathcal{K}_1 \leqslant_{FF} \mathcal{K}_2$. Of course, $\leqslant_{FF}$-reducibility is in general weaker than the $\leqslant_{EFF}$-reducibility, as the elementary example below illustrates. A much more insightful result separating Type II and Type I effective completeness will be given a bit later, in Theorem 8.2.25.

Example 8.2.16. Fix an abelian group $H \leqq \mathbb{Q}$ that has no computable presentation. For each $n \in \mathbb{N}$, consider the classes F_n of homogeneous c.d. groups of rank $\leqslant n$; there are groups of the form H^m, $m \leqslant n$, including $\{0\}$ for $m = 0$. Then each class has exactly one computable structure, this being $\{0\}$, and thus
$$F_n \equiv_{FF} F_m,$$
for all $m, n \in \mathbb{N}$, via the identity function. However, $F_{n+1} \nleqslant_{EFF} F_n$, by pigeonhole.

The exaggerated Example 8.2.16 above is rather unsatisfying. It works simply because the compared classes do not have "enough" computable members. A natural question arises as to whether the two notions of reducibility, $\leqslant_{EFF}$ and $\leqslant_{FF}$, can be distinguished using only subclasses $\mathcal{K}_1^c$ and $\mathcal{K}_2^c$ composed of the *computable* members of the compared classes.

We now explain why the answer is (essentially) negative, at least for all "natural" classes of structures that we have encountered in Proposition 7.1.1, and many more. We have already mentioned infinitary computable formulae in the introduction, but we delay the more detailed discussion until Chapter 10 (Definition 10.1.4). Here, we briefly explain the little that is needed to prove the next theorem.

A Π_2^c-*class of structures* is a class that is described by a single computable infinitary axiom of the form

$$\bigwedge_{i \in W} \forall \bar{x} \psi_i(\bar{x}),$$

where $(\psi_i(\bar{x}))_{i \in W}$ is a c.e. sequence of formulae (i.e., W is a c.e. set listing their Gödel numbers), each ψ_i being a c.e. disjunction of first-order existential formulae. For example, any class axiomatised by a computable collection of first-order $\forall\exists$-axioms is a Π_2^c-class. Every class from Proposition 7.1.1 is certainly like that. We do not need to know more about infinitary logic to prove the result below, which appears to be new.

All our classes have computable signatures, and we assume all structures have domain $\subseteq \omega$. Without loss of generality, we can also replace all operations with the relations representing their graphs. This will not violate the property of being a Π_2^c-class, but it will make the notion of "effective density" easier to define.

Definition 8.2.17. Computable structures are *effectively dense* in a class $\mathcal{K}$ if there is a uniformly computable collection of structures $(X_i)_{i \in \mathbb{N}}$ in $\mathcal{K}$ such that an arbitrarily large finite substructure of $A \in \mathcal{K}$ can be extended to some structure from the list.

All classes listed in Proposition 7.1.1 have this property, but this is not true for the ad-hoc classes described in Example 8.2.16.

Theorem 8.2.18. *Suppose $\mathcal{K}_1$ and $\mathcal{K}_2$ are Π_2^c-classes of algebraic structures in which furthermore computable structures are effectively dense. Then $\mathcal{K}_1 \leqslant_{FF} \mathcal{K}_2$ implies $\mathcal{K}_1^c \leqslant_{EFF} \mathcal{K}_2^c$, i.e., the FF-reduction can be witnessed by a Turing functional that is guaranteed to work on computable members of the classes.*

Proof of Theorem 8.2.18. In Exercise 2.4.34 we saw that $\mathcal{K}$ can be associated with $C(\mathcal{K}) \subseteq \omega^\omega$ so that computable structures are in a natural effective 1-1 correspondence with the computable points of this set. Computable structures are effectively dense in $\mathcal{K}$ if there is a uniformly computable sequence of structures in $\mathcal{K}$ so that the corresponding points are dense in $C(\mathcal{K})$.

Recall that in Exercise 4.2.88 we defined a subset of a computable Polish space to be effectively G_δ if it is an intersection of uniformly effectively open sets. The reader should convince themselves that all classes $\mathcal{K}$ from Proposition 7.1.1 correspond to effectively G_δ subsets $C(\mathcal{K})$ of ω^ω. In Exercise 10.1.67(1) we will see that, for any Π_2^c-class $\mathcal{K}$, $C(\mathcal{K}) \subseteq \omega^\omega$ is effectively G_δ. This is the only step when we need infinitary logic.

As we mentioned earlier, computable structures in $\mathcal{K}$ are in a uniform 1-1 correspondence with computable points in $C(\mathcal{K})$. (More generally, every computable structure in the language of $\mathcal{K}$ can be uniformly effectively associated with its code in ω^ω.) By Exercise 4.2.88, $C(\mathcal{K}_1)$ and $C(\mathcal{K}_2)$ admit computable complete metrics. Together with the respective dense sequences of computable (codes of) structures, $C(\mathcal{K}_1)$ and $C(\mathcal{K}_2)$ are computable Polish. Furthermore, we can uniformly effectively pass between computable points in the original metric on ω^ω and the new complete metric in $C(\mathcal{K}_i)$ $(i = 1, 2)$, provided that the points lie in $C(\mathcal{K}_i)$.

Recall that in Chapter 2 we proved Kreisel-Lacombe-Shoenfield-Markov Theorem 2.3.7 which says that every Markov (Type I) computable real function is Borel computable. That is, every Markov computable function is realised by a Turing functional that is guaranteed to work correctly when the inputs are computable reals. In Exercise 2.4.35 (Ceitin [82]) we saw that this result also holds for computable Polish spaces too via a relatively straightforward generalisation of Theorem 2.3.7.

A reduction $\mathcal{K}_1 \leqslant_{FF} \mathcal{K}_2$ induces a Markov computable function from $C(\mathcal{K}_1)$ to $C(\mathcal{K}_2)$. By Exercise 2.4.35, there is a computable functional that works on computable members of $C(\mathcal{K}_1)$ to $C(\mathcal{K}_2)$. This functional can be uniformly turned into a functional witnessing $\mathcal{K}_1^c \leqslant_{EFF} \mathcal{K}_2^c$. □

We remark that the effective density of computable structures in $\mathcal{K}_2$ can be dropped. Perhaps, the result can be extended to cover classes of separable structures as well. Indeed, we suspect that an even more general result holds, but we shall be satisfied with Theorem 8.2.18 as stated and leave it at that.

The FF-completeness of trees and torsion abelian groups

Similarly to what we had with EFF-completeness, any FF-complete class will have its isomorphism problem Σ_1^1-hard. However, the FF-completeness of $\mathcal{K}$ should not be confused with the usual Σ_1^1-completeness for the isomorphism problem for $\mathcal{K}$. Thus, the theorem below is *not* an elementary consequence of Theorem 8.1.5.

Theorem 8.2.19 (Fokina et al. [171]). *The class of computable trees is FF-complete.*

Before we prove the theorem, we briefly review some definitions introduced earlier in §8.1.2. Our trees are isomorphic to subtrees of $\omega^{<\omega}$. Let $S, T \subseteq \omega^{<\omega}$ be trees. Recall that we define the tree $S * T$ to consist of ordered pairs (a, τ), where $a \in S$ and $\tau \in T$. The successors of (σ, τ) are the pairs (σ', τ'), where σ' is a successor of σ in S, and τ' is a successor of τ in T. We also write $\mathrm{tr}(x)$ for the tree rank of a node x in T, and $\mathrm{tr}(T)$ for the tree rank of the root of T.

Proof of Theorem 8.2.19. We follow [171] closely. We will need a technical notion first introduced in [171].

Definition 8.2.20. A computable subtree T of $\omega^{<\omega}$ is *rank-saturated* provided that for all x in T:

1. If $\mathrm{tr}(x)$ is an ordinal α, then for all $\beta < \alpha$, x has infinitely many successors z such that $\mathrm{tr}(z) = \beta$;
2. If $\mathrm{tr}(x) = \infty$, then for all computable β, x has infinitely many successors z such that $\mathrm{tr}(z) = \beta$ and x has infinitely many successors z with $\mathrm{tr}(z) = \infty$.

The lemma below resembles Lemma 8.1.17.

Lemma 8.2.21. *There is a computable rank-saturated tree T^∞ such that $tr(T^\infty) = \infty$.*

Proof. In §8.1.2 we saw that there exists a computable linear order of order type $\omega_1^{CK}(1+\eta)$, the Harrison order $\mathcal{H}$. We let T^∞ be the set of finite sequences $((\alpha_0, k_0), \ldots, (\alpha_n, k_n))$, where $\alpha_0 > \cdots > \alpha_n$ in $\mathcal{H}$ and $k_0, \ldots, k_n \in \mathbb{N}$. If α_i corresponds to an ordinal α in ω_1^{CK}, then $\mathrm{tr}((\alpha_0, k_0), \ldots, (\alpha_i, k_i)) = \alpha$, and if α_i lies in the non-well-ordered part of $\mathcal{H}$, then $\mathrm{tr}((\alpha_0, k_0), \ldots, (\alpha_i, k_i)) = \infty$. □

Proposition 8.2.22. *If T is a computable tree, then $T * T^\infty$ is a computable rank-saturated tree and $tr(T) = tr(T * T^\infty)$.*

Proof. The tree rank of $T * T^\infty$ is clearly the same as the tree rank of T. For $x \in T * T^\infty$ of rank α and $\beta < \alpha$, we show that x has infinitely many successors of rank β. Say $x = (\sigma, \tau)$. Since $\mathrm{tr}(\tau) \geqslant \alpha$ and because T^∞ is rank-saturated, τ has infinitely many successors τ' of rank β. Also, $\mathrm{tr}(\sigma) \geqslant \alpha$, so σ has a successor σ' of rank at least β. Then for all such pairs (σ', τ'), $\mathrm{tr}(\sigma', \tau') = \beta$. □

We remark that, if E is a Σ^1_1-equivalence relation on a hyperarithmetical domain X, then we can always consider a new Σ^1_1 equivalence relation E_1 in which all elements in $\omega \backslash X$ are declared to be in one new equivalence class. If we can FF-reduce E to some other relation K, then the same reduction will work for E as well, by restriction.

We return to the proof of Theorem 8.2.19. Let E be a Σ^1_1 equivalence relation on ω. To prove that E is FF-reducible to the isomorphism relation on computable trees, we will build a computable sequence of computable trees $(T_n)_{n \in \mathbb{N}}$ such that for every $m, n \in \mathbb{N}$,

$$m\, E\, n \text{ if and only if } T_m \cong T_n.$$

The lemma below will in fact be more useful in the sequel than the theorem itself.

Lemma 8.2.23. *There is a computable sequence of trees $(T^*_{m,n})_{n,m \in \mathbb{N}}$ with the following properties:*

1. *if $m\, E\, n$, then $T^*_{m,n} \cong T^\infty$, where T^∞ is the rank-saturated tree with an infinite path;*

2. *if $\neg m\, E\, n$, then $T^*_{m,n} \cong T_\alpha$, where T_α is the rank-saturated tree of tree-rank α for a computable ordinal α such that for all $m' \in [m]_E$ and $n' \in [n]_E$, the relation $\neg m'\, E\, n'$ is witnessed by α.*

Proof. By Proposition 8.2.22, there exists a uniformly computable sequence of trees $\{T_{m,n}\}_{m,n \in \mathbb{N}}$ such that $\neg m\, E\, n$ iff $T_{m,n}$ is well-founded. We say that $\neg m\, E\, n$ is witnessed by α iff $T_{m,n}$ has tree-rank less than α. For every $m, n \in \mathbb{N}$, construct a new tree $T'_{m,n}$ in the following way. Let $\sigma_0, \sigma_1, \ldots$ be an enumeration of all finite sequences of natural numbers. Suppose $\sigma_s = (\sigma_0, \ldots, \sigma_k)$. Then, under the s-th node on level 1 (i.e., under the element of the form (s), $s \in \mathbb{N}$) of $T'_{m,n}$, we put the tree $P_s = T_{m,\sigma_0} * T_{\sigma_0,\sigma_1} * \cdots * T_{\sigma_k,n}$, identifying the top node of P_s with s. Then

$$\mathrm{tr}(T'_{m,n}) = \sup\{\mathrm{tr}(P_s) + 1 \mid s \in \mathbb{N}\}.$$

If $m\, E\, n$, then $T_{m,n}$ has an infinite path, i.e., $\mathrm{tr}(T_{m,n}) = \infty$. Thus, $\mathrm{tr}(T'_{m,n}) = \infty$. If $\neg m\, E\, n$, then for every $\sigma = (\sigma_0, \ldots, \sigma_k)$, $\mathrm{tr}(T_{m,\sigma_0} * \cdots * T_{\sigma_k,n})$ is a computable ordinal. Indeed, fix $m, n \in \mathbb{N}$ such that $\neg m\, E\, n$. For every finite sequence σ, consider the corresponding tree $P_s = T_{m,\sigma_0} * T_{\sigma_0,\sigma_1} * \cdots * T_{\sigma_k,n}$. Consider the function f from the set of finite sequences into WF (the indices of well-founded trees) such that $f(s)$ is the index of P_s. The function f is hyperarithmetical, and its domain is computable. By Fact 8.1.20, there is a computable bound on the range of f. Therefore, $T'_{m,n}$ has rank α for some computable α. Note that for all $m' \in [m]_E$ and $n' \in [n]_E$, we get the same bound α. Let $T^*_{m,n} = T'_{m,n} * T^\infty$. By Proposition 8.2.22, the tree $T^*_{m,n}$ is a computable rank-saturated tree, $\mathrm{tr}(T^*_{m,n}) = \mathrm{tr}(T'_{m,n})$; this is also uniform. □

Now we build the desired sequence $(T_n)_{n\in\mathbb{N}}$. Take the tree T consisting exactly of the sequences $(m, m, \ldots, m)$ of length $i \leqslant m$ for $m \in \mathbb{N}$. Now fix n and for every m, attach $T^*_{m,n}$ to the m-th leaf of T. The resulting tree is T_n. The sequence $(T_n)_{n\in\mathbb{N}}$ witnesses the reducibility: $m\,E\,n$ iff $T_m \cong T_n$.

Indeed, suppose $m\,E\,n$. For every $k \in [m]_E = [n]_E$,

$$\mathrm{tr}(T'_{k,m}) = \mathrm{tr}(T'_{k,n}) = \infty,$$

thus

$$T^*_{m,k} \cong T^*_{n,k} \cong T^\infty.$$

For every $k \notin [m]_E$,

$$\mathrm{tr}(T^*_{k,m}) = \mathrm{tr}(T^*_{k,n}) = \alpha,$$

and therefore

$$T^*_{m,k} \cong T^*_{n,k} \cong T_\alpha.$$

We conclude that $T_m \cong T_n$.

Suppose now that $\neg m\,E\,n$. Then $T^*_{m,m} \cong T^\infty$, while $T^*_{n,n} \cong T_\alpha$ for some computable α. Thus, $T_m \ncong T_n$. □

The proof of Theorem 8.2.19 implies the following:

Corollary 8.2.24 (Fokina et al. [171]). *The class of computable torsion abelian groups if FF-complete.*

The proof uses the technique of p-basic trees that will be explained in the next chapter; see Exercise 9.3.62 for an extended hint.

Friedman and Stanley [184] showed that there is no Borel embedding of the class of countable undirected graphs into the class of torsion abelian groups. (We cite [197] for a thorough explanation.) If there existed a Turing computable embedding taking graphs to torsion abelian groups, then this would be a Borel embedding. Combined with Corollary 8.2.24, this gives the following, somewhat unexpected:

Theorem 8.2.25 ([171, 169]). *The class of torsion abelian groups is FF-complete but not effectively complete.*

The theorem above contrasts with Exercise 8.2.31 and the theorems establishing effective completeness contained in §8.2.2 and §8.2.3. Also, it appears that the proof in [79] can be effectivised to illustrate the effective completeness (thus, FF-completeness) of Boolean algebras. In Theorem 8.3.10 we will prove that the class of computable torsion-free abelian groups is FF-complete. However, the proof contained in [426] appears to be effective, thus giving the EFF-completeness of this class as well.

Exercises

Exercise° 8.2.26 (Folklore; e.g., [166]). Show that structures with two unary function symbols are effectively complete with respect to degree spectra and computable dimension.

Exercise° 8.2.27 (Folklore; e.g., [123]). Show that structures in the language of one function with two arguments are effectively complete with respect to degree spectra and computable dimension.

Exercise° 8.2.28. Based on the material of [257], complete the proofs of Theorems 8.2.12 and 8.2.14.

Exercise 8.2.29 (Hirschfeldt, Khoussainov, Shore, and Slinko [257]). Prove that partial orderings are effectively complete, and indeed are complete with respect to degree spectra and effective dimension.

Exercise* 8.2.30 (Hirschfeldt, Khoussainov, Shore, and Slinko [257]). Same as above, but for lattices.

Exercise* 8.2.31 (Fokina et al. [171], based on Friedman and Stanley [184]). Show that linear orders form an effectively complete (EFF-complete) class.

Exercise* 8.2.32 (Essentially Friedman and Stanley [184]). Show that fields of characteristic zero form an effectively complete class. (Hint: Let $F^{\#}$ be the algebraic closure of $\mathbb{Q}(v_i : i \in \mathbb{N})$, where all v_i are algebraically independent. Given an undirected, irreflexive graph G upon vertices v_i, let $F(G)$ be the smallest subfield of $F^{\#}$ that contains the algebraic closures of $\mathbb{Q}(v_i)$ for all $i \in \mathbb{N}$ and the elements $\sqrt{v_i + v_j}$ for each $(v_i, v_j) \in E(G)$. Use the transformation $G \mapsto F(G)$.)

Exercise 8.2.33** (Kogabaev [309]). Show that the class of (pappian) projective planes are effectively complete with respect to degree spectra and computable dimension. (We omit the definitions.)

8.3 Torsion-free abelian groups and their connected duals

In this section we prove Theorem D that states that the homeomorphism problem for connected compact spaces is Σ_1^1-complete. Following the main pattern of the book, this is achieved by reducing the problem to a similar problem about discrete groups.

8.3.1 The isomorphism problem for torsion-free abelian groups

Whilst the class of completely decomposable torsion-free abelian groups is relatively well-behaved, the class of all torsion-free abelian groups is unclassifiable.

Theorem 8.3.1 (Downey and Montalbán [146]). *The isomorphism problem for computable torsion-free abelian groups is Σ_1^1-complete.*

Proof of Theorem 8.3.1. We follow [146] closely. By Theorem 8.1.31 and Remark 8.1.32, it is sufficient to prove the following

Proposition 8.3.2 (Downey and Montalbán [146]). *There is a computable operator G, that assigns to each tree T a torsion-free abelian group $G(T)$, in a way that*

1. *if $T_0 \cong T_1$, then $G(T_0) \cong G(T_1)$,*
2. *if T_0 is well-founded and T_1 is not, then $G(T_0) \not\cong G(T_1)$.*

We start by defining the operator G. Let T be a tree. Let $\mathbb{Q}^T$ the group whose elements are formal sums

$$\sum_{\sigma \in V} q_\sigma \sigma,$$

where V is a finite subset of T, $q_\sigma \in \mathbb{Q}$ and addition is computed componentwise. Note that if T is infinite, then $\mathbb{Q}^T$ is isomorphic to $\bigoplus_{i \in \mathbb{N}} \mathbb{Q}$. $G(T)$ will be a subgroup of $\mathbb{Q}^T$. We think of T as a subset of $\mathbb{Q}^T$. Let $\mathcal{P} = \{p_0, p_1, ...\}$ be the set of prime numbers, listed in increasing order. $G(T)$ is defined so that $\sigma \in T$ can be divided by all the powers of $p_{2|\sigma|}$, and if $|\sigma| > 0$, then $\sigma^- + \sigma$ can be divided by all the powers $p_{2|\sigma|-1}$, where σ^- is σ with its last element removed, i.e. $\sigma^- = \sigma \upharpoonright |\sigma| - 1$, where $\sigma \upharpoonright n$ denotes the restriction of σ to its prefix of length n. In other words, $G(T)$ is the subgroup of $\mathbb{Q}^T$ generated under addition by

$$\{\frac{1}{p_{2|\sigma|}^k} \sigma : \sigma \in T, k \in \mathbb{N}\} \cup \{\frac{1}{p_{2|\sigma|-1}^k} (\sigma^- + \sigma) : \sigma \in T, |\sigma| > 0, k \in \mathbb{N}\}.$$

Remark 8.3.3. For the reader familiar with Hjorth [261], we note that $G(T)$ is the *group eplag* corresponding to the *prime labeled graph* (V, E, f), where $V = T$, $E = \{(\sigma^-, \sigma) : \sigma \in T\}$, $f(\sigma) = p_{2|\sigma|}$, and $f((\sigma^-, \sigma)) = p_{2|\sigma|-1}$.

Note that the isomorphism type of $G(T)$ only depends on the isomorphism type of the tree T. This gives part (1) of Proposition 8.3.2. The second part follows immediately from the following lemma.

Lemma 8.3.4. *A tree T is non-well-founded iff in the group $G(T)$ there exists an infinite sequence $g_0, g_1, \ldots$ of elements such that for each i, g_i is divisible by all the powers of p_{2i} and $g_i + g_{i+1}$ is divisible by all the powers of p_{2i+1}.*

Before we prove this lemma, we need to prove some basic properties of $G(T)$. As before, for a set of prime numbers P, we use $\mathbb{Q}^{(P)}$ to denote the set of rational numbers whose denominators are products of powers of primes in P. For a finite set of primes $\{p_{i_0}, \ldots, p_{i_k}\}$, we write $\mathbb{Q}^{(p_{i_0},\ldots,p_{i_k})}$ instead of $\mathbb{Q}^{(\{p_{i_0},\ldots,p_{i_k}\})}$. Note that $\mathbb{Q}^{(\varnothing)} = \mathbb{Z}$. If P and R are sets of prime numbers then

$$\mathbb{Q}^{(P)} \cap \mathbb{Q}^{(R)} = \mathbb{Q}^{(P\cap R)} \quad \text{and} \quad \mathbb{Q}^{(P)} + \mathbb{Q}^{(R)} = \mathbb{Q}^{(P\cup R)}.$$

Lemma 8.3.5. *Let $h = \sum_{\sigma\in V} r_\sigma \sigma \in G(T)$ where $V \subseteq T$ and each $r_\sigma \neq 0$. If h is divisible by all the powers of p_{2n}, then $|\sigma| = n$ for every $\sigma \in V$.*

Proof. Multiply h by some integer and divide it by some power of p_{2n}, and obtain $g = \sum_{\sigma\in V} q_\sigma \sigma \in G(T)$ so that all the coefficients q_σ are of the form $\frac{m_\sigma}{p_{2n}^{i_\sigma}}$ for $m_\sigma \in \mathbb{Z}$, $i_\sigma \in \mathbb{Z}^+$, and $p_{2n} \not| m_\sigma$. In other words, all the coefficients of $g \in G(T)$ are in $\mathbb{Q}^{(p_{2n})}\backslash\mathbb{Z}$. By the definition of $G(T)$, every element of $G(T)$ can be written as follows:

$$g = \sum_{\tau\in W} a_\tau \tau + \sum_{(\tau^-,\tau)\in U} b_\tau(\tau^- + \tau),$$

where $W \subseteq T$, $U \subseteq \{(\tau^-,\tau) : \tau \in T\backslash\{\varnothing\}\}$, $a_\tau \in \mathbb{Q}^{(p_{2|\tau|})}$, and $b_\tau \in \mathbb{Q}^{(p_{2|\tau|-1})}$. Consider now $\sigma \in V$; we want to show that $|\sigma| = n$. We have that q_σ is equal to the coefficient of σ in the sum above. This coefficient is

$$a_\sigma + \left(\sum_{(\sigma,\tau)\in U} b_\tau\right) + b_\sigma,$$

where a_σ and b_σ might be 0. On the one hand we have that $q_\sigma \in \mathbb{Q}^{(p_{2n})}\backslash\mathbb{Z}$. On the other hand, the coefficient above belongs to $\mathbb{Q}^{(p_{2|\sigma|-1},p_{2|\sigma|},p_{2|\sigma|+1})}$. If $p_{2n} \neq p_{2|\sigma|}$, then $(\mathbb{Q}^{(p_{2n})}\backslash\mathbb{Z}) \cap \mathbb{Q}^{(p_{2|\sigma|-1},p_{2|\sigma|},p_{2|\sigma|+1})} = \varnothing$. Therefore $p_{2n} = p_{2|\sigma|}$ and $|\sigma| = n$ as wanted. □

Lemma 8.3.6. *Let $h = \sum_{\sigma\in V} r_\sigma \sigma \in G(T)$ where $V \subseteq T$ and each $r_\sigma \neq 0$. If h is divisible by all the powers p_{2n+1}, then, for every $\sigma \in V$ with $|\sigma| = n$, there exists $\tau \in V$ with $\sigma = \tau^-$*

Proof. As in the proof of the previous lemma, multiplying h by the right scalar, we obtain $g = \sum_{\sigma\in V} q_\sigma \sigma \in G(T)$ all of whose coefficients are in $\mathbb{Q}^{(p_{2n+1})}\backslash\mathbb{Z}$. Again, since $g \in G(T)$, we get that

$$g = \sum_{\tau\in W} a_\tau \tau + \sum_{(\tau^-,\tau)\in U} b_\tau(\tau^- + \tau),$$

where $W \subseteq T$, $U \subset \{(\tau^-,\tau) : \tau \in T\backslash\{\varnothing\}\}$, $a_\tau \in \mathbb{Q}^{(p_{2|\tau|})}$, and $b_\tau \in \mathbb{Q}^{(p_{2|\tau|-1})}$. Consider now $\sigma \in V$ with $|\sigma| = n$. We have that q_σ is equal to the coefficient of σ in the sum above. This coefficient is

$$a_\sigma + \left(\sum_{(\sigma,\tau)\in U} b_\tau\right) + b_\sigma,$$

where a_σ and b_σ might be 0. So, we have that $q_\sigma \in \mathbb{Q}^{(p_{2n+1})} \backslash \mathbb{Z}$ and that the coefficient above belongs to $\mathbb{Q}^{(p_{2n-1},p_{2n},p_{2n+1})}$. Therefore, the middle term, $\sum_{(\sigma,\tau)\in U} b_\tau$ has to be in $\mathbb{Q}^{(p_{2n+1})} \backslash \mathbb{Z}$: Because otherwise the coefficient above would belong to $\mathbb{Q}^{(p_{2n-1},p_{2n})}$, which has empty intersection with $\mathbb{Q}^{(p_{2n+1})} \backslash \mathbb{Z}$. So, there exists some $\tau \in T$ with $(\sigma, \tau) \in U$ and $b_\tau \in \mathbb{Q}^{(p_{2n+1})} \backslash \mathbb{Z}$. Pick one such τ. Note that $\sigma = \tau^-$. We claim that $\tau \in V$. Let us look at the coefficient of τ in g (which we want to show is not 0):

$$a_\tau + \sum_{(\tau,\delta)\in U} b_\delta + b_\tau.$$

The first two terms in this sum are in $\mathbb{Q}^{(p_{2n+2},p_{2n+3})}$, and the third one is in $\mathbb{Q}^{(p_{2n+1})} \backslash \mathbb{Z}$. Therefore this coefficient is not 0, and $\tau \in V$. □

Proof of Lemma 8.3.4. If T is not well-founded and X is an infinite path through T, then $\{g_i = X \upharpoonright i : i \in \mathbb{N}\} \subseteq T \subseteq G(T)$ is a sequence in $G(T)$ as wanted.

Suppose now that $\{g_i : i \in \omega\}$ is a sequence as in Lemma 8.3.4. Since g_i is divisible by all the powers of p_{2i}, by Lemma 8.3.5, we get that $g_i = \sum_{\sigma\in V_i} q_\sigma \sigma$, where V_i is a finite subset of $T \cap \omega^i$, and $q_\sigma \neq 0$. Since $g_i + g_{i+1} = \sum_{\sigma\in V_i \cap V_{i+1}} q_\sigma \sigma$ is divisible by all the powers of p_{2i+1}, then, by Lemma 8.3.6 we get that for every $\sigma \in V_i$, there exists $\tau \in V_{i+1}$ extending σ. Therefore, by induction we can choose a sequence $\sigma_i \in V_i$, for $i \in \mathbb{N}$, such that $\sigma_i \subset \sigma_{i+1}$. Hence T is not well-founded. □

The proof of Theorem 8.3.1 is complete. □

The proof above shows that there is a computable group such that the set of indices of computable groups which are isomorphic to it is Σ_1^1-complete. We state one more corollary. Let $T_0 = \{0^n : n \in \mathbb{N}\}$ where 0^n is the string with n many zeros $\langle 0, 0, ..., 0\rangle$. Let $G_0 = G(T_0)$. From the proof above we, we see that a tree T has an infinite path in and only if G_0 embeds in $G(T)$. Thus, we obtain

Corollary 8.3.7. *The index set*

$$\{i : G_i \text{ is torsion-free abelian and } G_0 \text{ embeds in } G_i\}$$

is Σ_1^1 complete.

8.3.2 Comparing integral homology and Čech cohomology

We remark that the Downey-Montalbán Theorem has applications to finitely presented groups. Baumslag, Dyer and Miller [31] proved that the c.e. presented groups are exactly the groups that appear in *integral homology sequences* of finitely presented groups. (We omit the complex definition.) Moreover, they proved that given any sequence $A_1, A_2, \ldots,$ of uniformly c.e. presented torsion-free abelian groups, where the first two are finitely generated, there exists a finitely presented group G whose integral homology is the given sequence. They also proved a similar result when the groups $A_1, A_2, \ldots$ are all computably presented. Baumslag, Dyer and Miller [31] left open whether c.e. presented torsion-free abelian groups always admit a computable presentation; this was settled by Khisamiev, see Theorem 5.1.41 which also appeared as Theorem A(3) in Part 1.

Observing that the construction in [31] of G from the sequence $A_1, A_2, \ldots,$ is effective, at least in the case when all A_i are 0 except for one, we get the following corollaries of Theorem 8.3.1.

Corollary 8.3.8 ([146],[184]). *Deciding whether two finitely presented groups have the same homology sequence is Σ^1_1-complete in general. Indeed, this is already true for the 3rd homology group in the sequence.*

There are many other results in the literature saying that some properties about finitely presented groups cannot be decided computably. For example, Lempp [334] showed that, for a finitely presented group, being torsion-free is Π^0_2-complete. See Miller [390] for a survey on decision problems for finitely presented groups. However, as far as we know, the only such property in the literature that would not be arithmetical is presented in Corollary 8.3.8 above. Note that deciding whether two finitely generated groups are isomorphic is merely Σ^0_3. The stark contrast with Σ^1_1-completeness in Corollary 8.3.8 perhaps partially explains why homological methods are not particularly easy to apply to finitely presented groups.

Recall that the isomorphism problem for computably compact spaces is Σ^1_1, and we have just showed it is Σ^1_1-complete.

Corollary 8.3.9. *Deciding whether the first Čech cohomology groups of given computably compact spaces are isomorphic is Σ^1_1-complete in general.*

Proof. The Čech cohomology groups are uniformly c.e. presented by Theorem 5.2.21. Thus, the upper bound is indeed Σ^1_1. To establish the Σ^1_1-completeness, use Theorem 5.2.1 to turn the groups witnessing Theorem 8.3.1 into connected compact spaces whose first Čech cohomology groups are isomorphic to the respective discrete groups in the sequence, by Theorem 5.2.24. □

While the Čech cohomology groups are certainly not the most straightforward invariants of a space, they are still quite useful in some situations. The corollary above might partially explain why this invariant is a balanced combination of complexity and utility.

8.3.3 The FF-completeness of torsion-free abelian groups

In this subsection, we prove the following stronger result, which implies Theorem 8.3.1 but is harder to prove. Recall Definition 8.2.15 which gives the natural Type I analogue of effective completeness.

Theorem 8.3.10 (Fokina et al. [171]). *The class of computable torsion-free abelian groups is FF-complete among Σ^1_1-equivalence relations.*

Theorem 8.3.10 predates (but of course does not imply) the recent result establishing the Borel completeness of this class of groups [426]. We remark that the proof contained in [426] appears to give the EFF-completeness of the class, thus implying Theorem 8.3.10 below. However, the class of torsion-free abelian groups is not effectively complete with respect to computable dimension; this follows from Corollary 10.3.25 and Theorem 10.3.3. It is not known whether it is effectively complete with respect to degree spectra.

Proof of Theorem 8.3.10

We shall need a class of trees slightly more general than the class of rank-saturated trees (Definition 8.2.20).

Definition 8.3.11 (Calvert, Knight, and Millar [78]). A tree $T \subseteq \omega^{<\omega}$ is *rank-homogeneous* provided that for all x at level n,

1. if $tr(x)$ is an ordinal, then for all y at level $n+1$ such that $tr(y) < tr(x)$, x has infinitely many successors z such that $tr(z) = tr(y)$,
2. if $tr(x) = \infty$, then for all y at level $n+1$, x has infinitely many successors z such that $tr(z) = tr(y)$.

It is not hard to see that every rank-saturated tree is rank-homogeneous. For a rank-homogeneous tree T, let $R(T)$ be the set of pairs (n, α) such that there is an element at level n of tree rank α (where α is an ordinal, not ∞). Note that the top node in T has rank ∞ just in case $R(T)$ has no pair of the form $(0, \alpha)$. Also note if T has a node of rank ∞, then the top node must have rank ∞, and if the top node has rank ∞, then there are nodes of rank ∞ at all levels. Thus, from the set of pairs $R(T)$ in which the second components are ordinals, we can deduce all of the information that would be given if we included pairs with second component ∞.

Proposition 8.3.12. *Suppose T, T' are rank-homogeneous trees. Then $T \cong T'$ iff $R(T) = R(T')$.*

Proof. Clearly, if $T \cong T'$, then $R(T) = R(T')$. Suppose $R(T) = R(T')$. To see that there is an isomorphism, we show that the set of finite partial rank-preserving isomorphisms between subtrees of T and T' has the back-and-forth property. The subtrees must be closed under predecessor in the large trees, and the finite partial isomorphisms must preserve all ranks, both ordinals and ∞. Given a finite subtree of one of the large trees, we can reach any further node by a finite sequence of steps in which the node being added is a successor of one already included. Therefore, it is enough to prove the following.

Claim: Let p be a rank-preserving isomorphism from the finite subtree τ of T onto the finite subtree τ' of T', and let $a \in T - \tau$ be a successor of $b \in \tau$. Suppose $b' = p(b)$. Then there exists a', a successor of b' in T', not already in $ran(p)$, such that a' and a have the same rank.

The rank of $p(b)$ is the same as that of b. If a has rank ∞, then b and b' also have rank ∞, and b' has infinitely many successors of rank ∞. If a has ordinal rank α, then b and b' have rank either ∞ or some $\beta > \alpha$. In either case, b' has infinitely many successors of rank α. We choose a' to be a successor of b', of the proper rank, not already in $ran(p)$. □

Let $T \mapsto G(T)$ be the transformation defined in Proposition 8.3.2. In view of Theorem 8.2.19, we claim that to establish Theorem 8.3.10 it is sufficient to prove the following:

Theorem 8.3.13 (Fokina et al. [169]). *For every two rank-homogeneous trees T and T', the groups $G(T)$ and $G(T')$ are isomorphic if, and only if, $T \cong T'$.*

We first explain how to derive Theorem 8.3.10 from Theorem 8.3.13, and then we prove Theorem 8.3.13. Given a Σ^1_1 equivalence relation E for every $n \in \mathbb{N}$, take the sequence of rank-saturated trees $(T^*_{m,n})_{m,n\in\mathbb{N}}$ as in Lemma 8.2.23. We explain how to pass effectively from the sequence to a group G_n such that $G_n \cong G_{n'}$ iff for all m, $T^*_{m,n} \cong T^*_{m,n'}$.

Let $(p_{m,k})_{k\in\mathbb{N}}$ be uniformly computable lists of distinct primes. For each m, consider $G_{m,n} = G(T_{m,n})$, but in the transformation we will be using the list of primes $(p_{m,k})_{k\in\mathbb{N}}$ (instead of just $(p_k)_{k\in\mathbb{N}}$ as defined in Proposition 8.3.2). By Theorem 8.3.13, we will have $G_{m,n} \cong G_{m',n'}$ if and only if $T^*_{m,n} \cong T^*_{m',n'}$. Let

$$G_n = \bigoplus_m G_{m,n}.$$

Since the sequences of primes are disjoint, we have $G_n \cong G_{n'}$ iff for all m, $G_{m,n} \cong G_{m,n'}$.

Proof of Theorem 8.3.13. We adopt the version of the proof that can be found in [367], with only minor further adjustments. Fix a tree T and the group $G(T)$. We do not distinguish between elements of the tree T and the corresponding elements of $G(T)$, which we call *vertex elements*. We also identify the empty set $\varnothing$ with the empty string λ, which is the root of T.

The following is taken from Hjorth [261] (Propositions 2.2 and 2.5).

Proposition 8.3.14. *Let φ be a homomorphism from $G(T)$ to $\mathbb{Q}$ such that $\varphi(v) = 1$ for $v \in T_n$ and $\varphi(v) = -1$ for $v \in T_{n+1}$. Let $h = \sum_{v\in V} c_v v + \sum_{u\in U} a_u u$, where $V \subseteq T_n$ and $U \subseteq T_{n+1}$. If $(p_{2n+1})^\infty | h$, then $\varphi(h) = 0$. Moreover, for each $v \in V$, if $h_v = c_v v + \sum_{u\in U_v} a_u u$, then $(p_{2n+1})^\infty | h_v$, and $\varphi(h_v) = 0$ (here $U_v \subseteq U$ contains all successors of v in U).*

Using Proposition 8.3.14, we obtain:

Lemma 8.3.15.

1. *Suppose $h = a_\varnothing \varnothing + \sum_{u\in U} a_u u$, where $U \subseteq T_1$. If $(p_1)^\infty | h$, then $a_\varnothing = \sum_{u\in U} a_u$.*
2. *Suppose $h = \sum_{v\in V} a_v v + \sum_{u\in U} b_u u$, where $U \subseteq T_{n+1}$, and V is the set of predecessors of these elements. For $v \in V$, let U_v be the set of successors of v. If $(p_{2n+1})^\infty | h$, then for each $v \in V$, $a_v = \sum_{u\in U_v} b_u$.*

Proof. For 1, we consider a homomorphism φ taking $\varnothing$ to 1 and taking elements at level 1 to -1. We have $\varphi(h) = 0 = a_\varnothing - \sum_{u\in U} a_u$. By Proposition 8.3.14, $a_\varnothing - \sum_{u\in U} a_u$. For 2, we consider a homomorphism φ taking all elements of V to 1 and all elements of U to -1. By Proposition 8.3.14, for each $v \in V$, $\varphi(a_v v + \sum_{u\in U_v} b_u) = 0 = a_v - \sum_{u\in U_v} b_u$. Therefore, $a_v = \sum_{u\in U_v} b_u$. □

Note that in Lemma 8.3.15, in Case 1, we may have $a_\varnothing = 0$ and $\sum_{u\in U} a_u = 0$, and in Case 2, we may have $a_v = 0$, and $\sum_{u\in U_v} b_u = 0$. We need a refinement of Lemma 8.3.5.

Lemma 8.3.16. *Suppose $(p_{2n})^\infty | h$.*

1. *If $n > 0$, then h can be expressed in the form $\sum_{v\in V} r_v v$, where $V \subseteq T_n$ and r_v is in $\mathbb{Q}^{(\{p_{2n}, p_{2n-1}\})}$.*
2. *If $n = 0$, then h has the form $r\varnothing$, where $r \in \mathbb{Q}^{(\{p_0\})}$.*

Proof. We consider the two cases separately.

Case 1: Suppose $n > 0$. By Lemma 8.3.5, h can be expressed in the form $\sum_{v \in V} r_v v$, where $V \subseteq T_n$, and $r_v \in \mathbb{Q}$. Just because $h \in G(T)$, we have $h = \sum_{u \in U} a_u u + \sum_{u \in W} b_u(u + u^-)$, where if $u \in T_k$, then $a_u \in \mathbb{Q}^{(\{p_{2k}\})}$, and $b_u \in \mathbb{Q}^{(\{p_{2k-1}\})}$. For u at level $k \neq n$, the coefficient of u in the expression for h must be 0. This coefficient has the form $a_u + (\sum_{w^-=u} b_w) + b_u$.

Claim: For all $k > n$, for u at level k (appearing in our decomposition), a_u and b_u are integers.

Proof of Claim. We work our way back from the largest $k > n$ with some u at level k that appears. For the greatest k, if u is at level k, and u appears, then no successor of u appears. We have $0 = a_u + b_u$, where $a_u \in \mathbb{Q}^{(\{p_{2k}\})}$ and $b_u \in \mathbb{Q}^{(\{p_{2k-1}\})}$. Then both a_u and b_u must be integers. Supposing that the claim holds for $k' > k$, where $k > n$, let u be an element at level k that appears. We have $0 = a_u + (\sum_{w^-=u} b_w) + b_u$, where $a_u \in \mathbb{Q}^{(\{p_{2k-1}\})}$, $b_u \in \mathbb{Q}^{(\{p_{2k-1}\})}$, and $\sum_{w^-=u} b_w \in Z$. Again a_u and b_u must be integers. □

Using the Claim, we can complete the proof for Case 1. For v at level n, the coefficient is $r_v = a_v + (\sum_{w^-=v} b_w) + b_v$, where $\sum_{w^-=v} b_w \in Z$, $a_v \in \mathbb{Q}^{(\{p_{2n}\})}$ and $b_v \in \mathbb{Q}^{(\{p_{2n-1}\})}$. Therefore, $r_v \in \mathbb{Q}^{(\{p_{2n}, p_{2n-1}\})}$.

Case 2: Suppose $n = 0$. Then the only possible v is $\varnothing$, so $h = r\varnothing$. Since there is no $\varnothing^-$, we have $r = a_\varnothing + \sum_{w^-=\varnothing} b_w$. By the argument above, $\sum_{w^-=\varnothing} b_w \in Z$. Since $a_\varnothing$ is in $\mathbb{Q}^{(\{p_0\})}$, r is also. □

A node in T_n has the feature that there is a successor chain of length n leading from $\varnothing$ to it. We try to describe this in the group $G(T)$. We define first the *pseudo-vertex-like* elements at level n, and then the *vertex-like* elements at level n. We start with the definition of pseudo-successor:

Definition 8.3.17 (pseudo-successor)**.** Suppose nonzero h and g are so that $p_{2n}^\infty | g$ and $p_{2n+2}^\infty | h$, for some $n \geqslant 0$. We say that h is a *pseudo-successor* of g if $(p_{2n+1})^\infty | (g + h)$.

Lemma 8.3.18. *There is an algebraic property $\Theta(x)$ such that for all $T \in RHT$ with $T_1 \neq \varnothing$, $\Theta(x)$ is satisfied just by $\varnothing$ and $-\varnothing$. (Indeed, $\Theta(x)$ a computable infinitary formula in the language of additive groups; computable infinitary formulae will to be defined in Chapter 10.)*

Proof. We let $\Theta(x)$ say the following:

1. $(p_0)^\infty | x$,
2. for primes $q \neq p_0$, $q \nmid x$,
3. x has a pseudo-successor,
4. $\dfrac{1}{p_0} x$ has no pseudo-successor.

It is not difficult to see that $\varnothing$ and $-\varnothing$ satisfy $\Theta(x)$. We must show that other elements do not. If x satisfies Condition 1, we can apply Part 2 of Lemma 8.3.16, to see that x has the form $r\varnothing$, where $r \in \mathbb{Q}^{(\{p_0\})}$. Then r has the form $\dfrac{z}{(p_0)^m}$, where $z \in Z$. Condition 2 implies that z is not divisible by any primes other than p_0. Therefore, x has the form $\pm p^k \varnothing$. Condition 3 says that x has a successor. Using this, we show that $k \geqslant 0$. Take y such that $(p_2)^\infty | y$. By Part 1 of Lemma 8.3.16, $y = \sum_{v \in V} s_v v$, where $V \subseteq T_1$ and $s_v \in \mathbb{Q}^{(\{p_2,p_1\})}$. If $(p_1)^\infty|(x + y)$, then by Lemma 8.3.15, $\pm(p_0)^k = \sum_{v \in V} s_v$. This implies that the right-hand side is an integer, and then the left-hand side is as well. Therefore, $x = \pm p_0^k$, where $k \geqslant 0$. Finally, we show that if x satisfies Condition 4, then k cannot be positive. If $k > 0$, then $\dfrac{1}{p_0}x = p_0^{k-1}\varnothing$. This satisfies Conditions 1 and 2. Moreover, if $v \in T_1$, then $p_0^{k-1}v$ is a successor of $\frac{1}{p_0}x$, contradicting Condition 4. Therefore, x must have the form $\pm\varnothing$. □

Definition 8.3.19 (pseudo-vertex-like)**.** An element $h \in G(T)$ is *pseudo-vertex-like*, or *p.v.l.*, *at level* n, if one of the following holds:

1. $n = 0$ and $\Theta(x)$ holds, or
2. $n > 0$ and
   (a) $p_{2n}^\infty|h$,
   (b) there exists a sequence $g_0, g_1, \ldots, g_n = h$, such that g_0 satisfies condition $\Theta(x)$ from Lemma 8.3.18, and for all $i < n$, we have $p_{2i}^\infty|g_i$ and $p_{2i+1}^\infty|(g_i + g_{i+1})$.

It is easy to see that all vertex elements are pseudo-vertex-like.

We define rank for p.v.l. elements by analogy with tree rank. We write $rk(h)$ for the rank of h in the group $G(T)$, and $tr(v)$ for the tree rank of v in the tree T.

Definition 8.3.20 (rank)**.** Let h be p.v.l. at level n.

1. $rk(h) = 0$ if h has no pseudo-successors,
2. for $\alpha > 0$, $rk(h) = \alpha$ if all pseudo-successors of h have ordinal rank, and α is the least ordinal greater than these ranks,
3. $rk(h) = \infty$ if h does not have ordinal rank.

We note that $rk(h) = \infty$ iff there is an infinite sequence $(g_i)_{i \in \omega}$ such that each g_i is p.v.l., $g_0 = h$ and g_{i+1} is a pseudo-successor of g_i.

Lemma 8.3.21. *Suppose h is p.v.l at level n, expressed in the form $\sum_{v \in V} r_v v$, where V is a finite subset of T_n and $r_v \neq 0$. Then for all v, $tr(v) \geqslant rk(h)$.*

Proof. We show by induction on α that if $rk(h) > \alpha$, then for all $v \in V$, $tr(v) \neq \alpha$. (We allow the possibility that $rk(h) = \infty$.) Let $rk(h) > 0$. Let g be a p.v.l. pseudo-successor for h. Then $(p_{2n+1})^\infty|(h + g)$. Say $g = \sum_{u \in U} s_u u$, where U is a set of vertex elements at level $n + 1$ and $s_u \neq 0$. By Lemma 8.3.6, for each $v \in V$, there is some $u \in U$ such that u is a successor of v. Therefore, $tr(v) \neq 0$.

Consider $\alpha > 0$, where the statement holds for $\beta < \alpha$. Suppose $rk(h) > \alpha$. Let g be a p.v.l. pseudo-successor of h such that $rk(g) \geqslant \alpha$. Say $g = \sum_{u \in U} s_u u$, where U is a set of vertex elements at level $n + 1$ and $s_u \neq 0$. By the Induction Hypothesis, $tr(u) \neq \beta$ for any $\beta < \alpha$, so $tr(u) \geqslant \alpha$. By Lemma 8.3.6, some $u \in U$ is a successor of v. Then $tr(v) \neq \alpha$. Finally, we show that if $rk(h) = \infty$, then for all $v \in V$, $tr(v) = \infty$. There must be an infinite sequence of p.v.l. elements $(g_k)_{k \in \mathbb{N}}$ such that $g_0 = h$ and g_{k+1} is a pseudo-successor of g_k. We have $g_k = \sum_{u \in U_k} s_u u$, where U_k is a set of vertex elements at level $n + k$, and $s_u \neq 0$. For each element of U_k, there is a successor in U_{k+1}. We obtain a chain of successors, starting with $v = v_0 \in U_0$, and choosing v_{k+1} a successor of v_k in U_{k+1}. Therefore, $tr(v) = \infty$. □

It is again helpful to consider an example.

Example: Let $v \in T_1$ and let u and u' be successors of v in T_2. Suppose that both u and u' have successors in T_3. Let $g = \frac{1}{p_4}u + \frac{p_4 - 1}{p_4}u'$. Since $p_4^\infty | u, u'$, we have $p_4^\infty | g$. Since $p_4(v + g) = (v + u) + (p_4 - 1)(v + u')$, we see that $p_3^\infty | (v + g)$. Therefore, g is p.v.l. and it is a pseudo-successor of v. We can show that g has no pseudo-successor, even though we have expressed it in terms of u and u', both of which have successors in T_3. Suppose that h is its a pseudo-successor at level 3. Then $h = \sum_{w \in W} r_w w$, where $W \subseteq T_3$ and $r_w \in Q$. By Lemma 8.3.16, we must have $r_w \in \mathbb{Q}^{(p_6, p_5)}$. We must have $p_5^\infty | (g + h)$. By Lemma 8.3.15, if $W_u, W_{u'}$ are, respectively, the sets of successors of u, u' in W, then $\sum_{w \in W_u} r_w = \frac{1}{p_4}$, and $\sum_{w \in W_{u'}} r_w = \frac{p_4 - 1}{p_4}$. This is a contradiction.

We strengthen the definition of p.v.l. element in order to rule out examples like the one above, in which g has no successor, but it has a decomposition in terms of elements all having successors.

Definition 8.3.22 (vertex-like). Let $g \in G(T)$. We say that g is *vertex-like*, or *v.l.*, if

1. g is p.v.l. at some level n, and
2. either
   (a) $rk(g) > 0$, or
   (b) $rk(g) = 0$ and for any decomposition $g = \sum_j r_j g_j$ such that all g_j are p.v.l. at level n, there exists j such that $rk(g_j) = 0$.

Lemma 8.3.23. *If v is a vertex element, then it is vertex-like.*

Proof. We already noted that a vertex element is p.v.l. Suppose v is at level n, and $rk(v) = 0$. Then v has no successors. We must show that if $v = \sum_j r_j g_j$, where each g_j is p.v.l. at level n, then for some j, $rk(g_i) = 0$. Suppose that for all j, $rk(g_j) \neq 0$. Say h_j is a p.v.l. pseudo-successor of g_j at level $n + 1$. By Lemma 8.3.5, each g_j has a decomposition in terms of tree elements at level n. Since $v = \sum_j r_j g_j$, v must appear with non-zero coefficient in the decomposition of some g_j. Then by Lemma 8.3.15, the corresponding h_j has a decomposition that involves successors of v with non-zero coefficients. This is a contradiction. □

We would like to show that if g is v.l. at level n, expressed in the form $\sum_{v \in V} r_v v$, where $V \subseteq T_n$ and $r_v \neq 0$, then $rk(g)$ is the minimum of $tr(v)$, for $v \in V$.

Lemma 8.3.24. *Suppose g is v.l. at level n. Say $g = \sum_{v \in V} r_v v$, where $V \subseteq T_n$. Then $rk(g) = 0$ iff there exists $v \in V$ such that $tr(v) = 0$.*

Proof. First, suppose there exists $v \in V$ such that $tr(v) = 0$. By Lemma 8.3.21, $tr(v) \geqslant rk(g)$, so $rk(g) = 0$. Next, suppose $rk(g) = 0$. The elements of V are p.v.l. and one of the decompositions of g is $\sum_{v \in V} r_v v$. By the definition of vertex-like, there is some v such that $rk(v) = 0$. Then v has no pseudo-successors, so v has no successors in T. Therefore, $tr(v) = 0$. □

Lemma 8.3.25. *If g is v.l. at level n and $rk(g) > 0$, then g has a decomposition $\sum_{v \in V} m_v v$ where all coefficients m_v are integers.*

Proof. By Lemma 8.3.16, g can be expressed in the form $\sum_{v \in V} r_v v$, where $V \subseteq T_n$ and $r_v \in \mathbb{Q}^{(\{p_{2n}, p_{2n-1}\})}$. Since $rk(g) > 0$, we have a p.v.l. pseudo-successor g', expressed in the form $\sum_{u \in U} s_u u$, where $U \subseteq T_{n+1}$ and $s_u \in \mathbb{Q}^{(\{p_{2n+2}, p_{2n+1}\})}$. Consider $h = g + g'$. Since $(p_{2n+1})^{\infty} | h$, we can apply Lemma 8.3.15. For each $v \in V$, let U_v be the set of successors of v in U. We have $r_v = \sum_{u \in U_v} s_u$. It follows that $\sum_{u \in U_v} s_u$ and r_v are integers. □

Suppose g is a v.l. element at level n. Recall that the definition of v.l. has two condidions, with the second split into two cases. If Condition 2 (a) holds for g, then Lemma 8.3.24 says that g can be expressed as a sum of vertex elements on level n with integer coefficients. If $rk(g) = 0$, then the decomposition of g involves some terminal vertex element.

Lemma 8.3.26. *Let g be v.l. at level n, with a decomposition $\sum_{v \in V} r_v v$, where $V \subseteq T_n$ and all coefficients r_v are non-zero. Then $rk(g) = min_{v \in V} tr(v)$.*

Proof. By Lemma 8.3.21, $tr(v) \geqslant rk(g)$ for all $v \in V$. We show by induction on α that if $tr(v) \geqslant \alpha$ for all $v \in V$, then $rk(g) \geqslant \alpha$. For $\alpha = 0$, the statement is trivially true. Suppose $\alpha > 0$, where the statement holds for all $\beta < \alpha$. If g satisfies Condition 2 (b) from the definition of v.l., then by Lemma 8.3.25, there is some $v \in V$ such that $tr(v) = 0$. Suppose $rk(g) = \beta$, where $0 < \beta < \alpha$. For all $v \in V$, $tr(v) > \beta$, so v has a successor u_v with $tr(u_v) \geqslant \beta$. By Lemma 8.3.25, we may suppose that all r_v are integers. We have a successor h of g, of the form $\sum_{v \in V} r_v u_v$. This h is vertex-like at level $n + 1$, and by the Induction Hypothesis, $rk(h) \geqslant \beta$. Then $rk(g)$ cannot be β after all. Finally, suppose $tr(v) = \infty$ for all $v \in V$. For each v, there is an infinite successor chain, and we can use these to form an infinite chain of successors of g, so $rk(g) = \infty$. □

We are ready to finish the proof of the theorem. Recall that for a tree T, $R(T)$ is the set of pairs (n, α) such that there is some $v \in T$ at level n with $tr(v) = \alpha$. Proposition 8.3.12 says that for rank-homogeneous trees T, T', $T \cong T'$ iff $R(T) = R(T')$.

Let T, T' be rank-homogeneous trees. We show that $T \cong T'$ iff $G(T) \cong G(T')$. We let $R(G(T))$ be the set of pairs (n, α) such that there is a v.l. $g \in G(T)$ at level n with $rk(g) = \alpha$. We can show that $R(T) = R(G(T))$. If $(n, \alpha) \in R(T)$, then there is a $v \in T$ and a corresponding vertex element v in $G(T)$ witnessing the rank. By Lemma 8.3.26, the group rank of v is equal to the tree rank of v. Therefore, $(n, \alpha) \in R(G(T))$. On the other hand, if $(n, \alpha) \in R(G(T))$, witnessed by $g = \sum_{v \in V} r_v v$, then by Lemma 8.3.26, there is some vertex element $v \in V_n$ such that $tr(v) = \alpha$. Therefore, $(n, \alpha) \in R(T)$. This completes the proof that $G : T \rightarrow G(T)$ is $1 - 1$ on isomorphism types. □

As explained earlier, this completes the proof of Theorem 8.3.10.

Exercises

See Exercise 10.1.68-10.1.70 for the Pullback Theorem (which relies on computable infinitary logic) and its consequences.

Exercise* 8.3.27 (Fokina et al. [169]). Show that there is a 1-1 effective transformation taking rank-homogeneous trees to Boolean algebras. (Hint: The idea is to attempt to code which sequences σ lie in T and which do not. While the coding won't be injective, it will preserve enough information about the tree ranks. Let Γ_n be any computable tree representing $\mathrm{Intalg}(\omega^{n+1} + \eta + 1)$. For an input tree T, let $\Gamma(T)$ consist of finite strings of the form $m_0 a_1 m_1 \ldots a_n m_n \sigma$, where

- $\langle a_1, \ldots, a_n \rangle \in T$,
- $m_i \in \mathbb{N}$, and
- $\sigma \in \Gamma_n$.

Set B_T equal to the Boolean algebra generated by the resulting tree $\Gamma(T)$.)

8.3.4 Compact spaces. Proof of Theorem D

Recall that Theorem D states that the homeomorphism problem for connected compact Polish spaces is Σ^1_1-complete.

Proof of Theorem D. Being a connected compact space is arithmetical (Remark 7.1.35), and by Proposition 8.1.34 the homeomorphism problem for compact spaces is Σ^1_1. Theorem 8.3.1 produces a uniformly computable sequence of pairs of computable torsion-free abelian groups; without loss of generality, all these groups can be assumed to be non-trivial.

As we explained in Remark 5.2.8, Effective Pontryagin Duality (Theorem 5.2.1) is uniform when passing from computable torsion-free abelian groups to their computably compact connected duals, provided that these groups are non-zero. (In fact, Theorem 8.3.1 produces a sequence of groups with computable bases which are given by the vertex-elements, so we can avoid the use of Dobrica's Theorem in this particular application of Theorem 5.2.1.) Since non-isomorphic torsion-free abelian groups correspond to non-homeomorphic connected compact spaces under the duality, this gives to the Σ^1_1-completeness of the homeomorphism problem for connected compact spaces. Indeed, it follows that there is a computably compact connected space so that its recognition problem is Σ^1_1-complete.

The proof of Theorem D is finished.

As a consequence of Theorem 8.3.10, Theorem 5.2.1, and Propositions 8.1.34, we actually obtain a stronger result:

Theorem 8.3.28. *The class of computably compact connected spaces is FF-complete among Σ^1_1-equivalence relations.*

Another interesting corollary is as follows.

Corollary 8.3.29. *The class of computable Banach spaces is FF-complete among Σ^1_1-equivalence relations.*

Proof. The upper bound is given by Proposition 8.1.35. The transformation $K \mapsto C(K; \mathbb{R})$ in Theorem 4.2.113 is uniform, injective on isomorphism types, and does not assume that K is totally disconnected. It remains to apply Theorem 8.3.28. ☐

We also note that, combined with Theorem 9.5.7, which will be established in the next chapter, Corollary 8.2.24 gives FF-completeness for computably compact profinite abelian groups. (The upper bound is given by Proposition 8.1.36.)

8.3.5 Exercises about degree spectra

The study of degree spectra of structures (Def. 8.3.30) is more related to the material of Part 1 of the book; e.g., see §3.2.3 and §4.1.7. However, there are tight technical connections between this topic and the machinery covered in Part 2 of the book so far. We compromise by stating some of these results as exercises here to inform the reader. For earlier exercises about degree spectra of structures, and of linear orders in particular, see Exercises 3.2.58-3.2.64.

We slightly adjust the approach taken in Richter [451] and define the *degree spectrum* of a countable algebraic structure A as follows.

Definition 8.3.30. The *degree spectrum* of a countable algebraic structure A is the set of Turing degrees that compute a presentation of the structure:

$$DSp(A) = \{\mathbf{a} : \mathbf{a} \text{ computes } B \cong A\},$$

where $\cong$ stands for algebraic isomorphism.

We shall need another standard definition.

Definition 8.3.31 (Jockusch)**.** If A is a countable structure, α is a computable ordinal, and $\mathbf{a} \geqslant \mathbf{0}^{(\alpha)}$ is a Turing degree, then A has α^{th} *jump degree* $\mathbf{a}$ if the set

$$\{\mathbf{d}^{(\alpha)} : \mathbf{d} \in DSp(A)\}$$

has $\mathbf{a}$ as its least element. In this case, the structure A is said to have α^{th} *jump degree.* If $\alpha = 0$ then we simply say that $\mathbf{a}$ is the *degree* of A. A structure A has *proper* α^{th} *jump degree* $\mathbf{a}$ if A has α^{th} jump degree $\mathbf{a}$ but not β^{th} jump degree for any $\beta < \alpha$.

Degree spectra of torsion-free abelian groups and their connected duals

The exercises below are not related to the statements of the results established in Section 8.3, but they are closely *technically* related to the material of the present section (specifically, to the proofs of Theorem D and Theorem 8.3.28). Another technically related Exercise 10.1.118 will be presented later.

Exercise° 8.3.32. Formulate the analogs of Definition 8.3.30 and Definition 8.3.31 for Polish spaces with respect to X-computable Polish presentations, and then with respect to X-computably compact presentations.

Exercise° 8.3.33 (Coles, Downey, and Slaman [97] for $n = 1$, Melnikov [365] and independently Calvert, Harizanov, and Shlapentokh [77] for $n > 1$)**.** Every torsion-free abelian group of finite rank n has first jump degree. Deduce consequences about their connected duals.

Exercise 8.3.34** (Andersen, Kach, Melnikov, and Solomon [12], Melnikov [365] for $\alpha = 2, 3$, folklore after Mal'cev for $\alpha = 1$)**.** Prove that for every computable ordinal $\alpha \geqslant 1$ and degree $\mathbf{a} > \mathbf{0}^{(\alpha)}$, there is a torsion-free abelian group having proper α^{th} jump degree $\mathbf{a}$.

Exercise 8.3.35** (Melnikov [371]). Show that for every computable ordinal β of the form $\delta + 2n + 1 > 1$, where δ is either zero or is a limit ordinal and $n \in \mathbb{N}$, there exists a torsion-free abelian group having an X-computable copy iff X is *non-low*$_\beta$. (The case when $\beta = 1$ is witnessed by a completely decomposable group; see Exercise 7.2.50.)

Exercise° 8.3.36 (Melnikov [374]). Use the two versions of Effective Pontryagin Duality established in §5.2.3 to transform the results stated in the three exercises above into results about Polish and compact Polish presentations of connected spaces (in the sense of Exercise 8.3.32).

Degree spectra in other standard classes*

Various exercises about *limitiwise monotonic spectra*, that have direct consequences about abelian p-groups and linear orders, can be found in the next chapter; see Exercises 9.1.27 – 9.1.36. See also Exercise 9.5.13 for degree spectra of profinite groups.

Exercise* 8.3.37 (Downey and Knight [134], Ash, Jockusch and Knight [19], Jockusch and Soare [273], Knight [304]). Prove the following:

1. If a linear order has a degree, it must be $\mathbf{0}$. If a linear order has first jump degree, it must be $\mathbf{0}'$ (Richter [450, 451]; this appeared earlier as Exercise 3.2.58).
2. For each computable ordinal $\alpha \geqslant 2$ and every Turing degree $\mathbf{a} \geqslant \mathbf{0}^{(\alpha)}$, there exists a linear order having proper α^{th} jump degree $\mathbf{a}$ (Def. 8.3.31).

Exercise 8.3.38** (Miller [393]). A Turing degree is is *hyperimmune* if it computes a function not dominated by any computable function. Show that there is a linear ordering which has a presentation of every nonzero Δ^0_2-degree, yet has no computable copy. Extend this argument to show that indeed the order has a copy in every hyperimmune degree.

Exercise 8.3.39 (Frolov, Harizanov, Kudinov, Kalimullin, Miller [186]). Show that for every $n > 1$ there is a linear ordering having an $\mathbf{x}$-computable copy iff $\mathbf{x}$ is non-low_n.

Exercise 8.3.40 (Melnikov [368]). Prove that for every $n > 2$, there is an ordered abelian group an $\mathbf{x}$-computable copy iff $\mathbf{x}$ is non-low_n.

Exercise* 8.3.41 (Jockusch and Soare [274]). Prove the following classical result. For $n \in \mathbb{N}$, if a Boolean algebra has n^{th} jump degree, then it is $\mathbf{0}^{(n)}$. In contrast, for each $\mathbf{a} \geqslant \mathbf{0}^{(\omega)}$, there exists a Boolean algebra with proper ω^{th} jump degree $\mathbf{a}$. (By Theorem 4.2.80, the same can be said about Stone spaces under homeomorphism.)

Exercise* 8.3.42 (Oates [419]). Show that for every computable α, there is a torsion abelian group having a proper α^{th} jump degree. (Hint: The proof exploits the rather well-understood algebra of countable abelian p-groups that will be presented in Chapter 9; see Section 9.3.).

Exercise* 8.3.43 (Kalimullin, Khoussainov, and Melnikov [281]). Show that there is a torsion abelian group computable in all hyperimmune degrees (Exercise 8.3.38) but which has no computable presentation.

Exercise 8.3.44 (Frolov, Kalimullin and Miller [187]). Show that every algebraic extension of a prime field (i.e., every algebraic field) has a jump degree, and that every upper cone of Turing degrees is a degree spectrum of an algebraic field.

Exercise* 8.3.45 (Marker and Miller [349]). We omit the definition of a deferentially closed filed and refer the reader to [349].

1. Show that every low differentially closed field has a computable presentation.
2. Let G be a countable symmetric irreflexive graph. Show that there exists a countable differentially closed field $K(G)$ of characteristic 0 such that

$$\mathrm{DSp}(K(G)) = \{\mathbf{d} : \mathbf{d}' \in DSp(G)\}.$$

In particular, there exists a differentially closed field that has an $\mathbf{x}$-computable copy iff $\mathbf{x}$ is non-*low*.

Exercise* 8.3.46 (Bazhenov, Frolov, Kalimullin, and Melnikov [44]). Show that there exists a distributive lattice with degree spectrum exactly the non-computable Turing degrees (cf. Exercise 3.2.64).

Exercise 8.3.47 (Dabkowska, Dabkowski, Harizanov, and Sikora [106]). Show that there exist centerless groups having and not having a Turing degree.

Degree spectra in general*

Beginning with the results of Slaman and Wehner (Exercise 3.2.64), there has been a line of investigation seeking to find various unexpected and complex degree spectra of structures (in general). Of course, all these results will also hold for any class that is complete with respect to degree spectra, including fields, 2-step nilpotent groups, lattices, and (discrete) metric spaces under isometry. We will not motivate these results and leave the judgement of their importance and significance to the reader. For more results of this sort, we cite the somewhat dated surveys [186, 167] and also the more recent papers [14, 160].

Exercise 8.3.48 (Knight (unpublished)). Show that thee non-trivial union of at most countably many ($\geqslant 2$) incomparable upper cones of Turing degrees cannot be realised as the degree spectrum of any structure. (For a proof that uses methods of computable topology, see [375].)

Exercise 8.3.49 (Goncharov et al. [212]). Prove that for every computable successor ordinal α there is a structure containing exactly non-low_α degrees. (The case when $\alpha = 0$ is Exercise 3.2.64.)

Exercise* 8.3.50 (Faizrahmanov and Kalimullin [160]). Show that there is a structure whose degree spectrum is the set of all non-low$_\omega$ degrees.

Exercise* 8.3.51 (Csima and Kalimullin [102]). Show that the collection of all hyperimmune degrees (Exercise 8.3.38) is a degree spectrum of a structure.

Exercise* 8.3.52 (Greenberg, Montalbán, and Slaman [221]). Prove that there exists a structure (indeed, a linear ordering) whose degree spectrum is the set of all non-hyperarithmetic degrees.

Exercise* 8.3.53 (Kalimullin [279]). Prove that the following collections of degrees can be realised as a degree spectrum of some structure.

1. For any low Turing degree d , the set $\{x \mid x \nleqslant_T d\}$.
2. For any Turing degrees c and d with $c', d' \leqslant_T \varnothing'$, the set $\{x \mid x \nleqslant_T c\} \cup \{x \mid x \nleqslant_T d\}$.

Exercise 8.3.54** (Kalimullin [280]). Show that there is a Turing degree $\mathbf{b} \leqslant \mathbf{0}''$ such that $\{\mathbf{x} : \mathbf{x} \nleqslant \mathbf{b}\}$ is *not* a degree spectrum of any structure.

8.3.6 Further related results*

There are a few more transformations that we did not mention in this chapter; see [257, 38, 294, 310]. When it comes to effective completeness, for more transformations see the cited earlier [171] and [307]. One recent notable result can be found in [239], where it is shown that the class of finitely generated groups is effectively universal among all finitely generated structures.

In the next chapter we will describe a transformation $G \leftrightarrow E_G$ between Ulm type 1 abelian p-groups and equivalence structures. It is (algebraically) quite elementary; however, it has surprisingly subtle computability-theoretic properties. For example, in Corollary 9.3.35 we will see that this transformation preserves degree spectra and computable dimension, but it does *not* preserve Δ^0_2-categoricity. In particular, the correspondence $G \leftrightarrow E_G$ is not witnessed by "computable functors", i.e., it does not transform isomorphisms to isomorphisms in a sufficiently nice uniform manner. Nonetheless, this effectively uniform correspondence will be immensely useful throughout the technically challenging Chapter 9.

There are some recent unexpected results in *primitive recursive* algebra that can be found in [123, 121]. In these papers, it was shown that for any reasonable definition of primitive recursive (PR-) universality, the class of graphs is *not* PR-universal. The same can be said about structures with two unary functions [121], which are also known to be computably universal.

When it comes to torsion-free abelian groups, several further results technically related to the Downey-Montalbán Theorem 8.3.1 can be found in [12, 371, 372]. As explained in [374], some of these results can be transformed into results about compact spaces using Effective Pontryagin Duality (Theorem 5.2.1).

A lot less is known about non-classification-type results for Polish spaces up to homeomorphism. Even less is known about compact spaces, with Theorem D and Corollary 8.3.9 being perhaps the strongest known results of this sort to date. We cite [139, 138] for some further results and references. We also cite [463, 108] for some closely related results concerning spaces that are not necessarily Polish.

8.4 What's next?

In the next chapter we will study Friedberg enumerations in more depth. We will use a $\mathbf{0}'''$-argument to produce such an enumeration for equivalence structures. We will then develop an effective algebraic apparatus sufficient to turn this computability-theoretic result into a Friedberg enumeration of abelian p-groups of finite Ulm type. Finally, another effective version of Pontryagin duality, this time between torsion discrete and profinite groups, will allow us to transfer this result to the topological setting.

Chapter 9

Enumerating structures without repetition

Recall that a *Friedberg enumeration of a class* is a uniformly computable list $(P_i)_{i\in\mathbb{N}}$ in which every member of the class is mentioned exactly once, up to isomorphism. The main goal of this chapter is to prove the following theorem.

Theorem E (Downey, Melnikov, and Ng [145]). For any fixed $n > 0$, there is a Friedberg enumeration of all computably compact pro-p abelian groups of pro-Ulm type $\leqslant n$.

For our proof, the usual infinite injury technique will no longer suffice. In this chapter we will see the first use of the $\mathbf{0}'''$-machinery which seems unavoidable in any proof of Theorem E. The chapter is organised as follows:

1. Section 9.1 presents the basics of computable equivalence structures and l.m. sets.
2. Section 9.2 contains a $\mathbf{0}'''$-construction that produces a Friedberg enumeration of all computable equivalence structures.
3. Section 9.3 presents a systematic exposition of computable abelian p-group theory sufficient to prove Theorem E.
4. Section 9.4 combines the main result of Section 9.2 with techniques of Section 9.3 to produce a Friedberg enumeration of all computable abelian p-groups of Ulm type $\leqslant n$ (for any fixed $n > 0$).
5. Section 9.5 contains another effective version of Pontryagin duality, this time between torsion discrete and profinite abelian groups. It is used to derive Theorem E from the result of Section 9.4.

This chapter is sufficiently independent from the rest of the material and can be read separately.

R. G. Downey, A. Melnikov, *Computable Structure Theory*, Theory and Applications of Computability, https://doi.org/10.1007/978-3-031-92433-0_9

9.1 Equivalence structures and limitwise monotonic sets

The goal of this section is to accumulate sufficient computability-theoretic techniques to derive our group-theoretic results. The seemingly elementary class of computable equivalence relations will be quite unexpectedly useful in such applications. It seems that almost every result about computable equivalence structures falls into one of two categories:

- an elementary exercise;
- a significant combinatorial challenge.

All results in this section are in the first category, but the main result of Section 9.2 falls into the second category.

9.1.1 Computable equivalence relations

Equivalence structures

For an equivalence relation $\sim$ on a set A, we will let $[x]_\sim = \{y : x \sim y\}$ denote the $\sim$-equivalence class of x. We will drop the subscript $_\sim$ where the meaning is clear. We will often identify an equivalence relation $\sim$ on $A \subseteq \omega$ with the respective *equivalence structure* $(A, \sim)$. An *equivalence structure* is an algebraic structure of the form $(A, \sim)$, where $\sim$ is an equivalence relation on A, and there are no further operations or relations on A.

In this chapter, all our equivalence structures will be at most countable.

Clearly, countable equivalence structures are fully described by the number of equivalence classes of each fixed size that occur in E, including the number of infinite classes.

Definition 9.1.1. Let $E = (A, \sim)$ be an equivalence structure.

1. The *character* of E is the set

$$\chi_E = \{\langle n, k\rangle : E \text{ has at least } k \text{ equivalence classes of size exactly } n\},$$

where $k, n \in \mathbb{N}$.

2. The *characteristic set* of E is

$$\#E = \{n : \langle n, 1\rangle \in \chi_E\}.$$

We say that E is *unbounded* if $\#E$ is infinite (i.e., E contains arbitrarily large finite classes); otherwise, E is *bounded.*

3. We write $R(E)$ to denote the equivalence structure obtained from E after removing all of its infinite classes. An equivalence structure is *reduced* if $R(E) = E$ (i.e., E has no infinite classes).

4. We let $D(E) = E - R(E)$ to be the restriction to E to its substructure that contains only infinite classes, and we write $rk\, D(E)$ to denote the number of (infinite) classes in $D(E)$ (and, thus, in E).

Observe that for every E, $\chi_{R(E)} = \chi_E$ and $\#R(E) = \#E$. Combined with the number of infinite classes $rk\, D(E)$, the χ_E describes E up to isomorphism.

Proposition 9.1.2. *Let E and K be two equivalence structures. Then $E \cong K$ iff $\chi_E = \chi_K$ and $rk\, D(E) = rk\, D(K)$.*

Our choice of notation and terminology may initially seem somewhat unnatural, but the reader will understand the rationale behind it when we discuss abelian p-groups in the next chapter.

Computable equivalence structures

An equivalence structure $(A, \sim)$ is computable if both the domain A and the relation $\sim$ are computable. The arithmetical hierarchy cannot capture the computable presentability of reduced unbounded equivalence structures.

Definition 9.1.3 (Khisamiev [290]). A (total, Δ^0_2) function $g : \omega \to \omega$ is *limitwise monotonic* (l.m.) if there is a computable approximation function $f(\cdot, \cdot)$ such that, for all x,

(i) $g(x) = \lim_s f(x, s)$, and

(ii) for all s, $f(x, s) \leqslant f(x, s + 1)$.

A set S is *l.m.* if it is the range of a l.m. function.

According to their definition, limitiwise monotonic sets are Σ^0_2. Clearly, every finite set is limitwise monotonic, and so is every c.e. and co-c.e. set. However, in Lemma 9.1.20 we will show that not every Σ^0_2-set is limitwise monotonic. The theorem below is folklore; it can be found in [75].

Theorem 9.1.4. *Let E be an equivalence structure.*

1. *If E has infinitely many infinite classes, then E is computably presented iff χ_E is Σ^0_2.*
2. *If E has finitely many infinite classes ($rk\, D(E) < \infty$), then E has a computable presentation iff χ_E is Σ^0_2 and $\#E$ is limitwise monotonic.*

Proof. We use the notation and terminology from Definition 9.1.1 throughout. We begin by disposing of the easy cases:

Fact 9.1.5. *If E is a computable equivalence structure, then χ_E is a Σ^0_2 set.*

Fact 9.1.6. *Let E be an equivalence structure that either has the sizes of all its finite classes bounded by some fixed $k \in \mathbb{N}$ or has infinitely many infinite classes. Then E is computably presentable iff χ_E is a Σ^0_2 set.*

Fact 9.1.7. *If E is an equivalence structure that has only finitely many infinite classes, then E is computably presentable iff $R(E)$ is computably presentable.*

The verification of these facts is left to Exercise 9.1.15. The only remaining case is when the structure is reduced and unbounded, i.e., $\#E$ is infinite and has no infinite classes (that is, $D(E) = \varnothing$).

Proposition 9.1.8. *Let E be a reduced and unbounded equivalence structure. Then E has a computable presentation if, and only if, χ_E is Σ_2^0 and $\#E = \{n : \langle n, 1\rangle \in \chi_E\}$ is l.m..*

Proof. First, suppose E is computable. By Fact 9.1.5, we only need to argue that $\#E$ is l.m.. We may assume the domain of E is ω, and let E_s be E restricted to $\{0, \ldots, s-1\}$. Define $f(i, s)$ to be the size of $[i]$ in E_s.

Conversely, suppose χ_E is Σ_2^0 and $\#E$ is l.m., and let $g = \lim_s f(\cdot, s)$ be such that $\#E = \mathrm{rng}(g)$. The set $\#E$ is infinite by assumption. Without loss of generality, we may assume g is injective; the proof of this fact is delayed until §9.1.3 (see Lemma 9.1.23).

Consider E' obtained from E by removing exactly one class of size k for every $k \in \#E$. We have that $\langle n, k\rangle \in \chi_{E'}$ iff $\langle n, k+1\rangle \in \chi_E$, and thus $\chi_{E'}$ is Σ_2^0. Clearly, E splits into the disjoint union

$$E = E' \sqcup E'',$$

where E'' has exactly one class of size k for each $k \in \#E$. If $[x]$ needs to be modified due to a change in the Σ_2^0-approximation of $\chi_{E'}$, search for i and s such that $f(i, s)$ is larger than any number mentioned so far. At later stages t, keep the size of $[x]$ equal to $f(i, t)$. Also, simultaneously build the E''-part of E, choosing $f(j, s)$ for the sizes of classes among those $j \leqslant s$ which have not yet been used for initialisation. □

This completes the proof of Theorem 9.1.4. □

Limitwise monotonicity captures the difference between c.e. and computably presented equivalence structures. In a c.e. presented structure, two points can become *equal* (not just equivalent) at a later stage.

Lemma 9.1.9. *E is c.e. presented iff χ_E is Σ_2^0.*

We leave the proof of the lemma to Exercise 9.1.17. Since there are Σ_2^0-sets that are not limitwise monotonic (delayed until Lemma 9.1.20), we arrive at the following corollary that resembles many similar results from Part 1 of the book.

Corollary 9.1.10 (Folklore). *There exists a c.e. presented equivalence structure that has no computable presentation.*

Computably categorical equivalence structures

Recall that an equivalence structure is bounded if all of its finite classes are uniformly bounded in size from above. A bounded equivalence structure can have infinite classes.

Theorem 9.1.11 (Calvert, Cenzer, Harizanov, and Morozov [75]). *An equivalence structure A is computably categorical iff almost all (i.e., all but finitely many) classes of A have the same size. (Note this includes the case when there are only finitely many classes in total.)*

Extended sketch. The easy implication is left to Exercise 9.1.18.

For the more difficult direction, assume that A is computably categorical but fails the algebraic property from the theorem. The proof is split into three cases which cover all possible isomorphism types of the structure.

Case 1. Assume that A has infinitely many classes of size k_0 and infinitely many classes of size k_1, where $k_0 < k_1$, allowing the possibility $k_1 = \omega$. Consider the structure E which is isomorphic to the substructure of A which contains all classes of A except for the classes of size k_0 and k_1. Since a limitwise monotonic (l.m.) set remains l.m. after removing finitely many elements, Theorem 9.1.4 guarantees that E has a computable presentation (which could be empty). Let also E_0 and E_1 be equivalence structures having infinitely many classes of the same size, with E_0 having all classes of size k_0 and E_1 having all its classes of size k_1. We will use E to build two computable presentations, B and C, of A that satisfy:

$$\mathcal{R}_e : \neg(B \cong_{\varphi_e} C),$$

for every e.

Let C be the computable presentation of A which is computably split into three disjoint computable substructures:

$$C = E \sqcup E_0 \sqcup E_1.$$

We define $B = E \sqcup U \sqcup E_1$, where $U \cong E_0 \sqcup E_1$ will be built in stages. (Note this guarantees $B \cong C \cong A$.)

The strategy is as follows. Initially, define every class in U to have size k_0. Use the $2e$-th class $[u_e]$ of U in B to diagonalise against φ_e, as follows:

1. Wait for $\varphi_e(u_e)$ to converge.
2. If $\varphi_e(u_e) \in E_0 \subset C$, then grow the class $[u_e]$ to have size $k_1 > k_0$.

It is clear that all requirements are met.

Case 2. Now assume that A has infinitely many infinite classes and infinitely many finite classes. The idea is similar to what we had in Case 1; however, we cannot always guarantee that the restriction of A to its finite classes, $R(A)$, has a computable presentation (see Theorem 9.1.4 and Lemma 9.1.20).

The idea is as follows. Each $\mathcal{R}_e$ will have a witness, u_e. As in Case 1, it is sufficient to make sure that the size of $\varphi_e(u_e)$ is different from the size of $[u_e]$ (if $\varphi_e(u_e)\downarrow$). The easiest way to achieve this is to declare $[\varphi_e(u_e)]$ to be infinite. However, we do not know whether the size of $[u_e]$ is finite, and therefore we may have to repeat the strategy for another u'_e, and then another one, and so on.

This time we build C, and we let $B = A$. In C, we will attempt to copy A, but sometimes some class $[x]$ in C will be declared infinite, and a new class will be introduced to C; this class will now copy the class of A that $[x]$ used to copy before it was declared infinite. (More formally, we define a partial embedding $\psi : A \to C$ in stages, so that all classes in $C \backslash \psi(A)$ are infinite. We initially define $\psi([z]) = [x]$, and later we may have to redefine $\psi([z]) = [x']$ for some x' with a fresh index that we put in C, together with all elements in its class $[x']$, specifically for this purpose. We will then argue that ψ is Δ^0_2. If we succeed, this will immediately imply that $C \cong B$.)

The strategy for $\mathcal{R}_e$ seeks the smallest index equivalence class in B that has appeared unchanged for sufficiently many stages.

Remark 9.1.12. Formally, let $g(e)$ be the Δ^0_2-function that outputs the smallest index finite class that has not yet been used by strategies of smaller indices. Using the Limit Lemma 3.1.3, fix a computable $\tilde{g}$ so that $g(e) = \lim_s \tilde{g}(e, s)$. To avoid interference between different strategies, we can assume $\tilde{g}(i, s) \leqslant \tilde{g}(e, t)$ whenever $i < e$. By the Recursion Theorem, we can assume we know the index of $\tilde{g}$. We omit all these formal details and focus on the key ideas instead. The idea is that searching for the next available finite class is clearly $\varnothing'$-effective, and after finitely many unsuccessful attempts, it will be located.

Let $[u_e]$ be this class. The strategy proceeds as follows:

1. Wait for $\varphi_e(u_e)$ to converge.
2. If the size of $[\varphi_e(u_e)]$ is currently equal to the size of $[u_e]$, then grow the class $[\varphi_e(u_e)]$ to be infinite.
3. Reintroduce the class that has been grown to be infinite, as explained earlier, by giving its new version a fresh large index in C.

If $[u_e]$ ever increases in size, then search for another available smallest index equivalence class in B, say $[u'_e]$, and repeat the strategy above with u'_e. If both $[u_e]$ and $[u'_e]$ ever increase in size, then define u''_e, and so on. Eventually, a class of finite size will be found in B, and the diagonalisation will be achieved.

In the presence of several $\mathcal{R}_e$, there is a potential conflict that needs to be resolved. Indeed, some class in C can be grown to be infinite and then reintroduced infinitely many times, thus potentially some finite class of A will never be copied into C. In other words, ψ may never achieve a stable value for some inputs.

To circumvent this difficulty, we simply do not allow $\mathcal{R}_j$ to grow classes in C that have already been shifted by some $\mathcal{R}_e$ for $e < j$. If φ_j converges to one such element "restrained" (in this sense) by a higher priority strategy, then we make $\mathcal{R}_j$ search for the next available witness. Since eventually each $\mathcal{R}_e$ is either forever inactive or is met, there will be only finitely many iterations until $\mathcal{R}_j$ finds a witness that can be used for diagonalisation. We leave the details to the reader.

The remaining case is as follows.

Case 3. The structure is unbounded but has at most finitely many infinite classes. By Theorem 9.1.4, in this case the set of finite sizes of classes that occur in the structure is a l.m. set. Let $g(x) = \lim_s l(x, s)$ be the l.m. function that witnesses this fact. In Lemma 9.1.23 we will show that g can be assumed to be injective.

We can non-uniformly fix representatives from the infinite classes (if there are any). Thus, in contrast with the previous case, just one witness is enough to diagonalise because we may assume our witnesses always come from finite classes. We keep the same notation as in the previous case. Let u_e be the witness assigned to $\mathcal{R}_e$.

1. Wait for $\varphi_e(u_e)$ to converge.

2. If $\varphi_e(u_e)$ belongs to one of the finitely many infinite classes in C, do nothing.

3. Otherwise, if the current size m of $[\varphi_e(u_e)]$ is equal to the current size of $[u_e]$, then search for x and s so that $l(x,s)$ is larger than any number used so far in the construction (including m), and grow the class $[\varphi_e(u_e)]$ to have size $l(x,s)$. (Unless a new x will be picked for e, eventually grow $[\varphi_e(u_e)]$ to have size $g(x)$.)

4. Introduce a fresh class in C (to replace $[\varphi_e(u_e)]$) having size m.

5. If at some later stage the size of $[u_e]$ becomes equal to that of $[\varphi_e(u_e)]$, repeat steps 3 and 4 using new large parameters.

To make sure that the right finite sizes are realised among the classes in C, follow the same strategy as in the proof of Proposition 9.1.8. Split C into two substructures, E' and E'', so that in E'', for every x there is exactly one class of size $g(x)$. We follow the proof of Proposition 9.1.8 with essentially only one modification. Specifically, at substage 3 of the strategy $\mathcal{R}_e$, we place $[\varphi_e(u_e)]$ in E'' and reintroduce a new class to replace the old version of $[\varphi_e(u_e)]$. We use the injectivity of g to carefully choose sizes of classes to be placed into E'' to avoid repetition. □

9.1.2 Beyond computable categoricity*

In spite of attacks by several authors, the following problem raised in [75] remains stubbornly open:

Problem 9.1.13. *Classify Δ^0_2-categorical (computable) equivalence structures.*

We have already seen that Δ^0_2-categorical homogeneous c.d. groups can be described in terms of semi-low sets. Also, the class of computable equivalence structures appears to be so incredibly tame that one might expect that Δ^0_2-categoricity in this class is much easier to capture. However, this intuition is misleading.

The only non-trivial case is when the structure is unbounded and has infinitely many infinite classes. It seems that $\mathbf{0}'$ is unable to detect how many classes of a fixed size are present in the structure. Thus, one might expect that repetitions of classes could be used to diagonalise against Δ^0_2-isomorphisms.

Using an unexpectedly complex argument, it was shown in [141] that Δ^0_2-categoricity of an equivalence structure is stable under removing repetitions from the sizes of finite classes. (In other words, if we keep only one class of each fixed finite size that is present in E, we obtain that the resulting structure is Δ^0_2-categorical iff the original structure was.) This means that the problem raised in [75] can be reduced to a problem about Σ^0_2-sets, as follows. For a Σ^0_2 set X, let $E(X)$ be the equivalence structure having infinitely many infinite classes, exactly one class of size n for each $n \in X$, and no other finite classes. Fact 9.1.6 guarantees that computable presentability of $E(X)$ is equivalent to X being Σ^0_2.

Question 9.1.14. *For which Σ^0_2 sets X is $E(X)$ Δ^0_2-categorical?*

It seems that no known property of a Σ^0_2 set captures Δ^0_2-categoricity of the respective equivalence structure; for more details, see [141].

Exercises

Exercise° 9.1.15. Prove Facts 9.1.5, 9.1.6 and 9.1.7.

Exercise° 9.1.16 (Folklore). Show that a computable equivalence structure has either 1 or ω-many computable presentations, up to computable isomorphism.

Exercise° 9.1.17. Prove Lemma 9.1.9.

Exercise° 9.1.18 (Calvert, Cenzer, Harizanov, and Morozov [75]). Prove that if an equivalence structure has almost all classes equal in size then it is computably categorical.

Exercise° 9.1.19. Give a complete formal proof of Theorem 9.1.11.

9.1.3 Calculus of limitwise monotonic sets and functions

The standard references for the material contained in this section are [133, 281].

Every c.e. set is clearly l.m., and so is every Π^0_1 set. However, not every Δ^0_2-set of the form $B \setminus C$, where B and C are c.e., is l.m., as the proof of the lemma below shows. (Differences of c.e. sets are called d-c.e. sets.)

Lemma 9.1.20 (Khisamiev [287]). *There is a d-c.e. set A which is not a l.m. set.*

Remark 9.1.21. Lemma 9.1.20 is usually attributed to Khisamiev [287], who applied it to characterise computably presented abelian p-groups of Ulm type 1 (to be discussed). It was independently (and slightly later) proven by Ash, Knight, and Oates (unpublished). Some 15 years later, it was rediscovered by Khoussainov, Nies, and Shore [295] in the context of computable model theory, and then shortly after also used to study computable linear orders [96] (see Exercise 9.1.32). It was recently re-rediscovered by Bosserhoff and Hertling [55] in the context of computable topology. See Exercise 9.1.36 for another recent application of l.m. sets in computable topology. The lemma is elementary; however, its history clearly indicates its importance.

Proof. We construct a set A satisfying the requirements:

$$\mathcal{R}_e : \text{The function } \varphi_e(\cdot,\cdot) \text{ does not witness that } A \text{ is a l.m. set.}$$

Initially, subdivide ω into subsequent disjoint intervals J_e such that each J_e has 2^{e+1}-many elements.

Towards meeting $\mathcal{R}_0$, we pick a witness $n_0 \in J_0$, put n_0 in A and keep $m < n_0$ out of A. Wait for a stage s so that for some x,

$$\varphi_0(x, s) = n_0.$$

If such an x is ever found, extract n_0 from A, and put $n_0 - 1$ in A. Declare this $x = x_0$ active.

The basic strategy for $\mathcal{R}_1$ is the same, mutatis mutandis, but the choice of the witness n_1 depends on the value of $\varphi_0(x_0, s)$, as follows.

The strategy for $\mathcal{R}_1$ when it will become active will pick its witness n_1 to be equal to the largest element of J_1. However, if at some stage t, $\varphi_0(x_0, t) = n_1$, then the strategy will be initialised by picking a fresh witness equal to the largest element in the lower half of J_1.

More generally, the strategy for $\mathcal{R}_e$ (when it becomes active at stage s) will pick a witness m_e so that $m_e \neq \varphi_i(x_i, s)$ for all $i < e$ (if these parameters are defined). For that, we divide J_e into two halves to avoid $\varphi_i(x_0, t)$, and then further subdivide each half further to avoid $\varphi_1(x_1, t)$, and so on. In the construction, we let the strategies act according to their instructions.

By induction, eventually (exactly) one element will be put in each interval J_e. Suppose $\varphi_e(\cdot,\cdot)$ indeed represents a l.m. set. If x_e is never defined, it means that no approximation ever enters J_e, and $\mathcal{R}_e$ is met. If x_e is defined, then the action of the strategy guarantee that no element of J_k for $k \geqslant e$ can possibly be equal to $\lim_t \varphi_e(x_e, t)$. □

Some of the statements that we presented in the previous subsection can be simplified using the combinatorial lemma below; we omit its proof.

Lemma 9.1.22 (Folklore; see, e.g., [133]). *An infinite Σ^0_2-set is l.m. iff it contains an infinite l.m. subset.*

The lemma below can often be convenient; we have already used it in the preceding subsection.

Lemma 9.1.23 (Harris [234]). *Suppose S is infinite and l.m.. Then there is an injective l.m. g with range S.*

Proof. Let $h(y) = \lim_x f(y, x)$ be a l.m. with range S. We have that the computable function $f(y, x)$ is non-decreasing in x for every $y \in \mathbb{N}$, so $\lim_x f(y, x) = \max_x f(y, x)$. Set $g(y, x) = 0$ if $x \leqslant y$. To define $g(y, x)$ for $x > y$, suppose that for every $y' < y$ and $x' < x$ the values of $g(y', x)$ and $g(y, x')$ have already been defined. Choose $\langle y_0, x_0\rangle$ least such that $x_0 > x$, $f(y_0, x_0) \geqslant \max_{x'<x} g(y, x')$ and $f(y_0, x_0) \notin \{g(y', x) \mid y' < y\}$. Set $g(y, x) = f(y_0, x_0)$.

It is not hard to check that $\max_x g(y, x)$ is total and injective. It is also evident that $\operatorname{rng}_y \max_x g(y, x) \subseteq S$, for every $n \in \mathbb{N}$. To see that

$$S \subseteq \operatorname{rng}_y \max_x g(y, x),$$

we use an inductive argument. The definition of g may be viewed as a construction, where at each stage we compute the value of g for exactly one new pair of arguments, and the value of f for a new pair of arguments as well. We denote our current guess about $\max_x g(y, x)$ at stage s by $[\max_x g(y, x)]_s$, and similarly for f. Suppose there are y and n such that $\max_x f(y, x) \notin \operatorname{rng}_y \max_x g(y, x)$, and y is least with this property.

Then there should be a stage s such that for every stage $t \geqslant s$ and every $y' < y$, $[\max_x f(y', x)]_s = [\max_x f(y', x)]_t \in \operatorname{rng}_y \max_x g(y, x)$. We may further assume that s satisfies $[\max_x f(y, x)]_s = \max_x f(y, x)$. By the definition of g, there should be a stage $t_0 \geqslant s$ and an argument y_0 such that $[\max_{x'<x} g(y_0, x')]_{t_0} = \max_x f(y, x)$, since we always start with $g(y, x) = 0$ for $x \leqslant y$. If there is no stage $t_1 \geqslant t_0$ and $y_1 < y_0$ such that $[\max_x g(y_1, x)]_{t_1} = [\max_x g(y_0, x)]_{t_0}$ then $\max_x g(y_0, x) = \max_x f(y, x)$. Therefore there should exist y_1 and a stage t_1 such that $[\max_x g(y_1, x)]_{t_1} = [\max_x g(y_0, x)]_{t_0}$ We can use the same argument to find y_2 and t_2 which play the same role for y_1 and t_1 as the latter arguments do for y and t. Since $y = y_0 > y_1 > \ldots$ we will find the least y_i in this sequence. But then $\max_x g(y_i, x) = \max_x f(y, x)$, contrary to the hypothesis. □

We leave the following corollary of the lemma above as an exercise.

Corollary 9.1.24. *Suppose that S is infinite and l.m.. Then there exists a strictly increasing l.m. function g so that $\operatorname{rng} g \subseteq S$. Indeed, we may assume that $x < g(x)$, for all x.*

Exercises

Exercise° 9.1.25. Prove Lemma 9.1.22.

Exercise° 9.1.26. Prove Corollary 9.1.24.

Exercise° 9.1.27 (Hirschfeldt, Miller and Podzorov [260]). There is a low Δ_2^0 set A which is not l.m..

Exercise° 9.1.28 (Downey, Kach, Turetsky [133], Wallbaum [500]). Show that for every nonzero Turing degree **a**, there is a set S which is limitwise monotonic in **a** but is not limitwise monotonic.

Exercise* 9.1.29 (Kalimullin, Khoussainov, and Melnikov [281]). Prove that there is a Σ_2^0 set S such that S is limitwise monotonic in every nonzero Δ_2^0-degree, but is not limitwise monotonic.

Exercise* 9.1.30 (Harris [234]). An η-representation of a set is a computable linear ordering of order-type $\eta + a_0 + \eta + a_1 + \eta + a_2 + \eta + \cdots$ where η is the order-type of the rationals, and $A = \{a_i : i \in \mathbb{N}\}$ (note we allow repetitions, and the sizes of the blocks a_i do not have to be non-decreasing in i). Show that A has an η-representation iff A is $\varnothing'$-l.m., i.e., is limitwise monotonic relative to $\varnothing'$.

Exercise 9.1.31 (Kach [275]). Recall that in the proof of Lemma 3.2.12 we defined the shuffle sum $\biguplus_i L_i$ of orders L_i to be the order obtained by placing L_i together densely. For a set $S \subseteq \omega + 1$, define $\sigma(S)$ to be $\biguplus_{\kappa \in S} \kappa$, where each κ is identified with its order-type. Show that for a set $S \subseteq \omega + 1$, the following are equivalent:

(1) The shuffle sum $\sigma(S)$ is computable.

(2) The set S is a limit infimum set, i.e., there is a total computable function $f(x, s)$ such that the function $F(x) = \liminf_s f(x, s)$ enumerates S.

(3) The set S is a limitwise monotonic set relative to $0'$ (allowing the limit of the monotonic approximation to be infinite).

Exercise* 9.1.32 (Coles, Downey, and Khoussainov [96]). Let η be the order-type of the rationals. A linear order is η-like if it is obtained from η by replacing every point by a finite block. The block set of a linear order L is the collection of the sizes of the finite blocks that occur in L.

(i) Show that the block set of a computably η-like linear order is $\varnothing'$-l.m..

(ii) Show that for any set S, $S \in \Sigma_3^0$ iff there is a computable linear order $L = A + B$ with A an η-like ordering such that the blocks of A have exactly the sizes of elements of S, and B has order type ω^*.

(iii) Deduce that there is a computable linear ordering with a Π_2^0 initial segment not isomorphic to a computable linear ordering.

Exercise* 9.1.33 (Hirschfeldt [255]). Using the concept of a limitiwise monotonic function, show that there is a complete theory of linear orderings with a computable model whose prime model does not have a computable presentation.

Exercise* 9.1.34 (Csima, Hirschfeldt, Knight and Soare [101]). A countable complete theory T is atomic if, for each formula $\phi(\bar{x})$ consistent with T, there is a principal type containing $\phi(\bar{x})$. Prove that for a set $X \leqslant_T \varnothing'$, the following are equivalent:

1. X is not low_2, i.e., $X'' >_T \varnothing''$;
2. every (infinite) Δ^0_2 set is limitiwise monotonic relative to X;
3. every complete atomic decidable theory has an X-decidable prime model.

Exercise* 9.1.35 (Khoussainov, Nies, and Shore [295]). A model $\mathcal{M}$ of a theory T is minimal if there is no formula $\phi(x)$ such that the sets $\{m \mid \mathcal{M} \models \phi(m)\}$ and $\{m \mid \mathcal{M} \models \neg\phi(m)\}$ are infinite. A theory T is *strongly minimal* if all models of T are minimal. A theory is *algebraically trivial* if the algebraic closure of every set X in every model of the theory equals to the union of the algebraic closures of elements of X.

Prove that for every given set S there exists an $\aleph_1$-categorical but not $\aleph_0$-categorical theory T_S with the following properties:

1. The theory T_S is strongly minimal and algebraically trivial,
2. Every (countable) non-prime model of T_S has a computable copy iff $S \in \Sigma^0_2$,
3. The prime model of T_S has a computable copy iff S is limitwise monotonic.

Exercise* 9.1.36 (Koh, Melnikov, and Ng [311]). A *k-star* is a topological space homeomorphic to k copies of the interval $[0,1]$ all joined at one end in a single point (the wedge sum of k copies of $[0,1]$ via 0). A 0-star is an isolated point. A *star-space* is the 1-point compactification of the disjoint union of stars. Prove the following:

1. Let M be a computably compact presentation of a star-space. Then the set $St(M) = \{n > 1 \mid M \text{ has an } n\text{-star component}\}$ is $0'$-limitwise monotonic.
2. For any Σ^0_3 set S, there is a star-space M c.e. closed in $[0,1]^2$ with $St(M) = S$.
3. Conclude that there is a c.e. subspace of $K = [0,1]^2$ not homeomorphic to any computably compact space. Consider $C(K;\mathbb{R})$ and note that it has a computable Banach presentation, thus witnessing that computable Banach–Stone duality (cf. Theorem 4.2.113) *fails* in general.

9.2 Enumerating equivalence structures

Recall that the Friedberg Enumeration Theorem 3.1.43 states that there exists a uniform enumeration of all c.e. sets in which every c.e. set is mentioned exactly once, up to equality. A Friedberg enumeration of a class $\mathcal{K}$ as a uniformly computable list $(F_i)_{i\in\mathbb{N}}$ that, up to isomorphism, contains exactly one instance of any computably presented structure in $\mathcal{K}$. The main goal of this section is to prove the following theorem established in [143]:

Theorem 9.2.1. *There is a Friedberg enumeration of all computable equivalence structures.*

The theorem solved a problem that was stated in [213]. It remained open for 15 years until it was partially solved in [330] and then completely settled in [143]. The reader should prepare themselves for a combinatorially involved proof. It may appear surprising that the most combinatorially challenging proof that we've seen so far in the book is about equivalence structures. However, recall that the isomorphism problem for equivalence structures in Π^0_4-complete (Proposition 7.1.5). Thus, eliminating repetitions will require a careful analysis of Π^0_4- and Σ^0_4-outcomes. The proof that we present here is a further polished and slightly reworked version of the proof from [143].

9.2.1 Preliminary analysis

Fix an effective listing $(E_n)_{n\in\mathbb{N}}$ of all computable equivalence structures, allowing repetitions. Let

$$E_n = \bigoplus_k \left[\!\left[\text{card } W_n^{[k]}\right]\!\right],$$

where W_n is the n-th c.e. set, and $[[\alpha]]$ is an equivalence class of size α. Our goal is to produce, using this sequence, another sequence $(U_n)_{n\in\mathbb{N}}$ in which isomorphism types are not repeated.

Notation 9.2.2. If E is a computable equivalence structure, then $\mathbf{e}_j$ will denote its j-th equivalence class. Equivalently, it is the j-th column under the enumeration above.

Since we are interested in isomorphism types, only the sizes of the $\mathbf{e}_j$ will matter.

Some elementary simplifications. Recall that in the proof of the Friedberg Enumeration Theorem 3.1.43, we adjoined ω to the list after the construction was finished. Also, the finite sets played the role of a "junk collector" in the proof of Theorem 3.1.43. We will use a similar technique for equivalence structures too, but there will be many more exceptional cases. Additionally, in the present proof, there will be two junk collectors: one composed of finite structures and the other of infinite structures of a special kind.

We assume that in our list all finite structures are non-empty. We will divide equivalence structures into five types:

(I) Finite equivalence structures.

(II) Finitely many finite classes and finitely many (with at least one) infinite classes.

(III) Finitely many finite classes and infinitely many infinite classes.

(IV) Infinitely many finite classes which are uniformly bounded in size, and any number of infinite classes.

(V) Infinitely many finite classes with arbitrarily large finite sizes, and any number of infinite classes. From now on, we will refer to a structure of type (V) as *unbounded.*

Our construction will produce a Friedberg enumeration of the isomorphism types in (I), (III), and (V). We will obtain the desired Friedberg enumeration by adjoining the missing isomorphism types (II) and (IV) to the list.

9.2.2 The setup

Recall $(E_n)_{n\in\mathbb{N}}$ is the effective listing of all computable equivalence structures induced by the standard enumeration of c.e. sets.

Requirements and preliminary remarks. In the construction, each computable equivalence structure E_i will have infinitely many strategies τ associated with it. The task of each τ is to produce a computable isomorphic copy $U_\tau \cong E_i$ unless E_i is bounded (not of type (V)) or E_i is isomorphic to some E_j with $j < i$. Thus, each such τ (associated with E_i and building U_τ) works towards meeting the requirement:

$$\bigwedge_{j<i} E_i \ncong E_j \text{ and } E_i \text{ unbounded } \implies U_\tau \cong E_i.$$

There will be *infinitely many strategies* τ associated with E_i. Recall that for every i it is Π^0_4 to tell if $E_i \cong E_j$ for some $j < i$ (Proposition 7.1.5). Let this predicate be $\forall k P(i,k)$, where $P(i,k)$ is Σ^0_3. For each k we will have a separate strategy τ guessing whether $P(i,k)$ holds. It is crucial that τ has to produce an isomorphic copy of E_i *only if* $P(i,k)$ fails.

We will see that the different τ's will collectively be able to take care of the *global requirement*, saying that the enumeration must be Friedberg, in particular, that $\forall \tau,\sigma\ [\tau \neq \sigma \mapsto U_\tau \ncong U_\sigma]$. We will also see that, depending on its true outcome, τ may produce a copy of the associated E_i, as well as several (perhaps, infinitely many) distinct finite structures, or a single structure of type (III); in the latter two cases, the structures will be placed into the *junk collector*, which will be operating outside the tree of strategies.

The junk collector will be further subdivided into a *finite junk collector* and an *infinite junk collector*, with infinite junk potentially transformable into finite junk.

Priority Tree. As we noted above, for every i, it is Π^0_4 to tell if $E_i \cong E_j$ for some $j < i$. Recall $P(i,k)$ is the Σ^0_3-predicate such that $\forall k P(i,k)$ holds iff $E_i \cong E_j$ for some $j < i$. Each node on the tree will be assigned the task of guessing $P(i,k)$ for some i and k, where i stands for the computable equivalence structure E_i. This is assigned in the usual way; a node τ is assigned $P(i,k)$, where $|\tau| = \langle i,k\rangle$. For convenience, we write E_τ for the equivalence structure associated with a strategy τ, and we write k_τ for the respective integer k.

The priority tree is a $1+\omega+2$-branching tree. The outcomes, ordered from left to right, are:

$$\mathtt{init} < \mathtt{pi}_2 0 < \mathtt{pi}_2 1 < \cdots < \mathtt{pi}_3 < \mathtt{wait}.$$

The leftmost outcome init is a Π^0_2 outcome, the $\mathtt{pi}_2 m$ are the Π^0_2 outcomes that collectively form the Σ^0_3 counterpart of $\mathtt{pi}_3$ (which is the Π^0_3 outcome). The waiting outcome wait is Σ^0_2. We will describe the role of each outcome later.

9.2.3 A single node in isolation

Fix τ on the construction tree. We now describe the actions and outcomes of τ, when considered in isolation. Assume τ is working for E_i and there are exactly $k_\tau - 1$ nodes $\sigma \subset \tau$ such that σ works for E_i and $\sigma * \mathtt{pi}_2 j \subseteq \tau$ for some j. If $\rho * o \subseteq \tau$ for some ρ that works for the same E_i and some other outcome o, we will simply assume τ is always inactive and does nothing when visited. The strategy τ will build a structure U_τ (we suppress τ and write U when the context is clear), which will be permanently abandoned when τ is initialised.

Initialisation. The strategy τ will be initialised if outcome $\mathtt{init}$ of τ is played. In isolation, this case corresponds to the scenario when E_τ has no finite classes (note that this can be detected in a Π^0_2 way), or if we move to the left of τ, unless we are moving from a $\sigma * \mathtt{pi}_3$ to a $\sigma * \mathtt{pi}_2 k$ outcome for some $\sigma \subset \tau$. In this case, the strategy abandons its current U_τ. If U_τ is abandoned, then it joins the *finite junk collector.*

We then restart with a new U_τ. After each initialisation, all parameters of the strategy are set to undefined. When τ is active again (if ever), it will set its U_τ equal to a structure with a single "large" finite class of size larger than any number seen so far in the construction.

The parameter fin. The node τ will also monitor the parameter $fin(\tau, j)$. When the context is clear, we write $fin(j)$ instead. This parameter will be used to copy E_τ into U_τ and guess whether E_τ is of the unbounded isomorphism type. Given the node τ and a stage s, define the sequence $f_1 < f_2 < \cdots < f_t$ of length at most s inductively by the following. Suppose f_{k-1} has been defined (for $k = 1$, take $f_{k-1} = 0$). Take f_k to be the least such that $f_k > f_{k-1}$ and the class $\mathbf{e}_{\tau, \mathbf{f_k}}$ of E_τ currently has size larger than k and is furthermore the oldest class in E_τ with index larger than f_{k-1} and with size larger than k. The age of a class is the number of stages it has not increased in size; so the oldest class is the one which has not increased in size for the longest time. If every class with index larger than f_{k-1} has size at most k, then f_k is not defined at stage s.

The parameter $fin(j)[s]$ is defined to be equal to the sequence $f_1 < f_2 < \cdots < f_{\langle \tau, j\rangle}$, if all the terms exist at stage s. Otherwise, we say that $fin(j)[s] \uparrow$. Note that at every stage of the construction, $fin(j)$ is a substring of $fin(j+1)$ (if they are both defined), and $fin(j)$ being undefined implies that $fin(j+1)$ is also undefined. We abuse our terminology and define the *range* of fin at stage s to be the string $f_1 < f_2 < \cdots < f_{\langle \tau, j\rangle}$ for j, the largest such that $fin(j)[s]$ is defined.

We explain the use of fin. It is not hard to see that the property of being unbounded is Π^0_3. In fact, the parameter fin is meant to guess whether E_τ is unbounded: the property of E_τ being unbounded is equivalent to the Π^0_3 property that $fin(j)$ holds a stable value for each j. This can be naturally incorporated into the outcomes of τ. The Π^0_2 outcome $\mathtt{pi}_2 j$ stands for the fact that j is the least such that $fin(j)$ does not have a limit, while the outcome $\mathtt{pi}_3$ stands for the fact that $fin(j)$ is stable for every j.

The reason for using $\langle \tau, j\rangle$ instead of $\langle \tau \rangle$ or j is to keep the "infinite junk" structures (of type (III)) produced by the different outcomes of different nodes non-isomorphic. To achieve this, we use the following modification of the pairing function throughout the rest of the proof:

We replace the standard pairing function $\langle \tau, j\rangle$ with $3^{\langle \tau, j\rangle}$.

As a consequence, the true outcome $\tau * \mathtt{pi}_2 j$ will produce an infinite junk structure (of type (III)) with somewhere between $\langle \tau, j\rangle$ and $2\langle \tau, j\rangle$ many finite classes, thus keeping structures produced by different outcomes (of various strategies) non-isomorphic.

The isomorphism ℓ. Let $\boldsymbol{e}_{\tau,i}$ and $\boldsymbol{u}_{\tau,i}$ be the ith classes in E_τ and U_τ, respectively. To assist us in organising the copying strategy, we will define a (potential) Δ^0_2 isomorphism $\ell_\tau : U_\tau \mapsto E_\tau$, which will be total only if $\tau * \mathtt{pi}_3$ is the true outcome of τ. (Note that the totality of a Δ^0_2 function is also Π^0_3.) Strictly speaking, ℓ will be a function mapping indices to indices, and we identify $\boldsymbol{u}_i$ with $\boldsymbol{e}_{\ell(i)}$. Initially, we set $\ell(i) \uparrow$ for all i. For convenience, whenever we wish to change the approximation to $\ell(i)$, we will first set $\ell(i) \uparrow$ before re-defining $\ell(i)$ at a later stage. The final limiting value of $\ell(i)$ is assigned the obvious meaning.

Action of τ. During the construction, whenever the node τ is visited, it will first check if there is a least (in the index) unmapped class $\boldsymbol{e}_{\tau,j}$ less than or equal to the largest element in the range of fin (unless we are waiting in step (i)), and if it exists, introduce a new class $\boldsymbol{u}_{\tau,i}$ to match it. This means that we will grow a new $\boldsymbol{u}_{\tau,i}$ to be equal in size to $\boldsymbol{e}_{\tau,j}$ and define $\ell_\tau(i) = j$. If (and only if) a new class is introduced in U_τ, we will also grow all classes $\boldsymbol{u}_{\tau,i}$ for which $\ell_\tau(i)$ is defined, to have size equal to $\boldsymbol{e}_{\tau,\ell(i)}$. Next, τ will process the following:

(i) For each nonempty class $\boldsymbol{u}_{\tau,i}$ such that $\ell_\tau(i) \uparrow$, we search for a corresponding $\boldsymbol{e}_{\tau,j}$ not yet mapped via ℓ_τ and with size larger than or equal to the size of $\boldsymbol{u}_{\tau,i}$.

  - If $\ell_\tau(i) \downarrow$ for every class $\boldsymbol{u}_{\tau,i}$ in U_τ, go to step (ii).
  - If $\ell_\tau(i) \uparrow$ for some class $\boldsymbol{u}_{\tau,i}$ in U_τ, and a large enough unmapped class in E_τ does not yet exist, play outcome `wait` and go to the next stage.
  - Otherwise, every class $\boldsymbol{u}_{\tau,i}$ with $\ell_\tau(i) \uparrow$ is able to find some image. We take the following actions, for each i. Define $\ell_\tau(i) = j$ (for the corresponding j). Play outcome `wait` and go to the next stage.

(ii) If we are here, it means that every nonempty class $\boldsymbol{u}_{\tau,i}$ has been mapped (i.e. $\ell_\tau(i) \downarrow$). Wait for $dom(fin)$ to increase beyond the previous maximum. If this is the first stage where $dom(fin)$ is longer than at any previous stage since the last τ-initialisation, we take outcome $\mathtt{pi}_3$ for this stage and go to (iii). Otherwise, play outcome `wait` at this stage.

(iii) Check if there is some $i < dom(fin)$ such that $fin(i)$ has changed, or the Π^0_2-predicate $Q(\tau, k_\tau, i)$, such that $P(\tau, k_\tau) = \exists i Q(\tau, k_\tau, i)$, has "fired". (Recall that a Π^0_2-predicate fires means that the Π^0_2 predicate appears true for one more stage in some computable approximation of the predicate; a Π^0_2-predicate holds iff it fires infinitely often). Without loss of generality, we assume that $Q(\tau, k_\tau, 0)$ never fires. Take i to be the least such, if it exists. Take the appropriate action below and go back to step (i).

  - If no such $i < dom(fin)$ exists, do nothing.
  - If $i = 0$, play outcome `init` and initialise τ.
  - If $i > 0$, play outcome $\mathtt{pi}_2(i-1)$. Preserve each class $\boldsymbol{u}_{\tau,k}$ for all $k < i$ as well as each class $\boldsymbol{u}_{\tau,k}$ such that $\ell_\tau(k)$ is in the tuple $fin(i-1)$ or $\ell_\tau(k) < i$. All other non-empty classes $\boldsymbol{u}_{\tau,m}$ we grow to size s (or some suitably large number, determined by the junk collector) and set $\ell_\tau(m) \uparrow$.

The strategy we describe here for τ assumes it acts in isolation. During the formal construction (§9.2.6), we will follow a slightly modified form of the strategy described here due to various technical reasons.

Analysis of the outcomes of τ. The true outcome of τ could be:

- *True outcome* wait: Since we only play outcomes to the left of wait finitely often, it is easy to see that in this case we must get stuck waiting forever in (i) or (ii). In either case, we will eventually stop adding new classes to U and stop growing existing classes of U. Thus, we end up producing a *finite structure* U.

- *True outcome* init: By convention, $Q(\tau, k_\tau, 0)$ never fires, so if init is the true outcome, then $fin(0)$ fails to hold a stable definition. In this case, we initialise τ infinitely often, and consequently, the strategy produces infinitely many finite structures. We will ensure that these structures are all pairwise non-isomorphic.

- *True outcome* $\mathtt{pi}_2 i$: In this case, either $Q(\tau, k_\tau, i+1)$ fires infinitely often, or $fin(i+1)$ fails to hold a stable definition. In the former case, as $P(\tau, k_\tau)$ holds, we have more evidence that E_τ is isomorphic to some E of higher priority, so we should not allow τ to copy E_τ into our list. In this case, U_τ will be an infinite junk structure with between $\langle \tau, i\rangle$ and $2i + \langle \tau, i\rangle < 2\langle \tau, i\rangle$ many finite classes.

- *True outcome* $\mathtt{pi}_3$: This means that all other outcomes to the left of $\mathtt{pi}_3$ are each visited only finitely often. Since outcome $\mathtt{pi}_3$ is played infinitely often, E_τ is not isomorphic to any higher-priority E, and E_τ is unbounded. We will argue in the verification that ℓ_τ is eventually total, stable at every input, and onto, and consequently witnesses that $U_\tau \cong E_\tau$.

9.2.4 Coordination between different τ

As the outcomes of the requirements are of order-type $1 + \omega + 2$, the true path of the construction will be $\varnothing'''$-computable. Since it is now possible to visit left of the true path infinitely often, we need to describe the effect each node has on the strategies on its right. If τ plays outcome init, then all nodes extending $\tau * o$ for an outcome $o \neq \mathtt{init}$ are initialised. If τ plays outcome $\mathtt{pi}_3$, then all nodes extending $\tau * \mathtt{wait}$ are initialised.

Now suppose τ plays outcome $\tau * \mathtt{pi}_2 i$. We will initialise every node extending $\tau * \mathtt{wait}$ or $\tau * \mathtt{pi}_2 j$ for $j > i$. However, we clearly do not wish to initialise a node $\sigma \supseteq \tau * \mathtt{pi}_3$, since σ could be on the true path. We ensure that the next time we visit σ again, we will *force* σ to play outcome $\mathtt{pi}_2 i$ (and take the corresponding actions) at least once, even though the basic strategy for σ does not require it to do so. This ensures that if $\tau * \mathtt{pi}_2 i$ is on the true path, then every $\sigma \supseteq \tau * \mathtt{pi}_3$ produces an infinite junk structure, unless σ is initialised infinitely often due to other reasons, and therefore does not copy any E into our list. On the other hand, if σ is on the true path, then for each i, σ is forced to play the outcome $\mathtt{pi}_2 i$ in this way only finitely often, and the true outcome of σ will still correctly reflect the outcomes of its basic strategy.

9.2.5 Coordination with junk collectors

We call finite structures that are produced by some node τ (or the $INFJUNK$ strategy) *finite junk*. Structures of type (III), which are built by some node τ, are called *infinite junk*.

In this subsection, we also explain several important modifications to the basic strategy of τ; these modifications are necessary for understanding the rest of the proof. The finite junk collector $FJUNK$ and the infinite junk collector $INFJUNK$ are strategies that act outside of the priority tree and will get to act at the end of every stage. Since the nodes on the construction tree will produce finite and infinite junk structures, the $FJUNK$ and $INFJUNK$ strategies are there to ensure that all isomorphism types of type (I) and (III) are listed. They will do so by adding to our list the missing structures of types (I) and (III) that are not produced as junk by nodes on the priority tree.

We split these actions into infinitely many substrategies, $FJUNK(F)$ and $INFJUNK(F)$, indexed by the isomorphism type F of a finite equivalence structure. These substrategies will seek to place a structure of isomorphism type F or $F \oplus I$, respectively, where I is the structure having infinitely many infinite classes.

Description of $FJUNK(F)$. Initially, when $FJUNK(F)$ is active for the first time, it checks whether there already exists a finite structure of isomorphism type F in the construction that is either permanently abandoned due to initialisation by some strategy on the tree, or corresponds to the outcome `wait` of some strategy and thus may (or may not) be truly abandoned. If none of the above possibilities occur, it introduces a new finite structure of type F. We say that this structure is the witness of $FJUNK(F)$. $FJUNK(F)$ has to ensure that *exactly* one copy of type F appears in our list. Thus, if $FJUNK(F)$ is already holding a witness structure, we need to ensure that *no* node on the construction tree can produce a structure of type F. There are three ways in which this may happen. We explain all three problematic situations and the modifications necessary to resolve the conflicts.

First, a finite structure being built by τ may be abandoned and permanently thrown into the finite junk pool after $FJUNK(F)$ had already chosen its witness. To avoid this conflict, we adopt the following modification to the basic strategy of τ:

Modification 1. If τ permanently abandons its structure due to an initialisation, we grow this abandoned structure F into a very large finite structure F'. Formally, given a finite structure F, we *enlarge* F by adding sufficiently many extra classes of size 1 to F to produce a structure F' that is different from any finite equivalence structure considered so far by the construction. Now, add the enlarged abandoned structure to the junk pool by assigning it as a witness for $FJUNK(F')$. Note that $FJUNK(F')$ has no current witness and has in fact never acted before in the construction.

Second, a strategy τ on the priority tree may be playing `wait` and might wait forever at construction steps (i) or (ii). Again, the isomorphism type of the structure being built by τ can be the same as F for some $FJUNK(F)$ that already has a witness.

Modification 2. When playing the `wait` outcome, τ will first enlarge its structure U_τ and wait with this enlarged finite structure. Since τ cares about copying E_τ only if it is unbounded, it does no harm to enlarge U_τ this way. Now, the strategy $FJUNK(U_\tau)$ (which has never acted before) is temporarily suspended, since τ is currently holding on to a structure of the same isomorphism type.

If later on τ finishes its wait, then U_τ will grow and $FJUNK(U_\tau)$ will then start a new structure as its witness. Since structures are always enlarged by adding a fresh number of new classes, $FJUNK(U_\tau)$ will be blocked in this way at most once.

Third, as we will see from the $INFJUNK$ strategy below, there might be a finite structure used as a witness by an $INFJUNK$ strategy which is permanently abandoned by the $INFJUNK$ strategy and added to the finite junk pool. In this case, we also make sure that the abandoned structure F is first enlarged and then added to $FJUNK(F')$ as a witness for the appropriate F'. Note that once a $FJUNK(F)$ strategy picks its follower, either on its own or is assigned its follower when a node τ or an $INFJUNK$ strategy abandons its current structure, this follower is permanently tied to $FJUNK(F)$ and no other strategy in the construction will produce a structure of the same type.

Description of $INFJUNK(F)$. We ensure that each strategy $INFJUNK(F)$ will produce in the limit a structure of the form $F \oplus I$, but at every finite stage, $INFJUNK(F)$ has a finite part of its intended structure. As in the case of $FJUNK(F)$, each $INFJUNK(F)$ will eventually choose its witness and will start growing it to a structure of isomorphism type $F \oplus I$.

The obvious conflict is that a node τ on the priority tree will also produce an infinite junk structure at the end if $\tau * \mathtt{pi}_2 i$ is its true outcome. Hence, we need to ensure that the corresponding $INFJUNK$ strategy does not duplicate this structure in our list.

To avoid repetition, $INFJUNK(F)$ must permanently abandon its current witness D every time τ is visited and makes more progress in constructing $F \oplus I$. The structure D will then be enlarged and placed into the finite junk collector. $INFJUNK(F)$ will then restart again with a new enlarged witness D' and start growing D' towards $F \oplus I$, until τ is again visited and makes more progress, if ever. In this way, either τ or $INFJUNK(F)$ will successfully construct $F \oplus I$ in our list. Note that given any F, there is at most one pair (τ, i) such that τ constructs $F \oplus I$ under outcome $\tau * \mathtt{pi}_2 i$.

We describe another possible conflict between infinite junk structures produced by the tree and $INFJUNK$ strategies. Consider the situation when τ is off the true path and is initialised infinitely often due to actions of other strategies. Then it may attempt to make progress in building $F \oplus I$ infinitely often, but in fact, it will produce only an infinite collection of finite structures due to being initialised. However, $INFJUNK(F)$ is also prevented from building $F \oplus I$ since its witness is also infinitely often reset due to us (wrongly) assuming that τ was making progress. This means that we will never produce a copy of $F \oplus I$ if we simply follow the instructions as described above. The difficulty goes away if we adopt the following modification to the basic module of τ:

Modification 3. In the definition of the parameter $fin(\tau, j)$, we will search for a longer sequence of integers $f_1 < f_2 < \cdots < f_{\langle \tau, \mathcal{I}_\tau, j \rangle}$, instead of $f_1 < f_2 < \cdots < f_{\langle \tau, j \rangle}$ as before. Here, $\mathcal{I}_\tau$ is the number of times τ has been initialised through the actions of another node. We do not count *self-initialisations*, where τ is initialised when playing outcome `init`. This modification causes τ to produce an infinite junk structure with between $\langle \tau, \mathcal{I}_\tau, i \rangle$ and $2i + \langle \tau, \mathcal{I}_\tau, i \rangle$ many finite classes, if τ was initialised exactly $\mathcal{I}_\tau$ times (by other nodes) and has true outcome $\tau * \mathtt{pi}_2 i$. After this modification, for each fixed F, the $INFJUNK(F)$ strategy only needs to worry about conflicts with a unique triple $(\tau, \mathcal{I}, i)$.

A final conflict between infinite junk structures produced by the tree and $INFJUNK$ strategies is more subtle. An $INFJUNK(F)$ strategy might have its witness structure reset infinitely many times because a strategy τ plays outcome $\mathtt{pi}_2 n$ infinitely often. However, the strategy τ might in fact have true outcome $\mathtt{pi}_2 m$ for $m < n$, and hence end up constructing an infinite junk structure that is not of type $F \oplus I$. This means that $F \oplus I$ is neither constructed by $INFJUNK(F)$ nor τ.

This problem can be fixed if we allow the $INFJUNK(F)$ strategy to pick finitely many witness structures $D_0, D_1, \ldots, D_n$ instead of a single witness, where $INFJUNK(F)$ is conflicted with outcome $\mathtt{pi}_2 n$ of τ. While $INFJUNK(F)$ detects no conflicts, it will grow D_n towards $F \oplus I$ and keep $D_0, \ldots, D_{n-1}$ finite. Whenever τ plays outcome $\mathtt{pi}_2 n$, $INFJUNK(F)$ will abandon D_n and begin with a new D_n (while keeping $D_0, \ldots, D_{n-1}$). When τ plays outcome $\mathtt{pi}_2 m$ for $m < n$, we will abandon $D_{m+1}, D_{m+2}, \ldots, D_n$ and restart these with new witness structures, while keeping $D_0, \ldots, D_m$. We also grow D_m for one more step towards making $D_m \cong F \oplus I$.

At the end, if τ has true outcome to the right of $\mathtt{pi}_2 n$, then $D_0, \ldots, D_{n-1}$ will finally stabilise at finite structures. We will have $D_n \cong F \oplus I$, but of course $U_\tau \not\cong F \oplus I$. If τ has true outcome $\mathtt{pi}_2 n$, then $D_0, \ldots, D_{n-1}$ are stable finite structures, while D_n will be infinitely often abandoned. Here, $D_\tau \cong F \oplus I$. Finally, if τ has true outcome $\mathtt{pi}_2 m$ for some $m < n$, then $D_0, \ldots, D_{m-1}$ are stable finite structures, while $D_{m+1}, \ldots, D_n$ are infinitely often abandoned. We finally make $D_m \cong F \oplus I$ and in this case, U_τ is an infinite junk structure not of type $F \oplus I$. Thus, either τ or one of the $INFJUNK(F)$ witnesses (and exactly one) will succeed in building $F \oplus I$.

9.2.6 Formal construction.

The basic strategies of τ and the junk collectors have been described above. We put it all together in this section. We adopt Modification 3 above in the definition of the parameter $fin(\tau, j)$. The construction will, at stage s, define the current approximation δ_s to the true path, where $|\delta_s| = s$. Suppose $\tau \subset \delta_s$ has been defined. We now need to describe the actions of τ and the outcome played at this stage.

Growing U_τ. The first thing we do is check if $\tau * \mathtt{pi}_3$ was played at the previous visit to τ, and if not, we do not grow U_τ and skip to the step "acting for τ" (to be explained after we finish with "growing U_τ"). Otherwise, suppose that $\tau * \mathtt{pi}_3$ was played at the previous visit to τ. Let $s_0 < s$ be the previous stage where U_τ last grew. Take the following actions:

- For each class $\boldsymbol{e}_{\tau,j}$ such that no $\ell_\tau(i)$ maps to it and where $j \leqslant$ the largest element in the current range of fin, introduce a new class $\boldsymbol{u}_{\tau,i}$ in U_τ to have the same size and set $\ell_\tau(i) = j$.

- For every class $\boldsymbol{u}_{\tau,i}$ in U_τ such that $\ell_\tau(i)$ exists, we grow $\boldsymbol{u}_{\tau,i}$ to have the same size as $\boldsymbol{e}_{\tau,\ell_\tau(i)}$.

- Enlarge U_τ by adding sufficiently many new $\boldsymbol{u}_\tau$ classes of size 1, so that the resulting finite structure has never been looked at by the construction.

If this is the first visit to τ since an initialisation (so that s_0 does not exist), we begin building a new structure U_τ by taking U_τ to be the enlargement of the empty structure.

Recall that $s_0 < s$ was the stage where we last grew U_τ; set $s_0 = 0$ if this does not exist. Check if one of the following holds:

- There exists a stage t such that $s_0 < t < s$, and some node α such that $\alpha * \mathtt{pi}_3 \subseteq \tau$ and $\alpha * \mathtt{pi}_2(i-1) \subseteq \delta_t$ for some i, or

- There is some $i < dom(fin)[s_0]$ such that $fin(i)$ has changed, or $Q(\tau, k_\tau, i)$ has fired between s_0 and s. (As before, we assume that $Q(\tau, k_\tau, 0)$ never fires.)

Pick the least i for which one of the above applies:

- $i = 0$: Play outcome `init` and initialise τ.

- $i > 0$: If the first alternative applies, we say that τ *is forced to play outcome* $\tau * \mathtt{pi}_2(i-1)$; if the second alternative applies, we say that τ *wants to play outcome* $\tau * \mathtt{pi}_2(i-1)$. In either case, we play outcome $\tau * \mathtt{pi}_2(i-1)$ and take the actions described under the basic strategy for τ in §9.2.3(iii) when outcome $\mathtt{pi}_2(i-1)$ is played.

- *No i found*: Play outcome `wait`.

In any case, go to Initialising other nodes (below).

Acting for τ. Suppose that we did not manage to grow U_τ. Now take the actions and outcome described under the basic strategy for τ in §9.2.3(i) or (ii), whichever applies. Proceed to Initialising other nodes (below).

Initialising other nodes. We have described the actions of τ and the outcome of τ at this stage. Now initialise all the nodes mentioned in §9.2.4. When a node τ is initialised, we will first enlarge the finite structure U_τ to U' and assign it as a witness to $FJUNK(U')$, reset all parameters associated with the node (except for the counter $\mathcal{I}_\tau$), and increase the value of $\mathcal{I}_\tau$ by 1, unless this is due to a self-initialisation.

This ends the description of the actions and the outcomes of τ. Suppose we have finished with the definition of δ_s of length s. Before we conclude the construction stage s, we shall act for the junk collector strategies. First, process the $INFJUNK$ strategies (below), followed by the $FJUNK$ strategies.

***INFJUNK* action.** For each F with a code number less than s, we act for $INFJUNK(F)$ as follows. (The actions for different F are independent.) First, let $\langle \tau, \mathcal{I}, n\rangle$ be the triple such that the number of classes of F is between $\langle \tau, \mathcal{I}, n\rangle$ and $2\langle \tau, \mathcal{I}, n\rangle$. This triple, if it exists, is unique. If this triple is not found, then $INFJUNK(F)$ will have no conflict with any structure built by the construction tree, and in this case, simply enumerate a witness $F \oplus I$ into our list.

Otherwise, fix this triple $\langle \tau, \mathcal{I}, n\rangle$. If $\mathcal{I}_\tau$ is currently not equal to $\mathcal{I}$, we will pick a new witness structure D for $INFJUNK(F)$ and begin growing D towards $F \oplus I$. If $\mathcal{I}_\tau$ later becomes equal to $\mathcal{I}$, we will abandon the previous witness structure and proceed as below. If $\mathcal{I}_\tau$ increases from $\mathcal{I}$ to $\mathcal{I}+1$, then we abandon all witness structures and pick a new D.

Otherwise, suppose that $\mathcal{I}_\tau = \mathcal{I}$. We begin by picking new witness structures $D_{-1}, D_0, D_1, \cdots, D_n$. At each stage, we will determine if $INFJUNK(F)$ has any conflicts with τ. If it is not the case that τ is visited and outcome $\texttt{pi}_2 m$ for some $m \leqslant n$ or outcome $\texttt{init}$ is taken, then there are no conflicts; at this stage, $INFJUNK(F)$ simply does nothing with $D_{-1}, D_0, \cdots, D_{n-1}$ and grows D_n for one more step towards making $D_n \cong F \oplus I$.

Suppose that τ is visited and outcome $\texttt{pi}_2 m$ for $m < n$ is taken. Then $INFJUNK(F)$ does nothing with $D_{-1}, D_0, \cdots, D_{m-1}$, and abandons and picks new witness structures for $D_{m+1}, \cdots, D_n$. It also grows D_m by one more step towards making $D_m \cong F \oplus I$.

Suppose that τ is visited and outcome $\texttt{init}$ is taken. Then $INFJUNK(F)$ abandons and picks new witness structures for $D_0, \cdots, D_n$. It also grows D_{-1} by one more step towards making $D_{-1} \cong F \oplus I$.

Finally, suppose that τ is visited and outcome $\texttt{pi}_2 n$ is taken. The action of τ at this stage was to restrain a set of classes G of U_τ and grow all the remaining ones. If $F \subseteq G$ and such that every class of $G - F$ is larger than the largest class in F, and has grown since the last time we considered this step, then a conflict occurs at this stage. $INFJUNK(F)$ will abandon its current witness and pick a new witness structure D_n and keep $D_{-1}, \cdots, D_{n-1}$. Otherwise, if there are no conflicts, then simply grow D_n.

In the strategy above, we need to ensure several things. First, every time we pick a new witness structure, we need to take it to be a suitable enlargement of the empty structure. Second, every time we grow an existing structure, we need to increase the size of one of its classes, as well as take a suitable enlargement of it. Third, if any witness structure is abandoned, $INFJUNK(F)$ will first enlarge it to a suitable finite structure D' and then assign it to $FJUNK(D')$ as a follower.

$FJUNK$ **action.** For each F with a code number less than s, we act for $FJUNK(F)$ as follows. If $FJUNK(F)$ has no current witness, and if there is no structure X currently in the list such that $X \cong F$, then $FJUNK(F)$ will introduce a witness structure of isomorphism type F. Otherwise, do nothing.

This ends the description of the construction.

9.2.7 Verification

We define the true path of the construction inductively, each time selecting the leftmost outcome which is visited infinitely often. It is easy to check that each node on the true path is initialised by another node only finitely often. Infinite self-initialisation is still possible.

We will need to argue two things: First, that all structures of type (I), (III), and (V) are represented in our list, and second, that our list does not contain repetitions.

We begin with an important lemma.

Lemma 9.2.3. *Let τ be a node on the construction tree, which is visited infinitely often by the construction, initialised by other nodes $\mathcal{I} < \infty$ times, and where $\tau * \mathtt{pi}_2 i$ is the leftmost outcome of τ visited infinitely often. Then the final structure U_τ built by τ is an infinite junk structure of type (III) with between $\langle \tau, \mathcal{I}, i\rangle$ and $2\langle \tau, \mathcal{I}, i\rangle$ many finite classes.*

Proof. Since $\mathtt{pi}_2 i$ is played only when we grow U_τ, this means that we grow U_τ infinitely often. This means that $\tau * \mathtt{pi}_3$ is also played infinitely often, and so $dom(fin)$ grows arbitrarily large. Since τ is visited infinitely often but outcomes to the left of $\mathtt{pi}_2 i$ are each played finitely often, this means that $fin(i)$ will eventually be defined and hold a stable value. After $fin(i)$ is stable, for each class $\boldsymbol{e}_{\tau,f}$ of $fin(i)$, we will be able to define $\ell_\tau^{-1}(f)$ and never again redefine $\ell_\tau^{-1}(f)$ (notice that these cannot be redefined by an outcome to the right of $\tau * \mathtt{pi}_2 i$). Since all classes in $fin(i)$ are finite, this means that there are at least $\langle \tau, \mathcal{I}, i\rangle$ many finite classes in U_τ.

Each time we play outcome $\mathtt{pi}_2 i$ we will preserve a fixed set of U_τ classes: these are $\boldsymbol{u}_{\tau,j}$ for $j < i+1$ and $\boldsymbol{u}_{\tau,\ell_\tau^{-1}(f)}$ for $f \in fin(i)$ or $f < i+1$. There are at most $2i + 2 + \langle \tau, \mathcal{I}, i\rangle < 2\langle \tau, \mathcal{I}, i\rangle$ many preserved classes. All other classes are increased in size. Thus U_τ will have at most $2\langle \tau, \mathcal{I}, i\rangle$ many finite classes.

Finally, each time we grow U_τ we enlarge it, so U_τ must contain infinitely many classes, and hence infinitely many infinite classes. Hence, U_τ is an infinite junk structure of type (III). Note that τ *does not* have to be on the true path in the statement of the lemma. □

Lemma 9.2.4. *Suppose τ is on the true path, and the true outcome of τ is $\mathtt{pi}_3$. Then $E_\tau \cong U_\tau$ is unbounded.*

Proof. Since $\tau * \mathtt{pi}_3$ is played infinitely often, $dom(fin)$ must grow arbitrarily large. Furthermore, $\tau * \mathtt{pi}_2 i$ is played finitely often for each i, which means that $fin(i)$ must eventually be defined and stable. Therefore, E_τ must be unbounded. Also, the unboundedness of E_τ implies that whenever ℓ_τ is set undefined for some class $\boldsymbol{u}_i$ in U_τ and later seeks a new image in E_τ, it will be able to find a suitable image (since E_τ contains arbitrarily large classes).

Similarly, since U_τ is grown at infinitely many stages and $dom(fin)$ grows arbitrarily large, whenever $\ell_\tau^{-1}(j)$ is set undefined for some class $\boldsymbol{e}_j$ in E_τ, we will always get a chance to redefine $\ell_\tau^{-1}(j)$. Furthermore, each $\ell_\tau(i)$ and $\ell_\tau^{-1}(j)$ will not be made undefined once $fin(i)$ or $fin(j)$ is stable; notice these cannot be made undefined by an outcome or node to the right of $\tau * \mathtt{pi}_2 i$ or $\tau * \mathtt{pi}_2 j$. Thus, ℓ_τ is total, and clearly bijective.

It is also easy to check that at every stage where $\ell_\tau(i)$ is defined, the current size of $\boldsymbol{u}_i$ is at most the current size of $\boldsymbol{e}_{\ell(i)}$: This is certainly true at the point when $\ell_\tau(i)$ receives a new definition, and after that, we only grow $\boldsymbol{u}_i$ to match the size of $\boldsymbol{e}_{\ell(i)}$ (unless under a $\mathtt{pi}_2 n$ outcome where we also make $\ell(i)$ undefined). Since there are infinitely many stages where we grow U_τ, we have in fact that the sizes of $\boldsymbol{u}_i$ and $\boldsymbol{e}_{\ell(i)}$ are equal. Therefore, $E_\tau \cong U_\tau$, and in fact, $E_\tau \cong_{\Delta_2^0} U_\tau$. □

Lemma 9.2.5. *All structures in our list are of type either (I), (III), or (V).*

Proof. Every time $FJUNK(F)$ receives or picks a witness structure, it is of type F and no further changes are made to this witness structure.

If $INFJUNK(F)$ picks a witness structure D, and if D is either later abandoned or is never grown again, then D ends up being finite. On the other hand, if D is never abandoned and is grown infinitely often, then we will be able to make it of type $F \oplus I$; the only issue is that we always enlarge a structure when growing, however, this action only adds classes of size 1, so it is compatible with extending it to $F \oplus I$.

Finally, if a structure U is introduced by a node τ, and is either later abandoned or is grown finitely often, then U ends up being finite. Otherwise, suppose U is grown infinitely often and never abandoned. This means that τ is visited infinitely often but initialised only finitely often, and the leftmost outcome of τ visited infinitely often has to be either $\mathtt{pi}_3$ or $\mathtt{pi}_2 n$ for some n. If it is $\mathtt{pi}_2 n$, then we apply Lemma 9.2.3. Otherwise, if it is $\mathtt{pi}_3$, then we note that τ has to be on the true path (otherwise, τ will be forced to play a $\tau * \mathtt{pi}_2 n$ outcome infinitely often for some n), and so we apply Lemma 9.2.4. □

Any activity on a structure X can be classified according to: X is first introduced to the list, X is grown, or X is abandoned. These are the only three ways in which X can be modified. It is easy to check the following fact:

Fact 9.2.6. *Any activity on a structure X must also be followed by an enlargement of the structure at the same time. The only exception is when a $FJUNK$ strategy picks a new witness structure.*

Lemma 9.2.7. *Each isomorphism type in (I), (III), and (V) is represented.*

Proof. Consider a finite type F, and let s be the stage where $FJUNK(F)$ is first active. If $FJUNK(F)$ ever picks a follower structure, it will be stable of type F, so let's assume it never gets to pick a follower. This means that at stage s there is at least one structure X currently in the list such that $X \cong F$. (In fact, there is at most one such X at any time, but so far we have not addressed uniqueness, and this will be done later. By Fact 9.2.6, no activity which occurs after stage s can produce a structure of type F. Thus, in order for $FJUNK(F)$ to remain blocked at every stage, it must be that one of these structures X already present at stage s is stable and of type F. So all structures in type (I) are represented.

Now fix F and consider $INFJUNK(F)$. If $INFJUNK(F)$ eventually detects no conflicts, then it will hold a stable witness structure of type $F \oplus I$. (Again, enlarging the structure while growing is compatible with making it of type $F \oplus I$). Otherwise, $INFJUNK(F)$ will infinitely often detect a conflict. This conflict must each time be with a node τ playing some outcome $m \leqslant n$ or $\mathtt{init}$, and during the time when $\mathcal{I}_\tau = \mathcal{I}$, for a unique triple $\langle \tau, \mathcal{I}, n \rangle$.

Let m be the least where the conflict happens infinitely often. If $m < n$, then D_m will hold a stable witness structure and we will make $D_m \cong F \oplus I$. Same for the case $\mathtt{init}$, where $D_{-1} \cong F \oplus I$.

On the other hand, if $m = n$ and a conflict is detected infinitely often, then $G - F$ must contain only infinite classes, and in fact, U_τ must contain only F as its set of finite classes. Hence, $U_\tau \cong F \oplus I$. (Note that U_τ is never abandoned unless $\mathcal{I}_\tau$ is increased or outcome `init` is played). So all structures in type (III) are represented.

Now finally we fix an unbounded E of type (V), and argue that it is represented. Let i be the least such that $E_i \cong E$; since i is least, there is some k such that $P(i,k)$ has Π^0_3-outcome. Let τ be along the true path assigned E_i and guessing $P(i,k)$. We claim that τ must have true outcome $\mathtt{pi}_3$. Since E_τ is unbounded, and τ being on the true path is initialised by another node only finitely often, this means that $\mathcal{I}_\tau$ is stable and hence $fin(\tau, n)$ eventually holds a stable value for each n. We cannot be stuck waiting under §9.2.3, step (i) for τ, as E_τ is unbounded, and so this means that outcome $\mathtt{pi}_3$ of τ is played infinitely often.

Suppose some outcome to the left of $\mathtt{pi}_3$ is played infinitely often. This cannot be `init` because $fin(0)$ holds a stable value. On the other hand, this cannot be outcome $\mathtt{pi}_2(n-1)$; τ cannot be forced to infinitely often play $\mathtt{pi}_2(n-1)$, otherwise the true path is to the left of τ, and τ cannot want to infinitely often play $\mathtt{pi}_2(n-1)$ since $fin(n)$ is eventually stable and $Q(\tau,k,n)$ eventually stops firing. Hence, $\tau * \mathtt{pi}_3$ is along the true path. By Lemma 9.2.4 we have $U_\tau \cong E_i$. Hence, all structures of type (V) are represented. □

Fix two structures X and Y in our list. We now argue that $X \not\cong Y$. Suppose X and Y are both finite. Then they each have a final activity before becoming stable; assume that the final activity for X occurs after the final activity for Y. The following are all the different possibilities:

- *Both X and Y are picked by $FJUNK$ strategies*: Since each $FJUNK$ strategy picks at most one witness structure, this means that $X \not\cong Y$.

- *X is not picked by a $FJUNK$ strategy*: By Fact 9.2.6, X is enlarged during its final activity, so $X \not\cong Y$.

- *X is picked by a $FJUNK$ strategy*: Since Y is already stable when X is picked, $X \not\cong Y$.

Now suppose that X and Y are both infinite; in particular, neither of them is ever abandoned. Again, the following are all the different possibilities:

- *Both X and Y are infinitely often grown by $INFJUNK$ strategies*: Since each single $INFJUNK(F)$ strategy produces at most one infinite structure at the end (which is necessarily of the type $F \oplus I$), this means that X and Y are witnesses of different $INFJUNK$ strategies; hence, $X \not\cong Y$.

- *Both X and Y are infinitely often grown by strategies on the priority tree*: Then there are nodes τ_x and τ_y responsible for X and Y respectively. Each node works on a single structure U_τ at any one time, so $\tau_x \neq \tau_y$. This means that τ_x and τ_y are initialised finitely often, visited infinitely often, and play a leftmost outcome $\mathtt{pi}_3$ or $\mathtt{pi}_2 n$ for some n infinitely often. If τ_x plays the leftmost outcome $\mathtt{pi}_2 n$, then apply Lemma 9.2.3 to see that X is an infinite junk structure with between $\langle \tau_x, \mathcal{I}_{\tau_x}, n\rangle$ and $2\langle \tau_x, \mathcal{I}_{\tau_x}, n\rangle$ many finite classes. If τ_x plays the leftmost outcome $\mathtt{pi}_3$, then apply Lemma 9.2.4 to see that $X \cong E_{\tau_x}$ is unbounded. The same applies to τ_y. Hence, it is clear that $X \not\cong Y$ except in the case where both τ_x and τ_y have the leftmost outcome $\mathtt{pi}_3$.

  If this is the case, then both τ_x and τ_y must lie along the true path. So, assume that $\tau_x * \mathtt{pi}_3 \subseteq \tau_y$. As τ_y is not deactivated, it means that $i_x \neq i_y$, where τ_x is assigned E_{i_x} and

τ_y is assigned E_{i_y}. Since *both* nodes τ_x and τ_y have the true outcome $\mathtt{pi}_3$, it must be that $E_{i_y} \not\cong E_{i_x}$ (note that it could be that $i_x > i_y$). Hence, $X \not\cong Y$.

- *X is infinitely often grown by τ and Y is infinitely often grown by $INFJUNK(F)$*: As in the analysis above, in order for τ to produce an infinite junk structure, it must be initialised finitely often, visited infinitely often, and play a leftmost outcome $\mathtt{pi}_2 n$ for some n infinitely often. Suppose that $X \cong Y \cong F \oplus I$. The number of finite classes in F must be between $\langle \tau, \mathcal{I}_\tau, n\rangle$ and $2\langle \tau, \mathcal{I}_\tau, n\rangle$. Since τ plays the leftmost outcome $\mathtt{pi}_2 n$, this means that $INFJUNK(F)$ will have stable finite witness structures $D_{-1}, \ldots, D_{n-1}$, while D_n gets abandoned infinitely often. This means that $INFJUNK(F)$ does not in fact produce an infinite structure, a contradiction.

This ends the proof.

Exercises

(See also Exercises 7.1.56-7.1.58.)

Exercise 9.2.8 (Goncharov, Lempp, and Solomon [214]). Show that there exists a Friedberg list of the family of all d-c.e. sets under equality, and also for any $n > 2$, of the family of all n-c.e. sets (sets in Σ^{-1}_n) for any $n > 2$. (A Δ^0_2 set is n-c.e. if it a we allow at most n changes in its Δ^0_2-approximation via the Limit Lemma 3.1.3.)

Exercise* 9.2.9 (Ospichev [424]). Prove a transfinite extension of the previous exercise to cover each class Σ^{-1}_α (for any computable α) in the Ershov hierarchy: There is a Friedberg list of the family of all sets Σ^{-1}_α, for any computable α. (We omit the definitions.)

Exercise 9.2.10 (Ospichev [425]). Prove that there is a Friedberg list of all partial computable functionals of any given finite type, including, e.g., Type III. (The general definition can be found in [155]. For example, a functional can take as inputs Type II objects and output Type I objects, making it a Type $\langle II, I\rangle$ object.)

Exercise 9.2.11 (Pour-El and Putnam [433]). A class K of c.e. sets is k-Friedberg if there exists a uniformly c.e. listing $A_0, A_1, \ldots$ of the class such that, for each $A \in K$, there exist at most k integers $i_1, \ldots, i_k$ such that $A = A_{i_1} = \cdots = A_{i_k}$. Prove:

1. Given any $k > 2$, there is a class K of finite sets which is k-Friedberg but not $(k-1)$-Friedberg.
2. There exists a class K of finite sets such that:
   (a) G is not k-Friedberg for any k,
   (b) there exists a uniform c.e. enumeration of the elements of K in which each element is repeated only finitely often.
3. There exists uniformly c.e. collection K of c.e. sets containing a single infinite set such that this infinite set is repeated infinitely often in every uniform computable enumeration of K.

Exercise 9.2.12** (Wehner [504]). Prove the following surprising and seemingly forgotten result: The index set uniformly computably enumerable families of c.e. sets with the property "K admits an enumeration with finite repetitions" (Exercise 9.2.11(2.b)) is Σ^0_6-*complete.*

9.3 Computable abelian p-groups

In this section, we present a systematic exposition of the theory of computable abelian p-groups sufficient to prove Theorem E. The main result of the section is as follows.

Theorem (Khisamiev [290] and [287] for $n = 1$). Suppose that A is a countable reduced abelian p-group of Ulm type $n < \omega$ ($n \geqslant 1$). Then the following conditions are equivalent:

1. A has a computable copy.
2. A has a computable p-basic tree representing it.
3. (a) For every $i < n$, the character χ_{A_i} is a Σ^0_{2i+2} set, and
   (b) for every $i < n$, the set
   $$\#A_i := \{m \mid (m, 1) \in \chi_{A_i}\}$$
   is $\mathbf{0}^{(2i)}$-limitwise monotonic.

(We will clarify the terminology in due course.) Khisamiev did not use the technique of p-basic trees, so the second item of the theorem was not known to him; it is due to Ash, Knight, and Oates (unpublished). We will need a modified version of the main technical proposition used in this theorem to produce a Friedberg enumeration of all abelian p-groups having Ulm type $\leqslant n$ when $n > 1$. This section also contains several other well-known theorems, including an algebraic characterisation of computably categorical abelian p-groups due to Goncharov and, independently, Smith.

9.3.1 Abelian p-groups basics

We refer the reader to §5.1.1 for the basic definitions of abelian group theory. We briefly discuss some further basic well-known algebraic facts directly related to the material of this chapter. The standard references are [194, 285] where all these facts can be found.

In this section all groups are additive and abelian. For a fixed prime p, a group is a p-group if for every x there is an $n > 0$ such that $p^n x = 0$. In particular, every p-group is torsion. On the other hand, given a torsion abelian group A, one can define the maximal p-subgroup

$$T_p(A) = \{a \in A \mid \text{order}(a) \text{ is a power of } p\}.$$

An elementary argument shows that, for a torsion abelian group A,

$$A = \bigoplus_p T_p(A),$$

where p ranges over all primes. This fact explains why, traditionally, the study of torsion abelian groups is often completely reduced to the investigation of p-groups. Also, for an abelian group A, define its *socle* $A[p]$ to be the subgroup of A consisting of 0 and all elements of A having order p. The socle can be viewed as a vector space over $\mathbb{Z}_p$, the cyclic group of order p which also happens to be a field.

Divisible groups

We have already encountered divisible (§5.1.1) and p-divisible (§5.2.1) torsion-free groups. Recall that a group A is *divisible* if for every $x \in A$ and every integer $m > 0$,

$$\exists g \in A \text{ such that } mg = x.$$

If we restrict m to powers of a fixed prime p, then we say that the group is *p-divisible.* That is, for every $x \in A$ and every integer $n > 0$,

$$\exists g \in A \text{ such that } p^n g = x.$$

Fact 9.3.1. *An abelian p-group is divisible iff it is p-divisible.*

We leave the proof to Exercise 9.3.13. One important example of a divisible abelian p-group is as follows.

Example 9.3.2. Fix a prime p. Consider the (abelian) group with generators $\{x_i : i \in \mathbb{N}\}$ defined by the relations

$$px_0 = 0, \quad px_{i+1} = x_i, \quad i \in \mathbb{N}.$$

It is clearly a p-group which is p-divisible. Thus, it is divisible by Fact 9.3.1. We will denote the resulting group by $\mathbb{Z}_{p^\infty}$. It is called the *Prüfer p-group,* and is sometimes also referred to as the *quasi-cyclic* p-group. For every fixed n, the subgroup generated by $x_0, \ldots, x_n$ is isomorphic to the cyclic group $\mathbb{Z}_{p^{n+1}}$. The Prüfer p-group is equal to the union (the "direct limit") of the sequence of its subgroups

$$\{0\} \subseteq \mathbb{Z}_p \subseteq \ldots \subseteq \mathbb{Z}_{p^n} \subseteq \mathbb{Z}_{p^{n+1}} \subseteq \ldots$$

under the natural inclusion.

A group is *reduced* if it has no divisible subgroups (other than $\{0\}$). We omit the proof of the following well-known theorem.

Theorem 9.3.3.
1. *If D is a divisible subgroup of an abelian group A, then there exists a subgroup B of A such that $D \oplus B = A$.*
2. *Every (non-zero) divisible abelian p-group is isomorphic to a direct sum of Prüfer p-groups.*
3. *Thus, every abelian p-group A splits into*

$$R(A) \oplus D(A),$$

where $R(A)$ is reduced, $D(A)$ is (maximal) divisible. If $D(A) \neq \{0\}$, then it further splits into the direct sum of Prüfer p-groups.

While $D(A)$ is uniquely defined (as a subset of A) to be the maximal divisible subgroup of A, $R(A) \cong A/D(A)$ is only unique up to isomorphism. The following elementary consequence of Theorem 9.3.3 was used in the previous chapter to establish that "being reduced" is Π^1_1-complete.

Corollary 9.3.4. *For an abelian p-group, being not reduced is equivalent to the existence of a sequence of non-zero elements $(x_i)_{i\in\mathbb{N}}$ such that for all i, $px_i = x_{i+1}$.*

Ulm factors

In a torsion-free group, if $\exists y\ my = x$ ($m \neq 0$) then there exists a unique such y. Thus, in the torsion-free case it makes sense to view the predicate

$$m|x \text{ if and only if } \exists y\ my = x$$

as a partial operation

$$div_m : x \mapsto \frac{x}{m},$$

if such an element exists in the group (and we set it undefined otherwise). Indeed, the predicate is always witnessed by at most one such y. Further, in the torsion-free case, if $my = x \neq 0$ for $m > 0$ and x is divisible by any n, then so is y. None of that is true for p-groups in general, as the next example illustrates.

Example 9.3.5. Fix a prime p. Consider the (abelian) group on the generators $\{x, y_i : i \in \mathbb{N}\}$ defined by the relations

$$px = 0,\ p^i y_i = x,\quad i \in \mathbb{N}.$$

It is clearly a p-group. For every i, $p^i \mid x$. However, it is not difficult to show that the group is not divisible. Indeed, only finitely many elements z of this group, namely $0, x, \ldots, (p-1)x$, have "infinite height", i.e., satisfy $\forall m > 0\ m \mid z$.

In an abelian p-group, an important property of elements is their *p-height*, or sometimes simply *height*, which is defined below.

Definition 9.3.6. Let A be an abelian p-group, and let $a \in A$. The p-height $h_p(a)$ of a is defined as follows:

$$h_p(a) = \begin{cases} \max\{m : p^m \mid a\} & \text{if this maximum exists,} \\ \infty & \text{otherwise.} \end{cases}$$

Define also

$$A' = \{a \in A \mid h_p(a) = \infty\},$$

which is easily seen to be a subgroup of A. (It will be always clear from the context that A' does *not* mean the Turing jump of A.) Iterate this process and define $A^{(0)} = A$, then $A^{(\alpha+1)} = (A^{(\alpha)})'$, and if α is a limit ordinal, $A^{(\alpha)} = \cap_{\beta<\alpha} A^{(\beta)}$.

Definition 9.3.7. Define the *Ulm factors* of A as follows:

$$A_\alpha = A^{(\alpha)}/A^{(\alpha+1)},$$

where α is an ordinal.

We are primarily interested in the case when A is countable. Since the sequence defined above is clearly nested under inclusion,

$$A \supseteq A' \supseteq A'' \supseteq \ldots \supseteq A^{(\alpha)} \supseteq \ldots,$$

for some (finite or countable) α, we must have $A^{(\alpha)} = A^{(\alpha+1)}$. The least such α is called *the Ulm type of A*. We remark that if A is computable, then $\alpha \leqslant \omega_1^{CK}$ (Exercise 10.1.63). It should be clear that, for this α, $A^{(\alpha)}$ is p-divisible and, thus, divisible. We say that A is *reduced* if $A^{(\alpha)} = \{0\}$, where α is the Ulm type of A; equivalently, if it has no non-trivial divisible subgroups. For some reduced R,

$$A \cong R \oplus A^{(\alpha)},$$

where $A^{(\alpha)}$ is divisible and α is equal to the Ulm type of A. Further, the Ulm type of R is equal to α as well. The subgroup R is not unique in general, but it is clearly unique up to isomorphism.

Groups of Ulm type 1

Clearly, for every α, $A_\alpha = A^{(\alpha)}/A^{(\alpha+1)}$ has no non-zero elements of infinite p-height.

Theorem 9.3.8 (Prüfer [436]). *Assume A is a countable abelian p-group that has no elements of infinite p-height. Then A splits into a direct sum of cyclic groups.*

We will also need the following fact that is easily derived from the preceding material, but it is rarely stated in the literature in the form in which we will need it.

Theorem 9.3.9. *Let A be an abelian p-group such that A' is divisible. Then A splits into a direct sum of cyclic and quasi-cyclic direct summands. Furthermore, the characteristic of A*

$$\chi_A = \{\langle n, k\rangle : A \textit{ has at least } k \textit{ direct summands isomorphic to } \mathbb{Z}_{p^n}\},$$

where $n, k \in \mathbb{N}$, together with the number of quasi-cyclic summands, fully describes A up to isomorphism.

Proof. Split A into its reduced part and its divisible part using Theorem 9.3.3. Further split it into the sum of cyclic and quasi-cyclic (Example 9.3.2) groups using Theorems 9.3.3 and 9.3.8. Uniqueness follows from considering dimensions of subspaces of the socle. In the divisible part, the number of the quasi-cyclic summands is equal to the dimension of its socle. In the reduced summand, for any fixed k, consider the dimension of the $\mathbb{Z}_p$-space

$$\{a \in A : pa = 0 \wedge h_p(a) \geqslant k\}/\{a \in A : pa = 0 \wedge h_p(a) > k\},$$

which is equal to the number of cyclic summands of order $\mathbb{Z}_{p^k}$. □

Note that χ_A is exactly the invariant of an equivalence structure E_A which has a class of size κ if and only if A has a cyclic summand of the form $\mathbb{Z}_{p^\kappa}$ ($\kappa \in \mathbb{N} \cup \{\infty\}$). In other words, Theorem 9.3.9 illustrates that (countable) abelian p-groups of Ulm type 1 and equivalence structures (Definition 9.1.1) can be described using the same invariants. We will soon see that this fact also holds effectively.

Groups of Ulm type $\alpha \geqslant 1$.

The following well-known extension of Prüfer's Theorem 9.3.8 is due to Ulm. One way to state the theorem is as follows.

Theorem 9.3.10 (Ulm [491]). *Let A be a countable reduced abelian p-group. Then the isomorphism types of its Ulm factors completely determine the isomorphism type of A.*

Recall that for every α, $A_\alpha = A^{(\alpha)}/A^{(\alpha+1)}$ satisfies the premises of Theorem 9.3.8. Thus, Ulm's Theorem says that a (countable, reduced) abelian p-group A is fully described by its *Ulm invariants*, which is the sequence

$$\{\chi_{A_\alpha} : \alpha \leqslant UT(A)\},$$

where $UT(A)$ is the Ulm type of A and $A_\alpha = A^{(\alpha)}/A^{(\alpha+1)}$ are its Ulm factors. If we do not want to restrict ourselves to reduced groups, then we can also add the number of quasi-cyclic summands of the divisible part to these invariants, as it was done in Theorem 9.3.9.

Any sequence of the form

$$\{\chi_\alpha : \alpha \leqslant \gamma\},$$

where perhaps only χ_γ is bounded, describes a group of bounded exponent, and can be an invariant of some reduced abelian p-group [285]. We will later prove a rather strong effective version of this fact, but only for groups of finite Ulm type.

A tree-representation of a group

In both Example 9.3.2 and Example 9.3.5, we used only relations of the form $px_i = x_j$ and $px_k = 0$ upon the free (abelian) group generated by the x_i. In each of these examples, the respective representation can be visualised as a tree in which 0 is the root and x_i is a successor of x_j if $px_i = x_j$. So the quasi-cyclic group in Example 9.3.2 is just an infinite isolated path, and the group in Example 9.3.5 has one immediate successor of 0 that has tree-rank ω. It is known that any countable abelian p-group can be represented this way, as we will see below. Thus, while groups of Ulm type 1 are very similar to equivalence structures, groups of higher Ulm type resemble trees.

Definition 9.3.11 (L. Rogers [455]). A *p-basic tree* is a set X together with a binary operation $p^n \cdot x$ of the sort $\{p^n \mid 0 < n < \omega\} \times X \to X$ such that:

1. There is a unique element $0 \in X$ for which $p \cdot 0 = 0$,
2. $p^k \cdot (p^m \cdot g) = p^{k+m} \cdot g$, for all $g \in X$ and $k, m \in \mathbb{N}$, and
3. For every element $x \in X$, there is a natural number n with $p^n \cdot x = 0$.

If a prime p is fixed, then we think of a p-basic tree as a rooted tree with 0 being the root. Given a p-basic tree X, one obtains a p-group $G(X)$ as follows. The set $X\backslash\{0\}$ is treated as the set of generators for $G(X)$, and we add $px = y$ into the collection of relations if $p \cdot x = y$ in X. Each element of the group $G(X)$ can be uniquely expressed as $\sum_{x \in X} m_x x$, where $m_x \in \{0, 1, \ldots, p-1\}$ (with almost all $m_x = 0$). We will usually deal with combinatorial trees which are subsets of $\omega^{<\omega}$, and each such tree can be interpreted as a p-basic tree. Note that the root must always be in the tree, for every group must contain at least the neutral element 0.

Theorem 9.3.12 (L. Rogers [455]). *Every countable abelian p-group is generated by some p-basic tree.*

Non-isomorphic trees can produce isomorphic p-groups, as we now explain. Suppose that T is a p-basic tree viewed as a rooted tree. We call a finite chain of nodes *simple* if it is isolated, i.e., every node along the chain has at most one successor. Consider the following procedure:

"Take a simple chain extending $v \in T$, detach it, and
attach this chain to the root of T."

The procedure is called *stripping*. If the tree rank of v does not change after stripping, then the stripped tree T_1 and the original tree T give rise to isomorphic p-groups: $G(T_1) \cong G(T)$. Note that $T_1 \not\cong T$ in general. This process can be iterated. The only restriction is that the tree ranks of nodes in the tree must be preserved under this transformation. We will take all these properties above for granted and refer to [455] for a detailed analysis.

Exercises

Exercise° 9.3.13. Prove Fact 9.3.1.

Exercise° 9.3.14. Describe the Ulm invariants of the group defined in Example 9.3.5. Draw the diagram of its Ulm tree that is obtained naturally from its definition. Give an example of a non-isomorphic tree that determines the same group. Produce an isomorphism of groups that witnesses it.

Exercise° 9.3.15. Let χ be the characteristic describing an equivalence structure having infinitely many classes of each fixed finite size. Produce a p-basic tree of the reduced abelian p-group A of Ulm type 2 whose Ulm invariants are $\chi_{A_0} = \chi_{A_1} = \chi$.

9.3.2 Computable abelian p-groups of Ulm type 1

Every (countable) abelian p-group of Ulm type 1 splits into a direct sum of cyclic and quasi-cyclic (Prüfer) summands $\mathbb{Z}_{p^\infty}$, by Theorem 9.3.8. Khisamiev [287] was the first to describe computably presentable abelian p-groups that split into a direct sum of cyclic groups. In this section, instead of giving a brute-force proof of this result, we reduce the problem of computable presentability of such a group to the problem of computable presentability of the equivalence structure that strongly resembles this group. The uniform procedure described in the proof of this result, Theorem 9.3.17, will be very useful in later subsections.

The transformation $G \mapsto E_G$

Recall that for an abelian p-group of Ulm type 1, its characteristic is defined to be

$$\chi_A = \{\langle n, k\rangle : A \text{ has at least } k \text{ direct summands isomorphic to } \mathbb{Z}_{p^n}\}.$$

Also, recall that $D(A)$ denotes the maximal divisible subgroup of A. If $D(A) \neq \{0\}$ then $D(A)$ further splits into a direct sum of Prüfer subgroups $\mathbb{Z}_{p^\infty}$. The number of the $\mathbb{Z}_{p^\infty}$-summands in $D(A)$ is equal to the $\mathbb{Z}_p$-rank of the *socle* of $D(A)$, which is the $\mathbb{Z}_p$-vector space $D(A)[p]$ generated by the elements of $D(A)$ whose order is equal to p:

$$rk\, D(A) = rk_{\mathbb{Z}_p} D(A)[p] = rk_{\mathbb{Z}_p}\{a \in D(A) : pa = 0\}.$$

These invariants are, of course, very similar to the respective invariants of an equivalence structure; see Definition 9.1.1.

Definition 9.3.16. An abelian p-group G of Ulm type 1, define E_G to be the equivalence structure such that $\chi_{E_G} = \chi_G$ and $rk\, D(E_G) = rk\, D(G)$. (In E_G, each summand of the form $\mathbb{Z}_{p^\kappa}$ is replaced with a class of size $\kappa \in \mathbb{N} \cup \{\omega\}$.)

Theorem 9.3.17 (Melnikov and Ng [377]). *An abelian p-group G of Ulm type 1 is computably presentable iff the equivalence structure E_G is computably presentable. Furthermore, this correspondence is uniform.*

Proof. We first clarify what we mean when we say that the correspondence is uniform. Suppose

$$G = \bigoplus_{i \in I} G_i,$$

where for each i the summand G_i is either a cyclic p-group or $\mathbb{Z}_{p^\infty}$. (We call such a decomposition *full.*) The value λ such that $\mathbb{Z}_\lambda \cong G_i$ is either a natural number n or the symbol ∞, and it will be denoted by $\#G_i$.

The definition of $G \mapsto E_G$. In the notation above, define E_G to be the equivalence structure in which the i-th equivalence class E_i has size exactly $\#G_i$. (We write $\#E_i$ to denote the size of E_i.)

By Theorem 9.3.9, the isomorphism type of E_G does not depend on the given full decomposition of G. We can pass from an equivalence structure to a group using the following transformation.

The definition of $E \mapsto G_E$. Given an equivalence structure $E = \sum_{i \in I} E_i$ (meaning that E_i is the i-th class of E), define

$$G_E = \bigoplus_{i \in I} G_i,$$

where G_i is either cyclic or quasi-cyclic and $\#G_i = \#E_i$ for each $i \in I$.

It follows that $G_{E_A} \cong A$ and $E_{G_U} \cong U$ for any equivalence structure U and any abelian p-group A of Ulm type 1.

Clearly, the transformation $E \mapsto G_E$ is *uniformly* computable, i.e., is witnessed by a computable functional whose inputs do not have to be computable. However, it is not immediately clear whether $G \mapsto E_G$ should also be uniformly computable.

Proposition 9.3.18 (Melnikov and Ng [377]). *The transformation $G \mapsto E_G$ defined above is uniformly effective. Furthermore, regardless of the Ulm type of the input abelian p-group G, the output of the uniform procedure is always an equivalence structure.*

Proof. The definition of the Turing functional representing $G \mapsto E_G$ is as follows. We work computably relative to the open diagram of G. Clearly, $\mathbb{Z}_p$-independence is decidable in the socle $G[p]$ of G, relative to the diagram. (As before, $G[p] = \{g \in G : pg = 0\}$.) First, initiate a uniformly effective enumeration of any basis $x_0, x_1, x_2, \ldots$ of the socle $G[p]$ of G. For each i such that x_i has been found, define

$$s_i = \sup_{m_0, \ldots, m_{i-1} \in \mathbb{Z}_p} h_p(x_i - \sum_{j=0}^{i-1} m_j x_j),$$

which can be G-effectively approximated from below, allowing the limit to be ∞. (If x_i has not yet been found or does not exist, then we can either set $s_i = -1$ or keep it undefined.) Initiate the enumeration of an equivalence structure U in which $\#U_i = s_i + 1$. Note that we never refer to the Ulm type of the input group.

We now check that the procedure described above satisfies the desired properties. Clearly, it is uniformly effective. Furthermore, regardless of the Ulm type of G, the function $i \mapsto s_i$ is monotonic, and thus U is well-defined. We claim that if the Ulm type of G is 1, then $U \cong E_G$. For this purpose, we define a full decomposition of G induced by the definition of s_i and the choice of the basis $x_0, x_1, \ldots$ of $G[p]$.

Evidently, $s_0 = h_p(x_0)$. Fix a (maximal) chain of p-divisions below x_0 that witnesses $h_p(x_0)$, and let C_0 be the subgroup of G generated by this chain. Then C_0 is either a pure cyclic or a quasi-cyclic subgroup of G. Since C_0 is either pure cyclic or divisible, it detaches as a direct summand of G,

$$G = C_0 \oplus A_1.$$

Fix the projection π_1 onto A_1. We claim that

$$h_p^{A_1}(\pi_1(x_1)) = h_p^{G/C_0}(x_1) = \sup_{m_0 \in \mathbb{Z}_p} h_p(x_1 - m_0 x_0) = s_1.$$

Fix a full decomposition of A_1, which clearly exists since the Ulm type of A_1 is 1. Then

$$x_1 = n_0 x_0 + \sum n_i y_i,$$

where the y_i come from distinct summands in the induced full decomposition of the socle of A_1. Note that the p-height of x_1 in G/C_0, $h_p^{G/C_0}(x_1)$, is equal to $h_p(\sum n_i y_i)$. Also, because the sum is direct,

$$h_p(m x_0 + \sum n_i y_i) \leqslant h_p(\sum n_i y_i)$$

for any m. Since in the definition of s_1 we take the supremum over all such m (including $m = n_0$), it follows that $h_p^{A_1}(\pi_1(x_1)) = s_1$, as claimed.

Fix a chain of p-divisions in A_1 that witnesses $h_p^{A_1}(\pi_1(x_1)) = s_1$, and let C_1 be the subgroup of A_1 generated by this chain. Similarly to C_0, it must be the case that C_1 detaches in A_1.

Suppose we have defined $C_0, \ldots, C_n$ and A_{n+1}, where the C_i are either cyclic or quasi-cyclic, and

$$G = \left(\bigoplus_{i=0}^{n} C_i \right) \oplus A_{n+1}.$$

As above, we can choose C_{n+1} that witnesses

$$h_p^{A_{n+1}}(\pi_{n+1}(x_{n+1})) = h_p^{G/\sum_i C_i}(x_{n+1}) = \sup_{m_0, \ldots, m_n \in \mathbb{Z}_p} h_p(x_{n+1} - \sum_{j=0}^{n} m_j x_j) = s_{n+1},$$

the proof of which is almost identical to the case $n = 1$ (here π_{n+1} is the projection onto A_{n+1}). As before, we get that C_{n+1} detaches within A_{n+1} to form A_{n+2}.

This way, we produce a subgroup B of G that satisfies the properties:

i. $B = \bigoplus_i C_i$, where the C_i are cyclic or quasi-cyclic subgroups,

ii. the socle of B is equal to the socle of G.

Property i. follows from the definition of $C_0, C_1, \ldots$. To see why ii. holds, recall that $x_0, x_1, \ldots$ is a basis of G_p, and

$$Span_{\mathbb{Z}_p}\{x_0,\ x_1 - n_{0,0}x_0,\ x_2 - n_{1,0}x_0 - n_{1,1}x_1,\ \ldots\} = Span_{\mathbb{Z}_p}\{x_0, x_1, \ldots\}$$

for any choice of $n_{i,j} \in \mathbb{Z}_p$. The generators of the socles of C_i are of the form $x_i - \sum_{j<i} n_{i,j}x_j$, thus ii. holds.

Recall that by our assumption G is a direct sum of cyclic and quasi-cyclic p-groups. We claim that i. and ii. together imply that $B = G$. Aiming for a contradiction, assume $\alpha \in G \backslash B$. Suppose also that $p^n\alpha \in B$ while $p^{n-1}\alpha \notin B$. Note such an n exists since $B[p] = G[p]$ (by ii. above). Without loss of generality, we can assume that $n = 1$. We arrive at

$$p\alpha = \sum_{i \leqslant k} d_i,$$

where $d_i \in C_i$ for each $i = 0, \ldots, k$. We may assume that each $d_i \neq 0$, otherwise we re-arrange the indexing of the C_i. This assumption is used throughout the proof of the claim below.

Claim 9.3.19. *In the notation as above, for each $i \leqslant k$ there exists $d'_i \in C_i$ such that $pd'_i = d_i$.*

Proof of Claim. Suppose such a d'_k does not exist (the case when $i = k$). But then the chain that generates C_k is not maximal in $G/(\sum_{j<k} C_j)$ as witnessed by the projection of a suitably chosen $\mathbb{Z}_p$-multiple of the coset of α. Thus, $d_k = pd'_k$ for some $d'_k \in C_k$.

To see why d'_{k-1} exists, consider $p(\alpha - d'_k) = p\alpha - d_k \in \bigoplus_{j<k} C_j$. As we had with α and d_k above, the C_{k-1}-projection of the element $(\alpha - d'_k)$ will witness the failure of maximality (in $G/\sum_{j<k-1} C_j$) of the chain used to define C_{k-1}, unless d'_{k-1} exists. We proceed in this manner to find $d'_{k-2}, \ldots, d'_0$. □

We conclude that $p\alpha = \sum_{i \leqslant k} pd'_i$. Then $p(\alpha - \sum_{i \leqslant k} d'_i) = 0$ and thus

$$\alpha - \sum_{i \leqslant k} d'_i \in G[p] = B[p] \subseteq B.$$

Together with $\sum_{i \leqslant k} d'_i \in B$ this gives $\alpha \in B$, contradicting the choice of α.

Finally, since all full decompositions of G are isomorphic (as decompositions), we have that $U \cong E_G$. □

The proof of Theorem 9.3.17 is finished. □

Consequences of Theorem 9.3.17

Recall that we write $R(A)$ for the reduced part of A consisting only of finite cyclic summands. Let $\#A = \#R(A)$ be the set of k such that A has at least one direct summand of the form $\mathbb{Z}_{p^k}$. As an immediate corollary of Theorem 9.3.17 and Theorem 9.1.4, we obtain:

Theorem 9.3.20 (Khisamiev [287]). *The following cases give a complete characterisation of all computably presented abelian p-groups of Ulm type 1:*

1. *If A has infinitely many quasi-cyclic summands, then A is computably presented iff* χ_A *is* Σ^0_2.

2. *If A has at most finitely many quasi-cyclic summands, then A has a computable presentation if and only if* $\chi_{R(A)}$ *is* Σ^0_2 *and* $\#A$ *is limitwise monotonic.*

Of course, Khisamiev proved Theorem 9.3.20 more directly, avoiding Proposition 9.1.8; see Exercise 9.3.39.

A computable abelian p-group of Ulm type 1 does not have to fully computably split into cyclic and quasi-cyclic summands. However, it follows from Theorem 9.3.17 that we can always uniformly produce a nice copy which does.

Corollary 9.3.21. *Given a computable p-group A of Ulm type 1, we can* uniformly *pass to a computable presentation H of A that admits an effective full decomposition into cyclic and quasi-cyclic subgroups.*

The next corollary contrasts greatly with the positive result about torsion-free abelian groups established in Part 1 (see Theorem A).

Corollary 9.3.22. *There exists a c.e. presented reduced abelian p-group that is not isomorphic to any computable group.*

Proof. In Corollary 9.1.10 we established that there is a c.e. presented equivalence structure without infinite classes that has no computable presentation. The transformation $E \mapsto G_E$ clearly turns c.e. presented structures having only finitely many classes into c.e. presented groups. To see why, whenever a class E_i of E decreases in size, declare all elements in the respective cyclic summand of G_E equal to zero, and introduce a new summand that will now copy E_i (in the sense that its order will be p^n, where n is the current size of E_i). See also Exercise 9.3.40. Apply Theorem 9.3.17 to G_E. □

Indeed, assuming Exercise 9.1.27 (which says that there are low Δ^0_2 sets that are not l.m.), we obtain:

Corollary 9.3.23. *There exists a low abelian p-group not isomorphic to any computable group, but isomorphic to a c.e. presented group.*

Another corollary concerning direct decompositions is as follows.

Corollary 9.3.24. *There exists a computable abelian p-group whose reduced part has no computable presentation.*

Proof. This follows from Theorem 9.3.20 and the existence of Σ^0_2 sets that are not l.m. (Lemma 9.1.20) □

Recall that one of the two main results of this chapter is the existence of a Friedberg enumeration of all abelian p-groups of Ulm type $\leqslant n$. In Theorem 9.2.1, we proved that such an enumeration exists for computable equivalence structures. Thus, Theorem 9.2.1 also implies the case when $n = 1$ for groups.

Corollary 9.3.25. *There is a Friedberg enumeration of all computable abelian p-groups of Ulm type* 1.

It follows from Theorem 9.3.17 and results in [330] and [213] that, in contrast, no such enumeration exists for all computable *reduced* abelian p-groups of Ulm type 1; we leave this to Exercise 9.3.41.

Computably categorical abelian p-groups*

It turns out that computably categorical abelian p-groups can be found only among groups of Ulm type 1, so we include their description here. The description of computably categorical abelian p-groups does not require any particularly deep methods. The theorem below is due to Goncharov [203] and Smith [474]. (The result is stated incorrectly in Smith [474].) Since this result will not be particularly important to us, we provide only an extended sketch that omits some of the standard details related to the priority method but explains the algebra.

Theorem 9.3.26. *A computable abelian p-group A is computably categorical iff A splits into a direct sum of cyclic and quasi-cyclic groups*

$$A = \bigoplus_i A_i,$$

where all but finitely many elementary summands A_i are isomorphic to some fixed $\mathbb{Z}_{p^\lambda}$, $\lambda \in \mathbb{N} \cup \{\infty\}$. (Note the direct sum can be finite.)

Proof. We first check that an abelian p-group of the form $A = \bigoplus_i A_i$, where almost all A_i are isomorphic to some fixed $\mathbb{Z}_{p^\lambda}$, is computably categorical. Unlike the analogous case for equivalence structures, it requires more effort to verify this claim. We split the verification into four lemmas that cover all possible cases. One important case is as follows.

Lemma 9.3.27. *Every divisible abelian p-group is computably categorical.*

Proof. Every divisible p-group splits into a sum of quasi-cyclic summands $\mathbb{Z}_{p^\infty}$. Consequently, there exists a "natural" copy of any such group where these summands form computable subgroups. Let G be such a "natural" copy. Given some other computable copy H, to show that it is computably isomorphic to G, it is sufficient to find a complete computable direct decomposition of H into quasi-cyclic summands. Once such a decomposition is obtained, we simply match the generators of the quasi-cyclic summands in H and G and extend this map to arbitrary elements to obtain a computable isomorphism.

Initially, pick any non-zero element $h_0 \in H$ in the socle $H[p]$ (i.e., $ph_0 = 0$). Initiate the enumeration of a sequence $(h_{0,i})_{i \in \mathbb{N}}$ with the properties:

$$h_{0,0} = h_0,\ ph_{0,1} = h_{0,0}, \ldots,\ ph_{0,i+1} = h_{0,i}, \ldots.$$

Such elements must exist and therefore will eventually be found; however, their choice is not unique. (We always pick the first found ones.) Let

$$H_0 = \langle h_{0,i} : i \in \mathbb{N} \rangle \cong \mathbb{Z}_{p^\infty}.$$

We claim that H_0 is a computable subgroup of H. Given any $h \in H$, calculate its order; suppose it is p^k. Then $h \in H_0$ iff $h \in \langle h_{0,k+1} \rangle$.

To define H_1, choose any $h_1 \in H[p]$ having the smallest index so that $h_1 \notin H_0$ (or, equivalently, such that $\langle h_1 \rangle \cap H_0[p] = \{0\}$). Define

$$h_{1,0} = h_1,\ ph_{1,1} = h_{1,0},\ \ldots,\ ph_{1,i+1} = h_{1,i}, \ldots,$$

and set $H_1 = \langle h_{1,i} : i \in \mathbb{N} \rangle$. We have $H_1[p] = \langle h_1 \rangle$ and, thus, $H_1[p] \cap H_0[p] = \{0\}$.

We claim that $H_1 \cap H_2 = \{0\}$. For suppose $h \in H_1 \cap H_2$ is non-zero. Then for some k, $p^k h = 0$ and $p^{k-1}h \neq 0$. But $p^{k-1}h \in H_0[p] \cap H_1[p] = \{0\}$, a contradiction. The subgroup H_1 is computable as well.

Then pick h_2 of order p outside $H_1 \oplus H_2$ to define H_3. As before, we obtain $H_3 \cap (H_1 \oplus H_2) = \{0\}$. We iterate this to define a subgroup of H of the form

$$D = \bigoplus_{i \geqslant 1} H_i,$$

where $H_i \cong \mathbb{Z}_{p^\infty}$ are uniformly computable. (We also include the case when this direct sum is finite, which corresponds to the case when the socle of H is finite and, thus, for some i we cannot find h_i.) Additionally, $D[p] = H[p]$. We claim that this implies $D = H$.

If there is some $d \in H \backslash D$, then it cannot have order p. For some $k > 0$, $p^k d$ has order p. Thus, for some $1 \leqslant m \leqslant k$, we must have that $p^m d \in D$ and $p^m d \neq 0$; let m be the smallest such. By divisibility of D, there exists a $u \in D$ such that $pu = p^m d$. Then $u - p^{m-1}d$ has order p, and thus $u - p^{m-1}d \in D$, which implies $p^{m-1}d \in D$, contradicting the minimality of m. □

Another important case is covered by the following lemma.

Lemma 9.3.28. *Suppose A is a direct sum of countably many copies of the cyclic group $\mathbb{Z}_{p^k}$, for some fixed k. Then A is computably categorical.*

Proof. This is the same as the previous lemma, but this time for every i, we define a finite sequence

$$h_{i,0} = h_i,\ ph_{i,1} = h_{i,0},\ \ldots,\ ph_{i,k-1} = h_{i,k-2},$$

where $ph_i = 0$ is taken outside of $\langle h_j : j < i \rangle$. The verification is essentially the same as in the proof of Lemma 9.3.27, including the very final step. There, the existence of u with $pu = p^m d$ follows from noting that the order of $p^m d$ has to be $< p^k$. □

Another case is as follows.

Lemma 9.3.29. *Suppose A splits into cyclic and quasi-cyclic p-groups, almost all of which are quasi-cyclic. (This includes the case when there are only finitely many summands in total.) Then A is computably categorical.*

Proof. As before, every such group has a "natural" presentation that computably splits into cyclic and quasi-cyclic summands. Given some other copy, it is sufficient to split it computably in a way that the isomorphism types of the summands can be decided.

If A is finite, then there is nothing to prove. If $A = F \oplus D$, where F is finite and D is divisible, then fix F non-uniformly. Let m be so large that $p^m F = 0$. Then repeat the proof of Lemma 9.3.27 by choosing the next element h_i to always have p-height at least m. □

The remaining case is covered by the next lemma.

Lemma 9.3.30. *Suppose A splits into cyclic and quasi-cyclic p-groups, almost all of which are cyclic and isomorphic to some $\mathbb{Z}_{p^k}$, for a fixed k. Then A is computably categorical.*

Proof. By the previous lemma, without loss of generality A splits into an *infinite* sum of cyclic and quasi-cyclic p-groups, all but finitely many of which are $\mathbb{Z}_{p^k}$. We non-uniformly fix the finitely many cyclic summands of exceptional sizes; let the subgroup generated by these summands be F.

If there are no quasi-cyclic summands, then proceed as in the proof of Lemma 9.3.28 by additionally requiring that the next h_i is outside the socle of F.

If there are several quasi-cyclic summands, then without loss of generality (after fixing the finitely many exceptional finite summands), assume that any element of order p^m (or larger) has to come from the finitely many exceptional quasi-cyclic summands (if there are any). Fix finitely many $\mathbb{Z}_p$-independent elements $d_{1,0}, \dots, d_{n,0}$ of the socle of A that have their p-height $\geqslant m$; where $n \geqslant 1$ is the largest such that an n-tuple exists.

These elements must come from different disjoint copies of $\mathbb{Z}_p$ (repeat the argument at the end of the proof of Lemma 9.3.27), and they must generate the divisible part of A. We then use these elements to define infinite chains of generators that together span the divisible part of the group, and additionally induce a direct decomposition of it into quasi-cyclic summands. For that, also fix $d_{1,m-1}, \dots, d_{n,m-1}$ such that $p^{m-1} d_{i,m-1} = d_{i,0}$, for all $1 \leqslant i \leqslant n$.

Repeat the process described in Lemma 9.3.27, but this time initiate the enumeration of $(d_{i,j})$ beginning with $j = m-1$, for each i. This results in a computable direct decomposition of the divisible part. Its socle is $\mathbb{Z}_p$-independent of the socles of the other cyclic summands that we define when we proceed as in the proof of Lemma 9.3.28. Thus, we obtain a computable direct decomposition of the entire group into cyclic and quasi-cyclic summands, in which, additionally, we know the isomorphism type of each summand. □

The lemmas above together give one implication of the theorem. The other implication is also split into several lemmas.

Lemma 9.3.31. *Suppose the orders of the cyclic summands in A are bounded, and suppose that for some $\lambda < \gamma$ in $\omega \cup \{\omega\}$, A has infinitely many summands of the form $\mathbb{Z}_{p^\lambda}$ and infinitely many summands of the form $\mathbb{Z}_{p^\gamma}$. Then A is not computably categorical.*

Proof. In this case, since the Ulm type is 1, the group splits into cyclic and quasi-cyclic summands, with the orders of all cyclic summands being uniformly bounded. By Theorem 9.3.17, we can further assume that it computably splits.

Proceed as in the proof of Theorem 9.1.11 (the case when there are infinitely many classes of size k_0 and infinitely many classes of size k_1, where $k_0 < k_1$, allowing the possibility $k_1 = \omega$). The ways in which this proof differs from the proof for equivalence structures are summarised below.

We use elements from the socles of the elementary summands $\mathbb{Z}_{p^\lambda}$ as witnesses (not just arbitrary elements of the summands). Also, a potential isomorphism φ_e can map a witness to a linear combination of the elements generating the socles of the summands in the other copy:

$$\varphi_e(w) = \sum_{i \leqslant k} m_i a_i, \qquad m_i \in \mathbb{Z}_p.$$

If $\varphi_e(w)$ is not of this form, then its order is not p, and thus we do not need to do anything to diagonalise against φ_e. If the p-height of $h_p(w)$ is currently declared unequal to $\min_{i \leqslant k} h_p(a_i)$, then we also do not need to act. Otherwise, if the p-heights look equal, we extend the $\mathbb{Z}_{p^\lambda}$-summand containing w to a $\mathbb{Z}_{p^\gamma}$-summand, thus increasing the p-height of the witness w.

As long as we keep infinitely many $\mathbb{Z}_{p^\lambda}$-summands, we will still end up with an isomorphic group. Thus, we simply never pick our witnesses from countably many such summands (this is also similar to the proof of Theorem 9.1.11). □

Lemma 9.3.32. *Suppose the reduced part of A contains elements of arbitrarily large orders. Then A is not computably categorical.*

Proof. Recall that every pure cyclic subgroup of A detaches as a direct summand in A. Thus, our assumption about A is equivalent to saying that A has arbitrarily large cyclic summands. Also, note that

$$\{m \mid A \text{ has a cyclic summand of order } p^m\}$$

is a Σ^0_2-set.

To see why, pick any element $a \in A$ of p-height 0. It generates a pure cyclic subgroup which detaches in A. Conversely, any cyclic summand has an element of p-height 0. Using $\mathbf{0}'$, we can decide whether the p-height of any element is 0, and thus $\mathbf{0}'$ can list all such elements. Further, for any given $a \in A$, $h_p(a)$ can be computably approximated from below.

If $A = \bigcup_s A_s$ is a computable group, then without loss of generality, we can assume that A_s is also given by its complete decomposition into cyclic p-groups (which of course do not have to be summands of the entire A). Furthermore, if a has p-height 0 in A_s, then $\langle a \rangle$ detaches in A_s.

Fact 9.3.33. *Let B be any finite subgroup of A. Then for every $m > 0$, there exists a cyclic subgroup C of A of order at least p^m such that*

$$A = C \oplus U,$$

where $B \subseteq U$.

Proof. We use the technique of p-basic trees explained in §9.3.1. Fix an arbitrary p-basic tree of A. Since A has arbitrarily large cyclic summands, without loss of generality, the root of the tree has arbitrarily long isolated finite chains attached to it. (To see why, use stripping. If there are no vertices of rank ω, and thus A' is divisible, then appeal to Prüfer's Theorem. Otherwise, if there is a node or tree rank ω, then strip infinitely many arbitrarily long finite chains below it and reattach them to the root of the tree.)

Since B is finite, every element is a linear combination of at most finitely many generators which label the vertices of the p-basic tree. There is a finite isolated chain of length at most m attached to the root that does not contain any of these generators. This chain generates C, and the rest of the vertices in the tree generate U; it is clear that $B \subseteq U$, $C \cap U = \{0\}$, and $C + U = A$. □

The idea is to use the fact above to imitate the strategy implemented in the harder cases in Theorem 9.1.11. Indeed, while A may not quite resemble an equivalence structure in the limit, it certainly resembles it at every stage.

We are given a computable $A = \bigcup_s A_s$, and we are building a computable $\bigcup_s B_s = B \cong A$ and meeting the requirements:

$$R_e : \varphi_e : A \to B \text{ is not an isomorphism,}$$

for every e. We also construct a Δ_2^0 map $\psi : B \to A$ so that at every stage s, ψ_s is an isomorphic embedding of B_s into A, and for every $i \in B$,

$$P_i : \psi(i) = \lim_s \psi_s(i) \text{ exists,}$$

and for each $j \in A$,

$$S_j : j \in \text{range}(\psi).$$

Most of the time ψ will simply copy A into B, in the following sense. If $\psi_s(B_s) \subseteq A$ is defined and $a \in A_{s+1} \backslash \psi_s(B_s)$, we introduce a fresh $b \in B_{s+1}$, set $\psi_{s+1}(b) = a$, and define the operation of $B_{s+1} = \langle B_s, b \rangle$ by retraction to copy the finite group $\langle \psi_s(B_s), a \rangle$; we also extend ψ to all elements of B_{s+1} naturally. This definition may have to be changed later due to an R_e action, as explained below. Whatever this potential change might be, to meet S_j, with priority of S_j we do not allow this change to involve $b \in B$ with $\psi(b) = j$, if such an element exists.

The strategy of meeting R_e is as follows. The strategy will have to *preserve* a finite subgroup $H = H_{e,s} \subseteq B_s$ that is controlled by higher priority requirements.

1. Monitor φ_e and preserve (with the priority of R_e) more and more of $\text{dom}(\varphi_e)$ from change if φ_e looks more like an isomorphism. It is Π_2^0 to say that φ_e is total and is a surjective isomorphism. Thus, preserve one more element if this predicate fires again. In this case, we say that the stage is e-expansionary.

2. Search for $w, v \in A$ and $u = \varphi_e(w) \in B$ with the properties:

   (a) $h_p(u) = h_p(w) = 0$ in A_s and B_s, respectively.

   (b) For some $U \subseteq H$, $B_s = U \oplus \langle w \rangle$.

   (c) $\psi_s(B_s) \cap \langle v \rangle = \{0\}$.

   (d) $\text{order}(v) > \text{order}(w)$.

   Additionally, we require that the index of $\langle u, v, w \rangle$ is the smallest possible.

3. Redefine ψ on B_s to map $\langle u \rangle$ to a subgroup of $\langle v \rangle$ and keep ψ unchanged on $U \supseteq H$. Protect the new definition of ψ with priority R_e.

If φ_e insists on being an isomorphism, then such u, v, w and U will eventually be found, by Fact 9.3.33 applied to B and A. Further, we will eventually find w whose p-height is 0 in A, not just in A_s.

Remark 9.3.34. Locating such an element w may require a few iterations. Also, as usual in such arguments, Fact 9.3.33 guarantees that such elements and decompositions must exist. However, we may find some decomposition and witnesses that work for the finite groups built so far; this decomposition does not have to extend to a decomposition of the entire group.

In this case, the change of ψ introduced by R_e will ensure that φ_e cannot possibly be an isomorphism, because any isomorphism must preserve p-height, and we made the p-height of the witness > 0 by embedding $\varphi_e(w) = u$ into a larger cyclic group that copies $\langle v \rangle$ under the new definition of ψ. We will then protect this definition with priority R_e. On the other hand, if no such triple is eventually found, then φ_e will eventually stop looking like an isomorphism, and thus the restraint imposed by R_e on ψ will be finite. In both cases, this restraint will contribute to what we denoted by H above, but for the lower priority requirements. Also, the S-requirements of higher priority will contribute to H, thus preventing ψ from changing even if R_e collectively do not have enough "expansionary stages" (however, introducing the S-requirements was indeed an overkill).

The rest of the proof is a standard finite injury argument of unbounded type (see, e.g., §3.1.6). Eventually, every restraint is finite, each R_e is met, and $\lim_s \psi_s$ exists and is an isomorphism of B onto A. □

Theorem 9.3.26 follows from the lemmas since they cover all possible cases. □

The transformations $E \mapsto G_E$ and $G \mapsto E_G$ viewed as $\leqslant_{EFF}$-reductions*

Let $\mathcal{P}_1$ be the class of abelian p-groups of Ulm type 1, and $\mathcal{E}$ be the class of equivalence structures. In the notation of Definition 8.2.1, we have that

$$\mathcal{P}_1 \equiv_{EFF} \mathcal{E},$$

which is perhaps not too surprising, however, it did require some effort to verify this fact.

It should be clear that degree spectra are preserved (Definition 8.2.3). Furthermore, Theorem 9.3.26 strongly resembles the similar description of computably categorical equivalence relations given in Theorem 9.1.11. However, the transformations are not as well-behaved as one might hope for.

Corollary 9.3.35. *The transformations $E \mapsto G_E$ and $G \mapsto E_G$ between equivalence structures and abelian p-groups of Ulm type 1 preserve degree spectra and computable dimension. However, $G \mapsto E_G$ does* not *preserve Δ^0_2-categoricity.*

The proof is left to Exercises 10.1.103. In fact, it follows from Exercise 10.1.102 that G can have exactly one summand isomorphic to $\mathbb{Z}_{p^\infty}$ and still not be Δ^0_2-categorical. In contrast, any equivalence structure with exactly one infinite class is clearly Δ^0_2-categorical. We see that $G \mapsto E_G$ cannot be modified to additionally transform isomorphisms to isomorphisms.

In Exercise 10.1.102 we will see that a certain Type I analogue of Δ^0_2-categoricity, weak uniform Δ^0_2-categoricity, is preserved under $G \leftrightarrow E_G$. Various notions of categoricity will be studied in detail in the next chapter.

Computably categorical torsion abelian groups*

According to Nazif Khisamiev, the problem below can be traced back to Mal'cev[1].

[1] Personal communication with A. Melnikov. This problem is often incorrectly attributed to Goncharov; see, e.g., Problem 12.17 in the "Kourovka notebook" (Issue 12, year 1992) published by Khukhro and Mazurov [296]. As of

Problem 9.3.36. *Describe computably categorical torsion abelian groups.*

Recall that every torsion abelian group A splits into the direct sum of its maximal p-subgroups $T_p(A)$; this is also clearly effective. It should be clear that in a computably categorical torsion A, each maximal p-component $T_p(A)$ is also necessarily computably categorical (Exercise 9.3.43). However, the converse fails; see Exercise 9.3.44. Using a rather involved combinatorial argument, Melnikov and Ng proved:

Theorem 9.3.37 (Melnikov and Ng [377])**.** *The index set of computably categorical torsion abelian groups is Π^0_4-complete.*

Proof. Omitted. Notably, the proof completely reduces Mal'cev's question to a technical problem about equivalence structures. □

Of course, whether Theorem 9.3.37 provides a *satisfactory* classification remains open to debate. (It certainly does not look like one.) However, at the very end of Section 10.1.1 (see Corollary 10.1.20), we shall argue that Theorem 9.3.37 could potentially be the best description of c.c. torsion abelian groups one can possibly hope for.

Exercises

Exercise° 9.3.38. Suppose χ is a characteristic describing an equivalence structure that contains only finite classes, and so that it contains arbitrarily large classes. Let T be any (p-basic) tree. Describe a (non-effective) procedure that builds a p-basic tree Γ so that $G(\Gamma)_0$ has characteristic χ and $G(\Gamma)' \cong G(T)$.

Exercise° 9.3.39. Prove Theorem 9.3.20 directly (not using equivalence relations).

Exercise° 9.3.40 (Khisamiev [289]; see also [370] for a sketch)**.** Suppose $A \cong \bigoplus_{k\in S} Z_{p^k}$. Then A is c.e. presentable if, and only if, S is Σ^0_2.

Exercise° 9.3.41. Prove that there is no Friedberg enumeration of all computable *reduced* abelian p-groups of Ulm type 1.

Exercise 9.3.42 (Goncharov [203])**.** Prove that every computable abelian p-group has either one or infinitely many computable presentations, up to computable isomorphism.

Exercise° 9.3.43. Suppose A is a torsion abelian group whose maximal p-subgroup is not c.c., for some p. Show that A is not c.c. either.

Exercise° 9.3.44 (Folklore; e.g. [377])**.** Give an example of a computable torsion abelian group that is not computably categorical, but so that all of its maximal p-subgroups are. (See Exercise 10.1.107 for a hint.)

2024, the "Kourovka notebook" is available online at https://kourovkanotebookorg.files.wordpress.com/2023/12/20tkt.pdf.

9.3.3 Groups of finite Ulm type > 1

By Theorem 9.3.20, computable abelian p-groups of Ulm type 1 have the same computability-theoretic and algebraic invariants as computable equivalence structures (Theorem 9.1.4). The case of arbitrary finite Ulm type is significantly more difficult. In this subsection, we prove Theorem 9.3, which we re-state below:

Theorem 9.3.45 (Ash, Knight, and Oates (unpublished); Khisamiev [290]). *Suppose that A is a (non-trivial) countable reduced p-group of Ulm type $n < \omega$, and let $A_0 = A/A', A_1, \dots, A_{n-1}$ be its Ulm factors. Then the following conditions are equivalent:*

1. *A has a computable copy.*
2. *A has a computable p-basic tree representing it.*
3. *(a) For every $i < n$, the character $\chi(A_i)$ is a Σ^0_{2i+2} set, and*

   (b) for every $i < n$, the set

$$\#A_i := \{m \colon (m, 1) \in \chi_{A_i}\}$$

   is $\mathbf{0}^{(2i)}$-l.m..

Proof. We prove the theorem by induction in the Ulm type n of the group.

Basic case ($n = 1$). Any group that computably splits into a direct sum of cyclic and quasi-cyclic summands can be represented by a computable p-basic tree; we leave this to Exercise 9.3.57. Thus, the basic case when $n = 1$ is given by Theorem 9.3.20 combined with Corollary 9.3.21.

Inductive step. The implication (2) → (1) is obvious. We first prove (1) → (3). We prove a slightly more general lemma that implies (1) → (3); it will be useful later.

Lemma 9.3.46. *Suppose A is an abelian p-group of Ulm type > 1 (which is not necessarily a reduced group). If A is computable then A' has a Δ^0_3-copy and A/A' has a computable copy.*

Proof. The subgroup A' consisting of all elements of G having infinite p-height is isolated by a Π^0_2-condition ("being infinitely p-divisible"), and therefore it has a $\mathbf{0}''$- computable presentation; we leave the details to Exercise 9.3.57.

This is almost the same as in the case when A is reduced. Since the Ulm type of A is at least 2, there must be an element $a \in A'$, $a \neq 0$, which is not divisible; equivalently, any p-basic tree of A' must have a non-trivial terminal node, for otherwise A' would be divisible and $A' = A''$, contradicting the assumption. This means that a has infinite p-height in A, but there is no x with the property $px = a$ which also has infinite p-height. Using a, define an l.m. function f, as follows. List all $x_1, x_2, \dots$ with the property $px_i = a$ and define $f(i) = h_p(x_i) + 1$, where $h_p(x_i)$ stands for the p-height of x_i.

We claim that the range of f is infinite and is contained in the collection of all n such that A/A' has a cyclic summand of order p^n. We verify this claim in the paragraph below.

It is clear that the range of f is infinite, for the p-height of a is infinite but it is not divisible. Since the heights of the x_i are unbounded, for each i there will be a j with $h_p(x_j) > h_p(x_i)$; this will imply $h_p(x_i - x_j) = h_p(x_i)$, because

$$h_p(x_i) = h_p(x_j + (x_i - x_j)) \geqslant \inf\{h_p(x_j), h_p(x_i - x_j)\}$$

and

$$h_p(x_i - x_j) \geqslant \inf\{h_p(x_i), h_p(x_j)\} = h_p(x_i).$$

Note that $p(x_i - x_j) = 0$, and for some c we have $p^{h_p(x_i - x_j)}c = (x_i - x_j)$. But this makes $\langle c \rangle$ a pure cyclic subgroup of A of order $h_p(x_i - x_j) = h_p(x_i)$ (that is, for each x in the subgroup its p-height in A is witnessed within the subgroup), and pure cyclic subgroups detach, so $A \cong B \oplus \langle c \rangle$. Since $(C \oplus D)' = (C' \oplus D')$ and $\langle c \rangle' = \langle c \rangle$, we have $A/A' \cong B' \oplus \langle c \rangle$, thus proving the claim.

Note that, essentially, we have just shown that if A has a cyclic summand of order p^n then A/A' also has a cyclic summand of order p^n. In fact, the converse implication is also true. To see why, suppose $\langle \alpha \rangle$ of order p^n detaches in A/A'. The coset α must contain an element a_0 such that $p^n a_0$ has infinite height, but $p^m a_0$ has finite height for each $m < n$. Also, if the p-height of a_0 in A was not zero then, for some $b \in A$, we would have $pb = a_0$, which would also hold modulo A'. So for some β we would have $p\beta = \alpha$, contradicting the choice of α. The same argument shows that the p-height of each $p^m a \in \langle a_0 \rangle$, $m < n$, is equal to the p-height of its coset in A/A' and is equal to m. Since the p-height of $x = p^n a_0$ is infinite, there exists some c with the property $p^n c = x$ and with $h_p(c) > 0$. Consider the element $y = a_0 - c$ and the cyclic subgroup $\langle y \rangle$ of A. Then $h_p(y) = 0$, for otherwise

$$h_p(a_0) = h_p(y + c) \geqslant \inf\{h_p(y), h_p(c)\} > 1$$

would contradict $h_p(a_0) = 0$. Similarly, for $m < n$, $h_p(p^m y) = m$; otherwise

$$h_p(p^m a_0) = h_p(p^m y + p^m c) \geqslant \inf\{h_p(p^m y), h_p(p^m c)\} > m$$

would contradict $h_p^A(p^m a_0) = h_p^{A/A'}(p^m \alpha) = m$. This shows that $\langle a_0 \rangle$ is pure in A and thus detaches as a direct summand of A.

So cyclic direct summands are the same in A and A/A'. This makes the characteristic

$$\chi_{A/A'} = \{\langle n, k \rangle : A/A' \text{ has at least } k \text{ cyclic summands of order } p^n\}$$

a Σ^0_2-set. Indeed, it is sufficient to search for $\mathbb{Z}_p$-independent $a_1, \ldots, a_m$ of order p such that, for each $i \leqslant m$, $h_p(a_i) = n$; the latter can be decided using $\mathbf{0}'$. Use Theorem 9.3.20 to produce a computable presentation of A/A'. □

By induction, Lemma 9.3.46 (combined with Theorem 9.3.20) gives (1) → (3).

We now prove (3) → (2). Observe that the proof of Lemma 9.3.46 illustrates that, if the Ulm type of A is > 1, then $A_1 = A/A'$ should satisfy the following property.

Definition 9.3.47. We say that an abelian p-group H is *proper* if it is reduced and have Ulm type 1 and is *unbounded*, i.e., contains elements of arbitrary large p-height.

Equivalently, H is proper iff $H' = 0$ and the sizes of the finite cyclic summands in its full direct decomposition are unbounded in size. The key technical step in the proof of (3) → (1) is the following proposition due to Ash, Knight, and Oates (unpublished).

Proposition 9.3.48. *There is a uniform procedure which, given a computable copy of a proper abelian p-group H and a Π^0_2 p-basic tree F, outputs a computable p-basic tree $T_H(F)$ with the properties $(T_H(F))' = F$ and $T_H(F)/(T_H(F))' \cong H$.*

We first explain how (3) → (2) of Theorem 9.3.45 follows from Proposition 9.3.48, and then we prove Proposition 9.3.48. By the inductive hypothesis, we have that A' is Δ^0_3-presented via a Δ^0_3-computable p-basic tree. Further, using a uniform procedure, we can turn any Δ^0_3-tree into a Π^0_2-subtree of $\omega^{<\omega}$, we leave this to Exercise 9.3.58. By Theorem 9.3.20 $H = A/A'$ has a computable presentation, which can be identified with the respective equivalence structure. Applying Proposition 9.3.48, we can produce a computable p-basic tree $T_H(F)$ representing A.

Proof of Proposition 9.3.48. We identify F with the group it represents. We also identify a proper abelian p-group H with the corresponding equivalence structure. Under this correspondence, an equivalence class of size n will represent a cyclic summand of order p^n. Recall that, by Theorem 9.3.18, this correspondence is also uniformly effective. Bearing in mind the uniform correspondence $H \leftrightarrow E_H$, we will abuse notation and write H for E_H. Since H is proper, the respective equivalence structure E_H will have only finite classes, but the sizes of these classes will be unbounded.

Fix a computable copy of $\omega^{<\omega}$ viewed as an infinitely branching tree with its root, the empty string e, located at its top. Identify F with a Π^0_2-subtree of $\omega^{<\omega}$ such that each $\sigma \in \omega^{<\omega}$ has infinitely many successors which do not belong to F; furthermore, we can assume that this set of successors of σ outside F has an infinite computable subset of nodes. This subset will be used to attach external chains whose sizes will be taken from H. These external chains will be called *auxiliary*. Each auxiliary chain will be associated with exactly one class/summand in $E_H \leftrightarrow H$ having size greater than or equal to the length of the auxiliary chain.

We also fix a computable predicate R such that $F = \{\sigma \mid \exists^\infty z\, R(\sigma, z)\}$. Whenever a new existential witness for z is found in R, we say that R "fires" on σ, or simply that σ "fires". We identify finite strings with their computable indices, and we assume that R firing on σ implies that R has also fired on every predecessor of σ at least once again at some previous stage. Without loss of generality, assume that at every stage exactly one node of $\omega^{<\omega}$ fires. Also recall that the empty string e belongs to F.

Construction. Initially, at stage 0, set $U = \varnothing$ and $T_0 = T_H(F)[0] = \{e\}$. At stage s, go through the four phases described below.

Phase 1: Updating ranks of nodes. Without loss of generality, at every stage exactly one node of $\omega^{<\omega}$ fires (recall $e \in F$). Suppose σ has fired. By our assumption, each initial segment τ of σ fired at least once again at some earlier stage. Let $U = \{u_1, \ldots, u_{n(s)}\}$ be the set of classes in H which are currently in the range of h, and let $\xi_1, \xi_2, \ldots, \xi_{n(s)}$ be simple auxiliary chains with $t(\xi_i) \in U$ and having indices $1, 2, \ldots, n(s)$, respectively.

Consider the subtree $T_{s-1} = T_H(F)[s-1]$ of $\omega^{<\omega}$ enumerated at the end of the previous stage $s-1$, and let K_s be the subtree of T_{s-1} rooted in σ (which has just fired).

1. Fix an m larger than any number mentioned so far and such that m is equal to the size of some class of H which is currently outside the range of t; if no such large class is seen in H at the stage, perform several extra steps in the enumeration of H until such a class is found.

2. Attach a new chain ξ_{s+1} of length $m - |\sigma|$ to σ.

3. If there is an auxiliary chain of length $n_j - |\sigma|$ associated with a size $n_j \in U$ and attached to σ, then enlarge this simple auxiliary chain to one of length n_j.

4. Suppose there is an auxiliary chain ξ attached to some τ extending σ, which is associated with some $n_k \in U$ but whose length is not equal to n_k. If there are no such chains, then do nothing. If $|\xi| = n_k - |\sigma|$ or longer, then again do nothing. Otherwise, suppose $|\xi| = n_k - d$, where $d < |\sigma|$. In this case, extend this auxiliary chain by adjoining $|\sigma| - d$ extra consequent nodes to the end of it.

Phase 2: Re-targeting. Suppose $i < s$ is the least such that $t(\xi_i)$ has grown in H since the previous stage. For ξ_i, add as many extra nodes as there are new points in $t(\xi_i)$. For each $j > i$ and which is not attached to the root e, update $t(\xi_j)$ and set it equal to the first found new class in H whose index is larger than the index of the current $t(\xi_j)$ and which is currently larger than $t(\xi_k)$ for every $k < j$; enumerate H until such a class is found.

Phase 3: Bookkeeping. Let u be the smallest among the classes that occur in H which is currently outside the range of h. Adjoin a simple auxiliary chain ξ of length $k = \mathrm{card}(u)[s]$ to the root e of T_{s-1}, set $h(\xi) = u$, and also enumerate ξ into U.

Verification. It is clear that the nodes which will end up having infinite rank are exactly the nodes of F; therefore, T' has the correct isomorphism type. Also, T is clearly a computably enumerable subtree of $\omega^{<\omega}$; it can be easily transformed into a computable tree. We must argue that $T/T' \sim H$.

By induction on a stage and on the index of a simple chain ξ, we can show that $h(\xi)$ is stable. Indeed, $h(\xi)$ has to be changed only if a chain of a smaller index has to be grown. Since all classes in H are finite and by the inductive hypothesis, there are only finitely many stages at which $h(\xi)$ has to be changed. Suppose $h(\xi)$ settled on some class u in H. Go to the stage at which the size of u reaches its final value k. After this stage we have $|\xi| \leqslant \mathrm{card}(u) = k$, and it may be shorter than k due to its position in T.

Phase 3 was responsible for making sure that no class of H is left without an auxiliary chain associated with it. Note that chains attached to the root e cannot be re-targeted again. We are guaranteed that e will have arbitrarily long finite chains attached to it, and therefore there is no need to worry about any stripping issues. We explicitly made sure that every class which could potentially be without an h-preimage will eventually be permanently associated with an auxiliary chain attached to e. Combined with the inductive argument above, this implies that every class in H will eventually be permanently associated with an auxiliary chain in T, and this correspondence is 1-1.

We now verify that the following conditions hold:

(P1) If a node $\sigma \in T_s$ is in F, then all finite auxiliary chains that are attached to σ in T_s, except for at most one (call it exceptional for σ at s), have their lengths equal to the sizes of classes that occur in H_s.

(P2) If ξ is an exceptional auxiliary chain for σ in (P1) at stage s, and $x \in F$, then there is a stage $t > s$ after which ξ is extended to a chain of length that is mentioned in H; after this stage, this auxiliary chain will never be exceptional for σ (or any other τ) again.

Condition (P1) is explicitly maintained at every stage. For a given σ, there is at most one exceptional ξ whose length is lagging behind the size of $h(\xi)$ according to the instructions in Phase

1. To see why (P2) holds, go to the stage at which the length of ξ reaches its final value. Since $\sigma \in F$, there is a longer chain which will eventually be attached to the same σ. Thus, according to the instructions at substage (3) of Phase 1, the length of ξ must be set equal to the size of $h(\xi)$.

It remains to consider what happens with nodes that are forever abandoned because they never fire again. Let σ be such a node, and assume its predecessor is in F. Then there are at most finitely many auxiliary chains attached to it or its successors. Go to the stage at which all of these chains reach their final value. The instructions of Phase 1 guarantee that after full stripping, this segment of the tree becomes a collection of disjoint simple chains having lengths equal to the sizes of the respective classes in H. Also, recall that Phase 3 guarantees that no classes are left without an h-preimage. Combined with (P1) and (P2), this shows that $T/T' \sim H$. □

This finishes the proof of Theorem 9.3.45. □

The following properties of the construction from the proof of Proposition 9.3.48 will be quite important later:

Property 9.3.49. Whenever a simple auxiliary chain obtains a new image in H, the chain grows in size.

Property 9.3.50. A chain is re-targeted only if some earlier introduced chain has grown.

Property 9.3.51. We re-introduce (the size of a) class in H that has been abandoned due to re-targeting, as follows. We attach a new simple auxiliary chain of the correct length to the root and associate it with the class. *The new simple chain will never be re-targeted again.*

Consequences of Proposition 9.3.48

Proposition 9.3.48 does not assume that F represents a reduced abelian group. This, for instance, implies:

Theorem 9.3.52. *Suppose A is an abelian p-group of Ulm type > 1 which is not necessarily a reduced group. Then the following are equivalent:*

1. *A has a computable copy;*
2. *A' has a Δ^0_3-copy and A/A' has a computable copy.*

Proof. (2) → (1). Recall A has Ulm type > 1, and therefore A/A' is infinite and furthermore the sizes of cyclic summands in A/A' are unbounded, for otherwise every element of infinite height in A would have to be divisible. We can therefore run the proof of Proposition 9.3.48, which does not require the p-basic tree for A' to be well-founded.

(1) → (2). This is Lemma 9.3.46. □

Remark 9.3.53. The transformation witnessing the proof of (2) → (1) is uniform if we guarantee that A/A' has only finite summands whose orders are *not* uniformly bounded.

Theorem 9.3.52 above *fails* for non-reduced groups of Ulm type 1; see Corollary 9.3.24. But the theorem also implies the following fact that should be compared with Khisamiev's Theorem A(3) about torsion-free abelian groups and with Corollary 9.3.22, which gives a counter-example for groups of Ulm type 1.

Corollary 9.3.54 (Melnikov [370]). *Every c.e. presented abelian p-group of Ulm type > 1 has a computable presentation.*

Sketch. If G is c.e. presented, then

$$G' = \{h \in G \mid h \neq 0 \,\&\, (\forall k)(\exists x)\, p^k x = h\}.$$

Thus, $\mathbf{0}''$ can list representatives of G'. Since the operation on G is computable and $=_G$ is $\mathbf{0}'$-computable, we conclude that G' has a Δ^0_3-presentation. Also, $\chi_{A/A'}$ remains a Σ^0_2 set. To see why, use $\mathbf{0}'$ to list all finite pure subgroups of G that must detach in G, and use their decompositions into cyclic summands to approximate $\chi_{A/A'}$. But additionally, we can also non-uniformly fix $a \neq 0$ of infinite p-height so that no x with the property $px = a$ has infinite p-height. We produce an l.m. function using the same procedure as described in the proof of Lemma 9.3.46. □

Our next task is to understand what happens when H in Proposition 9.3.48 is not necessarily reduced. This case will be necessary to produce a Friedberg enumeration of all computable abelian p-groups with Ulm types $\leqslant n$, where $n > 1$.

The modified Ash-Knight-Oates strategy

Suppose H is an equivalence structure. We identify H and the respective G_H, which is a direct sum of cyclic or quasi-cyclic p-groups. Recall that H is proper if it has only finite classes, but the sizes of the classes are not uniformly bounded. Recall the following definition.

Definition 9.3.55. Call a computable equivalence structure *an infinite junk* if it has infinitely many classes, almost all of which are infinite.

We abuse notation and write $T_H(F)$ for the *modified* Ash-Knight-Oates jump inversion, which is described in the lemma below.

Lemma 9.3.56. *There is a uniform procedure which, on input a computable copy of an equivalence structure H and a p-basic tree F represented as a Π^0_2-subtree of $\omega^{<\omega}$ with $e \in F$, outputs a computable p-basic tree $T_H(F)$ with the following properties:*

(1.) If H is proper, then $(T_H(F))' = F$ and $T_H(F)/(T_H(F))' \cong H$.

(2.) If H is an infinite junk, then $T_H(F) \cong H$.

(3.) If H is finite, then $T_H(F)$ is finite, and furthermore, its cardinality can be assumed arbitrarily large and with all possible uniformity[2].

Proof. We adopt the following modification to the original strategy of Ash, Knight, and Oates:

Modification 1. At every stage at which the Ash-Knight-Oates module initiates a new search through H or makes a change to its p-basic tree, adjoin a very long simple chain never seen so far to the root of the p-basic tree. Call this extra simple chain ***subsidiary***. If the subsidiary chain has just been introduced, then it does not have to copy any class in H. We also initiate a search for a new and large enough class in H that can be matched with the subsidiary chain in the future.

[2]Note that there are no assumptions on F apart from $e \in F$, which is equivalent to saying that 0 is in the subgroup generated by the p-basic tree F, and therefore this assumption is satisfied without any loss of generality.

The module will not act again until the search is finished (if ever). When the module acts again (if ever), the chain is handled as a standard auxiliary chain attached to e.

The module will later be associated with a node on the tree of strategies, and in particular, it may be initialised. We also attach a very long subsidiary chain to the root of the p-basic tree previously handled by the strategy if the strategy τ gets initialised. Since the old p-basic tree will be forever abandoned by the strategy, in this case there is no need to search for an image for the subsidiary chain in H (the image can be larger than the length of the chain).

(1.) Since H is proper, we will eventually succeed in finding a long enough class in H that can be matched to the subsidiary chain; the class in H may be (currently) larger than the chain. Once this is done, the chain becomes indistinguishable from the other many simple chains that we attach to the root 0 according to the non-modified instructions. There are no further interferences of the modification with the rest of the module. It follows that in the case when H is proper, the verification of the new module is almost literally the same as the verification contained in the previous section.

(2.) Here the modification plays no significant role either. However, the analysis of this scenario is new because the case of a non-reduced H has never been considered in the literature. Recall that the first few classes of H could be finite, but the rest of the classes are infinite, and there are infinitely many of them.

First, we claim that almost all auxiliary or subsidiary simple chains that we ever attach become infinite. Note that a simple chain may never find a stable pre-image among classes in H. However, at each intermediate step we always succeed in finding a long enough class in H to match with the chain. Whenever we switch, the chain itself must grow; see Property 9.3.49. Thus, we still grow the length of the chain to infinity, even though it may never find a stable image in H. Now consider those simple chains which do find a stable match in H. Almost all of these chains grow infinite by simply copying the respective stable class in H. The analysis also applies to the subsidiary simple chains from the modification. In particular, since we are never stuck at any intermediate step, there are infinitely many such infinite simple chains to be attached to the root. It follows that the divisible part of $T_H(F)$ has infinite rank.

There are at most finitely many exceptional chains that correspond to the finite classes in H. There may also be several finite configurations that become simple chains after stripping the tree. The latter corresponds to parts of the tree being forever abandoned in a Σ^0_2-outcome of the Π^0_2-approximation. Every individual simple chain, as well as each chain involved in an "abandoned" configuration, must grow whenever its image in H switches (Property 9.3.49). Thus, a chain or a configuration of chains can be finite only if each auxiliary chain involved in the configuration finds a stable image in H. There are only finitely many finite classes in H, and thus the reduced part of $T_H(F)$ must be finite. Furthermore, we may be forced to switch the image of a given chain only due to some currently shorter class of a smaller index having grown (Property 9.3.50).

If a finite class in H is skipped in the construction due to re-targeting, then it will be re-introduced again in the form of a simple chain attached to the root (Property 9.3.51). There are only finitely many classes having a smaller index than the index of the finite class. Therefore, by induction, each finite class will eventually find a stable image in the tree, which will be a simple chain of the correct length. It follows that the reduced part of $T_H(F)$ is isomorphic to the reduced part of H (viewed as a p-group).

(3.) This is obvious from the description of the modification because the subsidiary chain can be taken to be arbitrarily long. It is crucial that the chain does not have to copy any class in H at the stage when it is first introduced. □

Exercises

An earlier technically related exercise is Exercise 8.3.42.

Exercise° 9.3.57. Let A be a computable abelian group. Show that its first Ulm factor $A_1 = A/A'$ admits a Δ^0_3 presentation.

Exercise° 9.3.58. let $T \subseteq \omega^{<\omega}$ be a Δ^0_3 (combinatorial) tree. Show that it is isomorphic to a Π^0_2-subtree of $\omega^{<\omega}$, and this is uniform in the index of T provided that T is non-empty (thus, contains the root e of $\omega^{<\omega}$).

Exercise 9.3.59 (Calvert, Harisanov, Khight, and Miller [76]). Let K be the class of reduced abelian p-groups of Ulm type at most $m < \omega$, and let $A \in K$ be a computable group. Calculate the complexity of the index set $I(A) = \{i : A_i \cong A\}$. (Hint: The answer will depend on the isomorphism type of the last Ulm factor.)

Exercise 9.3.60 (Calvert [74]). Show that the isomorphism problem for computable reduced abelian p-groups of Ulm type $m \leqslant \omega$ is Π^0_{2m+1}-complete.

Exercise 9.3.61. Show that the isomorphism problem for (non necessarily reduced) computable abelian p-groups of Ulm type $m \leqslant \omega$ is Π^0_{2m+2}-complete.

Exercise 9.3.62. Use the techniques of p-basic trees to prove that computable torsion abelian groups form an FF-complete class; this appeared earlier as Corollary 8.2.24. (Hint: Fix the sequence of rank-saturated trees $T^*_{m,n}$ from the proof of Theorem 8.2.19 (see Lemma 8.2.23). Note that the isomorphism type of $T^*_{m,n}$ is fully determined by the tree rank of $T^*_{m,n}$; it is either T_α of rank α or T^∞ which is ill-founded. We may assume that the root of $T_{m,n}$ has exactly one immediate successor, and so the rank of the tree is always a successor ordinal. For each m, fix the mth prime p_m and consider the p_m-group $G_{m,n} = G_{p_m}(T^*_{m,n})$ generated by the p_m-basic tree $T^*_{m,n}$. If $T^*_{m,n}$ is well-founded, then $G_{m,n}$ is reducible, and the tree rank of $T^*_{m,n}$ uniquely determines the Ulm type of $G_{m,n}$ as well as the isomorphism type of the last Ulm factor of $G_{m,n}$. If $T^*_{m,n} \cong T^\infty$ then $G_{m,n}$ is divisible. Take $G_n = \bigoplus_{m\in\mathbb{N}} G_{m,n}$.)

Exercise* 9.3.63 (Smith [474]). We say that an abelian group A admits an effectively unique divisible closure if any two computable closures with the properties described in Exercise 5.1.36 computably isomorphic over A. Prove that for a computable abelian p-group A, having a unique divisible closure is equivalent to computability of the relation

$$p|a \text{ if and only if } \exists x \in A \; px = a$$

in A.

Exercise* 9.3.64 (Goncharov-Nurtazin [217], Harrington [231]). Whilst this is not about abelian groups, it is related to the previous exercise and to the material of §2.2.2. Harrington [231] used this result to show that every computable differential field has a computable differential closure, and that any decidable $\aleph_1$-categorical theory has a computable presentation.

Recall that a countable model A of a theory T is *prime* iff every finite tuple of A realises a principal type of T. Modify the construction of Theorem 2.2.28 to show that the following are equivalent.

(i) T has a decidable prime model.

(ii) T has a prime model and the set of principle types of T is computable.

Exercise 9.3.65. A group is said to be *primitive recursive* if its domain and the group operations $+$ and $-$ are primitive recursive. A primitive recursive group is *fully primitive recursive* or *punctual* if its domain is ω or an initial segment of ω.

1. Prove that every computable abelian p-group admits a punctual presentation ([282]).
2. Show that there is a computable torsion abelian group that is not isomorphic to any primitive recursive group ([86]).

9.4 Enumerating abelian p-groups

In this section, we prove the following:

Theorem 9.4.1 (Downey, Melnikov, and Ng [145]). *For each natural number $n > 0$ there is a Friedberg enumeration of all computable abelian p-groups of Ulm type $\leqslant n$.*

The case when $n = 1$ was established earlier in Corollary 9.3.25 and Theorem 9.2.1. It is clear that the groups in the list for $n = 1$ are uniformly represented by computable p-basic trees, which are inherited from the full decomposition induced by the corresponding equivalence structure.

Therefore, assume $n > 1$ throughout the rest of this subsection.

9.4.1 Plan of the proof

We use Theorem 9.3.17 and Proposition 9.3.18 throughout, sometimes with no explicit reference. In particular, we may occasionally identify E with the respective group G_E, and vice versa.

Fix a Friedberg enumeration $(E_j)_{j\in\mathbb{N}}$ of all infinite equivalence structures, and thus of infinite abelian p-groups of Ulm type 1 (with E identified with G_E). Inductively, fix a Friedberg enumeration $(F_i)_{i\in\mathbb{N}}$ of all isomorphism types of $\mathbf{0}''$-computable abelian p-groups of Ulm type $\leqslant n-1$. Furthermore, assume that they are represented by Π^0_2 p-basic trees whose indices $h(1), h(2), h(3), \ldots$ are given uniformly.

Remark 9.4.2. The function h is computable and not merely $\mathbf{0}''$-computable. It returns the index of the computable R_i such that

$$\sigma \in F_i \text{ if and only if } \exists^{\infty} z\, R_i(\sigma, z).$$

As we noted before in Exercise 9.3.58, it is well-known that there is a uniform procedure that transforms a non-empty Δ^0_3-tree into a Π^0_2-subtree of $\omega^{<\omega}$. Of course, $e \in F_i$ since it corresponds to 0.

Based on the Friedberg enumerations $(F_i)_{i\in\mathbb{N}}$ and $(E_j)_{j\in\mathbb{N}}$ described above, fix the effective listing $(F_i, E_j)_{i,j\in\mathbb{N}}$.

Proof idea. In the notation above, suppose F_i is well-founded and E_j has only finite but arbitrarily large classes; we call such F_i and E_j *true* and *proper*, respectively. Under these assumptions, we can uniformly produce a computable abelian p-group $T_{E_j}(F_i)$ of Ulm type at most n such that

$$(T_{E_j}(F_i))' \cong F_i \text{ and } T_{E_j}(F_i)/(T_{E_j}(F_i))' \cong E_j;$$

this is Theorem 9.3.52 and Remark 9.3.53. Since $(F_i)_{i\in\mathbb{N}}$ and $(E_j)_{j\in\mathbb{N}}$ are Friedberg, the Ulm classification theorem implies that unequal pairs correspond to non-isomorphic groups *provided that these pairs consist of true and proper members, respectively.* Furthermore, the Ulm classification theorem and Theorem 9.3.52 imply that each computable group of Ulm type k, $1 < k \leqslant n$, has the form $T_E(F)$ for some true F of type $< n$ and proper E having the correct complexities (Π^0_2 and computable, respectively).

The rough idea is as follows. Given (F_i, E_j), guess trueness and properness, and simultaneously attempt to enumerate $T_{E_j}(F_i)$. If *all* F_i and E_j in the list were true and proper, respectively, then $T_{E_j}(F_i), i, j \in \mathbb{N}$, would be a Friedberg enumeration of all computable abelian p-groups of Ulm type $1 < k \leqslant n$. Merging it with the Friedberg enumeration of all computable abelian p-groups of Ulm type 1, we would get the desired 1-1 list of all groups of types $\leqslant n$.

However, if F_i is not true or E_j is not proper, we cannot guarantee that $T_{E_j}(F_i)$ will have Ulm type > 1. This will conflict with the enumeration of all groups of type 1. Nonetheless, by carefully controlling the group produced in each of these two unpleasant outcomes, it is possible to incorporate this group of Ulm type 1 into the dynamic procedure of enumerating all type 1 groups (equivalent structures).

The global architecture of the proof. The construction will consist of three main modules.

The main module. On input F_i and E_j, it performs the following tasks:

- It measures whether F_i is true and E_j is proper. The combined complexity of these two guessing procedures is Σ^0_4 (to be verified), and it will be split into infinitely many Π^0_3-instances, one for each potential $\exists$-witness z in $\Sigma^0_4 = (\exists z)\Pi^0_3(z)$.

- It attempts to build $T_{E_j}(F_i)$. If F_i is true and E_j is proper, then, for exactly one z, exactly one submodule σ associated with (i, j, z) succeeds in building $T_{E_j}(F_i)$ of Ulm type > 1. This occurs only if the Π^0_3-predicate holds, and E is "true". The submodule σ also has several outcomes, which depend on the isomorphism type of E and also on how exactly the Π^0_3-predicate fails. Under these outcomes, either finite groups/structures or infinite junk structures (Definition 9.3.55) of Ulm type 1 are produced. They are placed into the junk collector; see below.

- The procedure associated with σ uniformly replaces E_j with a certain H_j and works with $T_{H_j}(F_i)$ instead of $T_{E_j}(F_i)$. The equivalence structure H_j has several convenient combinatorial properties (to be explained), and, of course, $H_j \cong E_j$ if the latter is proper. Thus, $T_{H_j}(F_i) \cong T_{E_j}(F_i)$ in the Π^0_3 outcome.

The module enumerating Ulm type 1 groups. This is literally the same as the one we used in the proof of Theorem 9.2.1, but with equivalence structures uniformly replaced by the respective Ulm type 1 abelian p-groups. We give a brief overview of this module here. Various sub-strategies are put together into a tree of strategies $\mathcal{T}$, in which the true path will be $\mathbf{0}'''$-computable. The tree $\mathcal{T}$ produces an enumeration of all equivalence structures, which mentions all structures having arbitrarily large finite classes exactly once; it also enumerates *some* isomorphism types of infinite junk and finite structures. The latter two are placed into the junk collector (see (3) below), which ensures all infinite junk and finite structures are mentioned exactly once up to isomorphism. The only missing isomorphism types are:

- Equivalence structures having finitely many classes and at least one of these being infinite.

- Equivalence structures which have infinitely many classes and are eventually bounded; that is, almost all classes are smaller than some fixed bound k specific to the structure.

The uniform Friedberg list of such structures can be easily produced independently and later adjoined to the Friedberg enumeration of the rest.

The junk collector. It is responsible for enumerating all infinite junk and finite equivalence structures/groups without repetition. Its actions are global. It handles the infinite junk and finite equivalence structures/groups produced by the two main modules as described above, and it also introduces its own to ensure that the enumeration is 1-1 and surjective on isomorphism types of infinite junk and finite equivalence structures/groups. The junk collector module has two submodules:

- *The infinite junk collector.* It is responsible for making sure that all computable isomorphism types of infinite junk structures/groups (Definition 9.3.55) are listed, without repetition. Its unsuccessful attempts result in abandoning a structure in the process; abandoned structures are permanently placed into the *finite junk collector.*
- *The finite junk collector.* Its task is to ensure all finite equivalence structures/abelian p-groups are mentioned in the list, and exactly once.

The construction will be described in §9.4.8. Here we give only a brief outline of the construction. One crucial observation is that, from the perspective of the junk collector, the products of Π^0_2- and Σ^0_2-outcomes of submodules of the main module are not really special when compared with similar outcomes of taken from the proof of Theorem 9.2.1. The construction will be split into three relatively independent phases.

1. Phase 1 is responsible for enumerating all Ulm type $k > 1$ ($k \leqslant n$) groups, all groups of Ulm type 1 having arbitrarily large finite cyclic summands, and *some* finite and infinite junk groups. At this phase of the construction, *the main module* and *the module enumerating Ulm type 1 groups* act simultaneously and independently according to their instructions. We ensure that there are no interactions between these two modules.
2. Phase 2 is responsible for expanding the output of Phase 1 so that the new enumeration also contains all isomorphism types of infinite junk structures. This is done using *the infinite junk collector.*
3. Phase 3 transforms the output of Phase 2 into an enumeration which additionally mentions every isomorphism type of a finite abelian p-group exactly once. This is done using *the finite junk collector.*

Finally, to get the desired Friedberg enumeration, we merge the output of Phase 3 with the Friedberg enumeration of all eventually bounded equivalence structures and all equivalence structures having finitely many classes, at least one of which is infinite; the latter, of course, are uniformly replaced with the respective abelian p-groups. This finishes the informal outline of the construction.

9.4.2 The basic strategy

Recall that the Ulm type of each F_i is at most $n-1$, and that each F_j is a Π^0_2 subtree of $\omega^{<\omega}$ whose index is given uniformly. Each E_j is a computable *infinite* equivalence structure, which can be viewed as an abelian p-group of Ulm type 1, in which a complete decomposition is known.

We identify E_i with the corresponding abelian p-group. According to our terminology, E_i is *proper* if it consists only of finite classes and the sizes of its classes are unbounded.

Definition 9.4.3. Let F be a p-basic tree. If F has a non-zero terminal node, then we say that F is *true*. Note that this is equivalent to saying that the reduced part of the corresponding p-group is non-trivial.

Lemma 9.4.4. *Let $(E_i)_{i\in\mathbb{N}}$ and $(F_i)_{i\in\mathbb{N}}$ be uniform enumerations of computable equivalence structures and Π^0_2 trees as defined above.*

1. *The property "E_i is proper" has complexity Π^0_3.*
2. *The property "F_i is true" has complexity Σ^0_4.*

Proof. For (1), just state that each class is finite (Π^0_3) and that there are arbitrarily large classes (Π^0_2). The statement "F_i is true" can be described by the formula:

$$(\exists x)\,[x \in F_i \;\wedge\; x \neq e \;\wedge\; (\forall y)(y \supset x \mapsto y \notin F_i)]$$

which gives an upper bound of Σ^0_4 for (2). □

It is not difficult to show that the bounds in the lemma above are optimal, and therefore the complexity of our guessing cannot be simplified.

Guessing trueness and properness. Given (F_i, E_j), we need to test whether F_i is true and E_j is proper. We suppress the subscripts in F_i and E_j and write (F, E) throughout the rest of this subsection. We begin with the simpler Π^0_3 guessing of properness for E. We index classes of a computable equivalence structure by natural numbers according to the order at which they appear in the enumeration of the equivalence structure. Write $[i]_E$ or simply $[i]$ for the i-th class of E.

Definition 9.4.5. An equivalence structure is *eventually bounded* if there is an $n \in \mathbb{N}$ such that all classes with indices greater than n are bounded in size by n.

Note that an eventually bounded structure may have infinite classes or finitely many classes.

Lemma 9.4.6. *For an equivalence structure E, eventual boundedness is a Σ^0_2-property.*

Proof. The property says:

$$(\exists n)(\forall i > n)\,\neg\left(\exists a_1,\ldots,a_{n+1} \in [i] \quad \bigwedge_{i\neq j, i,j\leqslant n+1} \quad a_i \neq a_j\right),$$

where the i-th class $[i]$ is *not necessarily the class containing the i-th element of* E; see the explanation preceding Definition 9.4.5. □

Guessing properness of E. The preliminary description of the outcomes of this guessing is:

$\Pi^0_2(j)$: E is *not* eventually bounded and the jth class in E is infinite.

Π^0_3: E is proper.

Σ^0_2: E is eventually bounded.

Since an equivalence structure is proper *iff* it is not eventually bounded and does not contain infinite classes, it is clear that the outcomes are exclusive and cover all possible cases.

Guessing trueness of F. Slicing Σ_4^0 into $(\Pi_3^0(z))_{z\in\mathbb{N}}$. Recall that the sentence saying that F is true has complexity Σ_4^0. We represent the respective Σ_4^0-predicate as $\exists z \Pi_3^0(z)$. As usual, we assume that the measured predicates satisfy the property of the uniqueness of existential witnesses. In particular, if $\exists z \Pi_3^0(z)$ holds, then there will be exactly one such z.

The outcomes of each $\Pi_3^0(z)$-guessing are:

$\Pi_2^0(j, z)$: This is a Π_2^0 outcome that says that j witnesses the failure of the $\Pi_3^0(z)$ predicate $\forall j \Sigma_2^0(j, z)$.

$\Pi_3^0(z)$: F is true with a Σ_4^0-witness z.

The collection of all $\Pi_2^0(j, z)$-outcomes can be viewed as the $\Sigma_3^0(z)$-complement of $\Pi_3^0(z)$.

The strategy for (F, E, z). Each triple (F, E, z) is associated with a strategy, in which z is interpreted as a potential existential witness for $\exists z \Pi_3^0(z)$ approximating the trueness of F. The strategy for one (F, E, z) in isolation relies on the guessing of F and E described above, and it also has the following two major tasks.

The first task: Building H. The strategy dynamically transforms the computable equivalence structure E into a computable equivalence structure H with the properties:

i. Assume E is not eventually bounded, and one of the two conditions holds:

(i.1) E has infinite classes, or

(i.2) F looks not true according to $\Sigma_3^0(z)$ (see the previous subsection).

In this case, H has infinitely many classes, with almost every class infinite. Furthermore, the number of finite classes in H produced by the strategy is specific to the node of the strategy and to the outcome of the strategy under which it is produced.

ii. If E is proper and F is true, then $H \cong E$.

iii. If E is eventually bounded, then H is finite.

Condition *i.* says that H is an infinite junk structure (Definition 9.3.55). We delay the detailed description of H and the verification of *i*-*iii* until §9.4.5. Also, a further minor adjustment to this transformation will be introduced in §9.4.6. For now, we take these properties for granted.

The second task: Building $T_H(F)$. The second task of the strategy is producing $T_H(F)$ based on the dynamic definition of H; here $T_H(F)$ stands for the *modified* version of the Ash-Knight-Oates operator defined in §9.3.3. As usual, we identify an equivalence structure with the direct sum of cyclic and quasi-cyclic p-groups in which cyclic summands $\mathbb{Z}_{p^n}$ naturally correspond to equivalence classes of size n. According to Lemma 9.3.56 and assuming the properties *i.*-*iii.* of H stated in the subsection above, we have the following different scenarios:

a. If H is infinite junk, then $T_H(F) \cong H$.

b. If H is proper, then $(T_H(F))' = F$ and $T_H(F)/(T_H(F))' \cong H$.

c. If H is finite, then so is $T_H(F)$.

Furthermore, by Lemma 9.3.56, the cardinality of the finite $T_H(F)$ in *c.* can be assumed as large as necessary.

9.4.3 Actions of the strategy for (F, E, z)

Whenever the strategy becomes active, it makes one more step in each of the two uniform procedures:

1. Approximate $T_H(F)$, where H is the uniformly modified version of E satisfying *i-iii* (see §9.4.5 for details) and $T_H(F)$ is the modified Ash-Knight-Oates operator satisfying *a-c* applied to F and H.

2. Monitor H and guess whether it has infinitely many classes, all of which are infinite. Since H is uniformly defined from E, this predicate is uniformly Π^0_2 in (the index for) E. If this predicate fires, then the basic module initialises itself by permanently abandoning its current $T_H(F)$. In this case, it creates a new version of $T_H(F)$, which is built from scratch. The new version will have a new index in the uniform enumeration of all type $\leqslant n$ abelian p-groups.

9.4.4 The outcomes

Assuming that H indeed satisfies the claimed properties *i-iii*, the strategy associated with (E, F, z) will have one of the following outcomes:

pi_0 : This is a Π^0_2 outcome which measures if all classes in H are infinite (and thus there are infinitely many such classes).

Every time it is played, the strategy is initialised, and its previous version of $T_H(F)$ is abandoned. Recall that the size of the abandoned $T_H(F)$ can be picked as large as necessary, according to Modification 1 from Section 9.3.3.

pi_j: $(j > 0)$ This is a Π^0_2 outcome which says that:

- E is not eventually bounded, i.e., it has arbitrarily large classes of arbitrarily large indices, and
- either the jth class in E is infinite, or F looks not true as witnessed by $\Pi^0_2(j, z)$.

By Lemma 9.3.56 and assuming properties *i-iii* of H, in this case, the strategy produces a computable $T_H(F) \cong H$ which can be identified with G_H, composed of at most finitely many cyclic and infinitely many quasi-cyclic direct summands. Furthermore, we will ensure that different strategies always produce non-isomorphic $T_H(F) \cong H$ under their Π^0_2-outcomes, and also different Π^0_2-outcomes of the same strategy give non-isomorphic $T_H(F) \cong H$. This will be clarified in §9.4.5. With extra care, we will make sure that these infinite junk structures/groups also differ from any infinite junk structure produced by the tree of strategies $\mathcal{T}$; see Section 9.4.6 for the description of $\mathcal{T}$ and §9.4.4 for the above-mentioned adjustment.

Π: This is a Π^0_3 outcome that says that E is proper and F is true.

In this case, by Lemma 9.3.56 and assuming properties *i-iii* of $E \mapsto H$, the strategy outputs a computable basic tree $T_H(F)$ with the properties $(T_H(F))' = F$ and $T_H(F)/(T_H(F))' \cong H \cong E$ (the latter two are identified with the respective groups). Furthermore, since $E \cong H$ is proper and F is true of type $< n$, the Ulm type of $T_H(F)$ is at least 2 and at most n.

fin: This is a Σ_2^0-outcome which says that E is eventually bounded.

In this case, $T_H(F)$ is finite. Furthermore, its cardinality can be controlled and made arbitrarily large, if necessary, according to Modification 1.

To finalise the description of the basic strategy, we must give a detailed description of the transformation $E \mapsto H$ and verify its claimed properties.

9.4.5 The description of $E \mapsto H$

First, we describe a transformation $E \mapsto \tilde{H}$ which takes care of most properties *i-iii* with the exception of the "furthermore" part of *iii*. Then, we further adjust $\tilde{H}$ and describe a transformation $\tilde{H} \mapsto H$ which also gives property *iii* in full. This modification is highly convenient in the general case of many strategies working together.

The definition of $\tilde{H}$.

Given an infinite computable equivalence structure E, the strategy produces a computable equivalence structure $\tilde{H}$ with the properties:

i. If E is not eventually bounded (Definition 9.4.5), and one of the two conditions holds:

(i.1) E has infinite classes, or

(i.2) F looks not true according to $\Pi_2^0(j, z)$ (see Guessing Trueness of F),

then $\tilde{H}$ has infinitely many classes with almost every class infinite.

ii. If E is proper and F is true, then $\tilde{H} \cong E$.

iii. If E is eventually bounded, then $\tilde{H}$ is finite.

We write $[m]_L$ for a class of an equivalence structure L with index m. Say that a stage s is expansionary if the parameter $\max\{\text{card}[i]_{E_s}, i \leqslant s\}$ has increased from the previous expansionary stage s'. The parameter measures whether the structure E has arbitrarily large classes with arbitrarily large indices. The simple construction below acts only at expansionary stages.

Construction. At every stage, each class in $\tilde{H}_s$ is matched with a class in E_s. Suppose at a stage $[n]_{\tilde{H}}$ is copying $[i]_E$. If $[i]_E$ has grown in E or the ith Π_2^0 instance of the Σ_3^0 predicate "F is not true (z)" has fired, then perform the following action. *Initialise* each class $[k]$ in $\tilde{H}$ that satisfies:

(1) $k > n$, and

(2) $[k]_{\tilde{H}}$ has been copying a class $[j]_E$ with $j \geqslant i$.

Each initialised class grows by one extra element and will be assigned to some large enough new class in E (if it exists). Until such large enough classes are found, the whole strategy (not just this simple procedure describing $\tilde{H}$) ceases its action. Then, once large enough classes are found, each currently abandoned class of E is assigned to a new class in $\tilde{H}$. This ends the construction.

The verification of i, ii and iii. To see why iii holds, recall that the procedure constructing $\tilde{H}$ acts only at expansionary stages. Since there are only finitely many such stages, $\tilde{H}$ remains finite. To check i and ii, note that each initialised class must grow. A class can be initialised only due to some larger index class growing or due to some higher-priority Π^0_2 instance of the predicate "F is not true (z)" firing; furthermore, in the former case, this larger $\tilde{H}$-index class must be copying a larger E-index class. There are only finitely many such classes. Thus, if all classes in E are finite and F looks not true according to instance z, then each class can be initialised only finitely often. Also, a class in E has to change its clone in $\tilde{H}$ only if a class with a smaller E-index grows. Therefore, ii follows by induction.

To check i, assume that $[j]$ is the left-most class of E – i.e., the one with the smallest index – that grows to infinity. Since all classes to the left of it are finite, there is a stage after which the class is stably assigned to a clone in $\tilde{H}$, call this clone $[k]$. There exist at most finitely many classes of $\tilde{H}$ to the right of $[k]$ that are controlled by classes in E having an index less than the index of $[j]$. All the rest are initialised infinitely often. Since E has arbitrarily large classes with arbitrarily big indices, every search for a new appropriate image for an initialised class is successful. In particular, E has infinitely many classes, and therefore so does $\tilde{H}$. Since each initialised class must grow, co-finitely many classes of $\tilde{H}$ are infinite.

The transformation from $\tilde{H}$ to H.

Fix a uniformly computable collection of non-intersecting intervals

$$I_0, I_1, \ldots, I_n \ldots$$

in ω which form its full partition, where the smallest number of I_n is equal to the largest number of I_{n-1} plus 1. We write $\max I_n$ for the largest number of I_n. (In the construction, we will also make sure that $\max I_i^\sigma \neq \max I_k^\tau$ for $\sigma \neq \tau$ and any strictly positive $i, k \in \mathbb{N}$.)

We are given $\tilde{H}$, which is either proper, an infinite junk, or finite (cf. i-iii). Recall that, according to our convention, every class of $\tilde{H}$ receives an index according to the stage at which it appears in the enumeration of $\tilde{H}$. The uniform definition of $\tilde{H}$ contained in the subsection above has the following property: If the size of $[i]$ in $\tilde{H}$ is infinite and has infinitely many classes, then so is $[k]$ for each class $[k]$ whose index is larger than the index for $[i]$. We must uniformly build a computable equivalence structure H and a map $\psi : H \to \tilde{H}$ by stages.

The idea is rather simple. We construct H so that it copies $\tilde{H}$, but the isomorphism ψ is defined not class-by-class but block-by-block. If some class in the k-th block of $\tilde{H}$ has grown, then in H we initialise all ψ-preimages of j-blocks for $j \geqslant k$. Whenever we initialise a block in H, each class in the block is increased in size.

We give formal details. At stage s, if a class of $\tilde{H}$ with index $j \in I_k$ has grown in size, then:

1. Declare ψ undefined for every class of $\tilde{H}_s$ with its index in I_m for some $m \geqslant k$.
2. Grow all classes of H_s which currently have no ψ-image to a size greater than any number mentioned so far.
3. Speed up the enumeration of $\tilde{H}$ and search for new, larger images for the finitely many classes in H_s for which ψ is currently undefined.

4. If (2) is ever finished, introduce new classes in H_s and match them with those classes of $\tilde{H}$ which currently have no ψ-preimages. Go to (1).

Lemma 9.4.7.

1. *If $\tilde{H}$ is proper, then $H \cong \tilde{H}$.*

2. *If $\tilde{H}$ is finite, then H is finite too.*

3. *If $\tilde{H}$ is an infinite junk, then so is H. Furthermore, either H has all classes infinite, or the total number of finite classes in H is equal to $\max I_k$ for some k.*

Proof. (1). By induction on the index i of a class $[k]_{\tilde{H}}$ and the index m of the block I_m such that $i \in I_m$, every class $[k]$ in $\tilde{H}$ eventually finds a stable ψ-preimage in H. Thus, in this case, ψ is a Δ_2^0-isomorphism of equivalence structures witnessing $H \cong \tilde{H}$.

(2). This is obvious.

(3). Let m be the smallest such that there is an infinite class in $\tilde{H}$ with index $j \in I_m$. Then the only classes which have stable ψ-preimages in H are the classes whose indices are in I_n for some $n < m$. If a class in H does not have a stable ψ-image, then its size is driven to infinity; indeed, since $\tilde{H}$ is an infinite junk, the search at (3) of the procedure describing H is always successful, and according to (2), whenever ψ is redefined, the class must be grown. If $m = 0$, then all classes in H end up infinite; otherwise, let $k = m - 1$. □

The lemma above and the properties of $E \mapsto \tilde{H}$ imply that the uniform transformation $E \mapsto \tilde{H} \mapsto H$ satisfies *i*-*iii* from "the first task" (building H), as desired.

9.4.6 The tree of strategies for Ulm type 1 groups

We will slightly adjust the notation from the proof of Theorem 9.2.1 to make it consistent with the other notation related to p-groups. We use the tree of strategies from Theorem 9.2.1 without any modification. The tree and various strategies associated with its nodes act independently from the rest of the construction, and the only interaction with the rest of the construction is via the junk collector. To successfully incorporate the construction from Theorem 9.2.1 into our proof, we will:

1. interpret equivalence structures as the respective Ulm type 1 groups, and

2. for every strategy associated with some σ along the tree, the infinite junk structures potentially produced by σ are non-isomorphic to any infinite junk structure produced by a strategy for (F_i, E_j, z) or by any other $\tau \neq \sigma$.

The first assertion is just a triviality, and the second is not really a modification either, for the construction in Theorem 9.2.1 already ensured that different nodes and different outcomes produce non-isomorphic infinite junk structures, and the precomputed bounds on the number of exceptional classes can be kept exactly the same as in the proof of Theorem 9.2.1. We will elaborate on this point at the very end of this subsection, where specifics will be spelled out.

The tree $\mathcal{T}$. Let $\mathcal{T}$ be the tree of strategies from the proof of Theorem 9.2.1. The order of the outcomes was:

$$\texttt{init} < \texttt{pi}_2 0 < \texttt{pi}_2 1 < \cdots < \texttt{pi}_3 < \texttt{wait}.$$

The tree $\mathcal{T}$ is composed according to this order; under the outcome $\texttt{pi}_3$ measuring $P(i, z)$, there is no other node working with some $z' > z$ in $P(i, z')$. If we view the tree of strategies $\mathcal{T}$ as one large module, its cumulative products can be classified as follows:

- *Equivalence structures having arbitrarily large finite classes.* All such structures are enumerated under the $\texttt{pi}_3$-outcomes along the true path, and without repetition (up to isomorphism).
- *Finite structures.* These come from true `init`- and `wait`-outcomes of various nodes in the tree, and are also produced due to initialisation. By making them larger than any number seen so far in the construction (see, e.g., Modification 1), we ensure there is no repetition among them, but we do not guarantee that all finite structures are produced by the tree.
- *Infinite junk structures produced by true* $\texttt{pi}_2$*-outcomes of various structures.* Note that some strategies off the true path can be forced to play their $\texttt{pi}_2$-outcomes. The number of sizes of exceptional classes is different for different nodes and below different outcomes of the same node. At every stage, the isomorphism type of the structure is guessed, with the guess eventually becoming correct if the outcome is played infinitely often.

  More specifically, if m is the number of times the strategy (call it τ) has been initialised, then the number of finite classes in U_τ produced under the true outcome $\texttt{pi}_2 j$ of τ should be between $\langle \tau, m, j\rangle$ and $2\langle \tau, m, j\rangle$, where the standard pairing function $\langle i, j\rangle$ is replaced with $3^{\langle i,j\rangle}$; this was verified in the proof of Theorem 9.2.1.

Separating the junk. Now, since we have explained the role of the intervals $[\langle \tau, m, j\rangle, 2\langle \tau, m, j\rangle]$, we are ready to introduce the following elementary but important adjustment to the basic strategy from §9.4.2.

Modification 2. We assume that for every strategy σ working with some (F_i, E_j, z), the parameters $\max I_k^\sigma = \max I_k$ described in §9.4.5 are taken from the complement of the set

$$\bigcup_{\tau \in \mathcal{T}} [\langle \tau, m, j\rangle, 2\langle \tau, m, j\rangle],$$

where $\mathcal{T}$ is the tree of strategies from Theorem 9.2.1. We furthermore assume that $\max I_k^\sigma \neq \max I_j^{\sigma'}$ whenever either $\sigma \neq \sigma'$ or $k \neq j$.

Infinite junk structures produced by various Π_2^0-outcomes $\texttt{pi}_2$ of different strategies are non-isomorphic. Thus, there is no conflict between Π_2^0-outcomes of different strategies, regardless of whether they live on the tree $\mathcal{T}$ or work with some triple (F_i, E_j, z). Any two distinct Π_2^0-outcomes of the same strategy (on the tree or working with a triple) produce non-isomorphic infinite junk structures as well.

9.4.7 The junk collector

The junk collector can be extracted from the proof of Theorem 9.2.1 *without any further modification*; see §9.2.5. We briefly go over the main ideas and discuss why Modification 1 and Modification 2 make managing the junk exactly the same as in the proof of Theorem 9.2.1. Recall that junk structures can be of two different kinds:

1. Finite junk. These are finite abelian p-groups/equivalence structures which are either produced due to initialisation or are built if the Σ^0_2-outcome is the true outcome. Because of Modification 1, the cardinalities of these finite groups may be assumed to be large and unseen at the stage when they are first introduced; see Lemma 9.3.56(3).

2. Infinite junk (see Definition 9.3.55). These are produced under various Π^0_2-outcomes, which are not their left-most Π^0_2-outcomes, of basic strategies either working with (F_i, E_j, z) or along the tree $\mathcal{T}$. According to Modification 2 in §9.4.6, the isomorphism type of the infinite junk structure produced by $\sigma\hat{}\xi$, where ξ is the Π^0_2-outcome of σ played infinitely often, will be uniquely determined by σ and ξ, regardless of the type of the strategy σ. At every stage at which the outcome is played, we will also have the current best guess on the isomorphism type of the structure.

The junk collector consists of two submodules working in coordination with each other.

The infinite junk collector. The task of this global strategy is to ensure that each isomorphism type of infinite junk structure H is represented in the global enumeration, and exactly once. The input of the infinite junk submodule is a uniform enumeration of abelian p-groups, some of which can be infinite junk. We write $L_0, L_1, \dots$ to denote these groups. At every stage, each L_i is finite and is identified with the respective equivalence structure E_{L_i} with all possible uniformity. This list is uniformly produced by sub-strategies of the main strategy and the tree $\mathcal{T}$, all working together, but the exact nature of this list is not important. We need only the following assumptions about this list.

(a1): We identify each L_i with its index, which is uniformly computable from i; without loss of generality, we may assume that the complement of the set of all these indices is an infinite computable set.

(a2): At every stage, at most one such $L = L_i$ can be declared *active*, which means that, in a Π^0_2-fashion, we have more evidence that L may end up being an infinite junk structure. In this case, the intended isomorphism type of L is also given in the form of a finite parameter describing the exceptional finite classes of L. At such a stage, L grows in size to a very large cardinality. If L is active infinitely often, then this parameter is the only one which appears as the best current guess infinitely many times.

(a3): Also, if $L_i \neq L_j$, then their parameters from (a2) above never describe the same isomorphism type of an infinite junk structure (cf. Modification 2).

Proposition 9.4.8. *Given a uniform list $(L_i)_{i\in\mathbb{N}}$ with properties (a1)-(a3) and which is injective on isomorphism types, the finite junk collector outputs a uniform list $(Z_i)_{i\in\mathbb{N}}$ which mentions each member of $(L_i)_{i\in\mathbb{N}}$ exactly once and mentions each isomorphism type of infinite junk structures exactly once. In addition to all infinite junk structures and members of $(L_i)_{i\in\mathbb{N}}$, it may only contain some isomorphism types of finite structures/groups, and also without repetition.*

This list $(Z_i)_{i\in\mathbb{N}}$ will serve as the input for the finite junk collector, which is briefly described below.

The finite junk collector. This global strategy must ensure that every isomorphism type of a finite abelian p-group is represented in the enumeration. As usual, we can identify such groups with finite equivalence relations. Recall also Modification 1.

The input is a uniform enumeration $(Z_i)_{i\in\mathbb{N}}$ of abelian p-groups. At every stage, each finite $Z_i[s]$ is identified with the respective equivalence structure $E_{Z_i[s]}$, with all possible uniformity. In the construction, this list is produced collectively by $\mathcal{T}$, sub-modules of the main module, and the infinite junk collector. We will need only the following dynamic properties of this list. These properties are immediate consequences of Modification 1 and the analysis contained in §9.4.6.

(b1): At every stage s, there is at most one i for which $Z_i[s+1]$ is larger than $Z_i[s]$. In this case, we also assume that the cardinality of $Z_i[s+1]$ is larger than any number mentioned so far in the construction. This applies to the case when $Z_i[s+1]$ is newly introduced too.

(b2): At every stage s, the finite list $(Z_i[s])_{i\leqslant s}$ contains no repetition up to isomorphism.

The work of the finite junk collector is summarised as follows.

Proposition 9.4.9. *On input of a uniform enumeration $(Z_i)_{i\in\mathbb{N}}$ which is injective on isomorphism types and satisfies (b1)-(b2), the finite junk collector produces a uniform enumeration $(B_i)_{i\in\mathbb{N}}$ which is injective on isomorphism types and mentions each isomorphism type from $(Z_i)_{i\in\mathbb{N}}$ and each finite isomorphism type.*

This finishes the description of the finite junk collector.

9.4.8 Construction

We are ready to put all the essential components together. The construction consists of three phases. The output of the first phase is the input of the second phase, and the output of the second is the input of the third. Apart from this obvious correlation via the input/output, there is no further interaction between the three phases.

Phase 1: Fix the effective enumeration $(F_i, E_j, z)_{i,j,z\in\mathbb{N}}$ which was defined at the beginning of §9.4.1. Also, fix the tree of strategies $\mathcal{T}$ defined in §9.4.6 and the strategies associated with its nodes.

In the first phase, we let all the basic strategies associated with each triple (F_i, E_j, z) and the strategies associated with $\mathcal{T}$ act according to their instructions; the instructions can be found in §§ 9.4.2 and 9.4.6, respectively.

Working together, these strategies produce a uniform enumeration $(L_i)_{i\in\mathbb{N}}$ of computable abelian groups which, as we shall argue, satisfy conditions $(a1)$-$(a3)$ from §9.4.7.

Phase 2: On input the enumeration $(L_i)_{i\in\mathbb{N}}$ listed at Phase 1, let the infinite junk collector act according to its instructions, as described in §9.4.7. Let $(Z_i)_{i\in\mathbb{N}}$ be the uniform enumeration produced as the result of these actions. We will argue that $(Z_i)_{i\in\mathbb{N}}$ will satisfy conditions $(b1)$-$(b2)$ from §9.4.7.

Phase 3: On input $(Z_i)_{i\in\mathbb{N}}$, let the finite junk collector act according to its instructions and produce a uniform enumeration $(B_i)_{i\in\mathbb{N}}$.

Finally, fix some Friedberg enumeration $(M_i)_{i\in\mathbb{N}}$ of all abelian p-groups of Ulm type 1, which correspond to equivalence structures having finitely many classes, at least one of which is infinite, and to eventually bounded equivalence structures having infinitely many classes. Merge $(B_i)_{i\in\mathbb{N}}$ with $(M_i)_{i\in\mathbb{N}}$ to produce an enumeration $(C_i)_{i\in\mathbb{N}}$. (For $i = 0, 1, \ldots$, set $C_{2i+1} = B_i$ and $C_{2i} = M_i$.)

We will argue that $(C_i)_{i\in\mathbb{N}}$ is a Friedberg enumeration of all computable abelian groups of Ulm type $\leqslant n$.

9.4.9 Verification

As usual, we identify equivalence structures and the respective Ulm type 1 groups throughout.

Lemma 9.4.10. *The enumeration $(L_i)_{i\in\mathbb{N}}$ produced at Phase 1 has the following properties:*

1. *It contains no repetition, up to isomorphism.*
2. *It includes all isomorphism types of computable abelian p-groups of Ulm types m, $1 < m < n$.*
3. *It mentions each isomorphism type of computable abelian p-groups having Ulm type 1 in which there are arbitrarily large finite cyclic summands.*
4. *It mentions some abelian p-groups of Ulm type 1 corresponding to infinite junk structures. In each of these cases, the respective group L_i in the list comes with an eventually stable sequence $(l^i_s)_{s\in\mathbb{N}}$ such that the number $l^i = \lim_s l^i_s$ describes the sizes of the finitely many exceptional classes in E_{L_i}. (If L_i is not infinite junk, then $(l^i_s)_{s\in\mathbb{N}}$ will be divergent.)*
5. *It includes some finite abelian p-groups.*
6. *Apart from the isomorphism types described in (2)-(4), no further isomorphism types will be enumerated.*
7. *It satisfies (a1)-(a3) from §9.4.7.*

Proof. (2): As we argued in §9.4.1, every computable abelian p-group A must have A' true and A/A' proper. Theorem 9.3.52 implies that, for some pair (F_i, E_j) we will have $F_i \cong A'$ and $A/A' \cong E_j$. The Σ^0_4-predicate described in §9.4.2 holds for this pair. In particular, for exactly one z the basic strategy working with (F_i, E_j, z) has a true Π^0_3-outcome; see §9.4.4 for the detailed analysis of the outcomes. Under this outcome, the strategy produces $T_{H_j}(F_i) \cong A$.

(3): See §9.4.6 for a detailed analysis of the structures produced by $\mathcal{T}$.

(4): This is explained in §9.4.6.

(5): This is merely an observation based on the descriptions of the strategies.

(6): This follows from the detailed analysis of the outcomes contained in §§ 9.4.4 and 9.4.6.

(7): Condition (a1) is a triviality, (a2) is a reformulation of (4) of this lemma, and (a3) is Modification 2 in §9.4.6.

(1): We use the same notation as in the proof of (2) of this lemma. Since the enumerations $(F_i)_{i\in\mathbb{N}}$ and $(E_j)_{j\in\mathbb{N}}$ are Friedberg and since we assumed uniqueness of existential witnesses throughout, the groups produced under true Π^0_3-outcomes corresponding to different pairs (F_i, E_j) are non-isomorphic, and there is at most one true Π^0_3-outcome for each such pair. The true Π^0_3-outcomes of strategies along the true path of $\mathcal{T}$ witness that the construction produces a complete list of all computable equivalence structures having arbitrarily large finite classes; see §9.4.6.

Modification 2 and the analysis contained in §9.4.6 imply that infinite junk structures produced by different strategies cannot be isomorphic. Finally, Modification 1 in the proof of Proposition 9.3.48 and the analysis contained in §9.4.6 guarantee that finite structures that appear in the list have no repetition, up to isomorphism; indeed, they all have distinct cardinalities. □

Lemma 9.4.11. *The enumeration $(Z_i)_{i\in\mathbb{N}}$ produced at Phase 2 has the following properties:*

1. *It contains no repetition, up to isomorphism.*

2. *It includes all isomorphism types which appear in* $(L_i)_{i \in \mathbb{N}}$.

3. *It includes all isomorphism types of infinite junk structures.*

4. *It satisfies* (b1) *and* (b2) *from 9.4.7.*

Proof. (1), (2), and (3) follow from Proposition 9.4.8, and (4) is an immediate consequence of Modification 1 and the analysis contained in §9.4.6; see also §9.4.7. □

Lemma 9.4.12. *The enumeration* $(B_i)_{i \in \mathbb{N}}$ *produced at Phase 3 has the following properties:*

1. *It contains no repetition, up to isomorphism.*

2. *It includes all isomorphism types which appear in* $(Z_i)_{i \in \mathbb{N}}$.

3. *It includes all isomorphism types of finite groups.*

Proof. This is a reformulation of Proposition 9.4.9. □

Combining the three lemmas above, we conclude that $(B_i)_{i \in \omega}$ is a Friedberg enumeration of almost all computable Ulm type $\leqslant n$ groups. This enumeration does not include the following special isomorphism classes of groups, namely:

1. abelian p-groups of Ulm type 1 which correspond to equivalence structures having finitely many classes, at least one of which is infinite, and

2. abelian p-groups corresponding to eventually bounded equivalence structures having infinitely many classes.

These two isomorphism classes have a combined uniformly computable Friedberg enumeration which we denote by $(M_i)_{i \in \mathbb{N}}$. By merging $(M_i)_{i \in \mathbb{N}}$ with $(B_i)_{i \in \omega}$, we obtain a computable Friedberg enumeration of all computable Ulm type $\leqslant n$ abelian p-groups, as desired.

The proof of Theorem 9.4.1 is complete.

Exercises

Exercise° 9.4.13. Show that there is a Friedberg enumeration of the following classes:

1. vector spaces over a fixed computable field;
2. algebraically closed fields;
3. well orderings of order-type less than a fixed computable ordinal α;
4. finitely generated abelian groups;
5. compact oriented surfaces (represented as finite simplicial complexes);
6. abelian p-groups of bounded order.

Exercise° 9.4.14. Show that there is no uniform list of finite presentations (given by their strong indices, as in Exercise 7.1.58) of f.p. groups in which every group is repeated at most finitely many times, up to isomorphism. (Hint: A *Markov property* P of finitely presentable groups is one for which:

1. P is preserved under group isomorphism.
2. There exists a finitely presentable group A_+ with property P.
3. There exists a finitely presentable group A_- that cannot be embedded as a subgroup in any finitely presentable group with property P.

The Adian–Rabin Theorem ([2, 439]) states: *Let P be a Markov property of finitely presentable groups. Then there is no algorithm that, given a finite presentation $\langle X \mid R\rangle$, decides whether or not the group defined by this presentation has property P.* Let P be the property "being the trivial group $T \cong \{e\}$", and suppose there is a list $V = (V_i)_{i\in\mathbb{N}}$ of finite presentations with the required properties. Let $V_{i_0}, \ldots, V_{i_k}$ be the only presentations of T in this list. To decide whether $\langle X \mid R\rangle \cong T$, wait for $\langle X \mid R\rangle \cong V_i$ for some i. This is c.e. by Exercise 7.1.58. See if $i = i_j$ for some $j \leqslant k$.)

Exercise* 9.4.15 (Lange, Miller, and Steiner [330]). There is a Friedberg enumeration of the family of computable algebraic fields.

Exercise* 9.4.16 (Hoyrup, Melnikov, and Ng [267]). The notion of a computable topological presentation was introduced in Definition 2.4.26. Recall that non-homeomorphic (Polish) spaces can share the same computable topological presentation (Exercise 2.4.27). Further, every Polish (more generally, countably based T_0) space admits a computable topological presentation (Exercise 4.2.105).

Show that there is a uniformly computable sequence of computable topological presentations $(T_i)_{i\in\mathbb{N}}$, so that every compact Polish space is represented by *exactly one* T_i from this sequence. The uniform sequence is given by parameters describing the cases (1)-(4):

(1) The space is finite.

(2) The space has a perfect kernel and has exactly $m \geqslant 0$ isolated points, $m \in \mathbb{N}$.

(3) Compact Polish spaces with infinitely many isolated points, in which isolated points are dense.

(4) Compact Polish spaces having infinitely many isolated points, but so that the isolated points are not dense.

For example, in (2) with $m = 0$, the presentation can be taken to be the standard computable topological presentation of 2^ω.

9.5 Computable profinite abelian groups

Under Pontryagin duality (see Section 5.2), discrete torsion abelian groups correspond to profinite abelian groups (and vice versa). The goal of this section is to show that this correspondence is uniformly computable, and furthermore, it preserves computable categoricity. As a consequence of this duality and the main result of the previous subsection (Theorem 9.4.1), we will obtain:

Theorem 9.5.1 (Downey, Melnikov, and Ng [145], Melnikov [373]). *For any fixed $n > 0$, there is a Friedberg enumeration of all computably compact pro-p abelian groups of pro-Ulm type $\leqslant n$.*

This result appeared in the Introduction as Theorem E. In the theorem, the pro-Ulm type of a pro-p group is the Ulm type of the Pontryagin dual of the group, which is a discrete abelian p-group (e.g., [297]). We will also derive a complete classification of computably categorical pro-p abelian groups.

9.5.1 Background

In Theorem 4.2.107 we showed that, for a profinite group P, P has a computably compact presentation *iff* it has a "recursive" presentation, in the following sense. There is a computable sequence of finite groups and surjective homomorphisms $0 \leftarrow_{f_0} A_0 \leftarrow_{f_1} A_1 \leftarrow_{f_2} \dots$ so that the group is homeomorphic to the projective (inverse) limit of this system. In other words, every element of the group is a sequence $(a_i)_{i \in \mathbb{N}}$, where $f_i(a_i) = a_{i-1}$ and $a_i \in A_i$; the operation is defined component-wise. All finite objects in such a presentation are given by their strong indices, i.e., as finite tuples.

A recursive profinite group can be viewed as the collection of (infinite) paths through a computably branching tree with no dead ends. In such a presentation, every (infinite) path represents an element of the group, and the operations are represented by computable operators acting on this totally disconnected topological space. We can define an (ultra)metric on this space, similarly to how it is done for Cantor space. With respect to this metric, the presentation becomes a computably compact Polish group. In this sense, a recursive profinite group is just a computably compact presentation of the group with some nice additional properties. Thus, for instance, we can define what we mean by a computable homeomorphism from a recursive profinite group to some other computable space or group. Furthermore, the non-trivial implication in Theorem 4.2.107 showed a bit more than was stated in the theorem:

Corollary 9.5.2. *Every computably compact profinite group is* computably *topologically isomorphic to a recursive profinite group.*

Proof. Exercise 9.5.11. □

It follows from the corollary above and the discussion before it that computably compact and recursive presentations of profinite groups are computably indistinguishable. Moreover, one can be replaced by the other with all possible effective uniformity. In particular, in the profinite case, the definition below can be restricted to recursive presentations without any loss of generality.

Definition 9.5.3. A (compact) Polish group is *computably categorical* if any two computably compact presentations of the group are computably topologically isomorphic.

Recall that we write $\widehat{A}$ for the Pontryagin dual of a compact or discrete abelian group A. To establish the effective Pontryagin duality between computable profinite and computable discrete torsion groups, we need several elementary properties of finite abelian groups and their duals.

Fact 9.5.4 (Folklore). *1. If C is cyclic, then $\widehat{C} \cong C$.*

2. If A, B are finite abelian, then $\widehat{A \oplus B} \cong \widehat{A} \oplus \widehat{B}$.

3. For every finite abelian A, $\widehat{A} \cong A$.

4. If $\phi : A \to B$ is a homomorphism of finite abelian groups, then $\widehat{\phi} : \widehat{B} \to \widehat{A}$ defined by the rule

$$\widehat{\phi}(\chi) = \chi \circ \phi$$

is a homomorphism between their duals.

5. In 4., ϕ is surjective iff *$\widehat{\phi}$ is injective.*

Proof. Exercise 9.5.12. □

Recall that the direct limit of a sequence of groups

$$0 \to A_0 \to A_1 \to \ldots,$$

where all arrows stand for injective embeddings, is essentially the union of the A_i, in which elements of A_i are identified with their images in A_{i+1} (etc.) throughout. (The embeddings do not have to be injective; but in this section, they will be such.) The lemma below is immediate.

Lemma 9.5.5. *Every (countable, discrete) torsion abelian group A can be viewed as a direct limit of finite groups and injective maps:*

$$0 \to A_0 \to A_1 \to \ldots.$$

Additionally, a torsion abelian group is computable iff *there is a direct system as above in which all finite groups and maps are given by their strong indices.*

The fact below essentially says that $\widehat{\varinjlim(A_i, \phi_i)} \cong \varprojlim(\widehat{A}_i, \widehat{\phi}_i)$ and $\widehat{\varprojlim(A_i, \phi_i)} \cong \varinjlim(\widehat{A}_i, \widehat{\phi}_i)$ for finite abelian groups A_i. It appears to be an old folklore, see the last chapter of [194]; we also cite [404]. A generalisation of this fact to locally compact A_i is the main result of [284].

Fact 9.5.6. *Suppose A is the inverse limit of a sequence*

$$0 \leftarrow_{f_0} A_0 \leftarrow_{f_1} A_1 \leftarrow_{f_2} \cdots$$

of finite abelian groups and surjective homomorphisms. Then $\widehat{A}$ is isomorphic to the direct limit of

$$0 \to_{\widehat{f_0}} \widehat{A_0} \to_{\widehat{f_1}} \widehat{A_1} \to_{\widehat{f_2}} \cdots,$$

where the dual maps are injective.

Conversely, suppose A is the direct limit of a sequence

$$0 \to_{f_0} A_0 \to_{f_1} A_1 \to_{f_2} \cdots$$

of finite abelian groups under injective embeddings. Then $\widehat{A}$ is isomorphic to the inverse limit of the sequence

$$0 \leftarrow_{\widehat{f_0}} \widehat{A_0} \leftarrow_{\widehat{f_1}} \widehat{A_1} \leftarrow_{\widehat{f_2}} \cdots,$$

where the dual maps are surjective.

Proof. Omitted. (This is essentially iterated Fact 9.5.4. See also the proof of Lemma 9.5.8.) □

For example, $\widehat{\mathbb{Z}_{p^\infty}} \cong J_p$, where J_p is the additive group of the p-adic integers. This is because the former is the direct limit of cyclic p-groups $\mathbb{Z}_{p^k}$, and the latter is their natural inverse limit.

9.5.2 Effective Pontryagin duality: the profinite case

We prove the following

Theorem 9.5.7 (Melnikov [373]). *Let P be a profinite abelian group.*

1. *P has a computably compact presentation iff $\widehat{P}$ is a computable presentation.*
2. *P is computably categorical iff $\widehat{P}$ is computably categorical.*

Proof. In the lemma below, all finite objects are given by their strong indices.

Lemma 9.5.8. *There is a computably uniform procedure which, on input finite abelian groups A, B and a homomorphism $\phi : A \to B$, outputs $\widehat{A}, \widehat{B}$ and the dual homomorphism $\widehat{\phi} : \widehat{B} \to \widehat{A}$.*

Proof. Using blind search, produce full direct decompositions of $A = \bigoplus_i A_i$ and $B = \bigoplus_j B_j$ into finitely many cyclic summands. Fix some generators a_i of b_j for these summands, one from each summand. Each $\chi : A \to \mathbb{T}$ is fully determined by the value of $\chi(a_i)$. The possible values of each fixed a_i under homomorphisms to $\mathbb{T}$ range over a finite set of rational numbers in $\mathbb{T}$ (viewed as $[0, 1]$ with 0 identified with 1), whose size is effectively and uniformly determined by the order of a_i. This gives a way of computing $\widehat{A}$ by going through all possibilities. In the resulting computable presentation of $\widehat{A}$, each $\chi \in \widehat{A}$ is additionally explicitly represented as a (strong index of the finite) map from A to $\mathbb{T} \cap \mathbb{Q}$. The same analysis applies to B and $\widehat{B}$.

We can use these representations of $\widehat{A}$ and $\widehat{B}$ to explicitly calculate $\widehat{\phi}(\chi) = \chi \circ \phi$, for each $\chi \in \widehat{B}$ (recall 4. of Fact 9.5.4). We then use a brute-force search to see which element of $\widehat{A}$ is equal to $\widehat{\phi}(\chi)$. □

Remark 9.5.9. In the preceding fact, $\widehat{A}$ can be additionally explicitly represented as a finite set of maps from A to $\mathbb{T} \cap \mathbb{Q}$ (i.e., the rational points in $\mathbb{T}$); the same is true for $\widehat{B}$.

We now turn to the proof of Theorem 9.5.7. By Theorem 4.2.107 and Lemma 9.5.5, to prove (1) of Theorem 9.5.7 it is sufficient to prove that, under Pontryagin duality, computable (surjective, linear) inverse systems of finite groups correspond to computable (injective, linear) direct systems of finite groups. But this is guaranteed by Lemma 9.5.8.

We prove (2). Unfortunately, we have to rely on some further well-known facts from the literature (e.g., [404]) which we will just state here without proof. It is well-known that, for a compact or discrete G, a topological homeomorphism witnessing $G \cong \widehat{\widehat{G}}$ can be chosen in the following canonical way:

$$g \to \langle \cdot, g \rangle,$$

where $\langle \chi, g \rangle = \chi(g)$ for any $\chi \in \widehat{G}$. But it follows from the proof of Lemma 9.5.8 (see Remark 9.5.9) that in the profinite/torsion discrete case, the presentation of $\widehat{G}$ obtained from G in (1) has the following property. It is effectively given by finite approximations of characters. In particular, the canonical isomorphism $i : G \to \widehat{\widehat{G}}$ is computable. When G is computable discrete, i^{-1} is also obviously computable. When G is computably compact profinite, i^{-1} is computable by Theorem 4.2.57.

Now let G be either computable discrete torsion or computably compact profinite, and assume $\widehat{G}$ is computably categorical. If A and B are effective presentations of G, calculate $\widehat{A}$ and $\widehat{B}$ using (1) of the theorem, and fix a computable isomorphism

$$\phi : \widehat{A} \to \widehat{B}.$$

It is well-known that 4. and 5. of Fact 9.5.4 can be extended to arbitrary locally compact groups. In particular, it follows that

$$\widehat{\phi}(\chi) = \chi \circ \phi,$$

where $\chi \in \widehat{\widehat{B}}$ and $\chi \circ \phi \in \widehat{\widehat{A}}$, is a topological group-isomorphism from $\widehat{\widehat{B}}$ onto $\widehat{\widehat{A}}$. Since ϕ is computable, $\widehat{\phi}$ is computable as well. (Exercise 9.5.15.) Let $i_0 : A \to \widehat{\widehat{A}}$ and $i_1 : B \to \widehat{\widehat{B}}$ be computable isomorphisms, existence of which was established earlier. Then $i_0^{-1} \circ \widehat{\phi} \circ i_1$ is the desired computable isomorphism from B onto A. □

9.5.3 Enumerating pro-p groups (Theorem E).

Theorem E (Theorem 9.5.1) states that for any fixed $n > 0$, there is a Friedberg enumeration of all computably compact pro-p abelian groups of pro-Ulm type $\leqslant n$.

Proof of Theorem E. The theorem is an immediate consequence of Theorem 9.4.1 and (1) of Theorem 9.5.7. □

Also, Theorem 9.3.26 combined with (2) of Theorem 9.5.7 gives:

Theorem 9.5.10 (Melnikov [373]). *A pro-p abelian group P is computably categorical iff it is homeomorphic to a (topological) direct product of cyclic p-groups and the group of p-adic integers J_p, in which all but finitely many factors are isomorphic to some fixed cyclic group or to J_p.*

With a bit of extra work, Theorem 9.3.37 implies that the index set of computably categorical profinite abelian groups is Π^0_4-complete ([373]). Since we omitted the proof of Theorem 9.3.37, we omit the proof of this fact as well. We, however, note that this Π^0_4-completeness result is likely the best possible description of computably categorical profinite abelian groups; see the discussion after Theorem 9.3.37.

Exercises

Exercise° 9.5.11. Prove Corollary 9.5.2.

Exercise° 9.5.12. Prove Fact 9.5.4.

Exercise° 9.5.13. Verify that the duality between torsion and profinite abelian groups preserves degree spectra. Use Exercise 8.3.43 and 8.3.42 to derive analogous results about profinite groups.

Exercise° 9.5.14 (Brodhead and Cenzer [65])**.** Prove that there is a Friedberg enumeration of all Π^0_1 classes (up to equality).

Exercise 9.5.15. Let $\widehat{\phi}(\chi) = \chi \circ \phi$ be the map defined in the last paragraph of the proof of Theorem 9.5.7 (2). Verify that this map is computable in the right sense, in both the profinite and the discrete cases. (Hint: Use the special properties of computable presentations of the duals that are produced in the proof of (1) of Theorem 9.5.7; i.e., Remark 9.5.9.)

Exercise 9.5.16 (Koh, Melnikov, and Ng [313])**.** The notion of an effectively continuous map can be extended naturally to right-c.e. spaces, as follows. Just repeat the definitions of c.e. names of open sets and require that $f^{-1}(W)$ is uniformly c.e. open for a c.e. open W. Using this notion, one can define what it means for a group to be right-c.e. presented. Show that there is a right-c.e. presented profinite group that is not topologically isomorphic to any computable Polish group.

Exercise 9.5.17 (Essentially [373])**.** Show that there exists a profinite group that has a computable Polish presentation but has no computably compact presentation. Conclude that there exists a computable Polish group that is approximable (Definition 2.4.8) but has no computably approximable presentation (Definition 2.4.9).

Exercise* 9.5.18. Prove Theorem 5.2.24 from Part 1 using an analogy of Fact 9.5.6 for torsion-free groups and their duals. [Note that the 1st cohomology of the unit circle $\mathbb{T}$ is $\mathbb{Z}$, and so is its dual. For a formal proof, see pp. 474–477 of [263].]

9.6 Further related results*

For a detailed exposition of the theory of computable abelian groups, we cite the surveys [291] and [370]. For a discussion of result related to (computable) profinite groups and totally disconnected locally compact groups, see §4.2.7. We remark that the effective Pontryagin duality for profinite and torsion groups (Theorem 9.5.7) established in this section enjoys many uniformly effective properties, essentially completely reducing the study of computable profinite groups to the theory of computably presented torsion abelian groups. As a consequence, most results about computable (discrete) torsion abelian groups that will appear throughout the rest of the book as exercises can be easily restated for profinite abelian groups as well; for example, see Exercises 10.1.116, 10.1.102, and 10.1.104. We will, however, usually omit the dual statements in these exercises.

Very little is known about computable abelian p-groups of Ulm type $\geqslant \omega$. We mention that in [74], Calvert gave sharp estimates for the isomorphism problem of reduced abelian p-groups of Ulm type $\leqslant \alpha$, where α is a computable ordinal (this covers all possible Ulm types by Exercise 10.1.63). The complexity of their index set is calculated in [76]. We omit the statements, but we note that these hyperarithmetical estimates of course increase monotonically α.

Here, we state only one further result that seems most related to the material of this chapter. Recall Khisamiev's characterisation of computably presented reduced abelian p-groups of finite Ulm type (Theorem 9.3). In this characterisation, χ_{A_i} is a Σ^0_{2i+2} set, and $\#A_i := \{n : (n,1) \in \chi_{A_i}\}$ has to be $\mathbf{0}^{(2i)}$-l.m.. It has been open for several decades whether Theorem 9.3 can be extended in any meaningful way to cover groups of Ulm type ω (and beyond). The obstacle is the non-uniformity of all known proofs of Theorem 9.3. If A is computable, then the index of the $\mathbf{0}^{(2i)}$-l.m. set is $\mathbf{0}^{(2i+3)}$-computable; note the extra three jumps required to find the index over the natural complexity of the function itself. Using the first known example of an iterated $\mathbf{0}'''$ argument, Downey, Melnikov, and Ng [140] proved:

Theorem 9.6.1. *The exists a computable reduced abelian p-group G of Ulm type ω such that the sequence of sets of indices for $\mathbf{0}^{(2i)}$-l.m. functions corresponding to $\#G_i$ is* not *uniformly Δ^0_{2i+3}.*

The theorem seems to suggest that some new idea is required to classify such computable groups, if a reasonable classification exists at all. Perhaps, no such classification should be anticipated. After all, the class of abelian p-groups of Ulm type ω is *not an arithmetical class.*

9.7 What's next?

In the next chapter, we present the foundations of the abstract theory of computably categorical structures. Interestingly, some of these results bear a strong resemblance to the results in Chapter 2, where we established the foundations of computable analysis. The connections with computable analysis in the next chapter go beyond mere analogy, as the main result of the subsequent chapter (Theorem F) is stated for Polish spaces up to isometry.

Chapter 10

Computable categoricity and computable dimension

In this section, we study computable structures and spaces up to computable isomorphism. The plan is as follows:

1. In Section 10.1, we establish a sequence of results that relate definability, categoricity, and decidability of structures. We also separate the notions of "plain" and relative computable categoricity for discrete algebraic structures.
2. In Section 10.2 we prove Theorem 10.2.9, which provides a syntactic characterisation of relative computable categoricity for Polish spaces viewed *up to isometry*. We apply this characterisation to the Urysohn space.
3. In Section 10.3, we investigate the problem of the number of computable presentations of a structure. We outline the proof of the well-known theorem of Goncharov, which states that there is a structure with exactly two computable presentations, up to computable isomorphism (Theorem 10.3.2). As far as we are aware, this is the first time Goncharov's proof is presented in book format. We also prove a powerful meta-theorem, Theorem 10.3.20, that generalises another well-known theorem of Goncharov (Theorem 10.3.2) to separable spaces *up to isometry*.

We finish the chapter, and the book, with a detailed proof of the following:

Theorem F (Melnikov and Ng [376]). The space $(C[0,1], d_{\mathrm{sup}})$ has infinitely many isometric, but not computably isometric, computable Polish presentations.

Theorem F follows from Theorem 10.3.20 combined with Theorem 2.4.20, which was the main result of Chapter 2.

R. G. Downey, A. Melnikov, *Computable Structure Theory*, Theory and Applications of Computability, https://doi.org/10.1007/978-3-031-92433-0_10

10.1 Relative computable categoricity for algebraic structures

In this section, we investigate the Type II version of computable categoricity, known as *relative computable categoricity*, focusing on discrete countable algebraic structures. The reader will notice parallels with our treatment of Type 2 computability in Chapter 2, particularly the Kreisel-Lacombe-Shoenfield-Markov Theorem 2.3.7 and Specker's Theorem 2.3.24. We will extend this general methodology to Polish spaces in the following section.

10.1.1 Relative computable categoricity

Many classifications of computably categorical algebraic structures exhibit a certain similarity. Computably categorical linear orderings are those with only a *finite* number of adjacencies (Theorem 3.2.2), computably categorical Boolean algebras are those with only a *finite* number of atoms (Theorem 4.1.17). In the class of torsion-free abelian groups, computable categoricity is captured by the *finiteness* of the rank of the group (Theorem 5.1.43), and the computable categoricity of abelian p-groups is also characterised by a certain *finite* parameter (Theorem 9.3.26). All these structures are not only computably categorical, but also possess the following apparently stronger property (as verified in Exercise 10.1.24).

Definition 10.1.1. We say that a computable structure A is *relatively* computably categorical if, for any isomorphic copy C of A, if C is $\mathbf{a}$-computable, then there exists an $\mathbf{a}$-computable isomorphism witnessing $A \cong C$.

Equivalently, if B_1 and B_2 are isomorphic copies of A, then we have

$$B_1 \cong_{\Delta_1^0(B_1 \oplus B_2)} B_2,$$

i.e., there is an isomorphism between B_1 and B_2 computable from the join of the open diagrams of B_1 and B_2. Now, it is not entirely clear whether this is a new notion. Perhaps all computably categorical computable structures are also relatively computably categorical. As we will see in Corollary 10.1.17, these two notions are indeed different. Whereas "plain" computable categoricity seems a purely computability-theoretic notion concerned with the intricacies of computation, relative computable categoricity links computability with the classical model-theoretic theme that relates definability to properties of structures.

The Scott Isomorphism Theorem

Fix a first-order language L, e.g., the language of groups or rings with identity. We have already mentioned infinitary formulae, but we will remind the reader that $L_{\omega_1\omega}$ is the logic with the usual first-order connectives and quantifiers, together with infinite countable disjunctions and conjunctions, and countably many variables.

To be more precise, we define Σ_0 and Π_0 formulae as the finitary (i.e., the usual first-order) quantifier-free ones. For $\omega_1 > \alpha > 0$, a Σ_α formula is one that is a countable disjunction of formulae of the form $\exists \overline{u}\, \psi(\overline{x}, \overline{u})$, where each ψ is a Π_β formula for some $\beta < \alpha$. The definition of Π_α is dual and uses a countable infinite conjunction of $\forall \overline{u}\, \psi(\overline{x}, \overline{u})$, where ψ is Σ_β for some $\beta < \alpha$. One striking property of infinitary sentences is that they can describe countable structures up to isomorphism.

Theorem 10.1.2 (Scott [461]). *Every countable structure has a Scott sentence, i.e., a sentence of $L_{\omega_1\omega}$ whose only models are ones isomorphic to A.*

Proof sketch. Given two tuples $\bar{a}, \bar{b} \in A$ of the same length, if they do not satisfy the same infinitary formulae, then let β be the least ordinal such that there is a Π_β-formula $\psi(\bar{x})$ which is true of $\bar{a}$ but not of $\bar{b}$, or vice versa. Since there are countably many finite tuples, there exists a countable ordinal α such that if tuples satisfy the same Π_α-formulae, then they satisfy the same $L_{\omega_1\omega}$-formulae. Without loss of generality, we can assume $\alpha > 1$.

Given a tuple $\bar{a} \in A$, let $\psi_{\bar{a}}(\bar{x})$ be the infinite conjunction of all Π_α-formulae true of $\bar{a}$ in A. We claim that the formula $\psi_{\bar{a}}(\bar{x})$ fully describes the automorphism type of $\bar{a}$ in A. This is because if $A \models \psi_{\bar{a}}(\bar{b})$ and $c \in A$, then $A \models \exists x \psi_{\bar{a},c}(\bar{b}, x)$, since $\bar{a}$ and $\bar{b}$ must satisfy the same infinitary formulae. Thus, one can use the family $\Lambda_A = \{\psi_{\bar{a}}\}_{\bar{a} \in A^{<\omega}}$ to construct an isomorphism between any two isomorphic copies of A using the usual back-and-forth method (to be detailed in the proof of Theorem 10.1.6).

The following sentence ensures that one can always pick an extension of $\bar{a}$ and find a formula in Λ_A that describes it:

$$\phi_{\bar{a}} = \forall \bar{x}(\psi_{\bar{a}}(\bar{x}) \to \bigwedge_{c \in A} \exists y \psi_{\bar{a},c}(\bar{a}, y)), \text{ and also } \phi_{\varnothing} = \bigwedge_{c \in A} \exists y \psi_c(y).$$

To guarantee that one can always find a suitable element realising a formula from the family, we use:

$$\eta_{\bar{a}} = \forall \bar{x}(\psi_{\bar{a}}(\bar{x}) \to \forall y \bigvee_{c \in A} \psi_{\bar{a},c}(\bar{a}, y)), \text{ and also } \eta_{\varnothing} = \forall y \bigvee_{c \in A} \psi_c(y).$$

The Scott sentence is

$$\bigwedge_{\bar{a} \in A^{<\omega}} \phi_{\bar{a}} \wedge \eta_{\bar{a}},$$

and it essentially says that Λ_A can be used to run a back-and-forth argument between any copies of A. □

Scott Families

The key step in the proof sketch of Theorem 10.1.2 was the use of a family Λ_A of infinitary formulae describing the automorphism orbits of tuples in A.

Definition 10.1.3. A *Scott family* for a countable structure A is a countable family Λ of $L_{\omega_1\omega}$ formulae with finitely many (fixed) parameters from A such that

(i) Each tuple $\overline{a}$ satisfies some $\psi \in \Lambda$.

(ii) If $|\overline{a}| = |\overline{b}|$ and they satisfy the same $\psi \in \Lambda$, then there is an automorphism of A taking $\overline{a} \mapsto \overline{b}$.

These observations suggest the following idea:

> Since infinitary formulae can be used to describe the automorphism orbits of tuples in countable structures, surely we can use *computable* ones to classify *computable* automorphism orbits in *computable* structures.

We will see that, at least in the case of *relative* computable categoricity, this intuition is correct and can lead to powerful results. (We also remark that the computable version of the Scott Isomorphism Theorem *fails*. We won't focus on this negative result and leave the exact statements to Exercise 10.1.120 and Exercise 10.3.19.)

Computable Scott families

Fix a computable language L. We define the class of *computable* infinitary formulae, $L^c_{\omega_1\omega}$, as follows.

Definition 10.1.4 (Computable infinitary formulae)**.** The Σ^c_0 and Π^c_0 formulas are the finitary first-order quantifier-free formulas. For $\alpha > 0$, a (notation for) a computable ordinal, a computable Σ_α formula is one that is a c.e. disjunction of formulas of the form $\exists \overline{u}\psi(\overline{x}, \overline{u})$, where each ψ is a computable Π_β formula for some computable $\beta < \alpha$. The dual class of computable Π_α formulae is defined similarly, but using computable conjunctions instead of disjunctions.

We will use the notation Σ^c_α and Π^c_α to emphasise that these are *computable* infinitary formulae. (Otherwise, we omit "c".) There are some technicalities regarding normal form (e.g., Exercise 10.1.49). Since we won't be too concerned with this, we leave the calculus of computable infinitary formulas to exercises (see §10.1.4). Some of the key properties of computable infinitary logic and results related to $L^c_{\omega_1\omega}$ will appear as exercises; see Exercise 10.1.49 onwards. We refer the reader to Ash and Knight [20] and Montalbán [401, 402] for a thorough exposition of this theory. Indeed, we will mainly focus on the case when α is finite, and in fact, we will rarely encounter Σ^c_α formulae with $\alpha > 2$.

The effective version of Definition 10.1.3 is as follows:

Definition 10.1.5. Let A be a countable (typically, computable) structure.

1. A Σ^0_1 *Scott family* (or *c.e. Scott family*) for A is one which is a c.e. set of (Gödel numbers of) first-order finitary existential formulae.
2. For a computable $\alpha > 1$, a Σ^0_α *Scott family* for A is one which is a c.e. set of (indices for) computable Σ^c_α formulae.

Note that in 1. we could have used computable infinitary Σ^c_1 formulae, but it would give the same notion. This is because $A \models \bigvee_{i\in\mathbb{N}} \exists\bar{x}\phi_i(\bar{x}, \bar{a})$ implies $\exists i\, A \models \exists\bar{x}\phi_i(\bar{x}, \bar{a})$, so we can just effectively search for such an i. Further, if two tuples satisfy the same infinite disjunctions, then they must be automorphic, and thus must satisfy the same disjunct $\exists\bar{x}\phi_i(\bar{x}, \bar{a})$.

Theorem 10.1.6 (Ventsov [498]). *Suppose that A is a computable structure. The following are equivalent.*

(i) A is relatively computably categorical.

(ii) A has a c.e. Scott family with a fixed tuple of constants $\overline{c}$.

Proof of Theorem 10.1.6. *(ii)* $\to$ *(i)*. Let Λ be the c.e. Scott family and suppose that B is a structure with $A \cong B$. Take $\overline{d}$ with $(A, \overline{c}) \cong (B, \overline{d})$. We construct the isomorphism f in stages, making the construction computable from B. So suppose that we have constructed f_s taking a finite part of A_s and a finite part of B_s, such that if $\overline{a} \mapsto \overline{b}$, then there exists $\varphi(\overline{c}, \overline{x}) \in \Lambda$, with

$$A \models \varphi(\overline{c}, \overline{a}) \wedge B \models \varphi(\overline{d}, \overline{b}).$$

Now at stage $s+1$, find the first $a \in A \backslash \{\overline{a}\}$ if $s+1$ is even, or symmetrically, the first $b \notin B \backslash \{\overline{b}\}$ if $s+1$ is odd, and in this latter case, find a pre-image of b. Assuming $s+1$ is even, find a formula $\theta(\overline{c}, \overline{a}, a) \in \Lambda$ and an element $b \in B$ with

$$A \models \theta(\overline{c}, \overline{a}, a) \wedge B \models \theta(\overline{d}, \overline{b}, b).$$

Define $f_{s+1}(a) = b$. Then $f = \cup_s f_s$ is a B-computable isomorphism taking A to B.

(i) $\to$ *(ii)*. Let A be a computable structure, and suppose it is relatively computably categorical. We shall replace all functions with their graphs in A and work with the open diagram $D(A)$ of A. We attempt to build B and an isomorphism $f : B \to A$ and diagonalise against all isomorphisms Φ_e^B from B to A. For some e, our diagonalisation attempt must fail, and this will allow us to produce a c.e. Scott family for the structure.

Without loss of generality, the domains of A and B will be ω. Then $D(B)$ will be defined using f by declaring $B \models \phi(\overline{b})$ if $A \models \phi(f(\overline{b}))$, for any atomic formula ϕ. We shall build B using the finite extension method (§3.1.2).

Notation. We identify a finite string $\sigma \in \omega^{<\omega}$ with the induced partial map. All our strings will be injective, meaning the respective maps are always assumed to be injective. We denote by $D(\sigma)$ the finite part of $D(B)$ restricted to the domain of σ.

We say that a finite partial injective map h *extends* another partial injective map g if $g \subseteq h$. We also say that h *properly extends* g if $g \subseteq h$, and if the domain of h contains the least $n \notin \mathrm{dom}(g)$ and the range of h contains the least $m \notin \mathrm{range}(g)$. We write $p \preceq q$ if p is an initial segment of q.

The definition below reflects that for q extending p, Φ_e looks a bit more like an extension of a potential isomorphism from B to A (assuming both p, q approximate f).

Definition 10.1.7. For two finite strings p, q and a functional Φ_e, we write $p \sqsubseteq_e q$ if $p \preceq q$ and the following conditions are satisfied:

1. $\Phi_e^{D(q)}$ is a partial isomorphism from (a subset of) $D(q) \subseteq B$ to A.
2. $\Phi_e^{D(q)}$ extends $\Phi_e^{D(p)}$.

If in 2., $\Phi_e^{D(q)}$ extends $\Phi_e^{D(p)}$ *properly*, then we write $p \sqsubset_e q$.

We now describe a diagonalisation attempt that must fail. Clearly, it is sufficient to attempt diagonalising against all Φ_e with $e > 0$.

Construction. Set $p_0 = \varnothing$.

Suppose p_{e-1} has been defined, and suppose $\Phi_1, \ldots, \Phi_{e-1}$ have been diagonalised against. To define p_e, check whether:

$$(\forall \tau \succeq p_{e-1}) \; [p_{e-1} \sqsubseteq_e \tau \rightarrow (\exists \rho) \; \tau \sqsubset_e \rho]. \tag{10.1}$$

If for some $\tau \succ p_{e-1}$ this property fails, then set p_e equal to any such τ. Declare Φ_e diagonalised.

Otherwise, proceed as follows. For all $j \geqslant e$, assuming p_{j-1} has been defined, pick $p_j \sqsupset_e p_{j-1}$. Declare that p_{e-1} *forces* Φ_e *to be an isomorphism*, written $p_{e-1} \Vdash \Phi_e$.

Set $f = \cup_e p_e$ and define B using pull-back via f, as explained earlier. This ends the construction of B.

Verification. We first observe that at every stage e we can pick an extension p_e of p_{e-1}. In particular, if $p_e \Vdash \Phi_e$, then this is guaranteed by (10.1). Thus, f and B are well-defined, and B is isomorphic to A via f, by construction.

We also claim that if Φ_e is declared diagonalised at stage e, then Φ_e is indeed not an isomorphism from B onto A. If $p_{e-1} \sqsubseteq_e \tau$ fails, then it must be because the Φ_e-computation along τ does not give a partial isomorphism to A. In this case, the assertion follows from the use principle. Otherwise, suppose $p_{e-1} \sqsubseteq_e \tau$ but there is no ρ with $\tau \sqsubset_e \rho$. In this case, Φ_e cannot be extended beyond $f \upharpoonright e = p_e = \tau$.

Since A is relatively computably categorical, for some $e > 0$ we must fail to diagonalise against Φ_e. This means that for $p_{e-1} : \{0, \ldots, e-1\} \mapsto \bar{c}$, we have that $p_{e-1} \Vdash \Phi_e$. In this case, we argue that *for any choice of* p_j^* ($j \geqslant e$) so that

$$p_e^* \sqsupset_e p_{e-1} \text{ and } p_j^* \sqsupset_e p_{j-1}^* \text{ for } j > e, \tag{10.2}$$

if we set $f^* = \sqcup_j p_j^*$ and let B^* be the resulting structure, then

$$\Phi_e^{D(B^*)} : B^* \rightarrow A$$

must be an isomorphism. Indeed, any such sequence of p_j^*-s witnesses that Φ_e is an isomorphism, according to Definition 10.1.7 (and by induction). This is, of course, also true about B in particular, but we shall need this stronger property of p_{e-1} to produce a c.e. Scott family. Note that we can computably enumerate all finite sequences $(p_j^*)_{e \leqslant j < k}$ that satisfy (10.2), because A is a computable structure, and thus we will eventually see whether conditions required by Definition 10.1.7 will be satisfied (note Φ_e has to be total on all such extensions of p_{e-1}). Finally, observe that any such finite sequence $(p_j^*)_{e \leqslant j < k}$ that satisfies (10.2) can be extended to an infinite such sequence.

Defining the Scott family Λ. Recall $p_{e-1} : \langle 0, \ldots, e-1 \rangle \mapsto \bar{c}$. The tuple $\bar{c}$ will be used as parameters for the family. To define Λ, proceed as follows.

List all finite sequences $(p_j^*)_{e \leqslant j < k}$, $k \in \mathbb{N}$, that satisfy (10.2), as well as all computations $\Phi_e^{D(p_j^*)}$ that define partial isomorphisms extending

$$p_{e-1} : \bar{c}' = \langle 0, \ldots, e-1 \rangle \mapsto \bar{c}.$$

In particular, each tuple $\bar{a}$ in A must eventually be in the range of one such $\Phi_e^{D(p_j^*)}$, say

$$\Phi_e^{D(p_j^*)} : \bar{b} \to \bar{a},$$

and assume this computation is the first found such. Let $\psi_{\bar{a}}(\bar{c}', \bar{d}, \bar{b})$ be the conjunction of the part of the (partial) diagram used in this computation. Set

$$\theta_{\bar{a}}(\bar{x}) = \exists \bar{y} \psi_{\bar{a}}(\bar{c}, \bar{y}, \bar{x}).$$

Define

$$\Lambda = \{\theta_{\bar{a}}(\bar{c}, \bar{x}) : \bar{a} \in A^{<\omega}\},$$

which is clearly c.e. and has one $\theta_{\bar{a}}(\bar{c}, \bar{x})$ for each $\bar{a} \in A^{<\omega}$.

Observe that for any B^* produced using extensions of p_j^* (fixed above for $\bar{a}$), $\Phi_e : B^* \to A$ isomorphically maps $\bar{c}'\bar{b}$ to $\bar{c}\bar{a}$. Since $\bar{c}'\bar{b}$ satisfies $\theta_{\bar{a}}(\bar{c}', \bar{b})$, then so does $\bar{c}\bar{a}$ (in A). This gives (i) of Definition 10.1.3 (for Λ).

It remains to verify (ii) of Definition 10.1.3 which, after adjusting notation, says that

$$A \models \theta_{\bar{a}}(\bar{c}, \bar{a}') \implies \bar{a} \text{ is automorphic to } \bar{a}' \text{ over } \bar{c}.$$

Note that $\bar{b}$ is just a tuple of natural numbers, and so is $\bar{c}' = \langle 0, \ldots, e-1 \rangle$. Consider B^{**} that extends p_{e-1} as follows. Consider p_j^{**} that assigns the exact same indices to the finite piece of the diagram corresponding to $\bar{c}, \bar{a}'$ (coded by $\psi_{\bar{a}}$) as p_j^* did for $\bar{c}, \bar{a}$. Since $p_{e-1} \Vdash \Phi_e$ and p_j^{**} with the described property still satisfies (10.2), it follows that we can build an isomorphism f^{**} and define B^{**} by running the construction above p_j^{**}.

We have that $\Phi_e : B^{**} \to A$ is an isomorphism that maps $\bar{c}'\bar{b}$ to $\bar{c}\bar{a}'$, and also, for the same Φ_e and via the exact same computation, $\Phi_e : B^* \to A$ isomorphically maps $\bar{c}'\bar{b}$ to $\bar{c}\bar{a}$.

Now, $\bar{c}'\bar{b}$ are the f^{**}-pre-images of $\bar{c}\bar{a}'$, and thus

$$\Phi_e^{-1} \circ f^{**} : A \to A$$

is an automorphism of A that fixes $\bar{c}$ and maps $\bar{a}$ to $\bar{a}'$, as required. □

Corollary 10.1.8 (Folklore). *The index set of relatively computably categorical structures is Σ_3^0-complete.*

Proof. Membership of Σ_3^0 follows from Theorem 10.1.6; simply state that there is a Σ_1^0 Scott family (see Definition 10.1.3). The completeness follows from the proof of Theorem 7.1.3 and Exercise 10.1.24. □

The next corollary justifies our description of relative computable categoricity as being a Type II analogue of computable categoricity.

Corollary 10.1.9 (Folklore). *If a computable structure is relatively computably categorical, then there is a single Turing functional (perhaps, with finitely many parameters) that witnesses this property.*

Proof. This functional is (essentially) given by the enumeration of the Scott family. □

When c.c. implies relative c.c.

Recall that a structure is n-decidable ($n \geqslant 0$) if we can decide the value of any first-order (finitary) Σ_n-formula about any tuple in the structure. For example, if a structure A is 2-decidable, we can decide arbitrary $\exists\forall$-statements about tuples in A. A 0-decidable structure is simply a computable structure. These notions differ for linear orders (Exercises 3.2.55 and 3.2.56) and for Boolean algebras as well (Exercises 4.1.41 and 4.1.43). Building on an earlier result of Nurtazin (see Exercise 10.1.26), Goncharov proved:

Theorem 10.1.10 (Goncharov [200]). *Suppose that A is a 2-decidable structure. Then the following are equivalent.*

(i) A is computably categorical.

(ii) A is relatively computably categorical.

We remark that, in general, "2-decidable" cannot be replaced with "1-decidable" in the premises of the theorem; this will appear as Exercise 10.1.30.

Proof of Theorem 10.1.10. We already have (ii) implying (i) in a stronger form since relative computable categoricity implies computable categoricity.

For the converse, Goncharov's construction works by trying to prove that A is not computably categorical, and in the end, we argue that the construction must fail, and from this, read off a Σ_1^0 Scott family.

We build a structure B and an isomorphism $f : A \to B$. We attempt to meet the requirements

$$R_e : \varphi_e \text{ is not an isomorphism witnessing } B \cong A.$$

The construction is finite injury. At stage s, we will consider finite partial injective functions p and $q \supseteq p$ approximating f^{-1}.

Let $p : \overline{d} \mapsto \overline{c}$ be from B to A. The goal is to show that if φ_e is a total and injective function from A onto B, then for some $\overline{b}$, the pre-images of $\overline{d}$ and $\overline{b}$ under φ_e and f do not satisfy the same existential formulae.

At stage s, suppose $\varphi_e \downarrow$ taking $\tilde{c}, \tilde{a}$ to $\overline{d}, \overline{b}$. Let $\delta(\overline{d}, \overline{b}, \overline{b_1})$ be the conjunction of all sentences (with constants naming elements of B) enumerated into the diagram of B by stage s. Suppose at stage s we have q with $p \subseteq q$ taking $\overline{d}, \overline{b}, \overline{b_1}$ to $\overline{c}, \overline{a}, \overline{a_1}$, where $p : \overline{d} \to \overline{c}$ is preserved by higher priority requirements. Note that q makes $A \models \delta(\overline{c}, \overline{a}, \overline{a_1})$.

Case 1. If $A \models \neg\exists\overline{y}\delta(\tilde{c}, \tilde{a}, \overline{y})$, then we can use q to satisfy R_e.

Case 2. Suppose $A \models \exists\overline{y}\delta(\tilde{c}, \tilde{a}, \overline{y})$. Then let $\overline{u}, \overline{x}$ be the variables corresponding to $\overline{d}, \overline{b}$ (and, thus, to $\tilde{c}, \tilde{a}$). Amongst the first s existential formulae, see if there is a $\theta(\overline{u}, \overline{x})$ with

$$A \models \theta(\overline{c}, \overline{a}) \leftrightarrow \neg\theta(\tilde{c}, \tilde{a}).$$

Then we can again use q to diagonalise R_e.

Case 3. Amongst the first s many existential formulae, there is some θ such that some of the tuples satisfying $\exists\overline{v}\delta(\overline{c},\overline{x},\overline{v})$ satisfy $\theta(\overline{c},\overline{x})$ and some do not. Take $\overline{n}$ and $\overline{m}$ with $A \models \delta(\overline{c},\overline{n},\overline{m})$ and

$$A \models \theta(\overline{c},\overline{n}) \leftrightarrow \neg\theta(\overline{\hat{c}},\overline{\hat{a}}).$$

Then we can satisfy R_e by replacing q with $q' \supseteq p$ where q' maps $\overline{d},\overline{b},\overline{\hat{b}}$ to $\overline{c},\overline{n},\overline{m}$.

Case 4. Suppose that $\xi(\overline{d},\overline{b},\overline{b'})$ is the first atomic sentence not decided in δ with $\overline{b'} \supseteq \overline{b_1}$. Let $\psi(\overline{u},\overline{x})$ denote $\exists\overline{v'}(\delta(\overline{u},\overline{x},\overline{v'}) \wedge \xi(\overline{u},\overline{x},\overline{v'}))$. Suppose that some of the tuples satisfying $\exists\overline{v'}\delta(\overline{u},\overline{x},\overline{v'})$ satisfy $\psi(\overline{c},\overline{x})$ and some do not. (Or this is true with $\neg\xi$ replacing ξ.) Then again we can use the same kind of strategy to meet R_e.

In any of the cases above, we will be able to meet R_e.

In the *construction*, we will choose the first $q \supseteq p$ to meet the highest priority R_j for $j \leqslant s$, if such a j exists. Once the diagonalisation is achieved via q, we will preserve the finite partial map q with priority of R_j. We also extend q to include constants from $\xi(\overline{d},\overline{b},\overline{n})$ so that q makes the sentence either true or false in A, and add the appropriate sentence to the diagram of B. We omit the usual details.

This ends (the description of) the construction.

Fix e so that for all $j < e$ we meet R_j (say by stage s_0), but R_e is never met. We claim that there is a Σ^0_1 Scott family. At stage s_0, we have a mapping p taking $\overline{d}$ to $\overline{c}$, with $\varphi_{e,s}$ having range $\overline{d}$. For each $s > s_0$, let $\varphi_{e,s}$ take $\overline{\hat{c}},\overline{\hat{a}_s}$ to $\overline{d},\overline{b_s}$. Let $\delta_s(\overline{d},\overline{b_s},\overline{b'_s})$ be the conjunction of all the sentences in the diagram by stage s.

Claim 10.1.11. *For any tuple $\overline{a}$ satisfying $\exists\overline{v}\delta_s(\overline{c},\overline{x},\overline{v})$ and a stage $t > s$, there exists $\overline{a'}$ such that $\overline{a},\overline{a'}$ satisfies $\exists\overline{v'}\delta_t(\overline{c},\overline{x},\overline{x'},\overline{v'})$. (The tuple $\overline{v'}$ has the same length as $\overline{b'_t}$ and $\overline{x}\overline{x'}$ essentially corresponds to $\overline{b_t}$.)*

We leave the proof of this as an exercise for the reader.

Defining the Scott family Λ. For each tuple $\overline{b} \in B$, define $\theta(\overline{c},\overline{x})$ as follows. Find the first stage $s > s_0$ where the range of $\varphi_{e,s}$ includes $\overline{d},\overline{b}$, and $q \supseteq p$ is defined on them. If q maps $\overline{d},\overline{b},\overline{b'}$ to $\overline{c},\overline{a},\overline{a'}$ at this stage, then let $\delta(\overline{d},\overline{b},\overline{b'})$ be the conjunction of all of the formulae enumerated into the diagram of B by s, and then define

$$\theta_{\overline{b}}(\overline{c},\overline{x}) = \exists\overline{v}\delta(\overline{c},\overline{x},\overline{v}).$$

Put $\theta_{\overline{b}}(\overline{c},\overline{x}) \in \Lambda$. Then Λ is of the correct form for a Σ^0_1 Scott family.

It is enough to show that each tuple satisfies some member of Λ, and if two tuples satisfy the same member, then they satisfy the same existential formulae $\theta(\overline{c},\overline{x})$ (Exercise 10.1.23). Since R_e is not met, we know that φ_e maps A bijectively and isomorphically onto B. Suppose that f maps $\overline{c},\overline{a}$ to $\overline{d},\overline{b}$. Then $\overline{a}$ must satisfy $\theta_{\overline{b}}(\overline{c},\overline{x})$. Note that at the stage when $\theta_{\overline{b}}(\overline{c},\overline{x})$ is determined, we might have q taking $\overline{b}$ to something other than $\overline{a}$. But it must be that $\overline{a}$ satisfies $\theta_{\overline{b}}(\overline{c},\overline{x})$, as when we change the function we preserve the diagram enumerated so far.

Let $\theta(\overline{c},\overline{x})$ be a (finitary) existential formula satisfied by some tuple that also satisfies $\theta_{\overline{b}}(\overline{c},\overline{x})$. Suppose that we determined $\theta_{\overline{b}}(\overline{c},\overline{x})$ at stage s and take $t \geqslant s$, where $\theta(\overline{c},\overline{x})$ is considered. Since we cannot use $\theta(\overline{c},\overline{x})$ to diagonalise φ_e at stage t, we know that if some tuple satisfies $\exists\overline{v'}\delta_t(\overline{c},\overline{x},\overline{x'},\overline{v'})$ and satisfies $\theta(\overline{c},\overline{x})$, then they all do. By Claim 10.1.11, we see

$$A \models \forall\overline{x}(\theta_{\overline{b}}(\overline{c},\overline{x}) \to \exists\overline{x'}\exists\overline{v'}\delta_t(\overline{c},\overline{x},\overline{x'},\overline{v'})).$$

That is

$$A \models \forall\overline{x}(\theta_{\overline{b}}(\overline{c},\overline{x}) \to \theta(\overline{c},\overline{x})).$$

It now follows from Exercise 10.1.23 that Λ is a Σ^0_1 Scott family. □

A transfinite extension of Theorem 10.1.6

A computable structure A is *relatively* Δ^0_α *categorical* if for any X-computable $B \cong A$, we have $B \cong_{\Delta^0_\alpha(X)} A$, where the class $\Delta^0_\alpha(X)$ stands for the relativisation of the class Δ^0_α defined in Section 8.1.

Theorem 10.1.12 (Ash [17]). *Fix a computable ordinal α. The following are equivalent for a computable structure S:*

(i) The structure S is relatively Δ^0_α-categorical.

(ii) The structure S has a Σ^c_α Scott family.

(iii) The structure S has a c.e. family Λ of Σ^c_α-formulae over some fixed $\overline{c} \in S$ such that each $\overline{a} \in S$ satisfies some $\theta \in \Lambda$, and if $\overline{a},\overline{b} \in S$ both satisfy the same θ then they satisfy the same Σ^c_α-formulae. (In other words, the Σ^c_α-types of S are effectively isolated by Σ^c_α-formulae.)

Proof. The implication $(ii) \to (iii)$ is trivial, and the proof of $(iii) \to (i)$ is not really different from the case of a c.e. family given by Exercise 10.1.23 and the proof of $(ii) \to (i)$ of Theorem 10.1.6. We focus on proving $(i) \to (ii)$. As sometimes happens, it will be easier to give a much more general construction, and then derive the theorem from this general machinery. The proof below is somewhat compressed since we will not really need to use $(i) \to (ii)$ in the sequel.

Building a generic copy of A. We reserve a computable collection of constants (that can be identified with ω) and, using these constants, we will build a bijection $f^* : B \to A$ using finite extensions. Our structure B^* upon the domain B will be defined using the pull-back from A via f^*.

In B^*, every existential quantifier $\exists x$ can be replaced with the infinite computable disjunction $\bigvee_{b\in B}$, and a first-order universal quantifier over elements of B can be replaced by an infinite computable disjunction over all elements of B. Thus, the following definition covers all computable sentences:

Definition 10.1.13. Let $p : B \to A$ be a finite partial injective map, and ϕ a computable infinitary sentence in the language of A augmented with constants from B. Define $p \Vdash \phi$ ("p forces ϕ") as follows:

1. If $\phi(\overline{b})$ is quantifier-free with parameters $\overline{b}$ in B, then $p \Vdash \phi(\overline{b})$ if $A \models \phi(p(\overline{b}))$ (this assumes $\overline{b}$ is in *dom* p).

2. $p \Vdash \bigvee_i \phi_i$ iff $\exists i\, p \Vdash \phi_i$.

3. $p \Vdash \bigwedge_i \phi_i$ iff $\forall i\, \forall p \geq q\ \exists \rho \geq p$ such that $\rho \Vdash \phi_i$.

Our formulae will be in normal form, unless otherwise specified. In particular, we identify $\neg\phi$ with its normal form. By induction on the complexity of formulae, we establish the following elementary but important lemma:

Lemma 10.1.14. *In the following, ϕ ranges over computable sentences (in the same language as before) and p, q, r are finite partial injective maps from B to A.*

1. *If $p \Vdash \phi$ and $q \geq p$, then $q \Vdash \phi$.*

2. *For each p, there exists a $q \geq p$ such that*
$$q \Vdash \phi \text{ or } q \Vdash \neg\phi.$$

3. *No p and ϕ can possibly satisfy*
$$p \Vdash \phi \text{ and } p \Vdash \neg\phi.$$

The proof is left to Exercise 10.1.53. Notice that the lemma tells us that every sentence is always "decided" by some extension of every given string. We adopt this self-descriptive terminology and say that q *decides* ϕ if $q \Vdash \phi$ or $q \Vdash \neg\phi$.

Before we proceed, we note that

$$F = \{p \,|\, p : B \to A \text{ finite partial injective map}\}$$

can be viewed as a subset of $\omega^{<\omega}$, which is dense in the closed $I \subseteq \omega^\omega$ that consists of injective maps. More generally, for a set of finite strings X, being dense is equivalent to saying that for each partial injective q there exists a $p \in X$ such that $p \geq q$; indeed, it is the same as saying that every basic open set $[q]$ intersects X. Note that, for each computable infinitary ϕ and each $a \in A$, the following sets are dense open in I :

$$D_\phi = \{p : p \text{ decides } \phi\}$$

and

$$D_a = \{p : \mathrm{rng}(p) \ni a\}.$$

This is a countable collection of dense open sets, so their intersection is also dense in I, by the Baire Category Theorem (see Exercise 4.2.38). Thus, there exists

$$f^* = \bigcup_{n \in \mathbb{N}} p_n \in I$$

where p_n "hits" the n-th dense open set in the sequence.

Any such f^* gives rise to a "generic" structure B^* via pullback from A. Indeed, any such f^* is clearly injective, and it is surjective because it is inside the set $D_{b=b}$ for every $b \in B$. The next lemma is perhaps the most useful property of such a generic B^*; it is also proven by induction on the complexity of ϕ:

Lemma 10.1.15. *In the following, ϕ ranges over computable sentences (in the language of A with constants from B).*

1. $B^* \models \phi$ *if and only if* $\exists n\ p_n \Vdash \phi$.

2. *For each ϕ and $\bar{b}$ in B, there is a computable formula $\psi_{\bar{b},\phi}$ of the same syntactic complexity as ϕ such that, for any $\bar{a}$ with $|\bar{a}| = |\bar{b}|$,*

$$A \models \psi_{\bar{b},\phi}(\bar{a}) \text{ if and only if } \bar{b} \mapsto \bar{a} \Vdash \phi.$$

We leave the proof to Exercise 10.1.54. Note that the B-constants are eliminated from ψ. Also, the inductive transformations used in the proof are uniformly effective.

The Scott family. We shall write $X^{(\alpha)}$ for the α-th jump of a set X, which is a (Turing) $\Delta^0_{1+\alpha}(X)$-complete set. To use α-th jump of the open diagram of B^*, $D(B^*)^{(\alpha)}$, we shall slightly adjust our notation and assume that the structure is relatively $\Delta^0_{1+\alpha}$-categorical, and so we need to produce a c.e. $\Sigma^c_{1+\alpha}$ Scott family.

The idea is to use Lemma 10.1.15 to turn the property

$$\Phi_e^{D(B^*)^{(\alpha)}} : A \to B \text{ is an isomorphism}$$

into a sequence of formulae that will be the Scott family for A. The key lemma is as follows:

Lemma 10.1.16. *There exist computable sentences (with constants in B) saying that:*

1. $\sigma \preceq (D(B^*))^{(\alpha)}$;

2. $\Phi^{(D(B^*))^{(\alpha)}} : A \to B^*$ *is total;*

3. $\Phi^{(D(B^*))^{(\alpha)}} : A \to B^*$ *is an isomorphism.*

In the sequel, we will only use the case when $\alpha = 1$, i.e., when the structure is relatively Δ^0_2-categorical. We will explain the Δ^0_2 case in a bit more detail, but the general case is actually not too different and can be derived by (effective) transfinite induction (cf. Exercise 10.1.51).

Sketch of (1), $\alpha = 1$. Identify a Turing functional Φ_e with the c.e. set $\{\langle n, \sigma, k\rangle : \Phi^\sigma_e(n) \downarrow = k\}$. Let X be a set. Then $\Phi^X_e(n) = k$ holds iff

$$\psi_{e,n,k}(X) \equiv \bigvee_{\langle n,\sigma,k\rangle \in W_e} \sigma \preceq X$$

holds. For a finite string σ, we have that $\sigma \preceq X'$ iff the property

$$\bigwedge_{\sigma(e)=1} \psi_{e,e,1}(X) \ \& \bigwedge_{\sigma(e)=0} \neg\psi_{e,e,1}(X)$$

holds. If X is taken to be the open diagram $D(B^*)$ of B^*, then $\psi_{e,n,k}(X)$ can be viewed as a computable Σ^c_1 formula:

$$\bigvee_{\langle n,\sigma,k\rangle \in W_e} D(\sigma),$$

where σ is (uniformly effectively in σ) interpreted as the truth values $D(\sigma)$ of the first few atomic facts about elements of B^* coded by σ. This allows one to express

$$\sigma \preceq D(B^*)'$$

as a conjunction of a computable Σ_1^c-sentence and a computable Π_1^c-sentence. □

Using similar manipulations, we can also demonstrate that there is a $\Sigma_{1+\alpha}^c$-sentence $\theta_{\bar{a},\bar{b}}$ saying that

$$\Phi_e^{D(B^*)^{(\alpha)}} : \bar{a} \to \bar{b},$$

where of course we also require $\Phi_e^{D(B^*)^{(\alpha)}}(a)\downarrow$, for all a in $\bar{a}$. (Note the elements in $\bar{a}$ are just natural numbers, and they only contribute to the indices for the c.e. sets used to take the disjunctions of B^*-sentences.)

Now, if B^* is a generic copy, then for some n, we have that

$$p_n \Vdash \text{“}\Phi_e^{D(B^*)^{(\alpha)}} : A \to B^* \text{ is an isomorphism”}.$$

Suppose p_n is $\bar{d} \mapsto \bar{c}$. For each $\bar{a}$ in A, there is a $q \succeq p_n$ and a tuple $\bar{a}'$ in A such that

$$q : \bar{d}, \bar{b} \mapsto \bar{c}, \bar{a}$$

and $q \Vdash \theta_{\bar{a}',\bar{b}}$. Notice that the choice of p_n and e guarantees that $\bar{a}'$ is automorphic to $\bar{a}$ over $\bar{c}$, via the composition of Φ_e and f^*, using *any* generic extension f^* of q.

We are almost done. To finish the proof, we shall invoke (2) of Lemma 10.1.15. For a tuple $\bar{a}'$ in A, let

$$\gamma_{\bar{a}'}(\bar{c}, \bar{x}) \equiv \bigvee_{\bar{b},\bar{b}_1 \in B} \exists \bar{u}_1 \psi_{\bar{d}\bar{b}\bar{b}_1\theta_{\bar{a}',\bar{b}}}(\bar{c}, \bar{x}, \bar{u}_1).$$

If $A \models \gamma_{\bar{a}'}(\bar{c}, \bar{a})$, then for some choice of parameters,

$$\bar{d}, \bar{b}, \bar{b}_1 \mapsto \bar{c}, \bar{a}, \bar{a}_1 \Vdash \text{“}\Phi_e^{D(B^*)^{(\alpha)}} : \bar{a}' \to \bar{b}\text{”},$$

which also makes any $\bar{a}'$ that satisfies this formula automorphic to $\bar{a}$ over $\bar{c}$ (as we already noted earlier). On the other hand, since Φ_e is an isomorphism between B^* and A, $\bar{a}'$ will satisfy some disjunct, and thus, $A \models \gamma_{\bar{a}'}(\bar{c}, \bar{a}')$.

Note that we can produce the formula uniformly in the tuple $\bar{a}$, e, and p_n. Since Lemma 10.1.15(2) preserves the syntactic complexity of formulae, we have that the formulae are $\Sigma_{1+\alpha}^c$, as required. □

A c.c. structure that is not relatively c.c.

Theorem 10.1.17 (Goncharov [205, Theorem 4]). *There is a computable structure A that is computably categorical but not relatively computably categorical.*

We remark that Goncharov derived his result from the earlier papers concerned with *numberings* (effective uniform enumerations) of sets of natural numbers independently obtained by Badaev [25] and Selivanov [462]. We adopt the modern version of the proof (e.g., [132]) that uses a tree of strategies similar to the one in the proof of the Minimal Pair Theorem 3.1.44.

Proof. Before discussing the formal details, we informally discuss the requisite ideas. The structure A will be a directed graph consisting of infinitely many finite connected components. Each component will consist of either two, three, or four cycles sharing only a single vertex v, termed the *root vertex.*

In order to prevent A from being relatively computably categorical, we diagonalise against all pairs $(\bar{c}, \Lambda)$, where $\bar{c}$ is a finite tuple of elements from the universe of A and Λ is a c.e. family of existential formulas with parameters from $\bar{c}$. We create vertices v_1 and v_2 such that v_1 and v_2 are not automorphic, but Λ cannot distinguish them.

In order to ensure A is computably categorical, we construct a partial computable map f_j from A to B_j (the j-th (partial) directed graph). If A and B_j are isomorphic, the map f_j will be an isomorphism.

More formally, we meet the following requirements to prevent relative computable categoricity:

$$\mathcal{R}_i : \text{The } i\text{-th pair } (\bar{c}_i, \Lambda_i) \text{ is not a Scott family for } A.$$

We meet the following requirements to ensure computable categoricity:

$$\mathcal{S}_j : \text{If } A \cong B_j, \text{ then } f_j : A \cong B \text{ is a computable isomorphism.}$$

Strategy for meeting $\mathcal{R}_i$ (in isolation). We take the following actions, being careful to use elements larger than those found in $\bar{c}_i$:

1. Fix a *large* number ℓ and create two new root vertices v_1 and v_2.
2. Attach a loop of length 2 and a loop of length 3ℓ to both v_1 and v_2 and a loop of length $3\ell + 1$ to v_1.
3. For every formula $\phi(x, \bar{c}_i) := (\exists \bar{y})[\psi(x, \bar{y}, \bar{c}_i)]$ in Λ_i, search for a tuple $\bar{a}_1 < s$ such that $A \models \psi(v_1, \bar{a}_1, \bar{c}_i)$.
4. If such a formula and tuple are found, attach a loop of length $3\ell + 2$ to v_1 and a loop of length $3\ell + 1$ to v_2.

These actions prevent $(\bar{c}_i, \Lambda_i)$ from witnessing that A is relatively computably categorical: If we never find a formula ϕ and tuple $\bar{a}_1$, then not every singleton satisfies some $\phi \in \Lambda_i$.

If we find a formula ϕ and tuple $\bar{a}_1$, let s be the stage at which these are found. Then by construction, the component of v_1 at stage s embeds into the component of v_2 at stage $s + 1$, and the component of v_2 at stage s embeds into the component of v_1 at stage $s+1$. This can be extended to an embedding $A_s \hookrightarrow A_{s+1}$ via the identity off these components, and notably this embedding maps v_1 to v_2 and fixes $\bar{c}_i$ elementwise.

Since ϕ is existential, we have

$$\begin{aligned} A_s \models \phi(v_1, \bar{c}) &\implies A_{s+1} \models \phi(v_2, \bar{c}) \\ &\implies A \models \phi(v_2, \bar{c}), \end{aligned}$$

but v_1 and v_2 are not automorphic.

Strategy for meeting $\mathcal{S}_j$ (in isolation). As the construction of A proceeds, we attempt to define f_j so that it maps components in A to components in B_j. Finding the image of a component in A is a two-step process: We identify root vertices in B_j as those vertices having out-degree at least two (this is the sole purpose of the loops of length two). While identifying root vertices in B_j, we also search for cycles emanating from already identified root vertices in B_j. When we find a component in B_j with the same lengths of cycles emanating from it as a component in A, we map the root vertex and cycles appropriately.

Conflicts between strategies and their resolution. Unfortunately, our action to defeat relative computable categoricity conflicts heavily with our action for computable categoricity. Trying to define a computable isomorphism between A and B_j, the naive approach would be to wait for the components to appear in A and B_j and to define the isomorphism appropriately. If and when the components grow in A or B_j, an opponent would have the opportunity to switch v_1 and v_2, killing our computable isomorphism f_j. As we need infinitely many pairs of components to defeat relative computable categoricity, an opponent would have sufficiently many opportunities to diagonalise against all computable functions.

The critical observation is that this opportunity to diagonalise can be prevented by slowing down the construction: For the finitely many higher-priority $\mathcal{R}_i$-strategies (which build finitely many finite components), the $\mathcal{S}_j$-strategy defines the computable isomorphism f_j nonuniformly. For the components built by lower-priority $\mathcal{R}_i$-strategies, we use the above-mentioned technique of *pushing on isomorphisms*: The $\mathcal{S}_j$-strategy will allow the lower-priority $\mathcal{R}_i$-strategy to extend its component in Step 4 only gradually as follows:

(4'a) Attach a loop of length $3\ell + 2$ to v_1.

(4'b) Wait until this loop appears in B_j for every $j < i$ for which $A \cong B_j$.

(4'c) Attach a loop of length $3\ell + 1$ to v_2.

In this way, the problem cannot occur: At any time, we will be able to distinguish v_1 and v_2 in B_j. Of course, it will likely be the case that $A \not\cong B_j$ for some $j < i$, in particular that some B_j with $j < i$ does not have a loop of length $3\ell + 2$. Hence, we may wait at Step 4'b unnecessarily (since we cannot effectively know whether $A \cong B_j$), causing Step 4'c not to be reached. This would cause our diagonalisation attempt against Λ_i to be unsuccessful.

The solution is to have $\mathcal{R}_i$-strategies guess the outcomes of higher priority $\mathcal{S}_j$-strategies via a priority tree. Each $\mathcal{R}_i$-strategy will have two outcomes: `wait`, (indicating that the strategy is still searching for a formula ϕ and a tuple $\overline{a}_1$) and `act` (indicating that the strategy has found the desired ϕ and $\overline{a}_1$). Each $\mathcal{S}_j$-strategy will have an infinite outcome ∞ (indicating that $\mathcal{S}_j$ believes $A \cong B_j$) and finite outcomes $\mathtt{k}$ for all $k \in \mathbb{N}$ (counting the number of times $\mathcal{S}_j$ has taken outcome ∞).

Full strategy for meeting $\mathcal{R}_i$. We take the following actions, always being careful to use elements larger than those found in $\overline{c}_i$:

1. Fix a *large* number ℓ and create two new root vertices v_1 and v_2.

2. Attach a loop of length 2 and a loop of length 3ℓ to both v_1 and v_2 and a loop of length $3\ell + 1$ to v_1.

3. For every formula $\phi(x) := (\exists \overline{y})[\psi(x, \overline{y}, \overline{c}_i)]$ in Λ_i, search for a tuple $\overline{a}_1 < s$ such that $A \models \psi(v_1, \overline{a}_1, \overline{c}_i)$.

4. If such a formula and tuple are found, attach a loop of length $3\ell + 2$ to v_1.

5. Wait until the next stage at which the strategy is accessible.

6. Attach a loop of length $3\ell + 1$ to v_2.

While the strategy is searching at Step 3, it has outcome `wait`. Once it has found a formula ϕ and a tuple $\overline{a}_1$, it has outcome `act`.

Full strategy for meeting $\mathcal{S}_j$. Let σ on the priority tree be the $\mathcal{S}_j$-strategy in question. Let s be the current stage. Let k be the number of stages less than s at which σ had outcome ∞.

We consider certain root vertices in A: For each $\tau < \sigma$ such that $\tau^\frown\texttt{wait} \preceq \sigma$, we consider the root vertices created by τ; for each $\tau < \sigma$ such that $\tau^\frown\texttt{act} \preceq \sigma$ and τ has reached Step 6, we consider the root vertices created by τ; and for each $\tau \not< \sigma$ with τ incomparable with $\sigma^\frown\texttt{k}$, we consider the root vertices created by τ.

For each root vertex v in A we are considering, if $f_j(v)$ is not yet defined, we search $B_{j,s}$ for a root vertex u with a component identical to the component of v and define $f_j(v) := u$ and then extend f_j to an isomorphism of the components. If $f_j(v)$ is defined, and the component of v appears identical to the component of $f_j(v)$ in $B_{j,s}$, we extend f_j to an isomorphism of the components, if it is not already.

After this action, if for every vertex v we are considering, $f_j(v)$ is defined and f_j is an isomorphism of the components of v and $f_j(v)$, then σ has outcome ∞ at stage s. Otherwise, it has outcome `k`.

Construction. For an $\mathcal{S}_j$-strategy, we order the outcomes as $\infty < \cdots < 2 < 1 < 0$. For an $\mathcal{R}_i$-strategy, we order the outcomes as `act` < `wait`. We create a priority tree by devoting each level to one requirement in some effective fashion. At stage s, we let all visited strategies of length at most s act in order of priority.

Verification. Define the true path through the priority tree in the usual fashion. We note the important fact that if the current path moves to the left of a node on the priority tree that has already been visited, that node can never be visited again.

It is immediate from the construction that A is a computable presentation. We verify that it is both computably categorical and not relatively computably categorical.

Claim 10.1.18. *The structure A is computably categorical.*

Proof. Fix an index j such that $A \cong B_j$, and let σ be the $\mathcal{S}_j$-strategy along the true path. By assumption, the presentation B_j contains a component isomorphic to every component of A, so σ will eventually define $f_j(v)$ for every vertex it considers. For the components built by $\tau < \sigma$, since σ is on the true path, these components will never grow once σ begins considering them, so f_j is correct on these.

For the components built by strategies τ incomparable with σ, τ can never be visited after σ begins considering them, and so they can never grow once they are considered. So f_j is correct on these.

For the components built by $\tau \succeq \sigma^\frown\infty$, if τ has final outcome `wait`, then the components never grow once σ begins considering them.

If τ adds the loop of length $3\ell + 2$ to v_1, then before τ added this loop, σ defined $f_j(v_1)$ to be an element of B_j with a loop of size $3\ell + 1$. After τ adds this loop, σ will never again have outcome ∞ unless a loop of length $3\ell + 2$ appears attached to $f_j(v_1)$, and if σ never again has outcome ∞, then v_1 is the unique vertex with a loop of size $3\ell + 1$. So the loop of length $3\ell + 2$ must appear on $f_j(v_1)$.

If τ adds the loop of length $3\ell + 1$ to v_2, σ must have outcome ∞ at some stage after τ attached the loop of length $3\ell + 2$ to v_1. So $f_j(v_1)$ has a loop of size $3\ell + 2$ and one of size 3ℓ, and $f_j(v_2)$ has a loop of size 3ℓ. Then there are only two loops of size 3ℓ in A, one with a loop of size $3\ell + 2$ and one without, so by elimination $f_j(v_2)$ must be the correct image of v_2. So the loop of length $3\ell + 1$ must appear on $f_j(v_2)$.

For the components built by $\tau \succeq \sigma^\frown \mathtt{k}$ for some k, if σ is considering this component at stage s, then it has had outcome ∞ more than k many times by stage s. So the components can never again grow once they are considered, so f_j is correct on these.

By the above, since f_j is correct on every component on which it is defined, and it will be defined on every component it considers, σ must have true outcome ∞. So by construction, every component is considered, and thus f_j is an isomorphism. □

Claim 10.1.19. *The structure A is not relatively computably categorical.*

Proof. Fix an index i and let σ be the $\mathcal{R}_i$-strategy along the true path. Then either σ will wait forever at Step 3, or it will reach Step 6. In the former case, the element v_1 fails to satisfy any $\phi \in \Lambda_i$. In the latter case, the nonautomorphic elements v_1 and v_2 satisfy $\phi \in \Lambda_i$. In either case, the family Λ_i is not a Scott family. □

This concludes the proof of Theorem 10.1.17. □

Divergence of plain and relative categoricity in non-universal classes

Can we find structures witnessing Theorem 10.1.17 in "natural" algebraic classes? Of course, it is only expected that any effectively universal class, such as the class of 2-step nilpotent groups (Theorem 8.2.14), can be used as a source of such examples. On the other hand, in all classes that are not effectively universal that we've studied in the book, "plain" and "relative" computable categoricity are equivalent for structures in the class (e.g., Exercise 10.1.24).

It is not difficult to see that the class of torsion abelian groups is not effectively universal (Exercise 10.4.7). Recall that Theorem 9.3.37 states that the index set of c.c. torsion abelian groups is Π^0_4-complete. Theorem 9.3.37 and Corollary 10.1.8 imply:

Corollary 10.1.20 (Melnikov and Ng [377]). *There exist computably categorical torsion abelian groups that are not relatively computably categorical.*

A closely related theorem is as follows:

Theorem 10.1.21 (Hirschfeldt, Kramer, Miller, and Shlapentokh [259]). *The index set of computably categorical algebraic fields is Π^0_4-complete. In particular, there exists an algebraic field that is computably categorical but not relatively computably categorical.*

As far as we know, in all algebraic classes in which computable categoricity is described by a purely algebraic property, "plain" and "relative" computable categoricity coincide. Thus, one should not expect an algebraic characterisation of computably categorical torsion abelian groups similar to that given in Theorem 9.3.26. Similarly, one should not expect to obtain a purely algebraic characterisation of computably categorical algebraic fields. Therefore, it could well be that Theorems 9.3.37 and 10.1.21 are the best possible classifications of computable categoricity in the respective classes.

Computable categoricity is unclassifiable in general

We also state without proof the following technically difficult result (it was stated earlier as Ex. 8.1.44):

Theorem 10.1.22 (Downey et al [131]). *The index set of computably categorical structures*

$$\{M_e : M_e \textit{ is a computably categorical graph}\}$$

is Π^1_1*-complete.*

The proof is omitted since it uses methods that were not covered in the book. Because of Corollary 10.1.8, Theorem 10.1.22 clearly implies Theorem 10.1.17. Also, it tells us that computably categorical structures are not classifiable in general, and also are not classifiable in any effectively universal class.

Exercises

Exercise° 10.1.23 (Folklore). Let A be a countable structure. Suppose that Λ is a family consisting of finitary existential formulas with parameters from a fixed finite tuple $\bar{a}$. Suppose that

(a) each $\bar{a}$ in A satisfies some formula in Λ, and

(b) if there is some formula φ in Λ satisfied by tuples $\bar{a}$ and $\bar{b}$, then the two tuples satisfy exactly the same existential formulas with parameters $\bar{a}$.

Then Λ is a Scott family for A.

Exercise° 10.1.24 (Folklore). Show that in the following classes, relative computable categoricity is equivalent to ("plain") computable categoricity:

1. Vector spaces over a fixed computable field.
2. Boolean algebras.
3. Linear orders.
4. Abelian p-groups.
5. Torsion-free abelian groups.

Exercise° 10.1.25. Prove the following. If a computable structure M is computably categorical, then its index set $\{i : M \cong M_i\}$ is Σ^0_3. Show that it can be Σ^0_3-complete even for a relatively c.c. structure.

Exercise 10.1.26 (Nurtazin [418]). We say that a decidable structure A is *decidably categorical* if, for all decidable $B \cong A$, $B \cong_{\Delta^0_1} A$; i.e., B is computably isomorphic to A. Prove that A is decidably categorical iff A has a Σ^0_1 Scott family.

Exercise° 10.1.27. Give a direct proof of Corollary 10.1.20 imitating the proof of Theorem 10.1.17.

Exercise° 10.1.28. Give an example of a decidably categorical structure (see Exercise 10.1.26) which is not computably categorical.

Exercise 10.1.29 (Millar [389]). Show that is A is computably categorical and 1-decidable, and $\overline{a} \in A$, then $(A, \overline{a})$ is also computably categorical[1].

Exercise* 10.1.30 (Kudinov [320]). There is a 1-decidable structure that is computably categorical but not relatively computably categorical.

Exercise* 10.1.31. A structure is *decidably categorical* (Exercise 10.1.26) if it has a unique decidable presentation, up to computable isomorphism.

1. Show that the index set of decidably categorical structures in a given computable signature lies in the class $\Sigma^0_{\omega+2}$.
2. Show that the index set of decidably categorical structures in each of the following classes is $\Sigma^0_{\omega+2}$-complete:
   (a) linear orders [209];
   (b) Boolean algebras [208].

Exercise 10.1.32 (Goncharov [205]). We say that a computable structure A is *computably stable* if, for all computable $B \cong A$, *every* isomorphism $f : B \to A$ is computable. For example $\mathbb{Q}$ is a computably categorical ordering which is not computably stable, and ω with successor function $s(x) = x+1$ is computably stable. Prove that A is computably stable iff for some tuple of constants $\overline{a} \in A$, $(A, \overline{a})$ is computably categorical and rigid. (That is, has no nontrivial automorphisms.)

Exercise 10.1.33 (Goncharov [205]). Suppose that A is a 1-decidable (thus, computable) structure. Show that A is computably stable (see Exercise 10.1.32) iff A has a c.e. *defining* family. That is a c.e. family Λ of finitary existential formulae defining all of the elements of A. This means that

(i) every element of A satisfies a formula in Λ.

(ii) no formula in Λ is satisfied by two distinct elements of A.

Exercise 10.1.34 (Ash and Nerode [22]). An m-ary relation U on a computable structure A is called *intrinsically c.e.* if for all computable $A \cong B$ via $f : A \to B$, then $f(U)$ is c.e.. Prove that if A is 2-decidable, then U is intrinsically c.e. iff it *formally c.e.*; meaning that there are a finite collection of parameters $a_1, \dots, a_k \in A$, and a c.e. sequence of existential formulae $\{\psi_n(x_1, \dots, x_m, y_1, \dots, y_k)\}$, such that

$$\forall x_1 \forall x_2 \dots \forall x_m U(x_1, \dots, x_m) \text{ iff } \vee_{n \in \mathbb{N}} \psi_n(x_1, \dots, x_m, a_1, \dots, a_k).$$

[1]This is not true if the hypothesis that A is 1-decidable is removed. This was shown by Cholak, Goncharov, Khoussainov and Shore [93], and then later Hirschfeldt, Khoussainov and Shore [258] showed that there is a computably categorical structure A and a constant $a \in A$ such that (A, a) has computable dimension ∞ (Definition 10.3.1).

10.1.2 Uniform computable categoricity

In this subsection, we compare different stronger notions of computable categoricity that naturally arise in algebraic examples.

Uniform versions of computable categoricity

There are at least two natural ways to define the concept of "uniform computable categoricity". In computability theory, uniformity is usually defined in terms of indices. From a model-theoretic viewpoint, however, it seems more natural to require the existence of an effective procedure for producing the desired isomorphisms that takes structures, rather than indices for structures, as input. For example, computable categoricity of the dense linear order $(\mathbb{Q}, <)$ is clearly witnessed by a Turing functional. These two different approaches are reflected in the following definition.

Definition 10.1.35 (Downey, Hirschfeldt, Khoussainov [126], Ventsov [497, 498]). Let A be a computable structure and let $M_0, M_1, \ldots$ be a computable list of all partial computable structures in the language of A.

1. A is *weakly uniformly computably categorical* (weakly u.c.c.) if there is a total computable f such that $M_e \cong A \Rightarrow \varphi_{f(e)} : M_e \cong A$.
2. A is *uniformly computably categorical* (u.c.c.) if there is a partial computable operator Ψ such that $M_e \cong A \Rightarrow \Psi(M_e) : M_e \cong A$ (that is, the function $x \mapsto \Psi(M_e; x)$ is an isomorphism from M_e to A).
3. A is *(weakly) u.c.c. with parameters* if there are finitely many elements $a_0, \ldots, a_n \in A$ such that $(A, a_0, \ldots, a_n)$ is (weakly) u.c.c..

Remark 10.1.36. For each partial computable g there exists a total computable f such that $g(e)\downarrow \Rightarrow \varphi_{f(e)} = \varphi_{g(e)}$. Thus for A to be weakly u.c.c. it is enough that there exist a partial computable g such that $M_e \cong A \Rightarrow g(e)\downarrow$ and $\varphi_{g(e)} : M_e \to A$.

The difference between the two notions defined above can be understood from a programming perspective. A computable structure A is weakly u.c.c. if there is a computable procedure that, given a program P outputting a structure B isomorphic to A, produces an isomorphism between B and A possibly using information about P. On the other hand, A is u.c.c. if there is a computable procedure that, given a computable structure B isomorphic to A, produces an isomorphism between B and A without knowledge of the particular program that is outputting B.

Relative vs. uniform categoricity with parameters

In the theorem below, by a Σ^0_1 Scott family we mean a c.e. Scott family consisting of first-order existential formulae.

Theorem 10.1.37 (Ventsov [497, 498]). *A computable structure is u.c.c. if and only if it has a Σ^0_1 Scott family without parameters, and is u.c.c. with parameters if and only if it has a Σ^0_1 Scott family with finitely many parameters.*

Proof. It is enough to prove the first part of the theorem, from which the second part follows immediately. Let A be a computable structure and let $M_0, M_1, \ldots$ be a computable list of all partial computable structures in the language of A.

If A is u.c.c. then let Ψ be as in Definition 10.1.35. Given $\overline{x} = (x_0, \ldots, x_k) \in A^{k+1}$, let $M_{e(\overline{x})}$ be a finite substructure of A such that, for every $i \leqslant k$, $x_i \in M_{e(\overline{x})}$ and $\Psi(M_{e(\overline{x})}; x_i)\downarrow$. Let $y_0, \ldots, y_n$ be the elements of $M_{e(\overline{x})}$ other than $x_0, \ldots, x_k$ and let $\delta(\overline{x}, y_0, \ldots, y_n)$ be the conjunction of the finitely many elements of the atomic diagram of $M_{e(\overline{x})}$. Define $\theta_{\overline{x}} \equiv \exists y_0, \ldots, y_n(\delta(\overline{x}, y_0, \ldots, y_n))$.

We claim that $\{\theta_{\overline{x}} : \overline{x} \in A^{<\omega}\}$ is a Scott family. It is enough to show that if $A \models \theta_{\overline{x}}(\overline{y})$ then $\overline{y}$ is in the orbit of $\overline{x}$. If $A \models \theta_{\overline{x}}(\overline{y})$ then there is an i and a g such that $g : A \cong M_i$, $g(\overline{y}) = \overline{x}$, and $M_{e(\overline{x})}$ is a substructure of M_i. It is easy to check that $(\Psi(A))^{-1} \circ \Psi(M_i)$ is an automorphism of A taking $\overline{y}$ to $\overline{x}$.

For the other direction, we can use the standard proof that a structure with a Scott family is computably categorical, since this proof produces computable isomorphisms uniformly. □

Combining the theorem above with Theorem 10.1.6, we obtain:

Corollary 10.1.38 (Downey, Hirschfeldt, Khoussainov [126]). *A computable structure is u.c.c. with parameters if and only if it is relatively computably categorical.*

In the second half of the proof of Theorem 10.1.37, there is no need for the input structure to be computable. This gives another way of proving Corollary 10.1.9 and allows us to conclude that relativising the notion of uniform computable categoricity does not make it any stronger. These results and observations should be compared with the Kreisel-Lacombe-Shoenfield-Markov Theorem 2.3.7 and with Theorem 2.3.18 about computable functions $[0, 1] \to \mathbb{R}$.

Type 1 vs Type 2 for uniform categoricity

We now prove a theorem of Kudinov which resembles Specker's Theorem 2.3.24 from Chapter 2. This theorem implies that there is a computably categorical structure that is not relatively computably categorical (Theorem 10.1.17). In [126], the theorem below is attributed to Kudinov [321]. The proof below is taken from [126].

Theorem 10.1.39. *There is a weakly u.c.c. structure which is not u.c.c. with parameters (equivalently, not relatively c.c., by Theorem 10.1.37).*

Proof. For each $n \in \mathbb{N}$, let $[n]$ be the directed graph consisting of $n+3$ many nodes $x_0, x_1, \ldots, x_{n+2}$ with an edge from x_0 to itself, an edge from x_{n+2} to x_1, and an edge from x_i to x_{i+1} for each $i \leqslant n+1$. We call x_0 the *top* of $[n]$.

For each $S \subset \omega$, the directed graph $[S]$ consists of one copy of $[s]$ for each $s \in S$, with all the tops identified.

Now let $D_0, D_1, \ldots$ be a standard list of all finite non-empty sets and let $B_0^n, B_1^n, \ldots$ be the nth uniformly c.e. (u.c.e.) collection of sets in some standard ordering of such collections. We will build a u.c.e. collection of finite sets $A_0, A_1, \ldots$ with the following properties.

1. There is no c.e. set W such that every A_i contains D_k for some $k \in W$ and $(k \in W \wedge D_k \subseteq A_i, A_j) \Rightarrow A_i = A_j$.

2. Let N be the set of all $i \in \mathbb{N}$ such that A_i is non-empty. There is a partial computable binary function h with the following property. For each $n \in \mathbb{N}$, if there is a 1–1 and onto map $g : \omega \to N$ such that, for all $i \in \mathbb{N}$, $B_i^n = A_{g(i)}$, then $h_n \equiv \lambda i(h(n, i))$ is such a map.

Suppose we have built $A_0, A_1, \ldots$ with these properties and let A be a computable directed graph consisting of the disjoint union of the graphs $[A_i]$, $i \in \mathbb{N}$. It is not hard to check that the first property above implies that A has no Scott family, while the second property implies that A is weakly u.c.c..

We now proceed with the construction of $A_0, A_1, \ldots$. We assume that if $s < \langle n, i \rangle$ then $B_i^n[s] = \varnothing$. Since we are only interested in collections of sets $B_0, B_1, \ldots$ for which there is a 1–1 and onto map $g : \omega \to N$ such that, for all $i \in \mathbb{N}$, $B_i = A_{g(i)}$, we also assume without loss of generality that for each $n, i, s \in \mathbb{N}$ there is a $j \in \mathbb{N}$ such that $B_i^n[s] \subseteq A_j[s]$. For each $e \in \mathbb{N}$, let $k(e)$ be such that $D_{k(e)} = \{2e\}$. We begin with $A_i = \varnothing$ for all $i \in \mathbb{N}$. At stage s, we proceed as follows.

1. Enumerate $2s$ into $A_{\langle s,0 \rangle}$.

2. Search for the least $e \leqslant s$ (if any) such that $k(e) \in M_e$ and $A_{\langle e,0 \rangle} = \{2e\}$. If such an e is found then enumerate $2\langle e, 0 \rangle + 1$ into $A_{\langle e,0 \rangle}$ and enumerate $2e$ and $2\langle e, 1 \rangle + 1$ into $A_{\langle e,1 \rangle}$. Say that e is *active* at stage s.

3. For each $e \leqslant s$, check if there exist $n, i \in \mathbb{N}$ such that $\langle n, i \rangle \leqslant s$, e has never caught-up for the sake of n before (defined below), $2e \in B_i^n[t]$, where $t \leqslant s$ is a stage at which e is active, and $B_i^n[s] \neq \{2e\}$. If so then proceed as follows. For each $i \leqslant s$ such that $2e \notin A_{\langle e,i \rangle}[s]$, enumerate $2e$ into $A_{\langle e,i \rangle}$. For each $i, j \leqslant s$ such that $2\langle e, j \rangle + 1 \notin A_{\langle e,i \rangle}[s]$, enumerate $2\langle e, j \rangle + 1$ into $A_{\langle e,i \rangle}$. Enumerate $2e$ and $2\langle e, s + 1 \rangle + 1$ into $A_{\langle e,s+1 \rangle}$. We say that e *catches-up for the sake of* n at stage s.

This completes the construction. We now check that $A_0, A_1, \ldots$ have the desired properties. Clearly, $A_0, A_1, \ldots$ are u.c.e.. Each $e \in \mathbb{N}$ can be active at most once, since e cannot be active unless $A_{\langle e,0 \rangle}$ is a singleton, and once e is active this is no longer the case. Furthermore, if e is active at stage s then e can only catch-up at most once for each n such that $2e \in B_i^n[s]$ for some $i \in \mathbb{N}$, which implies that e catches-up only finitely often. Since no numbers are enumerated into any $A_{\langle e,i \rangle}$, $i \in \mathbb{N}$, at any stage at which e is neither active nor catches-up, this means that each A_i is finite.

Fix $e \in \mathbb{N}$. If $k(e) \notin M_e$ then e is never active, so $A_{\langle e,0 \rangle} = \{2e\}$, which means that there is no $k \in M_e$ such that $D_k \subseteq A_{\langle e,0 \rangle}$. On the other hand, if $k(e) \in M_e$ then either e never catches-up, in which case both $A_{\langle e,0 \rangle}$ and $A_{\langle e,1 \rangle}$ contain $D_{k(e)}$ but $A_{\langle e,0 \rangle} \neq A_{\langle e,1 \rangle}$, or e catches-up for the last time at some stage s, in which case, both $A_{\langle e,0 \rangle}$ and $A_{\langle e,s+1 \rangle}$ contain $D_{k(e)}$ but $A_{\langle e,0 \rangle} \neq A_{\langle e,s+1 \rangle}$. Thus there is no c.e. set W such that every A_i contains D_k for some $k \in W$ and $(k \in W \wedge D_k \subseteq A_i, A_j) \Rightarrow A_i = A_j$.

Now define the map h as follows. Given $n, i \in \mathbb{N}$, assume that $h(n, j)$ has already been defined for all $j < i$. Wait until a stage s such that $B_i^n[s]$ is non-empty and equal to $A_k[s]$ for some $k \notin \text{range}(h \restriction i)$ and define $h(n, i) = k$. Clearly, h is partial computable.

Fix $n \in \mathbb{N}$ for which there is a 1–1 and onto map $g : \omega \to N$ such that, for all $i \in \mathbb{N}$, $B_i^n = A_{g(i)}$. (Note that this implies that each B_i^n is non-empty.) To complete the proof, we need to show that $h_n \equiv \lambda i(h(n, i))$ is a 1–1 map from ω onto N such that, for all $i \in \mathbb{N}$, $B_i = A_{h_n(i)}$.

Clearly, if h_n is defined for all $i \in \mathbb{N}$ then it is 1–1 and onto. (The surjectivity of h_n follows from the fact that, for each $e \in \mathbb{N}$, there are only finitely many $k \in \mathbb{N}$ such that $2e \in A_k$.) Suppose that, for all $j < i$, $h_n(j)$ is defined and $B_j^n = A_{h_n(j)}$. Then $h_n(i)$ must be defined, since otherwise there could be no 1–1 and onto map $g : \omega \to N$ such that, for all $l \in \mathbb{N}$, $B_l^n = A_{g(l)}$.

So we are left with showing that, for all $i \in \mathbb{N}$, $B_i^n = A_{h_n(i)}$. Fix $i \in \mathbb{N}$ and let $k = h_n(i)$. First suppose that s in the definition of $h(i)$ is such that $B_i^n[s] = A_k[s]$ is not a singleton. It is easy to check from the construction that if A_l and A_s share two elements then they are equal, so we have $B_i^n = A_{g(i)} = A_k$.

Now suppose that s in the definition of $h(i)$ is such that $B_i^n[s] = A_k[s]$ is a singleton. Then it must be the case that, for some $e \in \mathbb{N}$, $B_i^n[s] = \{2e\}$ and $k = \langle e, 0\rangle$. If e is never active then $B_i^n = \{2e\} = A_k$. Otherwise, every A_l that contains $2e$ has at least two elements, and thus so does B_i^n. But this means that, at some stage $t \geqslant s$, e catches-up for the sake of n. It is easy to check that this means that $A_l \supseteq B_i^n[t]$ if and only if $l = \langle e, u\rangle$, $u \leqslant t$. But it is also the case that $A_{\langle e,u\rangle} = A_{\langle e,v\rangle}$ for all $u, v \leqslant t$, so in fact $A_l = B_i^n[t]$ if and only if $l = \langle e, u\rangle$, $u \leqslant t$. In particular, $B_i^n = A_k$. □

We also remark that the results established so far in this section illustrate that the naive analogy of the Kreisel-Lacombe-Shoenfield-Markov Theorem 2.3.7 *fails* for structures, since one cannot hope to obtain a functional that would work *only* for computable structures.

Separating all notions of computable categoricity*

Consider the following statements about a computable structure A.

C: A is computably categorical.

W: A is weakly uniformly computably categorical.

WP: A is weakly uniformly computably categorical with parameters.

U: A is uniformly computably categorical.

UP: A is uniformly computably categorical with parameters.

R: A is relatively computably categorical.

The following theorem can be found in [126].

Theorem 10.1.40. *The following implications hold, and no other implications except the ones implied by transitivity hold in general.*

(i) $U \Rightarrow UP \Leftrightarrow R$.

(ii) $U \Rightarrow W$, $UP \Rightarrow WP$.

(iii) $W \Rightarrow WP \Rightarrow C$.

All the implications in this theorem are clear except for the equivalence of UP and R, which we have already established in Theorem 10.1.37. We have already proven WU $\not\Rightarrow$ UP in Theorem 10.1.39. We omit the verification that no other implications hold and refer the reader to Exercise 10.1.41 and [126].

Exercises

Exercise° 10.1.41. Prove that, in Theorem 10.1.40, WP does not imply W, and that UP does not imply U.

10.1.3 Relative Δ_2^0-categoricity*

The reader might wonder what happens if we weaken the hypothesis that the structure A is 2-decidable in Theorem 10.1.10; see also Exercise 10.1.30. In this subsection, we show that computable categoricity implies relative Δ_2^0-categoricity amongst 1-decidable structures. The crux of the proof is Lemma 10.1.43.

Definition 10.1.42. For a structure A and tuples $\overline{a}, \overline{p} \in A$, denote by $\Sigma_n\text{-tp}_{\overline{p}}(\overline{a})$ the set

$$\Sigma_n\text{-tp}_{\overline{p}}(\overline{a}) := \{\varphi(\overline{x}, \overline{y}) \in \Sigma_n : A \models \varphi(\overline{a}, \overline{p})\}$$

and denote by $\Sigma_n^c\text{-tp}_{\overline{p}}(\overline{a})$ the set

$$\Sigma_n^c\text{-tp}_{\overline{p}}(\overline{a}) := \{\varphi(\overline{x}, \overline{y}) \in \Sigma_n^c : A \models \varphi(\overline{a}, \overline{p})\},$$

where in both cases we consider only finitary (or computable infinitary, respectively) formulas in the language of the structure.

Lemma 10.1.43. *If A is computably categorical and 1-decidable, then there is a tuple $\overline{p} \in A$ such that distinct Σ_1-types over $\overline{p}$ are incomparable under inclusion, and for any $\overline{a}, \overline{a}' \in A$, if $\Sigma_1\text{-tp}_{\overline{p}}(\overline{a}) = \Sigma_1\text{-tp}_{\overline{p}}(\overline{a}')$, then $\Sigma_2^c\text{-tp}_{\overline{p}}(\overline{a}) = \Sigma_2^c\text{-tp}_{\overline{p}}(\overline{a}')$.*

Proof of Lemma 10.1.43. We build a computable presentation B isomorphic to A and attempt to make B not computably isomorphic to A. The amount of B constructed when we consider the computable function φ_e witnessing A and B being computably isomorphic will determine the parameter $\overline{p}$.

In order to ensure A and B are classically isomorphic, we build an isomorphism $F : B \to A$ in a Δ_2^0-manner. We build B by constructing its atomic diagram in stages. At each stage s, we enumerate the next atomic sentence true about A into the atomic diagram of B as determined by the isomorphism F (as approximated at stage s).

Let $\{\psi_i\}_{i \in \mathbb{N}}$ be a computable enumeration of all Σ_1-formulas in the language of A.

Strategy defeating ϕ_e. We fix a partial computable function $\phi_e : B \to A$ and seek to ensure that ϕ_e is not an isomorphism.

Let s_0 be the stage at which this strategy is initialised. This strategy takes no action until a stage $s_1 > s_0$ when $B_{s_0} \subseteq \text{dom} F_{s_1}$ and $A_{s_0} \subseteq \text{range} F_{s_1}$. We then let $\overline{b}_0 := B_{s_1}$ and restrain the strategy, in the sense that $F \upharpoonright \overline{b}_0$ cannot be changed by this strategy. At every stage s after becoming active, before we enumerate the next sentence into the atomic diagram of B, we look for an opportunity to change F in such a way that it still extends to an isomorphism, but such that $F \circ \phi_e^{-1}$ is guaranteed not to be an automorphism of A (ensuring that if F is an isomorphism, as it will be, then ϕ_e is not an isomorphism). We will find such opportunities if the types do not obey the conclusion of the lemma.

Before describing the strategy, we note the following. For any stage $t > s_1$, suppose $\overline{b}$ is a tuple from the domain of B_t, and let $\delta_t(\overline{b}_0, \overline{b}, \overline{f})$ be the atomic diagram of B_t, where $\overline{f} := B_t \backslash (\overline{b}_0 \cup \overline{b})$. Suppose $\overline{a} \in A$. At a stage $s > s_1$, we can redefine F to map $\overline{b}$ to $\overline{a}$ without changing $F \upharpoonright \overline{b}_0$ if and only if $A \models \exists \overline{x}\left[\delta_{s-1}(F_{s-1}(\overline{b}_0), \overline{a}, \overline{x})\right]$. Here, we consider δ_{s-1} instead of δ_s because when this strategy acts at stage s, we have not yet enumerated the next sentence into the atomic diagram of B.

At a stage $s > s_1$, we consider every triple $(\overline{b}, \overline{b}', \overline{d})$ with $\overline{b}, \overline{b}' \in \mathrm{dom}(\phi_{e,s})$ and $\overline{d} \in \mathrm{dom}(\phi_{e,s}) \backslash (\overline{b}_0 \cup \overline{b})$. If this is the first stage at which we have considered this triple, we use 1-decidability to determine if there is a tuple $\overline{c} \in A^{|\overline{d}|}$ such that

$$A \models \exists \overline{y}\left[\delta_{s-1}(F_{s-1}(\overline{b}_0), F_{s-1}(\overline{b}'), \overline{c}\,\overline{y})\right],$$

i.e., we ask whether we can redefine F by putting $F_s(\overline{b}) := F_{s-1}(\overline{b}')$ and $F_s(\overline{d}) := \overline{c}$ while respecting the restraint. If there is no such $\overline{c}$, we never consider this triple again (since we cannot redefine F, there is no point in considering it further). If there is such a $\overline{c}$, we search for one and assign it to this triple. When we consider this triple at future stages, this is the $\overline{c}$ to which we refer.

Then, for every triple $(\overline{b}, \overline{b}', \overline{d})$ being considered (along with its associated $\overline{c}$), we use 1-decidability to determine if

$$A \models \left[\psi_i(F_{s-1}(\overline{b}), F_{s-1}(\overline{d})) \iff \neg\psi_i(F_{s-1}(\overline{b}'), \overline{c})\right] \tag{10.3}$$

for some $i < s$, i.e., we ask whether redefining F by putting $F_s(\overline{b}) := F_{s-1}(\overline{b}')$ and $F_s(\overline{d}) := \overline{c}$ might be useful. If so, fix some triple and some $i_0 < s$ for which (10.3) holds. We use 1-decidability to determine whether

$$A \models \left[\psi_{i_0}(\phi_e(\overline{b}), \phi_e(\overline{d})) \iff \psi_{i_0}(F_{s-1}(\overline{b}), F_{s-1}(\overline{d}))\right], \tag{10.4}$$

i.e., we determine whether or not it is necessary to perform any action to prevent $F \circ \varphi_e^{-1}$ from being an automorphism. If (10.4) holds, we put $F_s(\overline{b}) := F_{s-1}(\overline{b}')$ and $F_s(\overline{d}) := \overline{c}$, and extend F_s such that $A_s \subseteq \mathrm{range}F_s$ and $B_s \subseteq \mathrm{dom}F_s$. If (10.4) fails, we put $F_s(\overline{b}) := F_{s-1}(\overline{b})$ and $F_s(\overline{d}) := F_{s-1}(\overline{d})$, and extend F_s such that $A_s \subseteq \mathrm{range}F_s$ and $B_s \subseteq \mathrm{dom}F_s$. Regardless of whether (10.4) holds or fails, we declare the strategy complete.

If (10.3) fails for all triples being considered and all $i < s$, we repeat the above process with $\exists \overline{y}\, \delta_s$ in place of ψ_i. That is, for every triple $(\overline{b}, \overline{b}', \overline{d})$ and associated $\overline{c}$ being considered, we use 1-decidability to determine if

$$A \models \neg\exists \overline{y}\left[\delta_s(F_{s-1}(\overline{b}_0), F_{s-1}(\overline{b}'), \overline{c}\,\overline{y})\right], \tag{10.5}$$

i.e., we ask whether we will lose the ability to redefine F after we enumerate the next atomic sentence into the diagram of B. If (10.5) fails for every triple, we will not lose the ability to redefine F, so we leave F alone and take no further action for this strategy at stage s.

If (10.5) holds for some triple, fix a triple for which it holds. We will lose the ability to redefine F, so we use 1-decidability to determine if

$$A \models \exists \overline{y}\left[\delta_s(\phi_e(\overline{b}_0), \phi_e(\overline{b}), \phi_e(\overline{d})\,\overline{y})\right], \tag{10.6}$$

i.e., we determine whether or not it is necessary to perform any action to prevent $F \circ \varphi_e^{-1}$ from being an automorphism. If (10.6) holds, we put $F_s(\overline{b}) := F_{s-1}(\overline{b}')$ and $F_s(\overline{d}) := \overline{c}$ and extend F_s such that $A_s \subseteq \mathrm{range}F_s$ and $B_s \subseteq \mathrm{dom}F_s$. If (10.6) fails, we put $F_s(\overline{b}) := F_{s-1}(\overline{b})$ and $F_s(\overline{d}) := F_{s-1}(\overline{d})$, and extend F_s such that $A_s \subseteq \mathrm{range}F_s$ and $B_s \subseteq \mathrm{dom}F_s$. Regardless of whether (10.6) holds or fails, we declare the strategy complete.

The strategy has two outcomes: `wait` and `stop`. Of course, these correspond to whether the strategy has been declared completed.

Construction. We put these strategies on a tree, performing a straightforward finite-injury argument in the usual manner. At each stage, the visited strategies on the tree act in priority order. After they have acted, if no strategy defined F_s, we define F_s by extending F_{s-1} to include A_s and B_s in the range and domain, respectively. Then the global strategy building B acts by taking the next atomic sentence $\theta_s(\overline{a})$ true about A and enumerating $\theta_s(F_s(\overline{a}))$ into the atomic diagram of B.

Verification. We verify $F := \lim_s F_s$ exists and is an isomorphism. Consequently, there will be a (least) k such that $\varphi_k : B \to A$ is a computable isomorphism. Let σ be the strategy for φ_k along the true path, and let $\overline{b}_0$ be the restraint of σ. We show the desired relationships between the types of tuples of A using $\overline{p} := F(\overline{b}_0)$.

Claim 10.1.44. *The function $F := \lim_s F_s$ exists and is an isomorphism.*

Proof. The existence of F follows from the fact that, if a strategy redefines F on an element (in either the domain or the range), then no lower-priority strategy can redefine F on that element. Thus, by induction, the function F can change only finitely many times on any element (in either the domain or the range).

By construction, the function F is surjective and respects atomic sentences. Thus, it is injective (as equality is an atomic sentence) and so an isomorphism. □

Claim 10.1.45. *If a strategy for defeating ϕ_e is along the true path and declared complete, then ϕ_e is not an isomorphism.*

Proof. Let $(\overline{b}, \overline{b}', \overline{d})$ be the triple we act for. Note that $F(\overline{b}) = F_s(\overline{b})$ and $F(\overline{d}) = F_s(\overline{d})$.

If we act because of some ψ_{i_0}, then regardless of whether (10.4) holds, we have

$$A \models \left[\psi_{i_0}(\phi_e(\overline{b}), \phi_e(\overline{d})) \iff \neg\psi_{i_0}(F_s(\overline{b}), F_s(\overline{d}))\right].$$

If we act because of δ_s, then regardless of whether (10.6) holds, we have

$$A \models \exists\overline{y}\left[\delta_s(\phi_e(\overline{b}_0), \phi_e(\overline{b}), \phi_e(\overline{d})\overline{y})\right] \iff \neg\exists\overline{y}\left[\delta_s(F_s(\overline{b}_0), F_s(\overline{b}), F_s(\overline{d})\overline{y})\right].$$

Thus, in either case, we have that $F \circ \phi_e^{-1}$ is not an automorphism. □

Claim 10.1.46. *The Σ_1-types over $\overline{p}$ are incomparable under inclusion.*

Proof. Towards a contradiction, we suppose that there are $\overline{a}, \overline{a}' \in A$ with

$$\Sigma_1\text{-tp}_{\overline{p}}(\overline{a}) \subsetneq \Sigma_1\text{-tp}_{\overline{p}}(\overline{a}'). \tag{10.7}$$

Consider any stage $s+1$ at which σ is visited such that $F_s(\overline{b}) = \overline{a}$ and $F_s(\overline{b}') = \overline{a}'$ for some $\overline{b}, \overline{b}' \in \text{dom}(\phi_{k,s})$. Then at such a stage, it will always be possible for σ to define $F_{s+1}(\overline{b}) = \overline{a}'$.

Note that (10.7) is equivalent to $\Sigma_1\text{-tp}(\overline{p}\,\overline{a}) \subsetneq \Sigma_1\text{-tp}(\overline{p}\,\overline{a}')$. Since $F \circ \phi_k^{-1}$ is an automorphism, we have

$$\Sigma_1\text{-tp}(\phi_k(\overline{b}_0)\,\phi_k(\overline{b})) \subsetneq \Sigma_1\text{-tp}(\overline{p}\,\overline{a}').$$

Fix a formula ψ_i true of $\overline{p}\,\overline{a}'$ but false of $\overline{p}\,\overline{a}$. Then at any stage $s > i$ when the strategy considers the triple $(\overline{b}_0\,\overline{b}, \overline{b}_0\,\overline{b}', \varnothing)$, it will redefine $F(\overline{b}) = \overline{a}'$ to defeat ϕ_k, contrary to our choice of k. □

Claim 10.1.47. *For any tuples $\overline{a}, \overline{a}' \in A$, if $\Sigma_1\text{-tp}_{\overline{p}}(\overline{a}) = \Sigma_1\text{-tp}_{\overline{p}}(\overline{a}')$ then $\Sigma_2^c\text{-tp}_{\overline{p}}(\overline{a}) = \Sigma_2^c\text{-tp}_{\overline{p}}(\overline{a}')$.*

Proof. Suppose $\Sigma_1\text{-tp}_{\overline{p}}(\overline{a}) = \Sigma_1\text{-tp}_{\overline{p}}(\overline{a}')$, or equivalently $\Sigma_1\text{-tp}(\overline{p}\,\overline{a}) = \Sigma_1\text{-tp}(\overline{p}\,\overline{a}')$. By symmetry, it suffices to show

$$\Sigma_2^c\text{-tp}(\overline{p}\,\overline{a}) \subseteq \Sigma_2^c\text{-tp}(\overline{p}\,\overline{a}').$$

Fix a formula $\exists \overline{x}\, \chi(\overline{p}\,\overline{a}, \overline{x}) \in \Sigma_2^c\text{-tp}(\overline{p}\,\overline{a})$ with $\chi \in \Pi_1^c$ and a witness $\overline{g} \in A$ so that $A \models \chi(\overline{p}\,\overline{a}, \overline{g})$. We show $\exists \overline{x}\, \chi(\overline{p}\,\overline{a}, \overline{x}) \in \Sigma_2^c\text{-tp}(\overline{p}\,\overline{a}')$.

Consider a stage $s > s_1$ when σ is visited, $F(\overline{b}) = \overline{a}, F(\overline{b}') = \overline{a}'$ and $F(\overline{d}) = \overline{g}$ have converged, and $\overline{b}_0, \overline{b}, \overline{b}', \overline{d} \in \mathrm{dom}(\phi_{k,s})$. Since

$$A \models \exists \overline{x}\, \delta_s(\overline{p}, \overline{a}, \overline{g}\,\overline{x}),$$

from $\Sigma_1\text{-tp}(\overline{p}\,\overline{a}) = \Sigma_1\text{-tp}(\overline{p}\,\overline{a}')$, we have

$$A \models \exists \overline{c}\, \exists \overline{y}\, \delta_s(\overline{p}, \overline{a}', \overline{c}\,\overline{y}).$$

Thus, there will be a $\overline{c}$ assigned to the triple $(\overline{b}_0\,\overline{b}, \overline{b}_0\,\overline{b}', \overline{d})$. Since σ is never declared complete (by Claim 10.1.45), there is never a stage $t > s$ when

$$A \models \neg \exists \overline{y}\left[\delta_t(\overline{p}, \overline{a}', \overline{c}\,\overline{y})\right].$$

Thus σ will never lose the ability to define $F(\overline{b}) = \overline{a}'$ and $F(\overline{d}) = \overline{c}$.

If there were some ψ_i such that

$$A \models \psi_i(\overline{p}\,\overline{a}', \overline{c}) \wedge \neg\psi_i(\overline{p}\,\overline{a}, \overline{g}),$$

then at some stage when we consider ψ_i, the strategy σ would be able to defeat ϕ_k, contrary to our choice of k.

Thus $A \models \chi(\overline{p}\,\overline{a}', \overline{c})$. We conclude $\exists \overline{x} \chi(\overline{p}\,\overline{a}', \overline{x}) \in \Sigma_2^c\text{-tp}(\overline{p}\,\overline{a}')$ as desired. □

This completes the proof of Lemma 10.1.43. □

We say that a computable structure A is relatively Δ_2^0-categorical if for any X-computable B, there is an X'-computable isomorphism $f : A \to B$. We are now ready to prove the main theorem of this section:

Theorem 10.1.48 (Downey, Kach, Lempp and Turetsky [132]). *Any 1-decidable, computably categorical structure A is relatively Δ_2^0-categorical.*

Proof. Fix the parameters $\overline{p}$ from the above lemma. For each $\overline{a} \in A$, let $\chi_{\overline{a}}(\overline{x})$ be the infinitary formula

$$\chi_{\overline{a}}(\overline{x}) := \bigwedge_{\substack{\psi \in \Pi_1(\overline{p}) \\ A \models \psi(\overline{a})}} \psi(\overline{x}),$$

i.e., the conjunction of all first-order Π_1-formulas (with parameters from $\overline{p}$) true of $\overline{a}$. As a consequence of 1-decidability, this is a Π^c_1-formula.

We show that the family of formulas $\{\chi_{\overline{a}}(\overline{x})\}_{\overline{a}\in A}$ constitutes a Scott family. By Theorem 10.1.12, it suffices to show that they isolate the Σ^c_2-types. We therefore suppose $A \models \chi_{\overline{a}}(\overline{a}')$ and show $\Sigma^c_2\text{-tp}_{\overline{p}}(\overline{a}') = \Sigma^c_2\text{-tp}_{\overline{p}}(\overline{a})$. If $A \models \chi_{\overline{a}}(\overline{a}')$, then every Π_1-fact true of $\overline{a}$ is true of $\overline{a}'$. Hence every Σ_1-fact true of $\overline{a}'$ is true of $\overline{a}$, i.e.,

$$\Sigma_1\text{-tp}_{\overline{p}}(\overline{a}') \subseteq \Sigma_1\text{-tp}_{\overline{p}}(\overline{a}).$$

By Lemma 10.1.43, it follows that $\Sigma_1\text{-tp}_{\overline{p}}(\overline{a}') = \Sigma_1\text{-tp}_{\overline{p}}(\overline{a})$. By Lemma 10.1.43 again, this implies that $\Sigma^c_2\text{-tp}_{\overline{p}}(\overline{a}') = \Sigma^c_2\text{-tp}_{\overline{p}}(\overline{a})$.

We conclude that the family of formulas $\{\chi_{\overline{a}}(\overline{x})\}_{\overline{a}\in A}$ constitutes a Scott family and so A is relatively Δ^0_2-categorical. □

We note that Exercise 10.1.105 implies that the result established in this subsection is, in some sense, optimal.

10.1.4 Exercises: Calculus of computable infinitary formulae

The following exercises are concerned with various properties of infinitary computable formulae. Most of these exercises are based on facts and theorems taken from [20], where the proofs and proof sketches can be found. We will generally avoid using infinitary logic in the book, with the only exceptions being Theorem 10.1.12 and Theorem 10.1.48. Nonetheless, the power of $L^c_{\omega_1\omega}$ in abstract computable structure theory should not be underestimated. For a smooth introduction to this theory, see [20]. Perhaps Exercises 10.1.65 and 10.1.68 are the only notable results in the sequence of these exercises that are directly related to the material of the book. However, including their complete and detailed proofs would significantly inflate the size of the already large book.

Unfortunately, it seems that the technique of infinitary computable formulae does not quite work for separable structures, especially when they are considered up to homeomorphism.

Exercises describing basic properties of infinitary formuale

In all exercises below, the language of structures in a class is always computable.

Exercise° 10.1.49. Prove the Normal Form Theorem for computable infinitary formulae.

(a) Given an index for a computable Σ^c_α (or Π^c_α) formula ϕ, we can find an index for a computable Π^c_α (or Σ^c_α, resp.) formula $\neg\phi$ that is logically equivalent to the negation of ϕ.

(b) Given indices for a pair of computable Σ^c_α (or Π^c_α) formulas ϕ and ψ, we can find indices for computable Σ^c_α (or Π^c_α, resp.) formulas logically equivalent to $(\phi \vee \psi)$ and $(\phi \wedge \psi)$.

(c) Given a formula θ that is a finite Boolean combination of computable Σ^c_β and Π^c_β formulas, for various $\beta < \alpha$, we can find a computable Σ^c_α formula and a computable Π^c_α formula, both logically equivalent to θ.

(d) Given a computable Σ^c_α formula ϕ, we can find a computable Σ^c_α formula logically equivalent to $\exists x\, \varphi$, and a computable $\Pi^c_{\alpha+1}$ formula logically equivalent to $\forall x\, \varphi$. (The case of a Π^c_α formula is dual.)

Exercise° 10.1.50. Let A be a computable structure, and ϕ a computable Σ^c_α formula in the language of A in free variables $\bar{x} \in A^n$. Prove that $\phi(A) = \{\bar{a} \in A^n : A \models \phi(\bar{a})\}$ is Σ^0_α.

Exercise° 10.1.51. ([20, Theorem 7.9].) For any hyperarithmetical set S, produce a sequence $(\varphi_n)_{n\in\mathbb{N}}$ of computable infinitary formulas, built up just out of $\top$ and $\bot$, such that φ_n is logically equivalent to

$$\varphi_n \equiv \begin{cases} \top & \text{if } n \in S, \\ \bot & \text{otherwise.} \end{cases}$$

Exercise 10.1.52. ([20, Theorem 7.15]). For any hyperarithmetical set F of computable infinitary propositional formulas, or computable infinitary predicate formulas with a fixed tuple of variables, there is a computable infinitary formula ϕ that is logically equivalent to the disjunction, and there is a computable infinitary formula ψ that is logically equivalent to the conjunction.

Exercises related to Theorem 10.1.12 and its proof

Exercise° 10.1.53. Prove Lemma 10.1.14.

Exercise° 10.1.54. Prove Lemma 10.1.15. (Hint: For the case when $\phi = \bigwedge_i \theta_i$ in (2), use

$$\bigwedge_{\bar{b}_1 \in B^{<\omega},\, i\in\mathbb{N}} \forall \bar{x}_1 \bigvee_{\bar{b}_2 \in B^{<\omega}} \psi_{\bar{b},\bar{b}_1,\bar{b}_2,\theta_i}(\bar{x},\bar{x}_1,\bar{x}_2),$$

which essentially says "for all i and all extensions of $\bar{b}$ that (say) map $\bar{b}_1$ to $\bar{x}_1$, there is a further extension mapping $\bar{b}_2$ to $\bar{x}_2$ that forces θ_i"; compare this to Definition 10.1.13(3).)

Exercise* 10.1.55 (Ash, Knight, Manasse and Slaman [21], Chisholm [89]). Prove the following: For a computable structure A with a further relation R, and α a computable ordinal, the following are equivalent:

(1) R is definable in A by a computable Σ^c_α formula.

(2) In all copies B of A, the image of R is Σ^0_α relative to (the open diagram $D(B)$ of) B.

(Hint: For the easier implication, use Ex. 10.1.50 relativised to $D(B)$. For the harder implication, use the method described in the proof of Theorem 10.1.12 to build a generic copy of A and consider p such that $p \Vdash$ "R is Σ^0_α".)

Exercise* 10.1.56 (Downey, Kach, Lempp, and Turetsky [132]). Let $\mathbf{d}$ be a Turing degree. A computable structure $\mathcal{S}$ is *relatively computably categorical above* $\mathbf{d}$ if between any two presentations A, $B \geqslant_T \mathbf{d}$ of $\mathcal{S}$, there is an isomorphism computable in $\deg(A) \cup \deg(B)$ (or $\Delta^0_\alpha(\deg(A) \cup \deg(B))$, respectively).

(i) Prove that for a computable structure $\mathcal{S}$, the following are equivalent:

  (a) The structure $\mathcal{S}$ is relatively Δ^0_α-categorical above $\mathbf{d}$.

  (b) Between any two presentations A and B of $\mathcal{S}$, there is an isomorphism computable in $\Delta^0_\alpha(\deg(A) \cup \deg(B) \cup \mathbf{d})$.

(ii) Prove that for any nonzero c.e. degree $\mathbf{d}$, there is a structure $\mathcal{S}$ that is relatively computably categorical above $\mathbf{d}$ but not computably categorical.

Exercises related to Barwise-Kreisel compactness for $L^c_{\omega_1\omega}$

Exercise° 10.1.57. ([20, Theorem 8.2]). Show the following. For any hyperarithmetical set F of computable infinitary sentences in a fixed computable language, there is a computable tree P (perhaps infinitely branching) such that F is consistent if and only if $[P] \neq \emptyset$, i.e., P has an infinite path. This is also uniform.

Exercise* 10.1.58. (Barwise-Kreisel Compactness for $L^c_{\omega_1\omega}$; [20, Theorem 8.3]). Fix a computable language L. Let F be a Π^1_1 set of computable infinitary sentences in $L^c_{\omega_1\omega}$ so that every Δ^1_1-subset of F is consistent. Using the previous exercise, show that F is consistent.

Exercise* 10.1.59. (Effective Compactness for $L^c_{\omega_1\omega}$; [20, Corollary 8.4]). Prove the computable version of the Barwise-Kreisel Compactness (Ex. 10.1.58): Let F be a Π^1_1 set of sentences in $L^c_{\omega_1\omega}$ so that every Δ^1_1-subset of F is has a computable model. Show that F has a computable model.

Exercise 10.1.60. ([20, Theorem 8.6]). Let A be a computable structure. Let F be a Π^1_1 set of sentences involving at least one extra new relation not in the language of A.

(a) If for every Σ^1_1 set $A \subseteq F$, A can be expanded to a model of A, then A can be expanded to a model of F.

(b) If for every Σ^1_1 set $A \subseteq F$, A can be expanded to a *computable* model of A, then A can be expanded to a *computable* model of F.

(Hint: Consider ϕ which is the conjunction of $D(A)$ (the atomic diagram of A) and the sentence $\bigvee_{a\in A} \exists x\, x = a$ and let $F^* = F \cup \{\phi\}$. Apply Ex. 10.1.58,10.1.59.)

Exercise 10.1.61. Prove that if two computable structures A and B (having the same computable language) satisfy the same computable infinitary sentences, then $A \cong B$. (Hint: Fix a partial map $p : A \to B$ that preserves satisfaction of computable infinitary formulas, say $\bar{a} \mapsto \bar{b}$. Note that by our assumption, we can choose p to be the empty map. Fix any $c \in A$ not in the domain of p, and let $\Gamma(\bar{b}, d)$ be the Π^1_1 set of computable infinitary sentences saying of the tuple $\bar{b}$ in B and a new constant d that $\bar{b}, d$ satisfies all the computable infinitary formulas that in A are true of $\bar{a}, d$. By Ex. 10.1.52, any Δ^1_1 subset of $\Gamma(\bar{b}, d)$ is equivalent to a single formula $\gamma(\bar{b}, d)$. Since $(\exists x)\gamma(\bar{a}, x)$ holds in A, $(\exists x)\gamma(\bar{b}, x)$ holds in B. By Ex. 10.1.60, B can be expanded to a model of all of $\Gamma(\bar{b}, d)$. Use this to build an isomorphism between A and B in the usual back-and-forth fashion.)

Exercise 10.1.62. ([20, Corollary 8.7]). Show that if A is a computable structure, then for any pair of tuples satisfying the same computable infinitary formulas in A there is an automorphism taking one to the other. (The same is true if A is a hyperarithmetical structure.)

Exercise 10.1.63. ([20, Theorem 8.17]). Let G be a computable abelian p-group. Use Ex. 10.1.60 to show that the Ulm type of G is $\leqslant \omega_1^{CK}$, and furthermore if G is reduced, then the Ulm type of G is $< \omega_1^{CK}$.

Exercises related to the classification themes

Exercise 10.1.64. ([20, Proposition 8.16]). Give a proof of Ex. 8.1.26 using Ex. 10.1.62 and working with the Boolean algebra directly rather than with the Stone space.

Exercise 10.1.65 (Goncharov and Knight [213]). Suppose that K is a class of structures, closed under isomorphism, and let K^c be the collection of its computable members. Then the following are equivalent:

(1) There is a computable infinitary sentence Ψ such that

$$K^c = \{A : A \models \Psi\}.$$

(2) The index set (the characterisation problem for K) $I(K) = \{i : A_i \in K^c\}$ is hyperarithmetical.

(Hint: The implication (1) $\to$ (2) is similar to Ex. 10.1.50 and is demonstrated by transfinite induction. For (2) $\to$ (1), first use Ex. 10.1.59 and Ex. 10.1.62 to show that there is a computable ordinal α such that for any computable structures $A \in K$ and $B \notin K$, there is a computable Π^c_α-sentence true in A and not in B. For each $A \in K$, use Ex. 10.1.52 to produce a computable infinitary sentence separating it from $B \notin K$, and then take the disjunction of such sentences over all $A \in K$.)

Exercise 10.1.66 (Knight and McCoy [305]). Recall that the class d-Σ^0_n consists of sets that can be expressed as $A \backslash B$, where $A, B \in \Sigma^0_n$. Define the class d-Σ^c_n of computable infinitary formulas to consist of all formulas of the form $\Psi \& \Theta$, where $\Psi \in \Sigma^c_n$ and $\Theta \in \Pi^c_n$. Observe that every structure M with a d-Σ^c_n Scott sentence has a d-Σ^0_n index set. Show that, however, there is a computable group $G \leqq (\mathbb{Q}, +)$ with the following properties:

1. The index set $I(G) = \{e : A_e \cong G\}$ is d-Σ^0_2-complete (i.e., m-complete d-Σ^0_2).
2. The group G does *not* have a computable d-Σ^c_2 Scott sentence.
3. However, G has a d-Σ_2 (non-computable) Scott sentence. Indeed, the sentence can be *low* d-Σ_2, in the sense that the infinite conjunctions and disjunctions are computable relative to a low oracle.

(Hint: Consider the subgroup of $(\mathbb{Q}, +)$ generated by fractions $\{\frac{1}{p_i^n} : i \in S, n \in \mathbb{N}\}$, where S is a non-computable low c.e. set given by Theorem 3.1.1.)

Exercise* 10.1.67 (Vanden Boom [494]). Suppose $\mathcal{K}$ is a class of structures in a computable language and closed under isomorphism, and let $C(\mathcal{K}) \subseteq \omega^\omega$ be the set of codes of structures in $\mathcal{K}$ from Exercise 2.4.34. In Exercise 4.2.88 we defined a set to be effectively G_δ if it is the intersection of effectively open sets.

1. Prove that the following are equivalent:

    (a) $\mathcal{K} = \{A : A \models \Psi\}$ for some Π^c_2 sentence Ψ;

    (b) $C(\mathcal{K}) \subseteq \omega^\omega$ is effectively G_δ.

2. Extend this result to arbitrary computable α and to arbitrary levels of the effective Borel hierarchy (we omit the definition). This is an effective version of the well-known theorem of Vaught [495].

Exercise* 10.1.68 (Knight, Miller, Vanden Boom [307]). Prove the Pullback Theorem: Suppose $K \leqslant_{EFF} K'$ via Φ (Def. 8.2.1). Then for any computable infinitary sentence ϕ in the language of K', we can effectively find a computable infinitary sentence ϕ^* in the language of K such that for $A \in K$,

$$A \models \phi^* \text{ if and only if } \Phi(A) \models \phi.$$

Moreover, if ϕ is computable Σ^c_α, or computable Π^c_α, then so is ϕ^*. (Hint: The basic case is essentially saying that the operator is effectively open; see the proof of Theorem 10.1.37. For the general case, generalise the construction from the proof of Theorem 10.1.12 to build a generic $A^* \in K$. Using A^*, given a Σ^c_α sentence ϕ in the language of K', produce a Σ^c_α-sentence ϕ^* in the language of K with the meaning "$p \Vdash \Phi(A^*) \models \phi$".)

Exercise° 10.1.69 (Knight, Miller, Vanden Boom [307]). Let VS and FLO be the classes of vector spaces over $\mathbb{Q}$ and finite linear orders, respectively. Define FVS to be the subclass of VS consisting of vector spaces of finite dimension. Let LO be the extension of FLO including infinite as well as finite linear orderings. Let PF be the class of finite prime fields. Use Ex. 10.1.68 to show that

$$PF <_{EFF} FLO <_{EFF} FVS <_{EFF} VS <_{EFF} LO.$$

Exercise 10.1.70 (Knight, Miller, Vanden Boom [307]). Let VS be the class of $\mathbb{Q}$-vector spaces. Show that $K \leqslant_{EFF} VS$ if and only if there exists a computable sequence $(\phi_n)_{n\in\mathbb{N}}$ of computable Σ^c_2 sentences in the language of K such that

1. for $A \in K$ and $m < n$, if $A \models \phi_n$ then $A \models \phi_m$, and
2. for $A, B \in K$, if $A \not\cong B$ then there is some n such that ϕ_n is true in only one of A, B.

10.1.5 Relativising computable categoricity to an oracle*

A structure is X-c.c. if any pair of X-computable copies of the structure are X-computably isomorphic. Theorem 10.1.17 illustrates that computable categoricity does not imply relative computable categoricity in general. If we take a c.c. structure A that is not relatively c.c., there must be some oracle X such that A is *not* X-c.c.. But what can we say about such X? The reader will note that the proofs of Theorems 10.1.10 and 10.1.6 relativise.

Proposition 10.1.71. *If A is computable and computably categorical relative to some degree $\mathbf{d} \geqslant 0''$, then A is c.c. relative to every degree above $\mathbf{0}''$.*

Proof. We argue that A has a $\mathbf{0}''$-computable Σ_1 Scott family, i.e., one consisting of first-order existential formulae that can be listed using $\mathbf{0}''$. We know that A is 2-decidable relative to $\mathbf{d}$, so by Theorem 10.1.10 it has a $\mathbf{d}$-computable Σ_1 Scott family. Now knowing that A has a Σ_1 Scott family, we will find a $\varnothing''$-computable such family. Given $\bar{a} \in A$ and an existential formula $\varphi(\bar{x})$ true of $\bar{a}$, ask whether for every tuple $\bar{b} \in A$ satisfying $\varphi(\bar{x})$ and every existential formula $\psi(\bar{x})$, $\psi(\bar{a})$ if and only if $\psi(\bar{b})$. If this is the case, then $\varphi(\bar{x})$ isolates the orbit of $\bar{a}$ and we can enumerate it into our $\varnothing''$-computable Scott family. □

Remark 10.1.72. The reader should not confuse relative categoricity with categoricity relative to a fixed oracle. There exist examples which are c.c. but not relatively Δ^0_α-categorical, where α can be an arbitrary large computable ordinal (follows from Theorem 10.1.22). However, Proposition 10.1.71

does not contradict these examples since a c.c. structure does not have to be c.c. relative to $\mathbf{d} \geqslant 0''$. The standard example of a structure which is c.c. but not relatively c.c. given in Theorem 10.1.17 has a $0'$-c.e. Σ_1 Scott family but no Σ_1^0 Scott family.

So suppose A is c.c. but not relatively c.c.. Proposition 10.1.71 implies that A perhaps goes from being c.c. to not c.c. relative to X for some $\varnothing <_T X <_T \varnothing''$ and then at degree $\mathbf{0}''$ the pathology disappears: either for all $\mathbf{d} \geqslant \mathbf{0}''$ we have that A is c.c. relative to $\mathbf{d}$, or for all $\mathbf{d} \geqslant \mathbf{0}''$ we have that A is not c.c. relative to $\mathbf{d}$. It is of course most natural to suspect that this process of switching from plain categoricity to the existence of a Scott family is monotonic, i.e., once the pathology disappears at a degree $\mathbf{a} > 0$, it never occurs again above $\mathbf{a}$. The result below was regarded as unexpected when it was proven.

Theorem 10.1.73 (Downey, Harrison-Trainor and Melnikov [124]). *There is a computable structure A and c.e. degrees $0 = Y_0 <_T X_0 <_T Y_1 <_T X_1 <_T \cdots$ such that*

1. *A is computably categorical relative to Y_i for each i,*
2. *A is not computably categorical relative to X_i for each i,*
3. *A is (relatively) computably categorical relative to $0'$.*

Proof. The structure A will be a directed graph consisting of infinitely many finite connected components. Each component will consist of either two, three, four, or five cycles sharing a single vertex, called the *root vertex* of the component. The root vertices can be identified as the only vertices of degree greater than two.

We build A stage-by-stage ensuring that it is computable. At the same time, we will build two sequences of uniformly c.e. sets $(X_i)_{i\in\mathbb{N}}$ and $(Y_i)_{i\in\mathbb{N}}$ and Turing reductions Ψ_k such that $B_k = \Psi_k^{X_k}$ is a X_k-computable copy of A which is not X_k-computably isomorphic to A. By enumerating elements into X_k, we can give Ψ_k permission to change B_k.

For every set Z, let $(M_i^Z)_{i\in\mathbb{N}}$ a uniformly Z-computable list of the (possibly partial) Z-computable structures. To ensure that A is computably categorical relative to each Y_k, we meet the requirements:

$$S_i^k : \text{If } M_i^{Y_k} \cong A, \text{ then } M_i^{Y_k} \text{ is } Y_k\text{-computably isomorphic to } A.$$

Recall that to make A not computably categorical relative to X_k, we build an X_k-computable structure $B_k = \Psi_k^{X_k}$ which is not X_k-computably isomorphic to A. To achieve this we meet the requirements:

$$R_i^k : \Phi_i^{X_k} : A \to B_k = \Psi_k^{X_k} \text{ is not an isomorphism.}$$

Note that for a fixed k, the R requirements share the same $B_k = \Psi_k^{X_k}$. We will build the X_i and Y_i by setting $Y_0 = \varnothing$, and

$$Y_{i+1} = X_i \oplus Y_{i+1}^* \qquad \text{and} \qquad X_i = Y_i \oplus X_i^*$$

where X_i^* and Y_i^* are c.e. sets to be defined by the construction. Thus we automatically have $\varnothing \equiv_T Y_0 \leqslant_T X_0 \leqslant_T Y_1 \leqslant_T X_1 \leqslant_T \cdots$.

Strategy for meeting R_i^k in isolation: Let $B = B_k = \Psi_k^{X_k}$. We take the following actions:

1. Choose a new large number ℓ and create two new root vertices a_1 and a_2 in A, and b_1 and b_2 in B.

2. Attach a loop of length 2 to a_1 and a_2 in A, and to b_1 and b_2 in B. Attach a loop of length $5\ell+1$ to a_1 in A and b_1 in B; and attach a loop of length $5\ell+2$ to a_2 in A and b_2 in B. The loop of length 2 is simply to identify a_1, a_2, b_1, and b_2 as root vertices.

$$\begin{array}{ll} a_1 : 2, 5\ell+1 & a_2 : 2, 5\ell+2 \\ b_1 : 2, 5\ell+1 & b_2 : 2, 5\ell+2 \end{array}$$

3. Wait for a stage s at which we see that $\Phi_i^{X_k} : A \to B$ maps $a_1 \mapsto b_1$ and $a_2 \mapsto b_2$. Let u be the use of this computation.

$$\begin{array}{l} a_1 : 2, 5\ell+1 \xrightarrow[\Phi]{} a_2 : 2, 5\ell+2 \\ b_1 : 2, 5\ell+1 \xrightarrow[\Phi]{} b_2 : 2, 5\ell+2 \end{array}$$

4. Choose a large number $v > u$. Attach loops of length $5\ell+3$ to a_1 in A and b_1 in B. Attach loops of length $5\ell+4$ to a_2 in A and b_2 in B with use $X_k[s] \upharpoonright 2v+2$ where s is the current stage. *Restrain* $X_k \upharpoonright 2v+2$ so that it may not be changed by another requirement of a lower priority (to be clarified). Enumerate ℓ into $Y_{k'}^*$ for $k' \geqslant k$; this will be used to meet the S requirements.

$$\begin{array}{l} a_1 : 2, 5\ell+1, 5\ell+3 \xrightarrow[\Phi]{} a_2 : 2, 5\ell+2, 5\ell+4 \\ b_1 : 2, 5\ell+1, 5\ell+3 \xrightarrow[\Phi]{} b_2 : 2, 5\ell+2, 5\ell+4 \end{array}$$

5. Attach loops of length $5\ell+2$ to a_1 in A and b_1 in B, and attach loops of length $5\ell+1$ to a_2 in A and b_2 in B.

$$\begin{array}{l} a_1 : 2, 5\ell+1, 5\ell+2, 5\ell+3 \xrightarrow[\Phi]{} a_2 : 2, 5\ell+1, 5\ell+2, 5\ell+4 \\ b_1 : 2, 5\ell+1, 5\ell+2, 5\ell+3 \xrightarrow[\Phi]{} b_2 : 2, 5\ell+1, 5\ell+2, 5\ell+4 \end{array}$$

6. Enumerate v into X_k^* (thus removing the loop of length $5\ell+3$ attached to b_1 and the loop of length $5\ell+4$ attached to b_2 in B). Attach a loop of length $5\ell+4$ to b_1 in B and a loop of length $5\ell+3$ to b_2 in B.

$$\begin{array}{l} a_1 : 2, 5\ell+1, 5\ell+2, 5\ell+3 \xrightarrow[\Phi]{} a_2 : 2, 5\ell+1, 5\ell+2, 5\ell+4 \\ b_1 : 2, 5\ell+1, 5\ell+2, 5\ell+4 \xrightarrow[\Phi]{} b_2 : 2, 5\ell+1, 5\ell+2, 5\ell+3 \end{array}$$

If $\Phi_i^{X_k} : A \to B_k$ is an isomorphism, it must be defined on a_1 and a_2 with some use u. In step (3), it must map $a_1 \mapsto b_1$ and $a_2 \mapsto b_2$ because a_1 and b_1 are the only elements with a loop of length $5\ell+1$, and a_2 and b_2 are the only elements with a loop of length $5\ell+2$. So at some stage s we see that $\Phi_i^{X_k[s] \upharpoonright u}$ maps $a_1 \mapsto b_1$ and $a_2 \mapsto b_2$. In steps (4), (5), and (6) we ensure that the elements have the following loops:

$$\begin{aligned} a_1 &: 2, 5\ell+1, 5\ell+2, 5\ell+3 \xrightarrow[\Phi]{} a_2 : 2, 5\ell+1, 5\ell+2, 5\ell+4 \\ b_1 &: 2, 5\ell+1, 5\ell+2, 5\ell+4 \xrightarrow[\Phi]{} b_2 : 2, 5\ell+1, 5\ell+2, 5\ell+3 \end{aligned}$$

At stage (6), we enumerated v into X_k^*, but we still have $X_k \upharpoonright u = X_k[s] \upharpoonright u$. So $\Phi_i^{X_k}$ still maps $a_1 \mapsto b_1$ and $a_2 \mapsto b_2$, and this does not extend to an isomorphism.

Injury and restraint between different requirements R_i^k*:* If R_i^k finds a computation $\Phi_i^{X_k}$ in step (3) with use u, it restrains $X_k \upharpoonright u$. Another requirement $R_{i'}^{k'}$ with $k' \leqslant k$ might have already chosen an element $v' < u$ in step (4), and want to enumerate v' into $X_{k'}$ in step (6). Since $k' \leqslant k$, this would enumerate an element into X_k as well, and potentially violating the restraint placed on X_k by R_i^k. We use the standard priority method; higher priority requirements $R_{i'}^{k'}$ are allowed to violate the restraint placed by R_i^k, in which case we say that R_i^k is *injured*; and when R_i^k places a restraint, it injures all lower priority requirements $R_{i'}^{k'}$ which then have to choose a value v' greater than the restraint placed by R_i^k. When a requirement is injured, it homogenises the elements a_1 and a_2 it has been working with, and also the elements b_1 and b_2:

$$\begin{aligned} a_1 &: 2, 5\ell+1, 5\ell+2, 5\ell+3, 5\ell+4 \qquad & a_2 &: 2, 5\ell+1, 5\ell+2, 5\ell+3, 5\ell+4 \\ b_1 &: 2, 5\ell+1, 5\ell+2, 5\ell+3, 5\ell+4 \qquad & b_2 &: 2, 5\ell+1, 5\ell+2, 5\ell+3, 5\ell+4 \end{aligned}$$

This means that any way of matching up a_1 and a_2 with b_1 and b_2 can be extended to an isomorphism. It then creates new elements a_1, a_2, b_1, and b_2 to work with.

Strategy for meeting S_i^k*:* The requirement S_i^k is responsible for building a Y_k-computable isomorphism $f = \Gamma^{Y_k}$ between A and $M_i^{Y_k}$. To define f, we must look at pairs of root nodes a_1, a_2 in A and decide on images for them in $M = M_i^{Y_k}$. Given a_1, a_2, let ℓ be such that each of a_1 and a_2 has a loop of length $5\ell + 1$ or $5\ell + 2$. We can Y_k-computably look for a pair of root nodes c_1, c_2 in M which also have such loops. Finally, identify the requirement $R_j^{k'}$ which was responsible for a_1, a_2. We have three cases in each of which we act differently:

- If $k' \leqslant k$: In step (4) of meeting the requirement $R_j^{k'}$, we enumerate ℓ into Y_k^*; so by checking whether $\ell \in Y_k^*$, we can determine whether $R_j^{k'}$ reached step (4) while working with a_1 and a_2. If it did not reach step (4), then exactly one of a_1, a_2 has a loop of length $5\ell + 1$, and the other has a loop of length $5\ell + 2$; so we can map whichever of a_1, a_2 has a loop of length $5\ell + 1$ to whichever of c_1, c_2 has a loop of the same size, and this will extend to an isomorphism. If $R_j^{k'}$ did reach step four, then exactly one of a_1, a_2 has a loop of length $5\ell + 3$, and the other has a loop of length $5\ell + 4$, and we can again match a_1, a_2 up with c_1, c_2.
- If $k' > k$ and $R_j^{k'}$ is of higher priority than S_i^k: S_i^k can non-uniformly know whether or not $R_j^{k'}$ ever reached step (4); the rest is similar to the previous case.
- If $k' > k$ and $R_j^{k'}$ is of lower priority than S_i^k: In this case Y_k does not know whether $R_j^{k'}$ reached step (4). Without loss of generality, we may assume that a_1 and c_1 have loops of length $5\ell + 1$ and a_2 and c_2 have loops of length $5\ell + 2$.

$$\begin{aligned} a_1 &: 2, 5\ell+1 \qquad & a_2 &: 2, 5\ell+2 \\ c_1 &: 2, 5\ell+1 \qquad & c_2 &: 2, 5\ell+2. \end{aligned}$$

Then have f map $a_1 \mapsto c_1$ and $a_2 \mapsto c_2$. If $R_j^{k'}$ never reaches step (4), then this extends to an isomorphism. The issue might be that $R_j^{k'}$ reaches step (4), but that M delays adding the loops from step (4) until $R_j^{k'}$ has reached step (6), so that A looks like

$$a_1 : 2, 5\ell + 1, 5\ell + 2, 5\ell + 3 \qquad\qquad a_2 : 2, 5\ell + 1, 5\ell + 2, 5\ell + 4.$$

M_j can now add loops so that it looks like:

$$c_1 : 2, 5\ell + 1, 5\ell + 2, 5\ell + 4 \qquad\qquad c_2 : 2, 5\ell + 1, 5\ell + 2, 5\ell + 3$$

This would defeat the isomorphism f.

The solution is to force R_i^k to wait for M_j to copy A after step (4) before preceding on to step (5). After step (4), A looks like

$$a_1 : 2, 5\ell + 1, 5\ell + 3 \qquad\qquad a_2 : 2, 5\ell + 2, 5\ell + 4$$

and this forces M_j, if it wants to be isomorphic to A, to add loops so that it looks like

$$c_1 : 2, 5\ell + 1, 5\ell + 3 \qquad\qquad c_2 : 2, 5\ell + 2, 5\ell + 4.$$

Thus f remains an isomorphism.

There is a complication in that M is a Y_k-computable structure, and we must build A computably. So we work with the stage-by-stage approximation to M given by the stage-by-stage approximation to Y_k, and whenever $R_j^{k'}$ sees more loops added to M (according to the current value of Y_k), it places a restraint on Y_k. If this restraint is ever violated by a higher priority requirement, we homogenise a_1 and a_2, so that the map f still extends to an isomorphism.

Of course, M does not have to copy A, and there will certainly be some j for which $M \ncong A$ and $R_j^{k'}$ gets stuck after step (4). To solve this we use a standard pressing strategy: the requirements $R_j^{k'}$ must guess at the outcomes of the higher priority requirements. While $R_j^{k'}$ is waiting for M to catch up, it starts a new instance guessing that M is not isomorphic to A; when M does catch up, this new instance is destroyed (and homogenised).

The action in the second case works even when $k' \leqslant k$, it is just that the first case also works as well. But the action in the third case does not work when $k' \leqslant k$, and it might be helpful to the reader, to aid in understanding the construction, to think about why this is true: In the infinitary outcome, the requirement $R_j^{k'}$ must be able to restrain Y_k, while still enumerating an element into $X_{k'}^*$; if $k' \leqslant k$, then enumerating an element into $X_{k'}^*$ would also enumerate an element into Y_k.

Priorities and guesses: We put a priority ordering on the requirements. A requirement S_i^k knows the outcome of the higher priority R requirements when it builds its isomorphism between $M_i^{Y_k}$ and A, and an R requirement must guess at the outcomes of the higher priority S requirements. The R requirements are the only requirements that are active during the construction and which can enumerate elements into the X_k and Y_k; the S requirements must be taken into account by the R requirements (e.g. by enumerating elements into the Y_k, or waiting to see loops in a structure $M_i^{Y_K}$), and the isomorphisms they ask for can be defined after the construction is finished.

Full strategy for meeting R_i^k: For simplicity we write $B = B_k$. Let F be a subset of the higher priority S requirements. For each such F, R_i^k can have a module which works under the assumption that the requirements in F are exactly the higher priority S requirements that R_i^k needs to wait for. We call this module a module of R_i^k with guess F. At certain stages in the module for working for F, we will have to wait to see some loops show up in a structure $M_j^{Y_{k'}}$ for some $S_j^{k'} \in F$; while waiting, we will start up a module for a set $G \subsetneq F$, which itself might start up another module, and so on.

When a module for R_i^k with guess F starts a new module with guess $G \subsetneq F$, we say that the module with guess F is the *parent* of the module with guess G, and the module with guess G is the *child* of the module with guess F. The module for R_i^k with F consisting of all of the higher priority S requirements is called the base module.

The module for R_i^k with guess F acts as follows:

1. Choose a new large number ℓ greater than the restraints of all higher priority requirements and also greater than the restraint of the parent modules of this module, and its parent, and so on. Create two new root vertices a_1 and a_2 in A, and b_1 and b_2 in B.

   Attach a loop of length 2 to a_1 and a_2 in A, and to b_1 and b_2 in B. Attach a loop of length $5\ell + 1$ to a_1 in A and b_1 in B; and attach a loop of length $5\ell + 2$ to a_2 in A and b_2 in B.

2. For each requirement $S_j^{k'}$ in F, with $k' < k$, wait until we see a pair of root nodes c_1 and c_2 in $M_j^{Y_{k'}}$ such that c_1 has a loop of length 2 and of length $5\ell + 1$, and c_2 has a loop of length 2 and of length $5\ell + 2$. If we ever see such elements, R_i^k puts a restraint r greater than the use of the oracle $Y_{k'}$ for the computation witnessing this. We say that such an $S_j^{k'}$ has *caught up*.

   If at any stage s we ever see any other elements connected to c_1 or c_2, R_i^k puts a restraint r greater than the use of the oracle $Y_{k'}$ for the computation witnessing this. (Just put this restraint once per structure $M_j^{Y_{k'}}$.) We say that such an $S_j^{k'}$ has been *killed*.

   Remark 10.1.74. We note that the strategy only cares about its current component (the elements a_1 and a_2 and the adjacent loops it is currently working with). Therefore, if if $S_j^{k'}$ has been declared killed then this status is *not* global, i.e., it is internal for this particular version of the R_i^k-strategy. We could of course make this status global and give the restraint some global priority (say, the priority of $S_j^{k'}$). However, this is not necessary. This is because the injury in the construction will be only finite. Thus, if some higher priority requirement changes the set below the use witnessing that $S_j^{k'}$ has been killed, we simply initialise R_i^k. In particular, for one fixed component the situation in which $S_j^{k'}$ was killed and then resurrected and then killed again etc. is impossible.

   While waiting, let G be the set of higher priority S requirements which have caught up but which have not been killed. Start a module of R_i^k with guess G. If we ever find that a new requirement has caught up, or that one which had caught up has now been killed, then the module with guess G must be homogenised (as described below), we reset G to be the new, larger, set of requirements which have caught up but which have not been killed, and start a module of R_i^k whose guess is the new G.

   If we ever see that every requirement in F has either caught up or been killed, move on to the next step.

3. Wait for a stage s at which we see that $\Phi_i^{X_k}\colon A \to B$ maps $a_1 \mapsto b_1$ and $a_2 \mapsto b_2$. Let u be the use of this computation. Enumerate ℓ into Y_k^*, injuring all lower priority requirements.

Choose a large number $v > u$ and set a restraint $r = v + 1$. Attach loops of length $5\ell + 3$ to a_1 in A and b_1 in B. Attach loops of length $5\ell + 4$ to a_2 in A and b_2 in B with use $X_k[s] \upharpoonright 2v + 2$ where s is the current stage.

4. For each requirement $S_j^{k'}$ in F, with $k' < k$, wait until we see loops of length $5\ell + 3$ and $5\ell + 4$ attached to c_1 and c_2 respectively. R_i^k puts a restraint r greater than the use of the oracle $Y_{k'}$ for the computation witnessing this, and we say that $S_j^{k'}$ has caught up.

   Again, if at any stage s we ever see any other elements connected to c_1 or c_2, R_i^k puts a restraint r greater than the use of the oracle $Y_{k'}$ for the computation witnessing this, and we say that $S_j^{k'}$ has been killed.

   While waiting, we start new modules of R_i^k with guesses $G \subsetneq F$ as in step (2).

   If we ever see that every requirement in F has either caught up or been killed, move on to the next step.

5. Attach loops of length $5\ell + 2$ to a_1 in A and b_1 in B, and attach loops of length $5\ell + 1$ to a_2 in A and b_2 in B.

   Enumerate v into X_k^* (thus removing the loop of length $5\ell + 3$ attached to b_1 and the loop of length $5\ell + 4$ attached to b_2 in B). Attach a loop of length $5\ell + 4$ to b_1 in B and a loop of length $5\ell + 3$ to b_2 in B.

For $k' \neq k$, whenever we do anything in A (e.g. creating the elements a_1 and a_2 or adding loops to them) do the same in $B_{k'}$ with no use.

Whenever a requirement increases its restraint, or enumerates an element, it injures all lower priority requirements. Each module of an injured requirement undergoes the homogenisation procedure described below, and then the requirement restarts with just the base module at the first step.

When a module is to be homogenised, we do the following: for any loop on a_1 for which there is no corresponding loop on a_2, we add a loop of that length to a_2, and vice versa. So a_1 and a_2 have loops of exactly the same lengths attached to them. Do the same for b_1 and b_2.

Construction

Recall that the S requirements do not take any action during the construction; we define the required isomorphisms after the construction.

At stage s, the first s R-requirements are allowed to act. For each of these requirements, in order from highest priority to lowest priority, do the following: First, if the requirement has never before acted or was injured, start the base module of the requirement with F being the set of all higher priority S requirements. Then, execute the base module until we end up at a step where we have to wait for a larger stage s. Then, if there is a child module, execute the child module until it has to wait, then any child of the child module, and so on.

Verification

Lemma 10.1.75. *Each requirement R_i^k is injured only finitely many times.*

Proof. We argue on induction that each R_i^k can injure the lower priority R requirements only finitely many times. To do this we suppose that a requirement R_i^k is never injured after some stage, and show that it injures the lower priority requirements only finitely many times.

An R_i^k module with guess F injures the lower priority requirements only finitely many times:

- once each time we find that a new requirement has caught up or been killed in step (2);
- once in step (3);
- once each time we find that a new requirement has caught up or been killed in step (4);
- once in step (5).

Now we argue that there are only finitely many R_i^k modules that ever run. We begin with the base module, say with guess F_0, and it has only one child at a time, with guess $F_1 \subsetneq F_0$; and its child module can have one child module, and so on. So at any one time, we have a chain of child modules with guesses $F_0 \supsetneq F_1 \supsetneq F_2 \supsetneq \cdots \supsetneq F_n$, and so $n \leqslant |F_0|$. A module can only homogenise its child module and start a new child module within the same step if a requirement has been found to have caught up or been killed, each of which can only happen once per requirement. The one exception to this is if the child module is homogenised in step (2) and the new child is started in step (4), but this can only happen once per module. So each module can have only finitely many child modules, each of which can have only finitely many child modules, and so on, and the depth is bounded by $|F_0|$. Thus there are only ever finitely many R_i^k modules running during the construction. □

Lemma 10.1.76. *Suppose that a module for R_i^k with guess G is never homogenised (which also means that R_i^k is never injured). Let $S_j^{k'}$ be a higher priority requirement and suppose that $M_j^{Y_{k'}} \cong A$. Then $S_j^{k'} \in G$.*

Proof. Suppose to the contrary that $S_j^{k'} \notin G$. Then, because $S_j^{k'}$ is contained in the guess by the base module, there must be some module with guess F containing $S_j^{k'}$, which has a child module G not containing $S_j^{k'}$, and neither module is ever homogenised.

So after some point the module with guess F must be stuck waiting at either stage (2) or (4), with G being exactly the set of higher priority requirements which have caught up but not been killed. But we will argue that since $M_j^{Y_{k'}} \cong A$, it must eventually catch up, and it can never be killed. This would cause the module with guess G to be homogenised, contradicting our initial assumption.

First, suppose that it is killed. Then at some stage s, we see in $M_j^{Y_{k'}}[s]$ that there is a root vertex such that no root vertex in A has loops of the same lengths; and we put a restraint on the use so that $M_j^{Y_{k'}}$ has such a vertex. (The restraint is never violated, or else R_i^k would be injured.) Thus we would have ensured that $M_j^{Y_{k'}} \not\cong A$, which is not the case.

It must also catch up, because $M_j^{Y_{k'}} \cong A$, and so whenever a vertex shows up in A with certain loops, it must show up in $M_j^{Y_{k'}}$; and so for some s, it must show up in $M_j^{Y_{k'}}[s]$. □

Lemma 10.1.77. *For each k, A is computably categorical relative to Y_k.*

Proof. We must show that each Y_k-computable structure copy of A is Y_k-computably isomorphic to A, i.e. that each requirement S_i^k is satisfied. Suppose that $A \cong M_j^{Y_k}$. For simplicity, write

$M = M_j^{Y_k}$. Let f be the Y_k-computable isomorphism defined in the strategy for meeting S_j^k. We argue that f is an isomorphism $M \to A$. It suffices to show that whenever we maps elements c_1, c_2 to a_1, a_2 respectively (as defined in the strategy for S_i^k), c_1 and a_1 have the same lengths of loops attached to them, and c_2 and a_2 have the same lengths of loops attached to them.

If the module that built a_1, a_2 was ever homogenised, then any way of mapping c_1, c_2 to a_1, a_2 extends to an isomorphism. So suppose that the module that built a_1, a_2 was never homogenised. We have three cases from the definition of f:

1. If $k' \leqslant k$: If $\ell \notin Y_k^*$ then the module that built a_1, a_2 did not reach step (3), then a_1 is the unique root vertex in A (and c_1 is the unique root vertex in M) with a loop of length $5\ell + 1$, and a_2 is the unique root vertex in A (and c_2 in M) with a loop of length $5\ell + 2$. So mapping c_1, c_2 to a_1, a_2 respectively extends to an isomorphism.

   If $\ell \in Y_k^*$ then the module that built a_1, a_2 reached step (3), and a_1 is the unique root vertex in A (and c_1 is the unique root vertex in M) with a loop of length $5\ell + 3$, and a_2 is the unique root vertex in A (and c_2 in M) with a loop of length $5\ell + 4$. So mapping c_1, c_2 to a_1, a_2 respectively extends to an isomorphism.

2. If $k' > k$ and $R_j^{k'}$ is of higher priority than S_i^k: Similar to above.

3. If $k' > k$ and $R_j^{k'}$ is of lower priority than S_i^k:

   Let F be the guess by the module of $R_j^{k'}$ that built a_1, a_2. By Lemma 10.1.76, since $M_i^{Y_k} \cong A$, and the module with guess F is not homogenised, we have $S_i^k \in F$.

   First of all, we argue that when we find a stage s and elements c_1 and c_2, such that in $M[s]$ there is a loop of length $5\ell + 1$ on c_1 and of length $5\ell + 2$ on c_2, that in $M[s]$ there is no loop of length $5\ell + 2$ on c_1 or $5\ell + 1$ on c_2. Otherwise, S_i^k would have been killed. (As in Lemma 10.1.76, if it is killed, then there is a restraint placed on Y_k which ensures that $M \ncong A$.)

   Then if the module with guess F does not make it to step (5), a_1 is the unique root vertex of A with a loop of length $5\ell + 1$, and a_2 is the unique root vertex with a loop of length $5\ell + 2$; so mapping c_1 to a_1 and c_2 to a_2 extends to an isomorphism.

   Then if the module with guess F does make it to step (5), a_1 is the unique root vertex of A with a loop of length $5\ell + 3$, and a_2 is the unique root vertex with a loop of length $5\ell + 4$; we must argue that c_1 gets a loop of length $5\ell + 3$ and c_2 gets a loop of length $5\ell + 4$. In step (2) we wait to see c_1 get a loop of length $5\ell + 1$ and c_2 get a loop of length $5\ell + 2$; and then in step (4) we wait to see c_1 get a loop of length $5\ell + 3$ and c_2 a loop of length $5\ell + 4$. We put a new restraint on Y_k every time we see a new loop. Moreover, if c_2 got the loop of length $5\ell + 3$, or c_1 got the loop of length $5\ell + 4$, then S_i^k would be killed. So it must be c_1 that gets the loop of length $5\ell + 3$ and c_2 that gets the loop of length $5\ell + 4$.

This completes the proof of the claim. □

Lemma 10.1.78. *For each k, B_k is isomorphic to A.*

Proof. The structure A consists entirely of pairs of root nodes a_1, a_2 produced by a module of an R requirement, and the elements forming the loops attached to these root nodes. Whenever we add elements a_1, a_2 to A for a requirement $R_i^{k'}$, $k' \neq k$, we add the same sort of elements to B_k (i.e., $B)_k$ just copies A. When we add elements a_1, a_2 to A for a requirement R_i^k, we add elements b_1, b_2

to B_k, and either a_1 has the same size loops as b_1 (and a_2 as b_2), or a_1 has the same size loops as b_2 (and a_2 and b_1); which case we are in depends on whether the requirement made it to step (5) or not. □

Lemma 10.1.79. *For each k, B_k is not X_k-computably isomorphic to A.*

Proof. We must argue that each requirement R_i^k is satisfied, so that no X_k-computable map $\Phi_i^{X_k}$ is an isomorphism between B_k and A. Fix some requirement R_i^k which we will show is satisfied. After some stage, it is no longer injured.

There is some module of R_i^k, say with guess F, which is never homogenised, and which either waits forever in step (3) or reaches step (5). (The other options for a particular module are that it waits forever in step (2) or step (4), and in each of these cases it has a child module; we know from the arguments in the previous lemma that each module must have some last child module, and that the depth of child modules is bounded, so that there must be some module with no child module.)

First suppose that the module waits forever in step (3). The either $\Phi_i^{X_k}$ is partial, or it maps a_1 to some element other than b_1. If it is is partial then it obviously cannot be an isomorphism. If it maps a_1 to some element other than b_1, then it cannot extend to an isomorphism, as a_1 and b_1 each have a loop of length $5\ell + 1$, and no other elements of A or B do, as no other requirement or module has the same value of ℓ.

Now suppose that the module waits forever in step (5). Since the module passed through step (3), we have that $\Phi_i^{X_k}$ maps $a_1 \mapsto b_1$ and $a_2 \mapsto b_2$. (The requirement R_i^k puts a restraint on the use of the computation found in step (3), which cannot be violated by any higher priority requirement; and if a lower priority requirement violated the restraint, R_i^k would have been injured.) But we ensure in step (5) that a_1 has a loop of length $5\ell + 3$ and that b_1 does not, so that $\Phi_i^{X_k}$ does not extend to an isomorphism. □

The proof is finished. □

10.1.6 Exercises: Beyond computable categoricity

Exercises about degrees of categoricity

All exercises in this paragraph use the following definition:

Definition 10.1.80 (Kalimullin, Fokina, Miller [172]). A Turing degree $\mathbf{x}$ is the *degree of categoricity for $\mathcal{S}$* if $\mathbf{x}$ is the least degree such that $\mathcal{S}$ is $\mathbf{x}$-computably categorical. A degree of categoricity $\mathbf{x}$ is said to be a *strong degree of categoricity* if and only if there are particular computable copies A and B of a structure with degree of categoricity $\mathbf{x}$ such that every isomorphism f between A and B computes $\mathbf{x}$.

For more background (and many more results), we refer the reader to the excellent survey [178].

Exercise 10.1.81 (Kalimullin, Fokina, Miller [172]). Prove that for each degree $\mathbf{d}$ satisfying at least one of the following properties, there exists a computable structure for which d is the degree of categoricity:

(1) $\exists m\,(\mathbf{0^{(m)}} \leqslant \mathbf{d} \wedge \mathbf{d}$ is c.e. in $\mathbf{0^{(m)}})$;

(2) $\mathbf{d}$ is d.c.e. in $\mathbf{0^{(m)}}$, for some $m \in \mathbb{N}$ (i.e., $\mathbf{d}$ is the difference of two $\mathbf{0^{(m)}}$-c.e. sets);

(3) $\mathbf{d} = \mathbf{0}^{(\omega)}$.

Exercise 10.1.82 (Miller [395]). Show that there exists a computable algebraic field having no degree of categoricity. Indeed, any computable algebraic field with a computable splitting algorithm which is not computably categorical has no degree of categoricity. (Use Ex. 4.2.61.)

Exercise 10.1.83 (Bazhenov [40]). Prove that every computable ordinal has a degree of categoricity.

Exercise* 10.1.84 (Csima, Franklin, and Shore [100]). Prove the following theorem. If α is a computable ordinal, then $0^{(\alpha)}$ is a strong degree of categoricity. If, in addition, α is a successor ordinal, then every degree that is c.e. or d.c.e. in and above $0^{(\alpha)}$ is a strong degree of categoricity.

Exercise 10.1.85 (Bazhenov [43]). Let $\alpha \geqslant 3$ be a computable ordinal, and consider a Turing degree $\mathbf{d}$ which is d.c.e. in and above $\mathbf{0}^{(\alpha+1)}$. Construct a computable linear order having degree of categoricity $\mathbf{d}$.

Exercise 10.1.86 (Anderson and Csima [13]). Show that there is a degree below $\mathbf{0}''$ that is not a degree of categoricity of any computable structure. (In fact, there is a Σ^0_2 degree that is not a degree of categoricity of any computable structure.)

Exercise 10.1.87 (Bazhenov [42]). Construct a Δ^0_4-categorical distributive lattice with no degree of categoricity.

Exercise 10.1.88 (Bazhenov [41]). Prove that every computable superatomic Boolean algebra has a degree of categoricity. Show that for every computable ordinal α, the α-th jump $\mathbf{0}^{(\alpha)}$ is the degree of categoricity for some computable Boolean algebra.

Exercise* 10.1.89 (Bazhenov [39]). Prove that every Δ^0_2-categorical Boolean algebra has degree of categoricity either $\mathbf{0}$ or $\mathbf{0}'$.

Exercise 10.1.90** (Csima and Ng [103]). Show that every Δ^0_2-degree is a strong degree of categoricity. (This solved a question left open in [172].)

Exercise* 10.1.91 (Csima and Rossegger [104]).
1. Show that every degree $\mathbf{d}$ with $\mathbf{0}' \leqslant \mathbf{d} \leqslant \mathbf{0}''$ is the strong degree of categoricity of a structure.
2. Show that every degree d with $\mathbf{0}^{(\alpha)} \leqslant \mathbf{d} \leqslant \mathbf{0}^{(\alpha+1)}$ for α a computable ordinal greater than 2 is the strong degree of categoricity of a rigid structure.

Exercises about relative Δ^0_2- and Δ^0_3-categoricity

Exercise° 10.1.92 (Cenzer, Harizanov, and Remmel [85]). An injection structure is a structure of the form (X, f), where f is an injection of the set X into itself. Show that every Δ^0_2-categorical injection structure is relatively Δ^0_2-categorical.

Exercise° 10.1.93 (Kach and Turetsky [277]). Prove that there is a Δ^0_2-categorical (computable) equivalence structure that is not relatively Δ^0_2-categorical.

Exercise* 10.1.94 (Fokina, Harizanov, and Turetsky [168]). Show that a computable homogeneous completely decomposable (c.d.) group of rank ω is relatively Δ^0_2-categorical iff it is isomorphic to $\bigoplus_{i\in\mathbb{N}}\mathbb{Q}^{(P)}$ for a computable set of primes, where $\mathbb{Q}^{(P)}$ is the subgroup of $\mathbb{Q}$ generated by $\{\frac{1}{p^n} : p \in P, n \in \mathbb{N}\}$. Conclude that there is a computable homogeneous c.d. group that is Δ^0_2-categorical but not relatively Δ^0_2-categorical.

Exercise 10.1.95 (McCoy [357]). Show that a computable Boolean algebra is relatively Δ^0_2-categorical if and only if it can be expressed as a finite direct sum $c_1 \vee \cdots \vee c_n$, where each c_i is either atomless, an atom, or a 1-atom. (Compare this with Exercise 4.1.57.)

Exercise 10.1.96 (Bazhenov [39]). Show relative Δ^0_2-categoricity and ("plain", "Type 1") Δ^0_2 categoricity are equivalent for Boolean algebras.

Exercise 10.1.97 (McCoy [357]). Show that a computable linear order L is relatively Δ^0_2-categorical iff L is a sum of finitely many intervals, each of type m (i.e., finite), $\omega, \omega^*, \mathbb{Z}$, or $n \cdot \eta$, such that each interval of type $n \cdot \eta$ has a supremum and an infimum.

Exercise 10.1.98 (Essentially Miller [395]). Show that every (computable) algebraic field is relatively Δ^0_3-categorical.

Exercise* 10.1.99 (McCoy [356]). Show that a computable Boolean algebra is relatively Δ^0_3-categorical if and only if it can be expressed as a finite direct sum of finitely many algebras which are atoms, atomless, 1-atoms, rank 1 atomic, or isomorphic to the interval algebra $\mathrm{Intalg}(\omega + \eta)$.

Exercise 10.1.100 (Moses [407]). We say that a 1-decidable structure is 1-*decidably categorical* if any two 1-decidable presentations of the structure are computably isomorphic. Show that a linear order is 1-decidably categorical iff it is relatively Δ^0_2-categorical (as described in Ex. 10.1.97).

Exercise* 10.1.101 (Bazhenov, Frolov, Kalimullin, and Melnikov [44]). Show that there exists a computably categorical (computable) distributive lattice that is not relatively Δ^0_2-categorical.

Exercise* 10.1.102 (Downey, Melnikov, and Ng [140]). Let G be a computable p-group of finite Ulm type n, such that:

- $G^{(n)} \cong \bigoplus_{j\leqslant m}\mathbb{Z}_{p^\infty}$ for some $m \in \mathbb{N}$, and
- the orders of the cyclic summands in G_{n-1} are not bounded.

Prove that G is *not* Δ^0_{2n}-categorical.

Exercise 10.1.103. Show that the transformation $G \leftrightarrow E_G$ between Ulm type 1 abelian p-groups and the respective equivalence structures has the following properties:

1. It preserves degree spectra and computable dimension;
2. It does *not* preserve Δ^0_2-categoricity.

(Hint: Use Exercises 9.1.16, 9.3.42, and 10.1.102.)

Exercise* 10.1.104 (Downey, Melnikov, and Ng [142]). Say that a computable algebraic structure is *weakly uniformly* Δ^0_2-*categorical* if there exists an effective procedure which, given indices i, j of computable copies M_i, M_j of the structure, produces a Δ^0_2-index e for a Δ^0_2-isomorphism from M_i onto M_j. Show that for (not necessarily reduced) computable abelian p-groups of Ulm type 1, the following hold:

1. The notion of weak uniform Δ_2^0-categoricity lies *strictly in-between* Δ_2^0- and relative Δ_2^0-categoricity. (In particular, all three notions are pairwise non-equivalent!)
2. The correspondence $G \leftrightarrow E_G$ between Ulm type 1 abelian p-groups and the respective equivalence structures preserves weak uniform Δ_2^0-categoricity.

Exercise 10.1.105 (Downey, Kach, Lempp, and Turetsky [132]). Prove that there is a 1-decidable, computably categorical structure M having a computable presentation A and a Δ_2^0-computable presentation B such that A and B are not Δ_2^0-isomorphic.

Exercise* 10.1.106 (Montalbán [400]). A countable structure A is *computably categorical on a cone* if there is a $Y \in 2^\omega$ such that A is X-computably categorical for all $X \geqslant_T Y$. Prove that he following are equivalent:

1. A is computably categorical on a cone.
2. A has a (not necessarily computable, $L_{\omega_1\omega}$) Σ_3 Scott sentence.

Exercise 10.1.107. Show that the most straightforward computable analogue of Exercise 10.1.106 fails. Indeed, there exists a computable torsion abelian group G (viewed in the language $\{+, -, 0\}$) with the properties:

1. G is not computably categorical;
2. G has a (computable) Π_2^c Scott sentence.

(Hint: Build $G = \bigoplus_{e\in\mathbb{N}} G_{p_e}$, where G_{p_e} is a p_e-group. We will have G_{p_e} isomorphic either to $\mathbb{Z}_{p_e^\infty} \oplus \mathbb{Z}_{p_e^\infty}$ or $\mathbb{Z}_{p_e^{n_e}} \oplus \mathbb{Z}_{p_e^{m_e}}$, for some $m_e, n_e \in \mathbb{N}$, $m_e > n_e$. We can easily build two copies of G and diagonalise against all potential isomorphisms φ_e between these copies, as follows. Wait for φ_e to halt on some non-zero $x_e \in G_{p_e}$ of order p_e. If this ever occurs, make sure $h_{p_e}(x) = n_e \neq m_e = h_p(\varphi_e(x_e))$. If φ_e never halts, proceed to building $G_{p_e} \cong \mathbb{Z}_{p_e^\infty} \oplus \mathbb{Z}_{p_e^\infty}$. The Π_2^c Scott sentence of G says that:

(i) G is a torsion abelian group.

(ii) For all i, G has exactly p_i^2 elements of order p_i (including 0).

(iii) For all i, monitor φ_e. At stage s, if diagonalisation has not yet occurred, let $\psi_{i,s}$ be a first-order $\exists$-sentence saying that $\mathbb{Z}_{p_e^s} \oplus \mathbb{Z}_{p_e^s}$ embeds in G. If the diagonalisation has already occurred and the parameters n_e, m_e have been already defined, then let $\psi_{i,s}$ be a sentence that says:

  (a) $\mathbb{Z}_{p_e^{m_e}} \oplus \mathbb{Z}_{p_e^{n_e}}$ isomorphically embeds in G,

  (b) neither $\mathbb{Z}_{p_e^{m_e}} \oplus \mathbb{Z}_{p_e^{n_e+1}}$ nor $\mathbb{Z}_{p_e^{m_e+1}}$ isomorphically embeds in G,

  which is a conjunction of an $\exists$- and a $\forall$-sentence.

Take the conjunction of (i), (ii), and

$$\bigwedge_{i,s\in\mathbb{N}} \psi_{i,s},$$

which were described in (iii). Appeal to the classification of finitely generated abelian groups and groups of Ulm type 1 to conclude that this is a Π_2^c Scott sentence for G.)

Exercises about (relative) Δ^0_α-categoricity for arbitrarily large α

Some of the exercises below will use the following convenient definition:

Definition 10.1.108. Fix a computable ordinal $\alpha > 1$. We say that a computable, relatively Δ^0_α-categorical structure is *optimally* relatively Δ^0_α-categorical if it is not Δ^0_β-categorical for $\beta < \alpha$.

Exercise 10.1.109 (Goncharov et al. [212]). Show that for every computable successor ordinal $\alpha > 1$, there is a Δ^0_α-categorical structure that is not relatively Δ^0_α-categorical.

Exercise 10.1.110 (Goncharov et al. [90]). Show that for every computable limit ordinal $\alpha \geqslant \omega$, there is a Δ^0_α-categorical structure that is not relatively Δ^0_α-categorical.

Exercise* 10.1.111 (Downey, Igusa and Melnikov [127]). Show that for every computable limit ordinal α, there is a structure $A = A_\alpha$, such that

(i) For each computable structure $B \cong A$, there is a $\beta < \alpha$ such that $B \cong_{\Delta^0_\beta} A$.

(ii) For each $\delta < \alpha$ there is a computable $C \cong A$, such that $C \ncong_{\Delta^0_\delta} A$.

Exercise* 10.1.112 (Downey, Melnikov, and Ng [144]). Produce a computable linear order with the property described in the previous exercise.

Exercise 10.1.113 (Folklore). Show that for each computable $\alpha > 1$ there exists an optimally relatively Δ^0_α-categorical structure (Definition 10.1.108).

Exercise* 10.1.114 (Ash [16]). Show that for each computable $\alpha > 1$ of the form $\delta + 2k$ (where δ is either 0 or a limit ordinal, and $k \in \omega$) there is an optimally relatively Δ^0_α-categorical well-order (Definition 10.1.108).

Exercise 10.1.115 (Melnikov [368])). Show that for each computable $\alpha > 1$ of the form $1 + \delta + 2k$ (where δ is either 0 or a limit ordinal, and $k \in \omega$) there is an optimally relatively Δ^0_α-categorical ordered abelian group (Definition 10.1.108).

Exercise 10.1.116 (Barker [30]). Prove that for each computable $\alpha > 1$, there is an optimally relatively Δ^0_α-categorical abelian p-group (Definition 10.1.108).

Exercise* 10.1.117 (Ocasio-Gonzalez [420]). Show that for each computable $\alpha > 1$ of the form $1 + \delta + 2k$ (where δ is either 0 or a limit ordinal, and $k \in \omega$) there is an optimally relatively Δ^0_α-categorical real closed field (Definition 10.1.108).

Exercise* 10.1.118 (Melnikov [372]). Prove** that for any computable successor ordinal of the form $\alpha = \delta + 2k$ (δ limit and $k \in \mathbb{N}$) there exists an optimally relatively Δ^0_α-categorical (Definition 10.1.108) computable torsion-free abelian group. (This solved a problem posed by Goncharov.)

Exercise* 10.1.119 (Montalbán [400]). A countable structure A is *Δ^0_α-categorical on a cone* (where α is a countable ordinal) if there is a $Y \in 2^\omega$ such that A is $\Delta^0_\alpha(X)$-computably categorical for all $X \geqslant_T Y$. (Without loss of generality, we may assume α is computable relative to Y.) Extend Exercise 10.1.106 to establish that the following are equivalent:

(C1) A is Δ^0_α-categorical on a cone.

(C2) A has a $\Sigma_{\alpha+2}$ Scott sentence.

Exercise* 10.1.120 (Alvir, Greenberg, Harrison-Trainor, and Turetsky [8]). The *Scott complexity* of a countable structure A is the lowest syntactic $L_{\omega_1\omega}$-complexity so that the structure has a Scott sentence of that complexity. Prove the following:

1. The Harrison order $\mathcal{H}$ has Scott complexity $\Pi_{\omega_1^{CK}+2}$.
2. There is a computable structure of Scott complexity $\Pi_{\omega_1^{CK}}$. (Essentially Knight and Millar [306].)
3. There exists a computable structure of Scott complexity $\Pi_{\omega_1^{CK}+1}$. (Essentially Harrison-Trainor, Igusa, and Knight [240].)
4. There exists a computable structure of Scott complexity $\Sigma_{\omega_1^{CK}+1}$.
5. There exists a computable structure of Scott complexity d-$\Sigma_{\omega_1^{CK}}$ (meaning that the sentence is a conjunction of a $\Sigma_{\omega_1^{CK}}$- and a $\Pi_{\omega_1^{CK}}$-sentence).

(See also Exercise 10.3.19.)

10.2 Computably isometric Polish spaces

In this section, we shall extend some of the methods and results accumulated for discrete structures in the previous section to separable structures. The main result of the subsection is Theorem 10.2.9, which describes relatively computably categorical spaces (up to isometry) in terms of *approximate Scott families*. We will see that even when spaces are viewed up to isometry, the situation becomes more subtle than in the discrete case. For example, the notion of a finite parameter will have to be adjusted. *To emphasise that our spaces are viewed up to isometry, we will often refer to a computable Polish presentation of a space as "a computable structure on the space".* This terminology is also consistent with the literature too, at least when it comes to isometric presentations.

10.2.1 Isometric computable categoricity

Of course, the notion of computable categoricity depends on the choice of the notion of an isomorphism. For example, if we view Polish spaces up to isometry, the natural notion is:

Definition 10.2.1 (Melnikov [369], Iljazović [268]). A Polish metric space is isometrically computably categorical if it has a unique computable Polish presentation up to computable surjective isometry.

From now on, we shall focus on *isometric computable categoricity.*

So "computably categorical" means "isometrically computably categorical" throughout the rest of this subsection.

Examples of isometrically c.c. spaces

Corollary 7.1.26 says that a computably compact metric space with at most finitely many self-isometries is isometrically computably categorical. The next example of a computably categorical metric space should be compared with Mal'cev's Theorem 2.2.16.

Theorem 10.2.2 (Melnikov [369], based on Pour-El and Richards [435]). *For any separable Hilbert space, the underlying Polish metric space is isometrically computably categorical.*

Proof. We begin with a lemma.

Lemma 10.2.3 (Pour-El and Richards [435]). *Let $\mathbb{H}$ be a separable Hilbert space. Then any two computable Banach presentations of the space are computably linearly isometric.*

Sketch. First, observe that the inner product $\langle \cdot, \cdot \rangle$ can be effectively reconstructed from the norm in a Hilbert space using the well-known formula

$$\langle u, v \rangle = \frac{1}{4}\left(\|u+v\|^2 - \|u-v\|^2\right).$$

Assume we are given two computable Banach space presentations A and B of the space. In each of these spaces, we use the inner product to produce a computable complete orthonormal system of vectors using (essentially) the iterated Gram–Schmidt process; see Exercise 10.2.26. We then use these orthonormal bases in A and B to compute a surjective linear isometry between the spaces, as follows. We first match the bases and then extend the map uniquely to the whole space using the projections of points onto the basic vectors. The latter can also be calculated using the inner product. □

Recall that all operations on a Hilbert space (including the inner product) can be effectively reconstructed from the norm (Theorem 2.4.19). The issue is that $||x|| = d(0, x)$, so it may seem that we need 0 to be a computable point in the Polish space. But since every affine shift of the space is a surjective self-isometry, we can take any special point in the underlying Polish metric space and "declare" it to be zero. We then use Theorem 2.4.19 and Lemma 2.4.17 to define new (automorphic) operations determined by the choice of zero. These operations turn the Polish space into an isomorphic (computable Banach) copy of our Hilbert space. Once the operations are computably defined, we can apply Lemma 10.2.3 to the resulting Hilbert space. □

We will also see that the Urysohn space is isometrically computably categorical (Proposition 10.2.14).

Recall that Theorem 2.4.20 states that there is an isometric computable Polish presentation of $(C[0,1], d_{sup})$ that computes 0 but dos not compute +. Theorem 2.4.20 implies:

Theorem 10.2.4 (Melnikov [369]). *The Polish space $(C[0,1], d_{sup})$ is not isometrically computably categorical.*

Proof. We use Lemma 2.4.17 throughout. To apply Theorem 2.4.20 we need the following elementary:

Claim 10.2.5. *Let $\mathbb{B}$ be a Banach space upon the domain B and d be the metric associated with the norm (i.e., $d(x, y) = ||x - y||$). Suppose X is a computable presentation of $(B, d, +)$, and suppose Y is a computable presentation of $(B, d, 0)$. If X and Y are computably isometrically isomorphic (as metric spaces), then + is computable in Y.*

Proof. Let $U : Y \to X$ be a surjective computable isometry. Recall that U^{-1} is also computable. By the theorem of Mazur and Ulam that we stated and discussed in §2.4.3, there exists a linear map $L : X \to Y$ such that $U(x) = L(x) + U(0)$, for every $x \in B$. We have that

$$U^{-1}(U(\beta) + U(\gamma) - U(0)) = U^{-1}(L(\beta + \gamma) + U(0)) = \beta + \gamma,$$

and therefore + is computable in Y. □

Now recall that the computable presentation of $C[0,1]$ constructed in Theorem 2.4.20 computes 0 but does not compute +. It follows from the claim above that this computable presentation cannot possibly be computably isometric to the "standard" one given in Example 2.4.18. □

10.2.2 Relative isometric computable categoricity

Definition 10.2.6 (Greenberg, Knight, Melnikov, and Turetsky [219]). A computable space $(M, d, (\alpha_i)_{i\in\mathbb{N}})$ is *relatively computably categorical* (r.c.c.) if any **a**-computable Polish space $(\beta_i)_{i\in\mathbb{N}}$ (whose completion is) isometric to M is **a**-computably isometrically isomorphic to $(\alpha_i)_{i\in\mathbb{N}}$.

For example, the separable Hilbert spaces of infinite dimension is relatively computably categorical.

In Theorem 10.1.6 we saw that for algebraic structures, this notion was equivalent to the existence of a c.e. Scott family. However, in their full generality, the theorems required finitely many parameters in the respective functionals and Scott families. What would this mean for a Polish space? Recall that in a Polish space, a point is no longer a finite parameter; it is not a Type I obejct. Further, it is not altogether clear what a "Scott family" would mean for an uncountable Polish space. We discuss these issues in detail next.

Approximate Scott families

To be able to adjust the methods developed for discrete algebraic structures, we will need our spaces to resemble algebraic structures. We view separable metric spaces as structures in the first-order language

$$L = \{d_{<r}(\cdot,\cdot), d_{>r}(\cdot,\cdot) : r \in \mathbb{Q}\},$$

where $M \models d_{<r}(x, y)$ iff $d_M(x, y) < r$, and similarly $M \models d_{>r}(x, y)$ iff $d_M(x, y) > r$. We do not allow $=$ and $\neg$ when we form L-formulae; thus, we restrict ourselves to *positive* atomic L-formulae with no equality. If X is a countable metric space, we write $D^+(X)$ for the positive open diagram of X in the language L, which consists of conjunctions and disjunctions of positive atomic L-formulae. Clearly, $(\alpha_i)_{i\in\mathbb{N}}$ is a computable Polish presentation of (a computable structure on) M iff $D^+((\alpha_i)_{i\in\mathbb{N}})$ is a computably enumerable set, under the standard Gödel numbering (we identify α_i with its index i).

An approximate Scott family. As we saw above, in countable structure theory, a Scott family of a structure A is a collection of formulae $\mathbb{S}$ such that:

(1) any finite tuple $\bar{a}$ in A satisfies some $\theta \in \mathbb{S}$, and

(2) if $\theta \in \mathbb{S}$ holds on both $\bar{a}$ and $\bar{b}$ in A, then there exists an automorphism of A taking $\bar{a}$ to $\bar{b}$.

A Σ^0_1 Scott family is a c.e. family consisting of first-order existential formulae.

Since perfect Polish metric spaces are uncountable, we cannot hope to just take the same definition literally. Instead, we allow both (1) and (2) to be true "up to ϵ". This can be viewed as a sequence of families, one family for each positive rational ϵ. Equivalently, we may assume that the family is indexed by rational numbers, so that Θ_ϵ is true with precision ϵ. The formal definitions are as follows.

Define a $\mathbb{Q}$-*indexing* of a family $\mathbb{F}$ of L-formulae to be a map $\mu : \mathbb{F} \to \mathbb{Q}^+$. If $\mathbb{F}$ admits a $\mathbb{Q}$-indexing μ, and there is no other indexing of $\mathbb{F}$ that is of any importance to us, we write Θ_ϵ for $\Theta \in \mathbb{F}$ to express that $\mu(\Theta) = \epsilon$.

In the following, tuples of M-points are viewed as elements of the corresponding direct product of M with the sup-metric inherited from d_M. We let $\mathrm{nbh}_\epsilon(\bar{x})$ denote the ϵ-ball centred in $\bar{x}$ in this metric.

Definition 10.2.7 (Greenberg, Melnikov, Knight, and Turetsky [219]). Let M be a Polish space. We say that a non-empty collection $\mathbb{S}$ of $\mathbb{Q}$-indexed first-order L-formulae is an *approximate Scott family* if the following two conditions hold:

1. For all tuples $\bar{x}$ and $\bar{x}'$ in M, if $M \models \Theta_\epsilon(\bar{x})$ and $M \models \Theta_\epsilon(\bar{x}')$ for some $\Theta \in \mathbb{S}$ and $\epsilon \in \mathbb{Q}^+$, then there exists an isometric automorphism of M taking $\bar{x}$ into $\mathrm{nbh}_\epsilon(\bar{x}')$.
2. For every tuple $\bar{x}$ in M and $\epsilon_1 \in \mathbb{Q}^+$, there exists a tuple $\bar{x}' \in \mathrm{nbh}_{\epsilon_1}\bar{x}$ and a formula Θ_{ϵ_2} with $\epsilon_2 < \epsilon_1$ in $\mathbb{S}$ such that $M \models \Theta_{\epsilon_2}(\bar{x}')$.

An approximate Scott family $\mathbb{S}$ is *c.e.* if there exists an effective listing of the Gödel numbers of $\mathbb{S}$, and the $\mathbb{Q}$-indexing is given by a (partial) computable function that halts on each member of $\mathbb{S}$. Without loss of generality, we may assume that $\mathbb{S}$ is actually indexed by rationals of the form 2^{-i}.

Parameters

We will allow our formulae an uniform procedures to use finitely many parameters. In a presentation of the structure, a list of parameters typically corresponds to a finite sequence of elements of ω. As a finite object, this list is trivially computable relative to the presentation. The situation is different in Polish metric spaces.

The idea is to allow a whole neighbourhood of a point to be a parameter. That is, any point from the neighbourhood can be put as a parameter, and this variation will not affect the desired properties. In this case we say that the parameter c is *stable*. So, for example, any tuple from the neighbourhood can be used in the computation of a functional witnessing uniform isometric categoricity with stable parameters. In the case of a Scott family, this requires a bit more care. We clarify this below.

Definition 10.2.8. We say that parameters $\bar{c} = (c_1, \ldots, c_n)$ of an approximate Scott family $\mathbb{S}$ are *stable* if there exists an open neighborhood $\overline{B} = B_1 \times \ldots \times B_k$ of $\bar{c}$ such that for any $\bar{c}' \in \overline{B}$, replacing $\bar{c}$ with $\bar{c}'$ in $\mathbb{S}$ gives an approximate Scott family of the space with parameters $\bar{c}'$ and furthermore:

(1)* If $M \models \Theta_\epsilon(\bar{c}, \bar{x})$ and $M \models \Theta_\epsilon(\bar{c}', \bar{x}')$, then there exists an automorphism of M taking $\bar{c}\bar{x}$ into $\mathrm{nbh}_\epsilon(\bar{c}'\bar{x}')$.

10.2.3 A characterisation of relatively c.c. Polish spaces

Recall that we view our Polish spaces up to isometry. So "computably categorical" means "isometrically computably categorical".

Theorem 10.2.9 (Greenberg, Knight, Melnikov, and Turetsky [219]). *Let M be a computable Polish space. The following are equivalent:*

1. *M is relatively computably categorical.*
2. *M possesses a c.e. approximate Scott family with stable parameters.*

Of course, the finitely many balls associated with the stable parameters contain special points. We also note that the result implies Theorem 10.1.6 for discrete structures. (To see why, use the effective universality of undirected irreflexive graphs and discrete metric spaces established in Chapter 8.) However, while at least one implication in Theorem 10.1.6 was essentially trivial, neither implication of Theorem 10.2.9 is particularly easy to establish. (The "easy" implication in Theorem 10.1.6 is no longer elementary here because an approximate Scott family allows us to merely *approximate* an isometry, and some extra care must be taken to ensure that, for example, the isometry is surjective.)

Proof of Theorem 10.2.9. (1) $\rightarrow$ (2) Fix a computable dense X in M. Without loss of generality, assume X has no repetition (Exercise 2.4.32). We identify elements of X with their indices from ω. We call such an X *a structure on* M rather than a computably isometric presentation of M. Consider the collection $\mathbb{P}$ of all quadruples (D, p, ℓ), where

(p1) D is a finite partial positive diagram upon an initial segment $0, \ldots, i$ of ω and in the language L.

(p2) $p : \{0, \ldots, i\} \rightarrow X$ is a finite partial embedding, i.e., D is true on the respective p-images of $1, \ldots, i$.

(p3) $\ell < i$ is such that each of the first ℓ-many elements of X are within 2^{-i} of some element of $range(p)$.

(p4) For each $k, j \leqslant i$ with $k \neq j$, there are $r, q \in \mathbb{Q}^+$ with $|r - q| < 2^{-i}$ such that D contains $d_{<r}(k, j)$ and $d_{>q}(k, j)$.

We write $\sigma, \tau, \rho \ldots$ to denote elements of $\mathbb{P}$, we assume $\sigma = (D_\sigma, p_\sigma, \ell_\sigma)$ and we write i_σ for $\sup dom(p_\sigma)$. (Recall $dom(p_\sigma)$ is an initial segment of ω.)

If ℓ_1, ℓ_2 are two finite strings in X (that is, functions from finite initial segments of ω to X), perhaps of different lengths, then let m be the length of the shortest of the two. Define

$$d(\ell_1, \ell_2) = \sup_{j<m} d(\ell_1(j), \ell_2(j)).$$

We say that τ is a refinement of σ and write $\tau \Supset \sigma$ if:

(e1) $dom(p_\tau) \supset dom(p_\sigma)$ and $D_\tau \supset D_\sigma$;

(e2) $\ell_\tau > \ell_\sigma$

(e3) $d(p_\sigma, p_\tau) < 2^{-i_\tau}$.

Define $\Subset$ accordingly. The relation $\Subset$ is a strict partial order on $\mathbb{P}$. Furthermore, each infinite $\Subset$-ascending chain $(D_i, p_i, s_i)_{i\in\mathbb{N}}$ corresponds to an isomorphic image N of M under the isometric surjective map $p = \lim_i p_i$. The positive diagram of N can be effectively enumerated given only positive information about $\cup_i D_i$, and without loss of generality we may identify it with $\cup_i D_i$. Conversely, given any structure Y on M and a finite partial diagram of any finite tuple $\bar{y}$ of special elements from Y, we can extend $\bar{y}$ to a $\bar{z}$ whose partial diagram σ satisfies $(p1) - (p4)$. We furthermore can represent Y as a sequence through $\mathbb{P}$ extending this σ.

Suppose $\sigma \in \mathbb{P}$. For any pair i, j in the domain of p_σ, pick r least and q largest such that $d_{<r}(i,j)$ and $d_{>q}(i,j)$ are mentioned in D_σ. Define $\overline{d}_\sigma(i,j) = r$ and $\underline{d}_\sigma(i,j) = q$.

Say that $f : \sigma \to M$ is σ-admissible if for all $i, j \in dom(p_\sigma)$

$$\underline{d}_\sigma(i,j) < d_M(f(i), f(j)) < \overline{d}_\sigma(i,j).$$

We also say that all extensions of f to maps with larger domains are σ-admissible.

Let Φ be an enumeration operator and $D = \bigcup_{\sigma_i} D_{\sigma_i}$ for some $\Subset$-ascending chain $(\sigma_i)_i$. For $\tau \in \mathbb{P}$ and D, $\lim_i \Phi_i^D$ is τ-admissible iff already for some finite subset D_{σ_i} and some i we can see this is the case (unless Φ^D lists sequences that are not fast Cauchy), which is a c.e. event. In this case, we say simply that $\Phi^{D_{\sigma_i}}$ is τ-admissible. Similarly, if Φ^D defines a collection of fast Cauchy names, then for every $x \in X$ and any $m \in \mathbb{N}$, a condition of the form $d_M(\lim_i \Phi_i^D(m), x) < 2^{-s}$ will be witnessed by some D_{σ_i} unless $(\Phi_k^D(m))_k$ is not fast Cauchy. In this case we write $d_M(\lim_k \Phi_k^{D_\sigma}(m), x) < 2^{-s}$.

We say that σ *forces* Φ *to be a computable isometry onto* $\bar{X}$ and write $\sigma \Vdash \Phi$ if for all $\tau \Supset \sigma$ there exist $\rho \Supset \tau$ with the properties:

1. Φ^{D_ρ} is τ-admissible;

2. for each $k, j \leqslant i_\tau$, we have $d_M(\Phi_j^{D_\rho}(k), \Phi_{j+1}^{D_\rho}(k)) < 2^{-j-1}$;

3. for each $j \leqslant i_\tau$, there exists $v \in dom(p_\rho)$ such that $d_M(\lim_k \Phi_k^{D_\rho}(v), j) < 2^{-i_\tau}$.

In (1) and (3) above we require all mentioned computations to halt. Notice that each of (1)-(3) above are naturally c.e. events.

According to our definitions, if (1) or (2) or (3) fails for a σ, then $\lim_i \Phi_i$ cannot possibly be a surjective isometry of the completion of any $D \in [\mathbb{P}]$ extending D_τ, for some τ above σ. If this is the case for τ, we write $\tau \Vdash \neg\Phi$. On the other hand, if $\sigma \Vdash \Phi$, then (2) ensures that Φ defines a sequence of D-computable points for any [infinite] $\Subset$-extension of σ. Also, if $\sigma \Vdash \Phi$, then (1) implies that $\lim_k \Phi_k$ is an isometric embedding, and (3) witnesses its completion contains X. Thus, the completion is a surjective isometry onto M.

In the construction, we list all enumeration operators Φ and build a path through $\mathbb{P}$ by stages. We try to ensure $\sigma \Vdash \neg\Phi$ (if we can) for each Φ, one-by-one, just as we did in the proof of Theorem 10.1.6. In particular, some σ must force Φ. Just as in the proof of Theorem 10.1.6, we shall proceed above σ using the extensions that witness that $\sigma \Vdash \Phi$; at no stage we shall be stuck.

Indeed, we again have that for *any* extension of the finite partial diagram σ to a (positive) diagram of a structure on M in the construction, as along as we choose the extensions that witness $\sigma \Vdash \Phi$, the operator Φ is a surjective computable isometry between the completion of the structure and $\overline{X}$. The desired stable tuple of parameters is the neighbourhood of (the range of) p_σ in M; it is determined by the precision with which the distances in D_σ are calculated. Recall $i_\sigma = \sup dom p_\sigma$, and let $\bar{i} = \langle 0, \ldots, i_\sigma \rangle$; will use $\bar{b}$ as the "stable parameters" representing $\bar{z}$.

To define the approximate Scott family, run the construction above σ (on *all* extensions witnessing $\sigma \Vdash \Phi$) and see for which tuples the operator halts. Assume $\Phi_i^{D_\tau}$ halts on a tuple $\bar{a}$ and some τ, has output $\bar{d}$, and suppose $\psi(\bar{i}, \bar{a}, \bar{c})$ is the part of the open positive diagram used in this computation. Put $\exists \bar{y} \psi(\bar{b}, \bar{x}, \bar{y})$ in $\mathbb{S}$ with index 2^{-i+2}; let this formula be $\theta_{2^{-i+1}}(\bar{b}, \bar{x})$. The intuition is that it describes $\bar{d}$ "up to 2^{-i+2}". For D, the final isometric Φ-image of $\bar{a}$ is at distance 2^{-i} from $\bar{d}$.

We use the same trick with renaming tuples to produce an extension D^* above σ as we used in Theorem 10.1.6. To argue that the operator Φ will isomorphically map any tuple that satisfies

$\exists\bar{y}\theta(\bar{b},\bar{x},\bar{y})$ into the 2^{-i+1}-nbhd of $\bar{d}$, produce such a D^* above σ that uses the same indices for this tuple as D_τ. This makes any pair of tuples that satisfy $\theta_{2^{-i+1}}(\bar{b},\bar{x})$ 2^{-i+2}-isometric, as required in (1) Definition 10.2.7. Condition (2) of Definition 10.2.7 follows from Φ being onto on any such D, so eventually every tuple will be ϵ-close to some Φ-output.

(2) $\rightarrow$ (1). Let $\mathbb{S}$ be an effective approximate Scott family with stable parameters $\bar{d}$ in $\overline{B}$. We fix a computable structure Y on M, and fix any X dense in M. We may assume that $\bar{d}$ are special in Y. In what will follow, elements of Y (and of X) are identified with the respective natural numbers, and for a set $Y \subseteq \mathbb{N}$, $Y \upharpoonright i$ denotes $Y \cap \{0, 1, \ldots, i-1\}$.

At the beginning of each step $i > 0$, we will have a rational $\epsilon_i = 2^{-i}$ as a precision parameter, a finite tuple $\bar{a}_i$ from Y and $\bar{b}_i$ from X, and a partial map $\bar{b}_i \mapsto \bar{a}_i$. We may assume that $\bar{b}_i$ is an initial segment of X, and that $\bar{b}_{i+1}$ extends $\bar{b}_i$ by one extra point.

At the end of step i, we will define $\epsilon_{i+1} = \epsilon_i/2$, choose $\bar{b}_{i+1}$ extending $\bar{b}_i$, and define $\bar{a}_{i+1}$ so that:

(a.) the initial segment of $\bar{a}_{i+1}$ of length the same as $\bar{a}_i$ is within $\mathrm{nbh}_{\epsilon_i}(\bar{a}_i)$;

(b.) $\mathrm{nbh}_{\epsilon_{i+1}/4}(\bar{a}_{i+1})$ contains an isomorphic image of $\bar{b}_{i+1}$;

(c.) If i is odd, then $Y\upharpoonright_i$ is within $\mathrm{nbh}_{\epsilon_{i-1}}$ of some substring of $\bar{a}_{i+1}$.

In (b.), an isomorphism is any surjective isometry of $\bar{X}$ onto $\bar{Y}$.

It is clear that a successful maintenance of (*a.*) implies that, for each n, the n-prefixes of $\bar{a}_i$ form a (uniformly) rapidly converging sequence in $\bar{Y}^n$. If we are successful in satisfying (*b.*), then the limit of these n-tuples will have the same distance matrix as the first n points of X. Therefore, we can naturally define an isometric embedding U as the limit of partial maps $\bar{a}_i \mapsto \bar{b}_i$. Finally, if (*c.*) is satisfied at the end of each stage, then each special point of Y is in the closure of the image $U(X)$, and thus (the completion of) U is surjective. Indeed, (*c.*) implies that for each y and for infinitely many i, there exists an x such that $U(x)$ is within $\mathrm{nbh}_{\epsilon_{i-1}+2\epsilon_{i+1}}(y)$. Since the sequence $(\epsilon_{i-1} + 2\epsilon_{i+1})_i$ converges to 0, we conclude that $U(X)$ is dense in $\bar{Y}$.

Step 0: Declare $\bar{a}_0 = \bar{d}$. We may assume that $\bar{d}$ lie in the centres of the respective balls in $\bar{B}$. Let r be the smallest radius among the balls witnessing stability of the parameters. Set the precision parameter $\epsilon_0 < r/16$. (This choice of ϵ_0 will imply that the initial segment of the sequence $\bar{a}_i$ will never leave $\bar{B}$.) Choose a formula $\Theta_{\epsilon_0/16}$ that holds on $\bar{d}$, and fix a special X-tuple $\bar{c}$ in the $\epsilon_0/16$-nbhd of an isomorphic image of $\bar{d}$ in the completion of X such that $X \models \Theta_{\epsilon_0/16}(\bar{c})$.

Step i alternates between two basic modules:

Extension (i is even).

1. Choose $z \in X$ least that is not yet among $\bar{b}_i$.

2. Let $\bar{x} \in \mathrm{nbh}_{\epsilon_i/16}(\bar{b}_i z)$ be the first found tuple of special points such that $\bar{x}$ satisfies some $\Theta_\delta \in \mathbb{S}$ labeled by $\delta < \epsilon_i/16$.

3. Pick some $\bar{y}$ in Y whose prefix is within $\epsilon_i/2$ of $\bar{a}_i$ and so that $\bar{y}$ satisfies Θ_δ.

4. Set $\bar{a}_{i+1} = \bar{y}$ and $\bar{b}_{i+1} = \bar{b}_i z$, define $\epsilon_{i+1} = \epsilon_i/2$, and proceed to the next step.

Onto (i is odd). For every point y among the first i special points of Y, do the following:

1. Pick a first found tuple v in X, a tuple $\bar{w}u$ in Y, and a formula Θ_δ in $\mathbb{S}$ labeled by $\delta < \epsilon_i/16$ such that:
   - $\bar{w}u \in \mathrm{nbh}_{\epsilon_i/4}(\bar{a}_i y)$;
   - $X \models \Theta_\delta(\bar{b}_i v)$ and $Y \models \Theta_\delta(\bar{w}u)$.
2. Define $\bar{b}_{i+1} = \bar{b}_i v$ and $\bar{a}_{i+1} = \bar{w}u$, and set $\epsilon_{i+1} = \epsilon_i/2$.

Verification. Recall that each $\Theta \in \mathbb{S}$ is an existential projection of an open positive formula. The witnesses of the latter form a (non-empty) open set in the corresponding power of M. From this, we obtain the following:

Fact 10.2.10. *Suppose Θ is a formula in $\mathbb{S}$, and let $\bar{m}$ be a tuple of points in M such that $M \models \Theta(\bar{m})$. Then for each $\epsilon > 0$, there exists a tuple of X-special points $\bar{x} \in nbh_\epsilon \bar{m}$ such that $X \models \Theta(\bar{x})$.*

As we explained above, it is sufficient to verify that the conditions $(a.)$ – $(c.)$ for every i, and that the substeps use only positive information about $D(X)$.

Extension. It follows from the definition of an approximate Scott family and Fact 10.2.10 that there exist Θ_δ and $\bar{x}$ with the desired properties. Thus, we will eventually find an atomic fact in $D(X)$ implying Θ_δ on such a tuple. By the I.H., there exists an isomorphic image of $\bar{b}_i$ within $\mathrm{nbh}_{\epsilon_i/4}(\bar{a}_i)$. In particular, the choice of the precision parameter guarantees that the initial segment of this isomorphic image will be within the nbhd witnessing stability of $\bar{d}$. Thus, we can apply property (1)* of an approximate Scott family with stable parameters. Consequently, by the choice of δ, we can find a $\bar{y} \in \mathrm{nbh}_{\epsilon_i/2}(\bar{a}_i)$ such that $\Theta_\delta(\bar{y})$ holds (this will be the true image of $\bar{b}_i$ or any tuple of special points sufficiently close to it). At some stage we will see a proof of that fact from $D(Y)$. It follows from the triangle inequality that $\mathrm{nbh}_{\epsilon_i/8}(\bar{y})$ contains an isomorphic image of $\bar{b}_{i+1} = \bar{b}_i z$. Both $(a.)$ and $(b.)$ hold for $i + 1$.

Onto. The verification of this step is similar. Using the properties of $\mathbb{S}$, Fact 10.2.10, and the I.H., we argue that such $\bar{w}u$ and $\bar{x}v$ can always be found. Indeed, $\mathrm{nbh}_{\epsilon_i/4}(\bar{a}_i)$ contains some isomorphic image of $\bar{b}_i$. Each neighborhood of the latter contains a tuple satisfying some Θ_δ labeled by a $\delta < \epsilon_i/16$, and whence its isomorphic image satisfies the formula as well. Note the initial segment of the tuple stays in $\bar{B}$. It is sufficient for us to wait for $D(X)$ and $D(Y)$ to enumerate facts implying what we need. It is clear that $(a.)$ – $(c.)$ hold for i. □

The following corollary is analogous to Corollary 10.1.9.

Corollary 10.2.11. *A relatively (isometrically) c.c. Polish space M is indeed* uniformly *relatively c.c., in the sense that there is a fixed functional Ψ which, given a structure B on M and after fixing finitely many stable parameters, outputs a surjective isometry between B and M.*

Proof. This follows from the existence of the approximate Scott family established in Theorem 10.2.9 and the proof of (2) → (1) of Theorem 10.2.9. Indeed, the procedure described there works for any isometric structure on the space. □

10.2.4 Uniform computable categoricity for Polish spaces

Definition 10.2.12 (Greenberg, Knight, Melnikov, and Turetsky [219]). A computable space $(M, d, (\alpha_i)_{i\in\mathbb{N}})$ is *uniformly computably categorical* (u.c.c.) if there exists a uniform procedure such that, given any *computable* Polish space $(\beta_i)_{i\in\mathbb{N}}$ (whose completion is) isometric to M, it produces a surjective isometry between the completions of $(\beta_i)_{i\in\mathbb{N}}$ and $(\alpha_i)_{i\in\mathbb{N}}$.

Our next theorem is analogous to Theorem 10.1.37.

Theorem 10.2.13 (Greenberg, Knight, Melnikov, and Turetsky [219]). *Let M be a computable Polish space. The following are equivalent:*

1. *M is relatively computably categorical.*
2. *M is uniformly computably categorical with stable parameters.*

Proof. (1) $\to$ (2) : This follows from Corollary 10.2.11.

We prove (2) $\to$ (1). Suppose uniform categoricity with stable parameters is witnessed by Φ, a tuple of balls $\bar{B}$, and $\bar{b}$ in $\bar{B}$. Let Y be a computable structure on M. We define a c.e. approximate Scott family as follows. We may assume $\bar{b}$ is special. Given a tuple $\bar{y}$ and a positive rational $\epsilon = 2^{-i} < 1$, compute $\Phi_{i+2}^{D(Y),\bar{b}}(\bar{y})$. Suppose the use of the computation is $D_{\bar{y},\epsilon}$, and let $\bar{z}$ be all points of Y mentioned in $D_{\bar{y},\epsilon}$ unequal to those among $\bar{y}$. Take the conjunction $\phi(\bar{b}\bar{z}\bar{y})$ of all formulae in $D_{\bar{y},\epsilon}$, and set $\Theta_{\epsilon,\bar{y}} = \exists\bar{z}\phi(\bar{B}\bar{z}\bar{w})$, where $\bar{w}$ are free variables. Define $\mathbb{S}$ to be the collection of $\Theta_{\epsilon,\bar{y}}$ where $\bar{y}$ and ϵ range over finite tuples of Y and rationals of the form 2^{-i}, respectively. Clearly, the definition of $\mathbb{S}$ is effective.

The density of Y in M implies that $\mathbb{S}$ satisfies (2) of Definition 10.2.7. We assume (for simplicity) that there are no parameters and show that $\mathbb{S}$ satisfies (1) of Definition 10.2.7. The proof of (1)* is almost literally the same. Suppose $\bar{x}'$ and $\bar{x}''$ both satisfy $\Theta_{2^{-i}}$. Since $\Theta_{2^{-i}}$ isolates an open set in M, there will be tuples $\bar{y}'$ and $\bar{y}''$ of Y-special points in the 2^{-i-2}-nbhds of $\bar{x}'$ and $\bar{x}''$, respectively, that satisfy $\Theta_{2^{-i}}$. Define new computable structures Y' and Y'' replacing $\bar{y}$ by $\bar{y}'$ and $\bar{y}''$, respectively, and also by re-naming the existential witnesses $\bar{z}'$ and $\bar{z}''$ corresponding to $\Theta_{2^{-i}}(\bar{y}')$ and $\Theta_{2^{-i}}(\bar{y}'')$ (taken to be Y-special) by the tuple $\bar{z}$ of Y-special points witnessing the existential quantifier in $\Theta_{2^{-i}}(\bar{y})$. Note the points stay "where they are" in M, but they get new names.

By the assumption, Φ_{i+2} is a 2^{i+2}-isometry from both Y' and Y'' onto a dense subset of $\bar{Y}$, and furthermore the computations of Φ_{i+2} are identical on $\bar{y}$ viewed either as a tuple in Y' or in Y'' or Y. Suppose this computation outputs a tuple $\bar{v}$ in Y. It follows that here exist automorphic images $\bar{w}, \bar{w}'$ and $\bar{w}''$ of $\bar{y}, \bar{y}'$ and $\bar{y}''$ (all viewed as elements of Y) which are at most 2^{-i-2}-far from $\bar{v}$ and thus are at most 2^{-i-1}-far apart. Recall that $\bar{y}'$ and $\bar{y}''$ were 2^{-i-2}-close to $\bar{x}'$ and $\bar{x}''$. Thus, 2^{-i-2}-nbhds of $\bar{w}'$ and $\bar{w}''$ must contain some automorphic images $\bar{a}'$ and $\bar{a}''$ of $\bar{x}'$ and $\bar{x}''$. It follows that $\bar{a}'$ and $\bar{a}''$ are at most 2^{-i}-far apart, as desired. $\square$

An application to the Urysohn space*

We assume that the reader is familiar with the basic properties of the Urysohn space, see [364] for a detailed exposition. Otherwise, if the reader is not interested in the Urysohn space, they can skip this subsection. This subsection will not be used in the rest of the book.

In effective algebra, using c.e. Scott families instead of uniform procedures is usually not particularly advantageous. In metric space theory, approximate Scott families allow us to separate a technical approximate back-and-forth construction from the rest of the argument. As an illustration, we give a significantly simplified proof of the following known fact first established in [369].

Proposition 10.2.14. *The Urysohn space is uniformly computably categorical (without parameters).*

Proof. We follow [219] very closely. Using the rational Urysohn space $\mathbb{UQ}$, we can produce an effective approximate Scott family for the Urysohn space $\mathbb{U}$, as follows.

For every tuple of points $\bar{u}$ in $\mathbb{UQ}$ and a rational $\epsilon > 0$, we effectively produce the distance matrix $D(\bar{u})$ that has rational entries. Using this matrix, for each u_1 and u_2 in $\bar{u}$ and $r = d(u_1, u_2)$, we take

$$\psi_{u_1,u_2}(x, y) = d_{<r+\epsilon/4}(x, y) \,\&\, d_{>r-\epsilon/4}(x, y),$$

and syntactically define the $\epsilon/2$-nbhd of the tuple

$$\Theta_\epsilon^{\bar{u}}(\bar{x}) = \bigwedge_{(u_1,u_2)\subseteq\bar{u}} \psi_{u_1,u_2}(x, y).$$

Each such formula describes the distance matrix with recision $\epsilon/4$. We then define $\mathbb{S}$ to be the collection of all such $\Theta_\epsilon^{\bar{u}}(\bar{x})$, where ϵ ranges over the positive rationals and $\bar{u}$ over finite tuples in $\mathbb{UQ}$.

Since distance in $\mathbb{UQ}$ is a computable function of two arguments ranging over positive rationals, the family and the labeling are effective. Definition 10.2.7(2) for $\mathbb{S}$ follows at once from the density of $\mathbb{UQ}$ in $\mathbb{U}$, and Definition 10.2.7(1) is a re-formulation of the *approximate extension property* [364, Definition 3.1] that holds in $\mathbb{U}$. □

Along these lines, approximate Scott families can be also used to simplify several known brute-force proofs including computable categoricity of separable Hilbert spaces that we saw earlier in Theorem 10.2.2, and also of Cantor space (Exercise 10.2.29).

10.2.5 Type II vs. Type I for isometric computable categoricity

The counterexamples established earlier for categoricity of algebraic structures (Theorem 10.1.39 and, more generally, Theorem 10.1.40) can be transferred to discrete Polish spaces (up to isometry) via Theorem 8.2.9. However, none of these counterexamples are quite satisfactory, as they are discrete. Restricting our notions to discrete spaces seems completely unnatural. We would like to have counterexamples that would not be discrete.

We have seen that computable functions between spaces can be viewed as Type II objects. However, stable parameters are not Type 0 objects, which makes the situation quite different from algebraic structures, unless the space is discrete. One way to address this is as follows.

Definition 10.2.15. We say that a point of a space is *intrinsically computable* if it is computable in all *computable* structures of the space.

Then such a point can be replaced with its index in any computable structure on the space. Then a function representing a computable isometry can use these indices as parameters, thus turning it into a Type I object. We arrive at the following version of isometric computable categoricity.

Definition 10.2.16 (Greenberg, Knight, Melnikov, and Turetsky [219]). A Polish space is uniformly (isometrically) computably categorical after fixing a finite tuple of intrinsically computable parameters, we say that the space is *weakly uniformly computably categorical* (weakly u.c.c.).

We now clarify why the notion above can be viewed as a Type I version of uniform isometric computable categoricity. An isometry $f : M \to N$ is fully determined by the images of the points in the dense sequence $(x_i)_{i \in \mathbb{N}}$ representing M. Thus, a functional computing an isometry can be replaced with the function which, on input $\langle i, n \rangle$, outputs j so that the special point of N indexed by j is 2^{-n}-close to $f(x_i)$. Definition 10.2.16 is equivalent to saying that there is a total computable h such that $\overline{M_e} \cong \overline{M} \Rightarrow \varphi_{h(e)} : \overline{M_e} \cong \overline{M}$, where $\varphi_{h(e)}$ is such a function which additionally takes the indices of intrinsically computable points as (non-uniform) parameters. Thus, Definition 10.2.16 is analogous to the notion of a weakly u.c.c. structure (with parameters) given in Definition 10.1.35.

It is known that any c.c. closed subspace of $\mathbb{R}^n$ is in fact weakly u.c.c. (Exercise 10.2.27). Interestingly enough, weak uniform categoricity does not imply relative computable categoricity even for computable compact subspaces of $\mathbb{R}^2$.

Theorem 10.2.17 (Greenberg, Knight, Melnikov, and Turetsky [219]). *There exists a (computable) weakly u.c.c. compact subspace of* $\mathbb{R}^2$ *that is not relatively c.c..*

Proof. Before we proceed to the proof of the theorem, we need to establish a proposition which we believe has some independent value.

In Theorem 10.2.9 the parameters could be chosen *relatively intrinsically computable*, in the following sense. We say that a point if *relatively intrinsically computable* (r.i.c.) if it is computable with respect to any given (isometric) structure on the space. If a Polish space is relatively (isometrically) computably categorical with stable parameters, then we can always pick these parameters special. Since the images of these points will be computable relative to the isometry witnessing relative computable categoricity of the space, we can replace the stable parameters with r.i.c. parameters in Theorems 10.2.9 and 10.2.13. We first prove a proposition that gives a gives a slightly different perspective on why r.i.c. parameters can be replaced by stable parameters in an approximate Scott family.

Definition 10.2.18. Let c be a a point in a Polish metric space M. A *formal name* of the point c is a sequence $(\Theta_i)_{i \in \mathbb{N}}$ of existential formulae in one free variable such that $U_i = \{x \in M : M \models \Theta_i(x)\}$ is an open set of diameter at most 2^{-i-1} that converges to c; i.e. $\bigcap_i U_i = \{c\}$.

Recall that our language allows neither equality nor negation. It follows that $\{x \in M : M \models \Theta(x)\}$ will be an open set for any existential formula Θ. We can relax the definition above and allow finitely many stable parameters $\bar{B}$ in the formulae $(\Theta_i)_{i \in \mathbb{N}}$. In this case we require that for any tuple $\bar{b}$ in $\bar{B}$ the formulae $\Theta_i(\bar{b}, x)$ should determine sets $U_i^{\bar{b}}$ of diameter at most 2^{-i-1} that converge to c, but different $\bar{b}$ may correspond to different sequences of open sets.

Proposition 10.2.19. *Let M be a computable Polish metric space. Then an (isometric) automorphis invariant point $a \in M$ is r.i.c. iff it admits a c.e. formal name with stable parameters.*

Proof sketch. First observe that computability of a point a w.r.t. X is equivalent to having an enumeration operator Φ such that $(\Phi_i^{D(X)})_{i\in\mathbb{N}}$ converges to the point. The first part of the proof is similar to the proof of (1) → (2) of Theorem 10.2.9 The definition of $\mathbb{P}$ is the same. Then take the first σ that forces some Φ to list open neighbourhoods in all extensions of σ.

Now, on any extension Y of σ that witnesses that $\sigma \Vdash \Phi$, the sequence $(\Phi_i^{D(Y)})_{i\in\mathbb{N}}$ computes exactly the same point on M, since the point is stable under automorphisms. For each i, let D_i be the use of the computation $\Phi_i^{D(Y)} = m$. Quantify over the points mentioned in D_i that are not the stable parameters and not M. A rather straightforward argument (similar to the proof of (1) → (2) in Theorem 10.2.9) shows that the constructed family of existential formulae is a formal name for x. □

We now prove Theorem 10.2.17. To prove the theorem, it is sufficient to construct a computable Polish metric space $M = (M, d, X)$ and a point $a \in M$ that satisfy:

1. M is rigid.
2. a is intrinsically computable but not relatively intrinsically computable.
3. There exists a uniform effective procedure that, given $D(Y)$ of a dense Y and a fast Cauchy name of a w.r.t. Y, produces a surjective isometry $\overline{Y} \to \overline{X}$.

The rigidity requirement will ensure that a is automorphism invariant, allowing us to use Proposition 10.2.19. Conditions (2) and (1) together imply that the space is not relatively computably categorical. Finally, (3) implies that the space is weakly uniformly categorical with parameter a.

A crude description of the space. The space will be a closed subset of the unit square $\mathbb{B}_2 = [0,1] \times [0,1]$. We describe the space by enumerating a set X of (rational) points in $\mathbb{B}_2$, and then we set $M = \overline{X}$. The set $X = \bigcup_s X_s$ will be enumerated by stages in the construction that will be described later. We explain several important features that X will have.

We will initially put $(0,1)$ into X_0. We will then set $a = (0,0)$; a will not be in X, but it will be an accumulation point of X and thus an element of M. We also fix a sequence of rationals $(\delta_n)_{n\in\mathbb{N}}$ that converges to 0 "very fast":

$$\sum_{i>n} \delta_i < \frac{1}{100}\delta_n$$

for every n. We also assume $\delta_0 < 1/10$. At the beginning of every stage s, we will choose a number n_s larger than any number previously mentioned in the construction and put $(0, \delta_{n_s})$ into X_s, thus making $(0,0)$ an accumulation point. In the construction, we will add more points to X. This will always be done by taking a finite initial segment $X_{s'}$ of the currently defined X_s and a (small enough) rational r and enumerating

$$X_{s'} + (r,0) = \{(x,y) + (r,0) : (x,y) \in X_{s'}\}$$

into X_{s+1}. We will call this operation the *r-shift* of $X_{s'}$. We may omit r and $X_{s'}$ and say simply *shift* if it is clear from the context which r and s' we use. As we will see, our r's will be chosen small enough to ensure $X \subseteq \mathbb{B}_2$. Indeed, if we number the r's in the order in which they are chosen, we will have

$$\sum_{i>n} r_i < \frac{1}{100} r_n$$

for every n, just as with the δ_n.

Basic properties of M. Already the crude description above allows us to make several conclusions about $M = \overline{X}$. All we need to know is that in the construction there will be infinitely many stages at which new points are introduced to X by shifting .

Claim 10.2.20. *M becomes uniformly categorical after fixing $a = (0, 0)$.*

Proof of Claim. First, note that the pair $\{(0,0),(0,1)\}$ is an automorphism base of M. Indeed, every point in M is completely described by its distances to $a = (0,0)$ and $(0,1)$. An easy exercise in Euclidean geometry shows that a better approximation of the $\{(0,0),(0,1)\}$-coordinates of a point x gives a better approximation to the point, with all possible effective uniformity.

Furthermore, since $\delta_0 < 1/10$, we see that $(0,1)$ is the unique point of M that satisfies $d(a,x) > 1/2$, and so a computes $(0,1)$ w.r.t. any structure on $\mathcal{M}$. Therefore, the space M becomes uniformly computably categorical after adjoining a to its signature. Indeed, an algorithm for an isometry between $(\mathcal{M}, a, X)$ and any $(\mathcal{M}, a, Y)$ can be sketched as follows:

1. Search for the element $b \in Y$ with $d(a,b) > 1/2$.
2. On input x, compute $d(x,a)$ and $d(x,(0,1))$ to a sufficiently high degree of precision.
3. Search for an element $y \in Y$ such that $d(y,a)$ and $d(y,b)$ are sufficiently close to $d(x,a)$ and $d(x,(0,1))$.
4. Output y.

Determining the required levels of precision is an exercise in Euclidean geometry and the triangle inequality. □

Claim 10.2.21. *M is rigid.*

Proof of Claim. Note that there exists an $m < 1$ such that

$$(x,1) \in M \Rightarrow x \leqslant m.$$

Indeed, we have $m = \sum_s r_s$, where the r_s range over all q such that q-shifts were ever performed in the construction. Since the r_i are small (in the same sense as δ_n), we will have $m < 1$. Furthermore, if we define m_y to be least such that

$$(x,y) \in M \Rightarrow x \leqslant m_y,$$

we see that $(m_{\delta_{n_s}})_{s \in \mathbb{N}}$ is non-increasing (recall that $(0, \delta_{n_s})$ is enumerated at stage s, and so $(0, \delta_{n_s})$ will not partake in shifts that occur before stage s). Finally,

$$(x,0) \in M \Rightarrow x = 0.$$

We conclude that the point a is the unique point satisfying

$$(\exists z, w)\ \left(d(a,z) = \sqrt{1+m^2}\ \wedge\ d(z,w) = m\ \wedge\ d(a,w) = 1\right),$$

and is thus automorphism invariant. As we have already noted above, $(0,1)$ is the unique point at distance 1 of $(0,0)$. Since $\{(0,0),(0,1)\}$ is an automorphism base of M, we conclude that M is rigid. □

The two claims above together with the analysis preceding the claims imply that it remains to make a intrinsically computable but not relatively intrinsically computable. Since M will be rigid, Proposition 10.2.19 implies that making a not r.i.c. is equivalent to diagonalising against all potential formal names for a. Fix an effective list $(Y_e)_{e\in\mathbb{N}}$ of all computably enumerable structures in the language $\mathcal{L}$. Fix also an effective listing of all c.e. sequences of existential $\mathcal{L}$-formulae $(\Theta_\epsilon(x,\bar{c}))_\epsilon$ effectively labeled by rationals, where the finitely many parameters $\bar{c}$ are special points from the computable structure X. Note that Proposition 10.2.19 ensures that parameters are stable, and since X is dense in M it will be sufficient to diagonalise against all families with parameters from X. Thus, we need to satisfy:

$$\mathcal{D}_{\Theta,\bar{c}} : (\Theta_\epsilon(x,\bar{c}))_\epsilon \text{ is not a formal name for } a = (0,0),$$

where Θ ranges over all c.e. families of existential $\mathcal{L}$-formulae effectively labeled by $\mathbb{Q}$ and $\bar{c}$ over X (equivalently, over ω). Also, for every $e \in \mathbb{N}$ we need to meet

$$\mathcal{I}_e : \text{If } \overline{Y}_e \cong M \text{ then } a = (0,0) \text{ is computable w.r.t. } Y_e.$$

Note that seeing that Y_e fails to determine a structure on a metric space is a c.e. event: Y_e fails to determine a structure on some metric space if it fails the triangle inequality – that is, if there are $i,j,k \in Y_e$ and $q_1,q_2 \in \mathbb{Q}^+$ with $Y_e \models d_{<q_1}(i,j) \wedge d_{<q_2}(j,k) \wedge d_{>(q_1+q_2)}(i,k)$.

The basic strategy for $\mathcal{D}_{\Theta,\bar{c}}$. At the first stage this strategy is visited, it chooses an n larger than any number previously mentioned in the construction and defines $\delta = \delta_n$. Let $\Theta_{\delta/4}(x,\bar{c}) = \exists\bar{z}\psi(x,\bar{c},\bar{z})$. The strategy searches for a special point b and a tuple of special points $\bar{y}$ such that

$$X_s \models \psi(b,\bar{c},\bar{y}) \text{ and } d(a,b) < \delta/4.$$

Since every positive atomic $\mathcal{L}$-formula isolates an open set in a metric space, we know that if $\exists\bar{z}\psi(a,\bar{c},\bar{z})$ is satisfied, there are such b and $\bar{y}$. If we ever see this happen at stage s, we perform the *δ-shift* of X_s by enumerating

$$X_s + (\delta,0) = \{(x,y) + (\delta,0) : (x,y) \in X_s\}$$

into X. After this is done, we will have $b' = b + (\delta,0)$ also satisfies $\Theta(x,\bar{c})$, since the witnesses $\bar{y}$ will also get shifted. But $d(a,b') \geqslant \delta$ contradicts $d(a,b') < \delta/4$ which must be the case if $(\Theta_c)_c$ is to be a formal name for a.

The strategy will perform only one shift; once it has enumerated points into X, it never performs another shift. The strategy has two outcomes, *win* and *wait*; it takes outcome *wait* while searching for the b and $\bar{y}$, and then takes outcome *win* after performing the δ-shift.

The full strategy for $\mathcal{D}_{\Theta,\bar{c}}$. The full strategy is only a small modification of the basic strategy. Suppose the strategy first sees the desired b and $\bar{y}$ at some stage s. It does not immediately perform the δ-shift. Instead, it takes outcome *wait* for one more stage. Then, at the next stage $t > s$ at which the strategy is visited on the priority tree, the strategy performs the δ-shift of X_s. That is, it enumerates all of $X_s + (\delta,0)$ into X_{t+1}. From this stage on, the strategy takes outcome *win*.

Note that, assuming the strategy is visited infinitely many times during the construction, this will suffice to show that the requirement is met.

The basic strategy for $\mathcal{I}_e$. This strategy will define a sequence $(w_k)_{k\in\mathbb{N}}$, which is intended to be a fast Cauchy name for a w.r.t. Y_e. It will also define an auxiliary sequence $(v_k)_{k\in\mathbb{N}}$.

The basic strategy in isolation is quite simple. Recall that $(\delta_n)_{n\in\mathbb{N}}$ denotes the sequence converging to 0 "very fast" (see the beginning of the proof). The idea is as follows: for any s, we know that there is a point $(0, \delta_{n_s})$ which is precisely distance δ_{n_s} from a (recall n_s from the crude description of the space). We will search for two points in Y_e which are approximately distance δ_{n_s} apart, and when we find such points, we will believe that they are close to a. In the absence of any r-shifts, we will be correct: one of the points must be δ_{n_s}, and the other must be either a or $(0, \delta_{n_t})$ for $t > s$.

Let k be least such that w_k is not yet defined. Let $n = n_k$ and $\epsilon = \delta_n$. If $k > 0$, let $m = n_{k-1}$ and $\tau = \delta_m$. The strategy searches for a pair of special points w, v in Y_e such that

$$Y_e \models .9 \cdot \epsilon < d(w, v) < 1.1 \cdot \epsilon.$$

If τ is defined, we also require that at least one of w and v is within $.1\tau$ of either w_{k-1} or v_{k-1}. If such a pair is ever found in Y_e, the strategy sets $w_k = w$ and $v_k = v_{k-1}$ and then repeats the process.

The outcomes are ∞ and $wait_k$ for each $k \in \mathbb{N}$. It takes outcome $wait_k$ while k is least with w_k undefined; when w_k is defined, the strategy takes outcome ∞ for a single stage and then begins taking outcome $wait_{k+1}$.

The full strategy for $\mathcal{I}_e$. We must primarily modify the choice of ϵ and τ. If s is the first stage at which σ is visited, σ takes no action at this stage. To simplify the later description, we declare that σ has outcome ∞ at stage s, even though the construction will not allow $\sigma^\frown$ to be visited at this stage.

If σ is visited at stage s, and this is not the first stage at which σ is visited, let $t_0 < s$ be the last stage at which σ took outcome ∞. Let $n = n_{t_0+1}$ and $\epsilon = \delta_n$ (recall n_{t_0+1} from the crude description of the space). If t_0 was also not the first stage at which σ was visited, let $t_1 < t_0$ be the last stage before t_0 at which σ took outcome ∞, let $m = n_{t_1+1}$ and $\tau = \delta_m$. The strategy now proceeds as the basic strategy.

Construction. The priority ordering and the tree of strategies are typical for infinite injury, so we skip the usual definitions.

At stage 0 we set $M_0 = \{(0, 1)\}$ and do nothing else. At every stage after 0, we begin by choosing a number n_s larger than any number previously mentioned in the construction and enumerating $(0, \delta_{n_s})$ into X_s. We then visit the root of the priority tree. We then proceed inductively, visiting strategies according to the outcome of the previous strategy, until some never before visited strategy is reached. Once that strategy has acted, we end the stage. This completes the description of the construction.

Finalising the proof. Much of the verification has already been done above. The infinite injury technique used in the construction is standard and contains no surprises, so we leave the usual inductive argument to the reader. It is clear, for instance, that all $\mathcal{D}$-requirements are met by strategies along the true path. It remains to argue that all $\mathcal{I}$-requirements are met along the true path.

Claim 10.2.22. *Suppose σ is an $\mathcal{I}_e$ strategy, Y_e is a structure on a metric space σ takes outcome ∞ infinitely many times (and thus $(w_k)_{k\in\mathbb{N}}$ is fully defined). For any $k_0 < k_1 < k_2$, let t be the stage at which w_{k_0} is defined, let $m = n_{t+1}$, and let $\delta = \delta_m$. Then $Y_e \models d_{<2\delta}(w_{k_1}, w_{k_2}) \wedge d_{<2\delta}(w_{k_1}, v_{k_2})$.*

Proof. Induction on the stage at which w_{k_2} is defined, using the fast convergence of $(\delta_n)_{n\in\mathbb{N}}$. Note that the τ of that stage is no larger than δ. □

It follows that the sequence $(w_k)_{k\in\mathbb{N}}$ defined by σ is a (possibly partial) fast Cauchy sequence.

Claim 10.2.23. *Suppose σ is an $\mathcal{I}_e$-strategy along the true path, and Y_e is a structure on $\mathcal{M}$. Then $d(w_k, a) < 2^{-k}$.*

Indeed, if ϵ was the value such that σ was searching for w and v satisfying $Y_e \models .9\cdot\epsilon < d(w,v) < 1.1\cdot\epsilon$ in order to define w_k, then one of w_k or v_k is within $.1\cdot\epsilon$ of a.

Proof. Let s_0 be the stage at which σ defined w_k. Let $t < s_0$ be the last stage at which σ took outcome ∞, and let $m = n_{t+1}$. Then $(0, \delta_m)$ was enumerated into X by our construction at the beginning of stage $t+1$, and $\epsilon = \delta_m$. Since w_{k-1} was defined at stage t, or $k = 0$, our construction will choose that $n_{t+1} > k$, and so $\delta_m < 2^{-k-1}$.

By construction, no $\mathcal{D}_{\Theta,\bar{c}}$-strategy will choose δ_m, as m was already chosen by the construction at the beginning of a stage. Thus, since the $(\delta_n)_{n\in\mathbb{N}}$ tend to 0 quickly, one of w_k or v_k must correspond to a point of the form $(x, 0)$ or (x, δ_ℓ), where $x = \sum_{n\in F}\delta_n$ for some possibly empty $F \subset \omega$ and $\ell < m$, while the other of w_k or v_k must correspond to a point of the form (x', δ_m), where $x' = \sum_{n\in F'}\delta_n$ for some possibly empty $F' \subset \omega$.

Since $Y_e \models d_{<1.1\cdot\delta}(w_k, v_k)$ and $(\delta_n)_{n\in\mathbb{N}}$ tends to 0 quickly, it must be that for every $n \leqslant m$, $n \in F$ if and only if $n \in F'$. Note that $\sum_{n>m}\delta_n$ is small compared to δ_m, so to complete the proof of the claim, it will suffice to show that there is no $n \in F$ with $n \leqslant m$. We do that now.

By construction, every $n \in F$ was chosen by some $\mathcal{D}_{\Theta,\bar{c}}$-strategy that performed its shift, so in particular $m \notin F$. Note also that any shift performed at or before stage t will not enumerate a translate of $(0, \delta_m)$, and so cannot contribute to F'. Since the strategy performing this shift has necessarily chosen an $n < m$, it also cannot contribute to F. Indeed, any $\mathcal{D}_{\Theta,\bar{c}}$ which sees its desired b and $\bar{y}$ at or before stage t will not enumerate a translate of $(0, \delta_m)$, and thus will not contribute to F or F'.

By the assumption that σ is along the true path, it follows that no $\mathcal{D}_{\Theta,\bar{c}}$-strategy which is an ancestor of σ contributes to F. Further, any $\mathcal{D}_{\Theta,\bar{c}}$-strategy which contributes to F and is incomparable with $\sigma^\frown$ must have been first visited after stage t. Thus the n such a strategy contributes was chosen larger than m.

Thus, it suffices to consider whether $n < m$ can be contributed by some $\mathcal{D}_{\Theta,\bar{c}}$-strategy $\rho \supseteq \sigma^\frown\infty$ which is first visited at or before stage t and which performs its shift at some stage $s_1 > t$. Therefore $s_1 \geqslant s_0$. Let $t_1 < s_1$ be the last stage before s_1 at which ρ was visited, so $t_1 \geqslant t$. Then when ρ acts at stage s_1, it enumerates $X_{t_1} + (\delta_n, 0)$ into X. Let $m_1 = n_{t_1+1}$.

Note that if $s_1 = s_0$, then $t_1 = t$ and $\delta_{m_1} = \delta_m$. Since X_{t_1} contains no pair of points which are distance δ_{m_1} apart, if $n \in F$, it cannot be that $s_1 = s_0$. So $s_1 > s_0$.

Take k_1 such that σ defined w_{k_1} at stage s_1. Since $s_1 > s$, $k_1 > k$. Since σ has outcome ∞ at stage s_1, $.9\delta_{m_1} < d(w_{k_1}, v_{k_1}) < 1.1\delta_{m_1}$. Since X_{t_1} contains no pair of points which are distance δ_{m_1} apart, if $n \in F$, then no pair of points within $\delta_n/2$ of w_k can satisfy this inequality. Since $\delta_n/2 > \delta_m$, this contradicts Claim 10.2.22. □

It follows that if the sequence $(w_k)_{k\in\mathbb{N}}$ is total, it converges to a.

Claim 10.2.24. *If σ is an $\mathcal{I}_e$-strategy along the true path, and Y_e is a structure on $\mathcal{M}$, then σ infinitely often takes outcome ∞, and so $(w_k)_{k \in \mathbb{N}}$ is fully defined.*

Proof. Suppose not. Let s_0 be the last stage at which σ takes outcome ∞. Let $n = n_{s_0+1}$. Eventually σ will find points w and v in Y_e which are arbitrarily close to a and the $(0, \delta_n)$, respectively. If s_0 is also the first stage at which σ is visited, this existence of such w and v contradicts our choice of s_0.

If s_0 is not the first stage at which σ is visited, let $t < s_0$ be the last stage before s_0 at which σ takes outcome ∞. Let $m = n_{t+1}$ and let k be such that σ defined w_k at stage s_0. Since one of w_k or v_k is strictly within $.1\delta_m$ of a by Claim 10.2.23, we can find w, v as above with w within $.1\delta_m$ of one of w_k, v_k, contradicting our choice of s_0. □

This completes the proof of Theorem 10.2.17. □

Corollary 10.2.25. *There exists an isometry-rigid computable $M \subseteq \mathbb{R}^2$ and a point $a \in M$ which is intrinsically computable but not relatively intrinsically computable.*

Exercises

Exercise° 10.2.26 (Pour-El and Richards [435]). Prove that every computable Hilbert space admits a complete computable orthonormal sequence.

Exercise 10.2.27 (Melnikov [369]). Prove that every isometrically c.c. closed subspace of $\mathbb{R}^n$ is weakly u.c.c..

Exercise° 10.2.28. Give a detailed proof of Proposition 10.2.19.

Exercise° 10.2.29 (Melnikov [369]). Show that Cantor space is isometrically computably categorical.

10.3 Computable dimension

The main definition of this section is as follows.

Definition 10.3.1. The *computable dimension* (the autodimension) of a computable structure is the number of its computable presentations, up to computable isomorphism.

Thus, having a computable dimension of 1 is synonymous with being computably categorical. Similarly, we can define the (isometric) computable dimension of a Polish space to be the number of its computable Polish presentations, up to computable isometry. Clearly, the computable dimension of a structure or a space is at most ω, in which case we say that it is infinite.

All algebraic structures from standard classes where computable categoricity has been described have a computable dimension of either 1 or ω. These classes include Boolean algebras, linear orders, abelian p-groups, and torsion-free abelian groups. A similar pattern is seen for natural Polish spaces; a non-trivial example, Theorem F, will be given in this section.

Goncharov [206] proved a powerful general result which often allows the complete separation of algebra from the combinatorics of computability theory. This theorem, which we refer to as Goncharov's Δ^0_2-Theorem, is as follows.

Theorem 10.3.2 (Goncharov [206])**.** *Suppose that A and B are computable structures with $A \cong_{\Delta^0_2} B$, and $A \not\cong_{\Delta^0_1} B$. Then the computable dimension of A is infinite.*

Consequences of Goncharov's Δ^0_2-Theorem include results about the computable dimension of structures in many further algebraic classes that are not covered in this book. These classes include differentially closed fields, difference closed fields, valued fields, and real closed fields; see [244, 236].

We will give a complete and detailed proofs of Theorem 10.3.2 and its generalisation Theorem 10.3.20 to Polish spaces. As we will see, the more general version remains applicable to discrete algebraic structures as well, as it implies Theorem 10.3.2. Using this generalised version, we will derive a consequence about the number of computable presentations of the space $(C[0,1], d_{\text{sup}})$ that was studied in Part I of the book; this is Theorem F.

However, before we prove this theorem, we must establish that it is not trivial. Indeed, perhaps any computable structure that is not computably categorical has computable dimension ω?

10.3.1 An algebraic structure of computable dimension 2

In this subsection, we give a detailed outline of the following well-known result of Goncharov.

Theorem 10.3.3 (Goncharov [205, 204])**.** *There is a computable structure A of computable dimension 2.*

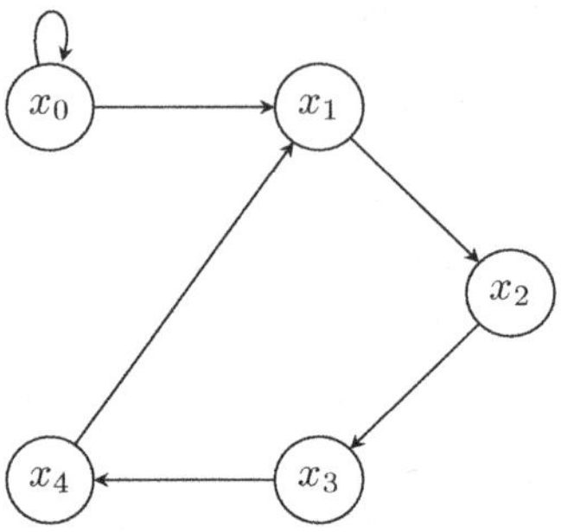

Figure 10.1: The component [3].

In the proof below, we will sketch (more or less) Goncharov's original technique, but using *loops* rather than Goncharov's families of sets. Since we will not actually need the result in the sequel, we provide only an extended sketch that emphasises the ideas related to the main strategy but omits most of the unpleasant combinatorics. We will keep our notation more or less consistent with that in [253], where all these details can be found[2].

Extended sketch. Let $(M_e)_{e\in\mathbb{N}}$ be the uniformly effective enumeration of all partially computable structures in the language of graphs.

We will build A_0, A_1 with $A_0 \cong A_1$, to meet the requirements below:

$$R_j : \varphi_e \text{ total} \Rightarrow A_0 \not\cong_{\varphi_e} A_1.$$

$$N_e : (M_e \cong A_0) \Rightarrow (M_e \cong_{\Delta_1^0} A_0) \vee (M_e \cong_{\Delta_1^0} A_1).$$

The loop-components. At every step, each A_i will consist of *directed* (tagged) "loops". Each such loop, denoted by $[n]$, consists of $n+3$ vertices $x_0, \ldots, x_{n+2}$, where:

1. $x_i x_{i+1}$ is a directed edge for each $0 \leqslant i < n+2$,
2. $x_{n+2}x_1$ is an edge, and
3. $x_0 x_0$ is an edge too (see Fig. 10.1 for the case when $n = 3$).

The element x_0 is called the *top* of C_n. The *n-cycle* is the portion $x_1, \ldots, x_{n+2}$. The element x_{n+2} is called the *bottom* of the loop and is where we plan to diagonalise. We can join two (or more) loops at the same top x_0 and would denote, for example, $[3,4]$ as a loop $x_0, \ldots, x_5$ joined at x_0 with $x_0, x_1', \ldots, x_6'$. If C_k is a k-loop and C_d is a d-loop then $C_k \cdot C_d$ is the loop $[k,d]$ thus formed. In the construction below, the numbers on the labels of such operations will be chosen to be fresh.

More generally, suppose U is a (finite, non-empty) set of natural numbers. Then $[U]$ will denote the graph that is obtained after joining $[v]$ for each $v \in U$. (Note that, unless $U \cap V = \varnothing$, $[V \cup U] \neq [V] \sqcup [U]$.) In the construction, each component will be of the form $[V]$ for some (finite, non-empty) set $V \subseteq \omega$. We define $[V] \cdot [U] = [V \cup U]$.

[2] As of 2024, the Ph.D. thesis is available at `https://www.math.uchicago.edu/~drh/Papers/dissertation.html`.

The transformations L and R. There will be infinitely many numbers set aside for new loops. If at stage s we have components $X_0, \ldots, X_n$, and $X_i = [S_i]$ for finite $S_i \subset \mathbb{N}$, the result of the *left transformation* $L(X_0, X_1, \ldots, X_n)$ is the set of components

$$X_0 \cdot X_1, X_1 \cdot X_2, \ldots, X_n \cdot X_0,$$

and the *right transformation* $R(X_0, X_1, \ldots, X_n)$ results in

$$X_0 \cdot X_n, X_1 \cdot X_0, \ldots, X_n \cdot X_{n-1}.$$

Note that if G is the structure with components $X_0, \ldots, X_n$, then if the structure G_1 is obtained by $L(X_0, X_1, \ldots, X_n)$, and if G_2 is obtained by $R(X_0, X_1, \ldots, X_n)$, then $G_1 \cong G_2$.

Meeting R_e. To satisfy R_e in isolation, we will have components Y_e^0, X_e^0, Z_e^0 for A_0 and similarly Y_e^1, X_e^1, Z_e^1 for A_1. (Note that no component will be chosen for more than one R_e.)

Let x_e^i be the bottom of X_e^i. We then wait for $\varphi_e(x_e^0) \downarrow= x_e^1$. Should this not occur, then φ_e cannot be an isomorphism. If this stage occurs, then we diagonalise by applying $L(X_e^0, Y_e^0, Z_e^0)$ and $R(X_e^1, Y_e^1, Z_e^1)$ to get $A_0[s+1]$ and $A_1[s+1]$, respectively. Note that R_e is satisfied since the components containing the x_e^i are no longer isomorphic. However, $L(X_e^0, Y_e^0, Z_e^0) \cong R(X_e^1, Y_e^1, Z_e^1)$.

The naive strategy for N_j (in presence of many R_e). To meet N_j, the structure M_j will be associated with a *special component* which will be denoted S_j. To ensure that $M_j \cong A_0$ implies $M_j \cong_{\Delta_1^0} A_0$ or $M_j \cong_{\Delta_1^0} A_1$, we place a copy S_j^i of S_j (which is initially a loop of a unique length assigned to N_j) into each A_i, $i = 0, 1$. We wait for M_j to respond by giving its version of S_j. We shall also incorporate S_j into the strategy for R_e, using $L(X_e^0, Y_e^0, Z_e^0, S_j^1)$ and $R(X_e^1, Y_e^1, Z_e^1, S_j^0)$ to diagonalise against φ_e.

The idea is as follows. After we finish our action for R_e, M_j will have to choose whether its version of S_j is extended to $S_j \cdot X_e$ or to $Z_e \cdot S_j$. If M_j never extends its version of S_j, then $M_j \not\cong A_0$. If M_j responds with $S_j \cdot X_e$, then we say that M_j *goes left*; in this case, we have made progress in demonstrating $M_j \cong_{\Delta_1^0} A_0$. Otherwise, if M_j responds by giving $Z_e \cdot S_j$, we say that M_j *goes right*, and in this case, we can make progress in demonstrating $M_j \cong_{\Delta_1^0} A_1$. In either case, the new larger component containing S_j will be declared the new special component for M_j. (For example, if M_j goes left and responds by extending $S_j = S_{j,s}$ to $S_{j,s} \cdot X_e$, then we shall set $S_{j,s+1} = S_{j,s} \cdot X_e$.)

The naive approach would be to repeat this strategy for other $X_{e'}^i, Y_{e'}^i, Z_{e'}^i$, and then for the next $X_{e''}^i, Y_{e''}^i, Z_{e''}^i$, and so on. For each such triple, M_j will have to choose between going either left or right. If (for example) M_j eventually always goes left, then $A_0 \cong_{\Delta_1^0} M_j$. Also, if M_j keeps switching between going left and right, then A_0 will have no infinite component, while the special component of M_j will grow infinite. In this case, we can certainly conclude that $M_j \not\cong A_0$.

However, there is a problem with this naive approach. For example, suppose M_j always goes left, so in A_0 we have:

$$S_j^0 \mapsto S_j^0 \cdot X_e^0 \mapsto S_j^0 \cdot X_e^0 \cdot X_{e'}^0 \mapsto S_j^0 \cdot X_e^0 \cdot X_{e'}^0 \cdot X_{e''}^0 \mapsto \ldots.$$

However, in A_1, we will have

$$S_j^1 \mapsto Z_e^1 \cdot S_j^1 \text{ and } X_e^1 \mapsto X_e^1 \cdot S_j^1 \cong S_j^1 \cdot X_e^1,$$

which results in the special component being shifted from the original copy of S_j^1 to the (now extended) component of X_e^1. The action for the requirement $R_{e'}$ will further shift the special component in A_1, and the same happens for $R_{e''}$, and so on. As a result, A_1 will have no infinite components, and this is obviously not desirable since we must keep $A_0 \cong A_1$.

The less naive strategy for N_j. To keep $A_0 \cong A_1$, we must find a way of reusing the old (abandoned) special components. We describe a brute-force attempt of reusing components and we explain what goes wrong with it. Then we shall finally get to the actual strategy for N_e that will finally work.

We modify the basic diagonalisation strategy for R_e as follows. Instead of using Y_e, X_e, Z_e and S_j, we shall incorporate two additional components, B and C. We will perform the transformation $L(X_e^0, Y_e^0, Z_e^0, B^0, S_j^0, C^0)$ in A_0 and $R(X_e^1, Y_e^1, Z_e^1, B^1, S_j^1, C^1)$ in A_1, where B^0, B^1, C^0, and C^1 will be chosen carefully. The idea is that one of these auxiliary components will be set equal to the old version of the special component (now abandoned). Of course, only one of A_0 or A_1 will currently require this "fix", depending on whether M_j goes left or right (since the definition of the special component of M_j depends on this).

It is perhaps best to illustrate this approach with an example. So, for example, suppose M_j goes left, i.e., it is currently following A_0. As discussed earlier, A_1 needs to be "fixed". To simplify our notation, assume $j = 0$. We suppress s in $S_{0,s}$, and we also replace e in X_e, Y_e, and Z_e with $0, 1, 2, \ldots$. (The exact value of e does not matter for N_0, as long as different indices correspond to different components.)

Suppose in A_1 we initially have:

$$Y_0^1 \quad X_0^1 \quad Z_0^1 \quad B_0^1 \quad \mathbf{S_0^1} \quad C_0^1$$

and after one application of the R-transformation, we obtain:

$$Y_0^1 \cdot C_0^1 \quad X_0^1 \cdot Y_0^1 \quad Z_0^1 \cdot X_0^1 \quad B_0^1 \cdot Z_0^1 \quad \underline{S_0^1 \cdot B_0^1} \quad \mathbf{C_0^1 \cdot S_0^1},$$

where the special component (in A_1) has shifted to $C_0^1 \cdot S_0^1$, as indicated by boldface, because $M = M_0$ responded by growing its copy of S_0 to $C_0 \cdot S_0$. The underlined component $S_0^1 \cdot B_0^1$ has now been temporarily abandoned but is set to be reused for isomorphism recovery.

To reuse $S_0^1 \cdot B_0^1$, we set:

$$Y_1^1 \quad X_1^1 \quad Z_1^1 \quad B_1^1 \quad \mathbf{S_1^1} = \mathbf{C_0^1} \cdot \mathbf{S_0^1} \quad C_1^1 = S_0^1 \cdot B_0^1$$

and after another iteration of the R-transformation, we arrive at:

$$Y_1^1 \cdot S_0^1 \cdot B_0^1 \quad X_1^1 \cdot Y_1^1 \quad Z_1^1 \cdot X_1^1 \quad B_1^1 \cdot Z_1^1 \quad \mathbf{C_0^1} \cdot \mathbf{S_0^1} \cdot \mathbf{B_1^1} \quad C_0^1 \cdot S_0^1 \cdot \cancel{S_0^1} \cdot B_0^1,$$

where $C_0^1 \cdot S_0^1 \cdot S_0^1 \cdot B_0^1 = [C_0^1 \cup S_0^1 \cup S_0^1 \cup B_0^1] = C_0^1 \cdot S_0^1 \cdot B_0^1$.

In A_0, we also apply the L-transformation to turn:

$$Y_1^0 \quad X_1^0 \quad Z_1^0 \quad B_1^0 \quad \mathbf{S_1^0} = \mathbf{C_0^0} \cdot \mathbf{S_0^0} \quad C_1^0 = S_0^0 \cdot B_0^0$$

into:

$$Y_1^0 \cdot X_1^0 \quad X_1^0 \cdot Z_1^0 \quad Z_1^0 \cdot B_1^0 \quad \mathbf{B_1^0} \cdot \mathbf{C_0^0} \cdot \mathbf{S_0^0} \quad C_0^0 \cdot S_0^0 \cdot \cancel{S_0^0} \cdot B_0^0 \quad S_0^0 \cdot B_0^0 \cdot Y_1^0$$

which is isomorphic to what we had in A_1. Of course, we also wait for M to respond, as before. In this example, we assume that M now chooses to follow A_1. In this case, A_0 will have to be recovered using a symmetric strategy involving $B_2 = S_0 \cdot C_0 \cdot B_0$. The old special component of A_1 that needed to be recovered (now turned into $Y_1^1 \cdot S_0^1 \cdot B_0^1$) will never have to be recovered again. Indeed, we shall now aim to recover $C_0^1 \cdot S_0^1 \cdot B_1^1$, and we will do so using new auxiliary components. If M switches between going left and going right infinitely many times, then no component of A_0 and no component of A_1 is infinite, while the special component of M has to be infinite. As before, if M eventually always follows A_0 (or A_1), then $M \cong_{\Delta_1^0} A_0$ (or $M \cong_{\Delta_1^0} A_1$, respectively).

However, there is still a *problem* with this strategy. Assume that M_0 always chooses to follow A_0. While it is true that, in A_1, the respective special component will grow infinite due to recycling, it will unfortunately grow one more infinite component, thus still resulting in $A_1 \not\cong A_0$. In the notation above, if $s = 0$ and $t = 1$ (etc.), both A_1 and A_0 will have the (intended) special component:

$$A_0 \quad : \quad S_0^0 \mapsto S_0^0 \cdot C_0^0 \mapsto S_0^0 \cdot C_0^0 \cdot B_0^0 \mapsto S_0^0 \cdot C_0^0 \cdot B_0^0 \cdot B_1^0 \mapsto \ldots;$$

$$A_1 \quad : \quad C_0^1 \mapsto S_0^1 \cdot C_0^1 \mapsto B_1^1 \cdot S_0^1 \cdot C_0^1 \mapsto B_0^1 \cdot B_1^1 \cdot S_0^1 \cdot C_0^1 \mapsto \ldots.$$

In this example, both will end up in an infinite component of the form $C_0 \cdot S_0 \cdot \bigodot_{i \in \mathbb{N}} B_i$, as in both A_0 and A_1 each B_i eventually finds its way into this component. However, in A_1, we will additionally have:

$$A_1 \quad : \quad S_0^1 \mapsto B_0^1 \cdot S_0^1 \mapsto C_0^1 \cdot B_0^1 \cdot S_0^1 \mapsto B_2^1 \cdot C_0^1 \cdot B_0^1 \cdot S_0^1 \mapsto \ldots,$$

which will also eventually grow into a component isomorphic to $C_0 \cdot S_0 \cdot \bigodot_{i \in \mathbb{N}} B_i$, by induction. However, it is problematic, as we must keep $A_0 \cong A_1$.

The actual strategy for N_j. To fix this, we shall not recover A_1 (or A_0) immediately. Instead, we shall allow one extra iteration of the L- or R-transformation in the respective structure, using fresh X_1, Y_1, Z_1, B_1, C_1, and only then will we re-use the abandoned component as C_2 to recover A_1 (we use B_2 to recover the A_0-side). This elementary variation completely fixes all issues and does not introduce any new ones. (As before, $j = 0$.) For example, in A_1, we will have:

Y_0^1	X_0^1	Z_0^1	B_0^1	S_0^1	C_0^1
$C_0^1 \cdot Y_0^1$	$Y_0^1 \cdot X_0^1$	$X_0^1 \cdot Z_0^1$	$Z_0^1 \cdot B_0^1$	$\underline{B_0^1 \cdot S_0^1}$	$\mathbf{S_0^1 \cdot C_0^1}$
Y_1^1	X_1^1	Z_1^1	B_1^1	$\mathbf{S_0^1 \cdot C_0^1}$	C_1^1
$C_1^1 \cdot Y_1^1$	$Y_1^1 \cdot X_1^1$	$X_1^1 \cdot Z_1^1$	$Z_1^1 \cdot B_1^1$	$\underline{\underline{B_1^1 \cdot S_0^1 \cdot C_0^1}}$	$\mathbf{S_0^1 \cdot C_0^1 \cdot C_1^1}$
Y_2^1	X_2^1	Z_2^1	B_2^1	$\mathbf{S_0^1 \cdot C_0^1 \cdot C_0^1}$	$\underline{B_0^1 \cdot S_0^1}$
$B_0^1 \cdot S_0^1 \cdot Y_2^1$	$Y_2^1 \cdot X_2^1$	$X_2^1 \cdot Z_2^1$	$Z_2^1 \cdot B_2^1$	$\underline{C_0^1 \cdot C_0^1 \cdot B_0^1 \cdot S_0^1}$	$\mathbf{B_0^1 \cdot S_0^1 \cdot C_0^1 \cdot C_1^1}$

The components in boldface are the special components at the respective stage. The underlined component is the one that we "recycle". The doubly-underlined component will never be extended. (This analysis assumes that M_0 is following A_0.) The reader should verify that this choice of components will guarantee that, in A_0, these steps indeed ensure isomorphism recovery. The case when we recover the A_0-side is symmetric, but it uses the B-components instead of C-components.

The outcomes of N_j. The outcomes of N_j are:

∞ : This Π_2^0 outcome measures whether M_j always eventually responds by revealing its versions of components.

wait : This is played in all other cases.

The Π_2^0 outcome can be further split into sub-outcomes: one saying that M_j goes left, another that M_j goes right, and a third one saying that M_j has switched again. Only under the first two outcomes do we promise to build an isomorphism from M_j to the respective A_i, and if the last one is the true outcome, then $M_j \not\cong A_0$. However, placing these sub-outcomes on the tree appears to be unnecessary, since this analysis can be done after the construction is finished.

Abandoning diagonalisation locations. Of course, M_j does not have to respond quickly, and this means that the weaker priority strategies may have to act before M_j responds. In this case, we shall not involve the special component of M_j in the diagonalisation process. If M_j eventually responds, we can *homogenise* all these previously used components of the lower-priority R_e. (This can be done, for example, by extending all of them to $X_e \cdot Y_e \cdot Z_e$ for the respective e.)

Then we shall put the resulting homogenised components in a queue, and we will require M_j to reveal a few more of such components the next time M_j is declared active. Such initialisation/homogenisation strategies are very typical in infinite injury arguments in computable structure theory, and there is certainly more than one way of implementing them.

The clones of the weaker priority strategies that believe M_j will never respond will, of course, be initialised, as this homogenisation contradicts their basic diagonalisation strategy. But this is fine, since these requirements will be met along the true path. The homogenisation will help in proving that the weaker priority strategies do not upset N_j, which is responsible for constructing a computable isomorphism to either A_0 or A_1; this is because N_j does not care which of the homogenised pieces M_j will reveal first, as they are all automorphic.

The general case. In the presence of many N-strategies, the combinatorics becomes heavier and much less transparent. We outline the key ideas.

In the construction, we can require that all N-nodes at the same level of the tree share the same strategy and the same special component. (In other words, placing the N-strategies on the tree appears to be unnecessary.) However, this is perhaps the only pleasant aspect of the general construction that we can think of.

This is because the L- and R-transformations will have to involve *all* the special components that are currently active in the construction. That is, all the components S_j corresponding to the N_j-strategies whose current outcome is ∞ should be involved.

The exact implementation of this is tedious. For example, we define

$$L(Y_0, \dots, Y_n; X; Z_0, \dots, Z_n; B_0, S_0, C_0; \dots; B_n, S_n, C_n)$$

to be

$$\begin{array}{lll} (Y_0, \dots, Y_n) \cdot X, & & X \cdot (Z_0, \dots, Z_n), \\ Z_0 \cdot B_0, & \dots, & Z_n \cdot B_n, \\ B_0 \cdot S_0, & \dots, & B_n \cdot S_n, \\ S_0 \cdot C_0, & \dots, & S_n \cdot C_n, \\ C_0 \cdot Y_0, & \dots, & C_n \cdot Y_n, \end{array}$$

where $(Y_0, \dots, Y_n) \cdot X$ is obtained from $Y_0, \dots, Y_n$ by creating a copy of X and putting an edge from the top node of X to the top node of each Y_i. The definition of $X \cdot (Z_0, \dots, Z_n)$ is similar, but this time $Z_0, \dots, Z_n$ are the new copies, while the edge still goes from the top of X to the top nodes of Z_i. (So, in particular, $(Y_0, \dots, Y_n) \cdot X \cong X \cdot (Y_0, \dots, Y_n)$.) The generalised R-transformation is defined dually, just as the basic R-transformation was defined to resemble the basic L-transformation; we omit this.

In the general construction, we shall use these generalised L- and R- transformations to incorporate all the special S-components of all currently active N-strategies. This is also true for the isomorphism recovery too, but we shall leave it at that. Modulo this combinatorics (which we omit), the rest of the construction is a relatively standard infinite injury argument.

If the reader is keen to see this combinatorics spelled out, we refer them to the cited earlier Ph.D. thesis [253] which has remained a standard reference for this technique for a couple of generations of computable structure theorists. There, a slight extension of the original Goncharov proof is presented in its gory detail. □

Consequences and generalisations of Theorem 10.3.3

It is well-known that for every $0 < n < \omega$ there is a computable structure of computable dimension n. The general case easily follows from the case of $n = 2$; see Exercise 10.3.8 and Exercise 10.3.9.

We remark that Theorem 10.3.3 is a consequence of the fact that there is a family of sets that has exactly two *Friedberg enumerations*, up to a certain natural reducibility (that is usually attributed to Kolmogorov) resembling m-reducibility [204]. Indeed, in our proof sketch of Theorem 10.3.3 we really only worked with families of sets that were represented by loops. In [205], Goncharov refers to [204] to derive Theorem 10.3.3 by turning a family of sets into a structure so that the equivalent uniform enumerations are in a 1-1 correspondence with computable isomorphisms between the respective structures. We see that there is an unexpected technical connection between the material of Chapter 9 and the theory of computable dimension.

As a consequence of results established in §8.2.2 and §8.2.3, we obtain:

Corollary 10.3.4. *The following classes contain examples of structures of computable dimension 2:*

1. *two-step nilpotent groups ([216], also follows from [257]);*
2. *fields ([397]);*
3. *Polish spaces under isometry.*

Goncharov's method has been widely used in the literature. For many applications of the method, see the exercises.

Open questions

In the proof of Theorem 10.3.3 that we outlined above, the isomorphism can be read off the true path, and hence $A_0 \cong_{\Delta^0_3} A_1$. For a long time, it was open how complex the isomorphism between A_0 and A_1 for the two representatives of a computable structure of computable dimension 2 could be. Using his new technique, Turetsky [486] has shown that it can be as complex as it could possibly be; not even hyperarithmetic (see Exercise 10.3.19). On the other hand, the use of infinite components is present in all known constructions of structures of finite computable dimension. The following question is open and seems difficult.

Question 10.3.5 (Hirschfeldt; see, e.g., [256]). *Is there a locally finite (directed) graph*[3] *of finite computable dimension* $n > 1$*?*

It follows from the Low Basis Theorem 4.2.47 that any pair of computable copies of a locally finite graph are isomorphic via an isomorphism that is low relative to (and above) $\varnothing'$, i.e., $X'' =_T \varnothing''$. Indeed, we do not know whether there exists any structure in which isomorphisms can be represented via a $\Pi^0_1(\varnothing')$-class that would have computable dimension 2.

The following question has been open for several decades:

Question 10.3.6 (Goncharov). *Is there an abelian group of finite computable dimension* $n > 1$*?*

Despite claims that can be found in the literature, the answer to this question is still unknown. A detailed discussion of this problem, and of the known cases, can be found in [377].

In Theorem 4.2.84 we established that effective Stone duality preserves computable categoricity, where computably compact Stone spaces are viewed up to (computable) homeomorphism. The same can be said about computable dimension as well. However, Boolean algebras cannot have non-trivial finite computable dimension; see Exercise 10.4.8.

Question 10.3.7. *Is there a compact Polish space having exactly two computably compact copies, up to computable* homeomorphism*?*

In the rest of the section we always view our spaces up to isometry.

Exercises

Exercise° 10.3.8 (Goncharov [205]). Prove that there exist structures of computable dimension 3. (Hint: Use the "symmetric" disjoint union of two copies of a structure having computable dimension 2. For that, consider the equivalence structure with two equivalence classes, one for each of the two domains.)

Exercise° 10.3.9 (Goncharov [205]). Prove that there exist structures of computable dimension $n > 2$. (Hint: Use the "symmetric" disjoint union of n-1-many copies of a structure having computable dimension 2.)

Exercise° 10.3.10 (McCoy [355], after Knight (unpublished)). McCoy attributes the following definition to Goncharov and Ventsov. A computable structure A has *relative computable dimension* $m < \omega$ if:

(i) there exist m copies (with universe ω) $B_1, \dots, B_m \cong A$ with no two of them pairwise $\Delta^0_1(D(B_1), \dots, D(B_m))$ isomorphic; and
(ii) for any $m+1$ copies (with universe ω) $B_1, \dots, B_{m+1} \cong A$, there are at least two that are $\Delta^0_1(D(B_1), \dots, D(B_{m+1}))$-isomorphic.

Prove that if a computable structure A has finite relative computable dimension, then it is relatively computably categorical. Thus, in particular, it has (relative) computable dimension 1. (Hint: Use the method of proof of Theorem 10.1.12 to produce a c.e. Scott family for A.)

Exercise° 10.3.11 (Nurtazin [418]). Define the *decidable dimension* of a structure to be the number of decidable presentations of the structure, up to decidable isomorphism. Prove that every (decidable) structure has decidable dimension either 1 or ω.

[3] A graph is locally finite if each vertex is adjacent to at most finitely many other vertices in the graph.

Exercise 10.3.12 (Goncharov [205]). Suppose that A is a 2-decidable computable structure which is not computably categorical. Show that the computable dimension of A is infinite.

Exercise* 10.3.13 (Hirschfeldt [254]). Assume R is a relation on a computable structure M. The *degree spectrum* of R is the set of all degrees of the isomorphic images of R in all computable structures classically isomorphic to M. Prove that for every c.e. degree $a > 0$, there exists an intrinsically c.e. relation R (see Exercise 10.1.34) on the domain of a computable structure so that the degree spectrum of R is $\{0, a\}$.

Exercise* 10.3.14 (Csima and Stephenson [105]). Prove that there is a computable rigid structure that has a degree of categoricity, but not a strong degree of categoricity (see Def. 10.1.80). (Hint: Produce a rigid structure of computable dimension 3 such that if $\mathbf{d_0}$, $\mathbf{d_1}$, and $\mathbf{d_2}$ are the degrees of isomorphisms between distinct representatives of the three computable equivalence classes, then each $\mathbf{d_i} < \mathbf{d_0} \oplus \mathbf{d_1} \oplus \mathbf{d_2}$.)

Exercise 10.3.15** (Hirschfeldt, Khoussainov, Shore [258]). Show that there exists a computably categorical structure whose expansion by a constant has infinite computable dimension.

Exercise 10.3.16** (Cholak, Goncharov, Khoussainov, and Shore [93]). Show that for any $n \in \mathbb{N}$ there exists a computably categorical structure A so that naming any element of the structure by a constant c turns A into a structure (A, c) having computable dimension n.

Exercise* 10.3.17 (Semukhin [464]). Show that there exists a structure of computable dimension two which is the prime model of its first-order theory.

Exercise 10.3.18.** A (countably infinite) structure is *punctual* or *fully primitive recursive* if its domain is ω and the operations and relations are uniformly primitive recursive. A function $f : \omega \to \omega$ is *punctual* if both f and f^{-1} are primitive recursive. The punctual dimension of a structure is the number of its punctual presentations, up to punctual isomorphism.

1. Prove that there exists a structure of punctual dimension 2 (Melnikov and Ng [379]).
2. Show that the hint given in Exercise 10.3.9 fails for punctual structures (Hammatt [228]).
3. Prove that for any finite $n > 2$ there exists a structure of punctual dimension n (Hammatt [228]).

Exercise 10.3.19** (Turetsky [486]).

1. Show that there is a structure A of computable dimension 2, so that the computable copies A_0 and A_1 witnessing the computable dimension of A are not hyperarithmetically isomorphic.
2. Show that the Scott complexity (Exercise 10.1.120) of such a structure can be $\Pi_{\omega_1^{CK}+2}$.

10.3.2 Goncharov's Δ_2^0-Theorem for discrete structures

In this subsection we prove Goncharov's Δ_2^0-Theorem 10.3.2: *If a structure has two computable presentations that are Δ_2^0 isomorphic but not computably isomorphic then its computable dimension is ω.* The particular version of the proof that we include below is due to Denis Hirschfeldt. We thank Denis for allowing us to include this proof in the book.

Proof of Theorem 10.3.2. It is enough to show that if $M_0, \ldots, M_n$, $n > 0$, are computable structures such that, for each $i \neq j \leqslant n$, M_i and M_j are Δ_2^0 isomorphic but not computably isomorphic, then there is a computable structure M_{n+1} that is Δ_2^0 isomorphic but not computably isomorphic to each M_i, $i \leqslant n$.

So let $M_0, \ldots, M_n$ be as above. We need to build a computable structure M_{n+1} to satisfy the requirements

$$\mathcal{R}_{e,i} : \Phi_e \text{ is not an isomorphism from } M_{n+1} \text{ to } M_i$$

for each $e \in \omega$, $i \leqslant n$, while at the same building a Δ_2^0 isomorphism g_i from M_{n+1} to M_i for each $i \leqslant n$.

The basic idea will be to ensure that if Φ_e is an isomorphism from M_{n+1} to M_i then g_j is computable for some $j \neq i$ (in the construction below, j will be $i + 1 \bmod n + 1$), from which it follows that M_i and M_j are computably isomorphic.

We need some notation to talk about "finite portions" of each M_i, so given a finite set $S \subset \omega$, let $M_i \upharpoonright S$ be the finite structure obtained from M_i by restricting the universe to S and restricting the language L of M_i to a finite language L_S by taking the first $|S|$ symbols of L, substituting all k-ary function symbols by $(k+1)$-ary relation symbols in the obvious way, and dropping any constants whose interpretation in M_i is not in S. This guarantees that $M_i \upharpoonright S$ is a structure in L_S, that checking whether $M_0 \upharpoonright S \cong N$ for a finite structure N in L_S is an effective procedure, and that if $g = \lim_s g_s$ is a Δ_2^0 function such that each g_s is an isomorphism from $M_i \upharpoonright S_s$ to some structure $\mathcal{N}_s$, $\bigcup_{s \in \omega} S_s = \omega$, and the N_s form a chain with limit N, then $g : M_i \cong N$ (where we think of N as a structure in L by converting the relation symbols in $\bigcup_{s \in \omega} L_{S_s}$ that correspond to function symbols in L back to function symbols).

For each $i \neq j \leqslant n$, let $f_{i,j}$ be a Δ_2^0 isomorphism from M_i to M_j. We assume without loss of generality that, for each $s \in \omega$, the stage s approximation $f_{i,j}[s]$ to $f_{i,j}$ is an isomorphism from $M_i \upharpoonright \operatorname{dom}(f_{i,j}[s])$ to $M_j \upharpoonright \operatorname{rng}(f_{i,j}[s])$ and $\operatorname{dom}(f_{i,j}[s+1]) \supsetneq \bigcup_{k \neq i \leqslant n} \operatorname{rng}(f_{k,i}[s])$.

A stage $s+1$ is *e-expansionary* if

$$\operatorname{dom}(\Phi_e[s]) \supsetneq \{0, \ldots, \sup(\operatorname{dom}(\Phi_e[t]))\}$$

and

$$\operatorname{rng}(\Phi_e[s]) \supsetneq \{0, \ldots, \sup(\operatorname{rng}(\Phi_e[t]))\},$$

where $t+1$ was the last e-expansionary stage before $s+1$ (or $t = 0$ if there have been none).

stage 0. Let $M_{n+1}[0] = \varnothing$. For each $i \leqslant n$, let $g_i[0] = \varnothing$. For each $e \in \omega$, $i \leqslant n$, let $r_{e,i}[0] = \langle e, i \rangle + 1$. (Here $\langle e, i \rangle$ will be $(n+1)e + i$.)

stage $s+1$. For each $e \in \omega$, $i \leqslant n$, if $\Phi_e[s]$ is not an isomorphism from $M_{n+1}[s] \upharpoonright \operatorname{dom}(\Phi_e[s])$ to $M_i \upharpoonright \operatorname{rng}(\Phi_e[s])$ then declare $\mathcal{R}_{e,i}$ to be satisfied.

Say that $\mathcal{R}_{e,i}$ requires attention if it is not satisfied and $s+1$ is an e-expansionary stage. Proceeding by recursion, define $r_{e,i}[s+1] = \max(\{\max(\operatorname{dom}(\Phi_e[s]))\} \cup \{r_{e',i'}[s+1] \mid \langle e', i' \rangle < \langle e, i \rangle\}) + 1$ if $\mathcal{R}_{e,i}$ requires attention and $r_{e,i}[s+1] = \max(\{r_{e',i'}[s+1] \mid \langle e', i' \rangle < \langle e, i \rangle\}) + 1$ otherwise.

Let m be the largest element of $M_{n+1}[s]$, and let e_0 and i_0 be such that r_{e_0,i_0} is the largest $r_{e,i}$ that is less than m. For convenience of notation, let $a_{-1} = 0$, $a_{\langle e,i \rangle} = r_{e,i}$ for $\langle e, i \rangle \leqslant \langle e_0, i_0 \rangle$, and $a_{\langle e_0, i_0 \rangle + 1} = m + 1$. We define $g_i[s+1]$, $i \leqslant n$, in two steps.

1. For each $k \leqslant \langle e_0, i_0 \rangle + 1$ in order, proceed as follows. Let e and i be such that $k = \langle e, i \rangle$ and let $j = i + 1 \bmod n + 1$. For each $a_{k-1} \leqslant x < a_k$, let $g_j(x)[s+1] = g_j(x)[s]$ and, for each $l \leqslant n$, $l \neq j$, let $g_l(x)[s+1] = (f_{j,l} \circ g_j)(x)[s]$.

   If now $g_l(x)[s+1] \neq g_l(x)[s]$ for some $l \leqslant n$ and $a_{k-1} \leqslant x < a_k$ then, for all $a_k \leqslant x \leqslant m$, let $g_j(x)[s+1] = g_j(x)[s]$ and, for each $l \leqslant n$, $l \neq j$, let $g_l(x)[s+1] = (f_{j,l} \circ g_j)(x)[s]$. In this case, proceed to step 2.

2. For $j = s \bmod n + 1$, let k be the first element of M_j that is not yet in the range of $g_j[s+1]$, define $g_j(m+1)[s+1] = k$ and, for each $l \leqslant n$, $l \neq j$, define $g_l(m+1)[s+1] = f_{j,l}(k)[s]$.

Notice that we have ensured that the images of $g_i[s+1]$ and $g_j[s+1]$ are isomorphic for each $i, j \leqslant n$, so we can define $M_{n+1}[s+1]$ to be the structure induced by $g_0^{-1}[s+1]$, where we think of this as a map from $M_0[s]$, and we then have that $M_{n+1}[s+1]$ extends $M_{n+1}[s]$ and, for each $i \leqslant n$, $g_i[s+1]$ is an isomorphism from $M_{n+1}[s+1]$ to $M_i[s]$.

This concludes the construction. Let M_{n+1} be the limit of the chain formed by the $M_{n+1}[s]$. Clearly, M_{n+1} is a computable structure and is isomorphic to each M_i, since each $g_i = \lim_s g_i[s]$ exists and is total and surjective.

To finish the proof, we need to show that each $\mathcal{R}_{e,i}$ eventually stops requiring attention. Assume for a contradiction that this is not the case and let $\langle e, i \rangle$ be the least pair such that $\mathcal{R}_{e,i}$ requires attention infinitely often. Note that this means that Φ_e is an isomorphism from M_{n+1} to M_i, since otherwise either there would be only finitely many e-expansionary stages or $\mathcal{R}_{e,i}$ would be satisfied at some point. Let $j = i + 1 \bmod n + 1$ and let s be a stage after which no $\mathcal{R}_{e',i'}$, $\langle e', i' \rangle < \langle e, i \rangle$, requires attention.

We claim that we have a computable isomorphism from M_i to M_j. To see this, first note that, for each $t > s$ and $\langle e', i' \rangle < \langle e, i \rangle$, $r_{e',i'}[t] = r_{e',i'}[s]$. Let $r = \max\{r_{e',i'}[s] \mid \langle e', i' \rangle < \langle e, i \rangle\}$. It follows from the fact that each $f_{k,l}$ is Δ^0_2 that there is a stage $t > s$ such that $g_k[t] \upharpoonright r = g_k \upharpoonright r$ for each $k \leqslant n$. For any $m > r$, if $u > t$ and $\Phi_e(m)[u] \downarrow$ then $r_{e,i}[v] > m$ for all $v \geqslant u$, so that $g_j(m)[v+1] = g_j(m)[v]$ for all $v \geqslant u$, and thus $g_j(m) = g_j(m)[u]$. But this means that g_j is computable, and hence that $g_j \circ \Phi_e^{-1}$ is a computable isomorphism from M_i to M_j, which is a contradiction. □

10.3.3 Goncharov's Δ^0_2-Theorem for Polish spaces

In this subsection we extend Goncharov's Δ^0_2-Theorem 10.3.2 to Polish spaces (up to isometry).

Definitions

A computable Polish presentation X of (M, d) is *rational-valued* if $d(x, y) \in \mathbb{Q}$ for every $x, y \in X$, and the distance d is represented by a computable *function* of two arguments mapping each pair of special points (x, y) to the corresponding rational number $d(x, y)$.

We also say that two computable Polish presentations L and L' of (M, d) are *limit equivalent* if there is a total computable function $g(x, s) : L \times \mathbb{N} \mapsto L'$ of two arguments such that the sequence $(g(x, s))_{s \in \mathbb{N}}$ is eventually stable on every x, and

$$f(x) = \lim_{s \to \infty} g(x, s)$$

is an isometric bijection of L onto L', where the limit is taken with respect to the standard discrete metric on $\mathbb{N}$.

Recall that in this chapter we view Polish spaces *up to isometry.*

Recall also that, to emphasise that our spaces are viewed up to isometry, we often use the term "computable structure" instead of "isometric computable Polish presentation". Further, rational-valued presentations can be viewed as algebraic structures in the language $\{D_r : r \in \mathbb{Q}\}$ (with equality), where $D_r(x, y) = 1$ iff $d(x, y) = r$. We view such structures up to isometry between their completions. (In the previous section we used the language $\{D_{<r}, D_{>r} : r \in \mathbb{Q}\}$ without equality.)

Theorem 10.3.20 (Melnikov and Ng [376]). *Suppose L and L' are two computable rational-valued structures on a Polish space (M, d) which are not computably isometric. If L and L' are limit equivalent, then (M, d) has infinitely many computable structures which are pairwise not computably isometric.*

As we already mentioned above, L and L' can be viewed as computable structures in the language $\{D_r : r \in \mathbb{Q}\}$ that are isomorphic relative to $0'$ but not computably isomorphic. By Goncherov's Δ^0_2-Theorem 10.3.2 there are infinitely many computable copies of the structure that are pairwise not computably isomorphic. However, these copies may be computably isometric as computable Polish spaces, just not by an isometry that takes special points to special points (recall we need to worry about their *completions*). Thus, the theorem above is *not* an elementary consequence of Goncharov's Δ^0_2-Theorem 10.3.2.

Proof of Theorem 10.3.20. The proof is organised as follows. First, we state the notations and the requirements. Next, we give an informal description which is followed by the formal construction and its verification.

Notation and conventions

We fix an effective listing $(\Psi_e)_{e \in \mathbb{N}}$ of all partial computable functions of two arguments, which includes all computable isometries from L to $\overline{L'}$. Here for each $S \subseteq M$, $\overline{S}$ stands for the completion of S in M. For every x and n such that $\Psi_e(x, n)\downarrow$, the number $\Psi_e(x, n)$ will be interpreted as an element of L'. The listing $(\Psi_e)_{e \in \mathbb{N}}$ satisfies the following conditions:

1. for every e, t, x, we have $d(\Psi_e(x, t), \Psi_e(x, t+1)) < 2^{-t-1}$, if $\Psi_e(x, t)$ and $\Psi_e(x, t+1)$ converge,
2. for every stage s and every e, t, x, we have $\Psi_{e,s}(x, t)\downarrow$ only if $\Psi_{e,s}(x, n)\downarrow$ for each $n \leqslant t$, and
3. if $\Theta : L \mapsto \overline{L'}$ is a computable isometry then there exists some e such that for every $x \in L$ we have $\Theta(x) = \lim_{n \to \infty} \Psi_e(x, n)$.

To see that $(\Psi_e)_{e \in \mathbb{N}}$ exists, we start with some universal listing of all partial computable functions of two variables, and limit ourselves to only those which satisfy (1) to (3). Since $d(\Psi_e(x, t), \Psi_e(x, t+1))$

is a computable fast converging sequence of rational numbers (in this case $d(\Psi_e(x,t), \Psi_e(x,t+1))$ is in fact rational), we will always be able to tell whenever $d(\Psi_e(x,t), \Psi_e(x,t+1)) < 2^{-t-1}$.

For every e and x, set $\Theta_e(x) = \lim_{n\to\infty} \Psi_e(x,n)$ if the limit exists (where the limit is taken with respect to the metric on M), and set $\Theta_e(x)\uparrow$ otherwise. The range of Θ_e, of course, does not have to be included in L'.

At stage s we set $\Theta_{e,s}(x)$ equal to $\Psi_{e,s}(x,m)$ if m is the largest such that $\Psi_{e,s}(x,m)\downarrow$, and we set $\Theta_{e,s}(x)$ undefined, otherwise. In the former case we let $\theta_{e,s}(x) = m$. Thus, $\Theta_{e,s}(x)$ is our stage s guess about $\Theta_e(x)$, and $\theta_{e,s}(x)$ indicates the error between $\Theta_{e,s}(x)$ and $\Theta_e(x)$. (This is essentially Notation 2.4.21.)

Let $f = \lim_s g_s$ be a Δ^0_2 permutation of natural numbers witnessing the limit equivalence of L and L'. As we have mentioned above, L and L' are essentially countable structures in the language $\{D_r : r \in \mathbb{Q}\}$. Thus we can safely assume that g_s is an isometry when restricted to the first s elements of its domain.

Requirements

We are going to produce a countably infinite family $\{A_m : m \in \mathbb{N}\}$ of computable structures on (M,d) which are pairwise not computably isometric. For every m, the structure A_m will be rational-valued.

We need to satisfy, for every $n > m$ and e, the following requirements:

$$N_{e,m,n} : \Theta_e \text{ does not induce an isomorphism from } \overline{A_m} \text{ onto } \overline{A_n},$$

and

$$R_m : A_m \text{ is isometric (in fact, limit equivalent) to } L \text{ and } L'.$$

To meet R_m we will construct surjective isometries between computable structures. Thus, A_m will be a rational-valued computable structure isomorphic to L and L' as a relational algebraic structure in the language $\{D_r : r \in \mathbb{Q}\}$.

Informal description

We first describe the strategy for R_m. To meet R_m we construct Δ^0_2 surjective isometric maps $\xi_m : A_m \mapsto L$ and $\eta_m : A_m \mapsto L'$. This is done via the approximations $\xi_{m,s}$ and $\eta_{m,s}$ where $\xi_m = \lim_s \xi_{m,s}$ and $\eta_m = \lim_s \eta_{m,s}$. Additionally we ensure that at each stage s, $g_s(\xi_{m,s}) = \eta_{m,s}$ on their domains:

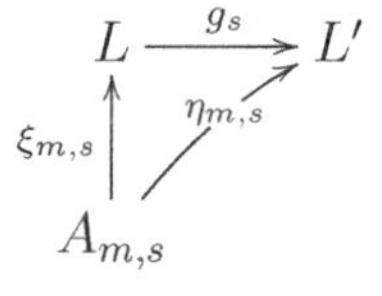

The main strategy of R_m is to copy either L or L', which is carried out via the surjective isometric maps ξ_m and η_m built by the strategy. The use of two maps rather than a single one will enable us to organise the activity of switching back and forth between copying L and copying L' during the construction. Since $g(x)$ may change several times before stabilising on a value, it may become necessary for us to redefine ξ_m and η_m during the construction in order to maintain the equality illustrated above. To be more specific, suppose at stage s we have defined $\xi_{m,s}(y)$ and

$\eta_{m,s}(y)$ such that $g_s(\xi_{m,s}(y)) = \eta_{m,s}(y)$. Suppose now that $g_{s+1}(x) \neq g_s(x)$ where $x = \xi_{m,s}(y)$. To keep the equality we have to do one of two things: either maintain $\xi_{m,s+1}(y) = \xi_{m,s}(y)$ and redefine $\eta_{m,s+1}(y) = g_{s+1}(x)$, or maintain $\eta_{m,s+1}(y) = \eta_{m,s}(y)$ and redefine $\xi_{m,s+1}(y) = z$ where $g_{s+1}(z) = g_s(x)$ (we speed up the approximation for g_s until such a z is found). In the former case we say that R_m *corrects via* η_m and in the latter case we say that R_m *corrects via* ξ_m. Each time R_m needs to correct, it will choose one of the two sides to preserve; this choice will be made so that the highest priority N-requirement with a current restraint larger than x is not injured. Since the approximation to $g_s(x)$ will eventually stabilise, at the end, η_m and ξ_m will be witnesses to the limitwise equivalence of A_m and L, and the limitwise equivalence of A_m and L', respectively.

An $N_{e,m,n}$-strategy in isolation will define a computable isometry between L and L' using the approximation $\Theta_{e,s} : A_m \to A_n$ and the maps ξ_m and η_n. Recall that $\Theta_e(x)$, if defined, is equal to the limit of the fast converging sequence $(\Psi_e(x,n))_{n\in\mathbb{N}}$ of points in L'. Recall also that for every s, $\Theta_{e,s}$ is a (partial) function from L to L', but the range of Θ_e itself may be outside L'.

If Θ_e is defined but does not induce an isometry, we will eventually see it because $\Theta_{e,s}$ will reflect it at some stage s. (This can only happen if for some $x, y \in L$, we have $d(x,y) \neq d(\Theta_e(x), \Theta_e(y))$.) The slightly more difficult case to handle is if Θ_e is not total, or Θ_e induces an isometry which is not onto. This, however, can be measured in a Π^0_2-way, and so to circumvent this difficulty, we use *expansionary stages* combined with a continuous version of the original Goncharov's preservation strategy, as follows: We call a stage s (e,m,n)-*expansionary*, if $\Theta_{e,s}$ "looks like an isometry from A_m to A_n with a certain precision" on a larger initial segment of its domain, with a better precision than at the previous expansionary stage, and with a further element of L' covered by a sufficiently small neighborhood of the range of $\Theta_{e,s}$. (The formal definition of an expansionary stage will be given later.) We will show that there are infinitely many (e,m,n)-expansionary stages iff Θ_e induces an onto isometry from $\overline{A_m}$ to $\overline{A_n}$.

We allow the strategy $N_{e,m,n}$ to act only at (e,m,n)-expansionary stages. At an (e,m,n)-expansionary stage s of the construction, $N_{e,m,n}$ will define the length of agreement between $\Theta_e(A_m)$ and A_n (this will be formally defined later) and will attempt to preserve $\xi_{m,s}$ on the domain of $\Theta_{e,s}$ and $\eta_{n,s}$ on the range of $\Theta_{e,t}$ for $t \leqslant s$. It is crucial that at every finite stage the domain and the (approximation to) range are both *finite sets.* If the restraint of this strategy $N_{e,m,n}$ eventually covers all of A_n and A_m then we would force both ξ_m and η_n to be computable functions. This would allow us to argue that L and L' are (contrary to the assumption of the theorem) computably isometric via the composition of ξ_m^{-1}, Θ_e and η_n.

The preservation strategy of $N_{e,m,n}$ described above potentially conflicts with the R_m-strategy when $g_s(\xi_m(x))$ changes value for x in the domain of of $\Theta_{e,s}$. Similarly, the preservation strategy of $N_{e,m,n}$ will potentially conflict with the R_n-strategy when $g_t(\xi_n(y))$ changes value for y in the range of $\Theta_{e,t}$. This is illustrated in the diagram below:

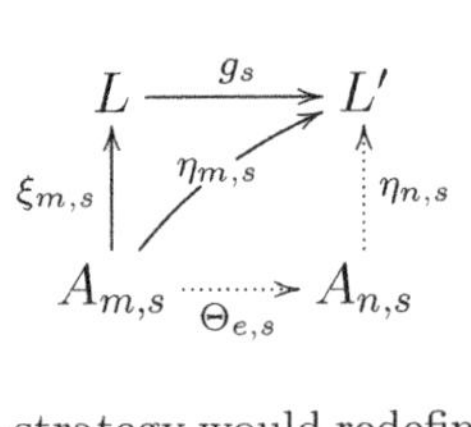

To prevent injuring $N_{e,m,n}$, the R_m-strategy would redefine η_m instead of ξ_m, while the R_n-strategy would redefine ξ_n instead of η_n. In this way the R_m- and R_n-strategies can maintain their equalities while not injuring $N_{e,m,n}$. Each $N_{e,m,n}$ eventually has finite restraint, and since the approximation $(g_s)_{s\in\mathbb{N}}$ will eventually settle on each finite subset of L, the overall construction involves only finite injury.

Definition 10.3.21 (*(e, m, n)-Expansionary stages*)**.** Recall that the elements of computable structures are identified with natural numbers. Hence in the following, $\Theta_{e,s} : A_m \to A_n$ is viewed as a map from $\mathbb{N}$ to $\mathbb{N}$. Given any stage s and e, m, n we let s^* be the largest $t < s$ such that t is an (e, m, n)-expansionary stage (set $s^* = 0$ if such a t does not exist).

We say that a stage s is *(e, m, n)-expansionary* if $s = 0$, or

1. the domain of $\Theta_{e,s}$ contains a longer initial segment of $\mathbb{N}$ since s^*, and for each $x \leqslant s^*$, $\theta_{e,s}(x) > s^*$;
2. for every $x, y \leqslant s^*$, $|d(x, y) - d(\Theta_{e,s}(x), \Theta_{e,s}(y))| \leqslant 2^{-s^*+1}$;
3. the 2^{-s^*}-neighborhood of the range of $\Theta_{e,s}$ contains the initial segment of $\mathbb{N}$ of length at least s^*.

Notice that every (e, m, n)-expansionary stage is associated with an initial segment of the domain of $\Theta_{e,s}$ (see (1)) and also with an initial segment of its range (see (3)). We denote these initial segments by $\sigma_{e,m,n,s}$ and $\tau_{e,m,n,s}$, respectively. To reduce cumbersome notation we drop e, m, n from the subscript when the context is clear.

Strategies

We describe the strategies for each requirement.

Strategy for $N_{e,m,n}$: If stage s is not (e, m, n)-expansionary then the strategy does nothing. Otherwise it sets the following restraints on the maps ξ_m and η_n until the next (e, m, n)-expansionary stage: Preserve the computation of $\xi_m(x)$ for every $x \leqslant \sigma_{e,m,n,s}$ and the computation of $\eta_n(y)$ for every

$$y \in L_{e,m,n,s} = \{\Theta_{e,t}(z) : z \leqslant |\sigma_{e,m,n,t}|, t \leqslant s\},$$

which is a finite set of points.

Strategy for R_m: At stage s of the construction, we define isometric partial maps $\xi_{m,s} : A_{m,s} \to L$ and $\eta_{m,s} : A_{m,s} \to L'$. By the choice of L and L', we can safely assume that g_s is an isometry when restricted to first s elements of its domain. We also assume that for every y mentioned before in the construction, there is an element x such that $g_s(x) = y$. The R_m-strategy does the following:

1. *Correction*: For each x such that $\xi_m(x)$ and $\eta_m(x)$ are currently defined, but $g_s(\xi_m(x)) \neq \eta_m(x)$, we correct via either (i) or (ii):

   (i) *Correction via η_m*: Maintain $\xi_m(x)$ and redefine $\eta_m(x) = g_s(\xi_m(x))$.

   (ii) *Correction via ξ_m*: Maintain $\eta_m(x)$ and redefine $\xi_m(x) = z$ where $g_s(z) = \eta_m(x)$.

   For each x where correction has to be done we pick the highest priority N-requirement such that $\xi_m(x)$ or $\eta_m(x)$ is restrained. We correct via η_m if N wants to restrain ξ_m, otherwise we correct via ξ_m. Initialise all lower priority N-strategies. (If no N-strategy restrains x then we correct via η_m.)

2. *Extension*: Let k be the least number which is not in the range of ξ_m. Find an element y in A_m such that $\xi_m(y)$ can be set equal k and $\eta_m(y)$ equal to $g_s(k)$ (i.e., we have to ensure that η_m, ξ_m are isometries of finite metric spaces). If such an element does not exist, introduce a new element y_0 to A_m and for each $y \in A_m$, declare the distances $d(y_0, y)$ correspondingly, i.e., set $d(y_0, y) = d(k, \xi_m(y)) = d(g_s(k), \eta_m(y))$. The extension substage is finished[4].

Construction

We fix an effective priority ordering of the N-strategies. The R-strategies are global strategies and are not assigned a priority, and will not be injured during the construction.

At stage 0 of the construction, initialise all N-strategies. At stage s, let the first s many N-strategies act according to their instructions described above. Next let the first s many R-strategies act.

Verification

We first show that each R_m is met, i.e., A_m is limitwise equivalent to L via ξ_m and to L' via η_m.

Lemma 10.3.22. *For every m, the maps $\xi_m = \lim_s \xi_{m,s}$ and $\eta_m = \lim_s \eta_{m,s}$ are well-defined, bijective, and isometric.*

Proof. The strategy for R_m cannot be injured. Fix an x, and we argue that $\lim_s \xi_{m,s}(x)$ and $\lim_s \eta_{m,s}(x)$ exists. Let N be the highest priority strategy that at some stage of the construction wants to preserve the computation of either ξ_m or η_m. Suppose N wishes to preserve the computation of ξ_m, say at some earliest stage s_0 (note that in this case N will never want to preserve the computation of η_m). The extension step in the construction ensures that when x is first enumerated in the structure A_m, we immediately define $\xi_m(x)$ and $\eta_m(x)$. Since this values are only redefined but never canceled, we have $\xi_{m,s_0}(x) \downarrow$ and $\eta_{m,s_0}(x) \downarrow$. Clearly $\xi_{m,s_0}(x)$ is never again redefined after s_0, since the correction step for x will always respect requirement N after s_0. Since $g_s(\xi_{m,s_0}(x))$ will be eventually stable, this means that $\eta_{m,s}(x)$ will be eventually stable. If N wishes to preserve η_m instead we proceed as above, but since g is onto we have that ${g_s}^{-1}(\eta_{m,s_0}(x))$ will be eventually stable.

The correction step ensures that $g_s(\xi_{m,s}(x)) = \eta_{m,s}(x)$ for every x and s. Hence this equality holds for the stable final values as well. Now it is easy to verify that since $g_s \upharpoonright s$ is an isometry of finite metric spaces for each s, the construction ensures that $\xi_{m,s}$ is an isometry of finite metric spaces at each step of the construction. Clearly ξ_m is injective because it is an isometry. Now the fact that ξ_m is onto follows easily from the fact that the $g_s(y)$ approximation is eventually stable, and by the action in the extension step. Since $\eta_m = g \circ \xi_m$ it follows that η_m is bijective and an isometry. □

Note that the lemma implies at once that all the sets A_m are computable structures on (isomorphic images of) M.

[4]Recall that we assume that g_s is an isometry on the first s elements, for every s. Therefore, we can always fix a $k \leqslant s$ and the corresponding g_s-image of k. The distances will agree, and we can safely set $d(y_0, y) = d(k, \xi_m(y))$. Obviously, since both L and L' are subsets of M, and in fact one is a permutation of the other, there is no further tension here. We also note that (due to other strategies acting) A_m may already have many elements outside the domain of ξ_m, and in this case we will not have to introduce new elements to A_m at this particular stage.

Lemma 10.3.23. *For every e, m, n, there are infinitely many (e, m, n)-expansionary stages iff Θ_e induces a computable isometry mapping A_m onto A_n.*

Proof. It is straightforward to check that the right to left direction holds. Suppose there are infinitely many (e, m, n)-expansionary stages. In this case condition (1) of Definition 10.3.21 ensures that for each x, $\Theta_e(x) = \lim_{t\to\infty} \Psi_e(x, t)$ exists. It suffices to check the following:

(i) For any x, y, we have $d(x, y) = \lim_s d(\Theta_{e,s}(x), \Theta_{e,s}(y))$, since the latter is the distance $d(\Theta_e(x), \Theta_e(y))$ in the closure $\overline{A_n}$.

(ii) For any y and any i, there exists some x and s such that $\theta_{e,s}(x) > i$ and $d(\Theta_{e,s}(x), y) < 2^{-i}$.

(i) above ensures that Θ_e induces an isometry in the closures, while (ii) ensures that Θ_e maps onto $\overline{A_n}$. It is easy to see that (i) and (ii) follow respectively from conditions (2) and (3) of Definition 10.3.21. □

Lemma 10.3.24. *For every e, m, n, $\limsup_t |\sigma_{e,m,n,t}| < \infty$ and $N_{e,m,n}$ is satisfied.*

Proof. We proceed by induction on $\langle e, m, n\rangle$. Suppose the lemma holds for all smaller indices. Hence there is a stage s_0 after which $N = N_{e,m,n}$ is never initialised, i.e. never injured by a higher priority requirement. Suppose that $\lim_{t>s_0} |\sigma_{e,m,n,t}| = \infty$. Then by Lemma 10.3.23, A_m and A_n are computably isometric. Since Θ_e induces an onto map, for each $z \in \mathbb{N}$ there is a first (e, m, n)-expansionary stage $\hat{s}_z > s_0$ such that $z \in L_{e,m,n,s}$ for every $s \geqslant \hat{s}_z$. This means that for each x, z, the first definition for $\xi_m(x)$ received after stage s_0 and the first definition for $\eta_n(z)$ received after stage $\hat{s}_z$ are stable and final. Hence ξ_m and η_n are computable functions. By Lemma 10.3.22 this means that L is computably isometric to A_m and L' is computably isometric to A_n, a contradiction. Since $\lim_t |\sigma_{e,m,n,t}| < \infty$, by Lemma 10.3.23, N is satisfied. □

The verification is finished, and the theorem is proved. □

Consequences of Theorem 10.3.20

First, we explain why Goncharov's Δ^0_2-Theorem 10.3.2 is a consequence of Theorem 10.3.20.

Proof of Theorem 10.3.2. The proofs of Theorems 8.2.7, 8.2.6, and 8.2.9 (combined) give an effective transformation $B \mapsto M(B)$ that turns a discrete algebraic structure B into a discrete metric space where, furthermore, the metric ranges over $\{0, 1, 2\}$. (The resulting space is viewed under isometry.) This combined transformation preserves computable dimension because it is preserved in Theorems 8.2.7, 8.2.6, and 8.2.9; see Exercise 8.2.8 and the discussion after the proof of Theorem 8.2.9. It is not difficult to see that the property of "being (or not being) Δ^0_2-isomorphic" is also preserved under the sequence of these transformations; see Exercise 8.2.10.

Furthermore, B can be effectively and uniformly reconstructed from $M(B)$ in the following sense. There is a uniformly effective transformation $M \mapsto S(M)$ such that whenever $M \cong M(B)$, we have that $S(M) \cong B$. Furthermore, this reverse transformation S also respects computable isomorphism classes; see Exercise 8.2.11.

So suppose $A \cong_{\Delta^0_2} B$ are two computable presentations of some (discrete) algebraic structure that are not computably isomorphic. Let $M(A)$ and $M(B)$ be the discrete, rational-valued spaces obtained from A and B after applying the transformation described above.

We have that $M(A) \cong_{\Delta^0_2} M(B)$ and $M(A) \not\cong_{\Delta^0_1} M(B)$. The sequence of spaces $(A_i)_{i\in\mathbb{N}}$ produced in the proof of Theorem 10.3.20 are pairwise computably non-isometric; note also that $\overline{A_n} = A_n$ for all n.

Let B_i be the algebraic structure that can be uniformly reconstructed from A_i by reverting the sequence of the transformations:

$$B_i = S(A_i), \quad i \in \mathbb{N}.$$

These structures witness the theorem. □

One well-known consequence of Goncharov's Δ^0_2-Theorem is the following generalisation of Nurtazin's Theorem 5.1.43.

Corollary 10.3.25 (Goncharov [203]). *Every computable torsion-free abelian group has computable dimension either* 1 *or* ω.

Proof. If the rank of the group is finite then the group is computably categorical. In the case of infinite rank, the strategy described in the proof of Nurtazin's Theorem 5.1.43 builds two computable copies of the group that satisfy the assumptions of Goncharov's Δ^0_2-Theorem 10.3.2. □

Similar results hold for many other classes, including ordered abelian groups, difference closed, real closed, differentially closed fields; see Exercises 10.4.5-10.4.17. (Not all of these results rely on Goncharov's Δ^0_2-Theorem 10.3.2.)

Corollary 10.3.26 (Downey, Harrison-Trainor and Melnikov [124]). *If a computable structure has finite computable dimension* > 1, *then it has computable dimension* ∞ *relative to* $0'$.

Proof. Let A be a computable structure. Relative to $0'$, it is 1-decidable. By Theorem 10.1.48, A has a $0'$-computable Σ^c_2 Scott family. If B is any other computable copy of A which is not computably isomorphic to A, A and B are $0''$-computably isomorphic. Thus A has computable dimension 1 or ∞ relative to $0'$. If it had computable dimension 1 relative to $0'$, then A and B would be $0'$-computably isomorphic, and so by Theorem 10.3.2 A would have computable dimension ∞. □

10.3.4 Proof of Theorem F

We now present an application of Theorem 10.3.20 to the space $C[0,1]$. It relies on a Theorem 2.4.24 proven at the very beginning of Part I of the book.

Theorem F (Melnikov and Ng [376]). The Polish space $(C[0,1], d_{\text{sup}})$ of continuous functions on the unit interval under the supremum metric d_{sup} has infinitely many isometric computable presentations that are pairwise not computably isometric.

Proof. We have already done most of the work; it remains to put all the pieces together. In Chapter 2, we proved Theorem 2.4.24 stating that there exist two limit-equivalent rational-valued computable structures on $(C[0,1], d_{\text{sup}})$ such that both compute the norm, but in one, $+$ is computable, and in the other, $+$ is not computable. These two computable Polish presentations are not computably isometric by Claim 10.2.5. Thus, Theorem F follows from Theorem 10.3.20. □

10.4 Further related results*

For a more thorough and detailed study of abstract computable structure theory for discrete structures, we cite [20], which has aged rather well, and also the recent books [401, 402]. The study of separable structures up to computable isometry was pioneered by Pour-El and Richards [435], which remains an excellent textbook.

Isometric and linearly isometric categoricity

The modern investigation of computably isometric spaces was initiated in [369, 268] and continued in many further works, most notably by McNicholl and his co-authors (e.g., [94, 360, 68]). Most of these results are concerned with either Banach spaces or compact spaces where the classification up to isometry seems more natural.

Banach spaces are typically viewed up to surjective *linear* isometry.

Definition 10.4.1 (Pour-El and Richards [435]). A Banach space is *linearly computably categorical* if it has a unique computable Banach (equivalently, Polish group) presentation, up to computable linear isometry.

The study of linearly computably categorical Banach spaces was pioneered by Pour-El and Richards [435]. They used different terminology and perhaps did not realise their notion was very similar to the analogous definition from effective algebra. We saw that $C[0,1]$ is not isometrically computably categorical, and it is known that it is not linearly computably categorical either [376]. For more about computably categorical Polish and Banach spaces, specifically the ones from Example 2.4.18, see [94, 376, 369, 268] and the exercises below. We leave open:

Question 10.4.2. *Is there a linearly computably categorical Banach space (Definition 10.4.1) such that the underlying Polish metric space is not isometrically computably categorical (Definition 10.2.1)?*

We do not know whether the restrictions on the computable Polish presentations in Theorem 10.3.20 can be dropped in general. However, in the (locally) compact case, we suspect the stabilisation of the Δ^0_2-process can be achieved for topological reasons.

We expect that Theorem 10.3.20 has a natural extension to the signatures of Banach spaces. However, it also appears that all known examples of Banach spaces that are not computably categorical have infinitely many computably non-isometric copies. We leave open:

Question 10.4.3. *Is there a Banach space of finite computable dimension $n > 1$?*

In the question above, we may interpret the dimension as the number of computable Banach presentations or computable Polish presentations. For Banach presentations, use computable linear isometry, and for Polish presentations just computable isometry (which, as we have seen, has to be affine). Both of these sub-questions are open.

Computable homeomorphic categoricity

Not much is known about computable *homeomorphic* categoricity of computably compact spaces or groups, with essentially the only known satisfactory results being Theorems 4.2.84 and 9.5.7. (Neither of the two results is particularly deep.) The topic is wide open.

Problem 10.4.4. *Extend the results of this chapter to computably compact spaces viewed* up to homeomorphism.

While we hope that Theorem 10.2.9 should not be too hard to extend to spaces viewed up to homeomorphism, extending deeper results to spaces up to homeomorphism will likely require significant new insights. Indeed, even up to isometry, the analogues of these deeper results have not yet been established. As we also stated earlier, it is not even known whether there is a compact Polish space having exactly two computably compact presentations, up to computable homeomorphism.

Computable group actions

One more related paper is [375], where many standard results of computable structure theory were proven in the more general setting of a computable Polish group acting on a computable Polish space. For example, [375] contains an analogue of Goncharov's Δ^0_2-Theorem 10.3.2 for computable group actions. Interestingly, some of the proofs of these more general results are simpler than their more specific analogues for structures. The paper has not yet found further applications; however, this may change in the future.

Exercises

Computable dimension and categoricity of discrete structures

See also Exercises 9.1.16 and 9.3.42

Exercise° 10.4.5 (Folklore). Show that in each of the following classes, the computable dimension of a computable structure is either 1 or ω:

1. Computable vector spaces (over a fixed computable field).
2. Algebraically closed fields.

Exercise 10.4.6 (Goncharov [203]). Prove that every computable abelian p-group has computable dimension either 1 or ω:

Exercise 10.4.7 (Melnikov and Ng [377]). Show that every torsion abelian group has computable dimension either 1 or ω.

Exercise° 10.4.8 (Dzgoev and Goncharov [210], Remmel [446]).

(i) Show that if A is a computable linear ordering which is not computably categorical, then it has computable dimension ∞.

(ii) Show that if B is a computable Boolean algebra which is not computably categorical then the computable dimension of B is ∞.

Exercise° 10.4.9 (Essentially Cenzer, Harizanov, and Remmel [85]). An injection structure is a structure of the form (A, f), where f is an injection on A. Prove that every computable injection structure has computable dimension either 1 or ω.

Exercise 10.4.10 (Levin [338]). Prove that every computable Archimedean ordered field has computable dimension either 1 or ω. (Archimedean means that there are no "infinitely large" elements; see the hint to Exercise 8.1.41.)

Exercise* 10.4.11 (Goncharov, Lempp, Solomon [215]). Show that every computable ordered abelian group has computable dimension either 1 or ω. (An ordered group is Archimedean if all non-zero elements of the group lie in one Archimedean class; see the definitions before Exercise 5.1.49. For a proof in the Archimedean case that relies on Goncharov's Δ_2^0-Theorem, see [244].)

Exercise* 10.4.12. We view a tree as a (strict) partial order. Prove the following:

1. Every tree if infinite height has computable dimension ω (Miller [394]).
2. Every tree of finite height has computable dimension either 1 or ω (Lempp, McCoy, Miller, and Solomon [335]).

Exercise* 10.4.13 (Alaev [5]). Show that the computable dimension of any (computable) Boolean algebra with finitely many distinguished ideals is either 1 or ω.

Exercise* 10.4.14 (Harrison-Trainor, Melnikov, and Montalbán [244]). Prove that every computable real closed field of infinite transcendence degree has computable dimension 1 or ω.

Exercise* 10.4.15 (Harrison-Trainor, Melnikov, and Montalbán [244]). Prove that every computable differentially closed field of infinite δ-degree (we omit the definitions, see [244]) has computable dimension 1 or ω.

Exercise* 10.4.16 (Harrison-Trainor, Melnikov, and Montalbán [244]). Show that every computable difference closed field of infinite transformal degree (we omit the definitions, see [244]) has computable dimension 1 or ω.

Exercise* 10.4.17 (Harrison-Trainor [236]). Prove that every computable algebraically closed valued field and every p-adically closed valued field of infinite transcendence degree has computable dimension 1 or ω.

Exercises about computably categorical Banach spaces

Exercise* 10.4.18 (McNicholl [360]). Show that (the real or complex) ℓ_p is linearly Δ_2^0-categorical (Definition 10.4.1), and it is linearly computably categorical if and only if $p = 2$.

Exercise* 10.4.19 (McNicholl and Stull [362]). Suppose p is a computable real such that $p \geqslant 1$. Define the linear isometry degree of a computable (complex or real) Banach presentation of ℓ_p to be the least powerful Turing degree d by which it is d-computably isometrically isomorphic to the standard presentation of ℓ_p. Show that this degree always exists and that when $p \neq 2$, these degrees are precisely the c.e. degrees.

Exercise* 10.4.20 (Clanin, McNicholl, and Stull [94]). Prove the following effective version of the Carathéodory Theorem: If Ω is a nonzero, nonatomic, and separable measure space, and if $p \geqslant 1$ is a computable real, then every computable presentation of $L^p(\Omega)$ is computably isomorphic to the standard computable presentation of $L^p[0,1]$.

Exercise* 10.4.21 (Brown and McNicholl [67]). We follow the notation and terminology of Ex. 10.4.19. Fix $p \neq 2$ and a computable Banach space $L^p(\Omega)$. Let be the least Turing degree that computes an isometric isomorphism between any two computable copies of $L^p(\Omega)$, if it exists. Prove that when Ω has only finitely many atoms this degree is $\mathbf{0}'$, and if it has infinitely many atoms, it is $\mathbf{0}''$.

Exercise 10.4.22** (Franklin et al. [179]). Show that the Banach space $C[0,1]$ is Δ_5^0-linearly isometrically categorical.

Primitive recursive dimension

The systematic study of primitive recursive analogues of computable algebraic results was initiated in [282]. We will not motivate these investigations here and refer the reader to the survey [32]. We shall include several results about the punctual dimension of structures here, to inform the reader. Most of these results are (somewhat unexpectedly) technically challenging.

We have already encountered primitive recursive and punctual structures earlier (e.g., Exercise 10.3.18). We repeat the main definitions here.

Definition 10.4.23.
1. A (countably infinite) structure is *punctual* or *fully primitive recursive* if its domain is ω and the operations and relations are uniformly primitive recursive.
2. A function $f : \omega \to \omega$ is *punctual* if both f and f^{-1} are primitive recursive.
3. The *punctual dimension* of a structure is the number of its punctual presentations, up to punctual isomorphism.
4. A structure is *punctually categorical* if it has punctual dimension 1.
5. If A, B are punctual structures, then A is *punctually reducible to* B, written $A \leqslant_{PR} B$, if there is a primitive recursive (but not necessarily punctual) surjective isomorphism $f : A \to B$.
6. $PR(A)$ is the *punctual degree structure* induced by the preorder $\leqslant_{PR}$ on the class of punctual presentations of A.

We note that both $|PR(A)|$ and the punctual dimension of a structure can be viewed as natural analogues of the computable dimension of A. While the punctual dimension certainly looks more familiar to a computable structure theorist, the partial order on $PR(A)$ carries more information about A. (Note that $|PR(A)| = \infty$ implies that the punctual dimension of A is also infinite. Surprisingly, not much is known about the relationship between these two notions of dimension beyond this observation and Exercise 10.4.30.)

Exercise° 10.4.24 (Dorzhieva et al. [116] based on [282])**.** Prove the following statements:

1. An equivalence structure S is punctually categorical iff it is either of the form $F \cup E$, where F is finite and E has only classes of size 1, or S has finitely many classes at most one of which is infinite.
2. A linear order is punctually categorical iff it is finite.
3. A Boolean algebra is punctually categorical iff it is finite.
4. An abelian p-group is punctually categorical iff it has the form $F \oplus \mathbb{V}$, where $p\mathbb{V} = \mathbf{0}$ and F is finite.
5. A torsion-free abelian group is punctually categorical iff it is the trivial group $\mathbf{0}$.

In each case, check that if the structure is not punctually categorical, then it has infinite punctual dimension.

Exercise* 10.4.25 (Dorzhieva et al. [115] based on [123])**.** Let G be an undirected graph. Prove that the following are equivalent:

1. G is punctually categorical.

2. G becomes a clique or an anti-clique (an independent set) after removing finitely many vertices $\bar{v} = v_0, \ldots, v_k$ with each v_i being either adjacent to all $x \in (G - \bar{v})$ or not adjacent to all $x \in (G - \bar{v})$.

Check that every punctual graph has punctual dimension either 1 or ω.

Exercise* 10.4.26 (Dorzhieva et al. [115] based on [121]). A *mono-unary structure* is a structure of the form (A, f), where f is a unary function symbol. Show that every mono-unary structure has computable dimension 1 or ω. (Note that a mono-unary structure can have computable dimension 2.)

Exercise* 10.4.27 (Kalimullin, Melnikov, and Ng [282]). Show that there exists a structure that is punctually categorical but not computably categorical.

Exercise 10.4.28** (Downey et al. [121]). Show that, for every computable α there is a punctually categorical structure that is not Δ^0_α-categorical. (Indeed, there is one that is not even Δ^1_1-categorical, as follows from the transformation defined in [121] and Exercise 10.3.19.)

Exercise* 10.4.29 (Melnikov and Ng [378]). Show that the punctual degrees of the random graph, the dense linear order, and the universal countable abelian p-group group $\bigoplus_{i \in \mathbb{N}} \mathbb{Z}_{p^\infty}$ are pairwise non-isomorphic (as partial orders).

Exercise 10.4.30** (Melnikov and Ng [378]). Show that for a graph G, the following are equivalent:

1. $|PR(G)| = 1$;
2. G is punctually categorical.

(We remark that it is still unknown where the equivalence holds for arbitrary structures.)

Exercise 10.4.31 (Bazhenov et al. [36]). Let A be an infinite finitely generated structure. Prove:

1. $PR(A)$ has a least element and is dense, i.e., for any punctual copies $A_0 <_{PR} A_1$ there is another punctual copy B with $A_0 <_{PR} B <_{PR} A_1$.

2*. $PR(A)$ can be infinite yet have a greatest element.

Exercise 10.4.32** (Greenberg et al. [218]). Show that there exists a structure A for which the partial order $PR(A)$ is not dense; that is, it has two punctual presentations $A_0 <_{PR} A_1$ so that the interval (A_0, A_1) in $PR(A)$ is empty.

Exercise 10.4.33** (Koh, Melnikov and Ng [312]). Show that the punctual degrees of the dense linear order are *not* dense.

Bibliography

[1] O. ABERTH, *Computable analysis*, McGraw-Hill International Book Company, 1980.

[2] S. ADYAN, *Algorithmic unsolvability of problems of recognition of certain properties of groups*, Doklady Akademii Nauk SSSR, 103 (1955), pp. 533–535. (in Russian).

[3] ——, *Finitely presented groups and algorithms*, Dokl. Akad. Nauk SSSR (N.S.), 117 (1957), pp. 9–12.

[4] ——, *Unsolvability of some algorithmic problems in the theory of groups*, Trudy Moskov. Mat. Obšč., 6 (1957), pp. 231–298.

[5] P. E. ALAEV, *Autostable I-algebras*, Algebra Logika, 43 (2004), pp. 511–550, 630.

[6] ——, *Strongly constructive Boolean algebras*, Algebra Logika, 44 (2005), pp. 3–23, 126.

[7] ——, *Categoricity for primitively recursive and polynomial Boolean algebras*, Algebra Logika, 57 (2018), pp. 389–426.

[8] R. ALVIR, N. GREENBERG, M. HARRISON-TRAINOR, AND D. TURETSKY, *Scott complexity of countable structures*, The Journal of Symbolic Logic, 86 (2021), pp. 1706–1720.

[9] K. AMBOS-SPIES, *Anti-mitotic recursively enumerable sets*, MLQ. Mathematical Logic Quarterly, 31 (1985), pp. 461–477.

[10] K. AMBOS-SPIES, K. WEIHRAUCH, AND X. ZHENG, *Weakly computable real numbers*, Journal of Complexity, 16 (2000), pp. 676–690.

[11] D. AMIR AND M. HOYRUP, *Strong computable type*, Computability, 12(3) (2023), pp. 227–269.

[12] B. ANDERSEN, A. KACH, A. MELNIKOV, AND R. SOLOMON, *Jump degrees of torsion-free abelian groups*, J. Symbolic Logic, 77 (2012), pp. 1067–1100.

[13] B. ANDERSON AND B. CSIMA, *Degrees that are not degrees of categoricity*, Notre Dame J. Form. Log., 57 (2016), pp. 389–398.

[14] U. ANDREWS, M. CAI, I. KALIMULLIN, S. LEMPP, J. S. MILLER, AND A. MONTALBÁN, *The complements of lower cones of degrees and the degree spectra of structures*, J. Symb. Log., 81 (2016), pp. 997–1006.

R. G. Downey, A. Melnikov, *Computable Structure Theory*, Theory and Applications of Computability, https://doi.org/10.1007/978-3-031-92433-0

[15] E. Artin and O. Schreier, *Algebraische konstruktion reeller körper*, Abh. Math. Sem. Univ., 5 (1927), pp. 85–89.

[16] C. Ash, *Recursive labeling systems and stability of recursive structures in hyperarithmetical degrees*, Trans. Amer. Math. Soc., 298 (1986), pp. 497–514.

[17] ——, *Categoricity in hyperarithmetical degrees*, Annals of Pure and Applied Logic, 34 (1987), pp. 1–14.

[18] ——, *A construction for recursive linear orderings*, J. Symbolic Logic, 56 (1991), pp. 673–683.

[19] C. Ash, C. Jockusch, and J. Knight, *Jumps of orderings*, Trans. Amer. Math. Soc., 319 (1990), pp. 573–599.

[20] C. Ash and J. Knight, *Computable structures and the hyperarithmetical hierarchy*, vol. 144 of Studies in Logic and the Foundations of Mathematics, North-Holland Publishing Co., Amsterdam, 2000.

[21] C. Ash, J. Knight, M. Manasse, and T. Slaman, *Generic copies of countable structures*, Ann. Pure Appl. Logic, 42 (1989), pp. 195–205.

[22] C. Ash and A. Nerode, *Intrinsically recursive relations*, in Aspects of effective algebra (Clayton, 1979), J. N. Crossley, ed., Upside Down A Book Co., Yarra Glen, Australia, 1981, pp. 26–41.

[23] M. Askes and R. Downey, *Online, computable and punctual structure theory*, Log. J. IGPL, 31 (2023), pp. 1251–1293.

[24] J. Avigad and V. Brattka, *Computability and analysis: the legacy of Alan Turing*, in Turing's legacy: developments from Turing's ideas in logic, vol. 42 of Lect. Notes Log., Assoc. Symbol. Logic, La Jolla, CA, 2014, pp. 1–47.

[25] S. A. Badaev, *Computable enumerations of families of general recursive functions*, Algebra i Logika, 16 (1977), pp. 129–148, 249.

[26] R. Baer, *Abelian groups without elements of finite order*, Duke Math. J., 3 (1937), pp. 68–122.

[27] R. Bagaviev, I. I. Batyrshin, N. Bazhenov, D. Bushtets, M. Dorzhieva, H. T. Koh, R. Kornev, A. Melnikov, and K. M. Ng, *Computably and punctually universal spaces*, Ann. Pure Appl. Logic, 176 (2025), pp. Paper No. 103491, 31.

[28] S. Banach, *Théorie des opérations linéaires*, Z subwencji Funduszu kultury narodowej, Warszawa, 1932.

[29] S. Banach and S. Mazur, *Sur les fonctions calculables*, Ann. Soc. Pol. de Math, 16 (1937), p. 402.

[30] E. Barker, *Back and forth relations for reduced abelian p-groups*, Ann. Pure Appl. Logic, 75 (1995), pp. 223–249.

[31] G. Baumslag, E. Dyer, and C. Miller, III, *On the integral homology of finitely presented groups*, Topology, 22 (1983), pp. 27–46.

[32] N. Bazhenov, R. Downey, I. Kalimullin, and A. Melnikov, *Foundations of online structure theory*, Bull. Symb. Log., 25 (2019), pp. 141–181.

[33] N. Bazhenov, S. Goncharov, and A. Melnikov, *Decompositions of decidable abelian groups*, Internat. J. Algebra Comput., 30 (2020), pp. 49–90.

[34] N. Bazhenov, M. Harrison-Trainor, I. Kalimullin, A. Melnikov, and K. M. Ng, *Automatic and polynomial-time algebraic structures*, J. Symb. Log., 84 (2019), pp. 1630–1669.

[35] N. Bazhenov, M. Harrison-Trainor, and A. Melnikov, *Computable Stone spaces*, Ann. Pure Appl. Logic, 174 (2023), pp. Paper No. 103304, 25.

[36] N. Bazhenov, I. Kalimullin, A. Melnikov, and K. M. Ng, *Online presentations of finitely generated structures*, Theoret. Comput. Sci., 844 (2020), pp. 195–216.

[37] N. Bazhenov, A. Melnikov, and K. M. Ng, *Every Δ^0_2 Polish space is computable topological*, Proc. Amer. Math. Soc., 152 (2024), pp. 3123–3136.

[38] N. Bazhenov and H.-C. Tsai, *On the effective universality of mereological theories*, MLQ Math. Log. Q., 68 (2022), pp. 48–66.

[39] N. A. Bazhenov, *Δ^0_2-categoricity of Boolean algebras*, Journal of Mathematical Sciences, 203 (2014), pp. 444–454.

[40] ——, *Degrees of autostability for linear orders and linearly ordered abelian groups*, Algebra and Logic, 55 (2016), pp. 257–273.

[41] ——, *Degrees of autostability relative to strong constructivizations for Boolean algebras*, Algebra and Logic, 55 (2016), pp. 87–102.

[42] ——, *Effective categoricity for distributive lattices and Heyting algebras*, Lobachevskii Journal of Mathematics, 38 (2017), pp. 600–614.

[43] ——, *Categoricity spectra of computable structures*, Journal of Mathematical Sciences, 256 (2021), pp. 34–50.

[44] N. A. Bazhenov, A. N. Frolov, I. Kalimullin, and A. Melnikov, *Computability of distributive lattices*, Sibirsk. Mat. Zh., 58 (2017), pp. 1236–1251.

[45] V. Becher and T. A. Slaman, *On the normality of numbers to different bases*, J. Lond. Math. Soc. (2), 90 (2014), pp. 472–494.

[46] I. Ben Yaacov and A. P. Pedersen, *A proof of completeness for continuous first-order logic*, J. Symbolic Logic, 75 (2010), pp. 168–190.

[47] I. Bilanovic, J. Chubb, and S. Roven, *Detecting properties from descriptions of groups*, Arch. Math. Logic, 59 (2020), pp. 293–312.

[48] E. Bishop, *Foundations of constructive analysis*, McGraw-Hill Book Co., New York-Toronto, Ont.-London, 1967.

[49] L. A. Bokut, *On a property of the Boone groups*, Algebra i Logika Sem., 5 (1966), pp. 5–23.

[50] ——, *On a property of the Boone groups. II*, Algebra i Logika Sem., 6 (1967), pp. 15–24.

[51] W. Boone, *The word problem*, Annals of Math, 70 (1959), pp. 207–265.

[52] W. Boone and H. Rogers, *On a problem of j. h. c. whitehead and a problem of alonzo church*, Journal of Symbolic Logic, 34 (1969), pp. 506–507.

[53] É. Borel, *Le calcul des intégral défines*, Journal de Mathématiques Pures and Appliquées, 8 (1912), pp. 159–210.

[54] V. Bosserhoff, *On the effective existence of Schauder bases*, J.UCS, 15 (2009), pp. 1145–1161.

[55] V. Bosserhoff and P. Hertling, *Effective subsets under homeomorphisms of* $\mathbb{R}^n$, Inform. and Comput., 245 (2015), pp. 197–212.

[56] V. Brattka, *Computability of Banach space principles*, FernUniversitat in Hagen, 2001.

[57] ——, *Computable versions of Baire's category theorem*, in Mathematical foundations of computer science, 2001 (Mariánské Lázně), vol. 2136 of Lecture Notes in Comput. Sci., Springer, Berlin, 2001, pp. 224–235.

[58] ——, *The inversion problem for computable linear operators*, in STACS 2003, vol. 2607 of Lecture Notes in Comput. Sci., Springer, Berlin, 2003, pp. 391–402.

[59] ——, *Borel complexity and computability of the Hahn-Banach theorem*, Arch. Math. Logic, 46 (2008), pp. 547–564.

[60] ——, *Plottable real number functions and the computable graph theorem*, SIAM J. Comput., 38 (2008), pp. 303–328.

[61] V. Brattka, M. de Brecht, and A. Pauly, *Closed choice and a uniform low basis theorem*, Ann. Pure Appl. Logic, 163 (2012), pp. 986–1008.

[62] V. Brattka and R. Dillhage, *Computability of compact operators on computable banach spaces with bases*, Mathematical Logic Quarterly, 53 (2007), pp. 345–364.

[63] V. Brattka and I. Kalantari, *A bibliography of recursive analysis and recursive topology*, in Handbook of recursive mathematics, Vol. 1, vol. 138 of Stud. Logic Found. Math., North-Holland, Amsterdam, 1998, pp. 583–620.

[64] V. Brattka and G. Presser, *Computability on subsets of metric spaces*, Theoretical Computer Science, 305 (2003), pp. 43–76.

[65] P. Brodhead and D. Cenzer, *Effectively closed sets and enumerations*, Arch. Math. Logic, 46 (2008), pp. 565–582.

[66] L. E. J. Brouwer, *Beweis, dass jede volle funktion gleichmässig stetig ist*, Koninklijke Nederlandse Akademie van Wetenschappen, Proc. Section of Sciences, 27 (1924), pp. 189–193.

[67] T. Brown and T. McNicholl, *Analytic computable structure theory and l^p-spaces, part 2*, Archive for Mathematical Logic, 59 (2020), pp. 427–443.

[68] T. Brown, T. McNicholl, and A. Melnikov, *On the complexity of classifying Lebesgue spaces*, J. Symb. Log., 85 (2020), pp. 1254–1288.

[69] B. Buchberger, *An Algorithm for Finding the Basis Elements of the Residue Class Ring of a Zero Dimensional Polynomial Ideal*, PhD thesis, University of Innsbruck, 1965.

[70] K. Burnik and Z. Iljazovic, *Computability of 1-manifolds*, Log. Methods Comput. Sci., 10 (2014), pp. 2:8, 28.

[71] P. Burton, C. J. Eagle, A. Fox, I. Goldbring, M. Harrison-Trainor, T. H. McNicholl, A. Melnikov, and T. Thewmorakot, *Computable gelfand duality*, 2024.

[72] W. Calvert, *The isomorphism problem for classes of computable fields*, Archive for Mathematical Logic, 43 (2004), pp. 327–336.

[73] ——, *Algebraic structure and computable structure*, ProQuest LLC, Ann Arbor, MI, 2005. Thesis (Ph.D.)–University of Notre Dame.

[74] ——, *The isomorphism problem for computable abelian p-groups of bounded length*, J. Symbolic Logic, 70 (2005), pp. 331–345.

[75] W. Calvert, D. Cenzer, V. Harizanov, and A. Morozov, *Effective categoricity of equivalence structures*, Ann. Pure Appl. Logic, 141 (2006), pp. 61–78.

[76] W. Calvert, V. Harizanov, J. Knight, and S. Miller, *Index sets of computable models*, Algebra Logika, 45 (2006), pp. 538–574, 631–632.

[77] W. Calvert, V. Harizanov, and A. Shlapentokh, *Turing degrees of isomorphism types of algebraic objects*, J. Lond. Math. Soc. (2), 75 (2007), pp. 273–286.

[78] W. Calvert, J. Knight, and J. Millar, *Computable trees of scott rank ω_1^{ck} , and computable approximation*, The Journal of Symbolic Logic, 71 (2006), pp. 283–298.

[79] R. Camerlo and S. Gao, *The completeness of the isomorphism relation for countable Boolean algebras*, Transactions of the American Mathematical Society, 353 (2001), pp. 491–518.

[80] J. Carson, V. Harizanov, J. Knight, K. Lange, C. McCoy, A. Morozov, S. Quinn, C. Safranski, and J. Wallbaum, *Describing free groups*, Trans. Amer. Math. Soc., 364 (2012), pp. 5715–5728.

[81] G. S. Ceitin, *On the theorem of cauchy in constructive analysis*, Uspehi Mat. Nauk., 10 (1955), pp. 207–209.

[82] ——, *Algorithmic operations in construcive complete separable metric spaces*, Doklady Akademii Nauk., 128 (1959), pp. 49–52.

[83] P. Cembranos and J. Mendoza, *Banach spaces of vector-valued functions*, vol. 1676 of Lecture Notes in Mathematics, Springer-Verlag, Berlin, 1997.

[84] D. CENZER, Π^0_1 *classes in computability theory*, in Handbook of Computability Theory, no. 140 in Studies in Logic and the Foundations of Mathematics, North-Holland, Amsterdam, 1999, pp. 37–85.

[85] D. CENZER, V. HARIZANOV, AND J. REMMEL, *Computability-theoretic properties of injection structures*, Algebra Logika, 53 (2014), pp. 60–108, 134–135, 137–138.

[86] D. CENZER AND J. REMMEL, *Polynomial-time abelian groups*, Ann. Pure Appl. Logic, 56 (1992), pp. 313–363.

[87] ——, *Index sets for* Π^0_1 *classes*, Annals of Pure and Applied Logic, 93 (1998), pp. 3–61. Computability Theory.

[88] ——, *Index sets for computable differential equations*, MLQ Math. Log. Q., 50 (2004), pp. 329–344.

[89] J. CHISHOLM, *Effective model theory vs. recursive model theory*, J. Symbolic Logic, 55 (1990), pp. 1168–1191.

[90] J. CHISHOLM, E. B. FOKINA, S. S. GONCHAROV, V. S. HARIZANOV, J. KNIGHT, AND S. QUINN, *Intrinsic bounds on complexity and definability at limit levels*, The Journal of Symbolic Logic, 74 (2009), pp. 1047–1060.

[91] J. CHISHOLM AND M. MOSES, *An undecidable linear order that is n-decidable for all n*, Notre Dame J. Formal Logic, 39 (1998), pp. 519–526.

[92] P. CHOLAK, R. DOWNEY, AND L. HARRINGTON, *On the orbits of computably enumerable sets*, J. Amer. Math. Soc., 21 (2008), pp. 1105–1135.

[93] P. CHOLAK, S. GONCHAROV, B. KHOUSSAINOV, AND R. SHORE, Journal of Symbolic Logic, 64 (1999), pp. 13–37.

[94] J. CLANIN, T. MCNICHOLL, AND D. STULL, *Analytic computable structure theory and* L^p *spaces*, Fund. Math., 244 (2019), pp. 255–285.

[95] C. R. J. CLAPHAM, *Finitely presented groups with word problems of arbitrary degrees of insolubility*, Proc. London Math. Soc. (3), 14 (1964), pp. 633–676.

[96] R. COLES, R. DOWNEY, AND B. KHOUSSAINOV, *On initial segments of computable linear orders*, Order, 14 (1997/98), pp. 107–124.

[97] R. COLES, R. DOWNEY, AND T. SLAMAN, *Every set has a least jump enumeration*, J. London Math. Soc. (2), 62 (2000), pp. 641–649.

[98] T. C. CRAVEN, *The topological space of orderings of a rational function field*, Duke Math. J., 41 (1974), pp. 339–347.

[99] ——, *The Boolean space of orderings of a field*, Trans. Amer. Math. Soc., 209 (1975), pp. 225–235.

[100] B. CSIMA, J. N. Y. FRANKLIN, AND R. A. SHORE, *Degrees of categoricity and the hyperarithmetic hierarchy*, Notre Dame Journal of Formal Logic, 54 (2013), pp. 215–231.

[101] B. Csima, D. Hirschfeldt, J. Knight, and R. Soare, *Bounding prime models*, J. Symbolic Logic, 69 (2004), pp. 1117–1142.

[102] B. Csima and I. Kalimullin, *Degree spectra and immunity properties*, Mathematical Logic Quarterly, 56 (2010), pp. 67–77.

[103] B. Csima and K. M. Ng, *Every Δ_2^0 degree is a strong degree of categoricity*, J. Math. Log., 22 (2022), pp. Paper No. 2250022, 18.

[104] B. Csima and D. Rossegger, *Degrees of categoricity and treeable degrees*, Journal of Mathematical Logic, 0 (2024), p. 2450002.

[105] B. Csima and J. Stephenson, *Finite computable dimension and degrees of categoricity*, Annals of Pure and Applied Logic, 170 (2019), pp. 58–94.

[106] M. Dabkowska, M. Dabkowski, V. Harizanov, and A. Sikora, *Turing degrees of nonabelian groups*, Proceedings of the American Mathematical Society, 135 (2007), pp. 3383–3391.

[107] A. Darbinyan, *Computability, orders, and solvable groups*, Journal of Symbolic Logic, 85 (2020), pp. 1588–1598.

[108] M. de Brecht, T. Kihara, and V. Selivanov, *Ideal presentations and numberings of some classes of effective quasi-polish spaces*, Computability, 13 (2024), pp. 1–24.

[109] M. Dehn, *Transformation der Kurven auf zweiseitigen Flächen*, Math. Ann., 72 (1912), pp. 413–421.

[110] J. C. E. Dekker, *Countable vector spaces with recursive operations. I*, J. Symbolic Logic, 34 (1969), pp. 363–387.

[111] ——, *Countable vector spaces with recursive operations. II*, J. Symbolic Logic, 36 (1971), pp. 477–493.

[112] ——, *Two notes on vector spaces with recursive operations*, Notre Dame J. Formal Logic, 12 (1971), pp. 329–334.

[113] V. Dobritsa, *Complexity of the index set of a constructive model*, Algebra i Logika, 22 (1983), pp. 372–381.

[114] ——, *Some constructivizations of abelian groups.*, (1983). Siberian Journal of Mathematics, 793 vol. 24,167–173 (in Russian).

[115] M. Dorzhieva, R. Downey, E. Hammatt, A. Melnikov, and K. M. Ng, *Punctually presented structures ii: comparing presentations*, Archive for Mathematical Logic, (2024).

[116] M. V. Dorzhieva, A. A. Issakhov, B. S. Kalmurzayev, R. A. Kornev, and M. V. Kotov, *Punctual dimension of algebraic structures in certain classes*, Lobachevskii J. Math., 42 (2021), pp. 716–725.

[117] R. Downey, *Bases of supermaximal subspaces and Steinitz systems. I*, J. Symbolic Logic, 49 (1984), pp. 1146–1159.

[118] ——, *Every recursive Boolean algebra is isomorphic to one with incomplete atoms*, Annals of Pure and Applied Logic, 60 (1993), pp. 193–206.

[119] ——, *Computability theory and linear orderings*, in Handbook of recursive mathematics, Vol. 2, vol. 139 of Stud. Logic Found. Math., North-Holland, Amsterdam, 1998, pp. 823–976.

[120] R. Downey, S. Goncharov, A. Kach, J. Knight, O. Kudinov, A. Melnikov, and D. Turetsky, *Decidability and computability of certain torsion-free abelian groups*, Notre Dame J. Form. Log., 51 (2010), pp. 85–96.

[121] R. Downey, N. Greenberg, A. Melnikov, K. M. Ng, and D. Turetsky, *Punctual categoricity and universality*, The Journal of Symbolic Logic, 85 (2020), pp. 1427–1466.

[122] R. Downey, N. Greenberg, and L. Qian, *Some open questions and recent results on computable Banach spaces*, in Twenty years of theoretical and practical synergies, vol. 14773 of Lecture Notes in Comput. Sci., Springer, Cham, [2024] ©2024, pp. 10–26.

[123] R. Downey, M. Harrison-Trainor, I. Kalimullin, A. Melnikov, and D. Turetsky, *Graphs are not universal for online computability*, Journal of Computer and System Sciences, 112 (2020), pp. 1–12.

[124] R. Downey, M. Harrison-Trainor, and A. Melnikov, *Relativizing computable categoricity*, Proceedings of the American Mathematical Society, 149 (2021), pp. 3999–4013.

[125] R. Downey and D. Hirschfeldt, *Algorithmic randomness and complexity*, Theory and Applications of Computability, Springer, New York, 2010.

[126] R. Downey, D. Hirschfeldt, and B. Khoussainov, *Uniformity in the theory of computable structures*, Algebra Logika, 42 (2003), pp. 566–593, 637.

[127] R. Downey, G. Igusa, and A. Melnikov, *On a question of Kalimullin*, Proc. Amer. Math. Soc., 146 (2018), pp. 3553–3563.

[128] R. Downey and C. Jockusch, *T-degrees, jump classes, and strong reducibilities*, Transactions of the American Mathematical Society, 301 (1987), pp. 103–136.

[129] ——, *Every low Boolean algebra is isomorphic to a recursive one*, Proceedings of the American Mathematical Society, 122 (1994), pp. 871–880.

[130] ——, *Effective presentability of Boolean algebras of Cantor-Bendixson rank* 1, J. Symbolic Logic, 64 (1999), pp. 45–52.

[131] R. Downey, A. Kach, S. Lempp, A. Lewis, A. Montalbán, and D. Turetsky, *The complexity of computable categoricity*, Advances in Mathematics, 268 (2015), pp. 423–466.

[132] R. Downey, A. Kach, S. Lempp, and D. Turetsky, *Computable categoricity versus relative computable categoricity*, Fundamentae Mathematica, 221 (2013), pp. 129–159.

[133] R. Downey, A. Kach, and D. Turetsky, *Limitwise monotonic functions and applications*, in Proceedings of STACS 2012, 2011, pp. 56–85.

[134] R. Downey and J. Knight, *Orderings with αth jump degree* $\mathbf{0}^{(\alpha)}$, Proc. Amer. Math. Soc., 114 (1992), pp. 545–552.

[135] R. Downey and S. Kurtz, *Recursion theory and ordered groups*, Ann. Pure Appl. Logic, 32 (1986), pp. 137–151.

[136] R. Downey and A. Melnikov, *Effectively categorical abelian groups*, J. Algebra, 373 (2013), pp. 223–248.

[137] ——, *Computable completely decomposable groups*, Trans. Amer. Math. Soc., 366 (2014), pp. 4243–4266.

[138] ——, *Computable analysis and classification problems*, in Beyond the Horizon of Computability - 16th Conference on Computability in Europe, CiE 2020, Fisciano, Italy, June 29 - July 3, 2020, Proceedings, M. Anselmo, G. D. Vedova, F. Manea, and A. Pauly, eds., vol. 12098 of Lecture Notes in Computer Science, Springer, 2020, pp. 100–111.

[139] ——, *Computably compact metric spaces*, The Bulletin of Symbolic Logic, 29 (2023), pp. 170–263.

[140] R. Downey, A. Melnikov, and K. M. Ng, *Iterated effective embeddings of abelian p-groups*, Internat. J. Algebra Comput., 24 (2014), pp. 1055–1084.

[141] ——, *On delta-2 categoricity of equivalence relations*, Ann. Pure Appl. Logic, 166 (2015), pp. 851–880.

[142] ——, *Abelian p-groups and the halting problem*, Ann. Pure Appl. Logic, 167 (2016), pp. 1123–1138.

[143] ——, *A Friedberg enumeration of equivalence structures*, J. Math. Log., 17 (2017), pp. 1750008, 28.

[144] ——, *Categorical linearly ordered structures*, Ann. Pure Appl. Logic, 170 (2019), pp. 1243–1255.

[145] ——, *Enumerating abelian p-groups*, Journal of Algebra, 560 (2020), pp. 745–790.

[146] R. Downey and A. Montalbán, *The isomorphism problem for torsion-free abelian groups is analytic complete*, J. Algebra, 320 (2008), pp. 2291–2300.

[147] R. Downey and M. F. Moses, *Recursive linear orders with incomplete successivities*, Transactions of the American Mathematical Society, 326 (1991), pp. 653–668.

[148] R. Downey and J. Remmel, *Computable algebras and closure systems: coding properties*, in Handbook of recursive mathematics, Vol. 2, vol. 139 of Stud. Logic Found. Math., North-Holland, Amsterdam, 1998, pp. 977–1039.

[149] R. Downey and M. Stob, *Splitting theorems in recursion theory*, Ann. Pure Appl. Logic, 65 (1993), p. 106.

[150] R. Downey and L. V. Welch, *Splitting properties of r.e. sets and degrees*, J. Symbolic Logic, 51 (1986), pp. 88–109.

[151] B. DUSHNIK AND E. MILLER, *Concerning similarity transformations of linearly ordered sets*, Bull. Amer. Math. Soc., 46 (1940), pp. 322–326.

[152] E. Z. DYMENT, *Recursive metrizability of enumerated topological spaces and bases of effective linear topological spaces*, Izv. Vyssh. Uchebn. Zaved. Mat., (1984), pp. 59–61.

[153] P. ENFLO, *A counterexample to the approximation problem in banach spaces*, Acta Mathematica, 130 (1973), pp. 309–317.

[154] Y. ERSHOV, *Numbered fields*, in Logic, Methodology and Philos. Sci. III (Proc. Third Internat. Congr., Amsterdam, 1967), North-Holland, Amsterdam, 1968, pp. 31–34.

[155] ——, *Computable functionals of finite types*, Algebra i Logika, 11 (1972), pp. 367–437, 496.

[156] ——, *The theory of numberings*, Library of the Department of Algebra and Mathematical Logic of Novosibirsk University, No. 13, Novosibirsk State Univ., Novosibirsk, 1974.

[157] ——, *Theorie der numierungen iii*, Z. Math. Logik, 23 (1977), pp. 289–371.

[158] ——, *Problems of solubility and constructive models [in russian]*, (1980). Nauka, Moscow (1980).

[159] Y. ERSHOV AND S. GONCHAROV, *Constructive models*, Siberian School of Algebra and Logic, Consultants Bureau, New York, 2000.

[160] M. FAIZRAHMANOV AND I. KALIMULLIN, *The enumeration spectrum hierarchy of n-families*, MLQ Math. Log. Q., 62 (2016), pp. 420–426.

[161] L. FEINER, *Orderings and Boolean algebras not isomorphic to recursive ones*, ProQuest LLC, Ann Arbor, MI, 1968. Thesis (Ph.D.)–Massachusetts Institute of Technology.

[162] ——, *Hiearchies of Boolean algebras*, J. Symbolic Logic, 35 (1970), pp. 365–374.

[163] ——, *Degrees of nonrecursive presentability*, Proc. Amer. Math. Soc., 38 (1973), pp. 621–624.

[164] S. FELLNER, *Recursive and finite axiomatizability of linear orderings*, PhD thesis, Rutgers, New Brundswick, NJ, USA, 1976.

[165] ——, *Recursiveness and finite axiomatizability of linear orderings*, ProQuest LLC, Ann Arbor, MI, 1976. Thesis (Ph.D.)–Rutgers The State University of New Jersey - New Brunswick.

[166] E. FOKINA, *Algorithmic properties of structures for the languages with two unary functional symbols*, Vestnik NGU, 8 (2008), pp. 90–101. (in Russian).

[167] E. FOKINA, V. HARIZANOV, AND A. MELNIKOV, *Computable model theory*, in Turing's legacy: developments from Turing's ideas in logic, vol. 42 of Lect. Notes Log., Assoc. Symbol. Logic, La Jolla, CA, 2014, pp. 124–194.

[168] E. FOKINA, V. HARIZANOV, AND D. TURETSKY, *Computability-theoretic categoricity and Scott families*, Ann. Pure Appl. Logic, 170 (2019), pp. 699–717.

[169] E. Fokina, J. Knight, A. Melnikov, S. Quinn, and C. Safranski, *Classes of Ulm type and coding rank-homogeneous trees in other structures*, J. Symbolic Logic, 76 (2011), pp. 846–869.

[170] E. B. Fokina and S.-D. Friedman, *Equivalence relations on classes of computable structures*, in Mathematical theory and computational practice, vol. 5635 of Lecture Notes in Comput. Sci., Springer, Berlin, 2009, pp. 198–207.

[171] E. B. Fokina, S.-D. Friedman, V. Harizanov, J. Knight, C. McCoy, and A. Montalbán, *Isomorphism relations on computable structures*, The Journal of Symbolic Logic, 77 (2012), pp. 122–132.

[172] E. B. Fokina, I. Kalimullin, and R. Miller, *Degrees of categoricity of computable structures*, Archive for Mathematical Logic, 49 (2010), pp. 51–67.

[173] G. B. Folland, *A course in abstract harmonic analysis*, Textbooks in Mathematics, CRC Press, Boca Raton, FL, second ed., 2016.

[174] D. Fowler and E. Robson, *Square root approximations in old babylonian mathematics: Ybc 7289 in context*, Historia Math, 25 (1998), pp. 366–378.

[175] A. Fox, *Computable presentations of C*-algebras*, The Journal of Symbolic Logic, (2023), pp. 1–26.

[176] ——, *Effective Metric Structure Theory of C-Star-Algebras*, ProQuest LLC, Ann Arbor, MI, 2023. Thesis (Ph.D.)–University of California, Irvine.

[177] A. Fox, I. Goldbring, and B. Hart, *Locally universal* C**-algebras with computable presentations*, J. Funct. Anal., 287 (2024), p. Paper No. 110652.

[178] J. Franklin, *Strength and weakness in computable structure theory*, in Computability and complexity, vol. 10010 of Lecture Notes in Comput. Sci., Springer, Cham, 2017, pp. 302–323.

[179] J. Franklin, R. Holzl, A. Melnikov, K. M. Ng, and D. Turetsky, *Computable classifications of continuous, transducer, and regular functions. Submitted*, 2024.

[180] R. Friedberg, *A criterion for completeness of degrees of unsolvability*, The Journal of Symbolic Logic, 22 (1957), pp. 159–160.

[181] ——, *Two recursively enumerable sets of incomparable degrees of unsolvability*, Proceedings of the National Academy of Sciences of the United States of America, 43 (1957), pp. 236–238.

[182] ——, *Three theorems on recursive enumeration*, The Journal of Symbolic Logic, 23 (1958), pp. 309–316.

[183] H. Friedman, S. Simpson, and R. Smith, *Countable algebra and set existence axioms*, Ann. Pure Appl. Logic, 25 (1983), pp. 141–181.

[184] H. Friedman and L. Stanley, *A Borel reducibility theory for classes of countable structures*, J. Symbolic Logic, 54 (1989), pp. 894–914.

[185] A. Fröhlich and J. Shepherdson, *Effective procedures in field theory*, Philos. Trans. Roy. Soc. London. Ser. A., 248 (1956), pp. 407–432.

[186] A. Frolov, I. Kalimullin, V. Harizanov, O. Kudinov, and R. Miller, *Spectra of high$_n$ and non-low$_n$ degrees*, J. Logic Comput., 22 (2012), pp. 755–777.

[187] A. Frolov, I. Kalimullin, and R. Miller, *Spectra of algebraic fields and subfields*, vol. 5635, 07 2009, pp. 232–241.

[188] A. Frolov and M. Zubkov, *The simplest low linear order with no computable copies*, The Journal of Symbolic Logic, 89 (2024), pp. 97–111.

[189] A. N. Frolov, *Δ_2^0-copies of linear orderings*, Algebra Logika, 45 (2006), pp. 354–370, 376.

[190] ——, *Linear orderings of low degree*, Sibirsk. Mat. Zh., 51 (2010), pp. 1147–1162.

[191] ——, *Computable presentability of countable linear orders*, in Proceedings of the seminar of the Kazan University Department of Algebra and Mathematical Logic (Russian), vol. 158 of Itogi Nauki Tekh. Ser. Sovrem. Mat. Prilozh. Temat. Obz., Vseross. Inst. Nauchn. i Tekhn. Inform. (VINITI), Moscow, 2018, pp. 81–115. Translation in J. Math. Sci. **256** (2021), no. 2, 199–233.

[192] L. Fuchs, *Abelian groups*, International Series of Monographs on Pure and Applied Mathematics, Pergamon Press, New York-Oxford-London-Paris, 1960.

[193] ——, *Partially ordered algebraic systems*, Pergamon Press, Oxford, 1963.

[194] ——, *Infinite abelian groups. Vol. I*, Pure and Applied Mathematics, Vol. 36, Academic Press, New York, 1970.

[195] ——, *Infinite abelian groups. Vol. II*, Academic Press, New York, 1973. Pure and Applied Mathematics. Vol. 36-II.

[196] D. J. Fuchs-Rabinowitsch, *Über eine Gruppe mit endlichvielen Erzeugenden und Relationen, die keine isomorphe Darstellung durch Matrizen von endlicher Ordnung zulässt*, C. R. (Doklady) Acad. Sci. URSS (N.S.), 27 (1940), pp. 425–426.

[197] S. Gao, *Invariant descriptive set theory*, vol. 293 of Pure and Applied Mathematics (Boca Raton), CRC Press, Boca Raton, FL, 2009.

[198] Z. Gao, S. Jain, B. Khoussainov, W. Li, A. Melnikov, K. Seidel, and F. Stephan, *Random subgroups of rationals*, in 44th International Symposium on Mathematical Foundations of Computer Science, vol. 138 of LIPIcs. Leibniz Int. Proc. Inform., Schloss Dagstuhl. Leibniz-Zent. Inform., Wadern, 2019, pp. Art. No. 25, 14.

[199] S. Goncharov, *The constructivizability of superatomic Boolean algebras*, Algebra i Logika, 12 (1973), pp. 31–40, 120.

[200] ——, *Autostability and computable families of constructivizations*, Algebra and Logic, 14 (1975), pp. 392–409.

[201] ——, *Certain properties of the constructivization of Boolean algebras*, Sibirsk. Mat. Ž., 16 (1975), pp. 264–278, 420. (loose errata).

[202] ——, *Bounded theories of constructive Boolean algebras*, Sibirsk. Mat. Z., 17 (1976), pp. 797–812.

[203] ——, *Autostability of models and abelian groups*, Algebra i Logika, 19 (1980), pp. 23–44, 132.

[204] ——, *Computable single-valued numerations*, Algebra and Logic, 19 (1980), pp. 325–356.

[205] ——, *The problem of the number of nonautoequivalent constructivizations*, Algebra i Logika, 19 (1980), pp. 621–639, 745.

[206] ——, *Limit equivalent constructivizations*, in Mathematical logic and the theory of algorithms, vol. 2 of Trudy Inst. Mat., "Nauka" Sibirsk. Otdel., Novosibirsk, 1982, pp. 4–12.

[207] ——, *Countable Boolean algebras and decidability*, Siberian School of Algebra and Logic, Consultants Bureau, New York, 1997.

[208] S. Goncharov, N. A. Bazhenov, and M. I. Marchuk, *The index set of Boolean algebras that are autostable relative to strong constructivizations*, Sibirsk. Mat. Zh., 56 (2015), pp. 498–512.

[209] ——, *Index set of linear orderings that are autostable relative to strong constructivizations*, J. Math. Sci. (N.Y.), 221 (2017), pp. 740–748.

[210] S. Goncharov and V. Dzgoev, *On constructivizations of some structures*, Akad. Nauk SSSR, Sibirsk. Otdel., Novosibirsk, 1978, (1978). (manuscript deposited at VINITI on July 26, 1978).

[211] ——, *Autostability of models*, Algebra i Logika, 19 (1980), pp. 45–58, 132.

[212] S. Goncharov, V. Harizanov, J. Knight, C. McCoy, R. Miller, and R. Solomon, *Enumerations in computable structure theory*, Annals of Pure and Applied Logic, 136 (2005), pp. 219–246.

[213] S. Goncharov and J. Knight, *Computable structure and antistructure theorems*, Algebra Logika, 41 (2002), pp. 639–681, 757.

[214] S. Goncharov, S. Lempp, and R. Solomon, *Friedberg numberings of families of n-computably enumerable sets*, Algebra and Logic, 41 (2002), pp. 81–86.

[215] ——, *The computable dimension of ordered abelian groups*, Adv. Math., 175 (2003), pp. 102–143.

[216] S. Goncharov, A. Molokov, and N. Romanovskii, *Nilpotent groups of finite algorithmic dimension*, Siberian Mathematical Journal, 30 (1989), pp. 63–68.

[217] S. Goncharov and A. Nurtazin, Algebra i Logika, 12 (1973), pp. 125–142.

[218] N. Greenberg, M. Harrison-Trainor, A. Melnikov, and D. Turetsky, *Non-density in punctual computability*, Ann. Pure Appl. Logic, 172 (2021), pp. Paper No. 102985, 17.

[219] N. Greenberg, J. F. Knight, A. Melnikov, and D. Turetsky, *Uniform procedures in uncountable structures*, J. Symb. Log., 83 (2018), pp. 529–550.

[220] N. Greenberg, A. Melnikov, A. Nies, and D. Turetsky, *Effectively closed subgroups of the infinite symmetric group*, Proceedings of the American Mathematical Society, 146 (2017), p. 1.

[221] N. Greenberg, A. Montalbán, and T. Slaman, *Relative to any non-hyperarithmetic set*, Journal of Mathematical Logic, 13 (2011).

[222] N. Greenberg, D. Turetsky, and L. Westrick, *Finding bases of uncountable free abelian groups is usually difficult*, Trans. Amer. Math. Soc., 370 (2018), pp. 4483–4508.

[223] V. Gregoriades, T. Kispéter, and A. Pauly, *A comparison of concepts from computable analysis and effective descriptive set theory*, Math. Structures Comput. Sci., 27 (2017), pp. 1414–1436.

[224] T. Grubba and K. Weihrauch, *On computable metrization*, Electron. Notes Theor. Comput. Sci., 167 (2007), pp. 345–364.

[225] A. Grzegorczyk, *Computable functionals*, Fund. Math., 42 (1955), pp. 168–202.

[226] ——, *On the definition of computable functionals*, Fund. Math., 42 (1955), pp. 232–239.

[227] ——, *On the definitions of computable real continuous functions*, Fund. Math., 44 (1957), pp. 61–71.

[228] E. Hammatt, *Structures of finite punctual dimension*, in Twenty Years of Theoretical and Practical Synergies. CiE 2024, P. E. G. L. M. F. Levy Patey, L., ed., vol. 14773 of Lecture Notes in Computer Science, Springer, Cham, 2024.

[229] V. Harizanov, *Pure computable model theory*, in Handbook of recursive mathematics, Vol. 1, vol. 138 of Stud. Logic Found. Math., North-Holland, Amsterdam, 1998, pp. 3–114.

[230] V. Harizanov, S. Lempp, C. McCoy, A. Morozov, and R. Solomon, *On the isomorphism problem for some classes of computable algebraic structures*, Arch. Math. Logic, 61 (2022), pp. 813–825.

[231] L. Harrington, *Recursively presentable prime models*, J. Symb. Log., 39 (1974), pp. 305–309.

[232] ——, *Mclaughlin's conjecture*. Handwritten notes, 1976.

[233] ——, *Building nonstandard models of peano arithmetic*. Handwritten notes, 1979.

[234] K. Harris, *η-representation of sets and degrees*, Journal of Symbolic Logic, 73 (2008), pp. 1097–1121.

[235] K. Harris and A. Montalbán, *Boolean algebra approximations*, Trans. Amer. Math. Soc., 366 (2014), pp. 5223–5256.

[236] M. Harrison-Trainor, *Computable valued fields*, Arch. Math. Logic, 57 (2018), pp. 473–495.

[237] ——, *There is no classification of the decidably presentable structures*, J. Math. Log., 18 (2018), pp. 1850010, 41.

[238] ——, *An introduction to the Scott complexity of countable structures and a survey of recent results*, Bull. Symb. Log., 28 (2022), pp. 71–103.

[239] M. Harrison-Trainor and M.-C. Ho, *Finitely generated groups are universal among finitely generated structures*, Annals of Pure and Applied Logic, 172 (2021), p. 102855.

[240] M. Harrison-Trainor, G. Igusa, and J. Knight, *Some new computable structures of high rank*, Proceedings of the American Mathematical Society, 146 (2018), pp. 3097–3109.

[241] M. Harrison-Trainor and A. Melnikov, *An arithmetic analysis of closed surfaces*, Trans. Amer. Math. Soc., 377 (2024), pp. 1543–1596.

[242] M. Harrison-Trainor, A. Melnikov, and R. Miller, *On computable field embeddings and difference closed fields*, Canad. J. Math., 69 (2017), pp. 1338–1363.

[243] M. Harrison-Trainor, A. Melnikov, R. Miller, and A. Montalbán, *Computable functors and effective interpretability*, The Journal of Symbolic Logic, 82 (2017), pp. 77–97.

[244] M. Harrison-Trainor, A. Melnikov, and A. Montalbán, *Independence in computable algebra*, J. Algebra, 443 (2015), pp. 441–468.

[245] M. Harrison-Trainor, A. Melnikov, and K. M. Ng, *Computability of Polish spaces up to homeomorphism*, The Journal of Symbolic Logic, (2020), pp. 1–25.

[246] K. Hatzikiriakou and S. Simpson, WKL_0 *and orderings of countable abelian groups*, in Logic and computation (Pittsburgh, PA, 1987), vol. 106 of Contemp. Math., Amer. Math. Soc., Providence, RI, 1990, pp. 177–180.

[247] G. Herrmann, *Die frage der endlich vielen schritte in der theorie der polynomideale*, Math. Ann., 95 (1926), pp. 736–788.

[248] P. Hertling, *A Banach-Mazur computable but not Markov computable function on the computable real numbers*, Ann. Pure Appl. Logic, 132 (2005), pp. 227–246.

[249] E. Hewitt, *The role of compactness in analysis*, The American Mathematical Monthly, 67 (1960), pp. 499–516.

[250] G. Higman, *Subgroups of finitely presented groups*, Proc. Roy. Soc. Ser. A, 262 (1961), pp. 455–475.

[251] D. Hilbert, *Über die theorie der algebraischen formen*, Mathematische Annalen, 36 (1890), pp. 473–534.

[252] ——, Bulletin of the American Mathematical Society, 8 (1902), pp. 437–479.

[253] D. Hirschfeldt, *Degree spectra of relations on computable structures*, ProQuest LLC, Ann Arbor, MI, 1999. Thesis (Ph.D.)–Cornell University.

[254] ——, *Degree spectra of intrinsically c.e. relations*, J. Symbolic Logic, 66 (2001), pp. 441–469.

[255] ——, *Prime models of theories of computable linear orderings*, Proc. Amer. Math. Soc., 129 (2001), pp. 3079–3083 (electronic).

[256] ——, *Some questions in computable mathematics*, in Computability and complexity, vol. 10010 of Lecture Notes in Comput. Sci., Springer, Cham, 2017, pp. 22–55.

[257] D. Hirschfeldt, B. Khoussainov, R. Shore, and A. Slinko, *Degree spectra and computable dimensions in algebraic structures*, Ann. Pure Appl. Logic, 115 (2002), pp. 71–113.

[258] D. Hirschfeldt, B. Khoussainov, and R. A. Shore, *A computably categorical structure whose expansion by a constant has infinite computable dimension*, J. Symbolic Logic, 68 (2003), pp. 1199–1241.

[259] D. Hirschfeldt, K. Kramer, R. Miller, and A. Shlapentokh, *Categoricity properties for computable algebraic fields*, Trans. Amer. Math. Soc., 367 (2015), pp. 3981–4017.

[260] D. Hirschfeldt, R. Miller, and S. Podzorov, *Order-computable sets*, Notre Dame J. Formal Logic, 48 (2007), pp. 317–347.

[261] G. Hjorth, *The isomorphism relation on countable torsion free abelian groups*, Fund. Math., 175 (2002), pp. 241–257.

[262] J. G. Hocking and G. S. Young, *Topology*, Dover Publications, Inc., New York, second ed., 1988.

[263] K. H. Hofmann and S. A. Morris, *The structure of compact groups—a primer for the student—a handbook for the expert*, vol. 25 of De Gruyter Studies in Mathematics, De Gruyter, Berlin, [2020] ©2020. Fourth edition [of 1646190].

[264] K. H. Hofmann and P. S. Mostert, *Cohomology theories for compact abelian groups*, Springer-Verlag, New York-Heidelberg; VEB Deutscher Verlag der Wissenschaften, Berlin, 1973. With an appendix by Eric C. Nummela.

[265] M. Hoyrup, *Genericity of weakly computable objects*, Theory of Computing Systems, 60 (2017), pp. 396–420.

[266] M. Hoyrup, T. Kihara, and V. Selivanov, *Degree spectra of homeomorphism types of Polish spaces*, Preprint., (2020).

[267] M. Hoyrup, A. Melnikov, and K. Ng, *Computable topological presentations*, (2025). To appear.

[268] Z. Iljazović, *Isometries and computability structures*, J.UCS, 16 (2010), pp. 2569–2596.

[269] Z. Iljazović, *Compact manifolds with computable boundaries*, Log. Methods Comput. Sci., 9 (2013), pp. 4:19, 22.

[270] Z. Iljazović and T. Kihara, *Computability of subsets of metric spaces*, in Handbook of computability and complexity in analysis, Theory Appl. Comput., Springer, Cham, [2021] ©2021, pp. 29–69.

[271] C. Jockusch and R. Shore, Π^0_1 *classes and degrees of theories*, Trans. Amer. Math. Soc., 173 (1972), pp. 33–56.

[272] ——, *Pseudo jump operators. I: The r.e. case*, Transactions of the American Mathematical Society, 275 (1983), pp. 599–609.

[273] ——, *Degrees of orderings not isomorphic to recursive linear orderings*, Ann. Pure Appl. Logic, 52 (1991), pp. 39–64. International Symposium on Mathematical Logic and its Applications (Nagoya, 1988).

[274] ——, *Boolean algebras, Stone spaces, and the iterated Turing jump*, J. Symbolic Logic, 59 (1994), pp. 1121–1138.

[275] A. Kach, *Computable shuffle sums of ordinals*, Arch. Math. Logic, 47 (2008), pp. 211–219.

[276] A. Kach, K. Lange, and R. Solomon, *Degrees of orders on torsion-free Abelian groups*, Ann. Pure Appl. Logic, 164 (2013), pp. 822–836.

[277] A. Kach and D. Turetsky, *Limitwise monotonic functions, sets, and degrees on computable domains*, J. Symbolic Logic, 75 (2010), pp. 131–154.

[278] I. Kalantari and G. Weitkamp, *Effective topological spaces. I. A definability theory*, Ann. Pure Appl. Logic, 29 (1985), pp. 1–27.

[279] I. Kalimullin, *Almost computably enumerable families of sets*, Mat. Sb., 199 (2008), pp. 33–40.

[280] ——, *Restrictions on the degree spectra of algebraic structures*, Siberian Mathematical Journal, 49 (2008), pp. 1034–1043.

[281] I. Kalimullin, B. Khoussainov, and A. Melnikov, *Limitwise monotonic sequences and degree spectra of structures*, Proc. Amer. Math. Soc., 141 (2013), pp. 3275–3289.

[282] I. Kalimullin, A. Melnikov, and K. M. Ng, *Algebraic structures computable without delay*, Theoret. Comput. Sci., 674 (2017), pp. 73–98.

[283] U. Kalvert, D. Kammins, D. F. Naĭt, and S. Miller, *Comparison of classes of finite structures*, Algebra Logika, 43 (2004), pp. 666–701, 759.

[284] S. Kaplan, *Extensions of the Pontrjagin duality. II. Direct and inverse sequences*, Duke Math. J., 17 (1950), pp. 419–435.

[285] I. Kaplansky, *Infinite abelian groups*, Revised edition, The University of Michigan Press, Ann Arbor, Mich., 1969.

[286] A. S. Kechris, *Classical descriptive set theory*, vol. 156 of Graduate Texts in Mathematics, Springer-Verlag, New York, 1995.

[287] N. Khisamiev, *Criterion for constructivizability of a direct sum of cyclic p-groups*, Izv. Akad. Nauk Kazakh. SSR Ser. Fiz.-Mat., (1981), pp. 51–55, 86.

[288] ——, *Hierarchies of torsion-free abelian groups*, Algebra i Logika, 25 (1986), pp. 205–226, 244.

[289] ——, *The arithmetic hierarchy of abelian groups*, Sibirsk. Mat. Zh., 29 (1988), pp. 144–159.

[290] ——, *Constructive abelian p-groups*, Siberian Adv. Math., 2 (1992), pp. 68–113.

[291] ——, *Constructive abelian groups*, in Handbook of recursive mathematics, Vol. 2, vol. 139 of Stud. Logic Found. Math., North-Holland, Amsterdam, 1998, pp. 1177–1231.

[292] ——, *On a class of strongly decomposable abelian groups*, Algebra Logika, 41 (2002), pp. 493–509, 511–512.

[293] N. Khisamiev and A. Krykpaeva, *Effectively completely decomposable abelian groups*, Sibirsk. Mat. Zh., 38 (1997), pp. 1410–1412, iv.

[294] B. Khoussainov and T. Kowalski, *Computable isomorphisms of Boolean algebras with operators*, Studia Logica, 100 (2012), pp. 481–496.

[295] B. Khoussainov, A. Nies, and R. Shore, *Computable models of theories with few models*, Notre Dame J. Formal Logic, 38 (1997), pp. 165–178.

[296] E. I. Khukhro and V. D. Mazurov, *Unsolved problems in group theory. The Kourovka Notebook*, 2023.

[297] J. A. Kiehlmann, *Classifications of countably-based abelian profinite groups*, J. Group Theory, 16 (2013), pp. 141–157.

[298] S. Kleene, *On notation for ordinal numbers*, The Journal of Symbolic Logic, 3 (1938), pp. 150–155.

[299] ——, *Recursive predicates and quantifiers*, Transactions of the American Math. Society, 53 (1943), pp. 41–73.

[300] ——, *On the forms of the predicates in the theory of constructive ordinals*, Amer. J. Math., 66 (1944), pp. 41–58.

[301] ——, *On the forms of the predicates in the theory of constructive ordinals. II*, Amer. J. Math., 77 (1955), pp. 405–428.

[302] ——, *Recursive functionals and quantifiers of finite types. I*, Trans. Amer. Math. Soc., 91 (1959), pp. 1–52.

[303] S. Kleene and E. Post, *The upper semi-lattice of degrees of recursive unsolvability*, Annals of Mathematics. Second Series, 59 (1954), pp. 379–407.

[304] J. Knight, *Degrees coded in jumps of orderings*, J. Symbolic Logic, 51 (1986), pp. 1034–1042.

[305] J. Knight and C. McCoy, *Index sets and Scott sentences*, Arch. Math. Logic, 53 (2014), pp. 519–524.

[306] J. Knight and J. Millar, *Computable structures of rank*, Journal of Mathematical Logic, 10 (2010), pp. 31–43.

[307] J. Knight, S. Miller, and M. Vanden Boom, *Turing computable embeddings*, The Journal of Symbolic Logic, 72 (2007), pp. 901–918.

[308] J. Knight and M. Stob, *Computable Boolean algebras*, J. Symbolic Logic, 65 (2000), pp. 1605–1623.

[309] N. T. Kogabaev, *The theory of projective planes is complete with respect to degree spectra and effective dimensions*, Algebra Logika, 54 (2015), pp. 599–627, 648, 650.

[310] ——, *Freely generated projective planes of finite computable dimension*, Algebra Logika, 55 (2016), pp. 704–737.

[311] H. T. Koh, A. Melnikov, and K. M. Ng, *Counterexamples in effective topology*. To appear.

[312] ——, *A non-density aspect of the rationals*. Submitted.

[313] H. T. Koh, A. Melnikov, and K. M. Ng, *Computable topological groups*, Journal of Symbolic Logic, (2023).

[314] A. Kokorin and V. Kopytov, *Fully ordered groups*, Halsted Press [John Wiley & Sons], New York-Toronto, Ont., 1974. Translated from the Russian by D. Louvish.

[315] M. Korovina and O. Kudinov, *Towards computability over effectively enumerable topological spaces*, in Proceedings of the Fifth International Conference on Computability and Complexity in Analysis (CCA 2008), vol. 221 of Electron. Notes Theor. Comput. Sci., Elsevier Sci. B. V., Amsterdam, 2008, pp. 115–125.

[316] ——, *The Rice-Shapiro theorem in computable topology*, Log. Methods Comput. Sci., 13 (2017), pp. Paper No. 30, 13.

[317] G. Kreisel, D. Lacombe, and J. R. Shoenfield, *Partial recursive functionals and effective operations*, in Constructivity in mathematics: Proceedings of the colloquium held at Amsterdam, 1957 (edited by A. Heyting), Studies in Logic and the Foundations of Mathematics, North-Holland Publishing Co., Amsterdam, 1957, pp. 290–297.

[318] L. Kronecker, *Grundzüge einer arithmetischen theorie der algebraischen grössen*, J. f. Math., 92 (1882), pp. 1–122.

[319] W. Krull, *Über polynomzerlegung mit endlich vielen schritten*, Math. Z., 59 (1953), pp. 57–60.

[320] O. V. Kudinov, *An autostable 1-decidable model without a computable Scott family of ∃-formulas*, Algebra and Logic, 35 (1996), pp. 255–260.

[321] ——, *Some properties of autostable models*, Algebra i Logika, 35 (1996), pp. 685–698, 752.

[322] G. Kuperberg, *Algorithmic homeomorphism of 3-manifolds as a corollary of geometrization*, Pacific J. Math., 301 (2019), pp. 189–241.

[323] A. Kurosh, *The theory of groups*, Chelsea Publishing Co., New York, 1960. Translated from the Russian and edited by K. A. Hirsch. 2nd English ed. 2 volumes.

[324] P. La Roche, *Effective Galois theory*, J. Symbolic Logic, 46 (1981), pp. 385–392.

[325] A. Lachlan, *Lower bounds for pairs of recursively enumerable degrees*, Proceedings of the London Mathematical Society, 16 (1966), pp. 537–569.

[326] D. Lacombe, *Extension de la notion de fonction recursive aux fonctions d'une ou plusieurs variables reelles. I*, C. R. Acad. Sci. Paris, 240 (1955), pp. 2478–2480.

[327] ——, *Extension de la notion de fonction recursive aux fonctions d'une ou plusieurs variables reelles. II, III*, C. R. Acad. Sci. Paris, 241 (1955), pp. 13–14, 151–153.

[328] R. E. Ladner, *Mitotic recursively enumerable sets*, Journal of Symbolic Logic, 38 (1973), pp. 199–211.

[329] S. Lang, *Algebra*, vol. 211 of Graduate Texts in Mathematics, Springer-Verlag, New York, third ed., 2002.

[330] K. Lange, R. Miller, and R. M. Steiner, *Classifications of computable structures*, Notre Dame J. Form. Log., 59 (2018), pp. 35–59.

[331] P. LaRoche, *Recursively presented Boolean algebras*, Notices AMS, 24 (1977), pp. 552–553.

[332] D. Lascar and B. Poizat, *An introduction to forking*, The Journal of Symbolic Logic, 44 (1979), pp. 330–350.

[333] S. Le Roux and M. Ziegler, *Singular coverings and non-uniform notions of closed set computability*, Mathematical Logic Quarterly, 54 (2008), pp. 545–560.

[334] S. Lempp, *The computational complexity of torsion-freeness of finitely presented groups*, Bull. Austral. Math. Soc., 56 (1997), pp. 273–277.

[335] S. Lempp, C. McCoy, R. Miller, and R. Solomon, *Computable categoricity of trees of finite height*, J. Symbolic Logic, 70 (2005), pp. 151–215.

[336] M. Lerman, *On recursive linear orderings*, in Logic Year 1979–80 (Proc. Seminars and Conf. Math. Logic, Univ. Connecticut, Storrs, Conn., 1979/80), vol. 859 of Lecture Notes in Math., Springer, Berlin, 1981, pp. 132 142.

[337] F. Levi. Habilitationsschrift, Leipzig, Teubner, 1917.

[338] O. Levin, *Computable dimension for ordered fields*, Arch. Math. Logic, 55 (2016), pp. 519–534.

[339] R. J. Lipton and Y. Zalcstein, *Word problems solvable in logspace*, J. Assoc. Comput. Mach., 24 (1977), pp. 522–526.

[340] P. Loth, *Classifications of abelian groups and Pontrjagin duality*, vol. 10 of Algebra, Logic and Applications, Gordon and Breach Science Publishers, Amsterdam, 1998.

[341] M. Lupini, A. Melnikov, and A. Nies, *Computable topological abelian groups*, J. Algebra, 615 (2023), pp. 278–327.

[342] N. Lusin and W. Sierpinski, *Sur un ensemble non measurable b*, Journal de Mathámatiques Pures et Appliquées, 9 (1923), pp. 53–72.

[343] R. C. LYNDON AND P. E. SCHUPP, *Combinatorial group theory*, Classics in Mathematics, Springer-Verlag, Berlin, 2001. Reprint of the 1977 edition.

[344] E. W. MADISON, *A note on computable real fields*, J. Symbolic Logic, 35 (1970), pp. 239–241.

[345] A. MAL'CEV, *Constructive algebras. I*, Uspehi Mat. Nauk, 16 (1961), pp. 3–60.

[346] ——, *On recursive Abelian groups*, Dokl. Akad. Nauk SSSR, 146 (1962), pp. 1009–1012.

[347] A. MARCONE AND M. VALENTI, *Effective aspects of Hausdorff and Fourier dimension*, Computability, 11 (2022), pp. 299–333.

[348] D. MARKER, *Model theory*, vol. 217 of Graduate Texts in Mathematics, Springer-Verlag, New York, 2002.

[349] D. MARKER AND R. MILLER, *Turing degree spectra of differentially closed fields*, J. Symb. Log., 82 (2017), pp. 1–25.

[350] A. MARKOV, *On the continuity of constructive functions*, Uspehi Matematicheskih Nauk, 9 (1954), pp. 226–230. (in Russian).

[351] ——, *On the continuity of constructive functions*, Uspekhi Matemicheskikh Nauk., 9 (1954), pp. 602–619.

[352] ——, *The insolubility of the problem of homeomorphy.*, Dokl. Akad. Nauk SSSR, 121 (1958), pp. 218–220.

[353] ——, *On constructive functions*, Proceedings of the Steklov Institute of Mathematics, LII (1958), pp. 315–348.

[354] D. MARTIN AND M. POUR-EL, *Axiomatizable theories with few axiomatizable extensions*, Journal of Symbolic Logic, 35 (1970), pp. 205–209.

[355] C. MCCOY, *Finite computable dimension does not relativize*, Archive for Mathematical Logic, 41 (2002), pp. 309–320.

[356] ——, *On Δ^0_3-categoricity for linear orders and Boolean algebras*, Algebra Logika, 41 (2002), pp. 531–552, 633.

[357] ——, *Δ^0_2-categoricity in Boolean algebras and linear orderings*, Ann. Pure Appl. Logic, 119 (2003), pp. 85–120.

[358] C. MCCOY AND J. WALLBAUM, *Describing free groups, Part II: Π^0_4 hardness and no Σ^0_2 basis*, Trans. Amer. Math. Soc., 364 (2012), pp. 5729–5734.

[359] W. MCCUNE, R. VEROFF, B. FITELSON, K. HARRIS, A. FEIST, AND L. WOS, *Short single axioms for Boolean algebra*, J. Automat. Reason., 29 (2002), pp. 1–16.

[360] T. MCNICHOLL, *Computable copies of ℓ^p*, Computability, 6 (2017), pp. 391–408.

[361] ——, *Computing the exponent of a Lebesgue space*, Journal of Logic and Analysis, 12 (2020), pp. Paper No. 7, 28 pp.

[362] T. McNicholl and D. M. Stull, *The isometry degree of a computable copy of ℓ^{p1}*, Computability, 8 (2019), pp. 179–189.

[363] T. H. McNicholl, *Evaluative presentations*, 2024.

[364] J. Melleray, *Some geometric and dynamical properties of the Urysohn space*, Topology Appl., 155 (2008), pp. 1531–1560.

[365] A. Melnikov, *Enumerations and completely decomposable torsion-free abelian groups*, Theory Comput. Syst., 45 (2009), pp. 897–916.

[366] ——, *Computable ordered abelian groups and fields*, in Programs, proofs, processes, vol. 6158 of Lecture Notes in Comput. Sci., Springer, Berlin, 2010, pp. 321–330.

[367] ——, *Computability and structure*, PhD thesis, The University of Auckland, 2012.

[368] ——, *Effective properties of completely decomposable abelian groups*, PhD thesis, Novosibirsk State University, Russian Federation, Novosibirsk, 2012.

[369] ——, *Computably isometric spaces*, J. Symbolic Logic, 78 (2013), pp. 1055–1085.

[370] ——, *Computable abelian groups*, The Bulletin of Symbolic Logic, 20 (2014), pp. 315–356.

[371] ——, *New degree spectra of abelian groups*, Notre Dame Journal of Formal Logic, 58 (2017).

[372] ——, *Torsion-free abelian groups with optimal Scott families*, Journal of Mathematical Logic, 18 (2017).

[373] ——, *Computable topological groups and Pontryagin duality*, Trans. Amer. Math. Soc., 370 (2018), pp. 8709–8737.

[374] ——, *New degree spectra of Polish spaces*, Sibirsk. Mat. Zh., (2021), p. 1091–1108.

[375] A. Melnikov and A. Montalbán, *Computable Polish group actions*, J. Symb. Log., 83 (2018), pp. 443–460.

[376] A. Melnikov and K. M. Ng, *Computable structures and operations on the space of continuous functions*, Fund. Math., 233 (2016), pp. 101–141.

[377] ——, *Computable torsion abelian groups*, Adv. Math., 325 (2018), pp. 864–907.

[378] ——, *The back-and-forth method and computability without delay*, Israel J. Math., 234 (2019), pp. 959–1000.

[379] ——, *A structure of punctual dimension two*, Proc. Amer. Math. Soc., 148 (2020), pp. 3113–3128.

[380] ——, *Separating notions in effective topology*, Internat. J. Algebra Comput., 33 (2023), pp. 1687–1711.

[381] A. Melnikov and A. Nies, *The classification problem for compact computable metric spaces*, in The nature of computation, vol. 7921 of Lecture Notes in Comput. Sci., Springer, Heidelberg, 2013, pp. 320–328.

[382] ——, *Computably locally compact totally disconnected groups*. Submitted., 2023.

[383] G. Metakides and A. Nerode, *Recursively enumerable vector spaces*, Ann. Math. Logic, 11 (1977), pp. 147–171.

[384] ——, *Effective content of field theory*, Ann. Math. Logic, 17 (1979), pp. 289–320.

[385] ——, *The introduction of nonrecursive methods into mathematics*, in The L. E. J. Brouwer Centenary Symposium (Noordwijkerhout, 1981), vol. 110 of Stud. Logic Found. Math., North-Holland, Amsterdam, 1982, pp. 319–335.

[386] G. Metakides, A. Nerode, and R. Shore, *Recursive limits on the Hahn-Banach theorem*, in Errett Bishop: reflections on him and his research (San Diego, Calif., 1983), vol. 39 of Contemp. Math., Amer. Math. Soc., Providence, RI, 1985, pp. 85–91.

[387] A. Meyer, *Program size in restricted programming languages*, Information and Control, 21 (1972), pp. 382–394.

[388] T. Millar, *Foundations of recursive model theory*, Ann. Math. Logic, 13 (1978), pp. 45–72.

[389] ——, *Recursive categoricity and persistence*, Journal of Symbolic Logic, 51 (1986), pp. 430–434.

[390] C. Miller, III, *Decision problems for groups—survey and reflections*, in Algorithms and classification in combinatorial group theory (Berkeley, CA, 1989), vol. 23 of Math. Sci. Res. Inst. Publ., Springer, New York, 1992, pp. 1–59.

[391] J. Miller, *Effectiveness for embedded spheres and balls*, Electron. Notes Theor. Comput. Sci., 66 (2002), pp. 127–138.

[392] ——, *Degrees of unsolvability of continuous functions*, J. Symbolic Logic, 69 (2004), pp. 555–584.

[393] R. Miller, *Computability, definability, categoricity, and automorphisms*, ProQuest LLC, Ann Arbor, MI, 2000. Thesis (Ph.D.)–The University of Chicago.

[394] ——, *The computable dimension of trees of infinite height*, J. Symbolic Logic, 70 (2005), pp. 111–141.

[395] ——, *d-computable categoricity for algebraic fields*, Journal of Symbolic Logic, 74 (2009), pp. 1325–1351.

[396] ——, Low_5 *Boolean subalgebras and computable copies*, J. Symbolic Logic, 76 (2011), pp. 1061–1074.

[397] R. Miller, B. Poonen, H. Schoutens, and A. Shlapentokh, *A computable functor from graphs to fields*, J. Symb. Log., 83 (2018), pp. 326–348.

[398] A. Montalbán, *On the triple jump of the set of atoms of a Boolean algebra*, Proc. Amer. Math. Soc., 136 (2008), pp. 2589–2595.

[399] ——, *Notes on the jump of a structure*, in Mathematical Theory and Computational Practice, K. Ambos-Spies, B. Löwe, and W. Merkle, eds., Berlin, Heidelberg, 2009, Springer Berlin Heidelberg, pp. 372–378.

[400] ——, *A robuster Scott rank*, Proc. Amer. Math. Soc., 143 (2015), pp. 5427–5436.

[401] ——, *Computable structure theory—within the arithmetic*, Perspectives in Logic, Cambridge University Press, Cambridge; Association for Symbolic Logic, Ithaca, NY, 2021.

[402] ——, *Computable structure theory – beyond the arithmetic*, To appear.

[403] T. Mori, Y. Tsujii, and M. Yasugi, *Computability structures on metric spaces*, in Combinatorics, Complexity, and Logic, D. S. Bridges, C. S. Calude, J. Gibbons, S. Reeves, and I. H. Witten, eds., vol. 1 of Discrete Mathematics and Theoretical Computer Science, Springer, Singapore, 1997, pp. 351–362.

[404] S. A. Morris, *Pontryagin duality and the structure of locally compact abelian groups*, Cambridge University Press, Cambridge-New York-Melbourne, 1977. London Mathematical Society Lecture Note Series, No. 29.

[405] Y. Moschovakis, *Recursive metric spaces*, Fund. Math., 55 (1964), pp. 215–238.

[406] ——, *Descriptive set theory*, vol. 155 of Mathematical Surveys and Monographs, American Mathematical Society, Providence, RI, second ed., 2009.

[407] M. Moses, *Recursive linear oderings with recursive successivities*, Annals of Pure and Applied Logic, 27 (1984), pp. 253–264.

[408] ——, *n-recursive linear orders without $(n+1)$-recursive copies*, in Logical methods. In honor of Anil Nerode's 60th birthday, Basel: Birkhäuser, 1993, pp. 572–592.

[409] A. A. Muchnik, *On the unsolvability of the problem of reducibility in the theory of algorithms*, Doklady Akademii Nauk SSSR (N.S.), 108 (1956), pp. 194–197.

[410] ——, *Solution of Post's reduction problem and certain other problems in the theory of algorithms*, Trud. Mosk. Math. Obsc., 7 (1958), pp. 391–401.

[411] J. R. Munkres, *Elements of algebraic topology*, Addison-Wesley Publishing Company, Menlo Park, CA, 1984.

[412] ——, *Topology*, Prentice Hall, Inc., Upper Saddle River, NJ, 2000. Second edition.

[413] ——, *Topology*, Prentice Hall, Inc., Upper Saddle River, NJ, second ed., 2000.

[414] J. Myhill, *A recursive function, defined on a compact interval and having a continuous derivative that is not recursive*, Michigan Math. J., 18 (1971), pp. 97–98.

[415] K. M. Ng, *Some properties of d.c.e. reals and their degrees*, 2005. M.Sc. thesis.

[416] P. Novikov, *On the algorithmic unsolvability of the word problem in group theory*, Trudy Mat. Inst. Steklov, 44 (1955), pp. 1–143.

[417] A. Nurtazin, *Computable classes and algebraic criteria of autostability*, Summary of Scientific Schools, Math. Inst., SB USSRAS, Novosibirsk, 1974.

[418] ——, *Strong and weak constructivizations, and enumerable families*, Algebra i Logika, 13 (1974), pp. 311–323, 364.

[419] S. Oates, *Jump Degrees of Groups*, PhD thesis, University of Notre Dame, 1989.

[420] V. Ocasio González, *Computability in the Class of Real Closed Fields*, PhD thesis, Notre Dame University, 2014.

[421] P. Odifreddi, *Classical Recursion Theory. The theory of functions and sets of natural numbers*, no. 125 in Studies in Logic and the Foundations of Mathematics, North-Holland Publishing Company, Amsterdam, 1990.

[422] S. P. Odintsov, *Generally constructive Boolean algebras*, in Handbook of recursive mathematics, Vol. 2, vol. 139 of Stud. Logic Found. Math., North-Holland, Amsterdam, 1998, pp. 1319–1354.

[423] S. P. Odintsov and V. L. Selivanov, *The arithmetical hierarchy and ideals of enumerated Boolean algebras*, Sibirsk. Mat. Zh., 30 (1989), pp. 140–149.

[424] S. Ospichev, *Friedberg numberings in the Ershov hierarchy*, Algebra and Logic, 54 (2015), pp. 283–295.

[425] ——, *Friedberg numberings of families of partially computable functionals*, Sib. Èlektron. Mat. Izv., 16 (2019), pp. 331–339.

[426] G. Paolini and S. Shelah, *Torsion-free abelian groups are Borel complete*, Ann. of Math. (2), 199 (2024), pp. 1077–1124.

[427] A. Pauly, D. Seon, and M. Ziegler, *Computing Haar Measures*, in 28th EACSL Annual Conference on Computer Science Logic (CSL 2020), M. Fernández and A. Muscholl, eds., vol. 152 of Leibniz International Proceedings in Informatics (LIPIcs), Dagstuhl, Germany, 2020, Schloss Dagstuhl–Leibniz-Zentrum fuer Informatik, pp. 34:1–34:17.

[428] E. N. Pavlovskii, *An estimate for the algorithmic complexity of classes of computable models*, Sibirsk. Mat. Zh., 49 (2008), pp. 635–649.

[429] L. Pontryagin, *Topological groups*, Translated from the second Russian edition by Arlen Brown, Gordon and Breach Science Publishers, Inc., New York-London-Paris, 1966.

[430] B. Poonen, *Undecidable problems: A sampler*, in Interpreting Gödel, Cambridge Univ. Press, Cambridge, 2014, pp. 211–241.

[431] E. Post, *Recursively enumerable sets of positive integers and their decision problems*, Bulletin of the American Mathematical Society, 50 (1944), pp. 284–316.

[432] M. Pour-El and J. Caldwell, *On a simple definition of computable function of a real variable—with applications to functions of a complex variable*, Z. Math. Logik Grundlagen Math., 21 (1975), pp. 1–19.

[433] M. Pour-El and H. Putnam, *Recursively enumerable classes and their application to recursive sequences of formal theories*, Archiv für mathematische Logik, 8 (1965), pp. 104–121.

[434] M. Pour-El and I. Richards, *Computability and noncomputability in classical analysis*, Trans. Amer. Math. Soc., 275 (1983), pp. 539–560.

[435] ——, *Computability in analysis and physics*, Perspectives in Mathematical Logic, Springer-Verlag, Berlin, 1989.

[436] H. Prüfer, *Unendliche Abelsche Gruppen von Elementen endlicher Ordnung*, PhD thesis, Humboldt-Universität zu Berlin, Berlin, Germany, 1921.

[437] L. Qian, *Computability-theoretic complexity of effective Banach spaces, M.Sc. thesis, Victoria University of Wellington*, 2021.

[438] M. Rabin, *Recursive unsolvability of group theoretic problems*, Ann. of Math. (2), 67 (1958), pp. 172–194.

[439] ——, *Recursive unsolvability of group theoretic problems*, Annals of Mathematics (2), 67 (1958), pp. 172–194.

[440] ——, *Computable algebra, general theory and theory of computable fields.*, Trans. Amer. Math. Soc., 95 (1960), pp. 341–360.

[441] R. Rado, *A proof of the basis theorem for finitely generated Abelian groups*, J. London Math. Soc., 26 (1951), pp. 74–75; erratum, 160.

[442] A. Raichev, *Relative randomness and real closed fields*, Journal of Symbolic Logic, 70 (2005), pp. 319–330.

[443] J. Remmel, *Recursively enumerable Boolean algebras*, Ann. Math. Logic, 14 (1978), pp. 75–107.

[444] ——, *Recursive Boolean algebras with recursive atoms*, J. Symb. Log., 46 (1981), pp. 595–616.

[445] ——, *Recursive isomorphism types of recursive Boolean algebras*, J. Symbolic Logic, 46 (1981), pp. 572–594.

[446] ——, *Recursively categorical linear orderings*, Proc. Amer. Math. Soc., 83 (1981), pp. 387–391.

[447] ——, *Recursive Boolean algebras*, in Handbook of Boolean algebras, Vol. 3, North-Holland, Amsterdam, 1989, pp. 1097–1165.

[448] H. Rice, *Classes of recursively enumerable sets and their decision problems*, Transactions of the American Mathematical Society, 74 (1953), pp. 358–366.

[449] ——, *Recursive real numbers*, Proceedings of the American Mathematical Society, 5 (1954), pp. 784–791.

[450] L. Richter, *Degrees of Unsolvability of Models*, PhD thesis, University of Illinois at Champaign-Urbana, 1977.

[451] ——, *Degrees of structures*, J. Symbolic Logic, 46 (1981), pp. 723–731.

[452] K. Riggs, *Computable Properties of Decomposable and Completely Decomposable Groups*, PhD thesis, Indiana University, Bloomington, Indiana, USA, 2014.

[453] R. W. Robinson, *Jump restricted interpolation in the recursively enumerable degrees*, Annals of Mathematics. Second Series, 93 (1971), pp. 586–596.

[454] H. Rogers, *Theory of recursive functions and effective computability*, MIT Press, Cambridge, MA, second ed., 1987.

[455] L. Rogers, *Ulm's theorem for partially ordered structures related to simply presented abelian p-groups*, Trans. Amer. Math. Soc., 227 (1977), pp. 333–343.

[456] J. G. Rosenstein, *Linear orderings*, vol. 98 of Pure and Applied Mathematics, Academic Press, Inc. [Harcourt Brace Jovanovich, Publishers], New York-London, 1982.

[457] G. Sacks, *On the degrees less than* $\mathbf{0}'$, Annals of Mathematics. Second Series, 77 (1963), pp. 211–231.

[458] ——, *Higher recursion theory*, Perspectives in Mathematical Logic, Springer-Verlag, Berlin, 1990.

[459] M. Schröder, *Effective metrization of regular spaces*, Computability and Complexity in Analysis, Informatik Berichte 235 (1998), pp. 63–80.

[460] S. Schwarz, *Recursive automorphisms of recursive linear orderings*, Annals of Pure and Applied Logic, 26 (1984), pp. 69–73.

[461] D. Scott, *Logic wit denumerably long formulas and finite strings of quantifiers*, in The Theory of Models, J. Addison, L. Henkin, and A. Tarski, eds., North-Holland, 1965, pp. 329–341.

[462] V. Selivanov, *The numerations of families of general recursive functions*, Algebra i Logika, 15 (1976), pp. 205–226, 246.

[463] ——, *Descriptive complexity of qcb0-spaces*, Theoretical Computer Science, 945 (2023), p. 113666.

[464] P. Semukhin, *Prime models of finite computable dimension*, The Journal of Symbolic Logic, 74 (2009), p. 336–348.

[465] S. Shelah, *Classification theory*, vol. 98 of Studies in Logic (London), College Publications, [London], second ed., [2023] ©2023. Mathematical Logic and Foundations.

[466] J. Shoenfield, *On degrees of unsolvability*, Annals of Mathematics. Second Series, 69 (1959), pp. 644–653.

[467] ——, *Degrees of Unsolvability*, no. 2 in North-Holland Mathematics Studies, North-Holland Publishing Company, Amsterdam-London, 1971.

[468] R. Shore, *Controlling the dependence degree of a recursively enumerable vector space*, J. Symbolic Logic, 43 (1978), pp. 13–22.

[469] W. SIERPINSKI, *General topology*, Mathematical Expositions, No. 7, University of Toronto Press, Toronto, 1952. Translated by C. Cecilia Krieger.

[470] S. G. SIMPSON, *Subsystems of second order arithmetic*, Perspectives in Mathematical Logic, Springer-Verlag, Berlin, 1999.

[471] T. SLAMAN, *Relative to any nonrecursive set*, Proc. Amer. Math. Soc., 126 (1998), pp. 2117–2122.

[472] R. SMITH, *The theory of profinite groups with effective presentations*, ProQuest LLC, Ann Arbor, MI, 1979. Thesis (Ph.D.)–The Pennsylvania State University.

[473] ——, *Effective aspects of profinite groups*, J. Symbolic Logic, 46 (1981), pp. 851–863.

[474] ——, *Two theorems on autostability in p-groups*, in Logic Year 1979–80 (Proc. Seminars and Conf. Math. Logic, Univ. Connecticut, Storrs, Conn., 1979/80), vol. 859 of Lecture Notes in Math., Springer, Berlin, 1981, pp. 302–311.

[475] R. SMULLYAN, *Theory of Formal Systems*, Princeton University Press, 1961.

[476] R. SOARE, *The Friedberg-Muchnik theorem re-examined*, Canadian Journal of Mathematics, 24 (1972), p. 1070–1078.

[477] ——, *Recursively enumerable sets and degrees*, Perspectives in Mathematical Logic, Springer-Verlag, Berlin, 1987. A study of computable functions and computably generated sets.

[478] ——, *Turing computability*, Theory and Applications of Computability, Springer-Verlag, Berlin, 2016. Theory and applications.

[479] R. SOLOMON, *Π^0_1 classes and orderable groups*, Ann. Pure Appl. Logic, 115 (2002), pp. 279–302.

[480] E. SPECKER, *Nicht konstrucktiv beweisbare sätze der analysis*, Journal of Symbolic Logic, 14 (1949), pp. 145–158.

[481] C. SPECTOR, *Recursive well-orderings*, J. Symbolic Logic, 20 (1955), pp. 151–163.

[482] D. SPREEN, *A characterization of effective topological spaces*, in Recursion theory week (Oberwolfach, 1989), vol. 1432 of Lecture Notes in Math., Springer, Berlin, 1990, pp. 363–387.

[483] E. STEINITZ, *Algebraische theorie der körper*, J. f. Math., 137 (1910), pp. 167–309.

[484] V. STOLTENBERG-HANSEN AND J. V. TUCKER, *Computable and continuous partial homomorphisms on metric partial algebras*, Bull. Symbolic Logic, 9 (2003), pp. 299–334.

[485] J. J. THURBER, *Degrees of Boolean algebras*, ProQuest LLC, Ann Arbor, MI, 1994. Thesis (Ph.D.)–University of Notre Dame.

[486] D. TURETSKY, *Coding in the automorphism group of a computably categorical structure*, Journal of Mathematical Logic, 20 (2020).

[487] A. TURING, *On computable numbers, with an application to the Entscheidungsproblem*, Proceedings of the London Mathematical Society, 42 (1936), pp. 230–265.

[488] ——, *On Computable Numbers, with an Application to the Entscheidungsproblem. A Correction*, Proceedings of the London Mathematical Society, 43 (1937), pp. 544–546.

[489] ——, *Finite approximations to Lie groups*, Annals of Mathematics, 39 (1938), pp. 105–111.

[490] ——, *Systems of logic based on ordinals*, Proceedings of the London Math. Society, 45 (1939), pp. 154–222.

[491] H. Ulm, *Zur theorie der abzählbar-unendlichen abelschen gruppen*, Math. Ann., 107 (1933), pp. 774–803.

[492] J. Vaisala, *A proof of the Mazur-Ulam theorem*, The American Mathematical Monthly, 110 (2003), pp. 633–635.

[493] B. van der Waerden, *Moderne Algebra. Parts I and II*, G. E. Stechert and Co., New York, 1943.

[494] M. Vanden Boom, *The effective Borel hierarchy*, Fund. Math., 195 (2007), pp. 269–289.

[495] R. Vaught, *Invariant sets in topology and logic*, Fund. Math., 82 (1974/75), pp. 269–294.

[496] R. L. Vaught, *Topics in the theory of arithmetical classes and Boolean algebras*, ProQuest LLC, Ann Arbor, MI, 1955. Thesis (Ph.D.)–University of California, Berkeley.

[497] Y. G. Ventsov, *Effective choice for relations and reducibilities in classes of constructive and positive models*, Algebra and Logic, 31 (1992), pp. 63–73.

[498] ——, *Effective choice operations on constructive and positive models*, Algebra and Logic, 32 (1993), pp. 23–28.

[499] L. Vietoris, *Über den höheren Zusammenhang kompakter Räume und eine Klasse von zusammenhangstreuen Abbildungen*, Math. Ann., 97 (1927), pp. 454–472.

[500] J. D. Wallbaum, *Computability of algebraic structures*, ProQuest LLC, Ann Arbor, MI, 2010. Thesis (Ph.D.)–University of Notre Dame.

[501] W. C. Waterhouse, *Profinite groups are Galois groups*, Proc. Amer. Math. Soc., 42 (1974), pp. 639–640.

[502] R. Watnick, *A generalization of Tennenbaum's theorem on effectively finite recursive linear orderings*, J. Symbolic Logic, 49 (1984), pp. 563–569.

[503] S. Wehner, *Enumerations, countable structures and Turing degrees*, Proc. Amer. Math. Soc., 126 (1998), pp. 2131–2139.

[504] ——, *On recursive enumerability with finite repetitions*, J. Symbolic Logic, 64 (1999), pp. 927–945.

[505] K. Weihrauch, *Computable analysis*, Texts in Theoretical Computer Science. An EATCS Series, Springer-Verlag, Berlin, 2000. An introduction.

[506] ——, *Computational complexity on computable metric spaces*, MLQ Math. Log. Q., 49 (2003), pp. 3–21.

[507] K. Weihrauch and X. Zheng, *Effectiveness of the global modulus of continuity on metric spaces*, Theoretical Computer Science, 219 (1999), p. 439 – 450.

[508] L. Westrick, *A lightface analysis of the differentiability rank*, J. Symb. Log., 79 (2014), pp. 240–265.

[509] W. M. White, *Characterizations for computable structures*, ProQuest LLC, Ann Arbor, MI, 2000. Thesis (Ph.D.)–Cornell University.

[510] R. F. Williams, *Expanding attractors*, Inst. Hautes Études Sci. Publ. Math., (1974), pp. 169–203.

[511] R. Xie, *Computability and Randomness*, PhD thesis, Victoria University of Wellington, 2024.

[512] Y. Xu and T. Grubba, *On computably locally compact Hausdorff spaces*, Mathematical Structures in Computer Science, 19 (2009), pp. 101 – 117.

[513] C. E. M. Yates, *A minimal pair of recursively enumerable degrees*, The Journal of Symbolic Logic, 31 (1966), pp. 159–168.

[514] I. D. Zaslavskii, *The refutation of some theorems in classical analysis in constructive analysis*, Uspehi Mat. Nauk., 10 (1955), pp. 209–1210.

[515] ——, *Differentiation and integration of constructive functions*, Dolk. Akad. Nauk. SSSR, 156 (1964), pp. 599–601.

Index

R. G. Downey, A. Melnikov, *Computable Structure Theory*, Theory and Applications of Computability, https://doi.org/10.1007/978-3-031-92433-0

The manufacturer's authorised representative in the EU is Springer Nature Customer Service Centre GmbH, Europaplatz 3, 69115 Heidelberg, Germany. If you have any concerns regarding our products, please contact ProductSafety@springernature.com

Printed and bound by CPI Group (UK) Ltd, Croydon, CR0 4YY
30/09/2026
02225570-0019